Theoretical and Mathematical Physics

The series founded in 1975 and formerly (until 2005) entitled *Texts and Monographs in Physics* (TMP) publishes high-level monographs in theoretical and mathematical physics. The change of title to *Theoretical and Mathematical Physics* (TMP) signals that the series is a suitable publication platform for both the mathematical and the theoretical physicist. The wider scope of the series is reflected by the composition of the editorial board, comprising both physicists and mathematicians.

The books, written in a didactic style and containing a certain amount of elementary background material, bridge the gap between advanced textbooks and research monographs. They can thus serve as basis for advanced studies, not only for lectures and seminars at graduate level, but also for scientists entering a field of research.

For other titles published in this series, go to
www.springer.com/series/720

Henk T.C. Stoof • Koos B. Gubbels •
Dennis B.M. Dickerscheid

Ultracold Quantum Fields

Springer

Henk T.C. Stoof
Utrecht University
Institute for Theoretical Physics
Leuvenlaan 4
3584 CE Utrecht
The Netherlands

Koos B. Gubbels
Utrecht University
Institute for Theoretical Physics
Leuvenlaan 4
3584 CE Utrecht
The Netherlands

Dennis B.M. Dickerscheid
Utrecht University
Institute for Theoretical Physics
Leuvenlaan 4
3584 CE Utrecht
The Netherlands

Library of Congress Control Number: 2008936953

ISSN 1864-5879
ISBN-13 978-1-4020-8762-2 (HB)
ISBN-13 978-1-4020-8763-9 (e-Book)

Published by Springer Science+Business Media B.V.
P.O. Box 17, 3300 AA Dordrecht, The Netherlands
In association with
Canopus Publishing Limited,
27 Queen Square, Bristol BS1 4ND, UK

www.springer.com and www.canopusbooks.com

Preface

On June 19th 1999, the European Ministers of Education signed the Bologna Declaration, with which they agreed that the European university education should be uniformized throughout Europe and based on the two-cycle bachelor-master's system. The Institute for Theoretical Physics at Utrecht University quickly responded to this new challenge and created an international master's programme in Theoretical Physics which started running in the summer of 2000. At present, the master's programme is a so-called prestige master at Utrecht University, and it aims at training motivated students to become sophisticated researchers in theoretical physics. The programme is built on the philosophy that modern theoretical physics is guided by universal principles that can be applied to any subfield of physics. As a result, the basis of the master's programme consists of the obligatory courses Statistical Field Theory and Quantum Field Theory. These focus in particular on the general concepts of quantum field theory, rather than on the wide variety of possible applications. These applications are left to optional courses that build upon the firm conceptual basis given in the obligatory courses. The subjects of these optional courses include, for instance, Strongly-Correlated Electrons, Spintronics, Bose-Einstein Condensation, The Standard Model, Cosmology, and String Theory. The master's programme in Theoretical Physics is preceded by a summer school that is organized in the last two weeks of August to help prospective students prepare for the intensive master's courses. Short courses are offered in quantum mechanics, electrodynamics, statistical physics and computational methods, and are aimed at overcoming possible deficiencies in any of these subjects.

The idea of writing this book came about during the period of 2000-2005, when one of us was teaching the course on Statistical Field Theory for the abovementioned master's programme in Theoretical Physics. The lecture notes used for this course were an extended version of the lecture notes for the Les Houches summer school on *Coherent Atomic Matter Waves* that took place in 1999. Although these lecture notes, in combination with the lectures and tutorials, were supposed to be self-contained, in practice students often expressed a desire for more calculational details, applications and background material.

It was also during this period that the research field of ultracold atomic gases, pushed in particular by the impressive experimental progress since the first observation of Bose-Einstein condensation in 1995, made rapid developments that helped shape the field as we know it today. Nowadays, many experimental groups around the world can routinely prepare quantum degenerate gases of bosons, fermions, and various mixtures thereof. Moreover, the microscopic details of these atomic gases are well known and can be controlled very accurately, leading to the exciting possibility of addressing fundamental questions about interacting quantum systems in unprecedented detail. Because of this, it is also possible to perform ab initio theoretical calculations that allow for a quantitative comparison with experiments, such that the connection between theory and experiment is particularly close in this field of physics. There are various ways to perform these calculations, but most research topics can be dealt with in a unified manner by using quantum field theoretical methods. Although there are several textbooks available on quantum field theory, to date there does not exist a textbook that applies advanced quantum field theory, and in particular its functional formulation, to ultracold atomic quantum gases.

The level of this textbook is geared to students beginning with their master's and to graduate students already working in the field of ultracold atoms. To overcome the differences in educational background between the various students, the book has been divided into three parts which can in principle be read independently of each other. The first part briefly introduces elementary concepts from mathematics, statistical physics, and quantum mechanics which are indispensable for a full understanding of the rest of the book. Various important concepts that return later in the language of quantum field theory are introduced here in a more familiar setting. At the end of each chapter, there are references to various excellent textbooks that provide more background on each of the discussed topics. This part of the book is particularly aimed at the Utrecht Summer School in Theoretical Physics and provides the participants with the appropriate background material for the obligatory field theory courses that form the basis of the master's programme in Theoretical Physics. The second part of the book is devoted to laying the conceptual basis of the functional formulation of quantum field theory from a condensed-matter point of view. This part forms the core of the above mentioned Statistical Field Theory course, in which also the canonical topics of superfluidity and superconductivity of interacting Bose and Fermi gases are treated. The third part of the book is then largely aimed at applications of the developed theoretical techniques to various aspects of ultracold quantum gases that are currently being explored, such that the chosen topics give an idea of the present status of the field. It is our hope that, after having read this part, students will be well prepared to enter this exciting field of physics and be able to start contributing themselves to the rapid developments that are taking place today.

The knowledge presented in this book has been acquired through many collaborations and interactions with our colleagues over the last two decades. Here, we would like to sincerely thank everybody involved for that. It is unfortunately impossible to give everybody the proper credit for their contribution. As a result, both in this short word of thanks, as well as in citing references throughout the book, subjective choices are made and important contributions left out. Our main aim in citing has been to provide students with interesting additional reading material, and not to give an exhaustive overview of the enormous amount of literature in the field of ultracold atoms. We hope to be forgiven for that. With this in mind, we thank the following persons together with the members of their groups, namely Immanuel Bloch, Georg Bruun, Keith Burnett, Eric Cornell, Peter Denteneer, Steve Girvin, Randy Hulet, Allan MacDonald, Cristiane Morais Smith, Guthrie Partridge, Chris Pethick, Subir Sachdev, Cass Sackett, Jörg Schmiedmayer, Kevin Strecker, Peter van der Straten, Stefan Vandoren, and Eugene Zaremba for the collaborations that have led to joint publications. We also thank the postdoctoral researchers Usama Al Khawaja, Jens Andersen, Behnam Farid, Masud Haque, Jani Martikainen, Pietro Massignan, and Nick Proukakis, and the graduate students Michel Bijlsma, Marianne Houbiers, Michiel Bijlsma, Rembert Duine, Dries van Oosten, Gianmaria Falco, Lih-King Lim, Mathijs Romans, Michiel Snoek, Arnaud Koetsier, and Jeroen Diederix of the Utrecht Quantum Fluids and Solids Group. In particular, we mention Usama Al Khawaja, Rembert Duine, Dries van Oosten, and Nick Proukakis for their direct contributions to the recent applications that are discussed in the third part of the book. We also thank our experimental colleagues Immanuel Bloch, Eric Cornell, Randy Hulet, Wolfgang Ketterle, and Wenhui Li, for kindly providing us with the experimental data that has allowed us to compare the theory to experiment in this book. We thank Rembert Duine for providing several exercises and for many helpful comments on the manuscript. Furthermore we express our gratitude to Tom Spicer from Canopus Publishing for all his effort in bringing forth this book. We are especially grateful to Randy Hulet for more than 15 years of friendship and fruitful collaboration, from which we benefitted greatly, both personally and professionally.

Finally, we wish to thank Jolanda, Maurice, Inèz, Joke, Harry, Winy, Theo, Roos, Hein, Paulien, Ryoko, Miguel, and the rest of our families and friends for all their unconditional support and for sharing the joy of life.

Utrecht, May 2008

Henk Stoof
Koos Gubbels
Dennis Dickerscheid

Contents

Chapter 1
Introduction

The field of many-body quantum physics has a long history of fundamental discoveries, many of which have gone far beyond our wildest imagination. These include the study of novel states of matter, the observation of previously unseen phase transitions, and the discovery of new macroscopic quantum effects which arise when the intriguing rules of quantum mechanics are no longer restricted to the subatomic world, but rather determine the collective behavior of systems that are observable with the naked eye. In the past, it has often been proven difficult to obtain the underlying theory that yields an accurate description of the collective quantum phenomenon on the microscopic level. A good example is the discovery of superfluidity in liquid ^{4}He by Pyotr Kapitsa, John Allen and Don Misener in 1938 [1, 2], where superfluidity refers to the fact that the liquid can flow without experiencing resistance, which leads for example to the spectacular fountain effect [3]. Since the atoms interact very strongly, the precise internal state of liquid helium is notoriously difficult to determine.

An exception to this rule, however, is the question of what happens to a noninteracting gas of bosons when it is cooled down to zero temperature. This question was already theoretically answered long before the discovery of superfluid helium. In fact, the answer was already obtained before the final formulation of quantum mechanics and before a good understanding of phase transitions was achieved. The question found its origin in the early 1920s, when Satyendra Bose introduced a different way of counting microstates than was usual in classical statistical mechanics [4]. In this way, he was able to rederive Planck's law for the energy spectrum of black-body radiation. Albert Einstein generalized this result in 1924 to the case of indistinguishable noninteracting massive bosons by including the effect of particle-number conservation, which led to the famous Bose-Einstein distribution [5]. Einstein also realized that a remarkable consequence of this Bose-Einstein distribution is that below a certain critical temperature

$$T_c = \frac{2\pi}{\zeta(3/2)^{2/3}} \frac{\hbar^2 n^{2/3}}{m k_B} , \tag{1.1}$$

it predicts that a macroscopic fraction of the bosons occupies the same one-particle quantum state. Here $\hbar$ is Dirac's constant, i.e. Planck's constant h divided by 2π, m is the mass of the particles, k_B is Boltzmann's constant, n is the particle density of the gas, and $\zeta(3/2) \simeq 2.612$. This promotes the wavefunction of that particular one-particle quantum state to the macroscopic level and gives rise to a new state of matter that is known as a Bose-Einstein condensate or BEC. It is believed that Bose-Einstein condensation is also the mechanism behind the superfluid behavior of liquid helium. However, in liquid helium the density is high and the interaction between the helium atoms is very strong, such that it is far from an ideal Bose gas. As a result, Einstein's theory needs to be modified considerably, and so far the properties of liquid helium have been impossible to determine analytically. Furthermore, the presence of a macroscopic occupation of a one-particle quantum state has never been directly observed in this system.

The microscopic theory for the phenomenon of superconductivity, which was discovered experimentally in 1911 by Heike Kamerlingh Onnes [6], also turned out to be an extremely challenging task. After superconductivity had been found it was studied experimentally in a wide variety of metals, leading to many important discoveries. A crucial example, known as the Meissner effect [7], reveals that a superconductor is a perfect diamagnet because any applied magnetic field is completely expelled from its interior. It took almost fifty years before John Bardeen, Leon Cooper, and Robert Schrieffer [8] finally realized that superconductivity is actually caused by a Bose-Einstein condensation of loosely bound fermion pairs. The Bardeen-Cooper-Schrieffer or BCS theory of superconductivity is based on the description of the electrons in a metal as a gas, where the electrons need an effectively attractive interaction to form stable Cooper pairs. Physically, this attractive interaction is the result of the rather subtle effect that the electrons can deform the positively charged ionic lattice that is present in the metal. It is perhaps ironic that if the theory was invented before the experimental discovery of Kamerlingh Onnes, physicists would probably have never started looking for superconductivity in metals, because electrons do not usually form pairs due to their strongly repulsive Coulomb interaction. In 1986, high-temperature superconductors were discovered in ceramic materials [9]. However, the precise microscopic mechanism governing these cuprates is still not clear today.

1.1 Ultracold Atomic Quantum Gases

From the moment that Bose-Einstein condensation was finally achieved in trapped dilute gases of bosonic alkali atoms in 1995 by the groups of Eric Cornell and Carl Wieman, Randy Hulet, and Wolfgang Ketterle [10, 11, 12], a completely new category of systems became available for studying macroscopic quantum effects. The most important ingredients for this accomplishment were the precooling of the atoms using laser cooling [13], the trapping of the atoms in a magnetic trap [14], the final cooling of the atoms using evaporative cooling [15], and the imaging of the gas either in situ or after expansion. In particular, the trapping of atoms and cooling by means of evaporation turned out to be crucial. The reason for this is hidden in the prediction of (1.1) that for the relevant low densities of $10^{12} - 10^{15}$ atoms per cubic centimeter, extremely low temperatures of $1 - 100$ nK are required to reach Bose-Einstein condensation. These are impossible to achieve if the gas is in contact with material walls. Once the atomic gas is magnetically or optically trapped, evaporative cooling can be relatively easily implemented by lowering the trap depth, so that only the most energetic atoms can escape from the trap and the remaining gas cools after re-thermalization. Because of their complete isolation these ultracold gases are, unlike solid-state systems, very clean in the sense that there are essentially no impurities unless deliberately added. Moreover, due to the low densities, interaction effects can be sufficiently small as to be treated with perturbation theory. As a result, it is possible to obtain an accurate microscopic description of these ultracold atomic quantum gases using advanced field-theoretical methods. This is one of the main goals of this book. Furthermore, these systems have also turned out to be very flexible, as the external trapping potential and the interatomic interaction are under complete experimental control. This allows for a systematic study of an enormous variety of interesting many-body systems, ranging from weakly interacting to strongly interacting, from one dimensional to three dimensional, from homogeneous to periodic, where the microscopic parameters are always precisely known and tunable.

Shortly after the achievement of Bose-Einstein condensation, it was predicted that the superfluid regime could also be reached in a dilute gas of fermionic atoms [16]. However, the realization of this intriguing possibility turned out to be even more difficult than reaching BEC. This comes about because the previously mentioned BCS theory for the condensation of fermion pairs predicts that the critical temperature is exponentially dependent on the inverse of the (negative) scattering length a, which describes the strength of the attractive interactions between the fermions. Namely, we have that

$$T_{\mathrm{c}} = \frac{4(9\pi)^{1/3}}{\mathrm{e}^{2-\gamma}} \frac{\hbar^2 n^{2/3}}{m k_{\mathrm{B}}} \exp\left\{ -\frac{\pi}{2k_{\mathrm{F}}|a|} \right\}, \tag{1.2}$$

where $\gamma \simeq 0.5772$ is Euler's constant and $k_{\mathrm{F}} = (3\pi^2 n)^{1/3}$ is the Fermi wavevector. This in general shifts the required temperature beyond the reach of experiments with ultracold gases, which are dilute and therefore usually characterized by $k_{\mathrm{F}}|a| \ll 1$. Also, it turns out that it is more difficult to obtain an experimental signature for the onset of the superfluid phase in the Fermi system than in the case of bosons. The use of Feshbach resonances, which were theoretically discovered in the alkalis by Eite Tiesinga, Boudewijn Verhaar and Henk Stoof [17], has fortunately solved both of these problems.

In a Feshbach-resonant atomic collision, two atoms collide and virtually form a long-lived molecule with a different spin configuration than the incoming two atoms, where the molecule ultimately decays into two atoms again. The scattering properties of the colliding atoms depend very sensitively on the energy difference of the molecular state with respect to the threshold of the two-atom continuum. This energy difference is known as the detuning and can be changed with an applied magnetic field, because the different spin states of the incoming atoms and the molecule lead to a different Zeeman shift. In particular, the Feshbach resonance allows for a precise tuning of the scattering length a, which opens up the exciting possibility of reaching the superfluid regime for fermionic atoms. Namely, the interactions can be made strongly attractive, i.e. $k_F|a| \gg 1$, which leads to a critical temperature comparable to that of an atomic Bose gas. This objective was ultimately achieved in a series of ground-breaking experiments by the group of Debbie Jin using ^{40}K [18] and the groups of Wolfgang Ketterle, John Thomas, Rudi Grimm, Christophe Salomon, and Randy Hulet using ^{6}Li [19, 20, 21, 22, 23]. A number of these experiments exploit the Feshbach resonance to its fullest by also using it to actually observe the Bose-Einstein condensate of Cooper pairs.

To understand the latter better, we realize that there is an intimate connection between the Bose-Einstein condensation of bosons and the Bose-Einstein condensation of loosely-bound fermionic Cooper pairs. Note that the first is responsible for the superfluidity of weakly-interacting Bose gases, while the latter is responsible for both the superfluidity of weakly-interacting Fermi gases and the superconductivity of metals. The connection between the two condensates might have already been anticipated from the fact that the critical temperatures in (1.1) and (1.2) are very similar in the strongly-interacting limit $k_F|a| \gg 1$. Moreover, the atomic Bose-Einstein condensation experiments make use of alkali atoms, which are hydrogen-like composite bosons that can be seen as an outer electron bond to an inner core consisting of the fermionic nucleus and a surrounding electron cloud with an even number of electrons. As a result, a condensate of bosonic atoms can also be seen as a Bose-Einstein condensate of tightly-bound fermion pairs. We may thus conclude that fermionic superconductivity and bosonic superfluidity are in fact two sides of the same coin, differing only in the strength of the attraction between the fermions. If the attraction is weak, the Cooper pairs are very weakly bound and their size is much larger than the average interparticle distance $n^{-1/3}$, which is also called the superconductivity or BCS limit. However, in the superfluidity or BEC limit, the attraction is strong and the pairs are much smaller than the average interparticle distance, such that they act as composite bosons.

With an atomic Fermi gas near a Feshbach resonance, we can now for the first time experimentally explore both sides of the coin in one and the same system, i.e. study the full physics of the BEC-BCS crossover as first envisaged by David Eagles and Tony Leggett [24, 25]. By changing the magnetic field we can go from a large positive detuning above the Feshbach resonance, in which case we have no stable molecular state and a weakly-attractive interaction, to a large negative detuning below the Feshbach resonance, in which case there exists a deeply bound molecular state. In this manner, we thus evolve from a condensate of loosely bound Cooper pairs to a condensate of tightly bound molecules. The evolution between these two extremes turns out to be a smooth crossover, such that the transition between diatomic molecules and Cooper pairs is continuous. As mentioned above, this feature has been used to detect the Bose-Einstein condensate of Cooper pairs, by conveniently converting it into a Bose-Einstein condensate of tightly bound bosonic molecules with the use of an adiabatic magnetic-field sweep across the Feshbach resonance. The reason why it is so easy to observe a condensate of ideal bosons can be readily understood from their macroscopic occupation of the same one-particle ground state, which has a minimal kinetic energy. As a result, the atoms or diatomic molecules hardly spread out upon releasing the condensate from the trap, which leads to a very distinct peak in the velocity distribution at low velocities. The first atomic Bose-Einstein condensate was observed [10] in exactly the same manner.

Presently, there are many exciting directions that are being explored with ultracold atomic gases. First of all, we remark that the fermionic atoms that form the pairs in the BEC-BCS crossover must have two different spin states due to the Pauli principle. As a result, this crossover physics is usually studied in a balanced Fermi mixture with an equal number of atoms in each of the two different spin states. At the moment, a hot topic is to explore what precisely happens to the gas when the Fermi mixture becomes imbalanced, so that it is impossible for all the atoms to pair up simultaneously. Understanding this problem may also shed light on the physics in the core of a neutron star, where an imbalanced mixture of free quarks with attractive interactions can exist. These quarks may then form what is known as a color superconductor [26].

Another important direction is associated with the possibility of creating an intense standing wave of light with counter-propagating laser beams. This gives rise to a periodic potential for the atoms due to the Stark effect, which is also known as an optical lattice [27]. These optical lattices are very interesting for various reasons. An important one is that they can be used to simulate ionic lattices, which offers the opportunity to explore various aspects of solid-state physics in the very controlled environment of ultracold atoms. A particularly exciting possibility in this respect is to study systematically the microscopic models that have been proposed to govern high-temperature superconductors. Moreover, optical lattices can also be used to create low-dimensional atomic gases. In particular, with a very deep two-dimensional optical lattice we can make a two-dimensional array of one-dimensional gases, whereas a one-dimensional optical lattice creates a one-dimensional stack of two-dimensional systems. Low-dimensional quantum gases are interesting, because they often give rise to intriguing strongly-correlated behavior that is very different from the three-dimensional case. In various cases, low-dimensional many-body systems even allow for exact theoretical solutions.

Also of much interest in current research is the use of Feshbach resonances between two different atomic species to create ultracold heteronuclear molecules. These kind of molecules can have a large electronic dipole moment, which leads to a strong anisotropic dipole-dipole interaction. Since this interaction has a long-range nature it can possibly be used in combination with an optical lattice to create a new kind of superfluid, first proposed by Geoffrey Chester in 1970 [28], called a supersolid. This unusual new state of matter, which shares the properties of both a solid and a superfluid, has recently drawn a lot of attention in the context of solid ^{4}He. However, these experiments appear to be inconclusive at present [29], such that ultracold atomic gases may be a better system to explore this intriguing possibility [30]. To conclude, we remark that many other directions are being explored at present, leading us to believe that ultracold atomic quantum gases will remain an exciting area of physics for many years to come.

1.2 Outline

To facilitate the use of this book, we end this introduction by presenting a short overview of its contents. The book is divided into three parts that to a large extent can be read independently of each other. Part I contains the introductory material that is necessary for understanding the formulation of the functional-integral approach to quantum many-body physics, which is the method of choice for most condensed-matter theorists active in research today. Part II is then the core of the book, where the functional formalism is constructed, developed and used to discuss the canonical topics of superfluidity in interacting Bose and Fermi gases. In Part III, we discuss various more recent applications of the many-body techniques that are developed in Part II in order to explain important experiments that have recently been performed in the field of ultracold quantum gases.

1.2.1 Part One

Part I consists of Chaps. 2 to 6. We start in Chap. 2 with the mathematical foundations that are needed to follow the calculations in the rest of the book. A consequence of using the functional-integral approach to quantum field theory is that we very often have to perform integrations over infinitely many variables. Most frequently used is the Gaussian integral, because it is one of the few functional integrals that we can solve exactly. In Chap. 2 we therefore discuss Gaussian integration over an arbitrary number of variables, where we not only consider real variables, but also complex variables and Grassmann variables. These last kind of variables change sign upon permutation, which is very convenient when describing indistinguishable fermions, whose antisymmetric behavior leads to precisely the same property. In Chap. 3, we briefly review the basics of quantum mechanics that are relevant to our purposes. In particular, we discuss the exact solution to the harmonic oscillator problem, which is important for two reasons. First of all, in order to perform experiments on ultracold atomic gases these gases are always trapped in space, and the trapping potential is typically well approximated by a harmonic potential. Second, the interacting many-body system described with quantum field theory turns out to be equivalent to an infinite number of interacting harmonic oscillators. As a result, we can already introduce various important concepts in the familiar setting of a single harmonic oscillator, where later these concepts are generalized to the more abstract language of quantum field theory. Examples are the coherent states and the use of perturbation theory, whose generalization is a way to describe interaction effects in many-body systems. Finally, we also consider some aspects of scattering theory, because for ultracold atomic gases it turns out that we can calculate many-body interaction effects from first principles if we know the two-atom scattering properties at low kinetic energy.

Chap. 4 is devoted to statistical physics. Since interacting quantum gases typically consist of at least millions of particles, an exact treatment of all microscopic degrees of freedom is unfeasible. However, we are usually only interested in the averaged macroscopic quantities, whose description actually becomes more convenient as the number of particles increases. This is the domain of statistical physics, which also tells us how to deal with the effects of thermal fluctuations. Since experiments with ultracold atomic gases are never performed at exactly zero temperature, it is usually not sufficient to consider only the many-body ground state. We then find from statistical physics that all macroscopic quantities can be directly obtained from the partition function of the gas, such that the main challenge of a many-body theoretical physicist is to determine this quantity in a sufficiently accurate approximation. For the ideal Bose and Fermi gases, this quantity can be computed exactly, and we find that these two systems behave very differently at low temperatures. In particular, the ideal Bose gas undergoes a phase transition to a new state of matter called a Bose-Einstein condensate, as was already mentioned in the discussion of (1.1). The precise knowledge of the noninteracting quantum gases is then a good starting point to discuss the effects of interactions, which is treated in the second part of the book.

In Chap. 5, we discuss quantum mechanics using Feynman's path-integral approach, which is rather different from the more familiar operator formalism of Chap. 3. Path integrals turn out to be very well suited for a generalization to quantum field theory, such that a thorough knowledge of them is very useful to fully understand all the calculations in the later chapters. Many subtleties of the functional-integral formalism already show up in this chapter, where we also immediately show how to deal with them. In particular, we derive the path-integral expression for the partition function of a single trapped atom. To also be able to derive the functional integral for the partition function of an interacting many-body system, we need to reformulate the quantum mechanics of a many-body system in a somewhat more convenient way. This is achieved in Chap. 6 via a procedure which is known as second quantization. In the second-quantized approach to many-body quantum theory, the particles are represented by creation and annihilation operators, which are conveniently constructed such that they automatically incorporate the quantum statistics of the particles. The eigenstates of these annihilation operators are called coherent states, and are the final ingredient needed to derive the functional formulation of quantum field theory.

1.2.2 Part Two

This is then achieved in Part II of the book, which consists of Chaps. 7 to 14. Part II forms the core of the book, in which we develop all the functional tools in quantum field theory that are needed to understand the equilibrium properties of ultracold atomic quantum gases. In fact, the introduced methods have a much wider range of applicability, such that they can actually be used as a starting point to tackle any quantum condensed-matter problem. However, in order to keep the book coherent, most applications we discuss are from the field of ultracold atoms. A reader with a good undergraduate education in quantum mechanics and statistical physics can most likely enter the discussion here, after a quick study of Gaussian integrals and Grassmann variables in Chap. 2 and the second-quantization formalism in Chap. 6. Part II starts off with Chap. 7, in which we derive the functional integral for the partition function of an interacting many-body system. We also reconsider the ideal quantum gases, for which the partition function reduces to a Gaussian functional integral such that it can be calculated exactly. We perform this calculation in three different ways to familiarize ourselves with functional integration, and to introduce various concepts that come back time after time throughout the rest of the book.

In Chap. 8 we discuss the effects of interactions between the particles, which in general leave the partition function unsolvable, such that we have to resort to appropriate approximation methods. A first way to systematically study interaction effects is by performing a perturbative expansion in the interaction. The general structure of the resulting perturbation theory is then very conveniently visualized with the use of Feynman diagrams. We also explain several features of the expansion that are valid up to any order in the interaction strength, and that are therefore especially useful for arriving at accurate approximations. In particular, we discuss the famous Hartree-Fock approximation, which is a selfconsistent approximation that sums an infinite number of Feynman diagrams and is used very often in condensed-matter physics to obtain a first understanding of the importance of interaction effects. We derive the Hartree-Fock theory by using a variational approach and by using a Hubbard-Stratonovich transformation. This exact transformation turns out to be a very versatile and powerful tool which comes back in many different guises throughout the book.

In Chap. 9 we discuss the Landau theory of phase transitions, where an important concept is the order parameter. This is the observable that distinguishes the two phases involved in the phase transition by quantifying the occurrence of order in the system. We then show that a nonzero value of the order parameter is often associated with a spontaneous breakdown of symmetry, which means that an equilibrium state of the system has less symmetry then the underlying microscopic Hamiltonian. The usefulness of the Hubbard-Stratonovich transformation introduced in the previous chapter becomes particularly obvious in the context of phase transitions, because it can be used to bring the order parameter exactly into the many-body theory. Moreover, we show that fluctuations of the order parameter field can become crucial close to the phase transition, such that they can even cause a breakdown of Landau theory. To go beyond Landau theory turns out to be an exceedingly difficult task and requires advanced field-theoretical methods to which we return in Chap. 14. In order to reach our goal of obtaining an ab initio microscopic description for the phase transition to the superfluid state in interacting atomic Bose and Fermi gases, we still need to understand some specific properties of the alkali atoms that are involved in the actual experiments. In particular, the spin structure of the atoms is important, because it affects the scattering properties of two atoms, where the resulting interaction strength is an input parameter for the quantum field theory of the trapped atomic quantum gas. Chap. 10 deals in more detail with both the spin structure and the scattering of atoms.

In Chap. 11 we apply the developed field-theorical machinery to discuss the famous Bogoliubov and Popov theories of Bose-Einstein condensation, leading, amongst others, to the equally famous Gross-Pitaevskii equation for the condensate wavefunction. The Bogoliubov theory is only valid for temperatures close to zero Kelvin, while the range of validity for the Popov theory is larger, because it takes into account fluctuation effects in a similar manner to the Hartree-Fock theory discussed in Chap. 8. The historically most important success of the Bogoliubov theory was the correct prediction for the vibrational eigenfrequencies of a fully Bose-Einstein condensed atomic cloud. In view of this success, we discuss these collective modes in some detail using a hydrodynamic-like approach. We also briefly discuss what happens when we try to bring the Bose-Einstein condensed gas into rotation, which leads to interesting properties due to the superfluid nature of the gas. Finally, we show that a condensate with effectively attractive interatomic interactions is metastable and ultimately collapses into a Bosenova. In Chap. 12, the Bose-Einstein condensation of Cooper pairs in an ultracold Fermi gas is discussed. In particular, we show how a Hubbard-Stratonovich transformation introduces the appropriate order parameter of the phase transition into the theory. This order parameter describes the condensate of Cooper pairs, which means that the superfluity of an atomic Fermi gas has the same physical origin as the superconductivity of metals. We also derive the critical temperature for the transition in mean-field theory, the result already announced in (1.2). Finally, we also give a more detailed discussion of the BEC-BCS crossover taking place in an atomic gas near a Feshbach resonance.

After having discussed these two explicit examples of phase transitions, we are ready for a more general discussion of the consequences of symmetries and symmetry breaking in quantum field theory. This is the topic of Chap. 13, which has a somewhat more formal nature than the two earlier chapters. However, its results are of much importance to practical calculations. We remember that in order to compare theory with experiments, we usually have to make approximations, because interacting quantum field theories are often too difficult to solve exactly. Obviously, we want to arrive at approximations that do not violate the underlying symmetries of the theory, which is particularly important in the discussion of phase transitions. This is because we need the corresponding symmetry breaking to occur spontaneously and not by the approximation that we make. It turns out that it is possible to derive identities, known as the Ward identities, that check if our approximations still preserve the underlying symmetries. We give a few explicit examples of these Ward identities, and discuss how they can be used in the calculation of certain directly measurable quantities in experiments. Another fundamental issue that we touch upon is the fact that spontaneous symmetry breaking can formally only occur for systems with an infinite number of particles, while realistic experiments always deal with a finite number of particles. We discuss how these two facts can be reconciled with each other for the specific case of superfluid atomic Bose and Fermi gases by discussing the phenomenon of phase diffusion.

In Chap. 14, we go beyond the Landau theory of phase transitions. This is necessary when critical fluctuations extend over the whole many-body system, giving rise to critical phenomena. Since the critical fluctuations now dominate at each length scale, the system is actually scale invariant, which we can use to describe it recursively at increasing wavelengths. This leads to the renormalization group theory of critical many-body systems which, amongst other results, provides the explanation for universality, i.e. the remarkable observation that very different microscopic systems have identical critical properties. We also apply the renormalization group approach to the imbalanced Fermi gas in the strongly-interacting regime, where we can compare the resulting homogeneous phase diagram with beautiful experimental results that were obtained recently.

1.2.3 Part Three

The last three chapters, 15 to 17, form Part III of the book, in which the functional formalism is applied to various recent topics in ultracold atomic gases. In Chap. 15 we discuss low-dimensional, i.e. one and two-dimensional, atomic Bose gases. An important challenge in this chapter is caused by the breakdown of Popov theory due to the enhanced importance of fluctuations in low dimensions. It is then explained in detail how the Popov theory can be modified in order to resolve these problems and, in particular, to describe the famous Kosterlitz-Thouless phase transition in two dimensions. The low-dimensional atomic gases are experimentally realizable with the use of optical lattices, which are the topic of Chap. 16. These lattices also give rise to interesting new physics in three dimensions, because they can be used to simulate solid-state-like periodic potentials, where the depth of the periodic potential is now tunable by varying the laser intensity. As a result, if a shallow optical lattice is loaded with a Bose-Einstein condensate of bosonic atoms, the superfluidity can be destroyed by increasing the lattice depth, which then leads to the Mott-insulator state with precisely one trapped atom at each lattice site. This phase transition happens at zero temperature, and is thus an example of a quantum phase transition, which was observed by Greiner et al. in 2002 [31]. The same experiment can also be performed with an ultracold Fermi mixture, which leads to the possibility of observing the Néel state, and hopefully eventually to new insights into high-temperature superconductors. Finally, we end the book in Chap. 17 with the theory for Feshbach resonances, which now have many important applications in ultracold atomic physics experiments. We start with the two-body atomic physics that causes the resonance, after which we also explain how this two-body physics can be accurately captured in a quantum field theory of atoms and molecules. As an application, we finally consider the coherent Josephson oscillations between a Bose-Einstein condensate of atoms and a Bose-Einstein condensate of molecules, where we also compare the results with some beautiful quantum-mechanical interference experiments that are the ultracold atomic analog of the neutrino oscillations known from high-energy physics.

Part I

Chapter 2
Gaussian Integrals

We must admit with humility that, while number is purely a product of our minds, space has a reality outside our minds, so that we cannot completely prescribe its properties a priori.
– Carl Friedrich Gauss.

In this chapter, we lay the mathematical foundations for the functional-integral formalism that we develop in later chapters. We start with introducing the Gaussian probability distribution together with the corresponding integrals over this distribution, called Gaussian integrals. These concepts are then generalized to higher dimensions, to the complex plane, and to what are called Grassmann variables. The multidimensional Gaussian integral is of great importance for the rest of this book. In Chap. 7, we show that it leads to an exact solution of noninteracting quantum gases, which then also forms the basis for a perturbative description of interacting quantum gases. The goal of this chapter is to highlight the practical use of several important mathematical results that are needed to understand the rest of the book. The chapter is not intended to be a full mathematical account of all the above-mentioned topics, meaning that proofs will often be omitted or replaced by illustrative examples. The more experienced reader who is already familiar with Gaussian integrals, complex analysis, and Grassmann algebras, can use this chapter for reference.

2.1 The Gaussian Integral over Real Variables

The Gaussian or normal probability distribution is the most common distribution in statistical physics. The main reason for this is that the probability distribution for the sum of N independent random variables, each with a finite variance, converges for large N to the Gaussian distribution. This is called the central limit theorem of probability theory. Famous physical examples of Gaussian distributions are the Maxwell distribution for the velocities of the atoms in a classical ideal gas, or the spatial distribution for an atom in the quantum-mechanical ground state of a harmonic trap. The Gaussian probability distribution is given by

$$P(x) = \sqrt{\frac{\alpha}{\pi}} \exp\left\{-\alpha x^2\right\}, \tag{2.1}$$

such that it is properly normalized to 1. This follows from

$$\int_{-\infty}^{+\infty} \mathrm{d}x\, \mathrm{e}^{-\alpha x^2} = \sqrt{\frac{\pi}{\alpha}}, \tag{2.2}$$

which is left as an exercise to the reader. The probability distribution of (2.1) has a maximum at $x = 0$, whereas in general the maximum could be at any arbitrary value x_0. Then, we have

$$P(x) = \sqrt{\frac{\alpha}{\pi}} \exp\{-\alpha(x-x_0)^2\}, \tag{2.3}$$

which corresponds, for example, to the probability distribution of the velocities in a thermal beam of atoms which is travelling at an average velocity x_0. The latter distribution has the property that the expectation value of the quantity x is equal to x_0, that is

$$\langle x \rangle \equiv \int_{-\infty}^{+\infty} \mathrm{d}x\, x\, P(x) = \sqrt{\frac{\alpha}{\pi}} \int_{-\infty}^{+\infty} \mathrm{d}x\, x\, \exp\{-\alpha(x-x_0)^2\} = x_0, \tag{2.4}$$

which is easily proven by performing the shift $x \to x + x_0$.

For our purposes, it is convenient to write the parameter α as $-G^{-1}/2 = -1/2G$, with $G < 0$. In the first instance, this looks overly complicated. However, it establishes a direct link with the notation used in later chapters for the Green's function in the functional-integral formalism. From now on, we also no longer explicitly denote the lower and upper limit of the integration when these are given by $-\infty$ and $+\infty$, respectively. With these changes, the Gaussian integral can be written as

$$\int \mathrm{d}x\, \exp\left\{\frac{1}{2}G^{-1}x^2\right\} = \sqrt{-2\pi G} = \sqrt{2\pi} \exp\left\{-\frac{1}{2}\log\left(-G^{-1}\right)\right\}. \tag{2.5}$$

2.1.1 Generating Function

By including a linear term Jx in the exponent, we introduce the generating function $Z(J)$ of the probability distribution. This is very useful because it allows us to calculate the expectation value of all the higher moments, i.e. the expectation values of x^n, by simply differentiating with respect to the current J. Specifically, we have for the Gaussian distribution

$$\begin{aligned} Z(J) &= \int \frac{\mathrm{d}x}{\sqrt{2\pi}} \exp\left\{\frac{1}{2}G^{-1}(x-x_0)^2 + Jx\right\} \\ &= \int \frac{\mathrm{d}x}{\sqrt{2\pi}} \exp\left\{\frac{1}{2}G^{-1}(x+GJ)^2 - \frac{1}{2}GJ^2 + Jx_0\right\} \\ &= \exp\left\{-\frac{1}{2}GJ^2 + Jx_0 - \frac{1}{2}\log\left(-G^{-1}\right)\right\}, \end{aligned} \tag{2.6}$$

where in the first step we performed the shift $x \to x + x_0$ before completing the square. Note that the additional factor $1/\sqrt{2\pi}$ conveniently cancels the factor $\sqrt{2\pi}$ coming from the Gaussian integral. The expectation value of x is now readily calculated from

$$\langle x \rangle = \frac{1}{Z(J)} \frac{\mathrm{d}}{\mathrm{d}J} Z(J) \bigg|_{J=0} = x_0, \tag{2.7}$$

and for $\langle x^2 \rangle$, we obtain

$$\langle x^2 \rangle = \frac{1}{Z(J)} \frac{\mathrm{d}^2}{\mathrm{d}J^2} Z(J) \bigg|_{J=0} = -G + x_0^2 = -G + \langle x \rangle^2. \tag{2.8}$$

Since we can always perform initially the shift $x \to x + x_0$, we consider from now on without loss of generality the case with $x_0 = 0$. A useful observation is that this leads to

$$\langle x^{2m+1} \rangle = 0, \tag{2.9}$$

where m is an integer. This is because the integrand of the integral

$$\int \mathrm{d}x\, x^{2m+1} \exp\left\{ \frac{1}{2} G^{-1} x^2 \right\}$$

is odd and the integral vanishes consequently. By repeatedly applying the derivative $\mathrm{d}/\mathrm{d}J$ an even number of times to the first line of (2.6) with $x_0 = 0$, we find that

$$\langle x^{2m} \rangle = \frac{1}{Z(J)} \frac{\mathrm{d}^{2m}}{\mathrm{d}J^{2m}} Z(J) \bigg|_{J=0}. \tag{2.10}$$

Explicitly calculating the right-hand side of (2.10), using the expression in the last line of (2.6), generates a large number of terms that vanish when we eventually take the limit $J \to 0$. To simplify the calculation, it is therefore convenient to realize that if we expand $Z(J)$ in powers of J only the terms proportional to J^{2m} contribute. In this manner, we find for $x_0 = 0$ that

$$\begin{aligned} \frac{1}{Z(J)} \frac{\mathrm{d}^{2m}}{\mathrm{d}J^{2m}} Z(J) \bigg|_{J=0} &= \frac{Z(0)}{Z(J)} \frac{\mathrm{d}^{2m}}{\mathrm{d}J^{2m}} \sum_{n=0}^{\infty} \frac{1}{n!} \left(-\frac{1}{2} G J^2 \right)^n \bigg|_{J=0} \\ &= \frac{(2m)!}{2^m m!} (-G)^m = (2m-1)!!(-G)^m, \end{aligned} \tag{2.11}$$

where $(2m-1)!! = (2m-1)(2m-3)(2m-5)\ldots 1$. Hence, we conclude that

$$\langle x^{2m} \rangle = (2m-1)!!(-G)^m. \tag{2.12}$$

It is important to realize that $(2m-1)!!$ is exactly the number of ways in which $2m$ numbers can be divided into m pairs. Thus, we have found that the expectation value of x^{2m} is equal to the sum of all possible ways in which $\langle x^{2m} \rangle$ can be factorized as

$\langle x^2 \rangle^m$. This last statement is the essence of the famous Wick's theorem that will turn out to be of great importance in later chapters.

2.1.2 Multi-Dimensional Gaussian Integral

The previous results can be immediately generalized to higher-dimensional integrals. Consider a diagonal $n \times n$ matrix $\mathbf{G}$,

$$\mathbf{G} = \begin{bmatrix} G_{11} & & & \\ & G_{22} & & \\ & & G_{33} & \\ & & & \ddots \end{bmatrix}, \tag{2.13}$$

with again $G_{jj} < 0$. Then, the inverse $\mathbf{G}^{-1}$ of $\mathbf{G}$ is clearly given by

$$\mathbf{G}^{-1} = \begin{bmatrix} \frac{1}{G_{11}} & & & \\ & \frac{1}{G_{22}} & & \\ & & \frac{1}{G_{33}} & \\ & & & \ddots \end{bmatrix}. \tag{2.14}$$

We want to evaluate the Gaussian integral

$$\int \left(\prod_{j=1}^{n} \mathrm{d}x_j \right) \exp\left\{ \frac{1}{2} \mathbf{x} \cdot \mathbf{G}^{-1} \cdot \mathbf{x} \right\} \equiv \int \mathrm{d}\mathbf{x} \exp\left\{ \frac{1}{2} \mathbf{x} \cdot \mathbf{G}^{-1} \cdot \mathbf{x} \right\}, \tag{2.15}$$

where $\mathbf{x}$ denotes the vector $(x_1, x_2, \ldots, x_n)$. Because the integral factorizes into a product of n one-dimensional integrals, we find that

$$\int \mathrm{d}\mathbf{x} \exp\left\{ \frac{1}{2} \mathbf{x} \cdot \mathbf{G}^{-1} \cdot \mathbf{x} \right\} = \frac{(2\pi)^{n/2}}{\sqrt{\prod_{j=1}^{n} (-G_{jj}^{-1})}} = \frac{(2\pi)^{n/2}}{\sqrt{\mathrm{Det}[-\mathbf{G}^{-1}]}}, \tag{2.16}$$

where $\mathrm{Det}[-\mathbf{G}^{-1}]$ denotes the determinant of the matrix $-\mathbf{G}^{-1}$. In the same way we find that (2.6) generalizes to

$$\begin{aligned} Z(\mathbf{J}) &= \int \frac{\mathrm{d}\mathbf{x}}{\sqrt{(2\pi)^n}} \exp\left\{ \frac{1}{2} \mathbf{x} \cdot \mathbf{G}^{-1} \cdot \mathbf{x} + \mathbf{J} \cdot \mathbf{x} \right\} \\ &= \exp\left\{ -\frac{1}{2} \mathbf{J} \cdot \mathbf{G} \cdot \mathbf{J} - \frac{1}{2} \mathrm{Tr}[\log(-\mathbf{G}^{-1})] \right\}, \end{aligned} \tag{2.17}$$

where we have taken again without loss of generality $\mathbf{x}_0 = \mathbf{0}$. Here $\mathrm{Tr}[\ldots]$ denotes the trace of a matrix, which is the sum of all diagonal elements. The n-th order correlation function $\langle x_{j_1} x_{j_2} \dots x_{j_n} \rangle$, given by the expectation value of the product of n coordinates x_j, is now easily calculated from

$$\langle x_{j_1} \dots x_{j_n} \dots \rangle = \frac{1}{Z(\mathbf{J})} \frac{\partial^n}{\partial J_{j_1} \dots \partial J_{j_n}} Z(\mathbf{J}) \bigg|_{\mathbf{J}=0}. \tag{2.18}$$

Example 2.1. Because $Z(\mathbf{J})$ depends quadratically on $\mathbf{J}$, it immediately follows that

$$\langle x_i \rangle = \frac{1}{Z(\mathbf{J})} \frac{\partial}{\partial J_i} Z(\mathbf{J}) \bigg|_{\mathbf{J}=0} = 0. \tag{2.19}$$

For the expectation value $\langle x_i x_j \rangle$, we find

$$\langle x_i x_j \rangle = \frac{1}{Z(\mathbf{J})} \frac{\partial^2}{\partial J_i \partial J_j} Z(\mathbf{J}) \bigg|_{\mathbf{J}=0} = -G_{ij}. \tag{2.20}$$

The above results were obtained for the specific case of a diagonal matrix. However, (2.17) is valid for any positive definite, symmetric matrix $-\mathbf{G}^{-1}$, where positive definite means that the matrix has only positive eigenvalues. First, note that $-\mathbf{G}^{-1}$ can always be assumed to be symmetric, because any antisymmetric part would give a vanishing contribution to the term $-\mathbf{x} \cdot \mathbf{G}^{-1} \cdot \mathbf{x}$. Then, a symmetric matrix can always be brought into diagonal form by a similarity transformation $\mathbf{S}$, which means that $\mathbf{S} \cdot \mathbf{G}^{-1} \cdot \mathbf{S}^{-1}$ is diagonal and $\mathbf{S}$ is orthonormal. Orthonormality implies that

$$|\mathrm{Det}[\mathbf{S}]| = 1, \tag{2.21}$$

such that the Jacobian of the coordinate transformation $\mathbf{x} = \mathbf{S}^{-1} \cdot \mathbf{x}'$ is equal to one. Applying the above considerations to (2.17), we have

$$\begin{aligned} Z(\mathbf{J}) &= \int \frac{d\mathbf{x}'}{\sqrt{(2\pi)^n}} \exp\left\{\frac{1}{2}\mathbf{x}' \cdot \mathbf{S} \cdot \mathbf{G}^{-1} \cdot \mathbf{S}^{-1} \cdot \mathbf{x}' + \mathbf{J} \cdot \mathbf{S}^{-1} \cdot \mathbf{x}'\right\} \\ &= \exp\left\{-\frac{1}{2}\mathbf{J} \cdot \mathbf{S}^{-1} \cdot \mathbf{S} \cdot \mathbf{G} \cdot \mathbf{S}^{-1} \cdot \mathbf{S} \cdot \mathbf{J}\right\} \frac{1}{\sqrt{\mathrm{Det}[-\mathbf{S} \cdot \mathbf{G}^{-1} \cdot \mathbf{S}^{-1}]}} \\ &= \exp\left\{-\frac{1}{2}\mathbf{J} \cdot \mathbf{G} \cdot \mathbf{J}\right\} \frac{1}{\sqrt{\mathrm{Det}[-\mathbf{G}^{-1}]}}, \end{aligned} \tag{2.22}$$

where we also used the property that for an orthogonal matrix the inverse matrix and the transposed matrix are the same. Thus, we find that (2.17) is valid for any positive definite matrix $-\mathbf{G}^{-1}$.

2.2 Complex Analysis

In the following, we generalize the results of the previous paragraph to Gaussian integrals over n complex variables z_j. Before doing so, we first review some concepts from elementary complex analysis. The complex plane is a two-dimensional linear space, meaning that any number in the complex plane can be written as $x+\mathrm{i}y$, where x and y are real. Instead of using x and y as the independent variables to parametrize the complex plane, it is more convenient for our purposes to make a coordinate transformation that maps x and y onto the independent variables z and z^* in the following way

$$z = x+\mathrm{i}y \qquad \text{and} \qquad z^* = x-\mathrm{i}y. \tag{2.23}$$

Here, $|z|^2 = z^* z = x^2 + y^2$ gives the square of the modulus of z, while the real and imaginary parts of z are given by $\mathrm{Re}[z] = (z+z^*)/2$ and $\mathrm{Im}[z] = (z-z^*)/2\mathrm{i}$. Instead of using the Cartesian coordinates x and y, it is also possible to introduce polar coordinates. In that case, complex numbers are written as

$$z = r\mathrm{e}^{\mathrm{i}\varphi}, \tag{2.24}$$

where $r = \sqrt{z^* z}$ is the complex modulus and $\varphi = \mathrm{Arg}[z]$ is the complex argument.

2.2.1 *Differentiation and Contour Integrals*

A general complex function $f(x,y)$ is a map from the complex plane to the complex plane and in general depends explicitly on both z and z^*. We write $f(x,y) = u(x,y) + \mathrm{i}v(x,y)$, where $u(x,y) = \mathrm{Re}[f(x,y)]$ and $v(x,y) = \mathrm{Im}[f(x,y)]$. In practise we will be dealing mostly with analytic functions, which turn out to depend only explicitly on $z = x+\mathrm{i}y$. Because such a function $f(x+\mathrm{i}y)$ or $f(z)$ only depends on z, we must have that $\mathrm{d}f/\mathrm{d}z = \partial f/\partial x = -\mathrm{i}\partial f/\partial y$ for an analytic function. Since

$$\frac{\partial f(x,y)}{\partial x} = \frac{\partial u(x,y)}{\partial x} + \mathrm{i}\frac{\partial v(x,y)}{\partial x} \tag{2.25}$$

and

$$-\mathrm{i}\frac{\partial f(x,y)}{\partial y} = -\mathrm{i}\frac{\partial u(x,y)}{\partial y} + \frac{\partial v(x,y)}{\partial y}, \tag{2.26}$$

we have that the functions u and v are not independent, but rather satisfy the following set of equations

$$\frac{\partial u(x,y)}{\partial x} = \frac{\partial v(x,y)}{\partial y} \qquad \text{and} \qquad \frac{\partial u(x,y)}{\partial y} = -\frac{\partial v(x,y)}{\partial x}. \tag{2.27}$$

These equations are known as the Cauchy-Riemann equations and satisfying them assures differentiability of the complex function.

Example 2.2. To explicitly demonstrate the use of the Cauchy-Riemann equations, we test for two simple complex functions whether or not they are analytic functions. First, consider the complex function $f(x,y) = x + \mathrm{i}y = z$, i.e. $u(x,y) = x$ and $v(x,y) = y$. Clearly it satisfies the Cauchy-Riemann equations, since $\partial u/\partial x = 1 = \partial v/\partial y$ and $\partial u/\partial y = 0 = -\partial v/\partial x$. However, the complex conjugate $f(x,y) = x - \mathrm{i}y = z^*$ is not analytic, because it does not satisfy the Cauchy-Riemann equations. Indeed, we have $\partial u/\partial x = 1 \neq -1 = \partial v/\partial y$. This illustrates the above statement that functions depending explicitly on z^* are not analytic.

Besides being able to take the derivative of a complex function we also want to be able to integrate it. In principle, the integral of a general complex function between two points in the complex plane depends on the specific path taken. However, if the function $f(z)$ satisfies the Cauchy-Riemann equations in all points enclosed by two different paths connecting z_i and z_f, then the integral $\int_{z_\mathrm{i}}^{z_\mathrm{f}} \mathrm{d}z f(z)$ gives the same result for each of the two paths. This leads directly to the Cauchy-Goursat theorem, stating that for any function f which is analytic on a closed contour C and at all points inside the contour, the integral along the contour vanishes, i.e.

$$\oint_C \mathrm{d}z\, f(z) = 0. \tag{2.28}$$

We will not prove this theorem here, but we give a simple demonstration in Example 2.3. With the Cauchy-Goursat theorem, it is then possible to prove the important Cauchy integral formula

$$f(z_0) = \frac{1}{2\pi\mathrm{i}} \oint_C \mathrm{d}z \frac{f(z)}{z - z_0}, \tag{2.29}$$

where the integration over the contour C is in a counterclockwise fashion. This will always be the convention for contour integration from now on.

Example 2.3. Consider the function $f(z) = z$, and let the contour C be the circle centered at $z = 0$ with radius R. The circle is parameterized by $z = R\mathrm{e}^{\mathrm{i}\phi}$, where φ runs counterclockwise from 0 to 2π. Hence, $\mathrm{d}z = \mathrm{i}R\mathrm{e}^{\mathrm{i}\varphi}\mathrm{d}\varphi$ and we find

$$\oint_C \mathrm{d}z\, z = \int_0^{2\pi} \mathrm{d}\varphi\, \mathrm{i}R^2\mathrm{e}^{2\mathrm{i}\varphi} = \frac{R^2}{2}\mathrm{e}^{2\mathrm{i}\varphi}\Big|_0^{2\pi} = 0. \tag{2.30}$$

This illustrates the Cauchy-Goursat theorem. Consider in (2.29) the function $f(z) = 1$ and take for the contour C the circle based at z_0 with radius R. We obtain

$$\frac{1}{2\pi \mathrm{i}}\oint_C \mathrm{d}z\, \frac{f(z)}{z-z_0} = \frac{1}{2\pi \mathrm{i}}\int_0^{2\pi} \mathrm{d}\varphi\, \frac{\mathrm{i}R\mathrm{e}^{\mathrm{i}\varphi}}{R\mathrm{e}^{\mathrm{i}\varphi}} = 1 = f(z_0). \tag{2.31}$$

This illustrates the Cauchy integral formula.

The Cauchy integral formula can be used to express the derivatives of a complex function in terms of a contour integral. By differentiating both sides of (2.29) n times with respect to z_0, we obtain

$$\frac{\mathrm{d}^n}{\mathrm{d}z_0^n} f(z_0) = \frac{n!}{2\pi \mathrm{i}}\oint_C \mathrm{d}z \frac{f(z)}{(z-z_0)^{n+1}}. \tag{2.32}$$

2.2.2 Laurent Series and the Residue Theorem

For a function $f(x)$ that depends on the real variable x, it is possible to make a Taylor series expansion around the nonsingular point x_0, i.e.

$$f(x) = \sum_{n=0}^{\infty} \frac{f^{(n)}(x_0)}{n!}(x-x_0)^n, \tag{2.33}$$

where $f^{(n)}(x) = \mathrm{d}^n f(x)/\mathrm{d}x^n$ and where we assumed that the sum on the right-hand side converges. A similar expansion holds for complex functions that are analytic throughout the interior of a circle centered at z_0 with radius R. In that case, the function can be written as

$$f(z) = \sum_{n=0}^{\infty} \frac{f^{(n)}(z_0)}{n!}(z-z_0)^n. \tag{2.34}$$

Now, suppose that we have a function that is singular at a single point z_0 that lies within a circle centered at z_0 and with radius R_1, as is illustrated in Fig. 2.1. We call S the region enclosed by the circle excluding the singular point z_0. For each point z that lies within S, the function $f(z)$ is given by

$$f(z) = \sum_{n=-\infty}^{\infty} a_n (z-z_0)^n, \tag{2.35}$$

where the coefficients are given by

$$a_n = \frac{1}{2\pi \mathrm{i}}\oint_C \mathrm{d}z\, \frac{f(z)}{(z-z_0)^{n+1}}, \tag{2.36}$$

and C is any contour that encloses z_0 and lies within S. This series expansion is also known as the Laurent series expansion. The coefficient a_{-1}, which is the integral of

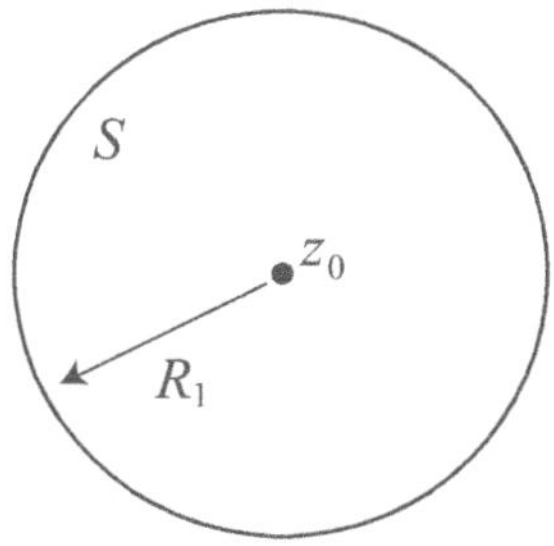

Fig. 2.1 Laurent series expansion. The region S is enclosed by a circle with radius R_1 that is centered at z_0, but excludes the point z_0 itself.

$f(z)$ along the contour C, is called the residue of f at the singular point z_0

$$a_{-1} = \frac{1}{2\pi \mathrm{i}} \oint_C f(z)\mathrm{d}z \equiv \mathrm{Res}[f(z_0)]. \tag{2.37}$$

For our purpose, analytic functions that have the following expansion in terms of a Laurent series are most relevant

$$f(z) = \sum_{n=-m}^{\infty} a_n (z - z_0)^n, \tag{2.38}$$

and the singularity at $z = z_0$ is called a pole of order m. For the residue, this leads to

$$\mathrm{Res}[f(z_0)] = \frac{1}{(m-1)!} \lim_{z \to z_0} \frac{\mathrm{d}^{m-1}}{\mathrm{d}z^{m-1}} \left((z - z_0)^m f(z)\right). \tag{2.39}$$

The concept of the residue allows for a generalization of Cauchy's integral formula of (2.29) to any contour enclosing a finite number of finite-order poles. This leads to the residue theorem, that is

$$\oint_C \mathrm{d}z\, f(z) = 2\pi \mathrm{i} \sum_j \mathrm{Res}[f(z_j)]. \tag{2.40}$$

Example 2.4. The function $f(z) = 1/((1 - \mathrm{i}z)(1 + \mathrm{i}z))$ is not analytic in $z = \pm \mathrm{i}$. To find the Laurent series expansion of $f(z)$ at $z = \mathrm{i}$, we start by writing

$$\frac{1}{1 + \mathrm{i}z} = -\mathrm{i}(z - \mathrm{i})^{-1}. \tag{2.41}$$

Moreover, the Taylor series of the term $1/(1 - \mathrm{i}z)$ is given by,

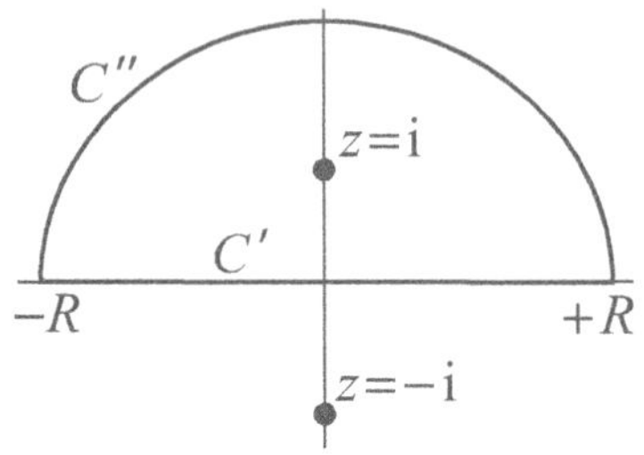

Fig. 2.2 Illustration of the contour $C = C' + C''$ used in Example 2.5.

$$\frac{1}{1-\mathrm{i}z} = \sum_{n=0}^{\infty} \frac{1}{2^{(n+1)}} \mathrm{i}^n (z-\mathrm{i})^n. \tag{2.42}$$

Multiplying the two terms gives the Laurent series of the function $f(z)$,

$$\begin{aligned} f(z) &= -\sum_{n=0}^{\infty} \left(\frac{\mathrm{i}}{2}\right)^{n+1} (z-\mathrm{i})^{n-1} \\ &= \sum_{n=-1}^{\infty} a_n (z-\mathrm{i})^n, \end{aligned} \tag{2.43}$$

where $a_n = -(\mathrm{i}/2)^{n+2}$. This shows that the residue is equal to $a_{-1} = -\mathrm{i}/2$.

Example 2.5. Suppose we want to calculate the following integral along the real axis

$$\int_{-\infty}^{\infty} \mathrm{d}x \, \frac{1}{1+x^2}.$$

This is a standard integral whose answer is known to be $\arctan x|_{-\infty}^{+\infty} = \pi$. Now we show how we can also obtain this result by making use of the residue theorem. We start by extending the function $f(x) = 1/(1+x^2)$ on the real axis to the function $f(z)$ in the complex plane, such that

$$f(z) = \frac{1}{(z+\mathrm{i})(z-\mathrm{i})}. \tag{2.44}$$

This function has poles in the complex plane at $z = \pm\mathrm{i}$. To be able to apply the residue theorem, we use the contour C shown in Fig. 2.2. It is the union of the line C', which is the part of the real axis from $-R$ to $+R$, and C'', which is the semicircle in the upper half-plane centered at zero with radius R. Our original integral can be obtained by taking the limit $R \to \infty$ and subtracting the integral along the path C''. Due to the residue theorem, we have that

$$\oint_C \mathrm{d}z \, f(z) = 2\pi\mathrm{i}\,\mathrm{Res}[f(z=\mathrm{i})] = \pi. \tag{2.45}$$

The contour integral along the semicircle C'' vanishes in the limit $R \to \infty$, since

$$\lim_{R\to\infty} \int_{C''} \mathrm{d}z \, \frac{1}{1+z^2} = \lim_{R\to\infty} \int_0^{\pi} \mathrm{d}\varphi \, \frac{\mathrm{i}}{R\mathrm{e}^{\mathrm{i}\varphi}} = 0. \tag{2.46}$$

As a result, we see that we retrieve the original answer, as desired.

2.3 Gaussian Integrals over Complex Variables

In this section, we generalize the results that we obtained for the Gaussian integral along the real axis in (2.5) to integrations over the complex plane. We assume that the complex number G^{-1} has a real part that is less than zero. We find

$$\begin{aligned}\int \mathrm{d}z^*\mathrm{d}z \, \exp\left\{G^{-1}z^*z\right\} &\equiv \int \frac{\partial(z^*,z)}{\partial(x,y)}\mathrm{d}x \, \mathrm{d}y \, \exp\left\{G^{-1}(x^2+y^2)\right\} \\ &= \int \mathrm{d}x \, \mathrm{d}y \, 2\mathrm{i}\exp\left\{G^{-1}x^2\right\}\exp\left\{G^{-1}y^2\right\} \\ &= -2\pi\mathrm{i}G, \end{aligned} \tag{2.47}$$

where the integral is over the full complex plane. The coordinate transformation of (2.23) that maps x and y onto z^* and z, also defines the measure $\mathrm{d}z^*\mathrm{d}z$ through the relation

$$\mathrm{d}z^*\mathrm{d}z \equiv \frac{\partial(z^*,z)}{\partial(x,y)}\mathrm{d}x \, \mathrm{d}y = 2\mathrm{i}\,\mathrm{d}x \, \mathrm{d}y, \tag{2.48}$$

where in the last step we explicitly calculated the Jacobian of the coordinate transformation. Just like in the real case, we can add a linear term $z^*J + J^*z$ to the exponent of the Gaussian integral and define the generating function

$$\begin{aligned} Z(J,J^*) &= \int \frac{\mathrm{d}z^*\mathrm{d}z}{2\pi\mathrm{i}} \, \exp\left\{G^{-1}z^*z + z^*J + J^*z\right\} \\ &= \exp\left\{-J^*GJ - \log\left(-G^{-1}\right)\right\}, \end{aligned} \tag{2.49}$$

which is shown by completing the square. As before, we can calculate all moments with this generating function, such that we have for example

$$\langle zz^*\rangle = \frac{1}{Z(J,J^*)}\frac{\mathrm{d}^2}{\mathrm{d}J^*\mathrm{d}J}Z(J,J^*)\bigg|_{J^*=J=0} = -G. \tag{2.50}$$

Next, consider the diagonal $n \times n$ matrix $\mathbf{G}^{-1}$, i.e.

$$\mathbf{G}^{-1} = \begin{bmatrix} \frac{1}{G_{11}} & & & \\ & \frac{1}{G_{22}} & & \\ & & \frac{1}{G_{33}} & \\ & & & \ddots \end{bmatrix}. \tag{2.51}$$

We want to evaluate the Gaussian integral

$$\int \left(\prod_{j=1}^{n} \frac{dz_j^* \, dz_j}{2\pi i} \right) \exp\left\{ \mathbf{z}^* \cdot \mathbf{G}^{-1} \cdot \mathbf{z} \right\} \equiv \int \frac{d\mathbf{z}^* d\mathbf{z}}{(2\pi i)^n} \exp\left\{ \mathbf{z}^* \cdot \mathbf{G}^{-1} \cdot \mathbf{z} \right\}, \tag{2.52}$$

where $\mathbf{z}$ is the complex vector $(z_1, \ldots, z_n)$. As before, the integral factorizes and we find that

$$\int \frac{d\mathbf{z}^* d\mathbf{z}}{(2\pi i)^n} \exp\left\{ \mathbf{z}^* \cdot \mathbf{G}^{-1} \cdot \mathbf{z} \right\} = \frac{1}{\prod_{j=1}^{n}(-G_{jj}^{-1})} = \frac{1}{\mathrm{Det}[-\mathbf{G}^{-1}]}. \tag{2.53}$$

Now, we can also generalize (2.49) to

$$\begin{aligned} Z(\mathbf{J}, \mathbf{J}^*) &= \int \frac{d\mathbf{z}^* d\mathbf{z}}{(2\pi i)^n} \exp\left\{ \mathbf{z}^* \cdot \mathbf{G}^{-1} \cdot \mathbf{z} + \mathbf{z}^* \cdot \mathbf{J} + \mathbf{J}^* \cdot \mathbf{z} \right\} \\ &= \exp\left\{ -\mathbf{J}^* \cdot \mathbf{G} \cdot \mathbf{J} - \mathrm{Tr}[\log(-\mathbf{G}^{-1})] \right\}. \end{aligned} \tag{2.54}$$

The above results were obtained for the specific case of a diagonal matrix. However, (2.54) is true for all positive definite hermitian matrices $-\mathbf{G}^{-1}$, because these can be diagonalized by a unitary transformation $\mathbf{U}$ with $|\mathrm{Det}[\mathbf{U}]| = 1$.

2.4 Grassmann Variables

To complete our discussion of Gaussian integrals we introduce another set of numbers, namely the set of anticommuting complex numbers or Grassmann numbers. These turn out to be very useful in setting up the functional-integral formalism for fermionic quantum gases. The reason for this is that, as we see later, fermionic behavior is mathematically expressed by anticommuting creation and annihilation operators. To illustrate this we note that if a fermionic creation operator anticommutes with itself then its square gives zero, which expresses the Pauli principle that two fermions cannot be created in the same quantum state. In order to consider eigenvalues of such anticommuting operators we need anticommuting numbers, i.e. the Grassmann numbers.

A Grassmann algebra is a set of Grassmann variables, which are called the generators of the algebra. They span a complex linear space by making linear combinations of them with complex coefficients. The simplest example that we can think of is the set $\{1, \phi\}$. By definition, we have for a Grassmann variable ϕ that its anticommutator vanishes, i.e. $[\phi, \phi]_+ = \phi\phi + \phi\phi = 0$. We can think of the elements

1 and ϕ as basis vectors of a linear space. However, in order to find a matrix representation of the algebra, it is actually more convenient to think of the elements also as operators. For instance, since we have that $1 \cdot 1 = 1$ and $1 \cdot \phi = \phi$, we see that 1 can be viewed as an operator on the Grassmann algebra that sends both 1 and ϕ to themselves. The element ϕ then sends the basis vector 1 to ϕ, while the other basis vector ϕ is mapped onto 0. In terms of matrices, the above mappings of basis vectors are readily found in matrix form as

$$1 \sim \begin{bmatrix} 1 & 0 \\ 0 & 1 \end{bmatrix} \quad \text{and} \quad \phi \sim \begin{bmatrix} 0 & 0 \\ 1 & 0 \end{bmatrix}. \tag{2.55}$$

Note that the above matrices automatically satisfy all rules imposed on the algebra. Furthermore, since $\phi^2 = 0$, the most general function of ϕ is simply $F(\phi) = f_1 + f_2\phi$.

The previous discussion is easily generalized to the set $\{1, \phi, \phi^*, \phi^*\phi\}$ of two such Grassmann variables, where ϕ and ϕ^* are independent variables. The two Grassmann variables anticommute with each other, giving

$$[\phi, \phi^*]_+ = \phi\phi^* + \phi^*\phi = 0. \tag{2.56}$$

As before, we also have that $\phi^2 = \phi^{*2} = 0$, such that the above set is complete. The complex conjugation in this algebra is defined by

$$(\phi)^* = \phi^*, \qquad (\phi^*)^* = \phi, \qquad (\phi^*\phi)^* = (\phi)^*(\phi^*)^* = \phi^*\phi, \tag{2.57}$$

and the most general function on this algebra yields

$$A(\phi^*, \phi) = a_{11} + a_{12}\phi + a_{21}\phi^* + a_{22}\phi^*\phi\ . \tag{2.58}$$

It is natural to define differentiation of Grassmann variables by

$$\frac{\partial}{\partial \phi} A(\phi^*, \phi) = a_{12} - a_{22}\phi^*, \tag{2.59}$$

where the minus sign occurs because we need to permute ϕ^* and ϕ before we can differentiate with respect to ϕ. The differentiation of (2.59) is called a left differentiation. Similarly, we have

$$\frac{\partial}{\partial \phi^*} A(\phi^*, \phi) = a_{21} + a_{22}\phi, \tag{2.60}$$

where this time the minus sign is absent, because now we do not have to permute the Grassmann variables. Furthermore, we find

$$\frac{\partial^2}{\partial \phi^* \partial \phi} A(\phi^*, \phi) = -\frac{\partial^2}{\partial \phi \partial \phi^*} A(\phi^*, \phi) = -a_{22}\ . \tag{2.61}$$

Next, we introduce integration over the Grassmann variables. Note that since $\phi^2 = 0$, we only have two possible integrals to consider, namely $\int \mathrm{d}\phi\ 1$ and $\int \mathrm{d}\phi\ \phi$. We define these by

$$\int \mathrm{d}\phi\ 1 = 0 \tag{2.62}$$

and

$$\int \mathrm{d}\phi\ \phi = 1\ . \tag{2.63}$$

This means that integration is equivalent to differentiation. The main reason for the above definitions is that we want the integration to obey the rules of partial integration. In particular, this implies that

$$\int \mathrm{d}\phi\ \frac{\partial F(\phi)}{\partial \phi} = 0\ , \tag{2.64}$$

for any function $F(\phi) = f_1 + f_2\phi$. Obviously, this condition requires that $\int \mathrm{d}\phi\ 1 = 0$. The result for $\int \mathrm{d}\phi\ \phi$ then turns out to be merely a question of normalization. The most general quadratic integral for the present Grassmann algebra thus yields

$$\int \mathrm{d}\phi^*\mathrm{d}\phi\ A(\phi^*,\phi) = \int \mathrm{d}\phi^*\mathrm{d}\phi\ (a_{11} + a_{12}\phi + a_{21}\phi^* + a_{22}\phi^*\phi) = -a_{22}\ . \tag{2.65}$$

All the above definitions are then readily further generalized to the Grassmann algebra generated by the Grassmann variables ϕ_j and ϕ_j^* with $j = 1,2,\ldots,n$. It is left as an exercise to show that with the above definitions, the Gaussian integral over $2n$ Grassmann variables leads to

$$\int \left(\prod_j \mathrm{d}\phi_j^*\mathrm{d}\phi_j\right)\ \exp\left\{\sum_{j,j'} \phi_j^* G_{j,j'}^{-1} \phi_{j'}\right\} = \mathrm{Det}[-\mathbf{G}^{-1}] = \mathrm{e}^{\mathrm{Tr}[\log(-\mathbf{G}^{-1})]}\ . \tag{2.66}$$

Note the difference with the result from (2.54), namely

$$\int \left(\prod_j \frac{\mathrm{d}\phi_j^*\mathrm{d}\phi_j}{2\pi\mathrm{i}}\right)\ \exp\left\{\sum_{j,j'} \phi_j^* G_{j,j'}^{-1} \phi_{j'}\right\} = \frac{1}{\mathrm{Det}[-\mathbf{G}^{-1}]} = \mathrm{e}^{-\mathrm{Tr}[\log(-\mathbf{G}^{-1})]}\ , \tag{2.67}$$

which is valid for ordinary complex variables. These last two results will be used extensively throughout the rest of the book.

2.5 Problems

Exercise 2.1. Prove the Gaussian integral in (2.2). To do so, consider

$$\left(\int \mathrm{d}x\ \mathrm{e}^{-\alpha x^2}\right)^2 = \int \mathrm{d}x\ \mathrm{d}y\ \mathrm{e}^{-\alpha(x^2+y^2)}, \tag{2.68}$$

and make use of the fact that the integrand is invariant under rotations, so that you can use polar coordinates to perform the integration.

Exercise 2.2. Consider a Gaussian probability distribution with nonzero average $\langle x \rangle = x_0$. Calculate $\langle x^3 \rangle$ and $\langle x^4 \rangle$ by transforming to the variable $x' = x - x_0$ that has a Gaussian probability distribution centered around zero, such that you can use (2.9) and (2.12). Also calculate $\langle x^3 \rangle$ and $\langle x^4 \rangle$ by making use of the generating function from (2.6), namely

$$Z(J) = \exp\left\{ -\frac{1}{2} G J^2 + x_0 J - \frac{1}{2} \log\left(-G^{-1}\right) \right\}. \tag{2.69}$$

Exercise 2.3. Observe that partial integration of the Gaussian integral leads to the following identity,

$$\int \frac{dx}{\sqrt{2\pi}} \exp\left\{ \frac{1}{2} G^{-1} x^2 \right\} = -\int \frac{dx}{\sqrt{2\pi}} G^{-1} x^2 \exp\left\{ \frac{1}{2} G^{-1} x^2 \right\}. \tag{2.70}$$

Prove now equation (2.12) by making repeated use of partial integration.

Exercise 2.4. Prove that

$$\oint_C dz \frac{1}{(z - z_0)^{n+1}} = 2\pi i \delta_{n,0} \tag{2.71}$$

by taking the contour C to be a circle with radius R around z_0, such that $z = z_0 + R e^{i\varphi}$ and the contour integral becomes an integral over φ.

Exercise 2.5. Contour integration
Using contour integration, show
(a) that the following one-dimensional integral yields

$$\int dq \frac{1}{E^+ - q^2/m} e^{iqx/\hbar} = -i\pi \sqrt{\frac{m}{E}} \exp\left\{ \frac{i|x|\sqrt{mE}}{\hbar} \right\}, \tag{2.72}$$

where $E^+ = E + i\eta$ with η an infinitesimally small positive number, and
(b) that the following three-dimensional integral yields

$$\int \frac{d\mathbf{k}}{(2\pi)^3} \frac{e^{-i\mathbf{k}\cdot\mathbf{r}}}{\alpha^2 + \gamma \mathbf{k}^2} = \frac{1}{4\pi\gamma} \frac{e^{-r/\xi}}{r}, \tag{2.73}$$

where $\xi = \sqrt{\gamma}/\alpha$ is also called the correlation length. Hint: use spherical coordinates $d\mathbf{k} = k^2 \sin(\vartheta)\, dk\, d\vartheta\, d\varphi$, such that $\mathbf{k} \cdot \mathbf{r} = kr\cos(\vartheta)$, and perform the integrations over the angles first.

Exercise 2.6. Find a matrix representation of the Grassmann algebra generated by ϕ and ϕ^*. Note that we need at least 4×4 matrices.

Exercise 2.7. Prove (2.66). To this end, it is instructive to first show that

$$\int \mathrm{d}\phi_1^* \, \mathrm{d}\phi_1 \, \mathrm{d}\phi_2^* \, \mathrm{d}\phi_2 \, \exp\{-\alpha_{11}\phi_1^*\phi_1 - \alpha_{12}\phi_1^*\phi_2 - \alpha_{21}\phi_2^*\phi_1 - \alpha_{22}\phi_2^*\phi_2\}$$
$$= \alpha_{11}\alpha_{22} - \alpha_{21}\alpha_{12}, \tag{2.74}$$

before considering the general case of an integral over $2n$ Grassmann variables.

Exercise 2.8. Hubbard-Stratonovich Transformation
Consider the following integral Z over the complex variables ϕ_j^* and ϕ_j,

$$Z = \int \left(\prod_{j=1}^{n} \frac{\mathrm{d}\phi_j^* \mathrm{d}\phi_j}{2\pi \mathrm{i}} \right) \exp\left\{ \sum_{j,j'=1}^{n} \left(\phi_j^* G_{0;j,j'}^{-1} \phi_{j'} - \frac{V_{j,j'}}{2} \phi_j^* \phi_{j'}^* \phi_{j'} \phi_j \right) \right\}, \tag{2.75}$$

where $\mathbf{G}_0^{-1}$ and $\mathbf{V}$ are invertible matrices with only negative eigenvalues, i.e.

$$\sum_{j''} V_{j,j''} V_{j'',j'}^{-1} = \sum_{j''} G_{0;j,j''}^{-1} G_{0;j'',j'} = \delta_{j,j'}. \tag{2.76}$$

Note that we cannot calculate the integral exactly, because it is not Gaussian, due to the quartic term in the exponential. However, we are going to perform a trick to transform the quartic term away.
(a) Show that the integral Z can be written as

$$Z = \int \left(\prod_{j=1}^{n} \frac{\mathrm{d}\phi_j^* \, \mathrm{d}\phi_j}{2\pi \mathrm{i}} \right) \left(\prod_{j=1}^{n} \frac{\mathrm{d}\eta_j}{\sqrt{2\pi}} \right) \exp\left\{ \frac{1}{2} \mathrm{Tr}[\log(-\mathbf{V})] \right\} \tag{2.77}$$
$$\times \exp\left\{ \sum_{j,j'} \left(\phi_j^* G_{0;j,j'}^{-1} \phi_{j'} + \frac{1}{2} \eta_j V_{j,j'} \eta_{j'} - \eta_j V_{j,j'} \phi_{j'}^* \phi_{j'} \right) \right\},$$

where η is a real variable. Note that there is no longer a quartic term, since we have transformed it away. This is the essence of the Hubbard-Stratonovich transformation, which we use many times when treating interacting quantum gases.
Hint: use the following identity

$$\int \left(\prod_{j=1}^{n} \frac{\mathrm{d}\eta_j}{\sqrt{2\pi}} \right) \exp\left\{ \frac{1}{2} \sum_{j,j'} (\eta_j - \phi_j^* \phi_j) V_{j,j'} \left(\eta_{j'} - \phi_{j'}^* \phi_{j'} \right) \right\} = \mathrm{e}^{-\mathrm{Tr}[\log(-\mathbf{V})]/2}.$$

(b) Show that Z can be written in the following way

$$Z = \mathrm{e}^{\mathrm{Tr}[\log(-\mathbf{V})]/2} \int \left(\prod_{j=1}^{n} \frac{\mathrm{d}\eta_j}{\sqrt{2\pi}} \right) \exp\left\{ \frac{1}{2} \sum_{j,j'} \eta_j V_{j,j'} \eta_{j'} - \mathrm{Tr}\left[\log\left(-\mathbf{G}_0^{-1} + \boldsymbol{\Sigma} \right) \right] \right\},$$

where we introduced the matrix $\Sigma_{j,j'} = \sum_{j''} \eta_{j''} V_{j'',j'} \delta_{j,j'}$.

Additional Reading

- S. Hassani, *Mathematical Physics, A Modern Introduction to Its Foundations*, (Springer-Verlag, Berlin, 1999).
- A mathematically concise textbook on many-particle systems is J. W. Negele and H. Orland *Quantum Many-Particle Systems*, (Westview Press, Boulder, 1998).
- For a thorough mathematical treatment of complex functions, S. Lang *Complex Analysis*, (Springer, Berlin, 1999).

Chapter 3
Quantum Mechanics

The quantum theory was born in 1900, with the twentieth century, and future centuries will list it among our own's most remarkable achievements. Designed to account for the puzzling behavior of matter at the submicroscopic scale of individual atoms, the theory has enjoyed phenomenal success. It has accounted in a quantitative way for atomic phenomena with numerical precision never before achieved in any field of science.
– N. David Mermin

At the end of the 19th century, the macroscopic world was understood in great detail. Newton's laws described mechanics ranging from the collisions of marbles to the motion of planets, Maxwell's equations explained electromagnetic phenomena and statistical physics was the underlying theory for thermodynamic observations. However, a few effects remained truly unexplained, such as the spectrum of black-body radiation and the photoelectric effect. When Planck first quantized the accessible energies for the modes inside a black body, thereby accurately reproducing the observed black-body spectra, he still regarded it as a dirty trick. Einstein was the first to take this idea more seriously when he used it to explain the photoelectric effect by introducing the quantum of light, nowadays called a photon. It was gradually realized that the microscopic world is governed by a set of rules that is completely different from the rules that we know in our everyday life. This set of rules is called quantum mechanics and its success in explaining the microscopic world has been enormous. An important example is Bohr's explanation for the discrete spectra of light that is absorbed and emitted by atoms. Initially quantum mechanics came in two seemingly different formulations, namely the one by Heisenberg now known as matrix mechanics, and the one by Schrödinger now known as wave mechanics. Soon these two pictures were shown to be equivalent, and the unified formulation of quantum mechanics was given by Dirac and Von Neumann. Later, yet another way of doing quantum mechanics was developed by Feynman with the use of path integrals. The latter formalism, which is the topic of Chap. 5, is generalized to quantum field theory in the second part of this book.

In this chapter we review the elementary concepts from quantum mechanics in the elegant formulation of Dirac, focusing on concepts that we need later on for the description of interacting quantum gases. We introduce various important concepts in a familiar setting, such as the number states and the coherent states, before generalizing them to the more abstract formalisms of second quantization and functional path integrals in later chapters.

3.1 Hilbert Spaces

In quantum mechanics, an isolated physical system is described by an abstract state vector $|\psi\rangle$ in an appropriate quantum-mechanical Hilbert space $\mathscr{H}$. This is a complex linear space that has an inner product and whose elements are the state vectors, denoted by $|\psi\rangle$, which are often called 'kets'. In a Hilbert space, adding up two different state vectors leads to a new state vector, which is known as the superposition principle. It is also possible to multiply a state vector with an arbitrary complex number, but this does not change the physical information contained in the vector. To be able to construct the inner product on the Hilbert space, we need to introduce the so-called 'bra' $\langle\psi|$, which is by definition the hermitian conjugate of the ket $|\psi\rangle$, i.e. $\langle\psi| = (|\psi\rangle)^\dagger$, where hermitian conjugation is a combination of both transposition and complex conjugation. The inner product of two vectors $|\psi\rangle$ and $|\psi'\rangle$ is then a complex number that we denote by the 'bracket' $\langle\psi|\psi'\rangle$. It is linear, i.e. $\langle\psi|(\alpha|\psi'\rangle + \beta|\psi''\rangle) = \alpha\langle\psi|\psi'\rangle + \beta\langle\psi|\psi''\rangle$, and satisfies $\langle\psi|\psi'\rangle = (\langle\psi'|\psi\rangle)^\dagger = \langle\psi'|\psi\rangle^*$. Well-known examples of inner products satisfying the above conditions include the standard inner product between two complex m-dimensional vectors, giving $\langle\psi|\psi'\rangle = \sum_{j=1}^m \psi_j^* \psi_j'$, and the inner product between two complex-valued functions f and g, for which we have $\langle f|g\rangle = \int \mathrm{d}x\, f^*(x)g(x)$.

Consider now a state vector characterized by a set of quantum numbers collectively denoted by ν. Then a set of such state vectors $\{|\nu\rangle\}$ is said to be linearly independent if the relation $\sum_\nu c_\nu |\nu\rangle = 0$ implies that the complex numbers c_ν are equal to zero for all ν. A set of independent vectors that span the whole vector space is called a basis. Furthermore, two vectors are said to be orthogonal if $\langle\nu|\nu'\rangle = 0$ and the length or norm of a vector is defined through the inner product as $\sqrt{\langle\nu|\nu\rangle}$. If all the vectors of a basis are mutually orthogonal and if the length of the basis vectors is also normalized to unity, then the basis is called orthonormal. In that case, we have $\langle\nu|\nu'\rangle = \delta_{\nu,\nu'}$, where $\delta_{\nu,\nu'}$ is the Kronecker delta that equals one when $\nu = \nu'$ and zero otherwise. An orthonormal set of vectors $|\nu\rangle$ is called complete if it satisfies the completeness relation

$$\sum_\nu |\nu\rangle\langle\nu| = \hat{1}. \tag{3.1}$$

The completeness relation ensures that an arbitrary vector can be expressed uniquely in terms of the orthonormal basis as

$$|\psi\rangle = \sum_\nu |\nu\rangle\langle\nu|\psi\rangle \equiv \sum_\nu c_\nu |\nu\rangle. \tag{3.2}$$

Demanding that the state vector $|\psi\rangle$ is properly normalized to unity, we have that

$$\langle\psi|\psi\rangle = \sum_{\nu,\nu'} c_\nu^* c_{\nu'} \langle\nu|\nu'\rangle = \sum_\nu |c_\nu|^2 = 1, \tag{3.3}$$

such that $|c_\nu|^2$ can be interpreted as the probability for the system to be in state $|\nu\rangle$. If $\{|\nu\rangle\}$ and $\{|\mu\rangle\}$ are both complete sets of orthonormal states, then we can always go from one basis to the other by inserting a completeness relation, i.e.

$$|\nu\rangle = \sum_{\mu} |\mu\rangle\langle\mu|\nu\rangle \tag{3.4}$$

and

$$|\mu\rangle = \sum_{\nu} |\nu\rangle\langle\nu|\mu\rangle. \tag{3.5}$$

Note that to generalize the above discussion to a continuous index ν, the sums over ν should be replaced by integrals.

Example 3.1. For a single atom on a line, typical examples of complete bases for its Hilbert space $\mathscr{H}$ are the momentum basis $|p\rangle$, whose states are labelled by the momentum of the particle, or the continuous position basis $|x\rangle$, whose states are labelled by the position of the particle. Since the position basis $|x\rangle$ is continuous, orthonormality is expressed by

$$\langle x|x'\rangle = \delta(x-x'), \tag{3.6}$$

where the appropriate generalization of the Kronecker delta is the Dirac delta function, such that the relation $\sum_{\nu} \delta_{\nu,\nu'} = 1$ generalizes to $\int \mathrm{d}x'\ \delta(x-x') = 1$. Moreover, completeness then implies that

$$\int \mathrm{d}x\ |x\rangle\langle x| = \hat{1}. \tag{3.7}$$

3.2 Observables

In quantum mechanics, dynamical variables are described by linear hermitian operators $\hat{O}$ that act on the Hilbert space and have a complete set of eigenstates. Such operators are also called observables. Linear means that $\hat{O}(c_1|\psi\rangle + c_2|\psi\rangle) = c_1\hat{O}|\psi\rangle + c_2\hat{O}|\psi\rangle$, while Hermitian means that the operator is its own Hermitian conjugate $\hat{O} = \hat{O}^{\dagger}$, defined by $\langle\psi'|\hat{O}^{\dagger}|\psi\rangle = (\hat{O}|\psi'\rangle)^{\dagger}|\psi\rangle = \langle\psi|\hat{O}|\psi'\rangle^{*}$. Examples of physical dynamical variables are the total energy, described by the Hamilton operator $\hat{H}$, the position, described by the position operator $\hat{x}$, and the momentum, described by the momentum operator $\hat{p}$. Important mathematical properties of Hermitian operators include that they are diagonalizable and that they have real eigenvalues. It is also convenient to introduce functions of observables. If $f(x)$ has a series expansion given by

$$f(x) = \sum_{i=0}^{\infty} a_i x^i, \tag{3.8}$$

then the operator function $f(\hat{O})$ is defined through

$$f(\hat{O}) = \sum_{i=0}^{\infty} a_i \hat{O}^i. \tag{3.9}$$

To perform quantum-mechanical calculations in practice, the position representation is often used. In this representation, the momentum operator has the familiar form

$$\hat{p} = \int \mathrm{d}x \, |x\rangle \left(-\mathrm{i}\hbar \frac{\mathrm{d}}{\mathrm{d}x}\right) \langle x| \equiv -\mathrm{i}\hbar \frac{\mathrm{d}}{\mathrm{d}\hat{x}}. \tag{3.10}$$

Then, we can substitute (3.10) into the eigenvalue equation

$$\hat{p}|p\rangle = p|p\rangle, \tag{3.11}$$

which leads to

$$\begin{aligned}\langle x|\hat{p}|p\rangle &= \int \mathrm{d}x' \, \langle x|x'\rangle \left(-\mathrm{i}\hbar \frac{\mathrm{d}}{\mathrm{d}x'}\right) \langle x'|p\rangle \\ &= -\mathrm{i}\hbar \int \mathrm{d}x' \, \delta(x-x') \frac{\mathrm{d}}{\mathrm{d}x'} \langle x'|p\rangle \\ &= -\mathrm{i}\hbar \frac{\mathrm{d}}{\mathrm{d}x} \langle x|p\rangle = p\langle x|p\rangle. \end{aligned} \tag{3.12}$$

Combining the above differential equation with the orthonormality condition

$$\langle p|p'\rangle = \int \mathrm{d}x \, \langle p|x\rangle\langle x|p'\rangle = \delta(p-p'), \tag{3.13}$$

we find that the position representation of the momentum states is given by

$$\langle x|p\rangle = \frac{1}{\sqrt{2\pi\hbar}} \exp\left\{\frac{\mathrm{i}px}{\hbar}\right\}. \tag{3.14}$$

Example 3.2. Consider a particle confined to a line of length L. The Hamiltonian is given by $\hat{H} = \hat{p}^2/2m$ with m the mass of the particle. The eigenstates of $\hat{H}$ we denote by $|n\rangle$ and the corresponding eigenvalues by ε_n. In the coordinate representation, the corresponding eigenvalue equation is given by

$$-\frac{\hbar^2}{2m} \frac{\mathrm{d}^2}{\mathrm{d}x^2} \chi_n(x) = \varepsilon_n \chi_n(x) \tag{3.15}$$

with $\chi_n(x) = \langle x|n\rangle$. The eigenvalue equation for a time-independent Hamiltonian is called the time-independent Schrödinger equation. Writing $p_n = \sqrt{2m\varepsilon_n}$, we see that the linearly independent solutions of the above Schrödinger equation are proportional to $\mathrm{e}^{\mathrm{i}p_n x/\hbar}$, where the normalization factor is determined below. Assuming periodic boundary conditions, i.e. $\chi_n(x+L) = \chi_n(x)$, we find that $p_n = 2\pi\hbar n/L$ with $n = 0, \pm 1, \pm 2, \ldots, \pm\infty$. As a result, the energy spectrum becomes discrete and

$$\varepsilon_n = \frac{2\pi^2\hbar^2}{mL^2} n^2. \tag{3.16}$$

The normalization of the wavefunctions $\chi_n(x)$ is fixed by

$$\int_{-L/2}^{L/2} \mathrm{d}x \, |\chi_n(x)|^2 = 1, \tag{3.17}$$

giving $\chi_n(x) = \mathrm{e}^{\mathrm{i}p_n x/\hbar}/\sqrt{L}$. Note also that the wavefunctions $\chi_n(x)$ are orthonormal, since

$$\int_{-L/2}^{L/2} \mathrm{d}x \, \chi_n^*(x)\chi_{n'}(x) = \int_{-L/2}^{L/2} \mathrm{d}x \, \frac{1}{L} \exp\left\{\frac{\mathrm{i}}{\hbar}(p_n - p_{n'})x\right\} = \delta_{n,n'}. \tag{3.18}$$

Example 3.3. Let $|\psi\rangle$ be a state vector for a particle confined to a line of length L and let $\psi(x) = \langle x|\psi\rangle$ be the corresponding wavefunction in the position representation. To obtain the wavefunction in the momentum representation, we have to perform a Fourier transform, because

$$\begin{aligned} \psi(p) = \langle p|\psi\rangle &= \int \mathrm{d}x \, \langle p|x\rangle\langle x|\psi\rangle \\ &= \int \frac{\mathrm{d}x}{\sqrt{2\pi\hbar}} \, \mathrm{e}^{-\mathrm{i}px/\hbar} \psi(x). \end{aligned} \tag{3.19}$$

This means in particular for the eigenstates of Example 3.2 that

$$\chi_n(p) = 2\hbar \frac{\sin((p-p_n)L/2\hbar)}{\sqrt{2\pi\hbar L}(p-p_n)}, \tag{3.20}$$

with $p_n = 2\pi\hbar n/L$.

Now, suppose that we have an ensemble of identical systems all in state $|\psi\rangle$ and we want to measure the dynamical variable O described by the observable $\hat{O}$, which has a complete set of eigenstates $|\nu\rangle$ with real eigenvalues ν. Then, a single measurement of the dynamical variable always yields one of the eigenvalues ν of the observable. However, if we perform a series of measurements on the ensemble and average over the result, then we experimentally determine the expectation value of the observable, which according to quantum mechanics is given by

$$\langle \hat{O} \rangle \equiv \langle \psi|\hat{O}|\psi\rangle = \sum_{\nu',\nu} c_{\nu'}^* c_\nu \langle \nu'|\nu|\nu\rangle = \sum_\nu \nu |c_\nu|^2, \tag{3.21}$$

where $|c_\nu|^2$ is thus to be interpreted as the probability of measuring the eigenvalue ν. Note that even for a perfect single measurement on a perfectly prepared system, we can in principle not be certain which of the eigenvalues ν we find as an outcome. The best we can do is to predict the probability $|c_\nu|^2$ of finding each eigenvalue ν, showing the intrinsic probabilistic character of quantum mechanics. Only in the special case when the system is in one of the eigenstates of the observable $\hat{O}$, then it is absolutely certain that a perfect measurement gives the corresponding eigenvalue $\nu = \langle \nu|\hat{O}|\nu\rangle$.

Example 3.4. For the eigenstates of Example 3.2, we have that

$$\langle \hat{p} \rangle = \int \mathrm{d}p \; p \, |\chi_n(p)|^2 = p_n. \tag{3.22}$$

Note that the momentum is not sharply defined in the eigenstates $\chi_n(p)$, since these states give rise to a width of order $\hbar/L$, which is a direct consequence of the Heisenberg uncertainty relation to which we come next. However, in the limit $L \to \infty$, this is resolved, because then $|\chi_n(p)|^2$ converges to the Dirac delta function such that the state has a definite momentum, as expected for an eigenstate of the free-particle Hamiltonian.

Upon considering two operators, such as for example the position operator $\hat{x}$ and the momentum operator $\hat{p}$, we can introduce their commutator, which is defined by

$$[\hat{x}, \hat{p}]_- \equiv \hat{x}\hat{p} - \hat{p}\hat{x}. \tag{3.23}$$

If two or more observables commute, i.e. when their commutator is zero, then it can be shown that they share a common complete set of eigenfunctions. However, a very fundamental postulate of quantum mechanics is that the position operator $\hat{x}$ and the momentum operator $\hat{p}$ actually do not commute and that they obey the commutation relation

$$[\hat{x}, \hat{p}]_- = \mathrm{i}\hbar. \tag{3.24}$$

This commutation relation then also gives rise to the expression of the momentum operator in position space, as given in (3.10). Furthermore, it is left as an exercise to show that the nonvanishing commutator of (3.24) directly leads to the famous Heisenberg uncertainty relation

$$\Delta x \Delta p \geq \frac{\hbar}{2}, \tag{3.25}$$

with the uncertainty or standard deviation given by $\Delta x \equiv \sqrt{\langle (\hat{x} - \langle \hat{x} \rangle)^2 \rangle}$.

3.3 Schrödinger vs. Heisenberg Picture

So far, we have not explicitly considered the time evolution of quantum-mechanical systems. The dynamics of a state vector $|\psi(t)\rangle$ is governed by the time-dependent Schrödinger equation, which in the basis-independent formulation is given by

$$i\hbar\frac{d}{dt}|\psi(t)\rangle = \hat{H}|\psi(t)\rangle, \tag{3.26}$$

where $\hat{H}$ is the Hamilton operator or Hamiltonian corresponding to the total energy of the system. If we expand the state vector in terms of the orthonormal basis $|\nu\rangle$, we find

$$|\psi(t)\rangle = \sum_{\nu} \psi_\nu(t)|\nu\rangle. \tag{3.27}$$

As before, we demand that the state $|\psi(t)\rangle$ is properly normalized, giving

$$\sum_{\nu} |\psi_\nu(t)|^2 = 1, \tag{3.28}$$

such that $|\psi_\nu(t)|^2$ represents the probability to be in state $|\nu\rangle$ at time t. Therefore, $\psi_\nu(t)$ is known as the probability amplitude or the wavefunction in the ν representation. Substituting the expansion of (3.27) into the Schrödinger equation and multiplying the resulting equation with $\langle\nu|$ gives

$$i\hbar\frac{d\psi_\nu(t)}{dt} = \sum_{\nu'} H_{\nu\nu'}\psi_{\nu'}(t), \tag{3.29}$$

with $H_{\nu\nu'} = \langle\nu|\hat{H}|\nu'\rangle$ the matrix elements of the Hamiltonian in the ν representation. By choosing the most appropriate representation for the specific quantum-mechanical problem, the Schrödinger equation is most conveniently solved in practice.

For a Hamiltonian that does not depend on time we can consider the time-independent Schrödinger equation, given by

$$\hat{H}|\psi_n\rangle = \varepsilon_n|\psi_n\rangle. \tag{3.30}$$

If the system is in an eigenstate of the time-independent Hamiltonian at a certain time t_0, then it will remain in this eigenstate and the time evolution of the state is given by

$$|\psi_n(t)\rangle = e^{-i\varepsilon_n(t-t_0)/\hbar}|\psi_n(t_0)\rangle, \tag{3.31}$$

which is easily verified by substituting the above solution in (3.26). In particular, we have for such an eigenstate $|\psi_n(t)\rangle$ that

$$\langle\psi_n(t)|\hat{H}|\psi_n(t)\rangle = \varepsilon_n \tag{3.32}$$

and the energy is conserved. To study the time evolution of an arbitrary wavefunction, we can either first expand it into eigenfunctions of the Hamiltonian, after which each of the eigenfunctions evolves according to (3.31), or we can formally solve the Schrödinger equation in (3.26) and find that the final state vector $|\psi(t_\mathrm{f})\rangle$ at time t_f depends on the initial state vector $|\psi(t_\mathrm{i})\rangle$ at time t_i through

$$|\psi(t_\mathrm{f})\rangle = \exp\left\{-\frac{\mathrm{i}}{\hbar}\hat{H}(t_\mathrm{f}-t_\mathrm{i})\right\}|\psi(t_\mathrm{i})\rangle\,. \tag{3.33}$$

As a result, it is natural to define the time-evolution operator as

$$\hat{U}(t,t_\mathrm{i}) = \exp\left\{-\frac{\mathrm{i}}{\hbar}\hat{H}(t-t_\mathrm{i})\right\}, \tag{3.34}$$

which describes the time evolution of the initial state $|\psi(t_\mathrm{i})\rangle$ and satisfies the differential operator equation

$$\mathrm{i}\hbar\frac{\mathrm{d}}{\mathrm{d}t}\hat{U}(t,t_\mathrm{i}) = \hat{H}\hat{U}(t,t_\mathrm{i})\,. \tag{3.35}$$

The time evolution operator is unitary, since

$$\hat{U}(t,t_\mathrm{i})\hat{U}^\dagger(t,t_\mathrm{i}) = \hat{U}^\dagger(t,t_\mathrm{i})\hat{U}(t,t_\mathrm{i}) = \hat{1}, \tag{3.36}$$

which is easily seen from the fact that $\hat{U}^\dagger(t,t_\mathrm{i}) = \hat{U}(t_\mathrm{i},t)$ and $\hat{U}(t,t_\mathrm{i})\hat{U}(t_\mathrm{i},t) = \hat{1}$. Since a unitary transformation preserves the length of a vector, we have that the time evolution conserves the total probability.

This formulation of quantum mechanics, where the state vectors depend on time and the operators are time independent, is known as the Schrödinger picture. A different, but equivalent way to formulate quantum mechanics is to use time-independent state vectors and operators that depend on time. This approach is known as the Heisenberg picture and can be obtained from the Schrödinger picture via a unitary transformation. The time evolution of the operator $\hat{O}(t)$ is determined by

$$\hat{O}(t) \equiv \hat{U}^\dagger(t,0)\hat{O}\hat{U}(t,0), \tag{3.37}$$

where the operator $\hat{O}$ on the right-hand-side is the time-independent operator from the Schrödinger picture. The equation of motion for the operator $\hat{O}$ is obtained by taking the time derivative of the above equation. Doing so, we find with the use of (3.35) the Heisenberg equation of motion

$$\mathrm{i}\hbar\frac{\mathrm{d}\hat{O}(t)}{\mathrm{d}t} = [\hat{O}(t),\hat{H}]_-. \tag{3.38}$$

The expectation value for any observable $\hat{O}$ is then the same in the two pictures, since

$$\langle\psi(t)|\hat{O}|\psi(t)\rangle = \langle\psi|\hat{U}^\dagger(t,0)\hat{O}\hat{U}(t,0)|\psi\rangle = \langle\psi|\hat{O}(t)|\psi\rangle, \tag{3.39}$$

showing that the two pictures are indeed equivalent.

3.4 Bosonic Harmonic Oscillator

One of the most common problems in quantum mechanics is that of a particle in a harmonic potential, whose importance can be understood as follows. Since a system near equilibrium finds itself close to the minimum of the potential energy, we can often make a series expansion around this minimum and only keep the quadratic term. This shows that many systems in or near equilibrium experience the harmonic potential. The simplest version is the one-dimensional harmonic oscillator, which is described by the Hamiltonian

$$\hat{H} = \frac{\hat{p}^2}{2m} + \frac{1}{2}m\omega^2\hat{x}^2, \tag{3.40}$$

where m is the mass of the particle and ω is the frequency that specifies the strength of the potential. The time-independent Schrödinger equation in the place representation thus becomes

$$\hat{H}\chi(x) = \left\{-\frac{\hbar^2}{2m}\frac{\mathrm{d}^2}{\mathrm{d}x^2} + \frac{1}{2}m\omega^2x^2\right\}\chi(x) = \varepsilon\chi(x), \tag{3.41}$$

which turns out to be analytically solvable. The solutions are given by

$$\chi_N(x) = \left(\frac{1}{l\sqrt{\pi}2^N N!}\right)^{1/2} \mathrm{e}^{-x^2/2l^2} H_N(x/l), \tag{3.42}$$

where $l = \sqrt{\hbar/m\omega}$ is the harmonic oscillator length and $H_N(x)$ are the Hermite polynomials. The corresponding eigenenergies are given by $\varepsilon_N = \hbar\omega(N+1/2)$ and the Hermite polynomials are defined through

$$H_N(x) = (-1)^N \mathrm{e}^{x^2}\frac{\mathrm{d}^N}{\mathrm{d}x^N}\mathrm{e}^{-x^2}, \tag{3.43}$$

leading to the property

$$\frac{\mathrm{d}}{\mathrm{d}x}H_N(x) = 2xH_N(x) - H_{N+1}(x). \tag{3.44}$$

It is left as an exercise to show that the eigenfunctions $\chi_N(x)$ satisfy

$$\sqrt{\frac{m\omega}{2\hbar}}\left(\hat{x} + \frac{\mathrm{i}}{m\omega}\hat{p}\right)\chi_N(x) = \sqrt{N}\chi_{N-1}(x), \tag{3.45}$$

$$\sqrt{\frac{m\omega}{2\hbar}}\left(\hat{x} - \frac{\mathrm{i}}{m\omega}\hat{p}\right)\chi_N(x) = \sqrt{N+1}\chi_{N+1}(x), \tag{3.46}$$

where the last equation shows that the creation operator

$$\hat{a}^\dagger \equiv \sqrt{\frac{m\omega}{2\hbar}}\left(\hat{x} - \frac{\mathrm{i}}{m\omega}\hat{p}\right) \tag{3.47}$$

effectively changes the eigenstate $\chi_N(x)$ to the eigenstate $\chi_{N+1}(x)$. The Hermitian conjugate operator $\hat{a}$ is known as the annihilation operator, and changes the eigenstate $\chi_N(x)$ to the eigenstate $\chi_{N-1}(x)$.

3.5 Creation and Annihilation Operators

The creation and annihilation operators defined in the previous section are convenient operators to represent the harmonic oscillator problem. We reconsider the Hamiltonian in (3.40) and write it in terms of the creation and annihilation operators as

$$\begin{aligned}\hat{H} &= \frac{1}{2}m\omega^2\left\{\frac{\hat{p}^2}{m^2\omega^2} + \hat{x}^2\right\} \\ &= \frac{1}{2}m\omega^2\left\{\left(\hat{x} - \frac{\mathrm{i}\hat{p}}{m\omega}\right)\left(\hat{x} + \frac{\mathrm{i}\hat{p}}{m\omega}\right) - \frac{\mathrm{i}}{m\omega}[\hat{x},\hat{p}]_-\right\} \\ &= \hbar\omega\left\{\hat{a}^\dagger\hat{a} + \frac{1}{2}\right\}.\end{aligned} \tag{3.48}$$

In terms of the creation and annihilation operators the Schrödinger equation becomes

$$\hat{H}|\chi\rangle = \hbar\omega\left(\hat{a}^\dagger\hat{a} + \frac{1}{2}\right)|\chi\rangle = \varepsilon|\chi\rangle. \tag{3.49}$$

Comparing the Hamiltonian in this form with the energy eigenvalues $\varepsilon_N = \hbar\omega(N + 1/2)$, we anticipate that the operator $\hat{a}^\dagger\hat{a}$ equals the operator $\hat{N}$, which has the natural numbers (including zero) as eigenvalues. Physically, the operator $\hat{N}$ counts the number of energy quanta that are stored in the harmonic oscillator and is appropriately called the number operator. The eigenstates of the number operator are denoted by $|N\rangle$. From the previous paragraph we have that

$$\hat{a}|N\rangle = \sqrt{N}|N-1\rangle, \tag{3.50}$$

and

$$\hat{a}^\dagger|N\rangle = \sqrt{1+N}|N+1\rangle, \tag{3.51}$$

showing indeed that $a^\dagger a|N\rangle = N|N\rangle$. We also have

$$[\hat{a},\hat{a}^\dagger]_- = 1, \tag{3.52}$$

and we can obtain any number state $|N\rangle$ by applying the creation operator successively N times to the ground state, giving

$$|N\rangle = \frac{(\hat{a}^\dagger)^N}{\sqrt{N!}}|0\rangle. \tag{3.53}$$

In the basis of eigenstates $|N\rangle$, it is easy to obtain the matrix representations of the creation and annihilation operators, respectively. We have, for instance, that

$$\hat{a} = \begin{bmatrix} 0 & \sqrt{1} & 0 & 0 & \dots \\ 0 & 0 & \sqrt{2} & 0 & \dots \\ 0 & 0 & 0 & \sqrt{3} & \dots \\ 0 & 0 & 0 & 0 & \dots \\ \vdots & \vdots & \vdots & \vdots & \ddots \end{bmatrix}, \tag{3.54}$$

and the matrix representation of the creation operator in this basis is given by the transpose of the above matrix.

3.6 Three-Dimensional Harmonic Oscillator

To generalize the one-dimensional harmonic oscillator to three dimensions, we consider the Hamiltonian

$$\hat{H} = \frac{\hat{\mathbf{p}}^2}{2m} + \frac{1}{2}m\omega^2\mathbf{r}^2 \tag{3.55}$$

with the corresponding time-independent Schrödinger equation given by

$$\left\{-\frac{\hbar^2\nabla^2}{2m} + \frac{1}{2}m\omega^2\mathbf{r}^2\right\}\psi(\mathbf{r}) = E\psi(\mathbf{r}). \tag{3.56}$$

In Cartesian coordinates, the eigenfunctions are simply given by the product of three one-dimensional harmonic oscillator wavefunctions, one for each direction (x, y, z), and the corresponding eigenvalues are $\varepsilon_{\mathbf{n}} = \hbar\omega(n_x + n_y + n_z + 3/2)$. However, to exploit the spherical symmetry of the system we might also use spherical coordinates (r, ϑ, φ), where ϑ is the zenith and φ the azimuth angle. For the kinetic term, this yields

$$-\frac{\hbar^2\nabla^2}{2m} = -\frac{\hbar^2}{2mr^2}\left\{\frac{\partial}{\partial r}\left(r^2\frac{\partial}{\partial r}\right) + \frac{1}{\sin\vartheta}\frac{\partial}{\partial\vartheta}\left(\sin\vartheta\frac{\partial}{\partial\vartheta}\right) + \frac{1}{\sin^2\vartheta}\frac{\partial^2}{\partial\varphi^2}\right\}. \tag{3.57}$$

Because the interaction potential is spherically symmetric, the Schrödinger equation has separable solutions of the form

$$\psi_{E\ell m}(\mathbf{r}) = R_{E\ell}(r)Y_{\ell m}(\vartheta, \varphi), \tag{3.58}$$

such that the radial and angular part of the wavefunction each satisfy their own differential equation. We have for the angular part

$$\left\{-\frac{1}{\sin\vartheta}\frac{\partial}{\partial\vartheta}\left(\sin\vartheta\frac{\partial}{\partial\vartheta}\right)-\frac{1}{\sin^2\vartheta}\frac{\partial^2}{\partial\varphi^2}\right\}Y_{\ell m}(\vartheta,\varphi)=\ell(\ell+1)Y_{\ell m}(\vartheta,\varphi), \quad (3.59)$$

which is analytically solvable and whose solutions are the well-known spherical harmonics

$$Y_{\ell m}(\vartheta,\varphi)=(-1)^m\sqrt{\frac{(2\ell+1)}{4\pi}\frac{(\ell-m)!}{(\ell+m)!}}P_\ell^m(\cos\vartheta)\mathrm{e}^{\mathrm{i}m\varphi}, \quad (3.60)$$

where the associated Legendre polynomials $P_\ell^m(\cos\theta)$ are determined by

$$P_\ell^m(x)=(1-x^2)^{|m|/2}\frac{1}{2^\ell\ell!}\frac{\mathrm{d}^{|m|}}{\mathrm{d}x^{|m|}}\frac{\mathrm{d}^\ell}{\mathrm{d}x^\ell}(x^2-1)^\ell, \quad (3.61)$$

with $\ell=0,1,2\ldots\infty$ and $m=-\ell,-\ell+1,\ldots,\ell-1,\ell$, where unfortunately the standard notation for this quantum number is identical to that for the mass of the particle. The spherical harmonics satisfy the following orthonormality relations

$$\int_0^{2\pi}\mathrm{d}\varphi\int_0^{\pi}\mathrm{d}\vartheta\,\sin\vartheta\,Y^*_{\ell' m'}(\vartheta,\varphi)Y_{\ell m}(\vartheta,\varphi)=\delta_{\ell,\ell'}\delta_{m,m'}, \quad (3.62)$$

which automatically means that the spherical harmonics are normalized,

$$\int_0^{2\pi}\mathrm{d}\varphi\int_0^{\pi}\mathrm{d}\vartheta\,\sin\vartheta\,|Y_{\ell m}(\vartheta,\varphi)|^2=1. \quad (3.63)$$

The radial part of the wavefunction satisfies the radial Schrödinger equation, which, after introducing $u_{E\ell}(r)=rR_{E\ell}(r)$, takes the form

$$\left\{-\frac{\hbar^2}{2m}\frac{\mathrm{d}^2}{\mathrm{d}r^2}+\frac{\hbar^2\ell(\ell+1)}{2mr^2}+\frac{1}{2}m\omega^2r^2-E\right\}u_{E\ell}(r)=0. \quad (3.64)$$

This radial differential equation can also be solved analytically and after multiplication with the spherical harmonics the final eigenstates are given by

$$\psi_{n\ell m}(r,\vartheta,\varphi)=\sqrt{\frac{2}{l^3}}\binom{n+\ell+1/2}{n}^{-1/2}\frac{1}{\sqrt{(\ell+1/2)!}}$$
$$\times\mathrm{e}^{-r^2/2l^2}L_n^{(1/2+\ell)}((r/l)^2)\left(\frac{r}{l}\right)^\ell Y_{\ell m}(\vartheta,\varphi), \quad (3.65)$$

where the associated Laguerre polynomials are defined through

$$L_n^{(1/2+\ell)}(x)=\frac{\mathrm{e}^x x^{-1/2-\ell}}{n!}\frac{\mathrm{d}^n}{\mathrm{d}x^n}(\mathrm{e}^{-x}x^{n+\ell+1/2}) \quad (3.66)$$

with the corresponding energy eigenvalues

$$\varepsilon_{n,\ell} = (2n + \ell + 3/2)\hbar\omega \,. \tag{3.67}$$

Note that the energy levels are degenerate in the quantum number m.

3.7 Coherent States

In this section, we introduce a very useful set of states in the context of the one-dimensional harmonic oscillator. This set consists of eigenstates of the annihilation operator, which are known as coherent states. Important physical examples of systems that are in a coherent state include lasers and Bose-Einstein condensates. Coherent states turn out to be of great importance in later chapters, when we set up the functional-integral formalism to describe interacting quantum gases. In this section, we study their relevant properties to serve this purpose. We start by looking at the creation operator, which cannot have eigenstates. To see this, we remember that an arbitrary state can always be expressed as a superposition of the number eigenstates $|N\rangle$. From (3.51) we see that, after applying the creation operator to an arbitrary state, the number eigenstate with the lowest eigenvalue of $\hat{N}$ has disappeared from the linear superposition. Thus, the new superposition can never be the same as the old one. However, the same argument does not hold for the annihilation operator. If we apply the annihilation operator to a superposition of number eigenstates, we find that

$$\hat{a} \sum_{N=0}^{M} \psi_N |N\rangle = \sum_{N=0}^{M-1} \psi_{N+1} \sqrt{N+1} |N\rangle. \tag{3.68}$$

For a finite number of terms, the number state with the highest eigenvalue of $\hat{N}$ would disappear from the superposition, which means that this state cannot be an eigenstate of the annihilation operator. However, if the superposition contains an infinite number of terms, we can circumvent this problem. Considering the state

$$|\phi\rangle = \exp\left\{\phi \hat{a}^\dagger\right\}|0\rangle \tag{3.69}$$

with ϕ a complex number, we note that application of the annihilation operator gives

$$\hat{a}|\phi\rangle = \hat{a}\, \exp\left\{\phi \hat{a}^\dagger\right\}|0\rangle = \hat{a} \sum_N \frac{\phi^N}{\sqrt{N!}} |N\rangle = \phi |\phi\rangle, \tag{3.70}$$

where in the last step we used (3.50). Thus, we see that the state $|\phi\rangle$ is indeed an eigenstate of the annihilation operator $\hat{a}$ with the eigenvalue ϕ.

It is important to realize that the coherent states are overcomplete and do not form an orthonormal set. This is easily shown by considering the inner product between two different coherent states

$$\langle\phi|\phi'\rangle = \sum_{N,N'} \langle N| \frac{(\phi^*)^N}{\sqrt{N!}} \frac{(\phi')^{N'}}{\sqrt{N'!}} |N'\rangle = \mathrm{e}^{\phi^*\phi'}. \tag{3.71}$$

However, the coherent states do obey a completeness or closure relation, that is

$$\int \frac{\mathrm{d}\phi^*\mathrm{d}\phi}{2\pi\mathrm{i}}\, \mathrm{e}^{-\phi^*\phi} |\phi\rangle\langle\phi| = \hat{1}\,. \tag{3.72}$$

To prove this important relation, we write the eigenvalue ϕ of the coherent state $|\phi\rangle$ as $\phi = \rho\mathrm{e}^{\mathrm{i}\theta}$, where both θ and ρ are real. Note that by making this change of variables, we pick up an additional factor of $2\mathrm{i}\rho$ coming from the Jacobian. Therefore, we find

$$\begin{aligned}\int \frac{\mathrm{d}\phi^*\mathrm{d}\phi}{2\pi\mathrm{i}}\, \mathrm{e}^{-\phi^*\phi} |\phi\rangle\langle\phi| &= \int \frac{\mathrm{d}\phi^*\mathrm{d}\phi}{2\pi\mathrm{i}} \sum_{N,N'} \mathrm{e}^{-\phi^*\phi} \frac{(\phi)^N}{\sqrt{N!}} |N\rangle\langle N'| \frac{(\phi^*)^{N'}}{\sqrt{N'!}} \\ &= \int_0^\infty \mathrm{d}\rho \int_0^{2\pi} \mathrm{d}\theta \sum_{N,N'} \frac{\rho}{\pi} \mathrm{e}^{-\rho^2} \frac{(\rho\mathrm{e}^{\mathrm{i}\theta})^N}{\sqrt{N!}} |N\rangle\langle N'| \frac{(\rho\mathrm{e}^{-\mathrm{i}\theta})^{N'}}{\sqrt{N'!}}\,.\end{aligned} \tag{3.73}$$

Since, the angular integration leads to $2\pi\delta_{N,N'}$, we are left with

$$\int \frac{\mathrm{d}\phi^*\mathrm{d}\phi}{2\pi\mathrm{i}}\, \mathrm{e}^{-\phi^*\phi} |\phi\rangle\langle\phi| = 2\sum_N \frac{|N\rangle\langle N|}{N!} \int_0^\infty \mathrm{d}\rho\; \rho^{2N+1}\mathrm{e}^{-\rho^2}. \tag{3.74}$$

This last integral is equal to $N!/2$, which follows from the change of variables $t = \rho^2$ and the integral form of the Gamma function $\Gamma(n)$, namely

$$\Gamma(n) = (n-1)! = \int_0^\infty \mathrm{d}t\; t^{n-1}\mathrm{e}^{-t}. \tag{3.75}$$

We then arrive at the desired result

$$\int \frac{\mathrm{d}\phi^*\mathrm{d}\phi}{2\pi\mathrm{i}}\, \mathrm{e}^{-\phi^*\phi} |\phi\rangle\langle\phi| = \sum_N |N\rangle\langle N| = \hat{1}. \tag{3.76}$$

The coherent states can also be used to perform a trace of an observable. Using a complete set of states $\{|\nu\rangle\}$, the trace of $\hat{O}$ is defined by

$$\mathrm{Tr}[\hat{O}] \equiv \sum_\nu \langle\nu|\hat{O}|\nu\rangle. \tag{3.77}$$

We can express this in terms of coherent states by inserting the completeness relation

$$\begin{aligned}\mathrm{Tr}[\hat{O}] &= \sum_\nu \langle\nu|\hat{O}|\nu\rangle = \int \frac{\mathrm{d}\phi^*\mathrm{d}\phi}{2\pi\mathrm{i}}\, \mathrm{e}^{-\phi^*\phi} \sum_\nu \langle\nu|\hat{O}|\phi\rangle\langle\phi|\nu\rangle \\ &= \int \frac{\mathrm{d}\phi^*\mathrm{d}\phi}{2\pi\mathrm{i}}\, \mathrm{e}^{-\phi^*\phi} \langle\phi|\hat{O}|\phi\rangle.\end{aligned} \tag{3.78}$$

3.8 Fermionic Harmonic Oscillator

To study a fermionic version of the one-dimensional bosonic harmonic oscillator, we demand that the energy quanta in the harmonic oscillator satisfy the Pauli exclusion principle. This means that a maximum of one fermionic quantum can be put into the oscillator, resulting in two possible number states, namely $|0\rangle$ and $|1\rangle$. To mimic such a system, we consider the following Hamiltonian consisting of two energy levels

$$\hat{H} = \begin{bmatrix} +\frac{\hbar\omega}{2} & 0 \\ 0 & -\frac{\hbar\omega}{2} \end{bmatrix}, \tag{3.79}$$

with eigenvalues $\pm\hbar\omega/2$. We denote the ground state $[0,1]$ by $|0\rangle$ and the excited state $[1,0]$ by $|1\rangle$. An arbitrary state is then given by the two-component vector $[\psi_\uparrow, \psi_\downarrow]$. In analogy with the bosonic case, we introduce creation and annihilation operators

$$\hat{b}^\dagger = \begin{bmatrix} 0 & 1 \\ 0 & 0 \end{bmatrix} \quad \text{and} \quad \hat{b} = \begin{bmatrix} 0 & 0 \\ 1 & 0 \end{bmatrix}. \tag{3.80}$$

These operators satisfy the anticommutation relation

$$[\hat{b}\ ,\hat{b}^\dagger]_+ \equiv \hat{b}\ \hat{b}^\dagger + \hat{b}^\dagger\hat{b}\ = \hat{1}, \tag{3.81}$$

as well as

$$[\hat{b},\hat{b}]_+ = [\hat{b}^\dagger,\hat{b}^\dagger]_+ = 0, \tag{3.82}$$

and have the following effect on the two number states

$$\hat{b}|0\rangle = 0, \qquad \hat{b}|1\rangle = |0\rangle, \tag{3.83}$$

and

$$\hat{b}^\dagger|0\rangle = |1\rangle, \qquad \hat{b}^\dagger|1\rangle = 0, \tag{3.84}$$

as required for fermionic creation and annihilation operators that satisfy the Pauli principle. This can also be written as

$$\hat{b}|M\rangle = \sqrt{M}|M-1\rangle \tag{3.85}$$

and

$$\hat{b}^\dagger|M\rangle = \sqrt{1-M}|M+1\rangle, \tag{3.86}$$

which is to be compared with (3.50) and (3.51) for the bosonic case. In terms of the creation and annihilation operators, the Hamiltonian has the form

$$\hat{H} = \hbar\omega\left(\hat{b}^\dagger\hat{b}\ - \frac{1}{2}\right). \tag{3.87}$$

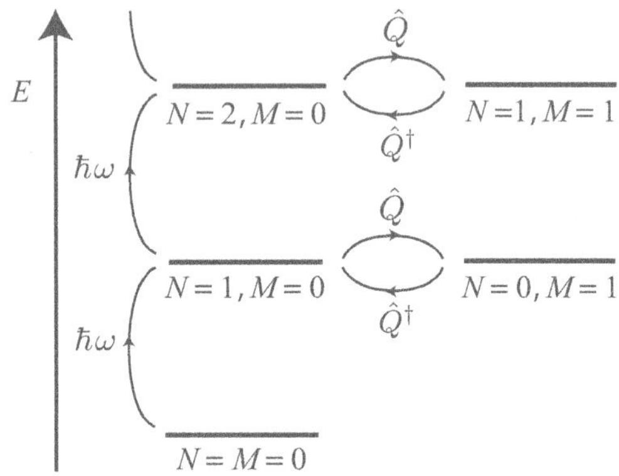

Fig. 3.1 Energy level structure of the supersymmetric one-dimensional harmonic oscillator. See Example 3.5.

Example 3.5. By adding the bosonic and the fermionic Hamiltonians in (3.49) and (3.87) we obtain

$$\hat{H} = \hbar\omega(\hat{a}^\dagger\hat{a} + \hat{b}^\dagger\hat{b}). \tag{3.88}$$

This Hamiltonian has a special kind of symmetry as we show now. The eigenvalues of this Hamiltonian are $\varepsilon_{N,M} = \hbar\omega(N+M)$ belonging to the eigenstates $|N,M\rangle$, where $N = 0,1,2,\ldots,\infty$ are the number of quanta in the bosonic harmonic oscillator and $M = 0,1$ are the number of quanta in the fermionic harmonic oscillator. We thus find that, while the ground state $|0,0\rangle$ is nondegenerate, every bosonic state $|N,0\rangle$, with $N \neq 0$, is degenerate with a fermionic state $|N-1,1\rangle$. This degeneracy is the result of a symmetry that is generated by the operator $\hat{Q} = \hat{b}^\dagger\hat{a}$, which changes a boson in a fermion and has the properties $\hat{Q}|N,0\rangle = \sqrt{N}|N-1,1\rangle$ and $[\hat{H},\hat{Q}]_- = 0$. The last relation is most easily seen by observing that the Hamiltonian can be written as

$$\hat{H} = \hbar\omega[\hat{Q}^\dagger,\hat{Q}]_+, \tag{3.89}$$

where we used the (anti)commutation relations for bosons and fermions, together with the the commutation of fermions with bosons. We also have that $\hat{Q}^2 = 0$, since $b^\dagger b^\dagger = 0$. This system exhibits a simple example of supersymmetry between fermions and bosons, resulting in a degeneracy between bosonic and fermionic states. The energy level structure of the supersymmetric harmonic oscillator is shown in Fig. 3.1.

3.9 Spin

As we show in this section, the spin-1/2 system is equivalent to the fermionic harmonic oscillator that was considered in the previous section. Furthermore, in the study of interacting quantum gases, the spin degree of freedom plays an important role both from a theoretical and an experimental point of view. For example, the atomic spin degree of freedom can be used by experimentalists to trap ultracold atoms by applying an external magnetic field. Later, we will also see an example where a fully spin-polarized atomic system behaves like a noninteracting ideal gas, while an atomic mixture of two spin states is strongly interacting. To appreciate such important effects, we look at the spin degree of freedom in a bit more detail.

Spin behaves quantum mechanically like an angular momentum, which means that we can describe it by an operator $\hat{\mathbf{S}}$, whose Cartesian components satisfy the commutation relations

$$[\hat{S}_x, \hat{S}_y]_- = \mathrm{i}\hbar\hat{S}_z, \quad [\hat{S}_y, \hat{S}_z]_- = \mathrm{i}\hbar\hat{S}_x, \quad [\hat{S}_z, \hat{S}_x]_- = \mathrm{i}\hbar\hat{S}_y. \tag{3.90}$$

From these commutation relations, it follows that the operators $\hat{\mathbf{S}}^2$ and $\hat{S}_z$ commute and can thus be diagonalized simultaneously. Their simultaneous eigenstates $|s, m_s\rangle$ satisfy

$$\hat{\mathbf{S}}^2|s, m_s\rangle = s(s+1)\hbar^2|s, m_s\rangle \tag{3.91}$$

and

$$\hat{S}_z|s, m_s\rangle = m_s\hbar|s, m_s\rangle. \tag{3.92}$$

For simplicity, we consider the case of a spin-$1/2$ atom, for which the z component of the spin angular momentum only takes on the values $\pm\hbar/2$. This implies that we have two basis vectors, $|1/2, 1/2\rangle$ and $|1/2, -1/2\rangle$ which we denote as $|\uparrow\rangle$ and $|\downarrow\rangle$, respectively. The spin angular momentum part of the wavefunction of the atom can be represented as the two-component vector

$$\begin{bmatrix} \psi_\uparrow(t) \\ \psi_\downarrow(t) \end{bmatrix},$$

where $\psi_\uparrow(t)$ is the amplitude for the atom to be in the $|\uparrow\rangle$ state and $\psi_\downarrow(t)$ is the amplitude for the atom to be in the $|\downarrow\rangle$ state, i.e.

$$|\psi(t)\rangle = \psi_\uparrow(t)|\uparrow\rangle + \psi_\downarrow(t)|\downarrow\rangle, \tag{3.93}$$

with $|\psi_\uparrow(t)|^2 + |\psi_\downarrow(t)|^2 = 1$. Because there are only two spin states, we can represent the operators $\hat{S}_i$ by 2×2 matrices, where $\hat{\mathbf{S}} = (\hbar/2)\hat{\boldsymbol{\sigma}}$ with $\hat{\sigma}_i$ represented by the Pauli matrices. In the basis of the eigenstates $|1/2, \pm 1/2\rangle$, the matrix representation of the Pauli matrices is given by

$$\boldsymbol{\sigma}_x = \begin{bmatrix} 0 & 1 \\ 1 & 0 \end{bmatrix}, \qquad \boldsymbol{\sigma}_y = \begin{bmatrix} 0 & -\mathrm{i} \\ \mathrm{i} & 0 \end{bmatrix}, \qquad \boldsymbol{\sigma}_z = \begin{bmatrix} 1 & 0 \\ 0 & -1 \end{bmatrix}. \tag{3.94}$$

The Hamiltonian that describes an atomic spin in a time-dependent magnetic field is given by the Zeeman interaction

$$\hat{H} = -\gamma \,\hat{\mathbf{S}} \cdot \mathbf{B}(t), \tag{3.95}$$

where γ determines the strength of the coupling of the spin to the magnetic field. Let us first consider the time evolution of the spin in a static magnetic field along the z axis. In this case, the Hamiltonian reduces to

$$\hat{H} = -\gamma \, B \, \hat{S}_z, \tag{3.96}$$

which is the fermionic harmonic oscillator of the previous paragraph with $\omega = -\gamma B$. The time evolution operator for this case is given by

$$U(t,0) = \exp\left\{ \frac{\mathrm{i}}{\hbar} \gamma B \, \hat{S}_z \, t \right\}. \tag{3.97}$$

Clearly, the eigenstates of the Hamiltonian are given by the eigenstates of $\hat{S}_z$, i.e. the states $|\uparrow\rangle$ and $|\downarrow\rangle$. For a state that at $t = 0$ is characterized by $|\psi(0)\rangle = \psi_\uparrow(0) \mid \uparrow\rangle + \psi_\downarrow(0) \mid \downarrow\rangle$, we have that

$$|\psi(t)\rangle = \psi_\uparrow(0)\mathrm{e}^{-\mathrm{i}\omega t/2} | \uparrow\rangle + \psi_\downarrow(0)\mathrm{e}^{\mathrm{i}\omega t/2} | \downarrow\rangle. \tag{3.98}$$

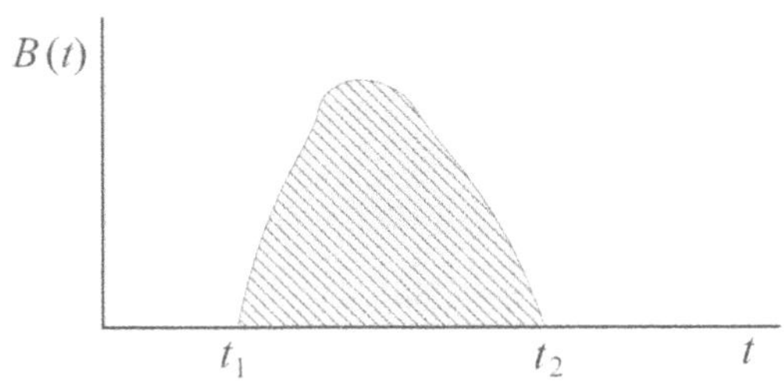

Fig. 3.2 Example of a time-dependent magnetic field, whose area determines the phase of a spin-1/2 atom.

We proceed with the more general case of a magnetic field whose magnitude depends on the time t, but whose direction is always along the x axis. Note that in deriving (3.34) for the time evolution operator $\hat{U}(t,t_\mathrm{i})$, it was explicitly used that the Hamiltonian was time independent. For the more general case that the Hamiltonian does depend on time, but that the Hamiltonians at different times commute, we have for the evolution operator

$$\hat{U}(t,t_{\rm i}) = \exp\left\{-\frac{\rm i}{\hbar}\int_{t_{\rm i}}^{t} {\rm d}t'\ H(t')\right\}. \tag{3.99}$$

In the case of a time-varying magnetic field, such as illustrated in Fig. 3.2, the time evolution of the spin states is determined by the area under the curve, because

$$\hat{U}(t,t_{\rm i}) = \exp\left\{\frac{\rm i}{\hbar}\gamma\hat{S}_x\int_{t_{\rm i}}^{t} {\rm d}t'\ B(t')\right\}. \tag{3.100}$$

From (3.94), we see that $\hat{\sigma}_x^2 = \hat{1}$, such that the time-evolution operator can be written as

$$\begin{aligned}
\hat{U}(t,t_{\rm i}) &= \exp\left\{\frac{\rm i}{\hbar}\gamma\hat{S}_x\int_{t_{\rm i}}^{t} {\rm d}t'\ B(t')\right\} = \exp\left\{-\frac{{\rm i}\theta(t)}{2}\hat{\sigma}_x\right\} \\
&= \sum_{n=0}^{\infty}\frac{1}{n!}\left(-\frac{{\rm i}\theta(t)}{2}\right)^n\hat{\sigma}_x^n \\
&= \sum_{n=0}^{\infty}\frac{(-1)^n}{(2n)!}\left(\frac{\theta(t)}{2}\right)^{2n}\hat{1} - {\rm i}\sum_{n=0}^{\infty}\frac{(-1)^n}{(2n+1)!}\left(\frac{\theta(t)}{2}\right)^{2n+1}\hat{\sigma}_x \\
&= \cos(\theta(t)/2)\hat{1} - {\rm i}\sin(\theta(t)/2)\hat{\sigma}_x,
\end{aligned} \tag{3.101}$$

where we introduce a shorthand notation for the integral in the argument of the exponent, namely

$$\theta(t) = -\gamma\int_{t_{\rm i}}^{t} {\rm d}t'\ B(t'). \tag{3.102}$$

This tells us that if an atom is initially in the state $|\uparrow\rangle$, then at a later time t it is in the state

$$|\psi(t)\rangle = \cos(\theta(t)/2)|\uparrow\rangle - {\rm i}\sin(\theta(t)/2)|\downarrow\rangle\,. \tag{3.103}$$

By manipulating the phase $\theta(t)$, which is determined by the area below the applied magnetic field pulse, we can control the final state. Three examples are of particular experimental importance. The first example is the application of a $\pi/2$ pulse, for which $\theta(t) = \pi/2$ after the pulse. The initial state $|\uparrow\rangle$ is then transferred into the state$(|\uparrow\rangle - {\rm i}|\downarrow\rangle)/\sqrt{2}$. The second example is the π pulse, for which $\theta(t) = \pi$ after the pulse, and the initial state $|\uparrow\rangle$ is transferred into the state $|\downarrow\rangle$. Finally, we can apply a constant magnetic field at all times after t_1. In that case

$$\theta(t) = -\gamma B(t - t_1) \tag{3.104}$$

and the probability amplitude for being in state $|\uparrow\rangle$ equals $P_\uparrow = \cos^2(\gamma B(t-t_1)/2)$, which is seen to oscillate with a period $T = 2\pi/\gamma B$. Such oscillations are known as Rabi oscillations.

3.10 Perturbation Theory

For most Hamiltonians, the time-independent Schrödinger equation cannot be solved exactly. However, if the Hamiltonian $\hat{H} = \hat{H}_0 + \hat{V}$ consists of two parts, one part $\hat{H}_0$ that is exactly solvable and a perturbation $\hat{V}$ that is weak, then we can use perturbation theory to obtain an approximate solution which is systematically improvable. The strategy is to expand both the eigenenergies E_ν and eigenstates $|\nu\rangle$ of $\hat{H}$ in powers of the perturbation

$$E_\nu = E_\nu^{(0)} + E_\nu^{(1)} + E_\nu^{(2)} + \ldots \tag{3.105}$$

and

$$|\nu\rangle = |\nu^{(0)}\rangle + |\nu^{(1)}\rangle + |\nu^{(2)}\rangle + \ldots, \tag{3.106}$$

where the superscript (j) denotes the contribution from the j-th order of the perturbative expansion in the interaction $\hat{V}$. For convenience, we assume that the set of eigenvalues is nondegenerate.

Since the unperturbed Hamiltonian $\hat{H}_0$ is exactly solvable, we have that

$$\hat{H}_0|\nu^{(0)}\rangle = E_\nu^{(0)}|\nu^{(0)}\rangle\,. \tag{3.107}$$

To find the corrections to these unperturbed eigenvalues and eigenstates, we introduce the operator

$$\hat{G}(E) = \frac{1}{E-\hat{H}} = \sum_\nu \frac{|\nu\rangle\langle\nu|}{E-E_\nu}\,, \tag{3.108}$$

which is seen to have the exact eigenstates $|\nu\rangle\langle\nu|$ as residues at the simple poles $E = E_\nu$. We may substitute the formal expansions from (3.105) and (3.106), and evaluate the resulting expansion up to first order in the interaction

$$\begin{aligned}\hat{G}(E) &= \sum_\nu \frac{|\nu^{(0)}\rangle\langle\nu^{(0)}| + |\nu^{(0)}\rangle\langle\nu^{(1)}| + |\nu^{(1)}\rangle\langle\nu^{(0)}| + \ldots}{E - E_\nu^{(0)} - E_\nu^{(1)} - \ldots} \\ &= \sum_\nu \left\{ \frac{|\nu^{(0)}\rangle\langle\nu^{(0)}|}{E-E_\nu^{(0)}} + \frac{|\nu^{(0)}\rangle\langle\nu^{(0)}|}{\left(E-E_\nu^{(0)}\right)^2} E_\nu^{(1)} + \frac{|\nu^{(0)}\rangle\langle\nu^{(1)}|}{E-E_\nu^{(0)}} + \frac{|\nu^{(1)}\rangle\langle\nu^{(0)}|}{E-E_\nu^{(0)}} + \ldots \right\}.\end{aligned} \tag{3.109}$$

To find the actual expressions for the first-order corrections, we consider the explicit expansion of $\hat{G}(E)$ in terms of $\hat{V}$. From (3.108), we have that

$$\hat{G}^{-1}(E) = E - \hat{H}_0 - \hat{V} = \hat{G}_0^{-1}(E) - \hat{V}, \tag{3.110}$$

where we also introduced $\hat{G}_0(E) = (E - \hat{H}_0)^{-1}$. The above equation can be rewritten as

$$\hat{G}(E) = \hat{G}_0(E) + \hat{G}_0(E)\hat{V}\hat{G}(E) = \hat{G}_0(E) + \hat{G}_0(E)\hat{V}\hat{G}_0(E) + \dots, \quad (3.111)$$

such that we can insert completeness relations of the unperturbed basis set $\{|\nu^{(0)}\rangle\}$ into the right-hand side of (3.111) to obtain

$$\hat{G}(E) = \sum_{\nu} \frac{|\nu^{(0)}\rangle\langle\nu^{(0)}|}{E - E_\nu^{(0)}} + \sum_{\nu,\nu'} \frac{|\nu'^{(0)}\rangle\langle\nu'^{(0)}|\hat{V}|\nu^{(0)}\rangle\langle\nu^{(0)}|}{\left(E - E_{\nu'}^{(0)}\right)\left(E - E_\nu^{(0)}\right)} + \dots . \quad (3.112)$$

Considering the case $\nu = \nu'$ in the second term of (3.112) and comparing it with (3.109), we find the first-order correction to the eigenenergy

$$E_\nu^{(1)} = \langle\nu^{(0)}|\hat{V}|\nu^{(0)}\rangle. \quad (3.113)$$

In the case that $\nu \neq \nu'$, the second term of (3.112) gives rise to the first-order correction $|\nu^{(1)}\rangle$ to the eigenstates. To see this more clearly, we split the fraction, giving

$$\sum_{\nu}\sum_{\nu'\neq\nu}\left\{\frac{|\nu'^{(0)}\rangle\langle\nu'^{(0)}|\hat{V}|\nu^{(0)}\rangle\langle\nu^{(0)}|}{(E_{\nu'}^{(0)} - E_\nu^{(0)})(E - E_{\nu'}^{(0)})} + \frac{|\nu'^{(0)}\rangle\langle\nu'^{(0)}|\hat{V}|\nu^{(0)}\rangle\langle\nu^{(0)}|}{(E_\nu^{(0)} - E_{\nu'}^{(0)})(E - E_\nu^{(0)})}\right\},$$

such that the second term of this expression can be directly compared with the fourth term of $\hat{G}^{-1}(E)$ in (3.109), while the first term corresponds to the third term in (3.109). As a result, we find for the first-order correction $|\nu^{(1)}\rangle$ to the eigenstates

$$|\nu^{(1)}\rangle = \sum_{\nu'\neq\nu} |\nu'^{(0)}\rangle \frac{\langle\nu'^{(0)}|\hat{V}|\nu^{(0)}\rangle}{E_\nu^{(0)} - E_{\nu'}^{(0)}}, \quad (3.114)$$

and the corresponding expression for the hermitian conjugate. In a similar way, we can derive higher-order corrections by looking at higher-order terms in the perturbative expansion. As an example, for the second-order correction to the energy we obtain

$$E_\nu^{(2)} = \sum_{\nu'\neq\nu} \frac{|\langle\nu'^{(0)}|\hat{V}|\nu^{(0)}\rangle|^2}{E_\nu^{(0)} - E_{\nu'}^{(0)}} . \quad (3.115)$$

Equations (3.113), (3.114) and (3.115) were first obtained by Schrödinger when he generalized a classical perturbative method developed by Lord Rayleigh to quantum mechanics. We note that the present way of deriving the perturbative corrections is more analogous to the methods used in quantum field theory, which is the topic of Chap. 8.

3.11 Scattering Theory

So far, we have mainly dealt with single-body quantum-mechanical problems. In this section, we look at the more general problem of two particles interacting via a

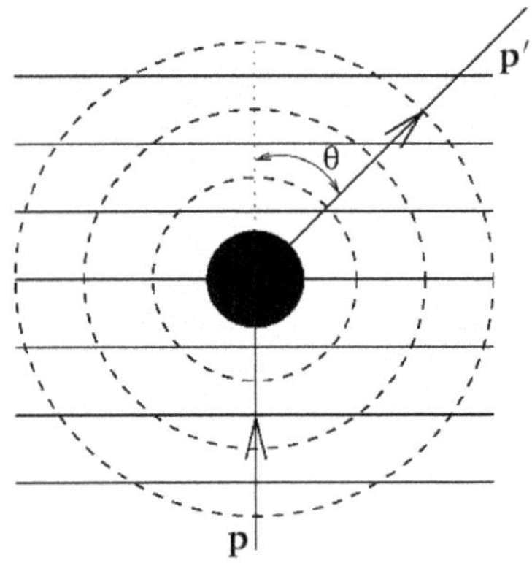

Fig. 3.3 Two-atom scattering in the center-of-mass reference frame. The atoms are initially in a plane-wave state with relative momentum $\mathbf{p}$, and scatter into the spherical wave with relative momentum $\mathbf{p}'$. Due to energy conservation we have that $|\mathbf{p}| = |\mathbf{p}'|$. The angle between $\mathbf{p}$ and $\mathbf{p}'$ is denoted by θ. The region where the interaction takes place is indicated by the black circle. Figure adapted from reference [32].

potential $V(\mathbf{r}_1 - \mathbf{r}_2)$ that only depends on the relative coordinate $\mathbf{r} = \mathbf{r}_1 - \mathbf{r}_2$ between the two particles. It is left as an exercise to show that the corresponding two-body Schrödinger equation separates into a part describing the center-of-mass motion and a part describing the relative motion, which then both effectively reduce to single-body problems. The center-of-mass part part behaves as a free particle with mass $M = m_1 + m_2$, which is equal to $2m$ for equal particle masses, whereas the relative part behaves as a single particle with reduced mass $\mu = m_1 m_2/(m_1 + m_2)$, equal to $m/2$ for equal masses. The relative wavefunction is then determined by the time-independent Schrödinger equation

$$\left\{-\frac{\hbar^2 \boldsymbol{\nabla}^2}{m} + V(\mathbf{r})\right\} \psi(\mathbf{r}) = \left\{\hat{H}_0 + V(\mathbf{r})\right\} \psi(\mathbf{r}) = E\psi(\mathbf{r}), \tag{3.116}$$

where we assumed equal masses and introduced $\hat{H}_0$ for the kinetic-energy operator. We can think of the above time-independent Schrödinger equation as describing a steady-state solution for a continuous stream of particles scattering from a central potential. We assume that the potential $V(\mathbf{r})$ is short ranged, which means that it becomes negligible at a certain distance.

If the separation between the atoms is large, such that the interaction potential can be neglected, then (3.116) becomes the Schrödinger equation for a free particle. In this region, the wavefunction is a superposition between the incoming plane-wave state with relative momentum $\mathbf{p}$ and the scattered state with relative momentum $\mathbf{p}'$, as also illustrated in Fig. 3.3. In the center-of-mass frame, the incoming plane wave corresponds to a kinetic energy $E = \mathbf{p}^2/m = 2\varepsilon_{\mathbf{p}}$. Since we are interested in elastic scattering processes, this energy is conserved. In scattering theory the relevant quantity to calculate is called the scattering amplitude, which is the probability amplitude for the initial plane-wave state with relative momentum $\mathbf{p}$ to scatter into the spherical wave with relative momentum $\mathbf{p}'$. In this section, we show that the scattered state is indeed a spherical wave.

From the above considerations, we have that the time-independent Schrödinger equation in basis-independent notation becomes

$$\{2\varepsilon_{\mathbf{p}} - \hat{H}_0\}\,|\psi_{\mathbf{p}}^{(+)}\rangle = \hat{V}|\psi_{\mathbf{p}}^{(+)}\rangle, \tag{3.117}$$

whose solutions $|\psi_{\mathbf{p}}^{(+)}\rangle$ can formally be obtained by applying $1/(2\varepsilon_{\mathbf{p}} - \hat{H}_0 + \mathrm{i}0)$ to both sides and realizing that the solution of the problem for $\hat{V} = 0$ is given by $|\mathbf{p}\rangle$, such that

$$|\psi_{\mathbf{p}}^{(+)}\rangle = |\mathbf{p}\rangle + \frac{1}{2\varepsilon_{\mathbf{p}} - \hat{H}_0 + \mathrm{i}0}\hat{V}|\psi_{\mathbf{p}}^{(+)}\rangle. \tag{3.118}$$

The notation i0 means $\mathrm{i}\varepsilon$ with $\varepsilon \downarrow 0$, where this limiting procedure is used to deal with the singular nature of the operator $1/(E - \hat{H}_0)$. The reason for choosing $+\mathrm{i}0$ and not $-\mathrm{i}0$ will become apparent when we obtain the solution for the scattering wavefunction $\langle \mathbf{r}|\psi_{\mathbf{p}}^{(+)}\rangle$. The equation for the scattering state, (3.118), is known as the Lippmann-Schwinger equation. To find the scattering wavefunction $\psi_{\mathbf{p}}^{(+)}(\mathbf{r})$, we multiply (3.118) with $\langle \mathbf{r}|$. Doing so, we obtain

$$\langle \mathbf{r}|\psi_{\mathbf{p}}^{(+)}\rangle = \langle \mathbf{r}|\mathbf{p}\rangle + \int \mathrm{d}\mathbf{r}'\, \langle \mathbf{r}|\frac{1}{2\varepsilon_{\mathbf{p}} - \hat{H}_0 + \mathrm{i}0}|\mathbf{r}'\rangle\langle \mathbf{r}'|\hat{V}|\psi_{\mathbf{p}}^{(+)}\rangle, \tag{3.119}$$

where the first term on the right-hand side corresponds to the incoming plane wave

$$\langle \mathbf{r}|\mathbf{p}\rangle = \frac{e^{\mathrm{i}\mathbf{p}\cdot\mathbf{r}/\hbar}}{(2\pi\hbar)^{3/2}}, \tag{3.120}$$

while the second term requires some more work, and gives

$$\begin{aligned}
&\int \mathrm{d}\mathbf{r}'\, \langle \mathbf{r}|\frac{1}{2\varepsilon_{\mathbf{p}} - \hat{H}_0 + \mathrm{i}0}|\mathbf{r}'\rangle\langle \mathbf{r}'|\hat{V}|\psi_{\mathbf{p}}^{(+)}\rangle \\
&= \int \mathrm{d}\mathbf{r}' \int \frac{\mathrm{d}\mathbf{p}'}{(2\pi\hbar)^3}\,\frac{\exp\{\mathrm{i}\mathbf{p}'\cdot(\mathbf{r}-\mathbf{r}')/\hbar\}}{2\varepsilon_{\mathbf{p}} - 2\varepsilon_{\mathbf{p}'} + \mathrm{i}0}\langle \mathbf{r}'|\hat{V}|\psi_{\mathbf{p}}^{(+)}\rangle \\
&= -\frac{m}{\hbar^2}\int \mathrm{d}\mathbf{r}'\,\frac{\exp\{\mathrm{i}p|\mathbf{r}-\mathbf{r}'|/\hbar\}}{4\pi|\mathbf{r}-\mathbf{r}'|}\langle \mathbf{r}'|\hat{V}|\psi_{\mathbf{p}}^{(+)}\rangle.
\end{aligned} \tag{3.121}$$

Here, we inserted in the first step the completeness relation of the momentum states and in the second step we used the known integral as shown in Table 3.1. Because the potential $V(\mathbf{r})$ is short ranged, we are primarily interested in the behavior of the scattering wavefunction at distances that are large with respect to this range. Therefore, we can expand the interatomic distance for $r \gg r'$ as

$$|\mathbf{r}-\mathbf{r}'| = \sqrt{r^2 - 2\mathbf{r}\cdot\mathbf{r}' + r'^2} = r\sqrt{1 - 2\frac{\mathbf{r}\cdot\mathbf{r}'}{r^2} + \frac{r'^2}{r^2}} \simeq r - \frac{\mathbf{r}}{r}\cdot\mathbf{r}', \tag{3.122}$$

which we substitute into (3.121), giving

$$\int d\mathbf{r}' \,\langle \mathbf{r}|\frac{1}{2\varepsilon_{\mathbf{p}} - \hat{H}_0 + i0}|\mathbf{r}'\rangle\langle \mathbf{r}'|\hat{V}|\psi_{\mathbf{p}}^{(+)}\rangle$$
$$= -\frac{m}{\hbar^2}\frac{e^{ipr/\hbar}}{4\pi r}\int d\mathbf{r}' \exp\left\{-ip\frac{\mathbf{r}}{r}\cdot\mathbf{r}'/\hbar\right\}\langle \mathbf{r}'|\hat{V}|\psi_{\mathbf{p}}^{(+)}\rangle. \tag{3.123}$$

Then, we define the vector $\mathbf{p}' = p\mathbf{r}/r$, which has the same length as the vector $\mathbf{p}$, but points in the direction $\mathbf{r}/r$. With this definition, we recognize that the integral on the right-hand side of (3.123) is nothing but a Fourier transform. By defining the scattering amplitude

$$f(\mathbf{p}',\mathbf{p}) = -\frac{1}{4\pi}(2\pi\hbar)^3\frac{m}{\hbar^2}\langle \mathbf{p}'|\hat{V}|\psi_{\mathbf{p}}^{(+)}\rangle, \tag{3.124}$$

we indeed find that at distances much larger than the range of the interaction, the total wavefunction can be written as the sum of an incoming plane wave and an outgoing spherical wave, that is

$$\psi_{\mathbf{p}}^{(+)}(\mathbf{r}) = \frac{1}{(2\pi\hbar)^{3/2}}\left\{e^{i\mathbf{p}\cdot\mathbf{r}/\hbar} + f(\mathbf{p}',\mathbf{p})\frac{e^{ipr/\hbar}}{r}\right\}. \tag{3.125}$$

Now, we can also understand the reason for adding the small positive imaginary part $+i0$ in (3.118). The wavefunction $\psi_{\mathbf{p}}^{(+)}(\mathbf{r})$ in (3.125) describes an incoming plane wave and an outgoing spherical wave, whereas a small negative imaginary part would have led to an an incoming spherical wave, i.e. a description of the time-reversed scattering process.

Table 3.1 The integral

$$I_d = \int \frac{d\mathbf{p}'}{(2\pi)^d}\frac{e^{i\mathbf{p}'\cdot(\mathbf{x}-\mathbf{x}')/\hbar}}{\mathbf{p}^2 - \mathbf{p}'^2 + i0},$$

where d is the dimension. The function $K_0(x)$ is the modified Bessel function of the second kind.

dimension d	I_d
1	$\frac{\exp\{ip\lvert x-x'\rvert/\hbar\}}{2ip}$
2	$\frac{K_0(-ip\lvert\mathbf{x}-\mathbf{x}'\rvert/\hbar)}{-2\pi}$
3	$\frac{\exp\{ip\lvert\mathbf{x}-\mathbf{x}'\rvert/\hbar\}}{-4\pi\lvert\mathbf{x}-\mathbf{x}'\rvert/\hbar}$

3.12 Many-particle Quantum Mechanics

If we want to go beyond two particles and describe a many-body system consisting of N identical particles, then we have to solve the N-body Schrödinger equation for the wavefunction $\Psi(\mathbf{x}_1,\ldots,\mathbf{x}_N,t)$, giving

$$\begin{aligned} &i\hbar\frac{\partial}{\partial t}\Psi(\mathbf{x}_1,\ldots,\mathbf{x}_N,t) \\ &= \left\{\sum_i\left(-\frac{\hbar^2\nabla_i^2}{2m}+V^{\text{ex}}(\mathbf{x}_i)\right)+\frac{1}{2}\sum_{i\neq j}V(\mathbf{x}_i-\mathbf{x}_j)\right\}\Psi(\mathbf{x}_1,\ldots,\mathbf{x}_N,t), \end{aligned} \tag{3.126}$$

where $V^{\text{ex}}(\mathbf{x}_i)$ is the external potential and $V(\mathbf{x}_i-\mathbf{x}_j)$ is the two-body interaction potential. In principle, there are also three-body interactions, which occur when the electronic clouds of three particles simultaneously overlap. However, for the interacting atomic quantum gases of interest in this book, these interactions almost never play a role, because under realistic experimental conditions the gases are very dilute. On the other hand, the external potential $V^{\text{ex}}(\mathbf{x}_i)$ is always present in a real experiment, because the atoms have to be prevented from heating by material walls. Typically, such an external trapping potential has a harmonic shape and can for example be created by a space-dependent magnetic field acting on the spin of the atom via the Zeeman interaction, i.e. $\hat{V}^{\text{ex}}(\mathbf{x})=-\gamma\,\mathbf{B}(\mathbf{x})\cdot\hat{\mathbf{S}}$. Another possibility is to trap the atoms optically by using the strong electric fields in a laser beam. This gives rise to an induced electric dipole moment of the atom, such that classically the trapping potential becomes $V^{\text{ex}}(\mathbf{x})=-\alpha E^2(\mathbf{x})$ with α the polarizability. This last kind of trapping is discussed more detailed later on, when we encounter optical lattices, which are periodic potentials created by counter-propagating laser beams.

Since the one-body Schrödinger equation is already often impossible to solve, a solution of the many-body Schrödinger equation with typically millions of interacting particles seems absolutely hopeless. However, the microscopic description in (3.126) actually contains far too much information, because typically we are only interested in a few macroscopic quantities such as the particle density n, the energy U, and the pressure p of the system. These macroscopic quantities are only weakly sensitive to the precise microscopic state the system, and can therefore be obtained by an appropriate average over the microscopic degrees of freedom. In the next chapter, we show how statistical physics is used to perform this averaging properly.

3.13 Problems

Exercise 3.1. Heisenberg Uncertainty Relation
(a) Calculate $\Delta x^2\Delta p^2$ for the coherent state in (3.70).
(b) Prove (3.25).

Exercise 3.2. Prove (3.45) and (3.46).

Exercise 3.3. Solve the one-dimensional Schrödinger equation for a particle that scatters of a hard-sphere potential, i.e.

$$V(x) = \begin{cases} \infty, & x < a. \\ 0, & x > a, \end{cases} \tag{3.127}$$

where a is positive.

Exercise 3.4. Two-particle Schrödinger equation
Consider the Schrödinger equation for two particles that interact through a potential that depends only on their relative coordinate $\mathbf{r}_1 - \mathbf{r}_2$, i.e.

$$\left\{ -\frac{\hbar^2 \nabla_1^2}{2m_1} - \frac{\hbar^2 \nabla_2^2}{2m_2} + V(\mathbf{r}_1 - \mathbf{r}_2) \right\} \psi(\mathbf{r}_1, \mathbf{r}_2) = E \psi(\mathbf{r}_1, \mathbf{r}_2). \tag{3.128}$$

(a) Show by introducing the relative coordinate $\mathbf{r} = \mathbf{r}_1 - \mathbf{r}_2$ and the center-of-mass coordinate $\mathbf{R} = (m_1 \mathbf{r}_1 + m_2 \mathbf{r}_2)/(m_1 + m_2)$ that

$$-\frac{\hbar^2 \nabla_1^2}{2m_1} - \frac{\hbar^2 \nabla_2^2}{2m_2} = -\frac{\hbar^2 \nabla_{\text{rel}}^2}{2\mu} - \frac{\hbar^2 \nabla_{\text{cm}}^2}{2M}, \tag{3.129}$$

where $M = m_1 + m_2$ is the total mass and $\mu = m_1 m_2/(m_1 + m_2)$ is the reduced mass.
(b) Show that the two-particle Schrödinger equation thus becomes two one-particle Schrödinger equations, where the center-of-mass part behaves as a free particle

$$\left\{ -\frac{\hbar^2 \nabla_{\text{cm}}^2}{2M} \right\} \psi_{\text{cm}}(\mathbf{R}) = E_{\text{cm}} \psi_{\text{cm}}(\mathbf{R}), \tag{3.130}$$

while the relative part behaves as a particle of mass μ in a central potential $V(\mathbf{r})$

$$\left\{ -\frac{\hbar^2 \nabla_{\text{rel}}^2}{2\mu} + V(\mathbf{r}) \right\} \psi_{\text{rel}}(\mathbf{r}) = E_{\text{rel}} \psi_{\text{rel}}(\mathbf{r}) \tag{3.131}$$

with $\psi(\mathbf{R}, \mathbf{r}) = \psi_{\text{cm}}(\mathbf{R}) \psi_{\text{rel}}(\mathbf{r})$ and $E = E_{\text{cm}} + E_{\text{rel}}$.

Additional Reading

- A classical work on quantum mechanics is P. A. M. Dirac, *Principles of Quantum Mechanics*, (Oxford University Press, London, 1958).
- There are also more recent textbooks on quantum mechanics, such as J. J. Sakurai, *Modern Quantum Mechanics*, (Addison-Wesley, Reading, 1994), and
- B. H. Bransden and C. J. Joachain, *Quantum Mechanics*, (Prentice Hall, New York, 2000).

Chapter 4
Statistical physics

There are four Laws. The third of them, the Second Law, was recognized first; the first, the Zeroth Law, was formulated last; the First Law was second; the Third Law might not even be a law in the same sense as the others.
– P.W. Atkins

Thermodynamics is a phenomenological theory for interacting many-body systems, whose empirical laws are taken from experiments. The aim of statistical physics is to derive these phenomenological laws starting from a true microscopic description of the many-body system. It is then needed to perform an appropriate average over the many microscopic degrees of freedom. The fundamental assumption of statistical physics is that every microscopic state that is accessible to the system is equally probable. Therefore, the problem of finding the correct probability distribution over which we have to average reduces to the problem of finding the total number of states of the system.

In this chapter, we discuss the basic concepts and techniques that are used in statistical physics to describe many-body systems. In particular, we briefly consider the three most frequently used statistical ensembles, namely the micro-canonical, the canonical and the grand-canonical ensemble. In the thermodynamic limit these ensembles become essentially equivalent, such that we can choose the ensemble that is most convenient to work with. In our case, this is nearly always the grand-canonical ensemble, which we then apply to the study of the ideal gases. We treat the classical gas that obeys Maxwell-Boltzmann statistics, the Bose gas that obeys Bose-Einstein statistics and the Fermi gas that obeys Fermi-Dirac statistics. At low temperatures, the ideal Bose gas undergoes a phase transition better known as Bose-Einstein condensation. A thorough knowledge of the ideal gases is important for understanding the interacting quantum gases, which is the topic of the second part of this book.

4.1 Legendre Transformations

The Legendre transformation is a convenient tool that is often used in thermodynamics, and in many other fields of physics. A familiar example comes from classical mechanics, which can be formulated in terms of the Lagrangian formalism or, equivalently, in terms of the Hamiltonian formalism. While the Lagrangian

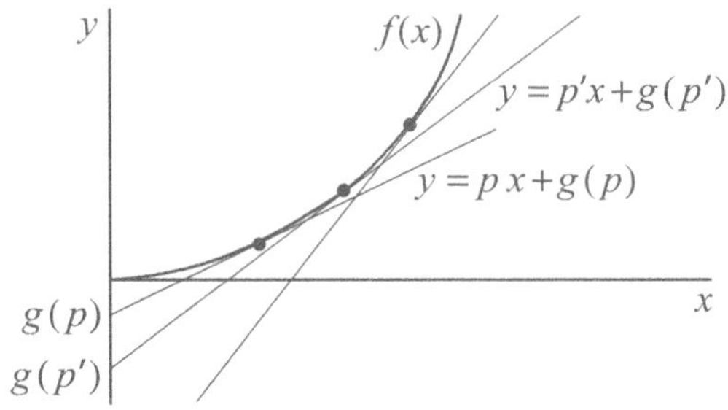

Fig. 4.1 Graphical illustration of the Legendre transform $g(p)$ for a convex function $f(x)$.

$L(x_i, \mathrm{d}x_i/\mathrm{d}t, t)$ is a function that depends explicitly on the positions x_i and the velocities $\mathrm{d}x_i/\mathrm{d}t$ of the particles, the Hamiltonian $H(x_i, p_i, t)$ depends on the generalized momenta $p_i = \partial L/\partial(\mathrm{d}x_i/\mathrm{d}t)$ rather than on the velocities. The relationship between the Hamiltonian and the Lagrangian of the system is given by

$$H(x_i, p_i, t) = \sum_i \frac{\mathrm{d}x_i}{\mathrm{d}t} p_i - L(x_i, \mathrm{d}x_i/\mathrm{d}t, t), \tag{4.1}$$

which is an example of a Legendre transformation. It is a change of variables that leads to an equivalent description of the system, such that it is often used in transforming to the most suitable variables for the situation of interest.

To study the Legendre transformation in somewhat more detail, we use a simple example, namely a convex function $f(x)$, such that $\mathrm{d}^2 f/\mathrm{d}x^2 > 0$. The function f determines a set of points (x, y) with $y = f(x)$. However, this set of points also gives rise to a set of tangent lines that contain exactly the same information, as is illustrated in Fig. 4.1. Each tangent line is uniquely determined by the slope of the function $f(x)$ at x and the intercept of the tangent with the y axis. Conversely, each tangent line determines a unique point on the curve (x, y). The corresponding Legendre transform $g(p)$ of $f(x)$ is defined as

$$g(p) = f(x(p)) - x(p)p, \tag{4.2}$$

where $x(p)$ is given by the root of the equation

$$p = \frac{\mathrm{d}f(x)}{\mathrm{d}x}. \tag{4.3}$$

Note that the convexity of the function $f(x)$ is required to make sure that this equation has a unique solution. In the following, we will encounter various examples of Legendre transformations in statistical physics.

4.2 Statistical Physics

The first law of thermodynamics, also known as the conservation of energy, states that if we apply an amount of heat $\mathrm{d}Q$ to a system, then this heat can be used either to increase the internal energy $\mathrm{d}U$ of the system or to let the system do an amount work $\mathrm{d}W$. Mathematically, this is expressed as

$$\mathrm{d}Q \equiv T\mathrm{d}S = \mathrm{d}U + \mathrm{d}W = \mathrm{d}U + p\mathrm{d}V - \mu\mathrm{d}N, \tag{4.4}$$

where T is the temperature, p the pressure, V the volume, μ the chemical potential, N the number of particles, and S the entropy of the system, defined by

$$S(U,V,N) \equiv k_{\mathrm{B}} \log g(U,V,N), \tag{4.5}$$

where $k_{\mathrm{B}} = 1.38 \cdot 10^{-23}$ J/K is Boltzmann's constant and $g(U,V,N)$ is the total number of states. The term $p\mathrm{d}V$ describes the mechanical work that the system can do by expanding, whereas the term $-\mu\mathrm{d}N$ describes the chemical work that the system can do by expelling particles. The first law of thermodynamics implies that

$$\left.\frac{\partial S}{\partial U}\right|_{V,N} = \frac{1}{T}, \quad \left.\frac{\partial S}{\partial V}\right|_{U,N} = \frac{p}{T}, \quad \left.\frac{\partial S}{\partial N}\right|_{U,V} = -\frac{\mu}{T}, \tag{4.6}$$

such that the entropy of the system $S(U,V,N)$ is everything we need to know for determining all relevant macroscopic variables. Instead of using the entropy $S(U,V,N)$, we can equally well use the internal energy $U(S,V,N)$, for which we have

$$\mathrm{d}U = T\mathrm{d}S - p\,\mathrm{d}V + \mu\mathrm{d}N, \tag{4.7}$$

leading to

$$T = \left.\frac{\partial U}{\partial S}\right|_{V,N}, \quad p = \left.\frac{\partial U}{\partial V}\right|_{S,N}, \quad \mu = \left.\frac{\partial U}{\partial N}\right|_{S,V}. \tag{4.8}$$

This last equation shows that the chemical potential is the energy needed to add a particle to the system at fixed entropy and volume.

4.2.1 Spin Chain

To make the above concepts more concrete, we study the one-dimensional Ising model, which is a spin chain with $N+1$ spin-$1/2$ particles. For later convenience, we take N to be even. Nearest neighboring spins interact through a coupling of strength J, such that the Hamiltonian becomes

$$\hat{H} = -\frac{4J}{\hbar^2}\sum_{j=1}^{N} \hat{S}_{z,j}\,\hat{S}_{z,j+1}, \tag{4.9}$$

where $\hat{S}_{z,j}$ denotes the z component of the spin operator $\hat{\mathbf{S}}$ at site j. Note that for positive J, the absolute ground state is reached when all the spins are aligned, which leads to ferromagnetic order. For negative J, the absolute ground state consists only of anti-aligned spins, which leads to antiferromagnetic order.

Next, we introduce what are known as bond variables

$$\hat{V}_j = \frac{1}{\hbar^2}(\hat{S}_{z,j+1} - \hat{S}_{z,j})^2, \tag{4.10}$$

which are zero for equal spins and one for opposite spins. The set $\{\hat{S}_{z,1},\ldots,\hat{S}_{z,N+1}\}$ is then equivalent to the set $\{\hat{S}_{z,1},\hat{V}_1,\ldots,\hat{V}_N\}$, since both sets can be used to characterize all possible states of the spin chain. Using the latter set, the Hamiltonian becomes

$$\hat{H} = \sum_{j=1}^{N}(2J\hat{V}_j - J) = -JN + 2J\sum_{j=1}^{N}\hat{V}_j, \tag{4.11}$$

and the model of interacting spins has been transformed to a model of noninteracting bonds, which allows for an exact solution. Note that since the Hamiltonian does not depend on $\hat{S}_{z,1}$, there is at least a two-fold degeneracy for each eigenvalue of the Hamiltonian. To identify these eigenvalues, we call the number of times that $\hat{V}_j$ is equal to zero $N/2 - s$, such that $-N/2 \leq s \leq N/2$ and $E = 2Js$. Calculating the number of states that give rise to this energy, we find

$$g(s,N+1) = 2\frac{N!}{(N/2+s)!(N/2-s)!}. \tag{4.12}$$

Then, we can use Stirling's formula

$$N! = \sqrt{2\pi N}N^N \mathrm{e}^{-N}\left(1 + \frac{1}{12N} + \ldots\right) \tag{4.13}$$

and consider the case $|s| \ll N/2$, such that

$$g(s,N+1) \simeq \sqrt{\frac{2}{\pi N}}2^{N+1}\exp\left(\frac{-2s^2}{N}\right), \tag{4.14}$$

where we used $(1-x/M)^M \simeq \mathrm{e}^{-x}$ for large M. Also note that $\int_{-\infty}^{\infty}\mathrm{d}s\; g(s,N+1) = 2^{N+1}$, showing that our approximation still reproduces the exact number of states.

An important observation is that $g(s,N+1)$ is a very sharply-peaked function of s/N with a spread of only $\mathcal{O}(1/\sqrt{N})$. To illustrate this point we also allow for fluctuations of the internal energy, due to for example heat exchange with another large system, called a heat bath. In the next section we show that the application of the fundamental assumption of statistical physics, which states that each microstate is equally probable, then leads to $P(s) \propto g(s,N+1)e^{-E(s)/k_BT}$ for the probability to measure an energy $E(s)$ of the spin chain. Considering now in particular such high temperatures that the contact with the heat bath allows for essentially arbitrary large energy fluctuations, then by far the most probable internal energy is given by $s = 0$. However, we realize that this last physical statement is actually rather weakly dependent on the fundamental assumption of statistical physics itself. This is because nearly all other initial probability distributions for the microstates would also lead to a very sharply-peaked distribution with a maximum at $s = 0$ after multiplication with the number of states $g(s,N+1)$. The fact that the physical predictions of the theory do not depend strongly on the fundamental assumption itself can be seen as a physical argument to explain why this assumption is valid. It is also important to realize that the very sharp behavior of the number of available states is not a specific feature of the one-dimensional Ising model, but is in fact inherent to any system in the thermodynamic limit, meaning that $N \longrightarrow \infty$. The crucial ingredient for the behavior is namely that the entropy is an extensive variable, which means that for large N the entropy is proportional to N, i.e. $S(U,V,N) = N\,S(U/N,V/N)$. As a result, we find

$$g(U,V,N) = e^{N\,S(U/N,V/N)/k_B}, \tag{4.15}$$

which for $N \longrightarrow \infty$ is indeed infinitely sharply peaked around the maximum of the entropy with only $\mathcal{O}(1/\sqrt{N})$ fluctuations in the intensive variables, such as for example the density $n = N/V$.

If the total internal energy of our system is fixed, which is the case for an isolated system, then thermodynamic equilibrium is reached when the system is at the maximum of its entropy. This is physically understood by noting that the maximum of the entropy corresponds to the maximum number of microstates, such that a system naturally tends to its most probable configuration. Describing a system at a fixed internal energy leads to the microcanonical ensemble, which is therefore particularly suitable for isolated systems. However, in practice, we are usually not dealing with an isolated system at a fixed total internal energy, but rather with a system at a fixed temperature in contact with its surroundings. Then, the system of interest together with its surroundings can be considered isolated and the microcanonical ensemble is applicable to the whole. The surroundings are usually called a heat bath or a reservoir, because their only physical relevance is to keep the temperature of the system of interest constant. Since the reservoir constantly exchanges heat, the internal energy of the system of interest fluctuates, such that it is not described by the microcanonical ensemble. Next, we derive the properties of such a system, which is said to be in the canonical ensemble. Another possibility we consider is that the reservoir not only exchanges heat, but also particles with the system of interest. Then, the chemical potential remains constant, whereas the number of particles in the system of interest fluctuates. This results in the grand-canonical ensemble. It is possible to show that in the thermodynamic limit the different ensembles become essentially equivalent. In this limit, we can therefore conveniently choose the ensemble that is most practical to work with, which is for our purposes usually the grand-canonical ensemble.

4.2.2 Canonical Ensemble

We define the (Helmholtz) free energy $F(T,V,N) = U - TS$ for an isolated system as the Legendre transform of U. From $\mathrm{d}F = \mathrm{d}U - T\,\mathrm{d}S - S\,\mathrm{d}T$, together with equation (4.4), we obtain

$$\mathrm{d}F = -S\,\mathrm{d}T - p\,\mathrm{d}V + \mu \mathrm{d}N. \tag{4.16}$$

Therefore, once we have obtained the Helmholtz free energy, the relevant thermodynamics variables are found from

$$S = -\left.\frac{\partial F}{\partial T}\right|_{V,N}, \quad p = -\left.\frac{\partial F}{\partial V}\right|_{T,N}, \quad \mu = \left.\frac{\partial F}{\partial N}\right|_{T,V}. \tag{4.17}$$

Next, we consider the total system to be split up into a system of interest and a large heat bath or reservoir. The two subsystems are coupled to each other such that the temperature of both remains constant. Then, the fundamental assumption of statistical physics tells us that the probability for the system of interest to be in a certain microstate ν with energy U_ν is given by

$$P(\nu) \propto g_{\mathrm{R}}(U_0 - U_\nu) = \mathrm{e}^{S_{\mathrm{R}}(U_0 - U_\nu)/k_{\mathrm{B}}}, \tag{4.18}$$

where U_0 is the total energy of the reservoir plus the system of interest, which we consider to be fixed, and where $g_{\mathrm{R}}(U_0 - U_\nu)$ is the number of microstates in the reservoir with an energy $U_0 - U_\nu$. In the case of a large reservoir, U_0 is much larger than U_ν, and we can expand the entropy of the reservoir S_{R} as

$$S_{\mathrm{R}}(U_0 - U_\nu) = S_{\mathrm{R}}(U_0) - U_\nu \frac{\mathrm{d}S_{\mathrm{R}}}{\mathrm{d}U_0} + \frac{1}{2} U_\nu^2 \frac{\mathrm{d}^2 S_{\mathrm{R}}}{\mathrm{d}U_0^2} + \ldots, \tag{4.19}$$

where the first term on the the right-hand side is of order N_{R}, the second of order N and the third of order N^2/N_{R}, which can be neglected if the number of particles in the reservoir N_{R} is much bigger than the number of particles N in the system of interest. Using (4.18) and $\mathrm{d}S_{\mathrm{R}}/\mathrm{d}U_0 = 1/T$, we find that $P(\nu) = \mathrm{e}^{-U_\nu/k_{\mathrm{B}}T}/Z$, where the normalization factor Z is determined by the condition that the total probability equals 1. The normalization factor is also called the canonical partition function and yields

$$Z = \sum_\nu \mathrm{e}^{-\beta U_\nu}, \tag{4.20}$$

where the sum is over all microstates of the system of interest and where we introduced $\beta = 1/k_{\mathrm{B}}T$. For a large system of interest, we can replace the sum over states by an integral over the internal energy U, which leads to

$$Z \simeq \int \mathrm{d}U\; g(U)\, \mathrm{e}^{-\beta U} = \int \mathrm{d}U\; \mathrm{e}^{-\beta(U - TS(U))} \simeq \mathrm{e}^{-\beta F}, \tag{4.21}$$

where we defined $F \equiv \langle U \rangle - TS(\langle U \rangle)$ with $\langle U \rangle$ minimizing $U - TS(U)$ and $S(U)$ the entropy corresponding to the system of interest. This definition generalizes the free energy to the canonical ensemble, in which the internal energy U is fluctuating. It can be shown that the minimized variable $F \equiv \langle U \rangle - TS(\langle U \rangle)$ corresponds to a maximized entropy for the entire system including reservoir. It is a general feature of the canonical ensemble that thermodynamic equilibrium is reached by minimizing the free energy. Furthermore, (4.21) shows that for large systems we have for the free energy $F = -k_{\mathrm{B}}T \log Z$, which we now also prove for any system size. Since the average energy $\langle U \rangle$ of the system in the canonical ensemble equals

$$\langle U \rangle \equiv \frac{1}{Z} \sum_{\nu} U_{\nu} \mathrm{e}^{-\beta U_{\nu}} = -\frac{\partial}{\partial \beta} \log Z, \tag{4.22}$$

and since by definition we have that $S = -\partial F / \partial T|_{V,N}$, the free energy obeys the differential equation

$$F = \langle U \rangle + T \left. \frac{\partial F}{\partial T} \right|_{V,N} . \tag{4.23}$$

Therefore, we find for the average energy

$$\langle U \rangle = F - T \left. \frac{\partial F}{\partial T} \right|_{V,N} = \left. \frac{\partial}{\partial \beta} (\beta F) \right|_{V,N} . \tag{4.24}$$

Together with equation (4.22), this shows that $F = -k_{\mathrm{B}}T \log Z + c k_{\mathrm{B}} T$. In the limit of zero temperature, Z will be dominated by the contribution corresponding to the lowest energy U_0, i.e. $Z \to g_0 \mathrm{e}^{-\beta U_0}$, where g_0 is the degeneracy of the ground state. This means that $F \to -k_{\mathrm{B}}T(\log g_0 + c) + U_0$. But since $-\partial F / \partial T = S \to -k_{\mathrm{B}} \log g_0$, we must take $c = 0$. This proves the important claim that

$$F(T,V,N) = -k_{\mathrm{B}}T \log Z(T,V,N) . \tag{4.25}$$

As a result, the general procedure to tackle a problem in the canonical ensemble is to calculate first the canonical partition function Z and from that the free energy F. Using (4.17), we then obtain all the relevant thermodynamical variables.

4.2.3 Grand-Canonical Ensemble

The calculation of the canonical partition function is often inconveniently complicated due to the fact that we need to sum over all microstates with a fixed number of particles. Instead, it is typically much simpler to relax the constraint of a fixed particle number and consider a system at a fixed chemical potential. This corresponds physically to a situation, where the system of interest is in contact with a reservoir with which it can exchange both heat and particles. The relevant thermodynamic variable in the grand-canonical ensemble is the thermodynamic potential Ω, which is the Legendre transform of the free energy F generalized to the case when the total particle number is fluctuating, that is

$$\Omega = F - \mu\langle N\rangle = \langle U\rangle - TS - \mu\langle N\rangle. \tag{4.26}$$

This means that the thermodynamic potential is an explicit function of $\Omega(T,V,\mu)$. As a result, we have

$$d\Omega = -S\,\mathrm{d}T - p\,\mathrm{d}V - \langle N\rangle \mathrm{d}\mu, \tag{4.27}$$

and once we know $\Omega(T,V,\mu)$, we can calculate the thermodynamical equilibrium properties of the system by

$$S = -\left.\frac{\partial\Omega}{\partial T}\right|_{V,\mu}, \quad p = -\left.\frac{\partial\Omega}{\partial V}\right|_{T,\mu}, \quad \langle N\rangle = -\left.\frac{\partial\Omega}{\partial\mu}\right|_{T,V}. \tag{4.28}$$

Using similar arguments as in the derivation for the canonical ensemble, we can show that the probability for the system of interest to be in a microstate ν with an internal energy U_ν and a number of particles N_ν is given by

$$P(\nu) = \frac{1}{Z}\exp\{-\beta(U_\nu - \mu N_\nu)\}, \tag{4.29}$$

where we introduced the grand-canonical partition function

$$Z = \sum_\nu \mathrm{e}^{-\beta(U_\nu - \mu N_\nu)}, \tag{4.30}$$

where the sum is over all microstates of the system of interest. We now also expect that as before

$$\Omega(T,V,\mu) = -k_{\mathrm{B}}T\log Z(T,V,\mu), \tag{4.31}$$

where to show this, we first observe that

$$\langle U\rangle - \mu\langle N\rangle \equiv \frac{1}{Z}\sum_\nu (U_\nu - \mu N_\nu)\mathrm{e}^{-\beta(U_\nu - \mu N_\nu)} = -\frac{\partial}{\partial\beta}\log Z. \tag{4.32}$$

From (4.27) we know that $S = -\partial\Omega/\partial T|_{V,\mu}$, and combining this with (4.26) yields

$$\langle U\rangle - \mu\langle N\rangle = \Omega - T\left.\frac{\partial\Omega}{\partial T}\right|_{V,\mu} = \left.\frac{\partial}{\partial\beta}(\beta\Omega)\right|_{V,\mu}. \tag{4.33}$$

We conclude that indeed $\Omega = -k_{\mathrm{B}}T\log Z$, where the integration constant can again be shown to equal zero by taking the limit $T \to 0$.

Furthermore, in the grand-canonical ensemble it is the minimum of the thermodynamic potential that leads to thermodynamic equilibrium, since it corresponds to the maximum of the entropy for the entire system including reservoir. Also note that the thermodynamic potential is not a completely new thermodynamic quantity, which we see as follows. Since the thermodynamic potential is extensive, meaning $\Omega(T,\lambda V,\mu) = \lambda\Omega(T,V,\mu)$, we have that $\Omega(T,V,\mu) = V\omega(T,\mu)$. However, we also know that

$$p = -\left.\frac{\partial \Omega}{\partial V}\right|_{T,\mu} = -\omega(T,\mu), \tag{4.34}$$

showing that the thermodynamic potential is directly related to the pressure by

$$\Omega(T,V,\mu) = -p(T,\mu)V. \tag{4.35}$$

4.3 Ideal Gases

The methods described in the previous section can be used to calculate the thermodynamic properties of the ideal Bose and Fermi gases in the grand-canonical ensemble. The ideal Bose (Fermi) gas consists of noninteracting identical bosons (fermions), for which in the homogeneous case, i.e. in the absence of an external potential, the single-particle Hamiltonian is simply given by $\hat{H} = \hat{\mathbf{p}}^2/2m$. We consider spinless bosons or fermions in a box of volume L^3, where we mean with spinless that there is only one spin degree of freedom, which is then irrelevant for the Hamiltonian. We impose periodic boundary conditions, such that the allowed momenta are given by $p_i = \hbar k_i = 2\pi\hbar n_i/L$, where i denotes the Cartesian direction x, y, z, and $n_i = 0, \pm 1, \pm 2, \ldots, \infty$, while the energy levels are given by $\varepsilon_{\mathbf{k}} = \hbar^2\mathbf{k}^2/2m$. For a system of N identical bosons or fermions, the state $\mathbf{N}$ of the system is uniquely specified by giving the occupation numbers $N_{\mathbf{k}}$ of the single-particle states with wavenumber $\mathbf{k}$. We have

$$N_{\mathbf{k}} = \begin{cases} 0,1,2,\ldots,\infty & \text{Bosons} \\ 0,1 & \text{Fermions,} \end{cases} \tag{4.36}$$

where due to the Pauli principle two identical fermions are forbidden to be in the same momentum state.

To calculate the grand-canonical partition function we have to sum the grand-canonical probability distribution over all microstates corresponding to a given number of particles, after which we also have to sum over all possible particle numbers. This means,

$$\begin{aligned} Z &= \sum_{N=0}^{\infty} \sum_{\mathbf{N}}{}' \exp\left\{-\beta\left(\sum_{\mathbf{k}} \varepsilon_{\mathbf{k}} N_{\mathbf{k}} - \mu N\right)\right\} \\ &= \sum_{N} e^{\beta\mu N} \sum_{\mathbf{N}}{}' \exp\left\{-\beta \sum_{\mathbf{k}} \varepsilon_{\mathbf{k}} N_{\mathbf{k}}\right\} = \sum_{N} e^{\beta\mu N} Z_N, \end{aligned} \tag{4.37}$$

where $\sum'$ indicates that the sum is constrained to all sets of occupation numbers that satisfy $N = \sum_{\mathbf{k}} N_{\mathbf{k}}$. We see that for each value of N, we have to calculate the canonical partition function Z_N, which is very cumbersome due to the constraint $N = \sum_{\mathbf{k}} N_{\mathbf{k}}$. However, the double sum can also be performed by summing for each momentum state $|\hbar\mathbf{k}\rangle$ over all possible values of the occupation numbers, i.e.

$$Z = \prod_{\mathbf{k}} \sum_{N_{\mathbf{k}}} \exp\{-\beta(\varepsilon_{\mathbf{k}} - \mu)N_{\mathbf{k}}\} = \prod_{\mathbf{k}} \left(1 \mp \mathrm{e}^{-\beta(\varepsilon_{\mathbf{k}}-\mu)}\right)^{\mp 1}, \tag{4.38}$$

where $\mp$ refers to bosons and fermions, respectively. The thermodynamic potential is readily calculated from (4.31) and is given by

$$\Omega = \pm k_{\mathrm{B}} T \sum_{\mathbf{k}} \log\left(1 \mp \mathrm{e}^{-\beta(\varepsilon_{\mathbf{k}}-\mu)}\right). \tag{4.39}$$

From the thermodynamic potential, we can calculate the average number of particles using $\langle N \rangle = -\partial\Omega/\partial\mu$ from (4.28), which leads to the average occupation numbers for the bosons N_{BE}

$$\langle N \rangle = \sum_{\mathbf{k}} \frac{1}{\exp\{\beta(\varepsilon_{\mathbf{k}} - \mu)\} - 1} \equiv \sum_{\mathbf{k}} N_{\mathrm{BE}}(\varepsilon_{\mathbf{k}}) \tag{4.40}$$

and for fermions N_{FD}

$$\langle N \rangle = \sum_{\mathbf{k}} \frac{1}{\exp\{\beta(\varepsilon_{\mathbf{k}} - \mu)\} + 1} \equiv \sum_{\mathbf{k}} N_{\mathrm{FD}}(\varepsilon_{\mathbf{k}}), \tag{4.41}$$

which are recognized as the famous Bose-Einstein and Fermi-Dirac distribution functions respectively. In the case of a negative chemical potential and in the limit $\beta|\mu| \gg 1$, the above distributions converge to the classical Maxwell-Boltzmann distribution, given by

$$\langle N \rangle = \sum_{\mathbf{k}} \exp\{-\beta(\varepsilon_{\mathbf{k}} - \mu)\} \equiv \sum_{\mathbf{k}} N_{\mathrm{MB}}(\varepsilon_{\mathbf{k}}). \tag{4.42}$$

This last point is also illustrated by the phase diagram for the ideal gases, shown in Fig. 4.2. Here, the three solid curves show for each distribution the chemical potential as a function of temperature for a fixed average number of particles $\langle N \rangle$. We see that left of the straight line given by $-\mu = 1/\beta$, that is when $\beta|\mu| > 1$, the three distributions indeed converge to the Maxwell-Boltzmann distribution. However, right of the straight line, they start to deviate significantly from each other. As a result, the left regime can be called classical, whereas the right regime can be called quantum mechanical, because here the difference in quantum statistics between bosons and fermions becomes clearly noticable.

4.3.1 Ideal Maxwell-Boltzmann Gas

To find an explicit expression for the relation between the total number of particles $\langle N \rangle$, the chemical potential and the temperature for the homogeneous Maxwell-Boltzmann gas, we have to perform the sum

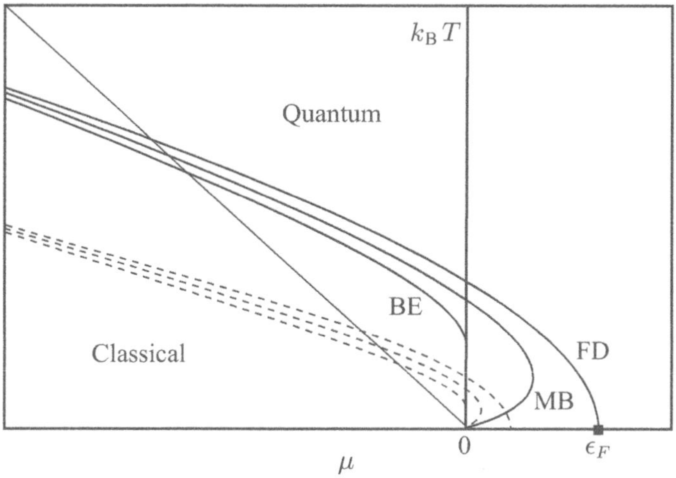

Fig. 4.2 The 'phase diagram' for the ideal gases. The solid lines show the behavior of the chemical potential μ as a function of the temperature T for a fixed average particle number, when the gas satisfies Bose-Einstein (BE), Fermi-Dirac (FD), or Maxwell-Boltzmann (MB) statistics. The dashed lines show the same, but for a smaller number of particles. The diagonal line $|\mu| = k_B T$ is the location of the crossover between the classical and quantum regimes.

$$\langle N \rangle = \sum_{\mathbf{k}} N_{\mathrm{MB}}(\varepsilon_{\mathbf{k}}) = \sum_{\mathbf{k}} \exp\{-\beta(\varepsilon_{\mathbf{k}} - \mu)\}, \tag{4.43}$$

whose exact result can be expressed in terms of elliptic theta functions. However, in practice it is often much more convenient to convert the sum over states into an integral. Since the ideal gas in a large box with periodic boundary conditions gives rise to quantized wavenumbers $\mathbf{k}$ according to $k_i = 2\pi n_i/L$ with $i = x, y, z$ and $n_i = 0, \pm 1, \pm 2, \ldots, \infty$, we have that each cube in $\mathbf{k}$ space with a volume $8\pi^3/L^3$ contains precisely one quantum state, or, conversely, that the number of quantum states in a volume $\mathrm{d}\mathbf{k} = \mathrm{d}k_x\ \mathrm{d}k_y\ \mathrm{d}k_z$ is on average equal to $L^3 \mathrm{d}\mathbf{k}/8\pi^3$. This can be used to convert a sum over momentum states into an integral, according to the substitution

$$\sum_{\mathbf{k}} \rightarrow \frac{L^3}{(2\pi)^3} \int \mathrm{d}\mathbf{k}, \tag{4.44}$$

which is allowed if the thermal energy or the chemical potential is much larger than the energy splittings between the quantum states. Note that it becomes exact in the continuum limit $L \longrightarrow \infty$. The above substitution is performed very often in practical calculations.

As a result, we obtain

$$\langle N \rangle = \sum_{\mathbf{k}} N_{\mathrm{MB}}(\varepsilon_{\mathbf{k}}) = \frac{V}{2\pi^2} \int_0^\infty \mathrm{e}^{-\beta(\varepsilon_k - \mu)} k^2 \mathrm{d}k, \tag{4.45}$$

where in the last step we exploited the spherical symmetry of the integrand by using spherical coordinates and performing the trivial angular integrals. We can rewrite the integral in terms of energies ε by using the substitution $\varepsilon = \hbar^2 k^2/2m$, which gives

$$\langle N \rangle = \frac{V}{(2\pi)^2} \left(\frac{2m}{\hbar^2} \right)^{3/2} \int_0^\infty \mathrm{e}^{-\beta(\varepsilon-\mu)} \sqrt{\varepsilon} \mathrm{d}\varepsilon. \tag{4.46}$$

From the above expression, we find the density of states $\mathscr{D}(\varepsilon)$, which is defined as the number of states in the interval $[\varepsilon, \varepsilon + \mathrm{d}\varepsilon]$. For the homogenous three-dimensional ideal gas, we have

$$\mathscr{D}(\varepsilon) = \frac{V}{(2\pi)^2} \left(\frac{2m}{\hbar^2} \right)^{3/2} \sqrt{\varepsilon}, \tag{4.47}$$

and we can write for the total number of atoms and the total energy

$$\langle N \rangle = \int_0^\infty \mathrm{d}\varepsilon \, \mathscr{D}(\varepsilon) N_{\mathrm{MB}}(\varepsilon) \quad \text{and} \quad \langle U \rangle = \int_0^\infty \mathrm{d}\varepsilon \, \varepsilon \, \mathscr{D}(\varepsilon) N_{\mathrm{MB}}(\varepsilon). \tag{4.48}$$

For the total number of particles, we find explicitly from (4.46) that

$$\langle N \rangle = V \left(\frac{m}{2\pi\hbar^2\beta} \right)^{3/2} \mathrm{e}^{\beta\mu}. \tag{4.49}$$

Solving the above equation for μ leads to

$$\mu = \frac{1}{\beta} \log(n\Lambda^3),$$

where $n = \langle N \rangle / V$ is the density of particles, and $\Lambda = (2\pi\hbar^2/mk_{\mathrm{B}}T)^{1/2}$ is the thermal de Broglie wavelength.

4.3.1.1 Inhomogeneous Maxwell-Boltzmann Gas

In a real experiment, an atomic gas is always subject to an external trapping potential, resulting in an inhomogeneous gas. The trapping potential is typically well approximated by a harmonic trapping potential, which leads to a different density of states compared to the homogenous gas. For an isotropic harmonic oscillator with frequency ω, i.e.

$$V^{\mathrm{ex}}(\mathbf{x}) = \frac{m}{2} \omega^2 \mathbf{x}^2, \tag{4.50}$$

we have that the single-particle eigenenergies are given by $\varepsilon_{\mathbf{n}} = \hbar\omega(n_x + n_y + n_z + 3/2)$ with $n_i = 0, 1, ..., \infty$. As a result, we find

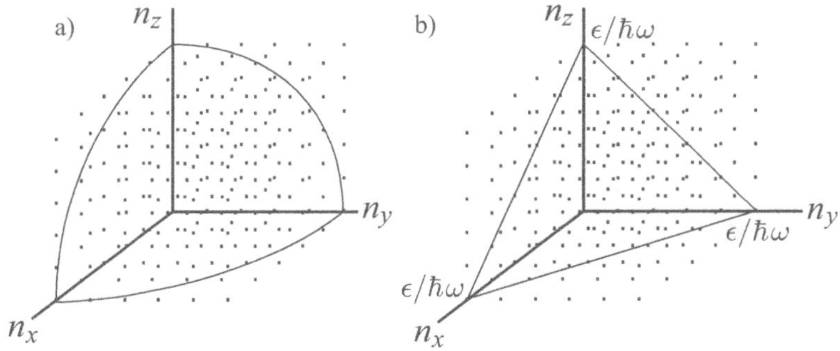

Fig. 4.3 a) For a homogeneous gas, the number of states $\mathscr{N}(\varepsilon)$ with an energy less than ε is given by the volume of the sphere in **n** space with radius $(mL^2\varepsilon/2\hbar^2\pi^2)^{1/2}$, of which the first octant is shown. The cubic lattice of dots represents the accessible single-particle quantum states b) For a harmonically trapped gas, $\mathscr{N}(\varepsilon)$ is given by the volume of the shown pyramid in the first octant. The density of states then follows from $\mathscr{D}(\varepsilon) = \mathrm{d}\mathscr{N}(\varepsilon)/\mathrm{d}\varepsilon$.

$$\langle N \rangle = \sum_{\mathbf{n}} N_{\mathrm{MB}}(\varepsilon_{\mathbf{n}}) = \sum_{\mathbf{n}} \exp\{-\beta(\varepsilon_{\mathbf{n}} - \mu)\} = \frac{\mathrm{e}^{\beta(\mu - 3\hbar\omega/2)}}{(1-\mathrm{e}^{-\beta\hbar\omega})^3}, \tag{4.51}$$

where we used $\sum_{n=0}^{\infty} \mathrm{e}^{-\alpha n} = 1/(1-\mathrm{e}^{-\alpha})$. Again, we can also obtain the total number of particles by converting the sum to an integral. First, we determine the number of states $\mathscr{N}(\varepsilon)$ with an energy less than ε. If $k_{\mathrm{B}}T$ is large compared to $\hbar\omega$, we may treat **n** as a continuous variable. If we also take $3\hbar\omega/2$ as our zero of energy, the number of states $\mathscr{N}(\varepsilon)$ is simply given by the volume of the triangular pyramid in **n**-space with the four edge points $(0,0,0), (\varepsilon/\hbar\omega,0,0), (0,\varepsilon/\hbar\omega,0), (0,0,\varepsilon/\hbar\omega)$, that is

$$\mathscr{N}(\varepsilon) = \frac{1}{6}\left(\frac{\varepsilon}{\hbar\omega}\right)^3, \tag{4.52}$$

which is also illustrated in Fig. 4.3. The density of states is then by definition equal to $\mathrm{d}\mathscr{N}(\varepsilon)/\mathrm{d}\varepsilon$, which yields

$$\mathscr{D}(\varepsilon) = \frac{1}{2(\hbar\omega)^3}\varepsilon^2. \tag{4.53}$$

We find for the total number of atoms that

$$\langle N \rangle = \int_0^{\infty} \mathrm{d}\varepsilon \, \frac{1}{2(\hbar\omega)^3}\varepsilon^2 \mathrm{e}^{-\beta(\varepsilon-\mu)} = \mathrm{e}^{\beta\mu}\left(\frac{k_{\mathrm{B}}T}{\hbar\omega}\right)^3, \tag{4.54}$$

which is seen to equal the result of (4.51), when $\beta\hbar\omega \ll 1$. Thus, we see that we may convert the sums over states into continuous integrals when the temperature is much larger than the energy spacing between states.

4.3.1.2 Local-Density Approximation

A third way to conveniently deal with the inhomogeneous Maxwell-Boltzmann gas is by applying the local-density approximation. The condition for this approximation is that the trapping potential varies slowly compared to the single-particle wavefunctions, such that the trap looks locally flat. Then, we are allowed to conveniently use the homogeneous density of states locally in the trap. Another way to formulate the above condition is that the thermal de Broglie wavelength of the particles should be small compared to the harmonic oscillator length of the trap. We are then allowed to simply absorb the trapping potential in the chemical potential $\mu(\mathbf{r}) = \mu - V^{\mathrm{ex}}(\mathbf{r}) = \mu - m\omega^2 r^2/2$ and use locally the theory for the homogeneous gas with the spatially varying chemical potential. With the use of (4.45), this leads for the local density of particles to

$$n(\mathbf{r}) = \frac{1}{2\pi^2} \int \mathrm{e}^{-\beta(\varepsilon_k - \mu(\mathbf{r}))} k^2 \mathrm{d}k = \mathrm{e}^{-\beta V^{\mathrm{ex}}(\mathbf{r})} \mathrm{e}^{\beta\mu} \left(\frac{m}{2\pi\beta\hbar^2} \right)^{3/2}, \tag{4.55}$$

using the result of (4.49). By integrating over space, we retrieve the total number of particles in the trap

$$\langle N \rangle = \int \mathrm{d}\mathbf{r}\, n(\mathbf{r}) = \mathrm{e}^{\beta\mu} \left(\frac{m}{2\pi\beta\hbar^2} \right)^{3/2} \int \mathrm{d}r\, r^2 \mathrm{e}^{-\beta m\omega^2 r^2/2} = \mathrm{e}^{\beta\mu} \left(\frac{k_{\mathrm{B}}T}{\hbar\omega} \right)^3, \tag{4.56}$$

showing that the local-density approximation gives exactly the same result as the continuum approximation, leading to (4.54). A big advantage of the local-density approximation is that we have also obtained an expression for the density profile, the density of atoms as a function of position in the trap $n(\mathbf{r})$, which is cumbersome to calculate in a different way.

4.3.2 *Ideal Bose Gas: Bose-Einstein Condensation*

For the homogeneous ideal Bose gas, we have

$$\langle N \rangle = \sum_{\mathbf{k}} \langle N_{\mathbf{k}} \rangle = \sum_{\mathbf{k}} N_{\mathrm{BE}}(\varepsilon_{\mathbf{k}}) = \sum_{\mathbf{k}} \frac{1}{\exp\{\beta(\varepsilon_{\mathbf{k}} - \mu)\} - 1}, \tag{4.57}$$

with $\langle N_{\mathbf{k}} \rangle$ the average number of particles in state $|\mathbf{k}\rangle$, which should therefore be larger or equal to zero. For the state with $\mathbf{k} = \mathbf{0}$, this means

$$\langle N_{\mathbf{0}} \rangle = \frac{1}{\exp\{-\beta\mu\} - 1} \geq 0, \tag{4.58}$$

from which we see that $\mu \leq 0$. Consider now an average number of particles $\langle N \rangle$, that we keep fixed. Upon lowering the temperature, we see from (4.57) that we have to increase the chemical potential to keep the number of particles constant, which can only be done until μ reaches zero. If we lower the temperature even further, then we can only keep the particle number constant by macroscopically occupying the ground state. As a result, we have for $\mu = 0$

$$\langle N \rangle = \langle N' \rangle + \langle N_{\mathbf{0}} \rangle = \sum_{\mathbf{k} \neq 0} \frac{1}{\exp\{\beta \varepsilon_{\mathbf{k}}\} - 1} + \langle N_{\mathbf{0}} \rangle, \tag{4.59}$$

with $\langle N' \rangle$ the number of particles in the excited states. The transition temperature T_c to a state with a macroscopic occupation of the ground state is given by the condition that $\langle N_{\mathbf{0}} \rangle$ is still negligible compared to $\langle N \rangle$, which means that all the particles can still be precisely accommodated in the excited states. This leads to the condition $\langle N \rangle = \langle N'(\mu = 0, T_c) \rangle$. The macroscopic occupation of the single-particle state with $\mathbf{k} = \mathbf{0}$ is called Bose-Einstein condensation, which, as we will see later in the book, gives rise to intriguing physical properties, such as for example superfluidity. Note that the Bose statistics, in which there is no limitation on the number of particles in a single quantum state, is crucial for the occurrence of the Bose-Einstein condensate. Indeed, in an ideal Fermi gas there is no condensation.

To find an explicit expression for the critical temperature, we consider

$$\langle N \rangle = \langle N'(\mu = 0, T_c) \rangle = \sum_{\mathbf{k} \neq \mathbf{0}} N_{\mathrm{BE}}(\varepsilon_{\mathbf{k}}) = \frac{V}{2\pi^2} \int_0^\infty \frac{k^2 \mathrm{d}k}{\exp\{\varepsilon_k / k_{\mathrm{B}} T_c\} - 1}, \tag{4.60}$$

which, as before, we can rewrite as

$$\langle N \rangle = \langle N'(\mu = 0, T_c) \rangle = \frac{V}{(2\pi)^2} \left(\frac{2m}{\hbar^2} \right)^{3/2} \int_0^\infty \frac{\sqrt{\varepsilon}\, \mathrm{d}\varepsilon}{\exp\{\varepsilon / k_{\mathrm{B}} T_c\} - 1}. \tag{4.61}$$

The integral of (4.61) can be performed exactly and yields

$$\langle N \rangle = \langle N'(\mu = 0, T_c) \rangle = V \zeta(3/2) \left(\frac{m k_{\mathrm{B}} T_c}{2\pi \hbar^2} \right)^{3/2}, \tag{4.62}$$

where

$$\zeta(\alpha) = \sum_{x=1}^{\infty} \frac{1}{x^\alpha} \tag{4.63}$$

is the Riemann zeta function and $\zeta(3/2) \simeq 2.612$. This equation can be inverted to give the critical temperature as a function of the density $n = \langle N \rangle / V$

$$T_c = \frac{2\pi \hbar^2}{k_{\mathrm{B}} m} \left(\frac{n}{\zeta(3/2)} \right)^{2/3}. \tag{4.64}$$

If we lower the temperature T below T_c, then the number of excited atoms becomes smaller, according to

$$\langle N'(\mu = 0, T)\rangle = V\zeta(3/2)\left(\frac{mk_{\mathrm{B}}T}{2\pi\hbar^2}\right)^{3/2} = \langle N\rangle (T/T_{\mathrm{c}})^{3/2}. \tag{4.65}$$

The rest of the atoms have to reside in the zero-momentum state and it follows that the number of condensed atoms is given by

$$N_0(T) = \langle N\rangle - \langle N'\rangle = \langle N\rangle \left\{1 - \left(\frac{T}{T_{\mathrm{c}}}\right)^{3/2}\right\}, \tag{4.66}$$

for $T \leq T_{\mathrm{c}}$. The behavior of the chemical potential as a function of temperature for the ideal Bose gas is given in Fig. 4.2.

4.3.2.1 Inhomogeneous Bose Gas

In a real experiment, the atomic Bose gas is trapped in an external trapping potential, which is typically harmonic and changes the density of states. Using the density of states obtained in (4.53), we find for the total number of atoms in the excited states when $\mu = 0$ that

$$\langle N'\rangle = \int_0^\infty \mathrm{d}\varepsilon \, \frac{1}{2(\hbar\omega)^3} \frac{\varepsilon^2}{\mathrm{e}^{\beta\varepsilon} - 1} = \zeta(3)\left(\frac{k_{\mathrm{B}}T}{\hbar\omega}\right)^3, \tag{4.67}$$

where $\zeta(3) \simeq 1.212$ Again, we have the critical condition for Bose-Einstein condensation $\langle N\rangle = \langle N'(\mu = 0, T_{\mathrm{c}})\rangle$, from which we obtain the critical temperature

$$T_{\mathrm{c}} = \frac{\hbar\omega}{k_{\mathrm{B}}}\left(\frac{\langle N\rangle}{\zeta(3)}\right)^{1/3}. \tag{4.68}$$

Below T_c, the number of atoms in the ground state of the harmonic oscillator is given by

$$\langle N_0(T)\rangle = \langle N\rangle - \langle N'\rangle = \langle N\rangle \left\{1 - \left(\frac{T}{T_{\mathrm{c}}}\right)^3\right\}. \tag{4.69}$$

The realization of Bose-Einstein condensation in a trapped dilute ultracold atomic gas was first realized by the group of Wieman and Cornell [10], and later by the groups of Hulet and Ketterle [11, 12] in 1995. For their achievements, Wieman, Cornell and Ketterle were awarded the Nobel prize in 2001. In Fig. 4.4, we see the experimental data as obtained in the first BEC experiment [10], which shows the velocity distribution of expanded clouds of ultracold rubidium atoms as a function of temperature. Let us see if we can understand this data from our present knowledge of the trapped ideal Bose gas. For N_0 noninteracting bosons in the ground state of a harmonic trap, the many-body wavefunction $\Psi(\mathbf{x}_1, \ldots, \mathbf{x}_{N_0})$, is simply a product of the single-particle states

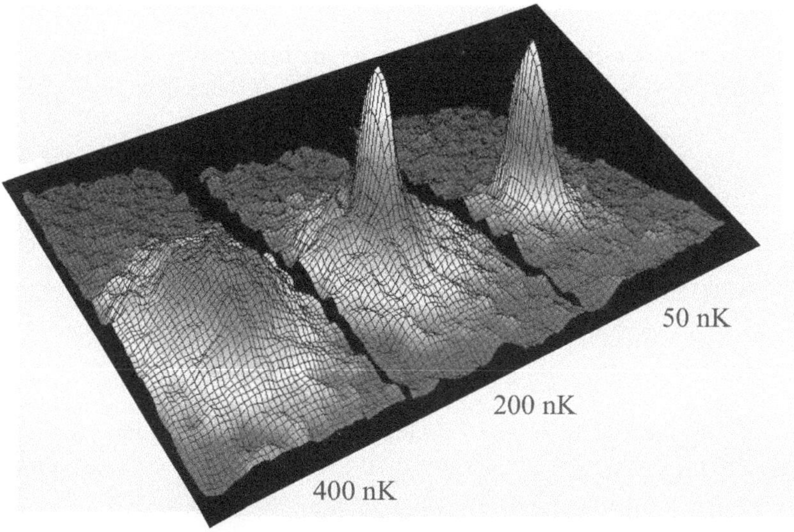

Fig. 4.4 Velocity distribution of rubidium-87 atom clouds at three different temperatures. The left image corresponds to a temperature just above the condensation temperature, the center image corresponds to a temperature at the onset of condensation and the right image shows a nearly pure Bose-Einstein condensate. Image by Mike Matthews, JILA [10].

$$\Psi(\mathbf{x}_1,\ldots,\mathbf{x}_{N_0}) = \prod_i \chi_0(\mathbf{x}_i), \tag{4.70}$$

where $\chi_0(\mathbf{x})$ is the wavefunction for the harmonic oscillator ground state. If we generalize our discussion to a general anisotropic harmonic potential $V^{\mathrm{ex}}(\mathbf{x}) = m(\omega_x^2x^2 + \omega_y^2y^2 + \omega_z^2z^2)/2$, which is actually the case for the experiment of Fig. 4.4, then we have for the ground state

$$\chi_0(\mathbf{x}) = \left(\frac{m\bar{\omega}}{\pi\hbar}\right)^{3/4} \exp\left\{-\frac{m}{2\hbar}(\omega_x x^2 + \omega_y y^2 + \omega_z z^2)\right\}, \tag{4.71}$$

with $\bar{\omega} = (\omega_x\omega_y\omega_z)^{1/3}$. The density distribution for the condensate is simply given by

$$n_0(\mathbf{x}) = N_0|\chi_0(\mathbf{x})|^2, \tag{4.72}$$

where we note that the spread in position for the above wavefunction is proportional to the harmonic oscillator lengths $l_i = \sqrt{\hbar/m\omega_i}$ with $i = x,y,z$ and is therefore independent of temperature. The same holds for the spread in the momenta, which can be obtained from the Fourier transform of (4.71), giving

$$\chi_0(\mathbf{p}) = \left(\frac{1}{\pi\hbar m\bar{\omega}}\right)^{3/4} \exp\left\{-\frac{1}{2m\hbar}(p_x^2/\omega_x + p_y^2/\omega_y + p_z^2/\omega_z)\right\}. \tag{4.73}$$

To obtain the density distribution of the excited atoms, we use the local-density approximation with the local homogeneous density of states from (4.47), such that

$$\langle N'(\mathbf{x})\rangle = \frac{V}{(2\pi)^2}\left(\frac{2m}{\hbar^2\beta}\right)^{3/2}\int_0^\infty \frac{\sqrt{u}\,\mathrm{d}u}{\mathrm{e}^u\mathrm{e}^{-\beta\mu(\mathbf{x})}-1}, \tag{4.74}$$

where we changed variables to $u=\beta\varepsilon$. The above integral can be performed exactly and yields

$$n'(\mathbf{x}) = \frac{\langle N'(\mathbf{x})\rangle}{V} = \left(\frac{m}{2\pi\hbar^2\beta}\right)^{3/2} g_{3/2}(\mathrm{e}^{\beta\mu(\mathbf{x})}) = \frac{g_{3/2}(\mathrm{e}^{\beta\mu(\mathbf{x})})}{\Lambda^3}, \tag{4.75}$$

where Λ is the thermal de Broglie wavelength and the polylogarithm $g_n(z)$ is defined by

$$g_n(z) = \sum_{x=1}^{\infty}\frac{z^x}{x^n}. \tag{4.76}$$

For $\beta|\mu| \gg 1$, the Bose distribution converges to the Maxwell-Boltzmann distribution, such that (4.75) gives the local-density Maxwell-Boltzmann result $n'(\mathbf{x}) = \mathrm{e}^{\beta\mu(\mathbf{x})}/\Lambda^3$. It is important to note that the spread in position for the excited particles depends on temperature and is given by $\sqrt{k_\mathrm{B}T/m\omega_i^2}$ for $i=x,y,z$. In a similar way, we can calculate the density of atoms in momentum space. By defining

$$\langle N'\rangle = \int\frac{\mathrm{d}\mathbf{k}}{(2\pi)^3}\int\mathrm{d}\mathbf{x}\frac{1}{\mathrm{e}^{\beta\varepsilon_\mathbf{k}}\mathrm{e}^{-\beta\mu(\mathbf{x})}-1} \equiv \int\mathrm{d}\mathbf{p}\; n'_\mathbf{p} \tag{4.77}$$

with the use of $\mathbf{p}=\hbar\mathbf{k}$, we have that

$$\begin{aligned} n'(\mathbf{p}) &= \frac{1}{(2\pi\hbar)^3}\int\frac{\mathrm{d}x\,\mathrm{d}y\,\mathrm{d}z}{\mathrm{e}^{\beta\varepsilon_\mathbf{p}}\mathrm{e}^{-\beta\mu(\mathbf{x})}-1} = \frac{1}{(2\pi\hbar)^3}\int\frac{\mathrm{d}x'\,\mathrm{d}y'\,\mathrm{d}z'}{\mathrm{e}^{\beta(\varepsilon_\mathbf{p}-\mu)}\mathrm{e}^{\beta V^{\mathrm{ex}}(\mathbf{x}')}-1} \\ &= \frac{1}{2\pi^2\hbar^3}\int_0^\infty\frac{r^2\,\mathrm{d}r}{\mathrm{e}^{\beta(\varepsilon_\mathbf{p}-\mu)}\mathrm{e}^{\beta m\bar{\omega}^2r^2/2}-1} \\ &= \frac{1}{\sqrt{2}\pi^2(\beta m\bar{\omega}^2)^{3/2}\hbar^3}\int_0^\infty\frac{\sqrt{u}\,\mathrm{d}u}{\mathrm{e}^{\beta(\varepsilon_\mathbf{p}-\mu)}\mathrm{e}^u-1} \\ &= \frac{1}{(2\pi\beta m\bar{\omega}^2)^{3/2}\hbar^3}g_{3/2}(\mathrm{e}^{-\beta(\varepsilon_\mathbf{p}-\mu)}), \end{aligned} \tag{4.78}$$

where in the first step we changed variables, for example $x=\bar{\omega}x'/\omega_x$, such that $V^{\mathrm{ex}}(\mathbf{x}') = m\bar{\omega}\mathbf{x}'^2/2$, which allows us to go to spherical coordinates. We also transformed to the variable $u=\beta m\bar{\omega}^2r^2/2$, which brings the integral to the form of (4.74). For $\beta|\mu|\gg 1$, the result of (4.78) converges to the Maxwell-Boltzmann distribution, which is given by $n'(\mathbf{p}) \propto \mathrm{e}^{-\beta(\varepsilon_\mathbf{p}-\mu)}$. It is important to note that the spread in momentum depends on temperature and is given by $\sqrt{k_\mathrm{B}T/m}$. The spread is thus the same in each direction, that is isotropic, even though the trap is anisotropic.

If we compare the spread in the momenta of the noncondensed particles with the spread of the condensed particles, then we see that the ratio is given by $\sqrt{2k_{\mathrm{B}}T/\hbar\omega_i}$, which in a typical experiment is much larger than unity. This follows from (4.68) and the fact that the number of atoms in the experiments obeys $\langle N\rangle \gg 1$. As a result, the momentum distribution of the condensate is much narrower than that of the thermal cloud. Furthermore, the spread of the condensate is anisotropic and independent of temperature, whereas for the thermal cloud it is isotropic and increases with temperature. Note that actually the Bose-Einstein condensate minimizes the spread in position and velocity with the minimum set by the Heisenberg uncertainty principle $\Delta x_i \Delta p_i = \hbar/2$, which is the general behavior for coherent states. We can compare these results with the experimental data for the velocity distributions above and below T_{c} as shown in Fig. 4.4. We see indeed that the condensate is much narrower and more anisotropic than the thermal cloud. However, the gas of rubidium atoms used in the experiment was not truly noninteracting. In Chap. 11, we consider in more detail Bose-Einstein condensation and take also interaction effects into account.

4.3.3 Ideal Fermi Gas

As mentioned before, for an ideal Fermi gas, we do not have Bose-Einstein condensation due to the Pauli principle which does not allow for a macroscopic occupation of a single quantum state. To determine the chemical potential for an ideal homogeneous Fermi gas as a function of temperature, we need to calculate

$$\langle N\rangle = \int_0^\infty d\varepsilon\, \mathscr{D}(\varepsilon) N_{\mathrm{FD}}(\varepsilon), \tag{4.79}$$

where we have that

$$N_{\mathrm{FD}}(\varepsilon) = \frac{1}{e^{(\varepsilon-\mu)/k_{\mathrm{B}}T} + 1}. \tag{4.80}$$

In the zero-temperature limit, the Fermi distribution becomes a step function and the fermions fill up all states up to a certain energy, called the Fermi energy ε_{F}. For an ideal Fermi gas, this Fermi energy is equal to the chemical potential μ at zero temperature and it can be calculated analytically in terms of the density from

$$\langle N\rangle = \int_0^\infty d\varepsilon\, \theta(\mu-\varepsilon)\mathscr{D}(\varepsilon) = \int_0^{\varepsilon_{\mathrm{F}}} d\varepsilon\, \mathscr{D}(\varepsilon), \tag{4.81}$$

with $\theta(x) = 1$ for $x > 0$ and zero otherwise. Using the density of states for the homogeneous ideal gas from (4.47), it follows that

$$n = \frac{\langle N\rangle}{V} = \frac{1}{6\pi^2}\left(\frac{2m\varepsilon_{\mathrm{F}}}{\hbar^2}\right)^{3/2}, \tag{4.82}$$

or inversely

$$\varepsilon_{\mathrm{F}} = \frac{\hbar^2}{2m}(6\pi^2 n)^{2/3}, \tag{4.83}$$

which we actually use as the definition for the Fermi energy in the case of a homogeneous spinless Fermi gas, also at nonzero temperatures. As a result, the Fermi energy is always determined by the density of particles and we can use it to rewrite the density of states as

$$\mathscr{D}(\varepsilon) = \frac{3}{2}\frac{\langle N\rangle}{\varepsilon_{\mathrm{F}}}\sqrt{\frac{\varepsilon}{\varepsilon_{\mathrm{F}}}}. \tag{4.84}$$

To treat the nonzero temperature case, where we look at temperatures $k_{\mathrm{B}}T \ll \varepsilon_{\mathrm{F}}$, we start by a partial integration, which yields

$$\langle N\rangle = \int_0^\infty \mathrm{d}\varepsilon\, \langle N\rangle \left(\frac{\varepsilon}{\varepsilon_{\mathrm{F}}}\right)^{3/2}\left(-\frac{\partial N_{\mathrm{FD}}(\varepsilon)}{\partial\varepsilon}\right). \tag{4.85}$$

The derivative of the Fermi distribution is given by

$$-\frac{\partial N_{\mathrm{FD}}(\varepsilon)}{\partial\varepsilon} = -\frac{\beta}{4\cosh^2(\beta(\varepsilon-\mu)/2)}, \tag{4.86}$$

which for low temperatures is a very sharply-peaked function around the chemical potential. In the zero-temperature limit, it becomes the Dirac delta function. As a result, the integral in the last equation can be performed by expanding the term $(\varepsilon/\varepsilon_{\mathrm{F}})^{3/2}$ around μ. Specifically, we have

$$\left(\frac{\varepsilon}{\varepsilon_{\mathrm{F}}}\right)^{3/2} \simeq \left(\frac{\mu}{\varepsilon_{\mathrm{F}}}\right)^{3/2} + \frac{3}{2}\left(\frac{\mu}{\varepsilon_{\mathrm{F}}}\right)^{1/2}\frac{\varepsilon-\mu}{\varepsilon_{\mathrm{F}}} + \frac{3}{8}\left(\frac{\varepsilon_{\mathrm{F}}}{\mu}\right)^{1/2}\frac{(\varepsilon-\mu)^2}{\varepsilon_{\mathrm{F}}^2} + \dots, \tag{4.87}$$

where substitution of this result into (4.85) gives

$$\begin{aligned}\langle N\rangle = &-\langle N\rangle\left(\frac{\mu}{\varepsilon_{\mathrm{F}}}\right)^{3/2}\int_{-\infty}^{+\infty}\mathrm{d}\varepsilon\,\frac{\partial N_{\mathrm{FD}}}{\partial\varepsilon} - \frac{3\langle N\rangle}{2\varepsilon_{\mathrm{F}}}\left(\frac{\mu}{\varepsilon_{\mathrm{F}}}\right)^{1/2}\int_{-\infty}^{+\infty}\mathrm{d}\varepsilon\,(\varepsilon-\mu)\frac{\partial N_{\mathrm{FD}}}{\partial\varepsilon}\\ &-\frac{3\langle N\rangle}{8\varepsilon_{\mathrm{F}}^2}\left(\frac{\varepsilon_{\mathrm{F}}}{\mu}\right)^{1/2}\int_{-\infty}^{+\infty}\mathrm{d}\varepsilon\,(\varepsilon-\mu)^2\frac{\partial N_{\mathrm{FD}}}{\partial\varepsilon} + \dots .\end{aligned} \tag{4.88}$$

Here, we extended the range of the integrals to $-\infty$, which is allowed, because $\partial N_{\mathrm{FD}}(\varepsilon)/\partial\varepsilon$ is exponentially suppressed in this regime. Performing the integrals, we obtain the Sommerfeld expansion, namely

$$\langle N\rangle = \langle N\rangle\left(\frac{\mu}{\varepsilon_{\mathrm{F}}}\right)^{3/2} + \langle N\rangle\frac{\pi^2}{8}\left(\frac{\varepsilon_{\mathrm{F}}}{\mu}\right)^{1/2}\left(\frac{k_{\mathrm{B}}T}{\varepsilon_{\mathrm{F}}}\right)^2 + \dots, \tag{4.89}$$

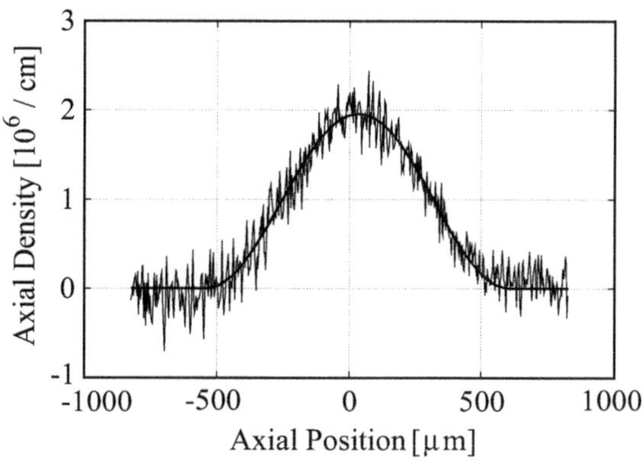

Fig. 4.5 Axial density profile of trapped fermionic Lithium-6 atoms in a single hyperfine state subject to an external magnetic field of $B = 967$ G. Also shown is the calculated axial density profile $n_z(z)$ for an ideal Fermi gas at zero temperature using the local-density approximation. Image by Wenhui Li and Randy Hulet, Rice University.

where we used that that the second term in (4.88) gives zero, because the integrand is odd, while the third term can be evaluated using the standard integral $\int du\, u^2/\cosh^2(u) = \pi^2/6$. We can now solve the above equation for μ, which leads for $k_B T \ll \varepsilon_F$ to

$$\mu = \varepsilon_F \left\{ 1 - \frac{\pi^2}{12} \left(\frac{k_B T}{\varepsilon_F} \right)^2 \right\}, \tag{4.90}$$

which means that for low temperatures the chemical potential decreases quadratically until it reaches the Fermi energy at zero temperature. This result is also shown in the phase diagram of Fig. 4.2.

4.3.3.1 Inhomogeneous Fermi Gas

To study the inhomogeneous ideal Fermi gas, we determine the density profile at zero temperature by applying the local-density approximation. Using the density of states of the homogeneous gas with a spatially varying chemical potential $\mu(\mathbf{r}) = \mu - V^{\text{ext}}(\mathbf{r})$, we find

$$\begin{aligned} n(\mathbf{r}) &= \frac{\langle N(\mathbf{r}) \rangle}{V} = \frac{1}{(2\pi)^2} \left(\frac{2m}{\hbar^2} \right)^{3/2} \int_0^\infty d\varepsilon\, \sqrt{\varepsilon}\, \theta(\mu(\mathbf{r}) - \varepsilon) \\ &= \frac{1}{6\pi^2} \left(\frac{m}{\hbar^2} \right)^{3/2} (2\mu - m\omega_\rho^2 \rho^2 - m\omega_z^2 z^2)^{3/2}, \end{aligned} \tag{4.91}$$

where we considered a cylindrically symmetric potential $V^{\text{ext}}(\mathbf{r}) = (m\omega_\rho^2\rho^2 - m\omega_z^2 z^2)/2$ with $\rho = (x^2+y^2)^{1/2}$ and where the above equation is valid for $\mu(\mathbf{r}) > 0$. Determining the integrated density $n_z(z)$ along the z axis, we find

$$\begin{aligned} n_z(z) &= \int_0^{2\pi} \mathrm{d}\varphi \int_0^{R'} \rho\, \mathrm{d}\rho\, \frac{1}{6\pi^2} \left(\frac{m}{\hbar^2}\right)^{3/2} (2\mu - m\omega_\rho^2\rho^2 - m\omega_z^2 z^2)^{3/2} \\ &= \frac{1}{15\pi m\omega_\rho^2} \left(\frac{m}{\hbar^2}\right)^{3/2} (2\mu - m\omega_z^2 z^2)^{5/2}, \end{aligned} \tag{4.92}$$

where the integration limit for ρ is given by $R' = \{(2\mu - m\omega_z^2 z^2)/m\omega_\rho^2\}^{1/2}$. Note that absorption measurements inevitably give rise to integrated density profiles. To show the validity of the local-density approximation for typical experimental circumstances in ultracold atomic Fermi gases, we then compare (4.92) with an actual absorption measurement that determines the axial density, which is shown in Fig. 4.5. The agreement is excellent.

The behavior of the ideal Bose gas and the ideal Fermi gas at low temperatures could not be more different. Whereas the Bose gas undergoes a phase transition to a Bose-Einstein condensate, the Fermi gas fills up all single-particle quantum states one by one until it reaches the Fermi energy. As a result, the states that are occupied in momentum space by the homogeneous Fermi gas form an incompressible sphere for temperatures below the Fermi temperature $T_{\text{F}} \equiv \varepsilon_{\text{F}}/k_{\text{B}}$. This incompressibility caused by the Pauli principle prevents for example white dwarfs and neutron stars from gravitational collapse. Note that the Fermi temperature usually represents a large energy scale in the system, because it is related to the total amount of particles. The striking difference between the Bose-Einstein and Fermi-Dirac statistics has been beautifully visualized in an experiment that has simultaneously trapped fermionic Lithium-6 and bosonic Lithium-7 gas clouds [33]. These isotopes are chemically the same, because they differ only by a single neutron in the nucleus. The results are shown in Fig. 4.6. At the initial temperature the two clouds are approximately the same size, but as the atom clouds are cooled down below the Fermi temperature the Bose gas contracts, whereas the Fermi gas cannot do so due to the pressure exerted by the exclusion principle. From (4.85), we see that the size of the fermionic cloud is at low temperatures given by $R_i = \sqrt{2\mu/m\omega_i^2}$ with $i = \rho, z$.

4.4 Density Matrix

Quantum mechanics is intrinsically probabilistic in the sense that preparing a system in a fully specified state $|\psi\rangle$ does not allow for a certain prediction for the outcome of an experiment measuring an observable $\hat{O}$. However, if we perform many measurements on exact copies of the same system, we can predict the average outcome of the experiment by calculating the expectation value $\langle\psi|\hat{O}|\psi\rangle$. Such a collection of copies of one and the same quantum state is called a pure ensemble. However,

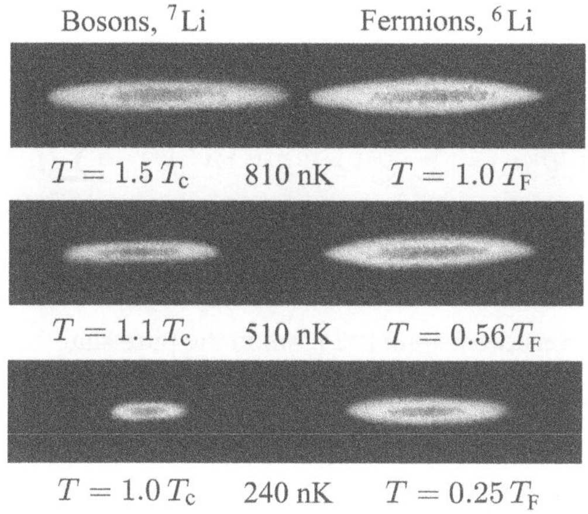

Fig. 4.6 Absorption images of the first quantum degenerate mixture of bosons and fermions. The images on the left are of bosonic ^{7}Li atoms, while those on the right are of fermionic ^{6}Li atoms, taken at progressively lower temperatures. For each temperature, also the ratio with the corresponding critical temperature for condensation T_c and the Fermi temperature T_F is given. Although the ^{6}Li and ^{7}Li atoms are simultaneously trapped in the same volume, the images are separated for clarity. The absorption is highest in the center of the clouds, where the atomic densities are largest. Adapted from A. G. Truscott, K. E. Strecker, W. I. McAlexander, G. Partridge, and R. G. Hulet, "Observation of Fermi Pressure in a Gas of Trapped Atoms", Science **291**, 2570 (2001). Reprinted with permission from AAAS.

in statistical physics we typically do not know precisely in which microstate the system is, but rather we know the probability for each microstate. A collection of copies of various quantum states, each with a certain probability, is called a mixed ensemble. Consider for example an unpolarized beam of spin-1/2 atoms, where half of the atoms have their spin up and half of the atoms have their spin down. Note that such a beam is not described by the state $|\psi\rangle = |\uparrow\rangle/\sqrt{2} + |\downarrow\rangle/\sqrt{2}$, since this would correspond to a beam of particles that are all in state $|\psi\rangle$ and therefore actually have a polarization in the x direction.

To correctly describe the consequences of a mixed ensemble of states $|\nu\rangle$, where each state has a probability p_ν, we introduce the density matrix as

$$\hat{\rho} = \sum_{\nu} |\nu\rangle p_\nu \langle\nu|, \tag{4.93}$$

where, since the total probability obeys $\sum_\nu p_\nu = 1$, we have that

$$\mathrm{Tr}[\hat{\rho}] = 1. \tag{4.94}$$

Note that the density matrix is by construction a semi-positive definite Hermitian operator, which means that its diagonal form has eigenvalues larger than or equal to zero. In the basis of the eigenstates $|\nu\rangle$ of the density matrix, the matrix elements are simply given by

$$\rho_{\nu\nu'} = \langle \nu|\hat{\rho}|\nu'\rangle = p_\nu \delta_{\nu\nu'}. \tag{4.95}$$

However, we may represent the density matrix in any other basis $|\mu\rangle$ via

$$\rho_{\mu\mu'} = \langle \mu|\hat{\rho}|\mu'\rangle = \sum_{\nu,\nu'} \langle \mu|\nu\rangle\langle \nu|\rho|\nu'\rangle\langle \nu'|\mu'\rangle. = \sum_{\nu} \langle \mu|\nu\rangle p_\nu \langle \nu|\mu'\rangle. \tag{4.96}$$

Example 4.1. We consider a spin-1/2 atom in the pure state

$$|\psi\rangle = \frac{1}{\sqrt{2}}(|\uparrow\rangle + |\downarrow\rangle). \tag{4.97}$$

As a result, in the basis $|1\rangle = (|\uparrow\rangle + |\downarrow\rangle)/\sqrt{2}$ and $|2\rangle = (|\uparrow\rangle - |\downarrow\rangle)/\sqrt{2}$, the density matrix is given by

$$\rho^{\text{pure}} = \begin{bmatrix} 1 & 0 \\ 0 & 0 \end{bmatrix}, \tag{4.98}$$

which is the characteristic diagonal form of a pure state. In the basis $|\uparrow\rangle$ and $|\downarrow\rangle$, the density matrix yields

$$\rho^{\text{pure}} = \frac{1}{2}\begin{bmatrix} 1 & 1 \\ 1 & 1 \end{bmatrix}. \tag{4.99}$$

In contrast, for a mixture of states where half of the spins is $|\uparrow\rangle$ and half is $|\downarrow\rangle$, we have in the basis $|\uparrow\rangle$ and $|\downarrow\rangle$

$$\rho^{\text{mixture}} = \frac{1}{2}\begin{bmatrix} 1 & 0 \\ 0 & 1 \end{bmatrix}, \tag{4.100}$$

which then does not correspond to a pure state. Whether or not we are dealing with a pure state can actually be most clearly distinguished by looking at the trace of $\hat{\rho}^2$, which only equals one for a pure state.

The expectation value or ensemble average of an operator $\hat{O}$ can now be calculated by using the density matrix. We define the average of an operator $\hat{O}$ over the mixed ensemble $|\nu\rangle$ as

$$\langle \hat{O}\rangle \equiv \sum_{\nu} p_\nu \langle \nu|\hat{O}|\nu\rangle = \sum_{\nu,\nu'} \langle \nu|\hat{\rho}|\nu'\rangle\langle \nu'|\hat{O}|\nu\rangle = \text{Tr}[\hat{\rho}\hat{O}], \tag{4.101}$$

where the trace has the convenient property that for two square matrices A and B we have $\text{Tr}[AB] = \text{Tr}[BA]$. As a result, $\text{Tr}[AB] = \text{Tr}[S^{-1}SAS^{-1}SB] = \text{Tr}[SAS^{-1}SBS^{-1}]$, and thus (4.101) also holds in any other basis representation. The time dependence of the density matrix follows from the Schrödinger equation for the states $|\nu(t)\rangle$, and is seen to obey

$$i\hbar \frac{d\hat{\rho}(t)}{dt} = [\hat{H}, \hat{\rho}(t)]_-, \tag{4.102}$$

which looks similar to the Heisenberg equation of motion (3.38) although the above relation is actually valid in the Schrödinger picture.

The density matrix that assigns the canonical probability to each energy eigenstate of a Hamiltonian reads

$$\hat{\rho} = \frac{e^{-\beta \hat{H}}}{Z}, \tag{4.103}$$

whereas the grand-canonical ensemble is described by

$$\hat{\rho} = \frac{e^{-\beta(\hat{H} - \mu \hat{N})}}{Z}. \tag{4.104}$$

It is the last density matrix that we use mostly throughout the rest of this book.

4.5 Problems

Exercise 4.1. One-dimensional Ising Model
(a) Show with the use of (4.14) that as a function of temperature the average energy $\langle U \rangle$ of the one-dimensional Ising model equals

$$\langle U \rangle = -N \frac{J^2}{k_B T}. \tag{4.105}$$

Why is this answer only valid for $k_B T \gg J$?
(b) Calculate the exact free energy F for the one-dimensional Ising model using the transformation to noninteracting bonds that yields (4.11). Show that for $k_B T \gg J$, you obtain for the energy the result of (a).
(c) Consider the one-dimensional Ising model with periodic boundary conditions. Write the partition function as

$$Z = \sum_{\{S_{z,i}\}} \langle S_{z,1}|\hat{T}|S_{z,2}\rangle \langle S_{z,2}|\hat{T}|S_{z,3}\rangle \dots \langle S_{z,N+1}|\hat{T}|S_{z,1}\rangle, \tag{4.106}$$

where $S_{z,i} = \pm\hbar/2$ and $\langle S_{z,i}|\hat{T}|S_{z,j}\rangle = e^{4JS_{z,i}S_{z,j}/\hbar^2 k_B T}$ is the transfer matrix, whose elements are the Boltzmann factors. Since we now have $Z = \mathrm{Tr}[\hat{T}^{N+1}]$, we can calculate Z if we know the eigenvalues of $\hat{T}$. Show that these eigenvalues are $e^{J/k_B T} + e^{-J/k_B T}$ and $e^{J/k_B T} - e^{-J/k_B T}$, respectively. Calculate the free energy and take the limit $N \to \infty$. Compare your results with (a) and (b).

Exercise 4.2. Show that in the grand-canonical ensemble we have that $P(\nu) = e^{-\beta(U_\nu - \mu N_\nu)}/Z$. To this end apply the same approach that was used to derive the probability distribution for the canonical ensemble. Note that in the present case you should expand the entropy of the reservoir both with respect to the internal energy and the number of particles.

Additional Reading

- C. Kittel and H. Kroemer, *Statistical Mechanics*, (W. H. Freeman and Company, New York, 1980).
- K. S. Huang, *Statistical Mechanics*, (Wiley, New York, 1987).
- A review of Bose-Einstein condensation in trapped gases is given by F. Dalfovo et al., Rev. Mod. Phys. **71**, 463 (1999), and by
- A. J. Leggett, Rev. Mod. Phys. **73**, 307 (2001)

Chapter 5
Path Integrals

It doesn't matter how beautiful your theory is, it doesn't matter how smart you are. If it doesn't agree with experiment, it's wrong.
– Richard Feynman.

In this chapter, we discuss the Feynman path-integral formulation of quantum mechanics. It is the generalization of the classical Lagrange and Hamilton formalisms to quantum mechanics. In the path-integral approach, the central object that determines the dynamics of the system is called the action. We start by considering the mathematical properties of an action, which is a functional, where we focus initially on functional differentiation. This allows us to determine the minimum of an action, which, according to the principle of least action, determines the classical motion. Having become familiar with actions, we go through the derivation of the path-integral expression for the quantum-mechanical transition amplitude using the time-slicing procedure. The same procedure returns in Chap. 7, when we derive the functional-integral formalism for quantum field theory. Having obtained the path integral, we discuss various ways of solving it and apply these methods to the free particle and to a particle in a potential. We also derive the path-integral expressions for matrix elements of operators and expectation values, where we see that the path-integral formalism gives rise to time-ordered expectation values. All these important concepts return many times in parts II and III of the book, when we use the generalization of the path-integral formalism to quantum field theory in the treatment of interacting quantum gases.

5.1 Functionals and Functional Derivatives

To fully appreciate the principle of least action, we need to carefully understand what an action actually is. Mathematically, the action is a functional, which is an object that assigns a number to a function $f(x)$. To illustrate this concept, we consider two simple, but important examples.

Example 5.1. A simple example of a functional is given by $F[f] = \int_a^b \mathrm{d}x\, f(x)$. In this case, the functional assigns to any function the value of the integral of that function between a and b, assuming that the integral exists. So, if $f(x) = 1$ and $g(x) = x$, we have for this example that $F[f] = (b-a)$ and $F[g] = (b^2 - a^2)/2$.

Another important example is given by the functional

$$\delta_x[f] = f(x), \tag{5.1}$$

which returns the value of the function at point x and defines the Dirac delta-function $\delta(x)$. We have

$$\delta_x[f] = \int \mathrm{d}x' \; \delta(x'-x) f(x') = f(x), \tag{5.2}$$

whereas the derivative of the Dirac delta-function follows from

$$\delta'_x[f] = \int \mathrm{d}x' f(x') \frac{\mathrm{d}}{\mathrm{d}x'} \delta(x'-x) = -\frac{\mathrm{d}}{\mathrm{d}x} f(x), \tag{5.3}$$

where we used partial integration.

To determine the least action, we need to be able to differentiate a functional. It is thus a natural question to ask what the change in $F[f]$ is as we vary $f(x)$. That is, we want to know the value $F[f+\delta f] - F[f]$, where $\delta f(x)$ is a small perturbation to the function $f(x)$. First, we note that the differential of an ordinary multivariate function $h(u_1, \ldots, u_n)$ is given by $\mathrm{d}h = \sum_i (\partial h/\partial u_i) \mathrm{d}u_i$. Since we may interpret the functional $F[f]$ as a infinitely multivariate function, where the value of f at each point x can be considered as an independent variable for the functional F, the natural generalization of ordinary differentiation to the functional derivative $\delta F[f]/\delta f(x)$ is given by

$$F[f+\delta f] - F[f] = \int \mathrm{d}x' \frac{\delta F[f]}{\delta f(x')} \delta f(x'). \tag{5.4}$$

The functional derivative can be expressed as the derivative in the direction of the Dirac delta-function, i.e.

$$\frac{\delta F[f]}{\delta f(x)} = \lim_{\varepsilon \to 0} \frac{F[f(x') + \varepsilon \delta(x'-x)] - F[f(x')]}{\varepsilon}, \tag{5.5}$$

which immediately follows from (5.4) by taking $\varepsilon \delta(x'-x)$ as a perturbation. If we have a functional of the form $F[f] = (f(x'))^n$, then the application of (5.5) is seen to yield

$$\frac{\delta (f(x'))^n}{\delta f(x)} = \lim_{\varepsilon \to 0} \frac{(f(x') + \varepsilon \delta(x'-x))^n - (f(x'))^n}{\varepsilon} = n(f(x'))^{n-1} \delta(x'-x), \tag{5.6}$$

which shows that the functional derivative obeys similar rules as ordinary differentiation.

Example 5.2. The functional derivative of the functional $F[f] = \int_a^b \mathrm{d}x'\ f(x')$ is given by

$$\frac{\delta F[f]}{\delta f(x)} = \int_a^b \mathrm{d}x'\ \delta(x'-x) = \begin{cases} 1 & \text{if} \quad a < x < b \\ 0 & \text{if} \quad x < a \quad \text{or} \quad x > b, \end{cases} \tag{5.7}$$

or, more generally, the functional derivative of the functional $F[f] = \int \mathrm{d}x'\ (f(x'))^n$ is given by

$$\begin{aligned} \frac{\delta F[f]}{\delta f(x)} &= \lim_{\varepsilon \to 0} \int \mathrm{d}x' \frac{(f(x') + \varepsilon\delta(x'-x))^n - (f(x'))^n}{\varepsilon} = \int \mathrm{d}x' n(f(x'))^{n-1}\delta(x'-x) \\ &= n(f(x))^{n-1}. \end{aligned} \tag{5.8}$$

5.2 Principle of Least Action

In classical mechanics, the trajectory of a particle between two space-time points $(\mathbf{x}_\mathrm{i}, t_\mathrm{i})$ and $(\mathbf{x}_\mathrm{f}, t_\mathrm{f})$ is very elegantly determined by the principle of least action. This principle states that the classical trajectory $\mathbf{x}_\mathrm{cl}(t)$ of the particle minimizes the functional $S[\mathbf{x}]$, which is known as the action and given by

$$S[\mathbf{x}] = \int_{t_\mathrm{i}}^{t_\mathrm{f}} \mathrm{d}t\ L(\mathbf{x}(t), \mathrm{d}\mathbf{x}(t)/\mathrm{d}t, t), \tag{5.9}$$

where $L(\mathbf{x}(t), \mathrm{d}\mathbf{x}(t)/\mathrm{d}t, t)$ is the Lagrangian for the system. For a particle with mass m and moving in a potential $V(\mathbf{x}, t)$, the Lagrangian is given by

$$L(\mathbf{x}(t), \mathrm{d}\mathbf{x}(t)/\mathrm{d}t, t) = \frac{1}{2} m \left(\frac{\mathrm{d}\mathbf{x}(t)}{\mathrm{d}t} \right)^2 - V(\mathbf{x}(t), t), \tag{5.10}$$

which is the kinetic energy minus the potential energy. The classical trajectory $\mathbf{x}_\mathrm{cl}(t)$ is determined by minimizing the action, where the end points remain fixed. Since $\mathbf{x}_\mathrm{cl}(t)$ is an extremum of the action $S[\mathbf{x}]$, we must have for the functional derivative of the action that

$$\left. \frac{\delta S[\mathbf{x}]}{\delta \mathbf{x}(t)} \right|_{\mathbf{x} = \mathbf{x}_\mathrm{cl}} = 0. \tag{5.11}$$

With the use of (5.8), we find that this implies for the classical path

$$\frac{\delta S[\mathbf{x}]}{\delta \mathbf{x}(t)} = \int_{t_\mathrm{i}}^{t_\mathrm{f}} \mathrm{d}t' \left\{ \frac{\partial L}{\partial \mathbf{x}} \delta(t'-t) + \frac{\partial L}{\partial (\mathrm{d}\mathbf{x}/\mathrm{d}t')} \frac{\mathrm{d}}{\mathrm{d}t'} \delta(t'-t) \right\} = 0. \tag{5.12}$$

Performing the integration, we arrive at

$$\frac{\partial L(\mathbf{x}(t),\mathrm{d}\mathbf{x}(t)/\mathrm{d}t,t)}{\partial \mathbf{x}(t)} = \frac{\mathrm{d}}{\mathrm{d}t}\frac{\partial L(\mathbf{x}(t),\mathrm{d}\mathbf{x}(t)/\mathrm{d}t,t)}{\partial(\mathrm{d}\mathbf{x}(t)/\mathrm{d}t)}, \tag{5.13}$$

which can be solved using the additional condition that the end points are fixed, i.e. $\mathbf{x}(t_\mathrm{i}) = \mathbf{x}_\mathrm{i}$ and $\mathbf{x}(t_\mathrm{f}) = \mathbf{x}_\mathrm{f}$. The above equation is known as the Euler-Lagrange equation. Upon substitution of (5.10), we find Newton's equation

$$m\frac{\mathrm{d}^2\mathbf{x}(t)}{\mathrm{d}t^2} = -\frac{\partial V(\mathbf{x}(t),t)}{\partial \mathbf{x}(t)}. \tag{5.14}$$

Example 5.3. For a free particle, we have

$$L = \frac{1}{2}m\left(\frac{\mathrm{d}\mathbf{x}(t)}{\mathrm{d}t}\right)^2, \tag{5.15}$$

such that the Euler-Lagrange equation leads to $m\mathrm{d}^2\mathbf{x}(t)/\mathrm{d}t^2 = 0$. Therefore, the classical path for a free particle satisfying the boundary conditions $\mathbf{x}_\mathrm{cl}(t_\mathrm{i}) = \mathbf{x}_\mathrm{i}$ and $\mathbf{x}_\mathrm{cl}(t_\mathrm{f}) = \mathbf{x}_\mathrm{f}$, is given by

$$\mathbf{x}_\mathrm{cl}(t) = \mathbf{x}_\mathrm{i} + (\mathbf{x}_\mathrm{f} - \mathbf{x}_\mathrm{i})\frac{t-t_\mathrm{i}}{t_\mathrm{f}-t_\mathrm{i}}. \tag{5.16}$$

As a result, we obtain for the extremal classical action in this case

$$S[\mathbf{x}_\mathrm{cl}] = \int_{t_\mathrm{i}}^{t_\mathrm{f}} \mathrm{d}t L(\mathbf{x}_\mathrm{cl}(t),\mathrm{d}\mathbf{x}_\mathrm{cl}(t)/\mathrm{d}t) = \frac{1}{2}m\frac{(\mathbf{x}_\mathrm{f}-\mathbf{x}_\mathrm{i})^2}{t_\mathrm{f}-t_\mathrm{i}}. \tag{5.17}$$

Example 5.4. For a particle in a harmonic potential, we have

$$L = \frac{1}{2}m\left(\frac{\mathrm{d}\mathbf{x}(t)}{\mathrm{d}t}\right)^2 - \frac{1}{2}m\omega^2\mathbf{x}^2(t) \tag{5.18}$$

and the Euler-Lagrange equation leads to $\mathrm{d}^2\mathbf{x}(t)/\mathrm{d}t^2 = -\omega^2\mathbf{x}(t)$. Therefore, the classical path for a particle in a harmonic potential satisfying the boundary conditions $\mathbf{x}_\mathrm{cl}(t_\mathrm{i}) = \mathbf{x}_\mathrm{i}$ and $\mathbf{x}_\mathrm{cl}(t_\mathrm{f}) = \mathbf{x}_\mathrm{f}$, is given by

$$\mathbf{x}_\mathrm{cl}(t) = \frac{\mathbf{x}_\mathrm{f}\sin(\omega(t-t_\mathrm{i})) - \mathbf{x}_\mathrm{i}\sin(\omega(t-t_\mathrm{f}))}{\sin(\omega(t_\mathrm{f}-t_\mathrm{i}))}. \tag{5.19}$$

As a result, we obtain for the classical action of the harmonic oscillator

$$S[\mathbf{x}_{\text{cl}}] = \int_{t_\text{i}}^{t_\text{f}} \mathrm{d}t L(\mathbf{x}_{\text{cl}}(t), \mathrm{d}\mathbf{x}_{\text{cl}}(t)/\mathrm{d}t)$$
$$= \frac{m\omega}{2\sin(\omega(t_\text{f}-t_\text{i}))}\left\{(\mathbf{x}_\text{i}^2+\mathbf{x}_\text{f}^2)\cos(\omega(t_\text{f}-t_\text{i})) - 2\mathbf{x}_\text{i}\cdot\mathbf{x}_\text{f}\right\}. \quad (5.20)$$

5.3 Phase-Space Representation

In the previous section, we formulated the principle of least action in terms of the Lagrangian $L(\mathbf{x}(t), \mathrm{d}\mathbf{x}(t)/\mathrm{d}t, t)$. However, in some situations we are more interested in the Hamiltonian, which is an explicit function of the coordinate $\mathbf{x}(t)$ and the momentum $\mathbf{p}(t)$. The Hamiltonian determines the total energy of the system. A good example is classical statistical mechanics, where we want to determine the canonical partition function, which is an integral of the Boltzmann weights $\mathrm{e}^{-\beta H(\mathbf{p},\mathbf{x})}$ over the total phase space. Thus, we would also like to formulate the principle of least action in terms of the Hamiltonian, which is related to the Lagrangian by means of a Legendre transformation

$$H(\mathbf{p}(t), \mathbf{x}(t), t) = \mathbf{p}(t)\cdot\frac{\mathrm{d}\mathbf{x}(t)}{\mathrm{d}t} - L(\mathbf{x}(t), \mathrm{d}\mathbf{x}(t)/\mathrm{d}t, t). \quad (5.21)$$

The velocity $\mathrm{d}\mathbf{x}(t)/\mathrm{d}t$ must be eliminated in favor of the momentum via

$$\mathbf{p}(t) = \frac{\partial L(\mathbf{x}(t), \mathrm{d}\mathbf{x}(t)/\mathrm{d}t, t)}{\partial(\mathrm{d}\mathbf{x}(t)/\mathrm{d}t)}, \quad (5.22)$$

such that action is given by the functional

$$S[\mathbf{p},\mathbf{x}] = \int_{t_\text{i}}^{t_\text{f}} \mathrm{d}t \left\{\mathbf{p}(t)\cdot\frac{\mathrm{d}\mathbf{x}(t)}{\mathrm{d}t} - H(\mathbf{p}(t), \mathbf{x}(t), t)\right\}. \quad (5.23)$$

The variation of the action has to vanish for the classical path $\mathbf{x}_{\text{cl}}(t)$ and $\mathbf{p}_{\text{cl}}(t)$, leading to the Hamilton equations of motion by demanding that the functional derivatives with respect to position and momentum are zero, giving

$$\frac{\mathrm{d}\mathbf{p}(t)}{\mathrm{d}t} = -\frac{\partial H(\mathbf{p}(t), \mathbf{x}(t), t)}{\partial \mathbf{x}(t)} \quad \text{and} \quad \frac{\mathrm{d}\mathbf{x}(t)}{\mathrm{d}t} = \frac{\partial H(\mathbf{p}(t), \mathbf{x}(t), t)}{\partial \mathbf{p}(t)}. \quad (5.24)$$

Example 5.5. Performing the Legendre transformation in (5.21) for the Lagrangian in (5.10), we find that

$$H(\mathbf{p}(t), \mathbf{x}(t)) = \frac{\mathbf{p}^2(t)}{2m} + V(\mathbf{x}(t), t), \quad (5.25)$$

where from the Hamilton equations of motion, we find again Newton's equations in the form

$$m\frac{\mathrm{d}\mathbf{x}(t)}{\mathrm{d}t} = \mathbf{p}(t) \quad \text{and} \quad \frac{\mathrm{d}\mathbf{p}(t)}{\mathrm{d}t} = -\frac{\partial V(\mathbf{x}(t),t)}{\partial \mathbf{x}(t)}. \tag{5.26}$$

5.4 The Feynman Path Integral

To generalize the above concepts to quantum mechanics, we wish to calculate the quantum-mechanical transition amplitude for a particle that is initially prepared at t_i with position $\mathbf{x}_\mathrm{i}$ to be found at a later time t_f at $\mathbf{x}_\mathrm{f}$, given by

$$W(\mathbf{x}_\mathrm{f},t_\mathrm{f};\mathbf{x}_\mathrm{i},t_\mathrm{i}) \equiv \langle \mathbf{x}_\mathrm{f},t_\mathrm{f}|\mathbf{x}_\mathrm{i},t_\mathrm{i}\rangle \equiv \langle \mathbf{x}_\mathrm{f}|\mathrm{e}^{-\mathrm{i}\hat{H}(t_\mathrm{f}-t_\mathrm{i})/\hbar}|\mathbf{x}_\mathrm{i}\rangle. \tag{5.27}$$

To calculate this matrix element, we proceed by dividing the finite time interval $t_\mathrm{f}-t_\mathrm{i}$ into M intervals by defining $\Delta t = (t_\mathrm{f}-t_\mathrm{i})/M$, where the intermediate times are denoted by $t_j = t_\mathrm{i} + j\Delta t$. In particular, this means that $t_0 = t_\mathrm{i}$ and $t_M = t_\mathrm{f}$. By inserting at every intermediate time a completeness relation, we find that the transition amplitude from (5.27) can be written as

$$\begin{aligned} W(\mathbf{x}_\mathrm{f},t_\mathrm{f};\mathbf{x}_\mathrm{i},t_\mathrm{i}) &= \int \left(\prod_{j=1}^{M-1} \mathrm{d}\mathbf{x}_j\right) \langle \mathbf{x}_M,t_M|\mathbf{x}_{M-1},t_{M-1}\rangle \\ &\quad \times \langle \mathbf{x}_{M-1},t_{M-1}|\mathbf{x}_{M-2},t_{M-2}\rangle \ldots \langle \mathbf{x}_1,t_1|\mathbf{x}_0,t_0\rangle \\ &= \int \left(\prod_{j=1}^{M-1} \mathrm{d}\mathbf{x}_j\right) \prod_{j=1}^{M} \langle \mathbf{x}_j,t_j|\mathbf{x}_{j-1},t_{j-1}\rangle. \end{aligned} \tag{5.28}$$

The interpretation of (5.28) is that quantum mechanically not just one, but many paths contribute to the transition amplitude, and that we have to sum over all of them. To illustrate this point, Fig. 5.1 shows several possible paths between the initial and final point. The Hamiltonian we consider here is a function of momentum and position $\hat{H} = \hat{H}(\hat{\mathbf{p}},\hat{\mathbf{x}})$, and is given by the sum of kinetic and potential energy,

$$\hat{H} = \frac{\hat{\mathbf{p}}^2}{2m} + V(\hat{\mathbf{x}}), \tag{5.29}$$

which leads to

$$\langle \mathbf{p}|\hat{H}|\mathbf{x}\rangle = \left\{\frac{\mathbf{p}^2}{2m} + V(\mathbf{x})\right\} \langle \mathbf{p}|\mathbf{x}\rangle \equiv H(\mathbf{p},\mathbf{x})\langle \mathbf{p}|\mathbf{x}\rangle. \tag{5.30}$$

Note that the crucial ingredient for obtaining (5.30) is that the Hamiltonian is normal ordered, which means that all the momentum operators are positioned to the left of

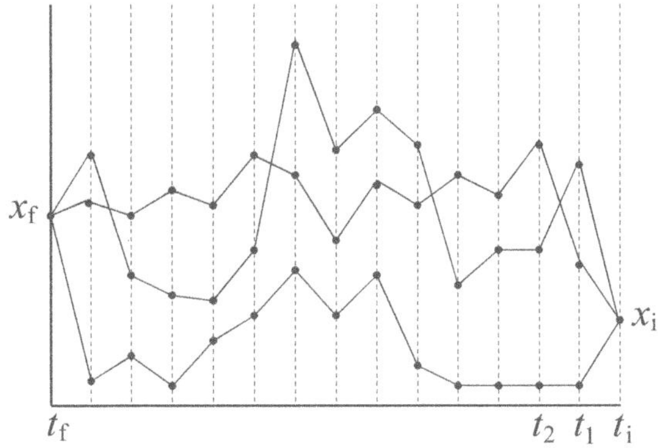

Fig. 5.1 Possible paths that contribute to the transition amplitude in the t vs. x plane.

the position operators. Upon taking the limit $M \longrightarrow \infty$, the time interval Δt becomes infinitesimally small. The matrix element $\langle \mathbf{x}_j t_j | \mathbf{x}_{j-1} t_{j-1} \rangle$ can then be evaluated by expanding the time-evolution operator to first order in Δt and neglecting all terms that are of order $(\Delta t)^2$ or higher. This is acceptable because it only leads to errors in the total transition amplitude that are of order $M\Delta t^2 = (t_\mathrm{f} - t_\mathrm{i})^2/M$, which vanish in the limit $M \longrightarrow \infty$. Thus, we have

$$\begin{aligned} \langle \mathbf{x}_j, t_j | \mathbf{x}_{j-1}, t_{j-1} \rangle &= \int \mathrm{d}\mathbf{p}_j \langle \mathbf{x}_j | \mathbf{p}_j \rangle \langle \mathbf{p}_j | \mathrm{e}^{-\mathrm{i}\hat{H}\Delta t/\hbar} | \mathbf{x}_{j-1} \rangle \\ &\simeq \int \mathrm{d}\mathbf{p}_j \langle \mathbf{x}_j | \mathbf{p}_j \rangle \langle \mathbf{p}_j | \left(1 - \mathrm{i}\hat{H}\Delta t/\hbar\right) | \mathbf{x}_{j-1} \rangle, \end{aligned} \tag{5.31}$$

which, after using (5.30) and writing the expansion as an exponential again, becomes

$$\langle \mathbf{x}_j, t_j | \mathbf{x}_{j-1}, t_{j-1} \rangle = \int \frac{\mathrm{d}\mathbf{p}_j}{(2\pi\hbar)^d} \, \mathrm{e}^{\mathrm{i}\mathbf{p}_j \cdot (\mathbf{x}_j - \mathbf{x}_{j-1})/\hbar - \mathrm{i}\Delta t H(\mathbf{p}_j, \mathbf{x}_{j-1})/\hbar}, \tag{5.32}$$

where we used $\langle \mathbf{x} | \mathbf{p} \rangle = \mathrm{e}^{\mathrm{i}\mathbf{p}\cdot\mathbf{x}/\hbar}/(2\pi\hbar)^{d/2}$ with d the number of spatial dimensions. Note that the re-exponentiation also leads only to errors of order $(\Delta t)^2$ and is therefore allowed. If we substitute this result into (5.28), we obtain for the transition amplitude

$$\begin{aligned} W(\mathbf{x}_\mathrm{f}, t_\mathrm{f}; \mathbf{x}_\mathrm{i}, t_\mathrm{i}) = \int &\left(\prod_{j=1}^{M-1} \mathrm{d}\mathbf{x}_j \right) \left(\prod_{j=1}^{M} \frac{\mathrm{d}\mathbf{p}_j}{(2\pi\hbar)^d} \right) \\ &\times \exp\left\{ \frac{\mathrm{i}}{\hbar} \sum_{i=1}^{M} \Delta t \left(\mathbf{p}_i \cdot \frac{(\mathbf{x}_i - \mathbf{x}_{i-1})}{\Delta t} - H(\mathbf{p}_i, \mathbf{x}_{i-1}) \right) \right\}. \end{aligned} \tag{5.33}$$

In the limit $M \longrightarrow \infty$, the first term in the argument of the above exponent becomes a derivative and the infinite sum becomes an integral. Then, we can write the transition amplitude as

$$W(\mathbf{x}_\mathrm{f},t_\mathrm{f};\mathbf{x}_\mathrm{i},t_\mathrm{i}) \tag{5.34}$$
$$= \int \mathrm{d}[\mathbf{p}] \int_{\mathbf{x}(t_\mathrm{i})=\mathbf{x}_\mathrm{i}}^{\mathbf{x}(t_\mathrm{f})=\mathbf{x}_\mathrm{f}} \mathrm{d}[\mathbf{x}] \exp\left\{\frac{\mathrm{i}}{\hbar}\int_{t_\mathrm{i}}^{t_\mathrm{f}} \mathrm{d}t \left(\mathbf{p}(t)\cdot\frac{\mathrm{d}\mathbf{x}(t)}{\mathrm{d}t} - H(\mathbf{p}(t),\mathbf{x}(t))\right)\right\},$$

where we have introduced the integration measures

$$\int \mathrm{d}[\mathbf{p}] \int_{\mathbf{x}(t_\mathrm{i})=\mathbf{x}_\mathrm{i}}^{\mathbf{x}(t_\mathrm{f})=\mathbf{x}_\mathrm{f}} \mathrm{d}[\mathbf{x}] \equiv \lim_{M\to\infty} \int \left(\prod_{j=1}^{M-1} \mathrm{d}\mathbf{x}_j\right)\left(\prod_{j=1}^{M} \frac{\mathrm{d}\mathbf{p}_j}{(2\pi\hbar)^d}\right). \tag{5.35}$$

In the case of a particle in a potential $\hat{H} = \hat{\mathbf{p}}^2/2m + V(\hat{\mathbf{x}})$, the Gaussian integral over momenta can be done exactly. Using the result of (2.17), we have that

$$\int \frac{\mathrm{d}\mathbf{p}_i}{(2\pi\hbar)^d} \exp\left\{\frac{\mathrm{i}}{\hbar}\Delta t\left(\mathbf{p}_i\cdot\frac{(\mathbf{x}_i-\mathbf{x}_{i-1})}{\Delta t} - \frac{\mathbf{p}_i^2}{2m}\right)\right\}$$
$$= \left(\sqrt{\frac{m}{2\pi\hbar\mathrm{i}\Delta t}}\right)^d \exp\left(\frac{\mathrm{i}m}{2\hbar\Delta t}(\mathbf{x}_i-\mathbf{x}_{i-1})^2\right), \tag{5.36}$$

which, after insertion in (5.34), leads to the following integral over the coordinates

$$W(\mathbf{x}_\mathrm{f},t_\mathrm{f};\mathbf{x}_\mathrm{i},t_\mathrm{i}) = \left(\sqrt{\frac{m}{2\pi\hbar\mathrm{i}\Delta t}}\right)^d \lim_{M\to\infty} \int \left(\prod_{j=1}^{M-1} \mathrm{d}\mathbf{x}_j \left(\sqrt{\frac{m}{2\pi\hbar\mathrm{i}\Delta t}}\right)^d\right)$$
$$\times \exp\left\{\frac{\mathrm{i}}{\hbar}\sum_{i=1}^{M}\Delta t\left(\frac{m}{2\Delta t^2}(\mathbf{x}_i-\mathbf{x}_{i-1})^2 - V(\mathbf{x}_{i-1})\right)\right\}. \tag{5.37}$$

Example 5.6. In the case of a free particle, the remaining integrals over the positions $\mathbf{x}_i$ are also Gaussian, such that they can be performed exactly. We first determine

$$\int \mathrm{d}\mathbf{x}_i \exp\left\{\frac{\mathrm{i}m}{2\hbar\Delta t}(\mathbf{x}_{i+1}-\mathbf{x}_i)^2\right\}\exp\left\{\frac{\mathrm{i}m}{2\hbar\Delta t}(\mathbf{x}_i-\mathbf{x}_{i-1})^2\right\}$$
$$= \exp\left\{\frac{\mathrm{i}m}{2\hbar\Delta t}(\mathbf{x}_{i+1}^2+\mathbf{x}_{i-1}^2)\right\}\int \mathrm{d}\mathbf{x}_i \exp\left\{\frac{\mathrm{i}m}{\hbar\Delta t}\mathbf{x}_i^2 - \mathrm{i}\frac{m}{\hbar\Delta t}\mathbf{x}_i\cdot(\mathbf{x}_{i+1}+\mathbf{x}_{i-1})\right\},$$
$$= \left(\sqrt{\frac{\mathrm{i}\pi\hbar\Delta t}{m}}\right)^d \exp\left\{\frac{\mathrm{i}m}{4\hbar\Delta t}(\mathbf{x}_{i+1}-\mathbf{x}_{i-1})^2\right\}, \tag{5.38}$$

where we again used (2.17). Substituting the result into (5.37), we finally obtain

$$\langle\mathbf{x}_\mathrm{f},t_\mathrm{f}|\mathbf{x}_\mathrm{i},t_\mathrm{i}\rangle = \left(\sqrt{\frac{m}{2\pi\mathrm{i}\hbar(t_\mathrm{f}-t_\mathrm{i})}}\right)^d \exp\left\{\frac{\mathrm{i}}{\hbar}\frac{m}{2}\frac{(\mathbf{x}_\mathrm{f}-\mathbf{x}_\mathrm{i})^2}{t_\mathrm{f}-t_\mathrm{i}}\right\}. \tag{5.39}$$

5.4.1 Continuum Limit and Fluctuation Expansion

In practice, it is not convenient to solve path integrals by going through the whole discretization procedure each time. Instead, we usually perform the integrals directly using continuum expressions. We have from (5.34) that

$$\begin{aligned} & W(\mathbf{x}_{\mathrm{f}},t_{\mathrm{f}};\mathbf{x}_{\mathrm{i}},t_{\mathrm{i}}) \\ &= \int_{\mathbf{x}(t_{\mathrm{i}})=\mathbf{x}_{\mathrm{i}}}^{\mathbf{x}(t_{\mathrm{f}})=\mathbf{x}_{\mathrm{f}}} \mathrm{d}[\mathbf{x}] \int \mathrm{d}[\mathbf{p}] \exp\left\{\frac{\mathrm{i}}{\hbar}\int_{t_{\mathrm{i}}}^{t_{\mathrm{f}}} \mathrm{d}t \left(\mathbf{p}(t)\cdot\frac{\mathrm{d}\mathbf{x}(t)}{\mathrm{d}t} - \frac{\mathbf{p}(t)^2}{2m} - V(\mathbf{x}(t))\right)\right\}, \end{aligned} \tag{5.40}$$

which is quadratic in momenta. In general, a convenient way to solve Gaussian integrals is by expanding the quadratic function $f(x) = G^{-1}x^2/2 + Jx$ around its maximum $x_0 = -GJ$, giving

$$\begin{aligned} \int \mathrm{d}x\, \mathrm{e}^{f(x)} &= \int \mathrm{d}x\, \exp\left\{ f(x_0) + \left.\frac{\mathrm{d}^2 f(x)}{\mathrm{d}x^2}\right|_{x=x_0} (x-x_0)^2 \right\} \\ &= \mathrm{e}^{f(x_0)}\sqrt{-2\pi G} = \mathrm{e}^{-GJ^2/2}\sqrt{-2\pi G}, \end{aligned} \tag{5.41}$$

which is valid, because the expansion up to second order is exact for a quadratic function and the linear term in $x - x_0$ is zero, if x_0 is the maximum of $f(x)$. The approach is particularly convenient if we are only interested in the behavior of the integral as a function of J, because all the dependence on J is contained in $f(x_0)$, which is obtained without even performing an integral. The rest of the expression is then merely a numerical prefactor independent of J. The above approach can be generalized to Gaussian functionals $F[p]$, giving

$$\begin{aligned} \int \mathrm{d}[p]\, \mathrm{e}^{F[p]} &= \mathrm{e}^{F[p_{\mathrm{cl}}]} \int \mathrm{d}[\pi]\, \exp\left\{ \int \mathrm{d}t \int \mathrm{d}t' \left.\frac{\delta^2 F[p]}{\delta p(t)\delta p(t')}\right|_{p=p_{\mathrm{cl}}} \pi(t)\pi(t') \right\} \\ &= \mathrm{e}^{F[p_{\mathrm{cl}}]}\mathcal{N}, \end{aligned} \tag{5.42}$$

where we introduced the momentum fluctuations $\pi(t) = p(t) - p_{\mathrm{cl}}(t)$ and $\mathcal{N}$ is a short-hand notation for the outcome of the functional integral on the right-hand side. If we apply this procedure to the momentum integral of (5.40), we find for the extremum of the functional in the exponent $\mathbf{p}_{\mathrm{cl}}(t) = m\mathrm{d}\mathbf{x}(t)/\mathrm{d}t$, leading to

$$\begin{aligned} & W(\mathbf{x}_{\mathrm{f}},t_{\mathrm{f}};\mathbf{x}_{\mathrm{i}},t_{\mathrm{i}}) = \\ & \mathcal{N} \int_{\mathbf{x}(t_{\mathrm{i}})=\mathbf{x}_{\mathrm{i}}}^{\mathbf{x}(t_{\mathrm{f}})=\mathbf{x}_{\mathrm{f}}} \mathrm{d}[\mathbf{x}] \exp\left\{\frac{\mathrm{i}}{\hbar}\int_{t_{\mathrm{i}}}^{t_{\mathrm{f}}} \mathrm{d}t \left(\frac{1}{2}m\left(\frac{\mathrm{d}\mathbf{x}(t)}{\mathrm{d}t}\right)^2 - V(\mathbf{x}(t))\right)\right\}, \end{aligned} \tag{5.43}$$

where the undetermined numerical prefactor $\mathcal{N}$ is independent of $\mathbf{x}(t)$. This last result illustrates the power of the fluctuation expansion and the direct use of continuum expressions, because we have obtained in one simple step the correct dependence of the path integral on $\mathbf{x}(t)$, leaving only a normalization factor undetermined. This normalization factor is often physically irrelevant or even drops out of calculations completely, as is for example the case for statistical expectation values seen in Sect. 5.7. Thus, we find that the transition amplitude is given by

$$W(\mathbf{x}_\mathrm{f},t_\mathrm{f};\mathbf{x}_\mathrm{i},t_\mathrm{i}) = \mathscr{N}\int_{\mathbf{x}(t_\mathrm{i})=\mathbf{x}_\mathrm{i}}^{\mathbf{x}(t_\mathrm{f})=\mathbf{x}_\mathrm{f}} \mathrm{d}[\mathbf{x}]\exp\left\{\frac{\mathrm{i}}{\hbar}\int_{t_\mathrm{i}}^{t_\mathrm{f}} \mathrm{d}t\; L(\mathbf{x}(t),\mathrm{d}\mathbf{x}(t)/\mathrm{d}t)\right\}$$
$$= \mathscr{N}\int_{\mathbf{x}(t_\mathrm{i})=\mathbf{x}_\mathrm{i}}^{\mathbf{x}(t_\mathrm{f})=\mathbf{x}_\mathrm{f}} \mathrm{d}[\mathbf{x}]\exp\left\{\frac{\mathrm{i}}{\hbar}S[\mathbf{x}]\right\}, \tag{5.44}$$

where comparison with (5.37) reveals that the correct normalization factor in the discretized case is actually given by $\mathscr{N} = (m/2\pi\mathrm{i}\hbar\Delta t)^{dM/2}$.

Example 5.7. In the case of a free particle, we can further evaluate (5.44). We start by doing another fluctuation expansion, $\mathbf{x}(t) = \mathbf{x}_\mathrm{cl}(t) + \boldsymbol{\xi}(t)$, where the classical path $\mathbf{x}_\mathrm{cl}(t)$ minimizes the action. The fluctuation $\boldsymbol{\xi}(t)$ is the deviation from the classical path and satisfies the boundary conditions $\boldsymbol{\xi}(t_\mathrm{f}) = \boldsymbol{\xi}(t_\mathrm{i}) = \mathbf{0}$. Hence, we find that the path integral in (5.44) can be written as

$$W(\mathbf{x}_\mathrm{f},t_\mathrm{f};\mathbf{x}_\mathrm{i},t_\mathrm{i}) = \exp\left\{\frac{\mathrm{i}}{\hbar}S[\mathbf{x}_\mathrm{cl}]\right\}\mathscr{N}\int_{\boldsymbol{\xi}(t_\mathrm{i})=\mathbf{0}}^{\boldsymbol{\xi}(t_\mathrm{f})=\mathbf{0}} \mathrm{d}[\boldsymbol{\xi}]\exp\left\{\frac{\mathrm{i}}{\hbar}\int_{t_\mathrm{i}}^{t_\mathrm{f}} \mathrm{d}t\;\frac{1}{2}m\left(\frac{\mathrm{d}\boldsymbol{\xi}(t)}{\mathrm{d}t}\right)^2\right\}$$
$$\equiv \exp\left\{\frac{\mathrm{i}}{\hbar}S[\mathbf{x}_\mathrm{cl}]\right\}f(t_\mathrm{f},t_\mathrm{i}), \tag{5.45}$$

where the linear terms in the fluctuations are zero by construction. Since the path integral on the right-hand side does not depend on the classical path, and in particular not on the initial and final positions, the fluctuation factor $f(t_\mathrm{f},t_\mathrm{i})$ only depends on the initial and final time. Actually, because the Lagrangian does not depend explicitly on time, the function $f(t_\mathrm{f},t_\mathrm{i})$ only depends on the difference $t_\mathrm{f}-t_\mathrm{i}$, and we have

$$f(t_\mathrm{f}-t_\mathrm{i}) = W(\mathbf{0},t_\mathrm{f};\mathbf{0},t_\mathrm{i}) = \langle\mathbf{0}|U(t_\mathrm{f},t_\mathrm{i})|\mathbf{0}\rangle = \langle\mathbf{0}|\mathrm{e}^{-\mathrm{i}\hat{H}(t_\mathrm{f}-t_\mathrm{i})/\hbar}|\mathbf{0}\rangle, \tag{5.46}$$

where the corresponding Hamiltonian for the free particle only depends on momentum. By inserting a full set of eigenstates of the Hamiltonian, we find that

$$\langle\mathbf{0}|U(t_\mathrm{f},t_\mathrm{i})|\mathbf{0}\rangle = \int\frac{\mathrm{d}\mathbf{p}}{(2\pi\hbar)^d}\exp\left\{-\frac{\mathrm{i}}{\hbar}\frac{\mathbf{p}^2}{2m}(t_\mathrm{f}-t_\mathrm{i})\right\}$$
$$= \left(\sqrt{\frac{m}{2\pi\mathrm{i}\hbar(t_\mathrm{f}-t_\mathrm{i})}}\right)^d, \tag{5.47}$$

so that (5.45) indeed agrees with the transition amplitude obtained in (5.39).

5.4.2 Gel'fand-Yaglom Method

In the previous section, we introduced the fluctuation expansion to calculate the transition amplitude for a free particle. In this section, we elaborate on this expansion and introduce a technique that also allows for an exact solution of the harmonic potential. We consider a system described by the action $S[\mathbf{x}] = \int \mathrm{d}t\; L(\mathbf{x}(t), \mathrm{d}\mathbf{x}(t)/\mathrm{d}t)$, where the potential $V(\mathbf{x})$ entering the Lagrangian only depends on the position $\mathbf{x}$. The classical path that follows from solving the Euler-Lagrange equations (5.13) we denote by $\mathbf{x}_{\mathrm{cl}}(t)$. A general path $\mathbf{x}(t)$ can always be written as $\mathbf{x}(t) = \mathbf{x}_{\mathrm{cl}}(t) + \boldsymbol{\xi}(t)$, where $\boldsymbol{\xi}(t)$ gives the deviation from the classical path and satisfies the boundary conditions $\boldsymbol{\xi}(t_{\mathrm{f}}) = \boldsymbol{\xi}(t_{\mathrm{i}}) = \mathbf{0}$. Inserting the fluctuation expansion in the action, gives

$$
\begin{aligned}
S[\mathbf{x}] = S[\mathbf{x}_{\mathrm{cl}} + \boldsymbol{\xi}] &= \int_{t_{\mathrm{i}}}^{t_{\mathrm{f}}} \mathrm{d}t \left\{ \frac{1}{2} m \left(\frac{\mathrm{d}\mathbf{x}_{\mathrm{cl}}(t)}{\mathrm{d}t} + \frac{\mathrm{d}\boldsymbol{\xi}(t)}{\mathrm{d}t} \right)^2 - V(\mathbf{x}_{\mathrm{cl}}(t) + \boldsymbol{\xi}(t)) \right\} \\
&= \int_{t_{\mathrm{i}}}^{t_{\mathrm{f}}} \mathrm{d}t \left\{ \frac{1}{2} m \left(\frac{\mathrm{d}\mathbf{x}_{\mathrm{cl}}(t)}{\mathrm{d}t} \right)^2 + m \frac{\mathrm{d}\mathbf{x}_{\mathrm{cl}}(t)}{\mathrm{d}t} \cdot \frac{\mathrm{d}\boldsymbol{\xi}(t)}{\mathrm{d}t} \right. \\
&\quad \left. + \frac{1}{2} m \left(\frac{\mathrm{d}\boldsymbol{\xi}(t)}{\mathrm{d}t} \right)^2 - V(\mathbf{x}_{\mathrm{cl}}(t) + \boldsymbol{\xi}(t)) \right\}.
\end{aligned} \tag{5.48}
$$

The second term on the right-hand side of the above equation can be rewritten as

$$
\begin{aligned}
\int_{t_{\mathrm{i}}}^{t_{\mathrm{f}}} \mathrm{d}t\; m \frac{\mathrm{d}\mathbf{x}_{\mathrm{cl}}(t)}{\mathrm{d}t} \cdot \frac{\mathrm{d}\boldsymbol{\xi}(t)}{\mathrm{d}t} &= \int_{t_{\mathrm{i}}}^{t_{\mathrm{f}}} \mathrm{d}t \left\{ m \frac{\mathrm{d}}{\mathrm{d}t} \left(\frac{\mathrm{d}\mathbf{x}_{\mathrm{cl}}(t)}{\mathrm{d}t} \cdot \boldsymbol{\xi}(t) \right) - m \frac{\mathrm{d}^2\mathbf{x}_{\mathrm{cl}}(t)}{\mathrm{d}t^2} \cdot \boldsymbol{\xi}(t) \right\} \\
&= m\, \frac{\mathrm{d}\mathbf{x}_{\mathrm{cl}}(t)}{\mathrm{d}t} \cdot \boldsymbol{\xi}(t) \bigg|_{t_{\mathrm{i}}}^{t_{\mathrm{f}}} - \int_{t_{\mathrm{i}}}^{t_{\mathrm{f}}} \mathrm{d}t\; m \frac{\mathrm{d}^2\mathbf{x}_{\mathrm{cl}}(t)}{\mathrm{d}t^2} \cdot \boldsymbol{\xi}(t),
\end{aligned} \tag{5.49}
$$

where the first term vanishes because of the boundary conditions for the fluctuations $\boldsymbol{\xi}(t)$. The second term can be rewritten using the fact that the classical path satisfies the Euler-Lagrange equations, giving

$$
m \frac{\mathrm{d}^2\mathbf{x}_{\mathrm{cl}}(t)}{\mathrm{d}t^2} = - \left. \frac{\mathrm{d}V(\mathbf{x})}{\mathrm{d}\mathbf{x}} \right|_{\mathbf{x}=\mathbf{x}_{\mathrm{cl}}}. \tag{5.50}
$$

As we show now explicitly, this term is actually cancelled by the first-order term in the series expansion of the potential around the classical solution. Note that this cancellation of the first-order terms in the fluctuations is required, because the classical path is by definition the minimum of the action. Explicitly, expanding the interaction up to second order in the fluctuations, we have

$$V(\mathbf{x}_{\rm cl}(t)+\boldsymbol{\xi}(t)) = V(\mathbf{x}_{\rm cl}(t)) + \left.\frac{\mathrm{d}V(\mathbf{x})}{\mathrm{d}\mathbf{x}}\right|_{\mathbf{x}=x_{\rm cl}} \cdot \boldsymbol{\xi}(t) + \frac{1}{2}\left.\frac{\mathrm{d}^2V(\mathbf{x})}{\mathrm{d}\mathbf{x}\,\mathrm{d}\mathbf{x}}\right|_{\mathbf{x}=\mathbf{x}_{\rm cl}} : \boldsymbol{\xi}(t)\boldsymbol{\xi}(t) + \ldots\,, \tag{5.51}$$

where the colon : defines the contraction of the tensor $\mathrm{d}^2V(\mathbf{x})/\mathrm{d}\mathbf{x}\,\mathrm{d}\mathbf{x}$ with the tensor $\boldsymbol{\xi}\boldsymbol{\xi}$, i.e.

$$\frac{1}{2}\left.\frac{\mathrm{d}^2V(\mathbf{x})}{\mathrm{d}\mathbf{x}\,\mathrm{d}\mathbf{x}}\right|_{\mathbf{x}=\mathbf{x}_{\rm cl}} : \boldsymbol{\xi}(t)\boldsymbol{\xi}(t) \equiv \sum_{i,j}\left.\frac{\mathrm{d}^2V(\mathbf{x})}{\mathrm{d}\mathbf{x}_i\,\mathrm{d}\mathbf{x}_j}\boldsymbol{\xi}_j(t)\boldsymbol{\xi}_i(t)\right|_{\mathbf{x}=\mathbf{x}_{\rm cl}}\,, \tag{5.52}$$

where the summations are over all spatial directions. Note that the expansion of (5.51) is exact for the harmonic potential. Combining all of the above, we find that the total action can be written as the sum of the classical action and terms that are quadratic in the fluctuations

$$\begin{aligned} S[\boldsymbol{\xi}] &= S[\mathbf{x}_{\rm cl}] + \frac{1}{2}\int_{t_{\rm i}}^{t_{\rm f}} \mathrm{d}t \left\{ m\left(\frac{\mathrm{d}\boldsymbol{\xi}(t)}{\mathrm{d}t}\right)^2 - \left.\frac{\mathrm{d}^2V(\mathbf{x})}{\mathrm{d}\mathbf{x}\,\mathrm{d}\mathbf{x}}\right|_{\mathbf{x}=\mathbf{x}_{\rm cl}} : \boldsymbol{\xi}(t)\boldsymbol{\xi}(t) + \ldots \right\} \\ &= S[\mathbf{x}_{\rm cl}] + \frac{1}{2}\int_{t_{\rm i}}^{t_{\rm f}} \mathrm{d}t \int_{t_{\rm i}}^{t_{\rm f}} \mathrm{d}t' \left.\frac{\delta^2 S[\mathbf{x}]}{\delta\mathbf{x}(t)\,\delta\mathbf{x}(t')}\right|_{\mathbf{x}=\mathbf{x}_{\rm cl}} : \boldsymbol{\xi}(t)\boldsymbol{\xi}(t') + \ldots\,, \end{aligned} \tag{5.53}$$

where the terms linear in the fluctuations $\boldsymbol{\xi}$ have indeed vanished.

We proceed by considering a potential for which the second-order term of the expansion in (5.51) gives rise to an harmonic oscillator frequency $\omega(t)$ that is the same in each of the d dimensions, that is

$$m\omega^2(t) \equiv \left.\frac{\mathrm{d}^2V(\mathbf{x})}{\mathrm{d}\mathbf{x}^2}\right|_{\mathbf{x}=\mathbf{x}_{\rm cl}}\,. \tag{5.54}$$

We note that the second term on the right-hand side of (5.53) does not depend on the classical path $\mathbf{x}_{\rm cl}$ and in particular not on the initial and final position. As a result, the transition amplitude can be written as a product of the classical amplitude and a fluctuation factor $f(t_{\rm f},t_{\rm i})$, i.e.

$$\begin{aligned} W(\mathbf{x}_{\rm f},t_{\rm f};\mathbf{x}_{\rm i},t_{\rm i}) &= \exp\left\{\frac{\mathrm{i}}{\hbar}S[\mathbf{x}_{\rm cl}]\right\} \\ &\quad\times\mathscr{N}\int_{\boldsymbol{\xi}(t_{\rm i})=\mathbf{0}}^{\boldsymbol{\xi}(t_{\rm f})=\mathbf{0}} \mathrm{d}[\boldsymbol{\xi}]\exp\left\{\frac{\mathrm{i}m}{2\hbar}\int_{t_{\rm i}}^{t_{\rm f}}\mathrm{d}t\left(\left(\frac{\mathrm{d}\boldsymbol{\xi}(t)}{\mathrm{d}t}\right)^2 - \omega^2(t)\boldsymbol{\xi}^2(t)\right)\right\} \\ &= \exp\left\{\frac{\mathrm{i}}{\hbar}S[\mathbf{x}_{\rm cl}]\right\}f(t_{\rm f},t_{\rm i}). \end{aligned} \tag{5.55}$$

We use the exact discretized-time expression of (5.37) for the fluctuation factor, i.e.

$$f(t_{\mathrm{f}},t_{\mathrm{i}}) = W(\mathbf{0},t_{\mathrm{f}};\mathbf{0},t_{\mathrm{i}}) = \left(\sqrt{\frac{m}{2\pi\hbar\mathrm{i}\Delta t}}\right)^{d} \lim_{M\to\infty}\int \left(\prod_{j=1}^{M-1} \mathrm{d}\boldsymbol{\xi}_j \left(\sqrt{\frac{m}{2\pi\hbar\mathrm{i}\Delta t}}\right)^{d}\right)$$
$$\times \exp\left\{\frac{\mathrm{i}m\Delta t}{2\hbar}\sum_{i=1}^{M}\left(\left(\frac{\boldsymbol{\xi}_i-\boldsymbol{\xi}_{i-1}}{\Delta t}\right)^2 - \omega_i^2\boldsymbol{\xi}_i^2\right)\right\}, \quad (5.56)$$

where we also used $\boldsymbol{\xi}_0 = \boldsymbol{\xi}_M = \mathbf{0}$.

The fluctuation factor can be further evaluated by making use of the method developed by Gel'fand and Yaglom [34]. We can immediately perform the Gaussian integrals over $\boldsymbol{\xi}$. Doing so, we find

$$f(t_{\mathrm{f}},t_{\mathrm{i}}) = \lim_{M\to\infty}\left(\sqrt{\frac{m}{2\pi\hbar\mathrm{i}\Delta t}}\frac{1}{\sqrt{\mathrm{Det}[A_{M-1}(\omega)]}}\right)^{d}, \quad (5.57)$$

where the matrix $A_{M-1}(\omega)$ is given by

$$A_{M-1}(\omega) = \begin{bmatrix} 2-(\Delta t)^2\omega_{M-1}^2 & -1 & 0 & 0 \\ -1 & 2-(\Delta t)^2\omega_{M-2}^2 & -1 & 0 \\ 0 & -1 & 2-(\Delta t)^2\omega_{M-3}^2 & -1 \\ 0 & 0 & -1 & \ddots \end{bmatrix}. \quad (5.58)$$

To calculate the determinant in the above expression, we introduce

$$\Psi_N = \Delta t\ \mathrm{Det}[A_N(\omega)], \quad (5.59)$$

where $A_N(\omega)$ is defined as the $N\times N$ matrix forming the lower right corner of $A_{M-1}(\omega)$. The functions Ψ_N and Ψ_{N-1} are related, as can be made explicitly clear by working out the determinants. We have that

$$\mathrm{Det}[A_N] = (2-(\Delta t)^2\omega_N^2)\mathrm{Det}[A_{N-1}] - \mathrm{Det}[A_{N-2}], \quad (5.60)$$

which implies

$$\frac{\Psi_N - 2\Psi_{N-1} + \Psi_{N-2}}{(\Delta t)^2} = \omega_N^2\Psi_{N-1}, \quad (5.61)$$

valid for $N>2$. The initial conditions are

$$\begin{aligned} A_1 &= 2-(\Delta t)^2\omega_1^2 \\ A_2 &= (2-(\Delta t)^2\omega_1^2)(2-(\Delta t)^2\omega_2^2) - 1, \end{aligned} \quad (5.62)$$

which means that

$$\Psi_1 = 2\Delta t - (\Delta t)^3\omega_1^2 \quad \text{and} \quad \frac{\Psi_2-\Psi_1}{\Delta t} = 1+\dots . \quad (5.63)$$

In the continuum limit $M \to \infty$, (5.61) results in a second-order differential equation for Ψ_N,

$$\frac{\mathrm{d}^2\Psi(t)}{\mathrm{d}t^2} = -\omega^2(t)\Psi(t) \tag{5.64}$$

with initial conditions

$$\Psi(t_\mathrm{i}) = \lim_{\Delta t \to 0} \Psi_1 = 0 \tag{5.65}$$

and

$$\left.\frac{\mathrm{d}\Psi(t)}{\mathrm{d}t}\right|_{t=t_\mathrm{i}} = \lim_{\Delta t \to 0} \frac{\Psi_2 - \Psi_1}{\Delta t} = 1\,. \tag{5.66}$$

Given the solution $\Psi(t)$, the fluctuation determinant of interest to us becomes

$$f(t_\mathrm{f}, t_\mathrm{i}) = \left(\sqrt{\frac{m}{2\pi \mathrm{i}\hbar\Psi(t_\mathrm{f})}}\right)^d . \tag{5.67}$$

Example 5.8. As a first application, we again consider the free particle. In this case, the differential equation in (5.64) becomes

$$\frac{\mathrm{d}^2\Psi(t)}{\mathrm{d}t^2} = 0. \tag{5.68}$$

The solution to this equation that satisfies the boundary conditions in (5.65) and (5.66) is given by

$$\Psi(t) = t - t_\mathrm{i}. \tag{5.69}$$

From this we see that

$$f(t_\mathrm{f}, t_\mathrm{i}) = \left(\sqrt{\frac{m}{2\pi \mathrm{i}\hbar(t_\mathrm{f} - t_\mathrm{i})}}\right)^d , \tag{5.70}$$

which is in agreement with (5.39).

As a second example, we calculate the transition amplitude for the case of a particle in a harmonic trapping potential $V(\mathbf{x}) = m\omega^2\mathbf{x}^2/2$, for which the Gel'fand Yaglom method is exact. The fluctuation factor is determined by solving the differential equation (5.64) with a constant frequency ω. We obtain

$$f(t_\mathrm{f}, t_\mathrm{i}) = \left(\sqrt{\frac{m}{2\pi \mathrm{i}\hbar}}\sqrt{\frac{\omega}{\sin\omega(t_\mathrm{f} - t_\mathrm{i})}}\right)^d . \tag{5.71}$$

Combining this with the contribution coming from the classical action given by (5.20), we find that

$$W(\mathbf{x}_\mathrm{f},t_\mathrm{f};\mathbf{x}_\mathrm{i},t_\mathrm{i}) = \left(\sqrt{\frac{m}{2\pi \mathrm{i}\hbar}}\sqrt{\frac{\omega}{\sin\omega(t_\mathrm{f}-t_\mathrm{i})}}\right)^d$$
$$\times \exp\left\{\frac{\mathrm{i}}{2\hbar}\frac{m\omega}{\sin\omega(t_\mathrm{f}-t_\mathrm{i})}\left\{(\mathbf{x}_\mathrm{f}^2+\mathbf{x}_\mathrm{i}^2)\cos(\omega(t_\mathrm{f}-t_\mathrm{i}))-2\mathbf{x}_\mathrm{f}\cdot\mathbf{x}_\mathrm{i}\right\}\right\}. \tag{5.72}$$

5.5 Matrix Elements and Time Ordering

We have now developed several methods to calculate quantum-mechanical transition amplitudes using path integrals. To further extend the path-integral formalism, we show next how it can be used to determine matrix elements of operators

$$\langle \mathbf{x}_\mathrm{f},t_\mathrm{f}|O(\hat{\mathbf{p}}(t),\hat{\mathbf{x}}(t),t)|\mathbf{x}_\mathrm{i},t_\mathrm{i}\rangle = \langle \mathbf{x}_\mathrm{f}|\mathrm{e}^{-\mathrm{i}\hat{H}(t_\mathrm{f}-t)}O(\hat{\mathbf{p}},\hat{\mathbf{x}},t)\mathrm{e}^{-\mathrm{i}\hat{H}(t-t_\mathrm{i})}|\mathbf{x}_\mathrm{i}\rangle, \tag{5.73}$$

where $O(\hat{\mathbf{p}}(t),\hat{\mathbf{x}}(t),t)$ is a function of the Heisenberg operators $\hat{\mathbf{p}}(t)$ and $\hat{\mathbf{x}}(t)$, evaluated at some time $t_\mathrm{i} < t < t_\mathrm{f}$. To derive the continuum limit of the corresponding path integral, we proceed as before by dividing the time interval into M subintervals, where t is given by one of the intermediate discrete times t_m. Moreover, we assume that $O(\hat{\mathbf{p}}(t),\hat{\mathbf{x}}(t),t)$ is in normal order, which means that all momentum operators are positioned to the left of the position operators. We note that a general operator can always be brought into normal form using the commutation relation of the position and momentum operator. For the matrix element of the operator, we then have

$$\langle \mathbf{x}_\mathrm{f},t_\mathrm{f}|O(\hat{\mathbf{p}}(t_m),\hat{\mathbf{x}}(t_m),t_m)|\mathbf{x}_\mathrm{i},t_\mathrm{i}\rangle = \int\left(\prod_{j=1}^{M-1}\mathrm{d}\mathbf{x}_j\right)\langle \mathbf{x}_M,t_M|\mathbf{x}_{M-1},t_{M-1}\rangle$$
$$\times\langle \mathbf{x}_{M-1},t_{M-1}|\mathbf{x}_{M-2},t_{M-2}\rangle\ldots\langle \mathbf{x}_m,t_m|O(\hat{\mathbf{p}}(t_m),\hat{\mathbf{x}}(t_m),t_m)|\mathbf{x}_{m-1},t_{m-1}\rangle\ldots$$
$$\times\langle \mathbf{x}_1,t_1|\mathbf{x}_0,t_0\rangle, \tag{5.74}$$

where we have seen in the previous paragraphs how to deal with the matrix elements $\langle \mathbf{x}_j,t_j|\mathbf{x}_{j-1},t_{j-1}\rangle$. However, the matrix element containing the operator function $\langle \mathbf{x}_m,t_m|O(\hat{\mathbf{p}}(t_m),\hat{\mathbf{x}}(t_m),t_m)|\mathbf{x}_{m-1},t_{m-1}\rangle$ is a little bit more complicated, such that by inserting two closure relations we obtain

$$\langle \mathbf{x}_m,t_m|O(\hat{\mathbf{p}}(t_m),\hat{\mathbf{x}}(t_m),t_m)|\mathbf{x}_{m-1},t_{m-1}\rangle \tag{5.75}$$
$$= \int \mathrm{d}\mathbf{x}'_m\int \mathrm{d}\mathbf{p}'_m\langle \mathbf{x}_m,t_m|\mathbf{p}'_m,t_m\rangle\langle \mathbf{p}'_m,t_m|O(\hat{\mathbf{p}},\hat{\mathbf{x}},t_m)|\mathbf{x}'_m,t_m\rangle\langle \mathbf{x}'_m,t_m|\mathbf{x}_{m-1},t_{m-1}\rangle$$
$$= \int \mathrm{d}\mathbf{x}'_m\int \mathrm{d}\mathbf{p}'_m\frac{1}{(2\pi\hbar)^d}\mathrm{e}^{\mathrm{i}\mathbf{p}'_m(\mathbf{x}_m-\mathbf{x}'_m)/\hbar}O(\mathbf{p}'_m,\mathbf{x}'_m,t_m)\langle \mathbf{x}'_m,t_m|\mathbf{x}_{m-1},t_{m-1}\rangle,$$

where we see that in general we thus need an extra integral over both position and momentum to calculate the above matrix element.

Example 5.9. For an operator function depending only on the position operator, we obtain

$$\begin{aligned}
&\langle \mathbf{x}_m, t_m | O(\hat{\mathbf{x}}(t_m), t_m) | \mathbf{x}_{m-1}, t_{m-1} \rangle \\
&= \int \mathrm{d}\mathbf{x}'_m \int \mathrm{d}\mathbf{p}'_m \frac{1}{(2\pi\hbar)^d} \mathrm{e}^{\mathrm{i}\mathbf{p}'_m(\mathbf{x}_m - \mathbf{x}'_m)/\hbar} O(\mathbf{x}'_m, t_m) \langle \mathbf{x}'_m, t_m | \mathbf{x}_{m-1}, t_{m-1} \rangle \\
&= O(\mathbf{x}_m, t_m) \langle \mathbf{x}_m, t_m | \mathbf{x}_{m-1}, t_{m-1} \rangle,
\end{aligned} \tag{5.76}$$

and the additional integrals over the primed coordinates are seen to vanish. For an operator function depending only on the momentum operator, we obtain

$$\begin{aligned}
&\langle \mathbf{x}_m, t_m | O(\hat{\mathbf{p}}(t_m), t_m) | \mathbf{x}_{m-1}, t_{m-1} \rangle \\
&= \int \mathrm{d}\mathbf{x}'_m \int \mathrm{d}\mathbf{p}'_m \int \mathrm{d}\mathbf{p}_m \frac{\mathrm{e}^{\mathrm{i}\mathbf{p}'_m(\mathbf{x}_m - \mathbf{x}'_m)/\hbar}}{(2\pi\hbar)^{2d}} O(\mathbf{p}'_m, t_m) \mathrm{e}^{\mathrm{i}(\mathbf{p}_m \cdot (\mathbf{x}'_m - \mathbf{x}_{m-1}) - \Delta t H(\mathbf{p}_m, \mathbf{x}_{m-1}))/\hbar} \\
&= \int \frac{\mathrm{d}\mathbf{p}_m}{(2\pi\hbar)^d} O(\mathbf{p}_m, t_m) \mathrm{e}^{\mathrm{i}(\mathbf{p}_m \cdot (\mathbf{x}_m - \mathbf{x}_{m-1}) - \Delta t H(\mathbf{p}_m, \mathbf{x}_{m-1}))/\hbar},
\end{aligned} \tag{5.77}$$

were we used (5.32) and (5.75). Again, we find that the additional integrals over the primed coordinates vanish.

Since the transition amplitudes for the infinitesimal time steps in equation (5.74) can be evaluated in exactly the same way as before, we finally obtain

$$\begin{aligned}
&\langle \mathbf{x}_\mathrm{f}, t_\mathrm{f} | O(\hat{\mathbf{p}}(t), \hat{\mathbf{x}}(t), t) | \mathbf{x}_\mathrm{i}, t_\mathrm{i} \rangle \\
&= \int \mathrm{d}[\mathbf{p}] \int_{\mathbf{x}(t_\mathrm{i}) = \mathbf{x}_\mathrm{i}}^{\mathbf{x}(t_\mathrm{f}) = \mathbf{x}_\mathrm{f}} \mathrm{d}[\mathbf{x}] O(\mathbf{p}(t), \mathbf{x}(t), t) \exp\left\{ \frac{\mathrm{i}}{\hbar} S[\mathbf{p}, \mathbf{x}] \right\},
\end{aligned} \tag{5.78}$$

where in principle, due to (5.75), the measure would now contain an extra integral over both position and momentum. However, as we saw in Example 5.9, these extra integrals vanish when the operator function depends only on the position operator or the momentum operator. Moreover, if we want to calculate the matrix element of (5.73) for a normally-ordered operator function of the form $O(\hat{\mathbf{p}}(t), \hat{\mathbf{x}}(t'), t, t')$ with $t > t'$, then we find that the momentum operators end up in a different time slice to the position operators, such that we can apply (5.76) and (5.77), and the measure in (5.78) actually does not contain any additional integrals. As a result, we should think of the path integral in the right-hand side of (5.78) as calculating $\langle \mathbf{x}_\mathrm{f}, t_\mathrm{f} | O(\hat{\mathbf{p}}(t^+), \hat{\mathbf{x}}(t), t) | \mathbf{x}_\mathrm{i}, t_\mathrm{i} \rangle$ with t^+ denoting the limiting procedure $\eta \downarrow 0$ of $t + \eta$.

From the above discussion, we realize that if we want to calculate the matrix element of equation (5.73) for any product of operators at different times with the path-integral approach, then we need to first order them in time to be able to apply the discretization procedure. This also holds the other way around. If we use the path-integral formulation to calculate matrix elements, then we automatically calculate the elements of time-ordered operators, given by

$$\mathrm{T}[\hat{A}(t)\hat{B}(t')] = \begin{cases} \hat{A}(t)\hat{B}(t') & t > t' \\ \hat{B}(t')\hat{A}(t) & t' > t \end{cases} . \tag{5.79}$$

A convenient way to calculate the matrix element of a time-ordered product of $\hat{\mathbf{x}}$ and $\hat{\mathbf{p}}$ operators in the path-integral formalism is to introduce an additional term

$$-\mathrm{i}\hbar \int_{t_\mathrm{i}}^{t_\mathrm{f}} \mathrm{d}t \; (\mathbf{J}(t)\cdot\mathbf{x}(t) + \mathbf{I}(t)\cdot\mathbf{p}(t))$$

to the action. Doing so, we have for the transition amplitude that

$$\begin{aligned} W_{I,J}(\mathbf{x}_\mathrm{f},t_\mathrm{f};\mathbf{x}_\mathrm{i},t_\mathrm{i}) = & \int \mathrm{d}[\mathbf{p}] \int_{\mathbf{x}(t_\mathrm{i})=\mathbf{x}_\mathrm{i}}^{\mathbf{x}(t_\mathrm{f})=\mathbf{x}_\mathrm{f}} \mathrm{d}[\mathbf{x}] \\ & \times \exp\left\{\frac{\mathrm{i}}{\hbar}\int_{t_\mathrm{i}}^{t_\mathrm{f}} \mathrm{d}t \; \left(\mathbf{p}(t)\cdot\frac{d\mathbf{x}(t)}{dt} - H(\mathbf{p}(t),\mathbf{x}(t),t)\right)\right\} \\ & \times \exp\left\{\int_{t_\mathrm{i}}^{t_\mathrm{f}} \mathrm{d}t \; (\mathbf{J}(t)\cdot\mathbf{x}(t) + \mathbf{I}(t)\cdot\mathbf{p}(t))\right\}, \end{aligned} \tag{5.80}$$

such that $W_{I,J}(\mathbf{x}_\mathrm{f},t_\mathrm{f};\mathbf{x}_\mathrm{i},t_\mathrm{i})$ becomes a generating functional. By differentiation of $W_{I,J}$ with respect to $\mathbf{J}$ and $\mathbf{I}$, we can immediately calculate the time-ordered matrix elements for any tensor product of operators. For instance, differentiating n times with respect to $\mathbf{J}$ yields

$$\begin{aligned} \langle \mathbf{x}_\mathrm{f},t_\mathrm{f}|T[\hat{\mathbf{x}}(t_1)\ldots\hat{\mathbf{x}}(t_n)]|\mathbf{x}_\mathrm{i},t_\mathrm{i}\rangle = & \left.\frac{\delta^n}{\delta\mathbf{J}(t_1)\ldots\delta\mathbf{J}(t_n)} W_{I,J}(\mathbf{x}_\mathrm{f},t_\mathrm{f};\mathbf{x}_\mathrm{i},t_\mathrm{i})\right|_{\mathbf{I}=\mathbf{J}=0} \\ = \int \mathrm{d}[\mathbf{p}] \int_{\mathbf{x}(t_\mathrm{i})=\mathbf{x}_\mathrm{i}}^{\mathbf{x}(t_\mathrm{f})=\mathbf{x}_\mathrm{f}} & \mathrm{d}[\mathbf{x}] \; \mathbf{x}(t_1)\ldots\mathbf{x}(t_n) \\ \times \exp\left\{\frac{\mathrm{i}}{\hbar}\int_{t_\mathrm{i}}^{t_\mathrm{f}} \mathrm{d}t \; \left(\mathbf{p}(t)\cdot\frac{\mathrm{d}\mathbf{x}(t)}{\mathrm{d}t} - H(\mathbf{p}(t),\mathbf{x}(t),t)\right)\right\}. \end{aligned} \tag{5.81}$$

For Hamiltonians of the form $\hat{H} = \hat{\mathbf{p}}^2/2m + V(\hat{\mathbf{x}})$, we can perform the Gaussian integral over the momenta in (5.80) exactly. Using the approach of Sect. 5.4.1, we find

$$\begin{aligned} W_{I,J}(\mathbf{x}_\mathrm{f},t_\mathrm{f};\mathbf{x}_\mathrm{i},t_\mathrm{i}) = \mathscr{N} \int_{\mathbf{x}(t_\mathrm{i})=\mathbf{x}_\mathrm{i}}^{\mathbf{x}(t_\mathrm{f})=\mathbf{x}_\mathrm{f}} \mathrm{d}[\mathbf{x}] \end{aligned} \tag{5.82}$$

$$\times \exp\left\{\frac{\mathrm{i}}{\hbar}\int_{t_\mathrm{i}}^{t_\mathrm{f}} \mathrm{d}t \; \left\{\frac{1}{2}m\left(-\mathrm{i}\hbar\mathbf{I}(t) + \frac{\mathrm{d}\mathbf{x}(t)}{\mathrm{d}t}\right)^2 - V(\mathbf{x}(t)) - \mathrm{i}\hbar\mathbf{J}(t)\cdot\mathbf{x}(t)\right\}\right\},$$

which is immediately rewritten as

$$W_{I,J}(\mathbf{x}_\mathrm{f},t_\mathrm{f};\mathbf{x}_\mathrm{i},t_\mathrm{i}) = \mathscr{N}\int_{\mathbf{x}(t_\mathrm{i})=\mathbf{x}_\mathrm{i}}^{\mathbf{x}(t_\mathrm{f})=\mathbf{x}_\mathrm{f}} \mathrm{d}[\mathbf{x}]\exp\left\{\frac{\mathrm{i}}{\hbar}S[\mathbf{x}]\right\}$$
$$\times\exp\left\{\frac{\mathrm{i}}{\hbar}\int_{t_\mathrm{i}}^{t_\mathrm{f}}\mathrm{d}t\left(-\frac{1}{2}m\hbar^2\mathbf{I}^2(t) - \mathrm{i}\hbar m\mathbf{I}(t)\cdot\frac{\mathrm{d}\mathbf{x}(t)}{\mathrm{d}t} - \mathrm{i}\hbar\mathbf{J}(t)\cdot\mathbf{x}(t)\right)\right\}. \quad (5.83)$$

Using the two equivalent expressions for $W_{I,J}(\mathbf{x}_\mathrm{f},t_\mathrm{f};\mathbf{x}_\mathrm{i},t_\mathrm{i})$, namely(5.80) and (5.83), we can find the connection between the matrix elements of the momentum $\mathbf{p}(t)$ and the velocity $\mathrm{d}\mathbf{x}(t)/\mathrm{d}t$. Differentiating both equations with respect to the currents, we find

$$\langle\mathbf{x}_\mathrm{f},t_\mathrm{f}|\mathbf{p}(t)\mathbf{x}(t')|\mathbf{x}_\mathrm{i},t_\mathrm{i}\rangle = \left.\frac{\delta^2 W_{I,J}}{\delta\mathbf{I}(t)\delta\mathbf{J}(t')}\right|_{\mathbf{I}=\mathbf{J}=0} = m\frac{\mathrm{d}}{\mathrm{d}t}\langle\mathbf{x}_\mathrm{f},t_\mathrm{f}|\mathbf{x}(t)\mathbf{x}(t')|\mathbf{x}_\mathrm{i},t_\mathrm{i}\rangle \quad (5.84)$$

and

$$\langle\mathbf{x}_\mathrm{f},t_\mathrm{f}|\mathbf{p}(t)\mathbf{p}(t')|\mathbf{x}_\mathrm{i},t_\mathrm{i}\rangle = \left.\frac{\delta^2 W_{I,J}}{\delta\mathbf{I}(t)\delta\mathbf{I}(t')}\right|_{\mathbf{I}=\mathbf{J}=0} = m^2\frac{\mathrm{d}^2}{\mathrm{d}t\,\mathrm{d}t'}\langle\mathbf{x}_\mathrm{f},t_\mathrm{f}|\mathbf{x}(t)\mathbf{x}(t')|\mathbf{x}_\mathrm{i},t_\mathrm{i}\rangle$$
$$- \mathrm{i}\hbar m\delta(t-t')\langle\mathbf{x}_\mathrm{f},t_\mathrm{f}|\mathbf{x}_\mathrm{i},t_\mathrm{i}\rangle. \quad (5.85)$$

The last term on the right-hand side of (5.85) might seem peculiar, but is actually present due to the canonical commutation relation between momentum and position. This can be made clear by explicitly considering the time derivatives of the time-ordered product of operators, for which we have

$$\begin{aligned} m^2\frac{\mathrm{d}}{\mathrm{d}t}\mathrm{T}\left[\hat{\mathbf{x}}(t)\hat{\mathbf{x}}(t')\right] &= m^2\frac{\mathrm{d}}{\mathrm{d}t}\left\{\theta(t-t')\hat{\mathbf{x}}(t)\hat{\mathbf{x}}(t') + \theta(t'-t)\hat{\mathbf{x}}(t')\hat{\mathbf{x}}(t)\right\} \\ &= m^2\left\{\delta(t-t')\hat{\mathbf{x}}(t)\hat{\mathbf{x}}(t') + \theta(t-t')\frac{\mathrm{d}\hat{\mathbf{x}}(t)}{\mathrm{d}t}\hat{\mathbf{x}}(t')\right. \\ &\quad \left. -\delta(t'-t)\hat{\mathbf{x}}(t')\hat{\mathbf{x}}(t) + \theta(t'-t)\hat{\mathbf{x}}(t')\frac{\mathrm{d}\hat{\mathbf{x}}(t)}{\mathrm{d}t}\right\} \\ &= m\mathrm{T}\left[\hat{\mathbf{p}}(t)\hat{\mathbf{x}}(t')\right] + m^2\delta(t-t')[\hat{\mathbf{x}}(t),\hat{\mathbf{x}}(t')]_-, \end{aligned} \quad (5.86)$$

where in the last step we used the Heisenberg equation of motion for the position operator, which for the Hamiltonian considered here yields

$$m\frac{\mathrm{d}\hat{\mathbf{x}}(t)}{\mathrm{d}t} = \hat{\mathbf{p}}(t). \quad (5.87)$$

Also note that the commutator on the right-hand side of (5.86) gives zero at equal times, such that this term vanishes. In the same manner, we now calculate the second derivative and find

$$m^2\frac{\mathrm{d}^2}{\mathrm{d}t\,\mathrm{d}t'}\mathrm{T}\left[\hat{\mathbf{x}}(t)\hat{\mathbf{x}}(t')\right] = \mathrm{T}\left[\hat{\mathbf{p}}(t)\hat{\mathbf{p}}(t')\right] - m\delta(t-t')[\hat{\mathbf{p}}(t),\hat{\mathbf{x}}(t')]_-\,, \quad (5.88)$$

where the commutator on the right-hand side yields $-i\hbar$ at equal times. As a result, we find indeed that the matrix elements of (5.85), computed conveniently in the path-integral formalism using the generating functional $W_{I,J}$, are in complete agreement with time-ordered products of operators.

5.6 Quantum-Mechanical Partition Function

So far, we have used path integrals to study the time evolution of dynamical quantum-mechanical systems. Next, we show how the path integral formalism can also be used to study equilibrium physics in quantum statistical mechanics. In Chap. 4, we found that all macroscopic thermodynamic variables of interest can be computed in equilibrium from the partition sum, which is in the canonical ensemble quantum mechanically given by

$$Z = \sum_{\nu} \langle \nu | \exp\{-\beta \hat{H}\} | \nu \rangle = \mathrm{Tr}\left[\exp\{-\beta \hat{H}\}\right], \tag{5.89}$$

where $\beta \equiv 1/k_{\mathrm{B}}T$ is the inverse temperature and $|\psi_n\rangle$ denotes the full set of eigenstates of the Hamiltonian. Since the trace is invariant under basis transformations, any complete set can be used to perform the trace. By identifying $t_{\mathrm{f}} = -i\hbar\beta$, we find that the canonical partition function can be related to the transition amplitude $W(\mathbf{x}_{\mathrm{f}}, t_{\mathrm{f}}; \mathbf{x}_{\mathrm{i}}, 0)$ as

$$Z = \int \mathrm{d}\mathbf{x}\, \langle \mathbf{x} | \mathrm{e}^{-\beta \hat{H}} | \mathbf{x} \rangle = \int \mathrm{d}\mathbf{x}\, \langle \mathbf{x} | \mathrm{e}^{-\mathrm{i}\hat{H}t_{\mathrm{f}}/\hbar} | \mathbf{x} \rangle = \int \mathrm{d}\mathbf{x}\, W(\mathbf{x}, -i\hbar\beta; \mathbf{x}, 0). \tag{5.90}$$

As a result, the partition function is expressed by the following path integral in phase space with the boundary condition $\mathbf{x}(0) = \mathbf{x}(t_{\mathrm{f}})$, giving

$$\begin{aligned} Z = & \int_{\mathbf{x}_M = \mathbf{x}_0} \left(\prod_{j=1}^{M} \mathrm{d}\mathbf{x}_j \right) \left(\prod_{j=1}^{M} \frac{\mathrm{d}\mathbf{p}_j}{(2\pi\hbar)^d} \right) \\ & \times \exp\left\{ \frac{\mathrm{i}}{\hbar} \int_0^{-i\hbar\beta} \mathrm{d}t \left(\mathbf{p}(t) \cdot \frac{\mathrm{d}\mathbf{x}(t)}{\mathrm{d}t} - H(\mathbf{p}(t), \mathbf{x}(t)) \right) \right\} \\ = & \int_{\mathbf{x}(\hbar\beta) = \mathbf{x}(0)} \mathrm{d}[\mathbf{x}] \int \mathrm{d}[\mathbf{p}] \\ & \times \exp\left\{ -\frac{1}{\hbar} \int_0^{\hbar\beta} \mathrm{d}\tau \left(-\mathrm{i}\mathbf{p}(\tau) \cdot \frac{\mathrm{d}\mathbf{x}(\tau)}{\mathrm{d}\tau} + H(\mathbf{p}(\tau), \mathbf{x}(\tau)) \right) \right\}, \end{aligned} \tag{5.91}$$

where in the second step we made a transformation to imaginary time $\tau = \mathrm{i}t$, also known as a Wick rotation. Since the trace leads to an extra integral over position, it restores the symmetry in the number of position and momentum integrals. However, note that we have not changed the notation of the measure, although there is now in principle one more integration compared with (5.34). This may seem like sloppy notation but, thinking in terms of continuous functional integration, the difference between the two expressions is actually incorporated by the boundary conditions of the integration. Put differently, by specifying each time the specific boundary conditions, it becomes clear which of the discrete integration measures is actually being used.

For the Hamiltonian $\hat{H} = \hat{\mathbf{p}}^2/2m + V(\hat{\mathbf{x}})$ we can integrate out the momentum, as explained in Sect. 5.4.1, such that we obtain

$$Z = \mathscr{N} \int_{\mathbf{x}(\hbar\beta)=\mathbf{x}(0)} \mathrm{d}[\mathbf{x}] \exp\left\{ -\frac{1}{\hbar} \int_0^{\hbar\beta} \mathrm{d}\tau \left(\frac{1}{2} m \left(\frac{\mathrm{d}\mathbf{x}(\tau)}{\mathrm{d}\tau} \right)^2 + V(\mathbf{x}(\tau)) \right) \right\}, \quad (5.92)$$

where we note that in imaginary time the Lagrangian actually becomes the Hamiltonian. Finally, the boundary condition in imaginary time $\mathbf{x}(\hbar\beta) = \mathbf{x}(0)$ can be enforced automatically in an elegant way by making a Fourier expansion, such that

$$\mathbf{x}(\tau) = \frac{1}{\sqrt{\hbar\beta}} \sum_{n=-\infty}^{\infty} \mathrm{e}^{-\mathrm{i}\omega_n \tau} \mathbf{x}_n, \quad (5.93)$$

with $\omega_n = 2\pi n/\hbar\beta$, which are called the bosonic Matsubara frequencies.

5.7 Expectation Values

The time evolution of a Heisenberg operator in imaginary time is defined by

$$O(\hat{\mathbf{p}}(\tau), \hat{\mathbf{x}}(\tau)) = \mathrm{e}^{\hat{H}\tau/\hbar} O(\hat{\mathbf{p}}, \hat{\mathbf{x}}) \mathrm{e}^{-\hat{H}\tau/\hbar}, \quad (5.94)$$

which follows from Wick rotation of the corresponding expression in ordinary time. The expectation value of $\hat{O}(\tau)$ in the canonical ensemble is then given by

$$\langle \hat{O}(\tau) \rangle = \frac{1}{Z} \sum_{\nu} \langle \nu | \hat{O}(\tau) \exp\{-\beta \hat{H}\} | \nu \rangle = \frac{1}{Z} \mathrm{Tr}\left[\hat{O} \exp\{-\beta \hat{H}\} \right], \quad (5.95)$$

where in the second step we used (5.94) and the cyclic invariance of the trace, such that the expectation value is seen to be independent of τ. Combining the discussions of Sects. 5.5 and 5.6, we see that the corresponding path integral is given by

$$\begin{aligned} \langle \hat{O} \rangle = & \frac{1}{Z} \int \mathrm{d}[\mathbf{x}] \int \mathrm{d}[\mathbf{p}]\, O(\mathbf{p}(\tau), \mathbf{x}(\tau)) \\ & \times \exp\left\{ -\frac{1}{\hbar} \int_0^{\hbar\beta} \mathrm{d}\tau \left(-\mathrm{i}\mathbf{p}(\tau) \cdot \frac{\mathrm{d}\mathbf{x}(\tau)}{\mathrm{d}\tau} + H(\mathbf{p}(\tau), \mathbf{x}(\tau)) \right) \right\}, \end{aligned} \quad (5.96)$$

with cyclic boundary conditions on the position. If the operator in question only depends on the position $\mathbf{x}$ we can integrate out the momenta, leading to

$$\langle \hat{O} \rangle = \tag{5.97}$$
$$\frac{1}{Z}\int d[\mathbf{x}]\, O(\mathbf{x}(\tau))\exp\left\{-\frac{1}{\hbar}\int_0^{\hbar\beta} d\tau \left\{\frac{1}{2}m\left(\frac{d\mathbf{x}(\tau)}{d\tau}\right)^2 + V(\mathbf{x}(\tau))\right\}\right\},$$

where we note that the numerical prefactor $\mathcal{N}$ coming from the momentum integrals cancels, since the same prefactor is also present in the path integral for the partition sum Z. Moreover, to calculate imaginary time-ordered expectation values explicitly, it is convenient to add source currents to the imaginary-time action in a manner completely analogous to the real-time case.

5.8 Hubbard-Stratonovich Transformation

The Hubbard-Stratonovich transformation is a powerful tool in rewriting path integrals. It makes use of (2.17), which in the continuum limit becomes

$$\int d[\kappa]\exp\left\{-\frac{1}{2\hbar}\int_0^{\hbar\beta} d\tau\, d\tau' \,(\kappa(\tau)-\kappa_0(\tau))M(\tau,\tau')\left(\kappa(\tau')-\kappa_0(\tau')\right)\right\}$$
$$\times\exp\left\{\frac{1}{2}\mathrm{Tr}[\log(M/\hbar)]\right\} = 1, \tag{5.98}$$

where we have interpreted the function $M(\tau,\tau')$ as a matrix with continuous indices τ and τ' over which we can take a trace in the following way

$$\mathrm{Tr}[M/\hbar] = \frac{1}{\hbar}\int_0^{\hbar\beta} d\tau\, M(\tau,\tau). \tag{5.99}$$

Since $\exp\{\mathrm{Tr}[\log(M/\hbar)]/2\}$ in (5.98) is merely a numerical prefactor, it is usually not of much importance.

Now, suppose that we want to calculate a partition sum using the path-integral formalism, but that we cannot solve it due to the presence of a certain term, such as for example a fourth-order term, that makes it impossible to deal with. As a result, we would like to have a method to transform such a term away. To this end, we notice that we can substitute the left hand-side of (5.98) into any path integral we wish to evaluate, since it is equal to unity. Moreover, (5.98) contains a lot of freedom in choosing the precise form of $M(\tau,\tau')$ and $\kappa_0(\tau)$, which we can use to cancel the term that we want to remove. To see precisely how this works, we look at the following example.

Example 5.10. Consider a single atom moving in a one-dimensional quartic potential, described by the Hamiltonian

$$\hat{H} = \frac{\hat{p}^2}{2m} + \alpha\hat{x}^2 + \frac{1}{2}\beta\hat{x}^4, \tag{5.100}$$

where β is not the inverse temperature, but corresponds to the rather standard notation for a fourth-order term, which comes from the Landau theory for phase transitions. This Landau theory is the topic of Chap. 9. Substituting the quartic potential into the path integral of (5.92), we find

$$Z = \mathcal{N}\int \mathrm{d}[\mathbf{x}]\exp\left\{-\frac{1}{\hbar}\int_0^{\hbar\beta}\mathrm{d}\tau\left(\frac{1}{2}m\left(\frac{\mathrm{d}x(\tau)}{\mathrm{d}\tau}\right)^2 + \alpha x^2(\tau) + \frac{1}{2}\beta x^4(\tau)\right)\right\}. \tag{5.101}$$

We have seen how to solve Gaussian path integrals, but how do we deal with the fourth-order term in the exponent? By choosing $\kappa_0(\tau) = x^2(\tau)$ and $M(\tau,\tau') = -\beta\delta(\tau-\tau')$ in (5.98) and substituting the result in (5.101), we find that the fourth-order term gets cancelled.

Although the Hubbard-Stratonovich transformation is extremely useful in removing an unwanted term in the original action, it goes at the cost of introducing an additional path integral over a new field $\kappa(\tau)$, which usually leaves the resulting path integral still unsolvable. In that sense, the transformation is only a formal rewriting of the original problem. However, in many situations the newly obtained action allows for various approximate solutions, such as for example by using a stationary phase or mean-field approximation to the newly introduced field $\kappa(\tau)$. This strategy is actually used several times in this book, since it allows for an elegant description of interacting quantum gases. It turns out that due to the flexibility of this technique, it finds applications ranging from the Hartree-Fock theory of an interacting normal gas to the Bardeen-Cooper-Schrieffer theory for a superfluid interacting Fermi gas. These issues are the topics of Chaps. 8 and 12.

5.9 Problems

Exercise 5.1. Derive the Euler-Lagrange equation by explicitly working out the variation

$$\delta S[\mathbf{x}_{\mathrm{cl}}] = S[\mathbf{x}_{\mathrm{cl}} + \delta\mathbf{x}] - S[\mathbf{x}_{\mathrm{cl}}] = 0. \tag{5.102}$$

Exercise 5.2. Show that the quantum-mechanical transition amplitude

$$W(\mathbf{x},t;\mathbf{x}_{\mathrm{i}},t_{\mathrm{i}}) \equiv \langle \mathbf{x},t|\mathbf{x}_{\mathrm{i}},t_{\mathrm{i}}\rangle \tag{5.103}$$

satisfies the time-dependent Schrödinger equation.
Hint: Make use of the complete set of eigenstates $\{|\nu\rangle\}$ of the Hamiltonian $\hat{H}$ with corresponding eigenvalues E_ν and show that the transition amplitude in (5.103) now

becomes

$$W(\mathbf{x},t;\mathbf{x}_\mathrm{i},t_\mathrm{i}) = \sum_\nu \chi_\nu^*(\mathbf{x})\chi_\nu(\mathbf{x}_\mathrm{i}) \exp\left\{-\frac{\mathrm{i}}{\hbar}E_\nu(t-t_\mathrm{i})\right\}. \quad (5.104)$$

Exercise 5.3. Transition Amplitudes
(a) Derive the path-integral expression for the amplitude $\langle \mathbf{p}_\mathrm{f} t_\mathrm{f} | \mathbf{p}_\mathrm{i} t_\mathrm{i} \rangle$.
(b) Derive the path-integral expression for the amplitude $\langle \mathbf{p}_\mathrm{f} t_\mathrm{f} | \mathbf{x}_\mathrm{i} t_\mathrm{i} \rangle$.

Exercise 5.4. Atom in a Quartic Potential
Consider a single atom moving in an external potential $V^\mathrm{ex}(\mathbf{x}) = \alpha\mathbf{x}^2 + \beta\mathbf{x}^4/2$.
(a) Give the path-integral expression for the matrix element $\langle \mathbf{x} | \mathrm{e}^{-\mathrm{i}\hat{H}t/\hbar} | \mathbf{x} \rangle$.
(b) Derive from this the path-integral expression for the partition function of the atom, by performing an analytic continuation to imaginary time. What are the boundary conditions for this path integral?
(c) Take the classical limit $\hbar\beta \downarrow 0$ and show that you have obtained the correct classical partition function for the atom.

Additional Reading

- R. P. Feynman, A. R. Hibbs, *Quantum Mechanics and Path Integrals*, (McGraw-Hill, New York, 1965).
- H. Kleinert, *Path integrals in Quantum Mechanics, Statistics, Polymer physics, and Financial Markets*, (World Scientific, Singapore, 2004).
- E. Zeidler, *Applied Functional Analysis, Applications to Mathematical Physics*, (Springer-Verlag, Berlin, 1995).

Chapter 6
Second Quantization

The fundamental laws necessary for the mathematical treatment of a large part of physics and the whole of chemistry are thus completely known, and the difficulty lies only in the fact that application of these laws leads to equations that are too complex to be solved.
– Paul Dirac

Quantization is the procedure of going from a classical theory to a quantum theory. An important example is the canonical quantization procedure for going from classical mechanics to quantum mechanics. It amounts to replacing the classical dynamical variables and their Poisson brackets by quantum mechanical operators and their commutators. Considering a system of particles in an external electromagnetic field, the above procedure leads to the original formulation of quantum mechanics where the motion of the particles is quantized, while the applied fields are still treated classically. However, it turns out that the canonical quantization procedure can also be extended to field theory, such that the classical fields are replaced by quantum-mechanical creation and annihilation operators. Since the resulting formulation allows for the quantization of the fields that were in the original quantum theory still treated classically, it is commonly referred to as second quantization. Another way to understand this name is by introducing a Lagrangian density that generates the Schrödinger equation upon applying the Euler-Lagrange equations. This means that the fields of the Lagrangian density become the wavefunctions in the Schrödinger equation. Upon applying the canonical quantization procedure to these fields, the wavefunctions actually become quantized themselves, and the corresponding operators are creation and annihilation operators of particles. It is this last way of applying second quantization that corresponds to the formalism developed in this chapter.

The reason for introducing the language of second quantization is that it turns out to be extremely convenient in the formulation of a quantum theory for many interacting particles. The starting point of this chapter is the more familiar first-quantized N-body Schrödinger equation in the place representation, where the Hamiltonian of interest is motivated from the study of ultracold atomic quantum gases. However, the resulting Hamiltonian is actually seen to be much more general, such that it also applies to a large class of condensed-matter problems. For identical particles, the resulting many-body wavefunction needs to be fully symmetric for bosons, whereas it needs to be fully antisymmetric for fermions. Since a fully (anti)symmetrized wavefunction consists of about $N!$ terms, we need to introduce a shorthand notation, because N is for our purposes typically a million or more. A convenient way to fully specify an (anti)symmetric wavefunction is in terms of the occupation numbers of the single-particle eigenstates from which the many-body state is constructed. This notation is appropriately called the occupation-number representation. Then, by relaxing the constraint of a fixed number of N particles, we introduce the Hilbert space of all (anti)symmetric many-body states, which is also known as Fock space. Since the number of particles in Fock space is not fixed, it is natural to define an annihilation operator, that can destroy a particle in a certain quantum state. From this definition, we immediately also obtain the creation operator, which is consequently used to construct all possible many-body states in Fock space. This procedure is then seen to incorporate automatically the statistics of the corresponding identical particles. By expressing also the many-body Hamiltonian in terms of the creation and annihilation operators, we arrive at our fully second-quantized many-body theory. Finally, we prove the complete equivalence between the old, more familiar formulation of the N-body Schrödinger equation in the place representation, also referred to as the first-quantized formalism, and the new formulation in terms of creation and annihilation operators, also referred to as the second-quantized formalism. This equivalence then ultimately validates all newly introduced definitions and expressions.

6.1 Many-Body Hamiltonian

Atoms form rich quantum systems with many internal degrees of freedom, such as the electronic and the nuclear spin. These internal degrees of freedom are important because they give rise to the hyperfine structure, the magnetic moment, and the electric polarizability of the atoms, which are used to manipulate them experimentally by applying magnetic and electric fields. In principle, calculating the effect of applied external fields on atoms can be a tedious task, just as deriving the full atomic interaction potential from first principles. In Chap. 10, where we focus on atomic physics, we address such questions in more detail. However, for our present goal only the outcome of such calculations are relevant, and we start from the point where the external potential and the interaction potential are given. Moreover, we restrict ourselves to atomic mixtures of at most two hyperfine states, which describes practically all important experiments that have been performed on ultracold atomic gases so far. We then add the remark that the following treatment is easily generalized to three hyperfine states or more. In this chapter we describe the hyperfine degrees of freedom by two effective spin states $|s,m_s\rangle$, whereas in Chap. 10 we overcome this simplification and treat the hyperfine structure in more detail.

Consider N identical atoms with mass m and effective spin $\mathbf{s}$ in a constant magnetic field $\mathbf{B}$ and an external potential $V^{\mathrm{ex}}(\mathbf{x})$, caused for example by a spatially varying electric field. As a result, the time-dependent Schrödinger equation we have to solve is

$$i\hbar\frac{\partial}{\partial t}|\Psi(t)\rangle = \hat{H}|\Psi(t)\rangle \tag{6.1}$$

with the Hamiltonian given by

$$\hat{H} = \sum_{i=1}^{N} \left\{ \frac{\hat{\mathbf{p}}_i^2}{2m} + V^{\mathrm{ex}}(\hat{\mathbf{x}}_i) - \gamma \hat{\mathbf{s}}_i \cdot \mathbf{B} \right\} + \frac{1}{2} \sum_{i \neq j=1}^{N} V(\hat{\mathbf{x}}_i - \hat{\mathbf{x}}_j) , \tag{6.2}$$

where we have that $[\hat{\mathbf{x}}_j, \hat{\mathbf{p}}_j]_- = i\hbar$, and all other commutators of the position and momentum operators vanish. The first term on the right-hand side is the sum of the one-particle Hamiltonians, including an effective Zeeman interaction $-\gamma \hat{\mathbf{s}}_i \cdot \mathbf{B}$ that accounts for a possible difference in the hyperfine energies of the two spin states. The second term describes the two-body interactions between the atoms. The factor $1/2$ in front of the interaction term makes sure that we count the potential-energy contribution of each pair of atoms only once. For notational simplicity, we assume in first instance that the interaction $V(\hat{\mathbf{x}}_i - \hat{\mathbf{x}}_j)$ is independent of the hyperfine states of atoms i and j. This is in general not the case for realistic atomic gases, and in later sections we overcome this approximation. Finally, we also have neglected possible three-body interactions. This is particularly valid for our goal of describing dilute interacting quantum gases, when it is highly improbable for three atoms to have an simultaneously overlapping electron cloud.

Although we have motivated the above Hamiltonian by atomic physics, it is in fact much more general. In particular, it is also the Hamiltonian describing the electron gas in metals and semiconductors. The external potential represents then the periodic potential provided by the ionic crystal and the interaction $V(\hat{\mathbf{x}}_i - \hat{\mathbf{x}}_j)$ is due to the Coulomb repulsion between the electrons. The effect of an external magnetic field is, however, not fully correctly incorporated in this case, because for charged particles the magnetic field does not only couple to the spin of the particles, but also to their momenta. To arrive at the correct Hamiltonian for an electron gas in a magnetic field, we need to perform the minimal-coupling substitution $\hat{\mathbf{p}}_i \to \hat{\mathbf{p}}_i + e\mathbf{A}(\hat{\mathbf{x}}_i)/c$, with $-e$ the electron charge and $\mathbf{A}(\mathbf{x})$ the vector potential that is related to the magnetic field by $\mathbf{B}(\mathbf{x}) = \boldsymbol{\nabla} \times \mathbf{A}(\mathbf{x})$. We come back to such orbital effects in Chap. 13, where they turn out to also be of importance for neutral atoms if the quantum gas is being rotated.

6.2 Fock Space

To tackle the many-body problem, we start by solving the one-particle time-independent Schrödinger equation. Assuming that we have done so, the eigenfunctions $\chi_{\mathbf{n}}(\mathbf{x}) \equiv \langle \mathbf{x} | \mathbf{n} \rangle$ that correspond to the eigenenergies $\varepsilon_{\mathbf{n}}$ satisfy

$$\left\{ -\frac{\hbar^2 \boldsymbol{\nabla}^2}{2m} + V^{\mathrm{ex}}(\mathbf{x}) - \varepsilon_{\mathbf{n}} \right\} \chi_{\mathbf{n}}(\mathbf{x}) = 0 , \tag{6.3}$$

where $\mathbf{n} = (n_x, n_y, n_z)$, such that the integers n_x, n_y, and n_z denote the three quantum numbers that are required to specify the one-particle eigenstates in the external potential. For the spin part, we introduce the shorthand notation $|\alpha\rangle = |s, m_s\rangle$, leading to an effective Zeeman energy ε_α

$$-\gamma \hat{\mathbf{s}} \cdot \mathbf{B} |\alpha\rangle = -\gamma m_s B |\alpha\rangle = \varepsilon_\alpha |\alpha\rangle, \tag{6.4}$$

where we conveniently took the quantization axis of the spin-angular momentum along the direction of the applied magnetic field. As a result, the total eigenenergy of the single-particle eigenstate $|\mathbf{n}, \alpha\rangle = |\mathbf{n}\rangle|\alpha\rangle$ is given by $\varepsilon_{\mathbf{n},\alpha} \equiv \varepsilon_\mathbf{n} + \varepsilon_\alpha$. In the absence of interactions, the eigenstates of the many-body Hamiltonian of (6.2) are given by the product states

$$\{|\mathbf{n}_1, \alpha_1\rangle_1 |\mathbf{n}_2, \alpha_2\rangle_2 \ldots |\mathbf{n}_N, \alpha_N\rangle_N\},$$

which form a basis for the N-particle Hilbert space $\mathscr{H}_N$,

$$\mathscr{H}_N = \mathscr{H}_1 \otimes \mathscr{H}_1 \otimes \cdots \otimes \mathscr{H}_1,$$

which is given by the tensor product of N single-particle Hilbert spaces. The inner product between two N particle product states is given by

$$\begin{aligned} &{}_N\langle \mathbf{n}_N, \alpha_N| \ldots {}_1\langle \mathbf{n}_1, \alpha_1||\mathbf{n}_1, \alpha_1\rangle_1 \ldots |\mathbf{n}_N, \alpha_N\rangle_N = \\ &{}_1\langle \mathbf{n}_1, \alpha_1|\mathbf{n}_1, \alpha_1\rangle_1 \ldots {}_N\langle \mathbf{n}_N, \alpha_N|\mathbf{n}_N, \alpha_N\rangle_N. \end{aligned} \tag{6.5}$$

A fundamental postulate of quantum mechanics is that identical particles are intrinsically indistinguishable. To study the consequences of this statement, we may introduce the permutation operator that interchanges two particles. Since the particles are indistinguishable, their interchange leads to the same physical state, such that the permutation operator only introduces a phase factor $\mathrm{e}^{\mathrm{i}\phi}$. Then, the eigenstates of the permutation operator with eigenvalue $\mathrm{e}^{\mathrm{i}\phi} = 1$ are symmetric states, and the particles obeying this symmetry are called bosons. The eigenstates with eigenvalue -1 are antisymmetric states, and the particles obeying this antisymmetry are called fermions. It turns out that, in one and two dimensions, particle-like excitations that obey $\mathrm{e}^{\mathrm{i}\phi} \neq \pm 1$ are possible. These exotic excitations are called anyons, and they give rise to fractional statistics. They also form the basis for the explanation of the fractional quantum Hall effect. However, so far all observed anyons in Nature have been exotic excitations of a low-dimensional system, whereas all elementary particles we know today are actually either bosons or fermions. Since any many-body wavefunction $|\Psi(t)\rangle$ describing identical (fermions) bosons has to be fully (anti)symmetric, it is actually very convenient to consider a Hilbert space that consists only of fully (anti)symmetrized states and that is therefore described by a fully (anti)symmetrized basis of product states.

To introduce a short-hand notation for this basis of fully (anti)symmetrized product states, we make the convention that we only specify the occupation numbers of the single-particle eigenstates $N_{\mathbf{n},\alpha}$ from which a basis state is formed. As a result, we use for the full set of basis states the shorthand notation $\{|\mathbf{N}\rangle\}$. It is then convenient to order the occupation numbers that make up the vector $\mathbf{N} = [\ldots, N_{\mathbf{n},\alpha}, \ldots]$ in a certain arbitrary, but fixed way. As an example, we may use an ordering in increasing single-particle eigenenergies, such that the number on the left corresponds to the ground state of the single-particle Hamiltonian, whereas upon moving to the right the occupation numbers correspond to excited single-particle states with monotonically increasing energies $\varepsilon_{\mathbf{n},\alpha}$. If degeneracies occur within the energy levels, then we need an additional arbitrary convention to order the single-particle states within a degenerate energy level. These concepts are made more concrete in the following example.

Example 6.1. Consider two particles in a one-dimensional harmonic oscillator with eigenvalues $\varepsilon_n = \hbar\omega(n+1/2)$ and corresponding single-particle eigenfunctions $|n\rangle$, as discussed in Sect. 3.5. Then, for two bosons the state $|1,1,0,\ldots\rangle$ corresponds to the fully symmetrized state in which they occupy the ground state and the first excited state of the single-particle Hamiltonian, that is

$$|1,1,0,\ldots\rangle = (|0\rangle_1|1\rangle_2 + |1\rangle_1|0\rangle_2)/\sqrt{2}, \tag{6.6}$$

whereas the state $|2,0,\ldots\rangle$ corresponds to the trivially symmetrized state in which both particles occupy the ground state

$$|2,0,\ldots\rangle = |0\rangle_1|0\rangle_2. \tag{6.7}$$

Then, for two fermions the state $|1,1,\ldots\rangle$ corresponds to the fully antisymmetrized state

$$|1,1,0,\ldots\rangle = (|0\rangle_1|1\rangle_2 - |1\rangle_1|0\rangle_2)/\sqrt{2}, \tag{6.8}$$

whereas the state $|2,0,\ldots\rangle$ is seen to give zero

$$|2,0,\ldots\rangle = |0\rangle_1|0\rangle_2 - |0\rangle_1|0\rangle_2 = 0. \tag{6.9}$$

The last expression shows that antisymmetrization of a state, in which two fermions occupy the same single-particle quantum state, immediately leads to zero. This statement is known as the Pauli exclusion principle. For fermions, we thus have only two possible occupation numbers, namely $N_{\mathbf{n},\alpha} = 0,1$. Note that any fermionic N-particle state $|\mathbf{N}\rangle$ can be written out explicitly in terms of one-particle states by using a Slater determinant. For bosons, we have $N_{\mathbf{n},\alpha} = 0,1,2,\ldots$, and any N-particle state $|\mathbf{N}\rangle$ can be written out explicitly using a permanent, which is a sign-less determinant.

The (anti)symmetrized basis set $\{|\mathbf{N}\rangle\}$ with the constraint $\sum_{\mathbf{n},\alpha} N_{\mathbf{n},\alpha} = N$ is a basis for the (anti)symmetrized N-particle Hilbert space. By relaxing this constraint, the set becomes a basis for the Hilbert space of all possible (anti)symmetrized many-body states. This space is also known as Fock space $\mathscr{F}$, and is defined by

$$\mathscr{F} = \mathscr{H}_0 \oplus \mathscr{H}_1 \oplus \mathscr{H}_2 \oplus \mathscr{H}_3 \oplus \ldots,$$

where $\mathscr{H}_0$ consists of the vacuum state with zero particles $|0\rangle$. The vacuum satisfies by definition $\langle 0|0\rangle = 1$ and its inner product with all other Fock states gives zero. Note that $|0\rangle$ should not be confused with the ground state of the single-particle Hamiltonian, which is a single-particle state and therefore orthonormal to the vacuum. Furthermore, the inner product between any two states having different number of particles is equal to zero, because two such states are by definition in different parts of the Fock space. We thus have that the complete set of basis states for the fermionic Fock space is given by $\{|0\rangle,|1,0,\ldots\rangle,|0,1,0,\ldots\rangle,\ldots,|1,1,0\ldots\rangle,|1,0,1,0,\ldots\rangle,\ldots,|1,1,1,0\ldots\rangle,\ldots\}$. For bosons, we can also have any number of particles in the same single-particle state, leading to additional Fock states, such as $|2,0,0,\ldots\rangle$,$|0,2,0,\ldots\rangle$, $|0,2,317,3,\ldots\rangle$ and so on.

Example 6.2. For a system of 5 bosonic atoms that are all in the ground state, we have

$$|\Psi\rangle = |5,0,0,\ldots\rangle, \tag{6.10}$$

whereas a filled Fermi-sea consisting of 5 fermionic atoms becomes

$$|\Psi\rangle = |1,1,1,1,1,0,0,\ldots\rangle. \tag{6.11}$$

6.3 Creation and Annihilation Operators

Since the (anti)symmetric set $\{|\mathbf{N}\rangle\}$ is by construction orthonormal, it can be conveniently used to represent any (anti)symmetric many-body wavefunction $|\Psi(t)\rangle$ as

$$|\Psi(t)\rangle = \sum_{\mathbf{N}} \Psi_{\mathbf{N}}(t)|\mathbf{N}\rangle \ , \tag{6.12}$$

where $\Psi_{\mathbf{N}}(t)$ is the amplitude for the many-body sytem to be in state $|\mathbf{N}\rangle$ at time t. In this basis, the Schrödinger equation becomes

$$i\hbar\frac{\partial}{\partial t}\Psi_{\mathbf{N}}(t) = \sum_{\mathbf{N}'} \langle \mathbf{N}|\hat{H}|\mathbf{N}'\rangle \ \Psi_{\mathbf{N}'}(t) \ . \tag{6.13}$$

This shows that we need to determine the matrix elements of the Hamiltonian between different states in Fock space. For the quantum theory of the harmonic oscillator in Sect. 3.5, where we encountered number states for the first time, it was very convenient to express all operators and number states in terms of creation and annihilation operators. In analogy with this treatment, we define the annihilation operator $\hat{\psi}_{\mathbf{n},\alpha}$ acting on Fock space by

$$\hat{\psi}_{\mathbf{n},\alpha}|\mathbf{N}\rangle = \hat{\psi}_{\mathbf{n},\alpha}|\ldots,N_{\mathbf{n},\alpha},\ldots\rangle = (\pm 1)^{M_{\mathbf{n},\alpha}}\sqrt{N_{\mathbf{n},\alpha}}|\ldots,N_{\mathbf{n},\alpha}-1,\ldots\rangle \ , \tag{6.14}$$

where the upper (lower) sign corresponds to bosons (fermions). The factor $M_{\mathbf{n},\alpha}$ gives for the many-body state $|\mathbf{N}\rangle = |\ldots,N_{\mathbf{n},\alpha},\ldots\rangle$ the total number of occupied single-particle states that are ordered to the left of the single-particle state corresponding to $N_{\mathbf{n},\alpha}$. If we use the ordering in increasing energy, then $M_{\mathbf{n},\alpha}$ thus counts the number of occupied states with an energy less then $\varepsilon_{\mathbf{n},\alpha}$. In the case that $\varepsilon_{\mathbf{n},\alpha}$ is degenerate, then $M_{\mathbf{n},\alpha}$ also includes the occupied states of the same energy that are placed to the left $\varepsilon_{\mathbf{n},\alpha}$ according to some arbitrary convention. The reason for the factor $M_{\mathbf{n},\alpha}$ in the case of fermions becomes clear in the example below. It is now left as an exercise to show that from the above definition for the annihilation operator, it automatically follows that the creation operator $\hat{\psi}^{\dagger}_{\mathbf{n},\alpha}$ is given by

$$\hat{\psi}^{\dagger}_{\mathbf{n},\alpha}|\ldots,N_{\mathbf{n},\alpha},\ldots\rangle = (\pm 1)^{M_{\mathbf{n},\alpha}}\sqrt{1 \pm N_{\mathbf{n},\alpha}}|\ldots,N_{\mathbf{n},\alpha}+1,\ldots\rangle\ . \tag{6.15}$$

Example 6.3. Consider a two-level system with ground-state energy ε_0 and excited state energy ε_1. The creation operator that creates a fermion in the ground state is denoted by $\hat{\psi}_0$, whereas the creation operator for the excited state is denoted by $\hat{\psi}_1$. Then, according to the above defintions, we have

$$\begin{aligned}
&\hat{\psi}^{\dagger}_0|0,0\rangle = |1,0\rangle, \quad \hat{\psi}^{\dagger}_1|0,0\rangle = |0,1\rangle, \\
&\hat{\psi}^{\dagger}_0\hat{\psi}^{\dagger}_1|0,0\rangle = \hat{\psi}^{\dagger}_0|0,1\rangle = |1,1\rangle, \\
&\hat{\psi}^{\dagger}_1\hat{\psi}^{\dagger}_0|0,0\rangle = \hat{\psi}^{\dagger}_1|1,0\rangle = -|1,1\rangle.
\end{aligned} \tag{6.16}$$

We thus find that due to the counting factor $(-1)^{M_{\mathbf{n},\alpha}}$ introduced for fermions, the permutation of two fermionic creation operators leads to a minus sign. This reveals that the antisymmetric nature of fermions upon interchange is now expressed by anticommuting fermionic creation and annihilation operators. Note that the particular way of ordering is merely a matter of convention, because any fixed ordering leads to $\hat{\psi}^{\dagger}_1\hat{\psi}^{\dagger}_0|0,0\rangle = -\hat{\psi}^{\dagger}_0\hat{\psi}^{\dagger}_1|0,0\rangle$, and is therefore equally appropriate. Ultimately, the anticommuting behavior of the fermionic operators is the defining property that matters.

From (6.14) and (6.15), it follows that the operator $\hat{\psi}^{\dagger}_{\mathbf{n},\alpha}\hat{\psi}_{\mathbf{n},\alpha}$ conveniently counts the number of particles in the single-particle state $|\mathbf{n},\alpha\rangle$, i.e.

$$\hat{\psi}^{\dagger}_{\mathbf{n},\alpha}\hat{\psi}_{\mathbf{n},\alpha}|\ldots,N_{\mathbf{n},\alpha},\ldots\rangle = N_{\mathbf{n},\alpha}|\ldots,N_{\mathbf{n},\alpha},\ldots\rangle\ . \tag{6.17}$$

It is left as another exercise to show that the creation and annihilation operators for bosons (fermions) satisfy the following set of (anti)commutation relations

$$\begin{aligned}
&[\hat{\psi}_{\mathbf{n},\alpha},\hat{\psi}_{\mathbf{n}',\alpha'}]_{\mp} = [\hat{\psi}^{\dagger}_{\mathbf{n},\alpha},\hat{\psi}^{\dagger}_{\mathbf{n}',\alpha'}]_{\mp} = 0, \\
&[\hat{\psi}_{\mathbf{n},\alpha},\hat{\psi}^{\dagger}_{\mathbf{n}',\alpha'}]_{\mp} = \delta_{\mathbf{n},\mathbf{n}'}\delta_{\alpha,\alpha'}\ ,
\end{aligned} \tag{6.18}$$

where again the upper (lower) sign refers to bosons (fermions). We use this notational convention always from now on. As a result, we have for fermions that $(\hat{\psi}^{\dagger}_{\mathbf{n},\alpha})^2 = 0$, expressing the Pauli principle that two particles cannot be created in the same state. From (6.15), it follows that any basis state $|\mathbf{N}\rangle$ in Fock space can be conveniently expressed as products of creation operators

$$|\mathbf{N}\rangle = \prod_{\mathbf{n},\alpha} \frac{(\hat{\psi}^{\dagger}_{\mathbf{n},\alpha})^{N_{\mathbf{n},\alpha}}}{\sqrt{N_{\mathbf{n},\alpha}!}} |0\rangle \,, \tag{6.19}$$

where for bosons the order of the creation operators does not matter, whereas the fermionic creation operators are ordered in the same way as the occupation numbers. For the ordering in increasing energy, this means that the creation operator corresponding to the lowest single-particle energy is positioned to the left.

Example 6.4. The creation and annihilation operators take a state in the Fock space and map it onto another state in the Fock space. If we consider a system that has only one quantum state available then the problem reduces to the bosonic or fermionic harmonic oscillator we considered in Chap. 3. Specifically, the matrix representation of the annihilation operator in the basis of Fock states is then for the bosonic case given by (3.54).

6.3.1 Second-Quantized Hamiltonian

Now that we have seen how the fully (anti)symmetrized Fock states are represented in terms of (fermionic) bosonic creation operators, we also want to express the relevant quantum mechanical operators, such as the external potential $V^{\text{ex}}(\hat{\mathbf{x}}_i)$ and the two-body interaction $V(\hat{\mathbf{x}}_i - \hat{\mathbf{x}}_j)$, in the language of second quantization. In this paragraph, we motivate physically the form of the relevant operators in terms of creation and annihilation operators acting on Fock space, where we postpone the proof of the full equivalence of the first and second-quantized formalism to Sect. 6.4. This proof then ultimately also validates the definition of the creation operation operator in (6.14), from which the annihilation operator of (6.15), the (anti)commutation relations of (6.18), and the basis representation of (6.19) immediately follow.

Without interactions, we have that the second-quantized form of the first-quantized N-body Hamiltonian in (6.2) is given by

$$\hat{H} = \sum_{\mathbf{n},\alpha} \varepsilon_{\mathbf{n},\alpha} \hat{\psi}^{\dagger}_{\mathbf{n},\alpha} \hat{\psi}_{\mathbf{n},\alpha} , \tag{6.20}$$

which is easily understood, since $\hat{\psi}^{\dagger}_{\mathbf{n},\alpha} \hat{\psi}_{\mathbf{n},\alpha}$ counts the number of particles in each single-particle state $|\mathbf{n},\alpha\rangle$, which is then multiplied by the single-particle eigenenergy $\varepsilon_{\mathbf{n},\alpha}$ to give the total energy of the noninteracting system. Including the two-body interaction term, we have that the second-quantized form of (6.2) becomes

$$\hat{H} = \sum_{\mathbf{n},\alpha} \varepsilon_{\mathbf{n},\alpha} \hat{\psi}^{\dagger}_{\mathbf{n},\alpha} \hat{\psi}_{\mathbf{n},\alpha} + \frac{1}{2} \sum_{\alpha,\alpha'} \sum_{\mathbf{n},\mathbf{n}',\mathbf{m},\mathbf{m}'} V_{\mathbf{n},\mathbf{n}';\mathbf{m},\mathbf{m}'} \hat{\psi}^{\dagger}_{\mathbf{n},\alpha} \hat{\psi}^{\dagger}_{\mathbf{n}',\alpha'} \hat{\psi}_{\mathbf{m}',\alpha'} \hat{\psi}_{\mathbf{m},\alpha}, \tag{6.21}$$

where the matrix elements of the two-body interaction are given by

$$V_{\mathbf{n},\mathbf{n}';\mathbf{m},\mathbf{m}'} = \int d\mathbf{x} \int d\mathbf{x}' \; \chi^*_{\mathbf{n}}(\mathbf{x}) \chi^*_{\mathbf{n}'}(\mathbf{x}') V(\mathbf{x}-\mathbf{x}') \chi_{\mathbf{m}}(\mathbf{x}) \chi_{\mathbf{m}'}(\mathbf{x}'). \tag{6.22}$$

The second-quantized expression for the interacting part of the Hamiltonian is also intuitively clear, because two particles that are initially in state $|\mathbf{m},\alpha\rangle$ and $|\mathbf{m}',\alpha'\rangle$ can scatter into the states $|\mathbf{n},\alpha\rangle$ and $|\mathbf{n}',\alpha'\rangle$ under the influence of the interaction potential $V(\mathbf{x}-\mathbf{x}')$. Moreover, the probability amplitude for this process to happen is given by $V_{\mathbf{n},\mathbf{n}';\mathbf{m},\mathbf{m}'}$. The corresponding scattering process is also shown schematically in Fig. 6.1(a).

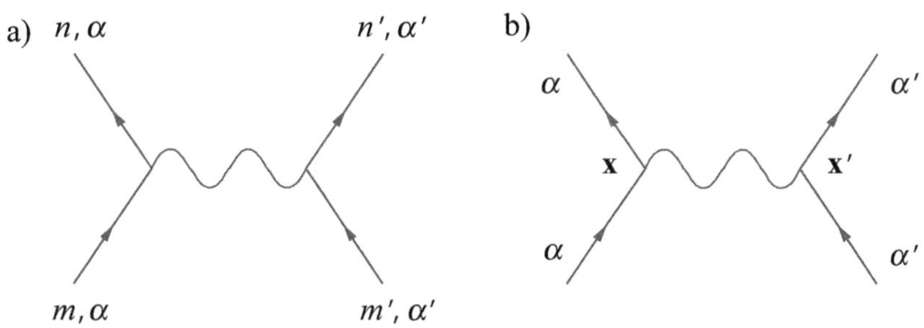

Fig. 6.1 Diagrammatic representation of the interaction terms in a) (6.21) and b) (6.29).

6.3.2 Field Operators

To be able to compare the first-quantized Hamiltonian in (6.2) and the second-quantized Hamiltonian in (6.21), it is most convenient to transform the creation and annihilation operators to real space. This is achieved by introducing the field operators

$$\hat{\psi}_{\alpha}(\mathbf{x}) = \sum_{\mathbf{n}} \hat{\psi}_{\mathbf{n},\alpha} \chi_{\mathbf{n}}(\mathbf{x}) \tag{6.23}$$

and

$$\hat{\psi}^{\dagger}_{\alpha}(\mathbf{x}) = \sum_{\mathbf{n}} \hat{\psi}^{\dagger}_{\mathbf{n},\alpha} \chi^{*}_{\mathbf{n}}(\mathbf{x}), \tag{6.24}$$

that annihilate and create a particle in the spin state $|\alpha\rangle$ at position $\mathbf{x}$, respectively. The inverse of the above relations is given by

$$\hat{\psi}_{\mathbf{n},\alpha} = \int \mathrm{d}\mathbf{x}\ \hat{\psi}_{\alpha}(\mathbf{x}) \chi^{*}_{\mathbf{n}}(\mathbf{x}) \tag{6.25}$$

and

$$\hat{\psi}^{\dagger}_{\mathbf{n},\alpha} = \int \mathrm{d}\mathbf{x}\ \hat{\psi}^{\dagger}_{\alpha}(\mathbf{x}) \chi_{\mathbf{n}}(\mathbf{x}). \tag{6.26}$$

As a result, we have for the (anti)commutation relations of the field operators

$$[\hat{\psi}_{\alpha}(\mathbf{x}), \hat{\psi}_{\alpha'}(\mathbf{x}')]_{\mp} = [\hat{\psi}^{\dagger}_{\alpha}(\mathbf{x}), \hat{\psi}^{\dagger}_{\alpha'}(\mathbf{x}')]_{\mp} = 0 \tag{6.27}$$

and

$$[\hat{\psi}_{\alpha}(\mathbf{x}), \hat{\psi}^{\dagger}_{\alpha'}(\mathbf{x}')]_{\mp} = \delta(\mathbf{x}-\mathbf{x}')\delta_{\alpha,\alpha'}, \tag{6.28}$$

where we used equation (6.18) and the completeness of the wavefunctions $\chi_{\mathbf{n}}(\mathbf{x})$.

With the use of the field operators, the second-quantized Hamiltonian of (6.21) can be written as

$$\begin{aligned} \hat{H} = & \sum_{\alpha} \int \mathrm{d}\mathbf{x}\ \hat{\psi}^{\dagger}_{\alpha}(\mathbf{x}) \left\{ -\frac{\hbar^2 \boldsymbol{\nabla}^2}{2m} + V^{\mathrm{ex}}(\mathbf{x}) + \varepsilon_{\alpha} \right\} \hat{\psi}_{\alpha}(\mathbf{x}) \\ & + \frac{1}{2} \sum_{\alpha,\alpha'} \int \mathrm{d}\mathbf{x} \int \mathrm{d}\mathbf{x}'\ \hat{\psi}^{\dagger}_{\alpha}(\mathbf{x}) \hat{\psi}^{\dagger}_{\alpha'}(\mathbf{x}') V(\mathbf{x}-\mathbf{x}') \hat{\psi}_{\alpha'}(\mathbf{x}') \hat{\psi}_{\alpha}(\mathbf{x}) , \end{aligned} \tag{6.29}$$

which is most easily shown by inserting (6.23) and (6.24) into (6.29) and using the orthonormality of the wavefunctions $\chi_{\mathbf{n}}(\mathbf{x})$. Furthermore, the number operator in terms of field operators becomes

$$\hat{N} = \sum_{\mathbf{n},\alpha} \hat{\psi}^{\dagger}_{\mathbf{n},\alpha} \hat{\psi}_{\mathbf{n},\alpha} = \sum_{\alpha} \int \mathrm{d}\mathbf{x}\ \hat{\psi}^{\dagger}_{\alpha}(\mathbf{x}) \hat{\psi}_{\alpha}(\mathbf{x}), \tag{6.30}$$

such that the density operator for particles in the state $|\alpha\rangle$ is given by $\hat{n}_{\alpha}(\mathbf{x}) = \hat{\psi}^{\dagger}_{\alpha}(\mathbf{x}) \hat{\psi}_{\alpha}(\mathbf{x})$. Similarly, the total effective spin operator $\hat{\mathbf{S}}$ is given by

$$\hat{\mathbf{S}} = \sum_{\mathbf{n},\alpha,\alpha'} \hat{\psi}^{\dagger}_{\mathbf{n},\alpha} \langle\alpha|\hat{\mathbf{s}}|\alpha'\rangle \hat{\psi}_{\mathbf{n},\alpha'} = \sum_{\alpha,\alpha'} \int \mathrm{d}\mathbf{x}\ \hat{\psi}^{\dagger}_{\alpha}(\mathbf{x}) \langle\alpha|\hat{\mathbf{s}}|\alpha'\rangle \hat{\psi}_{\alpha'}(\mathbf{x}) . \tag{6.31}$$

6.4 Equivalence of First and Second Quantization

To finally prove the equivalence of the first-quantization and second-quantization formalisms, we consider the general N-body state written in second quantization as

$$|\Psi(t)\rangle = \frac{1}{\sqrt{N!}}\sum_{\alpha_1\ldots\alpha_N}\int\left(\prod_{k=1}^{N}\mathrm{d}\mathbf{x}_k\right)\Psi_{\alpha_1\ldots\alpha_N}(\mathbf{x}_1,\ldots,\mathbf{x}_N;t)\hat{\psi}^{\dagger}_{\alpha_1}(\mathbf{x}_1)\ldots\hat{\psi}^{\dagger}_{\alpha_N}(\mathbf{x}_N)|0\rangle, \tag{6.32}$$

where we show that the expansion coefficients $\Psi_{\alpha_1\ldots\alpha_N}(\mathbf{x}_1,\ldots,\mathbf{x}_N;t)$ satisfy the first-quantized many-body Schrödinger equation. Note that because the Fock states are by construction (anti)symmetric, only the (anti)symmetric parts of the expansion coefficients survives the integration in (6.32). This is desired, because in first quantization the N-body wavefunction should indeed be (anti)symmetric in the case of (fermions) bosons. Next, we consider the N-body Schrödinger equation

$$\mathrm{i}\hbar\frac{\partial}{\partial t}|\Psi(t)\rangle = \hat{H}|\Psi(t)\rangle \tag{6.33}$$

with the second-quantized Hamiltonian as given in (6.29). We note that all the time dependence is taken into account by the expansion coefficients $\Psi_{\alpha_1\ldots\alpha_N}(\mathbf{x}_1,\ldots,\mathbf{x}_N;t)$. The left-hand side of (6.33) is then given by

$$\begin{aligned}\mathrm{i}\hbar\frac{\partial}{\partial t}|\Psi(t)\rangle = {} & \frac{1}{\sqrt{N!}}\sum_{\alpha_1\ldots\alpha_N}\int\left(\prod_{k=1}^{N}\mathrm{d}\mathbf{x}_k\right)\mathrm{i}\hbar\frac{\partial}{\partial t}\Psi_{\alpha_1\ldots\alpha_N}(\mathbf{x}_1,\ldots,\mathbf{x}_N;t)\\ & \times\hat{\psi}^{\dagger}_{\alpha_1}(\mathbf{x}_1)\ldots\hat{\psi}^{\dagger}_{\alpha_N}(\mathbf{x}_N)|0\rangle.\end{aligned} \tag{6.34}$$

To evaluate the right-hand side of (6.33) we first consider the contribution of the noninteracting part of the Hamiltonian, which yields

$$\begin{aligned}&\sum_{\alpha}\int\mathrm{d}\mathbf{x}\,\hat{\psi}^{\dagger}_{\alpha}(\mathbf{x})\left\{-\frac{\hbar^2\nabla^2}{2m}+V^{\mathrm{ex}}(\mathbf{x})+\varepsilon_\alpha\right\}\hat{\psi}_{\alpha}(\mathbf{x})\\ &\times\frac{1}{\sqrt{N!}}\sum_{\alpha_1\ldots\alpha_N}\int\left(\prod_{k=1}^{N}\mathrm{d}\mathbf{x}_k\right)\Psi_{\alpha_1\ldots\alpha_N}(\mathbf{x}_1,\ldots,\mathbf{x}_N;t)\prod_{l=1}^{N}\hat{\psi}^{\dagger}_{\alpha_l}(\mathbf{x}_l)|0\rangle =\\ &\sum_{i=1}^{N}\sum_{\alpha}\int\mathrm{d}\mathbf{x}\sum_{\alpha_1\ldots\alpha_N}\int\left(\prod_{k=1}^{N}\mathrm{d}\mathbf{x}_k\right)(\pm1)^{i+1}\,\hat{\psi}^{\dagger}_{\alpha}(\mathbf{x})\left\{-\frac{\hbar^2\nabla^2}{2m}+V^{\mathrm{ex}}(\mathbf{x})+\varepsilon_\alpha\right\}\\ &\times\frac{1}{\sqrt{N!}}\Psi_{\alpha_1\ldots\alpha_N}(\mathbf{x}_1,\ldots,\mathbf{x}_N;t)\,\delta(\mathbf{x}-\mathbf{x}_i)\delta_{\alpha,\alpha_i}\prod_{l\neq i=1}^{N}\hat{\psi}^{\dagger}_{\alpha_l}(\mathbf{x}_l)|0\rangle.\end{aligned} \tag{6.35}$$

Although this looks rather complicated, we actually only used

$$\hat{\psi}_{\alpha}(\mathbf{x})\prod_{l=1}^{N}\hat{\psi}^{\dagger}_{\alpha_l}(\mathbf{x}_l)|0\rangle=\sum_{i=1}^{N}(\pm 1)^{i+1}\delta(\mathbf{x}-\mathbf{x}_i)\delta_{\alpha,\alpha_i}\prod_{l\neq i=1}^{N}\hat{\psi}^{\dagger}_{\alpha_l}(\mathbf{x}_l)|0\rangle \tag{6.36}$$

which is proven by using the (anti)commutation relations of (6.28) to bring the annihilation operator all the way to the right, where it annihilates the vacuum, i.e. $\hat{\psi}_{\alpha}(\mathbf{x})|0\rangle=0$, and disappears. We then perform in (6.35) the integration over $\mathbf{x}$ and the summation over α, such that we retrieve the missing creation operator $\hat{\psi}^{\dagger}_{\alpha_i}(\mathbf{x}_i)$ which we consequently put back in the right place. Comparing (6.34) with (6.35) we then find that, in the absence of interactions, the expansion coefficients satisfy

$$\begin{aligned}
&\mathrm{i}\hbar\frac{\partial}{\partial t}\Psi_{\alpha_1\ldots\alpha_N}(\mathbf{x}_1,\ldots,\mathbf{x}_N;t)\\
&=\sum_{i=1}^{N}\left\{-\frac{\hbar^2\nabla^2}{2m}+V^{\mathrm{ex}}(\mathbf{x}_i)+\varepsilon_{\alpha_i}\right\}\Psi_{\alpha_1\ldots\alpha_N}(\mathbf{x}_1,\ldots,\mathbf{x}_N;t),
\end{aligned} \tag{6.37}$$

which is indeed the first-quantized many-body Schrödinger equation with the Hamiltonian of (6.2) in the absence of interactions.

To prove the equivalence of first and second quantization also in the presence of the two-body interactions, we need to extend the previous discussion with the second-quantized expression for the interaction, for which we need to show that

$$\begin{aligned}
&\frac{1}{2\sqrt{N!}}\sum_{\alpha_1\ldots\alpha_N}\int\left(\prod_{k=1}^{N}\mathrm{d}\mathbf{x}_k\right)\sum_{i\neq j=1}^{N}V(\mathbf{x}_i-\mathbf{x}_j)\Psi_{\alpha_1\ldots\alpha_N}(\mathbf{x}_1,\ldots,\mathbf{x}_N;t)\prod_{l=1}^{N}\hat{\psi}^{\dagger}_{\alpha_l}(\mathbf{x}_l)|0\rangle\\
&=\frac{1}{2}\sum_{\alpha,\alpha'}\int\mathrm{d}\mathbf{x}\int\mathrm{d}\mathbf{x}'\;\hat{\psi}^{\dagger}_{\alpha}(\mathbf{x})\hat{\psi}^{\dagger}_{\alpha'}(\mathbf{x}')V(\mathbf{x}-\mathbf{x}')\hat{\psi}_{\alpha'}(\mathbf{x}')\hat{\psi}_{\alpha}(\mathbf{x})\\
&\times\frac{1}{\sqrt{N!}}\sum_{\alpha_1\ldots\alpha_N}\int\left(\prod_{k=1}^{N}\mathrm{d}\mathbf{x}_k\right)\Psi_{\alpha_1\ldots\alpha_N}(\mathbf{x}_1,\ldots,\mathbf{x}_N;t)\prod_{l=1}^{N}\hat{\psi}^{\dagger}_{\alpha_l}(\mathbf{x}_l)|0\rangle.
\end{aligned} \tag{6.38}$$

The strategy is again to move the annihilation operators $\hat{\psi}_{\alpha'}(\mathbf{x}')$ and $\hat{\psi}_{\alpha}(\mathbf{x})$ through all the creation operators, giving

$$\begin{aligned}
&\hat{\psi}_{\alpha'}(\mathbf{x}')\hat{\psi}_{\alpha}(\mathbf{x})\prod_{l=1}^{N}\hat{\psi}^{\dagger}_{\alpha_l}(\mathbf{x}_l)|0\rangle\\
&=\sum_{i\neq j=1}^{N}(\pm 1)^{i+j+\theta(j-i)}\delta(\mathbf{x}'-\mathbf{x}_j)\delta_{\alpha',\alpha_j}\delta(\mathbf{x}-\mathbf{x}_i)\delta_{\alpha,\alpha_i}\prod_{\substack{l=1\\ l\neq i,j}}^{N}\hat{\psi}^{\dagger}_{\alpha_l}(\mathbf{x}_l)|0\rangle,
\end{aligned} \tag{6.39}$$

where θ is the step function, i.e. $\theta=0$ for $i>j$ and $\theta=1$ for $i<j$. Substituting (6.39) into the right-hand side of (6.38), we obtain

$$\frac{1}{2}\sum_{\alpha,\alpha'}\sum_{\alpha_1\ldots\alpha_N}\sum_{i\neq j=1}^{N}\int\left(\prod_{k=1}^{N}\mathrm{d}\mathbf{x}_k\right)\int\mathrm{d}\mathbf{x}\int\mathrm{d}\mathbf{x}'\ \hat{\psi}^\dagger_\alpha(\mathbf{x})\hat{\psi}^\dagger_{\alpha'}(\mathbf{x}')V(\mathbf{x}-\mathbf{x}')(\pm 1)^{i+j+\theta(j-i)}$$

$$\times\delta(\mathbf{x}'-\mathbf{x}_j)\delta_{\alpha',\alpha_j}\delta(\mathbf{x}-\mathbf{x}_i)\delta_{\alpha,\alpha_i}\frac{1}{\sqrt{N!}}\Psi_{\alpha_1\ldots\alpha_N}(\mathbf{x}_1,\ldots,\mathbf{x}_N;t)\prod_{\substack{l=1\\ l\neq i,j}}^{N}\hat{\psi}^\dagger_{\alpha_l}(\mathbf{x}_l)|0\rangle =$$

$$\frac{1}{2}\sum_{\alpha_1\ldots\alpha_N}\sum_{i\neq j=1}^{N}\int\left(\prod_{k=1}^{N}\mathrm{d}\mathbf{x}_k\right)\hat{\psi}^\dagger_{\alpha_i}(\mathbf{x}_i)\hat{\psi}^\dagger_{\alpha_j}(\mathbf{x}_j)V(\mathbf{x}_i-\mathbf{x}_j) \tag{6.40}$$

$$\times(\pm 1)^{i+j+\theta(j-i)}\frac{1}{\sqrt{N!}}\Psi_{\alpha_1\ldots\alpha_N}(\mathbf{x}_1,\ldots,\mathbf{x}_N;t)\prod_{\substack{l=1\\ l\neq i,j}}^{N}\hat{\psi}^\dagger_{\alpha_l}(\mathbf{x}_l)|0\rangle =$$

$$\frac{1}{\sqrt{N!}}\sum_{\alpha_1\ldots\alpha_N}\int\left(\prod_{k=1}^{N}\mathrm{d}\mathbf{x}_k\right)\sum_{i\neq j=1}^{N}\frac{1}{2}V(\mathbf{x}_i-\mathbf{x}_j)\Psi_{\alpha_1\ldots\alpha_N}(\mathbf{x}_1,\ldots,\mathbf{x}_N;t)\prod_{l=1}^{N}\hat{\psi}^\dagger_{\alpha_l}(\mathbf{x}_l)|0\rangle.$$

Comparison of the above equation with (6.33) (6.34) and (6.37) indeed shows that the coefficients $\Psi_{\alpha_1\ldots\alpha_N}(\mathbf{x}_1,\ldots,\mathbf{x}_N;t)$ satisfy the first-quantized N-body Schrödinger equation including interactions, namely

$$\begin{aligned}\mathrm{i}\hbar\frac{\partial}{\partial t}\Psi_{\alpha_1\ldots\alpha_N}(\mathbf{x}_1,\ldots,\mathbf{x}_N;t) &= \sum_{j=1}^{N}\left\{-\frac{\hbar^2\nabla^2}{2m}+V^{\mathrm{ex}}(\mathbf{x}_j)+\varepsilon_{\alpha_j}\right\}\Psi_{\alpha_1\ldots\alpha_N}(\mathbf{x}_1,\ldots,\mathbf{x}_N;t)\\ &\quad+\frac{1}{2}\sum_{i\neq j=1}^{N}V(\mathbf{x}_i-\mathbf{x}_j)\Psi_{\alpha_1\ldots\alpha_N}(\mathbf{x}_1,\ldots,\mathbf{x}_N;t),\end{aligned} \tag{6.41}$$

which is what we wanted to prove.

We have thus developed a second-quantized operator formalism with which we can study an interacting many-body quantum system of interest. In order to study its thermodynamic equilibrium properties, we need to determine the grand-canonical partition function, which can be derived from the time-evolution operator by going to imaginary time as explained in Sect. 5.6. Since we have expressed the experimentally relevant observables in terms of the field operators, the equilibrium properties of the system can now be studied through the imaginary-time evolution of the Heisenberg operator

$$\hat{\psi}(\mathbf{x},\tau)=\mathrm{e}^{(\hat{H}-\mu\hat{N})\tau/\hbar}\hat{\psi}_\alpha(\mathbf{x})\mathrm{e}^{-(\hat{H}-\mu\hat{N})\tau/\hbar} \tag{6.42}$$

at a fixed chemical potential μ [35]. Put differently, the desired quantum field theory would be defined by the Heisenberg equation of motion

$$\hbar\frac{\partial}{\partial\tau}\hat{\psi}_\alpha(\mathbf{x},\tau)=[\hat{H}-\mu\hat{N},\hat{\psi}_\alpha(\mathbf{x},\tau)]_-\ , \tag{6.43}$$

and we would need to solve this equation in a sufficiently accurate approximation.

However, the above operator equation turns out to be rather inconvenient to deal with in practice. Instead, we develop in the next chapters a functional integral formalism in which we merge the presently obtained many-body operator theory with Feynman's path-integral approach to quantum mechanics. In the functional approach, the field annihilation operators are represented by their eigenvalues. Since these eigenvalues can be manipulated algebraically, they are much more convenient to work with than the operators themselves. Furthermore, the functional-integral method allows for an elegant solution of the ideal gas, which is therefore an ideal starting point for incorporating interaction effects perturbatively. These are the topics of the next chapters. However, to derive the functional-integral formalism and treat the bosonic and fermionic quantum fluids in a unified manner, we need one more essential ingredient, namely the closure relation for the coherent states in Fock space. This is the topic of the next section.

6.5 Coherent States

In Sect. 3.7 the notion of the coherent state was introduced in the context of the one-dimensional harmonic oscillator, where it was defined as the eigenstate of the annihilation operator. This concept we may generalize to the Fock space for bosons, where we have that a general coherent state $|\phi\rangle$ is given by [36, 37]

$$|\phi\rangle = \exp\left\{\sum_{\mathbf{n},\alpha} \phi_{\mathbf{n},\alpha} \hat{\psi}^{\dagger}_{\mathbf{n},\alpha}\right\} |0\rangle , \tag{6.44}$$

which can be seen as follows. Using (6.18), we have that

$$\hat{\psi}_{\mathbf{n},\alpha}(\hat{\psi}^{\dagger}_{\mathbf{n},\alpha})^{n}|0\rangle = n(\hat{\psi}^{\dagger}_{\mathbf{n},\alpha})^{n-1}|0\rangle, \tag{6.45}$$

showing that $\hat{\psi}_{\mathbf{n},\alpha}$ acts as $\partial/\partial\hat{\psi}^{\dagger}_{\mathbf{n},\alpha}$ on the Fock states. As a result, we find

$$\hat{\psi}_{\mathbf{n},\alpha}|\phi\rangle = \phi_{\mathbf{n},\alpha} \exp\left\{\sum_{\mathbf{n},\alpha} \phi_{\mathbf{n},\alpha} \hat{\psi}^{\dagger}_{\mathbf{n},\alpha}\right\} |0\rangle , \tag{6.46}$$

showing that $|\phi\rangle$ is indeed a coherent state. We can also write equation (6.44) in terms of the field operators and obtain

$$|\phi\rangle = \exp\left\{\sum_{\alpha} \int d\mathbf{x}\ \phi_{\alpha}(\mathbf{x}) \hat{\psi}^{\dagger}_{\alpha}(\mathbf{x})\right\} |0\rangle , \tag{6.47}$$

where we introduced the field $\phi_{\alpha}(\mathbf{x}) = \sum_{\mathbf{n}} \phi_{\mathbf{n},\alpha}\chi_{\mathbf{n}}(\mathbf{x})$, such that we have $\hat{\psi}_{\alpha}(\mathbf{x})|\phi\rangle = \phi_{\alpha}(\mathbf{x})|\phi\rangle$. It is important to realize that these bosonic coherent states are not orthonormal. This was already shown in Sect. 3.7 for the coherent states in the context of the one-dimensional harmonic oscillator, whose number eigenstates are equivalent to a bosonic Fock space with only one single-particle quantum state with energy $\varepsilon_{\mathbf{n},\alpha} = \hbar\omega$. It is left as an exercise to generalize the expression for the overlap of the coherent states, as given by (3.71), to an arbitrary number of single-particle quantum states, leading to

$$\langle\phi|\phi'\rangle = \exp\left\{\sum_{\mathbf{n},\alpha} \phi^*_{\mathbf{n},\alpha}\phi'_{\mathbf{n},\alpha}\right\} = \exp\left\{\sum_{\alpha}\int \mathrm{d}\mathbf{x}\,\phi^*_{\alpha}(\mathbf{x})\phi'_{\alpha}(\mathbf{x})\right\} \equiv \mathrm{e}^{(\phi|\phi')}. \tag{6.48}$$

Furthermore, in Sect. 3.7 it was proven that the coherent states obey a closure relation, namely (3.72). It is left as another exercise to generalize the expression for the closure relation of the bosonic coherent states, as given by (3.72), to an arbitrary number of single-particle quantum states, leading to

$$\int \mathrm{d}[\phi^*]\mathrm{d}[\phi]\,\mathrm{e}^{-(\phi|\phi)}|\phi\rangle\langle\phi| = \hat{1}\,, \tag{6.49}$$

where the integration measure is defined by

$$\int \mathrm{d}[\phi^*]\mathrm{d}[\phi] \equiv \int \prod_{\mathbf{n},\alpha} \frac{\mathrm{d}\phi^*_{\mathbf{n},\alpha}\mathrm{d}\phi_{\mathbf{n},\alpha}}{2\pi\mathrm{i}}. \tag{6.50}$$

To extend the above discussion to fermions and consider also eigenstates of the fermionic annihilation operator, we need the Grassmann variables introduced in Sect. 2.4. Consider the state

$$|\phi_{\mathbf{n},\alpha}\rangle = \exp\left\{-\phi_{\mathbf{n},\alpha}\hat{\psi}^{\dagger}_{\mathbf{n},\alpha}\right\}|0\rangle = (1-\phi_{\mathbf{n},\alpha}\hat{\psi}^{\dagger}_{\mathbf{n},\alpha})|0\rangle\,, \tag{6.51}$$

where $\phi_{\mathbf{n},\alpha}$ is an anticommuting Grassmann variable that also anticommutes with all creation and annihilation operators in the Fock space, that is $[\phi_{\mathbf{n},\alpha}, \psi^{(\dagger)}_{\mathbf{n}',\alpha'}]_+ = 0$. Then, we have that

$$\begin{aligned}\hat{\psi}_{\mathbf{n},\alpha}|\phi_{\mathbf{n},\alpha}\rangle &= \hat{\psi}_{\mathbf{n},\alpha}|0\rangle + \phi_{\mathbf{n},\alpha}\hat{\psi}_{\mathbf{n},\alpha}\hat{\psi}^{\dagger}_{\mathbf{n},\alpha}|0\rangle \\ &= \phi_{\mathbf{n},\alpha}|0\rangle = \phi_{\mathbf{n},\alpha}(1-\phi_{\mathbf{n},\alpha}\hat{\psi}^{\dagger}_{\mathbf{n},\alpha})|0\rangle = \phi_{\mathbf{n},\alpha}|\phi_{\mathbf{n},\alpha}\rangle\,,\end{aligned} \tag{6.52}$$

such that $|\phi_{\mathbf{n},\alpha}\rangle$ is indeed an eigenstate of $\hat{\psi}_{\mathbf{n},\alpha}$ with the eigenvalue $\phi_{\mathbf{n},\alpha}$. More generally, we can now consider the states

$$|\phi\rangle = \exp\left\{-\sum_{\mathbf{n},\alpha}\phi_{\mathbf{n},\alpha}\hat{\psi}^{\dagger}_{\mathbf{n},\alpha}\right\}|0\rangle, \tag{6.53}$$

that also obey $\hat{\psi}_{\mathbf{n},\alpha}|\phi\rangle = \phi_{\mathbf{n},\alpha}|\phi\rangle$. Introducing the Grassmann-valued field $\phi_{\alpha}(\mathbf{x}) = \sum_{\mathbf{n}}\phi_{\mathbf{n},\alpha}\chi_{\mathbf{n}}(\mathbf{x})$, the latter two relations can be rewritten as

$$|\phi\rangle = \exp\left\{-\sum_{\alpha}\int \mathrm{d}\mathbf{x}\,\phi_{\alpha}(\mathbf{x})\hat{\psi}^{\dagger}_{\alpha}(\mathbf{x})\right\}|0\rangle \tag{6.54}$$

and $\hat{\psi}_\alpha(\mathbf{x})|\phi\rangle = \phi_\alpha(\mathbf{x})|\phi\rangle$.

It is important to note that the coherent states are not orthonormal. In contrast, we find that

$$\begin{aligned}\langle\phi|\phi'\rangle &= \langle 0|\prod_{\mathbf{n},\alpha}(1-\hat{\psi}_{\mathbf{n},\alpha}\phi^*_{\mathbf{n},\alpha})(1-\phi'_{\mathbf{n},\alpha}\hat{\psi}^\dagger_{\mathbf{n},\alpha})|0\rangle = \prod_{\mathbf{n},\alpha}(1+\phi^*_{\mathbf{n},\alpha}\phi'_{\mathbf{n},\alpha}) \\ &= \exp\left\{\sum_{\mathbf{n},\alpha}\phi^*_{\mathbf{n},\alpha}\phi'_{\mathbf{n},\alpha}\right\} = \exp\left\{\sum_\alpha\int d\mathbf{x}\,\phi^*_\alpha(\mathbf{x})\phi'_\alpha(\mathbf{x})\right\} \equiv e^{(\phi|\phi')}\,.\end{aligned} \tag{6.55}$$

Nevertheless, the coherent states do obey a closure relation, which can be shown as follows. First, consider the simplest case when the single-particle Hilbert space consists only of one quantum state $|\mathbf{n},\alpha\rangle$. Then, we have

$$\begin{aligned}\hat{1} &= |0\rangle\langle 0| + |1\rangle\langle 1| = |0\rangle\langle 0| + \hat{\psi}^\dagger_{\mathbf{n},\alpha}|0\rangle\langle 0|\hat{\psi}_{\mathbf{n},\alpha} \\ &= \int d\phi^*_{\mathbf{n},\alpha}d\phi_{\mathbf{n},\alpha}\,(1-\phi^*_{\mathbf{n},\alpha}\phi_{\mathbf{n},\alpha})(1-\phi_{\mathbf{n},\alpha}\hat{\psi}^\dagger_{\mathbf{n},\alpha})|0\rangle\langle 0|(1-\hat{\psi}_{\mathbf{n},\alpha}\phi^*_{\mathbf{n},\alpha}) \\ &= \int d\phi^*_{\mathbf{n},\alpha}d\phi_{\mathbf{n},\alpha}\,\exp\{-\phi^*_{\mathbf{n},\alpha}\phi_{\mathbf{n},\alpha}\}\,|\phi_{\mathbf{n},\alpha}\rangle\langle\phi_{\mathbf{n},\alpha}|,\end{aligned} \tag{6.56}$$

where we performed an integration over Grassmann variables as given by (2.65). This we can generalize to any number of single-particle states, for which we find

$$\begin{aligned}\hat{1} &= \sum_{\mathbf{N}}|\mathbf{N}\rangle\langle\mathbf{N}| \\ &= \int\prod_{\mathbf{n},\alpha}\left(d\phi^*_{\mathbf{n},\alpha}d\phi_{\mathbf{n},\alpha}\,(1-\phi^*_{\mathbf{n},\alpha}\phi_{\mathbf{n},\alpha})\right)\prod_{\mathbf{n},\alpha}(1-\phi_{\mathbf{n},\alpha}\hat{\psi}^\dagger_{\mathbf{n},\alpha})|0\rangle\langle 0|\prod_{\mathbf{n},\alpha}(1-\hat{\psi}_{\mathbf{n},\alpha}\phi^*_{\mathbf{n},\alpha}) \\ &= \int\left(\prod_{\mathbf{n},\alpha}d\phi^*_{\mathbf{n},\alpha}d\phi_{\mathbf{n},\alpha}\right)\exp\left\{-\sum_{\mathbf{n},\alpha}\phi^*_{\mathbf{n},\alpha}\phi_{\mathbf{n},\alpha}\right\}|\phi\rangle\langle\phi| \\ &\equiv \int d[\phi^*]d[\phi]\,e^{-(\phi|\phi)}|\phi\rangle\langle\phi|.\end{aligned} \tag{6.57}$$

Summarizing, we have thus found for bosons and fermions that

$$|\phi\rangle = \exp\left\{\pm\sum_\alpha\int d\mathbf{x}\,\phi_\alpha(\mathbf{x})\hat{\psi}^\dagger_\alpha(\mathbf{x})\right\}|0\rangle\,, \tag{6.58}$$

$$\langle\phi|\phi'\rangle = e^{(\phi|\phi')}\,, \tag{6.59}$$

$$\int d[\phi^*]d[\phi]\,e^{-(\phi|\phi)}|\phi\rangle\langle\phi| = \hat{1}\,. \tag{6.60}$$

Using the coherent states, we can express the trace of an operator $\hat{O}$ over the Fock space as

$$
\begin{aligned}
\mathrm{Tr}[\hat{O}] &= \sum_{\mathbf{N}} \langle \mathbf{N}|\hat{O}|\mathbf{N}\rangle \\
&= \sum_{\mathbf{N}} \int \mathrm{d}[\phi^*]\mathrm{d}[\phi]\ \mathrm{e}^{-(\phi|\phi)} \langle \mathbf{N}|\phi\rangle\langle\phi|\hat{O}|\mathbf{N}\rangle \\
&= \sum_{\mathbf{N}} \int \mathrm{d}[\phi^*]\mathrm{d}[\phi]\ \mathrm{e}^{-(\phi|\phi)} \langle \pm\phi|\hat{O}|\mathbf{N}\rangle\langle\mathbf{N}|\phi\rangle \\
&= \int \mathrm{d}[\phi^*]\mathrm{d}[\phi]\ \mathrm{e}^{-(\phi|\phi)} \langle \pm\phi|\hat{O}|\phi\rangle \ ,
\end{aligned}
\tag{6.61}
$$

where it is left as an exercise to show that due to the anticommuting nature of the Grassmann variables, we have in the fermionic case $\langle \mathbf{N}|\phi\rangle\langle\phi|\mathbf{N}\rangle = \langle -\phi|\mathbf{N}\rangle\langle\mathbf{N}|\phi\rangle$. We have now acquired all the mathematical tools that we need to construct a functional formalism for the unified treatment of bosonic and fermionic quantum fluids. This is the topic of the next chapter.

6.6 Problems

Exercise 6.1. Creation and Annihilation Operators
(a) Show that (6.15) follows from (6.14).
(b) Prove the various (anti)commutation relations from (6.18) for the (fermionic) bosonic creation and annihilation operators $\hat{\psi}^{\dagger}_{\mathbf{n},\alpha}$ and $\hat{\psi}_{\mathbf{n},\alpha}$, respectively.
(c) Show that the basis of the Fock space given in (6.19) is orthonormal.
(d) Prove the various (anti)commutation relations from (6.28) between the field operators $\hat{\psi}_{\alpha}(\mathbf{x})$ and $\hat{\psi}^{\dagger}_{\alpha}(\mathbf{x})$.

Exercise 6.2. Operators in Second Quantization
(a) For a single particle, the matrix elements of a general one-body operator $\hat{A}(\hat{\mathbf{r}})$ that depends on the coordinate $\hat{\mathbf{r}}$ are given by the first-quantized expression

$$
A_{\mathbf{n},\alpha;\mathbf{n}',\alpha'} = \langle \chi_{\mathbf{n},\alpha}|\hat{A}|\chi_{\mathbf{n}',\alpha'}\rangle = \int \mathrm{d}\mathbf{r}\ \chi^*_{\mathbf{n}}(\mathbf{r})\langle\alpha|\hat{A}(\mathbf{r})|\alpha'\rangle\chi_{\mathbf{n}'}(\mathbf{r}). \tag{6.62}
$$

Show, by considering these matrix elements, that in second quantization the operator $\hat{A}$ is given by

$$
\hat{A} = \sum_{\mathbf{n},\alpha,\mathbf{n}',\alpha'} A_{\mathbf{n},\alpha;\mathbf{n}',\alpha'} \hat{\psi}^{\dagger}_{\mathbf{n},\alpha}\hat{\psi}_{\mathbf{n}',\alpha'} . \tag{6.63}
$$

(b) Show that the factor 1/2 in front of the interaction term on the right-hand side of (6.21) is correct. Do this by considering the matrix elements of $V(\mathbf{x}-\mathbf{x}')$ for properly (anti)symmetrized and normalized two-particle states in the language of first quantization. Compare these to the same matrix elements calculated entirely with the use of creation and annihilation operators, i.e. in the language of second quantization.

Exercise 6.3. Heisenberg Equation of Motion
The goal is to derive and solve the Heisenberg equation of motion from (6.42) for the field operators $\hat{\psi}_\alpha(\mathbf{x},\tau)$ and $\hat{\psi}^\dagger_\alpha(\mathbf{x},\tau)$ in the case of an ideal quantum gas with no interactions. We assume that the single-particle Schrödinger equation has been solved and that its solution is given by (6.3).
(a) First, show that

$$[\hat{H}-\mu\hat{N},\hat{\psi}_\alpha(\mathbf{x},\tau)]_- = -\left\{-\frac{\hbar^2\nabla^2}{2m}+V^{\mathrm{ex}}(\mathbf{x})+\varepsilon_\alpha-\mu\right\}\hat{\psi}_\alpha(\mathbf{x},\tau). \quad (6.64)$$

both for bosons and fermions. Determine also the corresponding commutator for the creation operators.
(b) The resulting Heisenberg equation of motion takes a simple form by expanding the field operators in terms of the single-particle eigenstates $\chi_\mathbf{n}(\mathbf{x})$. Show that the solution becomes

$$\hat{\psi}_\alpha(\mathbf{x},\tau) = \sum_\mathbf{n} \hat{\psi}_{\mathbf{n},\alpha} \mathrm{e}^{-(\varepsilon_\mathbf{n}+\varepsilon_\alpha-\mu)\tau/\hbar}\chi_\mathbf{n}(\mathbf{x}) \quad (6.65)$$

and find the corresponding expression for $\hat{\psi}^\dagger_\alpha(\mathbf{x},\tau)$.

Exercise 6.4. Bosonic Coherent States
(a) Show that the coherent state in equation (6.54) is an eigenstate of the bosonic field operators $\hat{\psi}_\alpha(\mathbf{x})$. Do this by Taylor expanding the exponent and by explicitly calculating the effect of the annihilation operator on an arbitrary term in the expansion.
(b) Prove (6.48).
(c) Prove (6.49).

Exercise 6.5. Show that the minus sign in the right-hand side of (6.61) is indeed required for the fermionic case.

Exercise 6.6. Bardeen-Cooper-Schrieffer (BCS) Theory
In a Nobel prize-winning paper [8], Bardeen, Cooper and Schrieffer used a variational wavefunction to explain the superconducting state of metals at zero temperature. This exercise is mainly used as a training in second-quantized calculations, because presently not all the physics behind the wavefunction can be understood. However, it does serve as an excellent preparation for Chap. 12, where the concepts introduced in this exercise are extensively discussed, but then from the slightly different perspective of the functional-integral formalism. Consider the following second-quantized Hamiltonian for a two-component Fermi gas (the two components are labelled by a spin index $\alpha = \uparrow,\downarrow$) with an attractive interaction $V_0 < 0$ between opposite spins, i.e.

$$\hat{H} = \sum_{\mathbf{k},\alpha} \varepsilon_\mathbf{k} \hat{\psi}^\dagger_{\mathbf{k},\alpha}\hat{\psi}_{\mathbf{k},\alpha} + \frac{V_0}{V}\sum_{\mathbf{K},\mathbf{k},\mathbf{k}'} \hat{\psi}^\dagger_{\mathbf{K}/2+\mathbf{k}',\uparrow}\hat{\psi}^\dagger_{\mathbf{K}/2-\mathbf{k}',\downarrow}\hat{\psi}_{\mathbf{K}/2-\mathbf{k},\downarrow}\hat{\psi}_{\mathbf{K}/2+\mathbf{k},\uparrow}\,. \quad (6.66)$$

We determine the ground-state energy using the wavefunction

$$|\Psi_{\rm BCS}\rangle = \prod_{\mathbf{k}} \left(u_{\mathbf{k}} + v_{\mathbf{k}} \hat{\psi}^{\dagger}_{\mathbf{k},\uparrow} \hat{\psi}^{\dagger}_{-\mathbf{k},\downarrow} \right) |0\rangle \ , \tag{6.67}$$

where $u_{\mathbf{k}}, v_{\mathbf{k}}$ are variational parameters that we for simplicity take to be real. This wavefunction describes physically a Bose-Einstein condensate of fermion pairs with opposite spin and momentum, called Cooper pairs.
(a) Show that the BCS wavefunction leads to the following expecation values

$$\langle \Psi_{\rm BCS} | \hat{\psi}^{\dagger}_{\mathbf{k},\alpha} \hat{\psi}_{\mathbf{k},\alpha} | \Psi_{\rm BCS} \rangle = v^2_{\mathbf{k}}, \tag{6.68}$$

$$\langle \Psi_{\rm BCS} | \hat{\psi}_{\mathbf{k},\downarrow} \hat{\psi}_{-\mathbf{k},\uparrow} | \Psi_{\rm BCS} \rangle = u_{\mathbf{k}} v_{\mathbf{k}}. \tag{6.69}$$

(b) Show that $|\Psi_{\rm BCS}\rangle$ is normalized if $u^2_{\mathbf{k}} + v^2_{\mathbf{k}} = 1$.
(c) Show that the expectation value $\langle \Psi_{\rm BCS} | \hat{H} - \mu \hat{N} | \Psi_{\rm BCS} \rangle$ gives rise to the following terms, namely

$$\langle \Psi_{\rm BCS} | \hat{H} - \mu \hat{N} | \Psi_{\rm BCS} \rangle = 2 \sum_{\mathbf{k}} (\varepsilon_{\mathbf{k}} - \mu) v^2_{\mathbf{k}} + \frac{V_0}{V} \sum_{\mathbf{k},\mathbf{k}'} u_{\mathbf{k}} v_{\mathbf{k}} u_{\mathbf{k}'} v_{\mathbf{k}'} + \dots \ , \tag{6.70}$$

where we do not consider any other possible terms, $\hat{N}$ is the operator for the total number of atoms and μ is the chemical potential.
(d) Motivated by (b), we may write $u_{\mathbf{k}} = \sin\theta_{\mathbf{k}}$ and $v_{\mathbf{k}} = \cos\theta_{\mathbf{k}}$. Express $\langle \Psi_{\rm BCS} | \hat{H} - \mu \hat{N} | \Psi_{\rm BCS} \rangle$ as obtained from (6.70) in terms of $\theta_{\mathbf{k}}$, and minimize this result to find

$$0 = (\varepsilon_{\mathbf{k}} - \mu) \sin 2\theta_{\mathbf{k}} + \cos 2\theta_{\mathbf{k}} \Delta \ , \tag{6.71}$$

where Δ, which is also called the gap parameter, obeys the gap equation

$$\Delta = -\frac{V_0}{2V} \sum_{\mathbf{k}'} \sin 2\theta_{\mathbf{k}'} \ . \tag{6.72}$$

(e) Solve for $\theta_{\mathbf{k}}$ and show that the gap equation can be rewritten as

$$\frac{1}{V_0} = -\frac{1}{V} \sum_{\mathbf{k}} \frac{1}{2\sqrt{(\varepsilon_{\mathbf{k}} - \mu)^2 + \Delta^2}} \ . \tag{6.73}$$

The right-hand side of the last equation is not convergent. However, in Chap. 12, we learn how to deal with its ultraviolet divergence. So far, we have performed second-quantized calculations without actually realizing that they describe the superfluid state of attractively interacting fermions. In Chap. 12, we find that the gap Δ is interpreted as the energy needed to break up a Cooper pair, which means that it costs a certain amount of energy to excite the BCS ground state. As a result, an object moving through a condensate of Cooper pairs needs to have a minimum kinetic energy to transfer momentum, i.e. to experience friction.

Exercise 6.7. Itinerant Ferromagnetism
In a Fermi system, itinerant ferromagnetism is a result of the competition between the kinetic energy and repulsive interactions. The word "itinerant" reflects the fact that the magnetism is due to spins that are delocalized, as opposed to localized spins in, e.g., the Ising or Heisenberg models. Let us first convince ourselves of the fact that for the same total number of particles, the polarized Fermi gas has a larger kinetic energy than the unpolarized Fermi gas, where the Hamiltonian of the system is again given by (6.66), but now with $V_0 > 0$.
(a) Consider the unpolarized state $|\Psi_{\text{up}}\rangle = \prod'_{\mathbf{k}} \hat{\psi}^{\dagger}_{\mathbf{k},\uparrow} \hat{\psi}^{\dagger}_{\mathbf{k},\downarrow} |0\rangle$, where $\prod'$ denotes the restricted product of all states below the Fermi energy. Calculate its total kinetic energy in terms of the total density n. Note that for the total density we have $n = n_{\uparrow} + n_{\downarrow}$, where n_{α} is the density of atoms in spin state $|\alpha\rangle$.
(b) Next, consider the fully-polarized state $|\Psi_{\text{p}}\rangle = \prod'_{\mathbf{k}} \hat{\psi}^{\dagger}_{\mathbf{k},\uparrow} |0\rangle$. Calculate its kinetic energy in terms of the total density $n = n_{\uparrow}$. Note that the Fermi energy is now different from the previous exercise.
(c) We define the magnetization by $\hat{\mathbf{m}} \equiv \sum_{\mathbf{k},\alpha,\alpha'} \hat{\psi}^{\dagger}_{\mathbf{k},\alpha} \boldsymbol{\sigma}_{\alpha\alpha'} \hat{\psi}_{\mathbf{k},\alpha'}/V$, where $\boldsymbol{\sigma} = (\boldsymbol{\sigma}_x, \boldsymbol{\sigma}_y, \boldsymbol{\sigma}_z)$ is a vector of Pauli matrices. What is the magnetization for the above two states?
(d) Calculate the interaction energy for the polarized and unpolarized state and show that the polarized state has no interaction energy. Explain why.
(e) Consider now the state $|\Psi\rangle = \prod'_{\mathbf{k}} \hat{\psi}^{\dagger}_{\mathbf{k},\uparrow} \prod'_{\mathbf{k}'} \hat{\psi}^{\dagger}_{\mathbf{k}',\downarrow} |0\rangle$, where the products over $\mathbf{k}, \mathbf{k}'$ are restricted by the requirement that the density of $\uparrow$ atoms is $n_{\uparrow}$ and the density of $\downarrow$ atoms is $n_{\downarrow}$. Give the total energy as a function of $n_{\uparrow}$ and $n_{\downarrow}$.
(f) Using this result, show that the system becomes polarized if $V_0 \mathscr{D}(\varepsilon_{\text{F}}) > 1$, where $\mathscr{D}(\varepsilon_{\text{F}})$ is the density of states at the Fermi level. This is the Stoner criterion.
(g) Calculate the interaction energy of the state $|\Psi\rangle = \prod'_{\mathbf{k}} (\hat{\psi}^{\dagger}_{\mathbf{k},\uparrow} + e^{i\varphi} \hat{\psi}^{\dagger}_{\mathbf{k},\downarrow}) |0\rangle$, where φ is an arbitrary angle. In which direction is the magnetization of this state pointing?

Additional Reading

- A. L. Fetter and J. D. Walecka, *Quantum Theory of Many-Particle Systems*, (Dover, New York, 2003).
- H. Bruus and K. Flensberg, *Many-Body Quantum Theory in Condensed Matter Physics: An Introduction*, (Oxford University Press, Oxford, 2004).
- G.D. Mahan, *Many-Particle Physics*, (Plenum Press, New York, 1981).

Part II

Chapter 7
Functional Integrals

It is difficult for me to believe that quantum mechanics, working very well for currently practical set-ups, will nevertheless fail badly with improvements in counter efficiency.
– John S. Bell

In this first chapter of Part II, we introduce the functional-integral formalism that we use throughout the rest of the book to determine the equilibrium properties of a quantum fluid. Today, functional integrals are the preferred tool of researchers working on quantum many-body problems, in particular in condensed-matter physics and high-energy physics. Although this formulation of quantum field theory is of course fully equivalent to the operator formulation developed in the previous chapter, it is in practice more flexible to arrive at systematic approximation schemes and often also leads to a much simpler derivation of exact results. In this chapter, we derive the functional formulation of quantum field theory exactly along the same lines as we derived the path-integral approach to quantum mechanics in Chap. 5. To familiarize ourselves with this new method, we then discuss in detail the ideal quantum gases. The deep and fundamentally different consequences of interactions are then the topic of the rest of the book.

7.1 Grand-Canonical Partition Function

From statistical physics, we know that the equilibrium properties of an interacting many-body system follow from the grand-canonical partition sum

$$Z = \mathrm{Tr}\left[\mathrm{e}^{-\beta(\hat{H}-\mu\hat{N})}\right] , \tag{7.1}$$

where $\beta = 1/k_{\mathrm{B}}T$ and μ is the chemical potential. Our goal is to evaluate this quantity by combining many-body quantum-field theory with Feynman's path-integral approach to quantum mechanics. As we show next, our goal can be achieved by writing the partition function as a functional integral over time-dependent fields $\phi_\alpha(\mathbf{x}, \tau)$, in the same manner that we wrote the partition function for a single particle as a path integral over time-dependent paths $\mathbf{x}(\tau)$, which was achieved in Chap. 5. These fields are actually the eigenvalues of the coherent states that we introduced in the previous chapter. We start with (6.61) for the trace of an operator, evaluated as a functional integral over coherent states, giving

$$Z = \int d[\phi^*]d[\phi]\, e^{-(\phi|\phi)} \langle \pm\phi | e^{-\beta(\hat{H}-\mu\hat{N})} | \phi \rangle \,, \tag{7.2}$$

and observe that we are thus faced with the task of calculating the matrix elements $\langle \phi_M | e^{-\beta(\hat{H}-\mu\hat{N})} | \phi_0 \rangle$ with $\phi_{0,\alpha}(\mathbf{x}) = \phi_\alpha(\mathbf{x})$ and $\phi^*_{M,\alpha}(\mathbf{x}) = \pm\phi^*_\alpha(\mathbf{x})$. Just like in Sect. 5.6, we realize that the operator $e^{-\beta(\hat{H}-\mu\hat{N})}$ is identical to the quantum mechanical evolution operator $U(t,0) = e^{-i(\hat{H}-\mu\hat{N})t/\hbar}$ evaluated at $t = -i\hbar\beta$. Put differently, we need to calculate the matrix elements of the imaginary-time evolution operator $U(-i\tau,0)$ for $\tau = \hbar\beta$. To do so, we split the time interval $[0,\hbar\beta]$ into M pieces, with $\tau_j = j\hbar\beta/M$ and $j = 0,1,\ldots,M$, such that $\Delta\tau = \hbar\beta/M$. The procedure is summarized in Fig. 7.1.

Fig. 7.1 Illustration of the slicing of the imaginary time interval $[0,\hbar\beta]$ needed to derive Feynman's path-integral formulation of the partition function.

At each intermediate time τ_j, we then apply the closure relation of the coherent states, given by (6.60). This yields

$$\begin{aligned} &\langle \phi_M | e^{-\beta(\hat{H}-\mu\hat{N})} | \phi_0 \rangle \\ &= \int \left(\prod_{j=1}^{M-1} d[\phi_j^*]d[\phi_j]\, e^{-(\phi_j|\phi_j)} \right) \prod_{j=1}^{M} \langle \phi_j | e^{-\Delta\tau(\hat{H}-\mu\hat{N})/\hbar} | \phi_{j-1} \rangle \,. \end{aligned} \tag{7.3}$$

Now, we can use that in the limit $M \to \infty$ we only need to know the latter matrix elements up to order $\Delta\tau$, because terms of order $(\Delta\tau)^2$ lead to corrections only of order $M(\Delta\tau)^2 \propto 1/M$ to the total matrix element, which consequently vanish. Hence,

$$\begin{aligned} \langle \phi_j | e^{-\Delta\tau(\hat{H}-\mu\hat{N})/\hbar} | \phi_{j-1} \rangle &\simeq \langle \phi_j | 1 - \Delta\tau(\hat{H}-\mu\hat{N})/\hbar | \phi_{j-1} \rangle \\ &\equiv \langle \phi_j | \phi_{j-1} \rangle (1 - \Delta\tau H[\phi_j^*, \phi_{j-1}]/\hbar) \,, \end{aligned} \tag{7.4}$$

where we defined the grand-canonical Hamiltonian functional as

$$H[\phi^*,\phi] = \sum_{\alpha}\int d\mathbf{x}\ \phi^*_\alpha(\mathbf{x})\left\{-\frac{\hbar^2\nabla^2}{2m}+V^{\mathrm{ex}}(\mathbf{x})+\varepsilon_\alpha-\mu\right\}\phi_\alpha(\mathbf{x})$$
$$+\frac{1}{2}\sum_{\alpha,\alpha'}\int d\mathbf{x}\int d\mathbf{x}'\ \phi^*_\alpha(\mathbf{x})\phi^*_{\alpha'}(\mathbf{x}')V(\mathbf{x}-\mathbf{x}')\phi_{\alpha'}(\mathbf{x}')\phi_\alpha(\mathbf{x})\ , \quad (7.5)$$

which follows from the second-quantized form of the Hamiltonian and the number operator, given by (6.29) and (6.30), together with the defining property of the coherent states, namely $\hat{\psi}_\alpha(\mathbf{x})|\phi\rangle = \phi_\alpha(\mathbf{x})|\phi\rangle$ and $\langle\phi|\hat{\psi}^\dagger_\alpha(\mathbf{x}) = \langle\phi|\phi^*_\alpha(\mathbf{x})$.

Next, we can re-exponentiate the right-hand side of (7.4), leading again to errors only of order $(\Delta\tau)^2$, giving

$$\langle\phi_j|e^{-\Delta\tau(\hat{H}-\mu\hat{N})/\hbar}|\phi_{j-1}\rangle = e^{(\phi_j|\phi_{j-1})-\Delta\tau H[\phi^*_j,\phi_{j-1}]/\hbar}\ , \quad (7.6)$$

where we also used (6.59) for the overlap of the coherent states. As a result, the desired matrix element of the imaginary-time evolution operator becomes

$$\langle\phi_M|e^{-\beta(\hat{H}-\mu\hat{N})}|\phi_0\rangle = \int\left(\prod_{j=1}^{M-1} d[\phi^*_j]d[\phi_j]\ e^{-(\phi_j|\phi_j)}\right) \quad (7.7)$$
$$\times\exp\left\{\sum_{j=1}^{M}((\phi_j|\phi_{j-1})-\Delta\tau H[\phi^*_j,\phi_{j-1}]/\hbar)\right\},$$

which can then be manipulated into the suggestive form

$$\langle\phi_M|e^{-\beta(\hat{H}-\mu\hat{N})}|\phi_0\rangle = e^{(\phi_M|\phi_M)}\int\left(\prod_{j=1}^{M-1} d[\phi^*_j]d[\phi_j]\right)$$
$$\times\exp\left\{-\frac{1}{\hbar}\sum_{j=1}^{M}\Delta\tau\left(\hbar\frac{(\phi_j|\phi_j)-(\phi_j|\phi_{j-1})}{\Delta\tau}+H[\phi^*_j,\phi_{j-1}]\right)\right\}. \quad (7.8)$$

Taking the continuum limit $M\to\infty$ and putting $\phi_j\equiv\phi(\tau_j)$, we find that

$$\langle\phi_M|e^{-\beta(\hat{H}-\mu\hat{N})}|\phi_0\rangle$$
$$= e^{(\phi(\hbar\beta)|\phi(\hbar\beta))}\int_{\phi(0)=\phi_0}^{\phi^*(\hbar\beta)=\phi^*_M} d[\phi^*]d[\phi]\ e^{-S[\phi^*,\phi]/\hbar}\ , \quad (7.9)$$

with the Euclidean action given by

$$S[\phi^*,\phi] = \int_0^{\hbar\beta} d\tau\left\{\sum_\alpha\int d\mathbf{x}\ \phi^*_\alpha(\mathbf{x},\tau)\hbar\frac{\partial}{\partial\tau}\phi_\alpha(\mathbf{x},\tau)+H[\phi^*(\tau),\phi(\tau)]\right\}. \quad (7.10)$$

We have thus obtained the desired functional integral for the matrix element of (7.3), where we need to integrate over all fields $\phi_\alpha(\mathbf{x},\tau)$ that satisfy the boundary conditions $\phi_\alpha(\mathbf{x},0)=\phi_{0,\alpha}(\mathbf{x})$ and $\phi^*_\alpha(\mathbf{x},\hbar\beta)=\phi^*_{M,\alpha}(\mathbf{x})$. We note that for bosons the fields $\phi_\alpha(\mathbf{x},\tau)$ are ordinary complex functions, whereas for fermions they are

Grassmann valued functions, meaning that we pick up a minus sign each time we permute two such fields. Note that (7.9) is precisely the field theory analogue of the Feynman path integral for the transition amplitude obtained in (5.34). To calculate the grand-canonical partition function, we see from (7.2) that we need to put $\phi_{0,\alpha}(\mathbf{x})$ equal to $\pm\phi_{M,\alpha}(\mathbf{x})$ and perform a last functional integration over $\phi_{M,\alpha}(\mathbf{x})$ and $\phi^*_{M,\alpha}(\mathbf{x})$. Then, we finally obtain

$$Z = \int \mathrm{d}[\phi^*]\mathrm{d}[\phi]\ \mathrm{e}^{-S[\phi^*,\phi]/\hbar}, \tag{7.11}$$

where we need to integrate over all fields with the boundary conditions $\phi_\alpha(\mathbf{x},\hbar\beta) = \pm\phi_\alpha(\mathbf{x},0)$, i.e. the fields are periodic in $[0,\hbar\beta]$ for bosons and antiperiodic for fermions. The short-hand notation for the integration measure now means

$$\begin{aligned}\int \mathrm{d}[\phi^*]\mathrm{d}[\phi] &= \int \prod_{j=1}^{M} \mathrm{d}[\phi_j^*]\mathrm{d}[\phi_j] = \int \prod_{j=1}^{M}\prod_{\mathbf{n},\alpha} \frac{\mathrm{d}\phi^*_{j,\mathbf{n},\alpha}\mathrm{d}\phi_{j,\mathbf{n},\alpha}}{(2\pi\mathrm{i})^{(1\pm1)/2}} \\ &= \int \prod_{\mathbf{n},\alpha} \mathrm{d}[\phi^*_{\mathbf{n},\alpha}]\mathrm{d}[\phi_{\mathbf{n},\alpha}],\end{aligned} \tag{7.12}$$

where (anti)periodic boundary conditions were implied and we also used (6.50). Note that in (7.9) we have used the same notation for the integration measure as in (7.11), although there is in principle one more integration in the expression for the partition function. This may seem somewhat imprecise. However, thinking in terms of functional integration over all possible continuous functions $\phi_\alpha(\mathbf{x},\tau)$, we note that the difference between the two expressions is automatically accounted for by the different boundary conditions of the continuous integration. Put differently, by specifying each time the specific boundary conditions, it becomes clear which of the discrete integration measures is actually meant. From now on we also always imply (anti)periodic boundary conditions for (fermions) bosons, corresponding to the discrete measure of (7.12), unless stated otherwise.

Having arrived at an exact identity between the partition function and a functional integral, we are now going to familiarize ourselves with this identity. This means that in the next section we are going to treat various methods for performing functional integrals by considering the ideal quantum gases.

7.2 Ideal Quantum Gases

Since the partition functions Z_0 of the ideal quantum gases are known exactly, they are ideal test cases for our field-theoretical methods. Moreover, a thorough field-theoretical knowledge of the ideal quantum gases is important in its own right because it forms the basis for treating interacting quantum gases, in particular when the interaction effects are incorporated perturbatively. How such a perturbation theory is performed is discussed in detail in Chap. 8. For the noninteracting quantum gas, we have from (7.5) and (7.10) that

$$S_0[\phi^*,\phi] = \sum_\alpha \int_0^{\hbar\beta} d\tau \int d\mathbf{x} \times \phi^*_\alpha(\mathbf{x},\tau) \left\{ \hbar\frac{\partial}{\partial\tau} - \frac{\hbar^2\nabla^2}{2m} + V^{\text{ex}}(\mathbf{x}) + \varepsilon_\alpha - \mu \right\} \phi_\alpha(\mathbf{x},\tau) \tag{7.13}$$

and evaluating the partition function boils down to performing a Gaussian integral over the fields. It will be illustrative to evaluate this partition function in three different ways.

7.2.1 Semiclassical Method

The first method, which is called the semiclassical method, is the field-theoretical analogue of the way we solved the path integral for a free particle in the example of Sect. 5.4.1. We start by considering solely the matrix element $\langle \pm\phi| e^{-\beta(\hat{H}-\mu\hat{N})} |\phi\rangle$ and postpone the evaluation of the trace for the partition function to the end of the calculation. By using the full set of eigenstates $\chi_\mathbf{n}(\mathbf{x}) \equiv \langle \mathbf{x}|\mathbf{n}\rangle$ for the noninteracting single-particle Schrödinger equation, as introduced in (6.3), we can expand the fields as

$$\phi_\alpha(\mathbf{x},\tau) = \sum_\mathbf{n} \phi_{\mathbf{n},\alpha}(\tau)\chi_\mathbf{n}(\mathbf{x}). \tag{7.14}$$

As follows from (7.9), the matrix element $\langle \pm\phi| e^{-\beta(\hat{H}-\mu\hat{N})} |\phi\rangle$ is in the absence of interactions given by the following path integral

$$\int_{\phi(0)=\phi}^{\phi^*(\hbar\beta)=\pm\phi^*} \left(\prod_{\mathbf{n},\alpha} d[\phi^*_{\mathbf{n},\alpha}] d[\phi_{\mathbf{n},\alpha}] \right) \exp\left\{ \sum_{\mathbf{n},\alpha} \phi^*_{\mathbf{n},\alpha}(\hbar\beta)\phi_{\mathbf{n},\alpha}(\hbar\beta) \right\} \times \exp\left\{ -\frac{1}{\hbar} \int_0^{\hbar\beta} d\tau \sum_{\mathbf{n},\alpha} \phi^*_{\mathbf{n},\alpha}(\tau) \left(\hbar\frac{\partial}{\partial\tau} + \varepsilon_{\mathbf{n},\alpha} - \mu \right) \phi_{\mathbf{n},\alpha}(\tau) \right\},$$

which is the product for each $\mathbf{n}$ and α of the path integral

$$\int d[\phi^*_{\mathbf{n},\alpha}] d[\phi_{\mathbf{n},\alpha}] \times \exp\left\{ \phi^*_{\mathbf{n},\alpha}(\hbar\beta)\phi_{\mathbf{n},\alpha}(\hbar\beta) - \frac{1}{\hbar} \int_0^{\hbar\beta} d\tau\, \phi^*_{\mathbf{n},\alpha}(\tau) \left(\hbar\frac{\partial}{\partial\tau} + \varepsilon_{\mathbf{n},\alpha} - \mu \right) \phi_{\mathbf{n},\alpha}(\tau) \right\}$$

with the boundary conditions $\phi_{\mathbf{n},\alpha}(0) = \phi_{\mathbf{n},\alpha}$ and $\phi^*_{\mathbf{n},\alpha}(\hbar\beta) = \pm\phi^*_{\mathbf{n},\alpha}$. Note that this last path integral thus actually corresponds to the particular matrix element $\langle \pm\phi_{\mathbf{n},\alpha}| e^{-\beta(\varepsilon_{\mathbf{n},\alpha}-\mu)\hat{\psi}^\dagger_{\mathbf{n},\alpha}\hat{\psi}_{\mathbf{n},\alpha}} |\phi_{\mathbf{n},\alpha}\rangle$ for one specific value of $\mathbf{n}$ and α.

Then, we perform in analogy with Sect. 5.4.1 a fluctuation expansion, which amounts to a shift to the new integration variables $\xi_{\mathbf{n},\alpha}(\tau)$, such that $\phi_{\mathbf{n},\alpha}(\tau) = \phi_{\mathrm{cl};\mathbf{n},\alpha}(\tau) + \xi_{\mathbf{n},\alpha}(\tau)$ and $\phi^*_{\mathbf{n},\alpha}(\tau) = \phi^*_{\mathrm{cl};\mathbf{n},\alpha}(\tau) + \xi^*_{\mathbf{n},\alpha}(\tau)$, where $\phi_{\mathrm{cl};\mathbf{n},\alpha}(\tau)$ obeys the 'classical' equation of motion

$$\left.\frac{\delta S_0[\phi^*,\phi]}{\delta \phi^*_{\mathbf{n},\alpha}(\tau)}\right|_{\phi=\phi_{\mathrm{cl};\mathbf{n},\alpha}} = \left(\hbar\frac{\partial}{\partial\tau} + \varepsilon_{\mathbf{n},\alpha} - \mu\right)\phi_{\mathrm{cl};\mathbf{n},\alpha}(\tau) = 0 \tag{7.15}$$

and similarly for $\phi^*_{\mathrm{cl};\mathbf{n},\alpha}(\tau)$. This means that we fix the classical fields by requiring them to minimize the action. The solutions to (7.15) with the correct boundary values are

$$\phi_{\mathrm{cl};\mathbf{n},\alpha}(\tau) = \phi_{\mathbf{n},\alpha}\mathrm{e}^{-(\varepsilon_{\mathbf{n},\alpha}-\mu)\tau/\hbar} \quad \text{and} \quad \phi^*_{\mathrm{cl};\mathbf{n},\alpha}(\tau) = \pm\phi^*_{\mathbf{n},\alpha}\mathrm{e}^{(\varepsilon_{\mathbf{n},\alpha}-\mu)(\tau-\hbar\beta)/\hbar}, \tag{7.16}$$

such that we obtain

$$\exp\left\{\pm\mathrm{e}^{-\beta(\varepsilon_{\mathbf{n},\alpha}-\mu)}\phi^*_{\mathbf{n},\alpha}\phi_{\mathbf{n},\alpha}\right\}$$
$$\times\int \mathrm{d}[\xi^*_{\mathbf{n},\alpha}]\mathrm{d}[\xi_{\mathbf{n},\alpha}]\,\exp\left\{-\frac{1}{\hbar}\int_0^{\hbar\beta}\mathrm{d}\tau\,\xi^*_{\mathbf{n},\alpha}(\tau)\left(\hbar\frac{\partial}{\partial\tau}+\varepsilon_{\mathbf{n},\alpha}-\mu\right)\xi_{\mathbf{n},\alpha}(\tau)\right\}$$

with the boundary conditions $\xi^*_{\mathbf{n},\alpha}(\hbar\beta) = \xi_{\mathbf{n},\alpha}(0) = 0$. Note that in the last expression the path integral in the second line evaluates to 1, because it corresponds to the matrix element

$$\langle 0|\mathrm{e}^{-\beta(\varepsilon_{\mathbf{n},\alpha}-\mu)\hat{\psi}^\dagger_{\mathbf{n},\alpha}\hat{\psi}_{\mathbf{n},\alpha}}|0\rangle = \langle 0|(1-\beta(\varepsilon_{\mathbf{n},\alpha}-\mu)\hat{\psi}^\dagger_{\mathbf{n},\alpha}\hat{\psi}_{\mathbf{n},\alpha}+\ldots)|0\rangle = 1,$$

such that actually only the prefactor, namely $\exp\left\{\pm\mathrm{e}^{-\beta(\varepsilon_{\mathbf{n},\alpha}-\mu)}\phi^*_{\mathbf{n},\alpha}\phi_{\mathbf{n},\alpha}\right\}$, remains. Multiplying the obtained result for each value of $\mathbf{n}$ and α, we find the total matrix element

$$\langle\pm\phi|\mathrm{e}^{-\beta(\hat{H}-\mu\hat{N})}|\phi\rangle = \exp\left\{\pm\sum_{\mathbf{n},\alpha}\mathrm{e}^{-\beta(\varepsilon_{\mathbf{n},\alpha}-\mu)}\phi^*_{\mathbf{n},\alpha}\phi_{\mathbf{n},\alpha}\right\}, \tag{7.17}$$

which we then still have to integrate in order to evaluate the trace and find the partition function. Using (2.66) and (2.67), we find

$$Z_0 = \prod_{\mathbf{n},\alpha}\int\frac{\mathrm{d}\phi^*_{\mathbf{n},\alpha}\mathrm{d}\phi_{\mathbf{n},\alpha}}{(2\pi\mathrm{i})^{(1\pm1)/2}}\,\exp\left\{-\sum_{\mathbf{n},\alpha}\left(1\mp\mathrm{e}^{-\beta(\varepsilon_{\mathbf{n},\alpha}-\mu)}\right)\phi^*_{\mathbf{n},\alpha}\phi_{\mathbf{n},\alpha}\right\}$$
$$= \prod_{\mathbf{n},\alpha}(1\mp\mathrm{e}^{-\beta(\varepsilon_{\mathbf{n},\alpha}-\mu)})^{\mp1} = \exp\left\{\mp\sum_{\mathbf{n},\alpha}\log(1\mp\mathrm{e}^{-\beta(\varepsilon_{\mathbf{n},\alpha}-\mu)})\right\}. \tag{7.18}$$

This can be immediately compared with (4.38), where the only differences are that our present result is also valid in the presence of an external potential and that we have included the possibility of a spin degree of freedom. We have thus obtained the exact result for the ideal bosonic and fermionic quantum gases with the use of field-theoretical methods. The average total number of particles is obtained by using the thermodynamic identity of (4.28), namely

$$\langle \hat{N} \rangle_0 = \frac{1}{\beta} \frac{\partial \log Z_0}{\partial \mu}, \tag{7.19}$$

leading to

$$\langle \hat{N} \rangle_0 = \sum_{\mathbf{n},\alpha} \frac{1}{e^{\beta(\varepsilon_{\mathbf{n},\alpha}-\mu)} \mp 1}, \tag{7.20}$$

which is the summation over the previously obtained Bose-Einstein and Fermi-Dirac distributions of (4.40) and (4.41). Note that in the homogeneous case we have $\chi_{\mathbf{k}}(\mathbf{x}) = e^{i\mathbf{k}\cdot\mathbf{x}}/\sqrt{V}$, where $V = L^d$ is the volume of a d-dimensional box and the wavevector $\mathbf{k}$ is related to the momentum $\mathbf{p}$ through $\mathbf{p} = \hbar\mathbf{k}$. The general single-particle eigenenergies $\varepsilon_{\mathbf{n},\alpha}$ in (7.20), then result in

$$\varepsilon_{\mathbf{k},\alpha} = \frac{\hbar^2\mathbf{k}^2}{2m} + \varepsilon_\alpha, \tag{7.21}$$

which reduces to $\varepsilon_{\mathbf{p}} = \mathbf{p}^2/2m$ for spinless atoms.

7.2.2 Matsubara Expansion

Next, we evaluate the partition function for the noninteracting quantum gas again, but this time by writing the action first in a more convenient basis. We not only transform the space coordinate using the full set of eigenstates $\chi_{\mathbf{n}}(\mathbf{x}) \equiv \langle \mathbf{x}|\mathbf{n}\rangle$, but also the (imaginary) time coordinate by going to (imaginary) frequency space

$$\phi_\alpha(\mathbf{x},\tau) = \sum_{\mathbf{n}} \sum_{n=-\infty}^{\infty} \phi_{\mathbf{n},n,\alpha}\chi_{\mathbf{n}}(\mathbf{x}) \frac{e^{-i\omega_n\tau}}{\sqrt{\hbar\beta}}, \tag{7.22}$$

where $\omega_n = \pi(2n)/\hbar\beta$ for bosons and $\omega_n = \pi(2n+1)/\hbar\beta$ for fermions. These are known as the even and odd Matsubara frequencies, or as the bosonic and fermionic Matsubara frequencies, respectively. Note that the Matsubara frequencies conveniently incorporate the boundary conditions automatically, since they make the bosonic fields periodic in imaginary time, and the fermionic fields antiperiodic. Using the above expansion, we find for the action of (7.13) that

$$
\begin{aligned}
S_0[\phi^*,\phi] &= \sum_\alpha \int_0^{\hbar\beta} d\tau \int d\mathbf{x}\, \phi^*_\alpha(\mathbf{x},\tau) \\
&\times \left\{ \hbar\frac{\partial}{\partial\tau} - \frac{\hbar^2\nabla^2}{2m} + V^{\mathrm{ex}}(\mathbf{x}) + \varepsilon_\alpha - \mu \right\} \phi_\alpha(\mathbf{x},\tau) \\
&= \sum_\alpha \int_0^{\hbar\beta} d\tau \int d\mathbf{x} \sum_{\mathbf{n},n} \sum_{\mathbf{n}',n'} \phi^*_{\mathbf{n},n,\alpha} \chi^*_{\mathbf{n}}(\mathbf{x}) \frac{e^{i\omega_n\tau}}{\sqrt{\hbar\beta}} \\
&\times (-i\hbar\omega_{n'} + \varepsilon_{\mathbf{n}'} + \varepsilon_\alpha - \mu)\, \phi_{\mathbf{n}',n',\alpha} \chi_{\mathbf{n}'}(\mathbf{x}) \frac{e^{-i\omega_{n'}\tau}}{\sqrt{\hbar\beta}} \\
&= \sum_{\mathbf{n},n,\alpha} \phi^*_{\mathbf{n},n,\alpha} (-i\hbar\omega_n + \varepsilon_{\mathbf{n},\alpha} - \mu)\, \phi_{\mathbf{n},n,\alpha},
\end{aligned}
\tag{7.23}
$$

where we used the orthonormality of the single-particle eigenstates and

$$
\int_0^{\hbar\beta} d\tau \frac{e^{i(\omega_n - \omega_{n'})\tau}}{\hbar\beta} = \delta_{n,n'} . \tag{7.24}
$$

Example 7.1. Note that each time we perform a transformation on the integration variables, we in principle pick up a corresponding Jacobian. In this respect there is a subtlety with the continuum limit of the functional integral that we for completeness' sake would like to point out here, although it turns out to be unimportant in practice and can therefore be safely ignored. If we consider the discrete version of the above Fourier transform

$$
\phi_{j,\mathbf{n},\alpha} = \sum_{n=0}^{M-1} \phi_{\mathbf{n},n,\alpha} \frac{e^{-i\omega_n\tau_j}}{\sqrt{\Delta\tau M}}, \tag{7.25}
$$

with $\Delta\tau = \hbar\beta/M$, we find that

$$
\sum_j \phi^*_{j,\mathbf{n},\alpha}\phi_{j,\mathbf{n},\alpha} = \sum_n \frac{1}{\Delta\tau} \phi^*_{\mathbf{n},n,\alpha}\phi_{\mathbf{n},n,\alpha}, \tag{7.26}
$$

which means that the Jacobian for the transformation from discrete time slices to the Matsubara frequencies is given by

$$
\frac{\partial(\{\phi^*_{j,\mathbf{n},\alpha}, \phi_{j,\mathbf{n},\alpha}\})}{\partial(\{\phi^*_{\mathbf{n},n,\alpha}, \phi_{\mathbf{n},n,\alpha}\})} = (\Delta\tau)^{\mp M} = \left(\frac{M}{\hbar\beta}\right)^{\pm M} . \tag{7.27}
$$

Note that the difference between the Jacobians for bosons and fermions is a consequence of the fact that for Grassmann variables, we have that

$$
\int d\phi\; c\,\phi = c = \int d(c\phi)\; c\,(c\phi),
$$

whereas for ordinary complex integration, we would use

$$\int d\phi \; c\,\phi = \int d(c\phi)\;\frac{1}{c}\;(c\phi).$$

This also implies that for Grassmann variables $d\phi$ does not have the same physical dimension as ϕ. Indeed, their dimensionalities are precisely inverse to each other. Both observations are easily understood from the fact that integration is equivalent to differentiation for Grassmann variables. Although the factor $M^{\pm M}$ from (7.27) is present when working with the discrete expression, it has no obvious continuum limit. As a result, the Jacobian is usually taken to be just $(\hbar\beta)^{\mp M}$. The difference amounts to a numerical prefactor in the partition sum, which cancels when calculating expectation values. Moreover, we show below that the prefactor actually also plays no role in obtaining the exact answer for the partition sum of the ideal quantum gas directly in the continuum limit.

For the grand-canonical partition function of the gas, we thus find from (7.23) that

$$\begin{aligned} Z_0 = &\int \left(\prod_{\mathbf{n},n,\alpha} \frac{d\phi^*_{\mathbf{n},n,\alpha} d\phi_{\mathbf{n},n,\alpha}}{(2\pi i)^{(1\pm 1)/2}} \frac{1}{(\hbar\beta)^{\pm 1}} \right) \\ &\times \exp\left\{ -\frac{1}{\hbar} \sum_{\mathbf{n},n,\alpha} \phi^*_{\mathbf{n},n,\alpha}(-i\hbar\omega_n + \varepsilon_{\mathbf{n},\alpha} - \mu)\phi_{\mathbf{n},n,\alpha} \right\} , \end{aligned} \tag{7.28}$$

where we included the factor $1/(\hbar\beta)^{\pm 1}$ coming from the Jacobian of the coordinate transformation as explained in the example. Note that we can immediately perform the Gaussian integrals to obtain

$$\begin{aligned} Z_0 &= \prod_{\mathbf{n},n,\alpha} (\beta(-i\hbar\omega_n + \varepsilon_{\mathbf{n},\alpha} - \mu))^{\mp 1} \\ &= \exp\left\{ \mp \sum_{\mathbf{n},n,\alpha} \log(\beta(-i\hbar\omega_n + \varepsilon_{\mathbf{n},\alpha} - \mu)) \right\} , \end{aligned} \tag{7.29}$$

where we used (2.66) and (2.67). To evaluate the resulting sum over Matsubara frequencies, it is useful to add a convergence factor $e^{i\omega_n\eta}$ and finally take the limit $\eta \downarrow 0$. The precise reason for this procedure is explained in Example 7.2 at the end of this section. Comparing the above expression with (7.18), we thus need to prove

$$\lim_{\eta\downarrow 0} \sum_n \log(\beta(-i\hbar\omega_n + \varepsilon - \mu))e^{i\omega_n\eta} = \log(1 \mp e^{-\beta(\varepsilon-\mu)}) . \tag{7.30}$$

In first instance, we can show that this is correct up to a constant by differentiating the latter equation with respect to $\beta\mu$. We then obtain

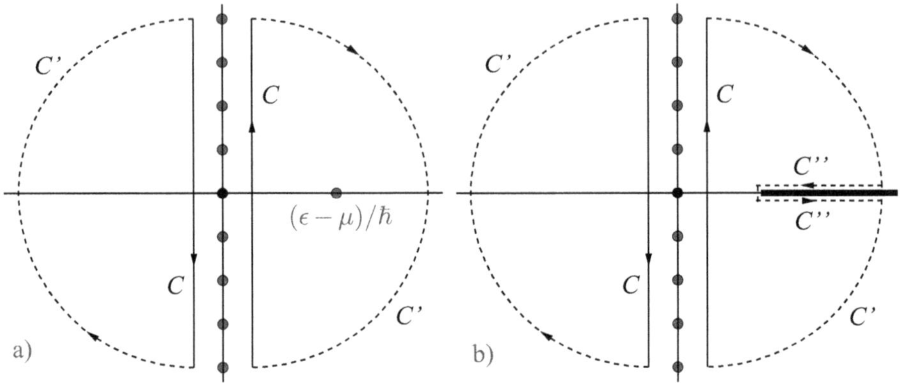

Fig. 7.2 Illustration of the contour integrations that are required to perform the summation over the Matsubara frequencies. a) The black dots indicate the position of the poles in the integrand from (7.32). b) The thick line indicates branch cut of the integrand from (7.34).

$$\lim_{\eta \downarrow 0} \frac{1}{\hbar\beta} \sum_n \frac{e^{i\omega_n \eta}}{i\omega_n - (\varepsilon-\mu)/\hbar} = \mp \frac{1}{e^{\beta(\varepsilon-\mu)} \mp 1}, \tag{7.31}$$

which can be shown by using contour integration in the following way. First note that the function $\hbar\beta/(e^{\hbar\beta z} \mp 1)$ has simple poles at the even and odd Matsubara frequencies, such that its residue at the poles is given by ± 1. Hence, by applying the residue theorem from (2.40), we find that

$$\lim_{\eta \downarrow 0} \frac{1}{2\pi i} \int_C dz \frac{e^{\eta z}}{z-(\varepsilon-\mu)/\hbar} \frac{\pm 1}{e^{\hbar\beta z} \mp 1} = \lim_{\eta \downarrow 0} \frac{1}{\hbar\beta} \sum_n \frac{e^{i\omega_n \eta}}{i\omega_n - (\varepsilon-\mu)/\hbar} \tag{7.32}$$

with C a contour that fully encloses the imaginary axis as shown in Fig. 7.2a. Next, we note that we can freely add the infinite arcs C' to the contour C, because the additional integration along the arcs C' on the left-hand side of (7.32) vanishes. The reason for this is that the integrand behaves as $\pm e^{-(\hbar\beta-\eta)\mathrm{Re}(z)}/|z|$ for $\mathrm{Re}(z) \to \infty$ and as $-e^{\eta \mathrm{Re}(z)}/|z|$ for $\mathrm{Re}(z) \to -\infty$. Thus, for any $0 < \eta < \hbar\beta$, the integrand always vanishes much faster than $1/|z|$ on the contour C', namely exponentially fast, such that the corresponding integral indeed gives no contribution. By considering now the closed contours formed by C and C', we see that the poles at the Matsubara frequencies are no longer inside the contour. Now, we only have the simple pole at $z = (\varepsilon-\mu)/\hbar$ inside the contour and application of the residue theorem leads to

$$\lim_{\eta \downarrow 0} \frac{1}{2\pi i} \int_{C+C'} dz \frac{e^{\eta z}}{z-(\varepsilon-\mu)/\hbar} \frac{\pm 1}{e^{\hbar\beta z} \mp 1} = \mp \frac{1}{e^{\beta(\varepsilon-\mu)} \mp 1}, \tag{7.33}$$

which is then seen to prove (7.31). As a result, we have also shown the correctness of (7.30) up to a constant independent of $\beta\mu$.

However, we can also prove (7.30) directly using contour integration, which we show explicitly for the bosonic case, while the fermionic case is left as an exercise. We can start again by writing the sum over bosonic Matsubara frequencies as a contour integral along the contour C, which gives

$$\lim_{\eta\downarrow 0}\sum_n \log(\beta(-\mathrm{i}\hbar\omega_n+\varepsilon-\mu))\mathrm{e}^{\mathrm{i}\omega_n\eta} = \lim_{\eta\downarrow 0}\frac{1}{2\pi\mathrm{i}}\int_C \mathrm{d}z\, \log(\beta(-\hbar z+\varepsilon-\mu))\,\frac{\mathrm{e}^{\eta z}}{\mathrm{e}^{\hbar\beta z}-1}. \tag{7.34}$$

We can also add the arc C' at no cost, because its contribution evaluates to zero. However, this time we need to be careful near the positive real axis, because for $z > (\varepsilon-\mu)/\hbar$ the logarithm in the integrand of (7.34) has a branch cut. As a result, we need to integrate around this branch cut, using C'' as a part of the contour which is shown in Fig. 7.2b. Since now the contour $C+C'+C''$ does not contain any poles, application of the residue theorem simply yields

$$\lim_{\eta\downarrow 0}\frac{1}{2\pi\mathrm{i}}\int_{C+C'+C''} \mathrm{d}z\, \log(\beta(-\hbar z+\varepsilon-\mu))\,\frac{\mathrm{e}^{\eta z}}{\mathrm{e}^{\hbar\beta z}-1} = 0. \tag{7.35}$$

This leads to

$$\begin{aligned}\lim_{\eta\downarrow 0}\int_C \frac{\mathrm{d}z}{2\pi\mathrm{i}}\log(\beta(-\hbar z+\varepsilon-\mu))\frac{\mathrm{e}^{\eta z}}{\mathrm{e}^{\hbar\beta z}-1} &= \int_{C''}\frac{\mathrm{d}z}{2\pi\mathrm{i}}\log(\beta(-\hbar z+\varepsilon-\mu))\frac{-1}{\mathrm{e}^{\hbar\beta z}-1}\\ &= \int_{(\varepsilon-\mu)/\hbar}^{\infty}\mathrm{d}z\frac{-1}{\mathrm{e}^{\hbar\beta z}-1}\\ &= \log(1-\mathrm{e}^{-\beta(\varepsilon-\mu)}),\end{aligned} \tag{7.36}$$

where we used that $\log(x+\mathrm{i}\varepsilon)-\log(x-\mathrm{i}\varepsilon) = 2\pi\mathrm{i}$ for $x<0$ and $\varepsilon\downarrow 0$. Comparing (7.34) with (7.36) shows that we have indeed proven (7.30), which leads to the conclusion that the Matsubara expansion gives the same exact solution for the ideal quantum gas as the semiclassical method of the previous paragraph.

Example 7.2. Consider the following grand-canonical partition sum for bosonic particles that have only access to one single-particle quantum state with energy ε

$$Z = \mathrm{Tr}[\mathrm{e}^{-\beta(\varepsilon-\mu)\hat{\psi}^\dagger\hat{\psi}}], \tag{7.37}$$

where $\hat{\psi}^\dagger$ creates a single boson in this single-particle quantum state. Then, the slicing procedure of Sect. 7.1 leads to the following path integral for the partition sum,

$$\begin{aligned}Z(\eta) &= \int \mathrm{d}[\phi^*]\mathrm{d}[\phi]\exp\left\{-\frac{1}{\hbar}\int_0^{\hbar\beta}\mathrm{d}\tau\phi^*(\tau+\eta)\left(\hbar\frac{\partial}{\partial\tau}+\varepsilon-\mu\right)\phi(\tau)\right\}\\ &= \int \mathrm{d}[\phi^*]\mathrm{d}[\phi]\exp\left\{-\frac{1}{\hbar}\sum_n \phi_n^*(-\mathrm{i}\hbar\omega_n+\varepsilon-\mu)\,\mathrm{e}^{\mathrm{i}\eta\omega_n}\phi_n\right\},\end{aligned} \tag{7.38}$$

Note that the slicing procedure from Sect. 7.1 shows that the fields $\phi^*(\tau+\eta)$ and $\phi(\tau)$ come from operators $\hat{\psi}^\dagger$ and $\hat{\psi}$ acting on the imaginary time slices $\Delta\tau(j+1) = \tau+\Delta\tau$ and $\Delta\tau j = \tau$, respectively. Indeed, (7.4) shows that for a normal-ordered Hamiltonian, $\hat{\psi}^\dagger$ acts on slice $j+1$ when $\hat{\psi}$ acts on slice j, which then explains why we should take η to be infinitesimally positive in the limit $\Delta\tau \downarrow 0$.

It is left as an exercise to show that if we take η to be infinitesimally negative, we find for the partition function

$$\lim_{\eta\uparrow 0} Z(\eta) = \left(e^{\beta(\varepsilon-\mu)} - 1\right)^{-1} = e^{-\beta(\varepsilon-\mu)}\left(1 - e^{-\beta(\varepsilon-\mu)}\right)^{-1} = e^{-\beta(\varepsilon-\mu)} \lim_{\eta\downarrow 0} Z(\eta)$$
$$= e^{-\beta(\varepsilon-\mu)}\mathrm{Tr}[e^{-\beta(\varepsilon-\mu)\hat{\psi}^\dagger\hat{\psi}}] = \mathrm{Tr}[e^{-\beta(\varepsilon-\mu)\hat{\psi}\hat{\psi}^\dagger}], \tag{7.39}$$

which shows that there is a discontinuity of $e^{-\beta(\varepsilon-\mu)}$ in the partition sum of (7.38) as a function of η. Moreover, this discontinuity precisely represents the equal-time commutator in the operator formalism. We will see more examples of this feature in the next chapters.

7.2.3 Green's Function Method

The third method turns out to be the most convenient and versatile technique, and it is also the most easily extended to interacting quantum gases. Therefore, this is the method that is actually used most frequently throughout the book. The method is based on explicitly introducing the (inverse) noninteracting Green's function $G^{-1}_{0;\alpha,\alpha'}(\mathbf{x},\tau;\mathbf{x}',\tau')$, such that the partition sum of the noninteracting quantum gas can be written as

$$Z_0 = \int d[\phi^*]d[\phi] \exp\left\{\sum_\alpha \int_0^{\hbar\beta} d\tau \int d\mathbf{x} \right.$$
$$\left. \times \sum_{\alpha'} \int_0^{\hbar\beta} d\tau' \int d\mathbf{x}'\ \phi^*_\alpha(\mathbf{x},\tau) G^{-1}_{0;\alpha,\alpha'}(\mathbf{x},\tau;\mathbf{x}',\tau')\phi_{\alpha'}(\mathbf{x}',\tau')\right\}, \tag{7.40}$$

where the specific function $G^{-1}_{0;\alpha,\alpha'}(\mathbf{x},\tau;\mathbf{x}',\tau')$ leading to this expression still has to be determined. However, note that we may interpret G_0^{-1} as a 'matrix' both in spin space and in coordinate space with matrix elements $G^{-1}_{0;\alpha,\alpha'}(\mathbf{x},\tau;\mathbf{x}',\tau')$. The indices of this matrix are discrete in the spin coordinates, but continuous in the space and imaginary-time coordinates. In analogy with (2.66) and (2.67), valid for discrete matrices, we can then write

$$Z_0 = \exp\left\{\mp\mathrm{Tr}[\log(-G_0^{-1})]\right\} . \tag{7.41}$$

To be able to work practically with the above expression, we need to introduce concepts such as matrix multiplication and taking a trace in coordinate space. For matrix multiplication of the matrices M and M', we have in analogy with ordinary matrix multiplication

$$[M\cdot M']_{\alpha,\alpha''}(\mathbf{x},\tau;\mathbf{x}'',\tau'')$$
$$=\sum_{\alpha'}\int_0^{\hbar\beta}\mathrm{d}\tau'\int\mathrm{d}\mathbf{x}' M_{\alpha,\alpha'}(\mathbf{x},\tau;\mathbf{x}',\tau')M'_{\alpha',\alpha''}(\mathbf{x}',\tau';\mathbf{x}'',\tau''), \tag{7.42}$$

where also the trace has its familiar form of adding diagonal elements, namely

$$\mathrm{Tr}[M]=\sum_{\alpha}\int_0^{\hbar\beta}\mathrm{d}\tau\int\mathrm{d}\mathbf{x}\, M_{\alpha,\alpha}(\mathbf{x},\tau;\mathbf{x},\tau). \tag{7.43}$$

The definition of the inverse of a matrix is given by

$$\delta_{\alpha,\alpha''}\delta(\mathbf{x}-\mathbf{x}'')\delta(\tau-\tau'')$$
$$=\sum_{\alpha'}\int_0^{\hbar\beta}\mathrm{d}\tau'\int\mathrm{d}\mathbf{x}' M_{\alpha,\alpha'}(\mathbf{x},\tau;\mathbf{x}',\tau')M^{-1}_{\alpha',\alpha''}(\mathbf{x}',\tau';\mathbf{x}'',\tau''), \tag{7.44}$$

such that the Dirac delta function is the unity matrix in coordinate space, as follows also from

$$[M\cdot 1]_{\alpha,\alpha''}(\mathbf{x},\tau;\mathbf{x}'',\tau'')$$
$$=\sum_{\alpha'}\int_0^{\hbar\beta}\mathrm{d}\tau'\int\mathrm{d}\mathbf{x}' M_{\alpha,\alpha'}(\mathbf{x},\tau;\mathbf{x}',\tau')\delta_{\alpha',\alpha''}\delta(\mathbf{x}'-\mathbf{x}'')\delta(\tau'-\tau'')$$
$$=M_{\alpha,\alpha''}(\mathbf{x},\tau;\mathbf{x}'',\tau''). \tag{7.45}$$

Finally, we also introduce the following convenient short-hand notation

$$(\phi|M|\phi) \tag{7.46}$$
$$=\sum_{\alpha,\alpha'}\int\mathrm{d}\tau\int\mathrm{d}\mathbf{x}\int\mathrm{d}\tau'\int\mathrm{d}\mathbf{x}'\,\phi^*_{\alpha}(\mathbf{x},\tau)M_{\alpha,\alpha'}(\mathbf{x},\tau;\mathbf{x}',\tau')\phi_{\alpha'}(\mathbf{x}',\tau').$$

Note that the notation $(\phi|M|\phi)$ is basis independent and upon basis transformation this inner product can be evaluated in any other basis. As a result, (7.40) becomes in basis-independent notation

$$Z_0=\int\mathrm{d}[\phi^*]\mathrm{d}[\phi]\,\exp\left\{(\phi|G_0^{-1}|\phi)\right\}=\exp\left\{\mp\mathrm{Tr}[\log(-G_0^{-1})]\right\}. \tag{7.47}$$

Comparing (7.40) with (7.11) and (7.13), we find that the inverse Green's function is given by

$$G^{-1}_{0;\alpha,\alpha'}(\mathbf{x},\tau;\mathbf{x}',\tau') \tag{7.48}$$
$$= -\frac{1}{\hbar}\left\{\hbar\frac{\partial}{\partial\tau} - \frac{\hbar^2\nabla^2}{2m} + V^{\mathrm{ex}}(\mathbf{x}) + \varepsilon_\alpha - \mu\right\}\delta(\mathbf{x}-\mathbf{x}')\delta(\tau-\tau')\delta_{\alpha,\alpha'},$$

which is easily shown by substituting the above expression in (7.40) and integrating and summing over the primed coordinates. Using the definition of the inverse, we have that

$$\delta_{\alpha,\alpha'}\delta(\mathbf{x}-\mathbf{x}'')\delta(\tau-\tau'')$$
$$= \sum_{\alpha'}\int_0^{\hbar\beta} \mathrm{d}\tau' \int \mathrm{d}\mathbf{x}'\, G^{-1}_{0;\alpha,\alpha'}(\mathbf{x},\tau;\mathbf{x}',\tau')G_{0;\alpha',\alpha''}(\mathbf{x}',\tau';\mathbf{x}'',\tau''), \tag{7.49}$$

such that we can multiply (7.48) on both sides with $G_{0;\alpha',\alpha''}(\mathbf{x}',\tau';\mathbf{x}'',\tau'')$, which after integration over the singly-primed coordinates leads to

$$-\hbar\delta(\mathbf{x}-\mathbf{x}')\delta(\tau-\tau')\delta_{\alpha,\alpha'}$$
$$= \left\{\hbar\frac{\partial}{\partial\tau} - \frac{\hbar^2\nabla^2}{2m} + V^{\mathrm{ex}}(\mathbf{x}) + \varepsilon_\alpha - \mu\right\} G_{0;\alpha,\alpha'}(\mathbf{x},\tau;\mathbf{x}',\tau')\,, \tag{7.50}$$

where we renamed the doubly-primed coordinates as singly-primed coordinates again. This shows that $G_{0;\alpha,\alpha'}(\mathbf{x},\tau;\mathbf{x}',\tau')$ is indeed mathematically a Green's function, and its physical meaning will be extensively studied in the rest of the book. Also note that $G_{0;\alpha,\alpha'}(\mathbf{x},\tau;\mathbf{x}',\tau')$ is not an operator, but rather the solution to (7.50), given by

$$G_{0;\alpha,\alpha'}(\mathbf{x},\tau;\mathbf{x}',\tau') = \delta_{\alpha,\alpha'}\sum_{\mathbf{n},n}\frac{-\hbar}{-\mathrm{i}\hbar\omega_n + \varepsilon_{\mathbf{n},\alpha} - \mu}\chi_{\mathbf{n}}(\mathbf{x})\chi^*_{\mathbf{n}}(\mathbf{x}')\frac{\mathrm{e}^{-\mathrm{i}\omega_n(\tau-\tau')}}{\hbar\beta}, \tag{7.51}$$

which can be readily checked by substitution in (7.50).

In general, we can expand any matrix in terms of Matsubara modes and single-particle eigenfunctions using

$$M_{\alpha,\alpha'}(\mathbf{n},i\omega_n;\mathbf{n}',\mathrm{i}\omega_{n'})$$
$$\equiv \int_0^{\hbar\beta}\mathrm{d}\tau\,\mathrm{d}\tau'\int\mathrm{d}\mathbf{x}\,\mathrm{d}\mathbf{x}'\,M_{\alpha,\alpha'}(\mathbf{x},\tau;\mathbf{x}',\tau')\chi^*_{\mathbf{n}}(\mathbf{x})\chi_{\mathbf{n}'}(\mathbf{x}')\frac{\mathrm{e}^{\mathrm{i}\omega_n\tau-\mathrm{i}\omega_{n'}\tau'}}{\hbar\beta}. \tag{7.52}$$

For the noninteracting Green's function of (7.51), this becomes

$$G_{0;\alpha,\alpha'}(\mathbf{n},i\omega_n;\mathbf{n}',\mathrm{i}\omega_{n'}) = \delta_{\mathbf{n},\mathbf{n}'}\delta_{n,n'}\delta_{\alpha,\alpha'}\frac{-\hbar}{-\mathrm{i}\hbar\omega_n + \varepsilon_{\mathbf{n},\alpha} - \mu}\,, \tag{7.53}$$

where we used the orthonormality of the single-particle wavefunctions and the Matsubara modes to perform the integrals over the space and imaginary-time coordinates. It explicitly shows that the noninteracting Green's function is a diagonal matrix in the expansion coefficients, which in a less superfluous notation becomes

$$G_{0;\alpha}(\mathbf{n},\mathrm{i}\omega_n) = \frac{-\hbar}{-\mathrm{i}\hbar\omega_n + \varepsilon_{\mathbf{n},\alpha} - \mu}, \tag{7.54}$$

which means that the inverse noninteracting Green's function is now easily found as

$$G^{-1}_{0;\alpha}(\mathbf{n},\mathrm{i}\omega_n) = \frac{-\mathrm{i}\hbar\omega_n + \varepsilon_{\mathbf{n},\alpha} - \mu}{-\hbar}. \tag{7.55}$$

Example 7.3. As an example, we consider the homogeneous case, for which we have that $\chi_{\mathbf{k}}(\mathbf{x}) = \mathrm{e}^{\mathrm{i}\mathbf{k}\cdot\mathbf{x}}/\sqrt{V}$. Substituting this into (7.51), we find for the noninteracting homogeneous Green's function that

$$\begin{aligned} &G_{0;\alpha,\alpha'}(\mathbf{x},\tau;\mathbf{x}',\tau') \\ &= \delta_{\alpha,\alpha'} \frac{1}{\hbar\beta V} \sum_{\mathbf{k},n} \frac{-\hbar}{-\mathrm{i}\hbar\omega_n + \varepsilon_{\mathbf{k},\alpha} - \mu} \mathrm{e}^{\mathrm{i}\mathbf{k}\cdot(\mathbf{x}-\mathbf{x}')}\mathrm{e}^{-\mathrm{i}\omega_n(\tau-\tau')}. \end{aligned} \tag{7.56}$$

Note that the right-hand side of this equation explicitly shows that the Green's function only depends on the difference in the coordinates $G_{0;\alpha,\alpha'}(\mathbf{x}-\mathbf{x}',\tau-\tau')$. This is the direct consequence of the fact that the homogeneous ideal gas is both translationally invariant in position and imaginary time. The equation also shows that the Fourier transform of the noninteracting Green's function is given in diagonal form by

$$G_{0;\alpha,\alpha'}(\mathbf{k},\mathrm{i}\omega_n) = \frac{-\hbar}{-\mathrm{i}\hbar\omega_n + \varepsilon_{\mathbf{k},\alpha} - \mu}\delta_{\alpha,\alpha'}. \tag{7.57}$$

Going back to (7.41) for the partition sum of the noninteracting quantum gas, we find upon substitution of the inverse Green's function of (7.54), that

$$\begin{aligned} Z_0 &= \exp\left\{ \mp \sum_{\mathbf{n},n,\alpha} \log(\beta(-\mathrm{i}\hbar\omega_n + \varepsilon_{\mathbf{n},\alpha} - \mu)) \right\} \\ &= \exp\left\{ \mp \sum_{\mathbf{n},\alpha} \log(1 \mp \mathrm{e}^{-\beta(\varepsilon_{\mathbf{n},\alpha}-\mu)}) \right\}, \end{aligned} \tag{7.58}$$

where the logarithm in the exponent contains an additional factor of $\hbar\beta$, which comes from the Jacobian of the transformation from imaginary time to the Matsubara frequencies. Although this factor should be included for dimensional reasons, it was actually already seen to be irrelevant for the Matsubara sum in proving (7.30), i.e. in obtaining the final answer of (7.36).

7.3 Wick's Theorem

Next, we elaborate on the physical meaning of the Green's function introduced in the previous section. To this end, we first consider the time-ordered expectation value

$$\begin{aligned}&\langle T[\hat{\psi}_\alpha(\mathbf{x},\tau)\hat{\psi}^\dagger_{\alpha'}(\mathbf{x}',\tau')]\rangle\\&\equiv\theta(\tau-\tau')\langle\hat{\psi}_\alpha(\mathbf{x},\tau)\hat{\psi}^\dagger_{\alpha'}(\mathbf{x}',\tau')\rangle\pm\theta(\tau'-\tau)\langle\hat{\psi}^\dagger_{\alpha'}(\mathbf{x}',\tau')\hat{\psi}_\alpha(\mathbf{x},\tau)\rangle\ ,\end{aligned} \tag{7.59}$$

where $\langle\hat{\psi}_\alpha(\mathbf{x},\tau)\hat{\psi}^\dagger_{\alpha'}(\mathbf{x}',\tau')\rangle$ gives the probability for a single particle that was added to the system at position $\mathbf{x}'$ and imaginary time τ' to be found at a different position $\mathbf{x}$ at another imaginary time τ. Since the removal of a particle can be seen as the creation of a hole, the expectation value $\langle\hat{\psi}^\dagger_{\alpha'}(\mathbf{x}',\tau')\hat{\psi}_\alpha(\mathbf{x},\tau)\rangle$ actually describes the propagation of a hole in the reversed direction. In this section, we prove that the Green's function of the previous section is directly related to the time-ordered expectation value of (7.59), such that this Green's function is also commonly referred to as the one-particle propagator or correlation function. Note that although (7.59) is formulated in imaginary time, we can also retrieve real-time information by performing a Wick rotation. Furthermore, since the expectation value is determined in the grand-canonical ensemble, we have that the imaginary time Heisenberg operator $\hat{\psi}_\alpha(\mathbf{x},\tau)$ is defined through

$$\hat{\psi}_\alpha(\mathbf{x},\tau)=\mathrm{e}^{(\hat{H}-\mu\hat{N})\tau/\hbar}\hat{\psi}_\alpha(\mathbf{x})\mathrm{e}^{-(\hat{H}-\mu\hat{N})\tau/\hbar} \tag{7.60}$$

and therefore obeys the Heisenberg equation of motion

$$\hbar\partial_\tau\hat{\psi}_\alpha(\mathbf{x},\tau)=[\hat{H}-\mu\hat{N},\hat{\psi}_\alpha(\mathbf{x},\tau)]_-. \tag{7.61}$$

For the noninteracting case it reads

$$\hbar\frac{\partial}{\partial\tau}\hat{\psi}_\alpha(\mathbf{x},\tau)=\left\{\frac{\hbar^2\boldsymbol{\nabla}^2}{2m}-V^{\mathrm{ex}}(\mathbf{x})-\varepsilon_\alpha+\mu\right\}\hat{\psi}_\alpha(\mathbf{x},\tau)\ , \tag{7.62}$$

which in combination with (7.59) leads to

$$\begin{aligned}&\hbar\frac{\partial}{\partial\tau}\langle T[\hat{\psi}_\alpha(\mathbf{x},\tau)\hat{\psi}^\dagger_{\alpha'}(\mathbf{x}',\tau')]\rangle_0=\hbar\delta(\tau-\tau')\langle[\hat{\psi}_\alpha(\mathbf{x},\tau),\hat{\psi}^\dagger_{\alpha'}(\mathbf{x}',\tau')]_\mp\rangle_0\\&+\left\{\frac{\hbar^2\boldsymbol{\nabla}^2}{2m}-V^{\mathrm{ex}}(\mathbf{x})-\varepsilon_\alpha+\mu\right\}\langle T[\hat{\psi}_\alpha(\mathbf{x},\tau)\hat{\psi}^\dagger_{\alpha'}(\mathbf{x}',\tau')]\rangle_0\ .\end{aligned} \tag{7.63}$$

Substituting the equal-time (anti)commutation relations of (6.28), we find

$$[\hat{\psi}_\alpha(\mathbf{x},\tau),\hat{\psi}^\dagger_{\alpha'}(\mathbf{x}',\tau)]_\mp=\delta(\mathbf{x}-\mathbf{x}')\delta_{\alpha,\alpha'}, \tag{7.64}$$

which suggests that

$$G_{0;\alpha,\alpha'}(\mathbf{x},\tau;\mathbf{x}',\tau') = -\langle T[\hat{\psi}_\alpha(\mathbf{x},\tau)\hat{\psi}^\dagger_{\alpha'}(\mathbf{x}',\tau')]\rangle_0 \,. \tag{7.65}$$

We can provide a more careful proof of this important relation, which bridges the gap between the functional formulation of quantum field theory and the more familiar operator formalism. First of all, from the slicing procedure used to derive the functional-integral formalism in Sect. 7, it follows that with path integrals we always automatically calculate time-ordered expectation values. This was actually more carefully explained in Sect. 5.5. Therefore, we have that

$$\begin{aligned}\langle T[\hat{\psi}_\alpha(\mathbf{x},\tau)\hat{\psi}^\dagger_{\alpha'}(\mathbf{x}',\tau')]\rangle_0 &= \frac{1}{Z_0}\int d[\phi^*]d[\phi]\;\phi_\alpha(\mathbf{x},\tau)\phi^*_{\alpha'}(\mathbf{x}',\tau')e^{-S_0[\phi^*,\phi]/\hbar} \\ &= \langle \phi_\alpha(\mathbf{x},\tau)\phi^*_{\alpha'}(\mathbf{x}',\tau')\rangle_0 \,. \end{aligned} \tag{7.66}$$

As also explained in Sect. 5.5, it is most convenient to calculate expectation values in the path-integral approach by adding external currents $J_\alpha(\mathbf{x},\tau)$ and $J^*_\alpha(\mathbf{x},\tau)$ to the partition function, where in the fermionic case these currents are Grassmann variables. Then, the partition function generalizes to a generating functional, given by

$$\begin{aligned} Z_0[J,J^*] = \int d[\phi^*]d[\phi]\;\exp\Bigg\{ &-\frac{1}{\hbar}S_0[\phi^*,\phi] \\ &+\sum_\alpha \int_0^{\hbar\beta} d\tau \int d\mathbf{x}\,\left(\phi^*_\alpha(\mathbf{x},\tau)J_\alpha(\mathbf{x},\tau)+J^*_\alpha(\mathbf{x},\tau)\phi_\alpha(\mathbf{x},\tau)\right)\Bigg\} \,. \end{aligned} \tag{7.67}$$

Since the right-hand side of (7.65), as given by (7.66), can be conveniently expressed as a second derivative of the generating functional to the currents, we have that proving (7.65) actually boils down to showing that

$$-G_{0;\alpha,\alpha'}(\mathbf{x},\tau;\mathbf{x}',\tau') = \frac{\pm 1}{Z_0}\left.\frac{\delta^2 Z_0[J,J^*]}{\delta J^*_\alpha(\mathbf{x},\tau)\delta J_{\alpha'}(\mathbf{x}',\tau')}\right|_{J,J^*=0} \,. \tag{7.68}$$

Using the short-hand notation introduced in (7.46), we have

$$Z_0[J,J^*] = \int d[\phi^*]d[\phi]\;\exp\{(\phi|G_0^{-1}|\phi)+(\phi|J)+(J|\phi)\} \,. \tag{7.69}$$

The terms in the exponent can be rewritten as $(\phi+G_0J|G_0^{-1}|\phi+G_0J)-(J|G_0|J)$, which is usually called completing the square. Note that here the short-hand notation G_0J actually corresponds to

$$[G_0\cdot J]_\alpha(\mathbf{x},\tau) = \sum_{\alpha'}\int d\mathbf{x}'\,d\tau'\;G_{0;\alpha,\alpha'}(\mathbf{x},\tau;\mathbf{x}',\tau')J_{\alpha'}(\mathbf{x}',\tau'). \tag{7.70}$$

Performing a shift in the integration variables $\phi+G_0J\rightarrow\phi$ then leads to

$$Z_0[J,J^*] = Z_0[0,0]\mathrm{e}^{-(J|G_0|J)} = Z_0[0,0]\exp\left\{-\sum_\alpha \int_0^{\hbar\beta} \mathrm{d}\tau \int \mathrm{d}\mathbf{x} \right. \tag{7.71}$$
$$\left. \times \sum_{\alpha'} \int_0^{\hbar\beta} \mathrm{d}\tau' \int \mathrm{d}\mathbf{x}' \; J^*_\alpha(\mathbf{x},\tau) G_{0;\alpha,\alpha'}(\mathbf{x},\tau;\mathbf{x}',\tau') J_{\alpha'}(\mathbf{x}',\tau') \right\} ,$$

which indeed proves (7.68) after performing the corresponding functional differentiation.

The generating functional $Z_0[J,J^*]$ is in fact very convenient for calculating the expectation value of the time-ordered product of any number of operators. With the expressions for $Z_0[J,J^*]$ from (7.67) and (7.71), we can easily show that this results in the sum of all possible products of time-ordered expectation values of two operators, which is better known as Wick's theorem. For instance, we have that

$$\begin{aligned} &\langle \phi^*_\alpha(\mathbf{x},\tau)\phi^*_{\alpha'}(\mathbf{x}',\tau')\phi_{\alpha''}(\mathbf{x}'',\tau'')\phi_{\alpha'''}(\mathbf{x}''',\tau''')\rangle_0 \\ &= \langle \phi^*_\alpha(\mathbf{x},\tau)\phi_{\alpha'''}(\mathbf{x}''',\tau''')\rangle_0 \langle \phi^*_{\alpha'}(\mathbf{x}',\tau')\phi_{\alpha''}(\mathbf{x}'',\tau'')\rangle_0 \\ &\pm \langle \phi^*_\alpha(\mathbf{x},\tau)\phi_{\alpha''}(\mathbf{x}'',\tau'')\rangle_0 \langle \phi^*_{\alpha'}(\mathbf{x}',\tau')\phi_{\alpha'''}(\mathbf{x}''',\tau''')\rangle_0 , \end{aligned} \tag{7.72}$$

where we obtained the left-hand side of the above equation by differentiating (7.67) four times with respect to the appropriate currents, while we obtained the right-hand side by differentiating (7.71) with respect to the same currents. Wick's theorem plays a crucial role in the next section, where we start with the treatment of interaction effects and find that due to this theorem we can systematically set up a perturbation theory to take such effects into account.

Example 7.4. Note that for the ideal quantum gas, we immediately find

$$\langle \phi_\alpha(\mathbf{x},\tau)\phi_{\alpha'}(\mathbf{x}',\tau')\rangle_0 = \frac{1}{Z_0} \left. \frac{\delta^2 Z_0[J,J^*]}{\delta J^*_\alpha(\mathbf{x},\tau)\delta J^*_{\alpha'}(\mathbf{x}',\tau')} \right|_{J,J^*=0} = 0, \tag{7.73}$$

where in the first step we used (7.67), while in the second step we used (7.71). In later chapters, we see examples of situations where these anomalous expectation values are not equal to zero. This is then found to be intimately connected to phenomena as Bose-Einstein condensation and superfluidity. For the ideal gas, the expectation value for any odd number of operators evaluates to zero. In the case of three operators, we have for example that

$$\begin{aligned} \langle \phi_\alpha(\mathbf{x},\tau)\phi^*_{\alpha'}(\mathbf{x}',\tau')\phi_{\alpha''}(\mathbf{x}'',\tau'')\rangle_0 &= \frac{\pm 1}{Z_0} \left. \frac{\delta^2 Z_0[J,J^*]}{\delta J^*_\alpha(\mathbf{x},\tau)\delta J_{\alpha'}(\mathbf{x}',\tau')\delta J^*_{\alpha''}(\mathbf{x}'',\tau'')} \right|_{J,J^*=0} \\ &= 0, \end{aligned} \tag{7.74}$$

where again in the first step we used (7.67), while in the second step we used (7.71).

7.4 Problems

Exercise 7.1. Ideal-Gas Partition Function
In this exercise, we solve the ideal quantum gas in a fourth exact manner, namely by applying the Gaussian integral formulas from (2.66) and (2.67) directly to the expression of the partition sum following from (7.2) and (7.8), which is indeed quadratic in the time-sliced fields $\phi_{j,\alpha}(\mathbf{x})$.
(a) First, expand the fields $\phi_{j,\alpha}(\mathbf{x})$ in terms the eigenstates $\chi_{\mathbf{n}}(\mathbf{x})$ of the single-particle Hamiltonian to show that the partition function can be written as

$$Z = \int \prod_{j=1,}^{M} \prod_{\mathbf{n},\alpha} \left(\frac{\mathrm{d}\phi^*_{j,\mathbf{n},\alpha} \mathrm{d}\phi_{j,\mathbf{n},\alpha}}{(2\pi\mathrm{i})^{(1\pm1)/2}} \right) \exp\left\{ \sum_{j=1}^{M} \sum_{\mathbf{n},\alpha} \left(\phi^*_{j,\mathbf{n},\alpha} \phi_{j,\mathbf{n},\alpha} \right.\right.$$
$$\left.\left. + \left[1 - (\varepsilon_{\mathbf{n},\alpha} - \mu) \frac{\Delta\tau}{\hbar} \right] \phi^*_{j,\mathbf{n},\alpha} \phi_{j-1,\mathbf{n},\alpha} \right) \right\} \equiv \int \mathrm{d}[\phi^*] \mathrm{d}[\phi] \mathrm{e}^{-(\phi|A|\phi)}. \quad (7.75)$$

(b) Realizing that the matrix $A_{\mathbf{n},\alpha,j,j'}$ is diagonal in $\mathbf{n}$ and α, give its explicit form in terms of j and j' for a specific value of $\mathbf{n}$ and α. Show that the determinant of this matrix is given by

$$\mathrm{Det}[A_{\mathbf{n},\alpha,j,j'}] = 1 \mp (1 - (\varepsilon_{\mathbf{n},\alpha} - \mu)\beta/M)^M, \quad (7.76)$$

Hint: for the correct form of $A_{\mathbf{n},\alpha,j,j'}$, the (anti)periodic boundary relation between the fields at $j = 0$ and $j = M$ is crucial.
(c) Using the result from (b) and taking the limit $M \to \infty$, show that the partition function for the ideal quantum gas is indeed given by (7.18).
Hint: use $\lim_{M\to\infty}(1 - x/M)^M = \mathrm{e}^{-x}$.

Exercise 7.2. Matsubara Summation and Contour Integration
(a) Show by contour integration, in a similar way as for (7.31), that

$$\lim_{\eta \downarrow 0} \frac{1}{\hbar\beta} \sum_n \frac{\mathrm{e}^{\mathrm{i}\omega_n \eta}}{-\mathrm{i}\omega_n - (\varepsilon - \mu)/\hbar} = \mp \frac{1}{\mathrm{e}^{\beta(\varepsilon-\mu)} \mp 1} - 1\,. \quad (7.77)$$

(b) Show by contour integration, in a similar way as for (7.30), that for bosons we have

$$\lim_{\eta \downarrow 0} \sum_n \log(\beta(-\mathrm{i}\hbar\omega_n + \varepsilon - \mu)) \mathrm{e}^{-\mathrm{i}\omega_n \eta} = \log(\mathrm{e}^{\beta(\varepsilon-\mu)} - 1)\,. \quad (7.78)$$

(c) Show by contour integration that for fermions we have

$$\lim_{\eta \downarrow 0} \sum_n \log(\beta(-\mathrm{i}\hbar\omega_n + \varepsilon - \mu)) \mathrm{e}^{\mathrm{i}\omega_n \eta} = \log(1 + \mathrm{e}^{-\beta(\varepsilon-\mu)})\,, \quad (7.79)$$

where we note that for the ideal Fermi gas we can also have $\mu > 0$, whereas for ideal bosons we only have $\mu \leq 0$

Exercise 7.3. Prove (7.72) explicitly by an appropriate differentiation of $Z_0[J, J^*]$.

Chapter 8
Interactions and Feynman Diagrams

The bottom line for mathematicians is that the architecture has to be right. In all the mathematics that I did, the essential point was to find the right architecture. It's like building a bridge. Once the main lines of the structure are right, then the details miraculously fit. The problem is the overall design.
– Freeman Dyson.

In the previous chapter, we introduced the functional-integral formalism of quantum field theory for the treatment of many-body systems. In particular, we calculated the partition sum for the ideal quantum gas in three different ways. The last method involved the noninteracting Green's function, which was seen to form the bridge between the more familiar operator formalism and the newly obtained functional-integral formalism. In this chapter we extend the notion of the Green's function to interacting systems, starting with the Lehmann representation of the interacting Green's function. This representation is exact and shows that in general the poles of the interacting Green's function correspond to the single-particle excitations of the many-body system, which are also called the quasiparticle excitations. Realizing that the interacting Green's function gives us both the elementary excitations of the many-body system, as well as the expectation value of the one-particle observables, the question arises how to determine this important quantity in practice.

Since it is in general not possible to determine the interacting Green's function exactly, we need to develop approximate methods to take interaction effects into account. A systematic way to do so is by performing a perturbation theory in powers of the interaction, which is the many-body analogue of the perturbation theory known from quantum mechanics. The rather cumbersome expressions resulting from this expansion are then elegantly represented in terms of Feynman diagrams, which make it possible to see the general structure of the expansion. To lowest order in the interaction, the noninteracting Green's function is then found to be modified by the Hartree and Fock diagrams which consequently can be used to construct a self-consistent Hartree-Fock theory. The name of the resulting theory comes from the full analogy with the nonperturbative Hartree-Fock theory for many electrons in an atom. This theory is then not only studied diagrammatically, but also variationally and finally with the Hubbard-Stratonovich transformation. The latter is the technique most frequently used in the following chapters when we are interested in phase transitions occurring in interacting quantum gases.

8.1 Quasiparticles

In a homogeneous noninteracting quantum gas, the energies of the single-particle eigenstates for a particle with mass m are given by

$$\hbar\omega_{\mathbf{k}} = \varepsilon_{\mathbf{k}} - \mu = \frac{\hbar^2\mathbf{k}^2}{2m} - \mu, \tag{8.1}$$

where the chemical potential μ implies that we work in the grand-canonical ensemble. This relation is also called the dispersion or the energy-momentum relation for the particles. Consider, as an example, the ground state for an ideal Fermi gas, which has all single-particle states filled up to the chemical potential. We note that this ground state can be excited by either adding a particle outside the Fermi sea, or by removing a particle from the Fermi sea, i.e. by creating a hole in the Fermi sea. These kinds of excitation are therefore appropriately called the single-particle excitations of the many-body system. In the case of the ideal Bose gas, we can of course also make excitations by either adding or removing particles. If the quantum gas becomes interacting then the particles are not separate objects anymore, but also feel their surrounding medium. As a result, the interactions between the particles change the dispersion of (8.1), where the physical reason for this change is that a particle travelling through the gas now has to move other particles out of the way or temporarily drag neighboring particles along. As it turns out, the composite object of the particle and its surrounding cloud often behaves very much like a particle again, such that it is then suggestively called a quasiparticle. The resulting single-particle excitation spectrum in the presence of the interacting medium is therefore also called the quasiparticle excitation spectrum. Typically, after the dressing with the surrounding cloud has been taken into account, the quasiparticles interact only weakly with each other such that it takes a long time before their momentum is changed by interactions with other quasiparticles. To study these quasiparticles and their dispersion experimentally, we can perform tunnelling experiments that either add or remove a single particle from the system. A famous example of a theory where the quasiparticle picture plays an important role is Landau's Fermi liquid theory for liquid helium-3 [38], which explains why this strongly-interacting quantum liquid shares so many similarities with the noninteracting Fermi gas. In the same way, Fermi liquid theory is an underlying reason why the model of a free electron gas works so surprisingly well in describing the properties of metals.

To give the quasiparticle picture a solid foundation we consider the interacting Green's function, which is the extension to the interacting case of the noninteracting Green's function that we studied extensively in the previous chapter. The Green's function, or one-particle propagator $G_{\alpha,\alpha'}(\mathbf{x},\tau;\mathbf{x}',\tau')$, is a very important quantity to determine theoretically because, as we will see, it gives us both the quasiparticle excitation spectrum for our system of interest and the expectation value of any one-particle observable. As an example of the last, we see that the average density of particles in the spin state $|\alpha\rangle$ for an ideal gas is given by

$$\langle \hat{\psi}^\dagger_\alpha(\mathbf{x},\tau)\hat{\psi}_\alpha(\mathbf{x},\tau)\rangle_0 = \mp G_{0;\alpha,\alpha}(\mathbf{x},\tau;\mathbf{x},\tau^+) \tag{8.2}$$
$$= \lim_{\eta\downarrow 0}\sum_{\mathbf{n},n} \frac{\pm e^{i\omega_n\eta}}{\beta(-i\hbar\omega_n+\varepsilon_{\mathbf{n},\alpha}-\mu)}|\chi_\mathbf{n}(\mathbf{x})|^2 = \sum_\mathbf{n}\frac{1}{e^{\beta(\varepsilon_{\mathbf{n},\alpha}-\mu)}\mp 1}|\chi_\mathbf{n}(\mathbf{x})|^2 ,$$

where we used (7.31), (7.51) and (7.65). Moreover, we used the notation τ^+ for the limit $\eta \downarrow 0$ of $\tau+\eta$. From the time-ordered definition of the Green's function in (7.65), we see that this is the correct limiting procedure to get the operators in the correct order for calculating the average density. This type of ordering, where the creation operators are placed to the left of the annihilation operators, is also known as normal ordering. Note that all the second-quantized operators that were introduced in Sect. 6.3 are normal ordered.

From the argument of the Bose or Fermi distribution function in (8.2), we see that the single-particle or elementary excitations have an energy of $\varepsilon_{\mathbf{n},\alpha}-\mu$. Considering the Fourier transform of the Green's function to the Matsubara frequencies,

$$G_{\alpha,\alpha'}(\mathbf{x},\mathbf{x}';i\omega_n) = \delta_{\alpha,\alpha'}\sum_\mathbf{n}\frac{-\hbar}{-i\hbar\omega_n+\varepsilon_{\mathbf{n},\alpha}-\mu}\chi_\mathbf{n}(\mathbf{x})\chi^*_\mathbf{n}(\mathbf{x}') , \tag{8.3}$$

and replacing $i\omega_n$ by ω, we see that $G_{\alpha,\alpha'}(\mathbf{x},\mathbf{x}';\omega)$ has poles at precisely $\hbar\omega = \varepsilon_{\mathbf{n},\alpha}-\mu$. As we see in the next subsection, this important result turns out to be very general, such that the poles in the (non)interacting Green's function $G_{\alpha,\alpha'}(\mathbf{x},\mathbf{x}';\omega)$ correspond to the energies of the elementary or quasiparticle excitations of the (non)interacting many-body system. Moreover, these quasiparticle excitation energies can in general also have a negative imaginary component, which then results in a finite lifetime of the excitation.

8.1.1 The Lehmann Representation

Next, we study in detail the analytic structure of the one-body correlation function for an interacting many-body system, i.e. the interacting Green's function, and show in more detail how it is related to the quasiparticle spectrum. It is instructive to consider the expansion coefficients $G_{\alpha,\alpha'}(\mathbf{n},\tau;\mathbf{n}',\tau')$ of the interacting Green's function $G_{\alpha,\alpha'}(\mathbf{x},\tau;\mathbf{x}',\tau')$ with respect to the complete set of single-particle eigenstates $\chi_\mathbf{n}(\mathbf{x})$, such that we obtain

$$G_{\alpha,\alpha'}(\mathbf{x},\tau;\mathbf{x}',\tau') = \sum_{\mathbf{n},\mathbf{n}'}\chi_\mathbf{n}(\mathbf{x})\chi^*_{\mathbf{n}'}(\mathbf{x}')G_{\alpha,\alpha'}(\mathbf{n},\tau;\mathbf{n}',\tau'), \tag{8.4}$$

where by definition

$$G_{\alpha,\alpha'}(\mathbf{n},\tau;\mathbf{n}',\tau') \equiv -\langle T[\hat{\psi}_{\mathbf{n},\alpha}(\tau)\hat{\psi}^\dagger_{\mathbf{n}',\alpha'}(\tau')]\rangle . \tag{8.5}$$

To understand the physical content of the above expansion coefficients, we consider a system with on average $N \gg 1$ particles described by the grand-canonical Hamiltonian $\hat{H} - \mu\hat{N}$. We write $|0;N\rangle$ for the ground state of this system, such that at zero temperature, we have

$$G_{\alpha,\alpha'}(\mathbf{n},\tau;\mathbf{n}',\tau') = -\theta(\tau-\tau')\langle 0;N|\hat{\psi}_{\mathbf{n},\alpha}(\tau)\hat{\psi}^{\dagger}_{\mathbf{n}',\alpha'}(\tau')|0;N\rangle \\ \mp\theta(\tau'-\tau)\langle 0;N|\hat{\psi}^{\dagger}_{\mathbf{n}',\alpha'}(\tau')\hat{\psi}_{\mathbf{n},\alpha}(\tau)|0;N\rangle. \quad (8.6)$$

Denoting the full set of eigenstates and corresponding eigenenergies of the N-particle Hamiltonian $\hat{H}$ by $\{|\nu;N\rangle\}$ and $\{E^N_\nu\}$, we insert

$$\sum_\nu |\nu;N\pm 1\rangle\langle\nu;N\pm 1| = \hat{1} \quad (8.7)$$

into (8.6) and obtain

$$G_{\alpha,\alpha'}(\mathbf{n},\tau;\mathbf{n}',\tau') = \\ -\theta(\tau-\tau')\sum_{\nu''}\langle 0;N|\hat{\psi}_{\mathbf{n},\alpha}(\tau)|\nu'';N+1\rangle\langle\nu'';N+1|\hat{\psi}^{\dagger}_{\mathbf{n}',\alpha'}(\tau')|0;N\rangle\ . \\ \mp\theta(\tau'-\tau)\sum_{\nu''}\langle 0;N|\hat{\psi}^{\dagger}_{\mathbf{n}',\alpha'}(\tau')|\nu'';N-1\rangle\langle\nu'';N-1|\hat{\psi}_{\mathbf{n},\alpha}(\tau)|0;N\rangle. \quad (8.8)$$

Explicitly substituting the imaginary-time dependence of the creation and annihilation operators as given by (7.60), we obtain

$$G_{\alpha,\alpha'}(\mathbf{n},\tau;\mathbf{n}',\tau') = -\theta(\tau-\tau')\sum_{\nu''}\langle 0;N|\hat{\psi}_{\mathbf{n},\alpha}|\nu'';N+1\rangle\langle\nu'',N+1|\hat{\psi}^{\dagger}_{\mathbf{n}',\alpha'}|0;N\rangle \\ \times\exp\{-(E^{N+1}_{\nu''}-E^N_0-\mu)(\tau-\tau')/\hbar\} \\ \mp\theta(\tau'-\tau)\sum_{\nu''}\langle 0;N|\hat{\psi}^{\dagger}_{\mathbf{n}',\alpha'}|\nu'';N-1\rangle\langle\nu'';N-1|\hat{\psi}_{\mathbf{n},\alpha}|0;N\rangle \\ \times\exp\{-(E^{N-1}_{\nu''}-E^N_0+\mu)(\tau'-\tau)/\hbar\}. \quad (8.9)$$

Note that the matrix element $\langle\nu'';N+1|\hat{\psi}^{\dagger}_{\mathbf{n}',\alpha'}|0;N\rangle$ gives the overlap of the exact $N+1$ particle state $|\nu'';N+1\rangle$ and the N particle ground state $|0;N\rangle$ to which we add one single atom with quantum numbers $\mathbf{n}'$ and α'. Similarly, the matrix element $\langle\nu'';N-1|\hat{\psi}_{\mathbf{n},\alpha}|0;N\rangle$ gives the overlap of the exact $N-1$ particle state $|\nu'';N-1\rangle$ and the N particle ground state $|0;N\rangle$ from which we removed one single atom with quantum numbers $\mathbf{n}$ and α. To make the physical meaning of the difference in energies $E^{N+1}_{\nu''}-E^N_0$ more clear, we add and subtract from it the ground-state energy for the $N+1$ particle system, i.e.

$$E^{N+1}_{\nu''}-E^N_0 = (E^{N+1}_{\nu''}-E^{N+1}_0)+(E^{N+1}_0-E^N_0). \quad (8.10)$$

The quantities $E^{N+1}_{\nu''}-E^{N+1}_0 = \hbar\omega_{\nu''}\geq 0$ give the energies of all the possible excitation energies above the many-body ground-state, where the ground state contributions themselves give zero. The difference $E^{N+1}_0-E^N_0=\mu$ is the chemical potential because it tells us what the energy cost is to add a particle to the system.. Similarly, the difference $E_0^{N-1} - E_0^N = -\mu$ tells us what the energy gain is to remove a particle from the ground state of the system. Note that we have implicitly used $N \gg 1$, i.e. we consider the thermodynamic limit, such that the magnitude of the latter two quantities are equal and the excitation energies do not depend on N.

As a result, we can now determine the Fourier transform of (8.9), which boils down to Fourier transforming exponentially decaying functions for all excited states and the step function for the ground state contributions. Explicitly, we find for the Fourier transform with respect to the imaginary time difference $\tau - \tau'$, where we may set τ' to zero, that

$$\begin{aligned} G_{\alpha,\alpha'}(\mathbf{n},\mathbf{n}';\mathrm{i}\omega_n) &= \int_{-\infty}^{\infty} \mathrm{d}\tau \, \exp(\mathrm{i}\omega_n\tau) G_{\alpha,\alpha'}(\mathbf{n},\tau;\mathbf{n}',0) \\ &= \sum_{\nu''} \frac{\langle 0;N|\hat{\psi}_{\mathbf{n},\alpha}|\nu'';N+1\rangle\langle\nu'',N+1|\hat{\psi}^\dagger_{\mathbf{n}',\alpha'}|0;N\rangle}{\mathrm{i}\omega_n^+ - \omega_{\nu''}} \\ &\mp \sum_{\nu''} \frac{\langle 0;N|\hat{\psi}^\dagger_{\mathbf{n}',\alpha'}|\nu'';N-1\rangle\langle\nu'';N-1|\hat{\psi}_{\mathbf{n},\alpha}|0;N\rangle}{\mathrm{i}\omega_n^- + \omega_{\nu''}}, \end{aligned} \tag{8.11}$$

where we introduced the convention $\omega_n^\pm = \lim_{\eta\downarrow 0}(\omega_n \pm \mathrm{i}\eta)$. Note that the Fourier transform of $1/\mathrm{i}\omega_n^\pm$ can be used as a definition for the step function $\mp\theta(\pm\tau)$. Also note that because we work at zero temperature, the Matsubara frequencies $\mathrm{i}\omega_n$ have become continuous variables. The representation of the Green's function from (8.11) is known as the Lehmann representation and it brings out the analytic structure as a function of $\mathrm{i}\omega_n$ explicitly. Note that the above calculation is exact and that it is thus valid for any interacting many-body system at zero temperature. The final form of (8.11) indeed shows that in general the poles of $G_{\alpha,\alpha'}(\mathbf{n},\mathbf{n}';\omega)$, corresponds to the exact elementary or single-particle excitations of the many-body system. Although we have considered here for calculational convenience only the zero-temperature case, the same calculation can also be performed at nonzero temperatures. Then, we should take the trace over the density matrix in (8.6), which turns out to result in additional Boltzmann factors but does not change the analytic structure of the Lehmann representation. As a result, we may conclude that the poles of the exact interacting Green's function always correspond to the exact quasiparticle excitations.

Example 8.1. We have seen in the previous chapter (see Example 7.3) that the Fourier transform of the homogeneous noninteracting Green's function is given by

$$G_{0;\alpha,\alpha'}(\mathbf{k},\mathrm{i}\omega_n) = \frac{-\hbar}{-\mathrm{i}\hbar\omega_n + \varepsilon_{\mathbf{k},\alpha} - \mu}\delta_{\alpha,\alpha'}. \tag{8.12}$$

After the analytic continuation, i.e. $\mathrm{i}\omega_n \to \omega$, we see that this Green's function has a pole when the energy $\hbar\omega$ is precisely equal to $\varepsilon_{\mathbf{k},\alpha} - \mu$. The same expression for the noninteracting Green's function also follows from (8.11), which we show here explicitly for the Fermi case. To this end, we note that the excitation energies indeed become $\varepsilon_{\mathbf{k},\alpha} - \mu$, while the matrix elements yield

$$\langle \mathbf{k}'', \alpha'', N+1 | \hat{\psi}^{\dagger}_{\mathbf{k}',\alpha'} | 0; N \rangle = \delta_{\alpha',\alpha''} \delta_{\mathbf{k}',\mathbf{k}''} \theta(\varepsilon_{\mathbf{k}',\alpha'} - \mu) \tag{8.13}$$

and

$$\langle \mathbf{k}'', \alpha''; N-1 | \hat{\psi}_{\mathbf{k}',\alpha'} | 0; N \rangle = \delta_{\alpha',\alpha''} \delta_{\mathbf{k}',\mathbf{k}''} \theta(\mu - \varepsilon_{\mathbf{k}',\alpha'}). \tag{8.14}$$

It is then left as an exercise to show that the Lehmann representation also leads to the correct noninteracting Green's function in the Bose case.

8.1.2 The Spectral-Weight Function

In the homogeneous case, we can extend the discussion of the Lehmann representation by exploiting the diagonality of the Green's function in the momentum $\hbar\mathbf{k}$. In the previous subsection, we found that the interacting Green's function $G_{\alpha,\alpha'}(\mathbf{k},\omega)$ in general only has poles on the real frequency axis. To be able to deal with these poles analytically, we add a infinitesimally small imaginary part to the frequency, such that the Green's function now has a nonzero imaginary part. As an example, we then obtain from (7.57) for the noninteracting case

$$\mathrm{Im}[G_{0;\alpha,\alpha'}(\mathbf{k},\omega^+)] = -\hbar\pi\delta(\hbar\omega - \varepsilon_{\mathbf{k},\alpha} + \mu)\delta_{\alpha,\alpha'}, \tag{8.15}$$

where $G_{0;\alpha,\alpha'}(\mathbf{k},\omega^+)$ is called the retarded noninteracting Green's function. Note that in obtaining (8.15), we have actually performed the analytic continuation $\mathrm{i}\omega_n \to \omega^+$, which is mathematically allowed because (8.11) shows that the Green's function is analytic on the entire upper half of the complex plane plane. If we study (8.15) as a function of $\hbar\omega$, we thus find an infinitely sharp line at the energy $\varepsilon_{\mathbf{k},\alpha} - \mu$. In spectroscopy such a line is known to correspond to a state with an infinite lifetime, because any decay mechanism yields a broadening of the line. This is understood from Heisenberg uncertainty relation, where the longer the lifetime $\tau_{\mathbf{k},\alpha}$ of the state, the more accurately we can determine its energy according to $\Delta E \geq \hbar/2\tau_{\mathbf{k},\alpha}$. Note that the infinite lifetime of the single-particle state $|\mathbf{k},\alpha\rangle$ is expected for the noninteracting quantum gas, because there is no mechanism that can remove a particle from its state. These observations lead naturally to the general definition of the spectral-weight function, which is given by

$$\rho_\alpha(\mathbf{k},\omega) = -\frac{1}{\pi\hbar}\mathrm{Im}[G_{\alpha,\alpha}(\mathbf{k},\omega^+)]. \tag{8.16}$$

The spectral function $\rho_\alpha(\mathbf{k},\omega)$ gives the energies, the amplitudes and the lifetimes of the states accessible to a particle with momentum $\mathbf{k}$ and spin $|\alpha\rangle$ in the presence of a medium, such that the spectral function can also be interpreted as a single-particle density of states.

Applying the above definition of the spectral-weight function to the Green's function in the Lehmann representation of (8.11), we find that

$$\begin{aligned}\rho_\alpha(\mathbf{k},\omega) &= -\frac{1}{\pi\hbar}\mathrm{Im}[G_{\alpha,\alpha}(\mathbf{k},\omega^+)] = \sum_{\nu''}\left|\langle\nu'';N+1|\hat{\psi}^\dagger_{\mathbf{k},\alpha}|0;N\rangle\right|^2\delta\left(\hbar\omega-\hbar\omega_{\nu''}\right)\\ &\mp\sum_{\nu''}\left|\langle\nu'';N-1|\hat{\psi}_{\mathbf{k},\alpha}|0;N\rangle\right|^2\delta\left(\hbar\omega+\hbar\omega_{\nu''}\right),\end{aligned} \tag{8.17}$$

where we note that only states with $|\nu'',N\pm1\rangle = |\lambda'',\mathbf{k},\alpha,N\pm1\rangle$ give rise to a nonzero overlap in the above expression, because $\mathbf{k}$ and α are good quantum numbers for the homogeneous quantum gas. Using the above result and the completeness of the exact eigenstates $|\nu'',N\pm1\rangle$ of the many-body Hamiltonian, we find the following important identity that the spectral-weight function satisfies

$$\int \mathrm{d}(\hbar\omega)\;\rho_\alpha(\mathbf{k},\omega) = \langle 0;N|\hat{\psi}_{\mathbf{k},\alpha}\hat{\psi}^\dagger_{\mathbf{k},\alpha}|0;N\rangle \mp \langle 0;N|\hat{\psi}^\dagger_{\mathbf{k},\alpha}\hat{\psi}_{\mathbf{k},\alpha}|0;N\rangle = 1, \tag{8.18}$$

which is also known as the frequency sum-rule. Using the interpretation of the spectral function as a single-particle density of states in the presence of a medium, it simply means that one particle occupies precisely the equivalent of one quantum state.

In obtaining the Lehmann representation of the exact interacting Green's function in the previous subsection, we made use of the exact eigenstates and eigenenergies of the many-body Hamiltonian. This made the discussion rather formal, because in general it turns out to be impossible to determine the exact eigenstates and eigenenergies when dealing with an interacting system of a large number of particles. As a result, it is our task to find manageable approximations for the Green's function that describe the physics of the many-body system of interest. In the next section, we show that the Green's function for a homogeneous system with spin-independent interactions can always be written in the form

$$G_{\alpha,\alpha'}(\mathbf{k},\mathrm{i}\omega_n) = \frac{-\hbar}{-\mathrm{i}\hbar\omega_n+\varepsilon_{\mathbf{k}}-\mu+\hbar\Sigma_\alpha(\mathbf{k},\mathrm{i}\omega_n)}\delta_{\alpha,\alpha'}, \tag{8.19}$$

where the selfenergy $\hbar\Sigma_\alpha(\mathbf{k},i\omega_n)$ is a complex-valued function that describes the effects of the interactions between the atoms. As a result, our task of finding manageable approximations for the Green's function amounts to finding manageable approximations to the selfenergy. This is the topic of the rest of this chapter. Note already that due to the presence of a complex-valued selfenergy, the spectral-weight function for the interacting system may change significantly its shape.

We have seen that for the noninteracting case the spectral-weight function $\rho_\alpha(\mathbf{k},\omega)$ is a delta function for every wavenumber. This implies that a particle with momentum $\hbar\mathbf{k}$ has a well-defined energy and therefore an infinite lifetime. In the interacting case, it is sometimes possible to approximate the selfenergy by

$\hbar\Sigma_\alpha(\mathbf{k}, i\omega_n) \simeq \Delta\varepsilon_{\mathbf{k},\alpha} - i\hbar/2\tau_{\mathbf{k},\alpha}$. Then, the spectral-weight function becomes the Lorentz profile

$$\rho_\alpha(\mathbf{k}, \omega) = \frac{\hbar}{2\pi\tau_{\mathbf{k},\alpha}} \frac{1}{(\hbar\omega - \varepsilon_{\mathbf{k},\alpha} - \Delta\varepsilon_{\mathbf{k},\alpha} + \mu)^2 + (\hbar/2\tau_{\mathbf{k},\alpha})^2}, \tag{8.20}$$

such that the energy of the atom is no longer well defined and the corresponding lifetime has become finite. More precisely, the Green's function now has a pole at $\varepsilon_{\mathbf{k},\alpha} + \Delta\varepsilon_{\mathbf{k},\alpha} - \mu - i\hbar/2\tau_{\mathbf{k},\alpha}$, which implies that the time evolution of the corresponding wavefunction is given by

$$e^{-i(\varepsilon_{\mathbf{k},\alpha}+\Delta\varepsilon_{\mathbf{k},\alpha}-\mu)t/\hbar} e^{-t/2\tau_{\mathbf{k},\alpha}} .$$

We see that due to the interactions with other atoms, the energy of an atom with momentum $\hbar\mathbf{k}$ and spin $|\alpha\rangle$ is shifted with $\Delta\varepsilon_{\mathbf{k},\alpha}$, while its lifetime has become equal to $\tau_{\mathbf{k},\alpha}$, which describes an exponential decay $e^{-t/\tau_{\mathbf{k},\alpha}}$ of the probability for the atom to be precisely in state $|\mathbf{k}, \alpha\rangle$.

8.1.3 Collective Excitations

To end this discussion on excitations, we note that there are also other ways to excite a system than by taking out or adding a particle. In particular, we can have excitations that do not change the number of particles, such as when we kick a certain particle out of one single-particle quantum state into another single-particle quantum state. Note that in the case of a filled Fermi sea, such a procedure would lead to a particle outside the sea and a hole inside the sea, also called a particle-hole pair. Just as the single-particle excitations are described by the poles of the single-particle Green's function, the particle-hole excitations are described by the poles of the two-particle Green's function, which is given by

$$\langle T[\hat{\psi}^\dagger_\alpha(\mathbf{x}, \tau)\hat{\psi}_{\alpha'}(\mathbf{x}', \tau')\hat{\psi}^\dagger_{\alpha''}(\mathbf{x}'', \tau'')\hat{\psi}_{\alpha'''}(\mathbf{x}''', \tau''')]\rangle .$$

These kind of excitations are called collective excitations, because they have to do with the behavior of the system as a whole. To illustrate this, we note for example that $\hat{\psi}^\dagger_\alpha(\mathbf{x}, \tau)\hat{\psi}_\alpha(\mathbf{x}, \tau)$ is the density operator, such that the above correlation function is able to describe density correlations rather than merely single-particle correlations. A well-known example of a collective excitation is a phonon in a crystal, which is a vibration of the lattice propagating through the system. These kinds of density fluctuations are also called sound waves, because they are the same kind of compression waves that are detectable by the ear in air. Another example of a collective excitation is a density fluctuation of electrons propagating through a metal, in which case we speak of a plasmon. These kind of collective fluctuations return in the last section of the chapter, when we discuss the Jellium model for electrons in a metal.

Using Wick's theorem from (7.72), we find for the noninteracting gas that

$$\begin{aligned}&\langle\phi^*_\alpha(\mathbf{x},\tau)\phi^*_{\alpha'}(\mathbf{x}',\tau')\phi_{\alpha''}(\mathbf{x}'',\tau'')\phi_{\alpha'''}(\mathbf{x}''',\tau''')\rangle_0\\&=\langle\phi^*_\alpha(\mathbf{x},\tau)\phi_{\alpha'''}(\mathbf{x}''',\tau''')\rangle_0\langle\phi^*_{\alpha'}(\mathbf{x}',\tau')\phi_{\alpha''}(\mathbf{x}'',\tau'')\rangle_0\\&\pm\langle\phi^*_\alpha(\mathbf{x},\tau)\phi_{\alpha''}(\mathbf{x}'',\tau'')\rangle_0\langle\phi^*_{\alpha'}(\mathbf{x}',\tau')\phi_{\alpha'''}(\mathbf{x}''',\tau''')\rangle_0\,,\end{aligned}\tag{8.21}$$

such that in this case the two-particle Green's function does not give rise to additional poles compared to the single-particle Green's function. This shows that collective excitations are truly interaction effects. We remark that collective excitations can also be studied experimentally by using probes that do not change the number of particles and that are sensitive to, for example, the particle density. This can be achieved by scattering experiments, where scattering of photons is often used for the atomic gases, while in condensed-matter physics more techniques are available in general, such as scattering of electrons, ions, and neutrons. As a last remark, we note that theoretically the collective degrees of freedom are particularly convenient to introduce into the many-body theory with the Hubbard-Stratonovich transformation, which we already briefly encountered in Sect. 5.8. We generalize this technique to quantum field theory later in this chapter and apply it, amongst others, to the collective excitations of the Jellium model in Sect. 8.7.

8.2 Perturbation Theory

Having discussed the meaning and the relevance of the Green's function, the question arises as to how to determine it in practice for an interacting quantum system. In the previous chapter, we found that the functional formalism conveniently allowed for a variety of methods to solve the ideal quantum gas. Next, we show that this formalism also accommodates a whole set of tools to tackle interacting many-body systems. The action for the interacting quantum gas of interest has the form

$$S[\phi^*,\phi]=S_0[\phi^*,\phi]+S_{\text{int}}[\phi^*,\phi],\tag{8.22}$$

where the noninteracting part is given by

$$\begin{aligned}&S_0[\phi^*,\phi]\\&=\sum_\alpha\int_0^{\hbar\beta}\mathrm{d}\tau\int\mathrm{d}\mathbf{x}\;\phi^*_\alpha(\mathbf{x},\tau)\left\{\hbar\frac{\partial}{\partial\tau}-\frac{\hbar^2\nabla^2}{2m}+V^{\text{ex}}(\mathbf{x})+\varepsilon_\alpha-\mu\right\}\phi_\alpha(\mathbf{x},\tau)\end{aligned}\tag{8.23}$$

and the interactions are described by

$$\begin{aligned}&S_{\text{int}}[\phi^*,\phi]=\\&\frac{1}{2}\sum_{\alpha,\alpha'}\int_0^{\hbar\beta}\mathrm{d}\tau\int\mathrm{d}\mathbf{x}\int\mathrm{d}\mathbf{x}'\;\phi^*_\alpha(\mathbf{x},\tau)\phi^*_{\alpha'}(\mathbf{x}',\tau)V(\mathbf{x}-\mathbf{x}')\phi_{\alpha'}(\mathbf{x}',\tau)\phi_\alpha(\mathbf{x},\tau)\,,\end{aligned}\tag{8.24}$$

as follows from (7.5) and (7.10). Note that only the symmetric part of the spin-independent interaction survives the real-space integrals, which is easily shown by interchanging the integration variables. As a result, we can take the spinless interaction to be symmetric in the coordinates. We wish to calculate the interacting Green's function, whose expression in the functional formalism is given by

$$\begin{aligned} G_{\alpha,\alpha'}(\mathbf{x},\tau;\mathbf{x}',\tau') &= -\langle \phi_\alpha(\mathbf{x},\tau)\phi^*_{\alpha'}(\mathbf{x}',\tau')\rangle \\ &= -\frac{1}{Z}\int \mathrm{d}[\phi^*]\mathrm{d}[\phi]\ \phi_\alpha(\mathbf{x},\tau)\phi^*_{\alpha'}(\mathbf{x}',\tau')\mathrm{e}^{-S[\phi^*,\phi]/\hbar}, \end{aligned} \tag{8.25}$$

where, due to the presence of the fourth-order interaction term, the functional integral cannot be performed exactly anymore. As a result, we need to develop approximate methods to take the interaction effects into account.

The first approach we discuss is based on the systematic expansion of the functional integral in powers of the interaction. It is the field-theoretical analogue of the perturbation theory known from quantum mechanics, which was discussed in Sect. 3.10. We start with expanding the exponent containing the interaction in both the numerator and the denominator of (8.25). Up to first order, we find for the partition function in the denominator

$$\begin{aligned} Z &= \int \mathrm{d}[\phi^*]\mathrm{d}[\phi]\ \mathrm{e}^{-S_0[\phi^*,\phi]/\hbar}\mathrm{e}^{-S_{\mathrm{int}}[\phi^*,\phi]/\hbar} \\ &= \int \mathrm{d}[\phi^*]\mathrm{d}[\phi]\ \mathrm{e}^{-S_0[\phi^*,\phi]/\hbar}\left(1-\frac{S_{\mathrm{int}}[\phi^*,\phi]}{\hbar}\right) \equiv Z_0\left(1-\frac{1}{\hbar}\langle S_{\mathrm{int}}[\phi^*,\phi]\rangle_0\right). \end{aligned} \tag{8.26}$$

Using Wick's theorem, (7.72), we thus have

$$\begin{aligned} -\frac{1}{\hbar}\langle S_{\mathrm{int}}[\phi^*,\phi]\rangle_0 = \frac{1}{2}\sum_{\alpha,\alpha'}\int_0^{\hbar\beta}\mathrm{d}\tau\int \mathrm{d}\mathbf{x}\int \mathrm{d}\mathbf{x}'\ & G_{0;\alpha,\alpha}(\mathbf{x},\tau;\mathbf{x},\tau^+) \\ & \times\frac{-V(\mathbf{x}-\mathbf{x}')}{\hbar}G_{0;\alpha',\alpha'}(\mathbf{x}',\tau;\mathbf{x}',\tau^+) \\ \pm\frac{1}{2}\sum_{\alpha}\int_0^{\hbar\beta}\mathrm{d}\tau\int \mathrm{d}\mathbf{x}\int \mathrm{d}\mathbf{x}'\ & G_{0;\alpha,\alpha}(\mathbf{x}',\tau;\mathbf{x},\tau^+) \\ & \times\frac{-V(\mathbf{x}-\mathbf{x}')}{\hbar}G_{0;\alpha,\alpha}(\mathbf{x},\tau;\mathbf{x}',\tau^+), \end{aligned} \tag{8.27}$$

where the noninteracting Green's function $G_{0;\alpha,\alpha}(\mathbf{x},\tau;\mathbf{x}',\tau')$ was studied extensively in the previous chapter. We again used the convention $\tau^+ = \lim_{\eta\downarrow 0}(\tau+\eta)$, and we note the difference with the corresponding notation for the frequency ω^+. From now on, the convention of this notation will always be that for imaginary times τ^+ we add an infinitesimal real part, while for frequencies ω^+ we add an infinitesimal imaginary part. The necessity of the limiting procedure $G_{0;\alpha,\alpha}(\mathbf{x},\tau;\mathbf{x},\tau^+)$ can be understood as follows. It was already explained in Sects. 5.5 and 7.3 that the expectation values calculated with the functional approach always correspond to time-ordered expectation values of operators, such that we automatically have

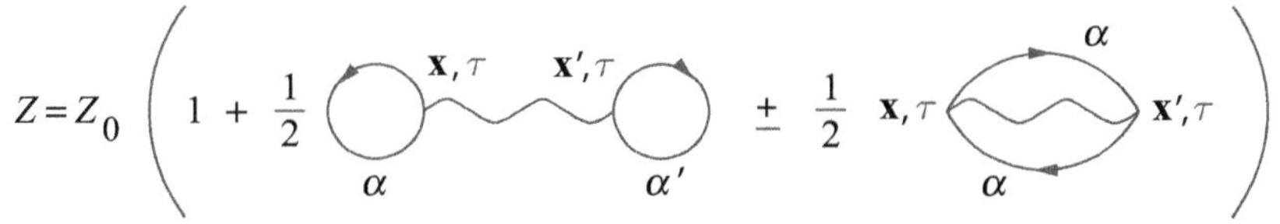

Fig. 8.1 Diagrammatic representation of the partition function up to first order in the interaction.

$$\langle S_{\text{int}}[\phi^*,\phi]\rangle_0 = \sum_{\alpha,\alpha'} \int_0^{\hbar\beta} d\tau \int d\mathbf{x} \int d\mathbf{x}' \, V(\mathbf{x}-\mathbf{x}') \\ \times \left\langle T\left[\hat{\psi}^\dagger_\alpha(\mathbf{x},\tau)\hat{\psi}^\dagger_{\alpha'}(\mathbf{x}',\tau)\hat{\psi}_{\alpha'}(\mathbf{x}',\tau)\hat{\psi}_\alpha(\mathbf{x},\tau)\right]\right\rangle_0 . \tag{8.28}$$

But, since all of the above operators are at equal time, how do we know which of them comes first? Remember that the interaction part of the action comes from the Hamiltonian of (6.29), which is normal-ordered at equal time, such that all the creation operators are to the left of the annihilation operators. To incorporate this equal-time normal ordering into the functional-integral formalism, we give the creation operators an infinitesimal increase in time $\hat{\psi}^\dagger_\alpha(\mathbf{x}',\tau^+)$ such that time-ordering automatically places them to the left of the annihilation operators. From (7.65) we see that the imaginary-time coordinate of the creation operator actually corresponds to the second imaginary-time argument of the Green's function, which therefore explains the expression found in (8.27). Note that we already encountered this procedure in (8.2), where we indeed found that $G_{0;\alpha,\alpha}(\mathbf{x},\tau;\mathbf{x},\tau^+)$ corresponds to the expectation value of the normal-ordered operators at equal time $\langle\hat{\psi}^\dagger_\alpha(\mathbf{x},\tau)\hat{\psi}_\alpha(\mathbf{x},\tau)\rangle_0$ such that it describes the density of atoms for an ideal gas.

From (8.27) it then becomes clear that, if we want to calculate higher-order terms in the perturbative expansions of (8.25) and (8.26), we soon find that the expressions get very cumbersome to write out explicitly. Therefore, it is convenient to introduce a short-hand notation that makes it possible to easily keep track of all the expanded terms and that allows us to understand the general structure of the expansion. Such a notation is obtained in terms of Feynman diagrams, where we represent the factor $-V(\mathbf{x}-\mathbf{x}')/\hbar$ by a wiggly line and the noninteracting Green's function $G_{0;\alpha,\alpha}(\mathbf{x},\tau;\mathbf{x}',\tau')$ by a thin arrowed line pointing from $(\mathbf{x}',\tau')$ to $(\mathbf{x},\tau)$. The result of this procedure for (8.26) and (8.27) is shown in Fig. 8.1 where we have, for clarity's sake, also explicitly indicated the various coordinates and spin degrees of freedom that are integrated or summed over in (8.27). In practice, however, these integrated or summed degrees of freedom are rarely explicitly specified in the corresponding Feynman diagrams. For the numerator of (8.25), we obtain up to first order in the interaction

$$-\int d[\phi^*]d[\phi]\;\phi_\alpha(\mathbf{x},\tau)\phi^*_{\alpha'}(\mathbf{x}',\tau')e^{-S_0[\phi^*,\phi]/\hbar}\left(1-\frac{S_{\text{int}}[\phi^*,\phi]}{\hbar}\right) \\ = Z_0\left(-\langle\phi_\alpha(\mathbf{x},\tau)\phi^*_{\alpha'}(\mathbf{x}',\tau')\rangle_0 + \frac{1}{\hbar}\langle\phi_\alpha(\mathbf{x},\tau)\phi^*_{\alpha'}(\mathbf{x}',\tau')S_{\text{int}}[\phi^*,\phi]\rangle_0\right), \tag{8.29}$$

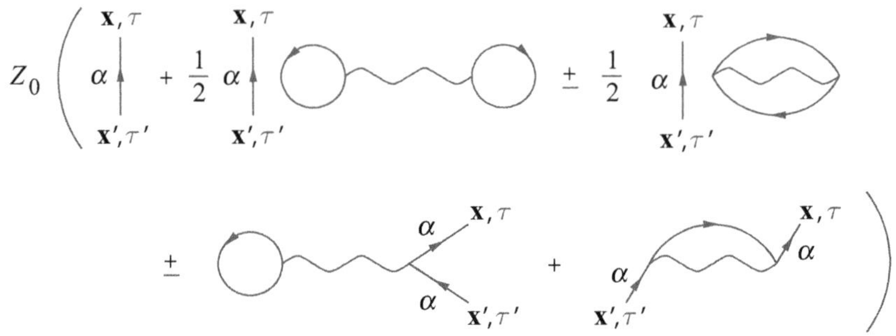

Fig. 8.2 Diagrammatic representation of the numerator of (8.25) up to first order in the interaction.

where as an example we will work out in more detail the second expectation value in the above equation.

Example 8.2. To evaluate $\langle\phi_\alpha(\mathbf{x},\tau)\phi^*_{\alpha'}(\mathbf{x}',\tau')S_{\rm int}[\phi^*,\phi]\rangle_0/\hbar$, we first note that for unequal spins α,α' it gives zero. This is seen by using Wick's theorem to factorize the expectation value into noninteracting Green's functions, resulting always in one of them having unequal spins and thus yielding zero. Also note that due to Wick's theorem, a part of the factorization has the form

$$\langle\phi_\alpha(\mathbf{x},\tau)\phi^*_{\alpha'}(\mathbf{x}',\tau')\rangle_0\langle S_{\rm int}[\phi^*,\phi]\rangle_0/\hbar\,,$$

which is the product of the noninteracting expectation value of the interaction, for which the Feynman diagrams are shown in Fig. 8.1, with the noninteracting Green's function, drawn as a single line. This simple product is then represented by the disconnected diagrams shown in Fig. 8.2. The rest of the terms obtained from Wick's theorem are given by

$$\begin{aligned}-\frac{1}{\hbar}\sum_{\alpha'}\int_0^{\hbar\beta}d\tau''\int d\mathbf{x}''\int d\mathbf{x}'''\,V(\mathbf{x}''-\mathbf{x}''')\\ \times\Big(\pm G_{0;\alpha}(\mathbf{x},\tau;\mathbf{x}'',\tau'')G_{0;\alpha}(\mathbf{x}'',\tau'';\mathbf{x}',\tau')G_{0;\alpha'}(\mathbf{x}''',\tau'';\mathbf{x}''',\tau''^+)\\ +\,\delta_{\alpha,\alpha'}G_{0;\alpha}(\mathbf{x},\tau;\mathbf{x}'',\tau'')G_{0;\alpha}(\mathbf{x}''',\tau'';\mathbf{x}',\tau')G_{0;\alpha'}(\mathbf{x}'',\tau'';\mathbf{x}''',\tau''^+)\Big)\,,\end{aligned}$$

where we explicitly used that the spinless interaction potential can be taken symmetric in the coordinates. The above expression is represented by the last two connected diagrams in Fig. 8.2. In general, if we want to evaluate the noninteracting expectation value of $2n$ fields, consisting of n conjugated pairs, then we get $n!$ terms due to Wick's theorem. However, after integration over coordinates several terms may be

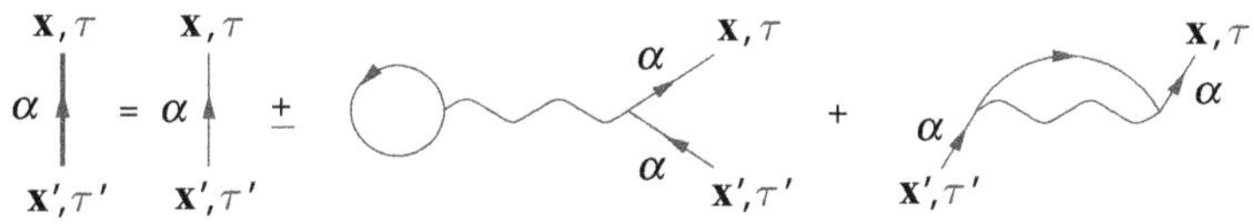

Fig. 8.3 Interacting Green's function, represented by the thick line, up to first order in the interaction.

Combining the results of (8.26) and (8.29), we find for the Green's function of (8.25) in the first instance

$$-\langle \phi_\alpha(\mathbf{x},\tau)\phi^*_{\alpha'}(\mathbf{x}',\tau')\rangle$$
$$= \frac{-\langle \phi_\alpha(\mathbf{x},\tau)\phi^*_{\alpha'}(\mathbf{x}',\tau')\rangle_0 + \frac{1}{\hbar}\langle \phi_\alpha(\mathbf{x},\tau)\phi^*_{\alpha'}(\mathbf{x}',\tau')S_{\text{int}}[\phi^*,\phi]\rangle_0}{1-\frac{1}{\hbar}\langle S_{\text{int}}[\phi^*,\phi]\rangle_0}. \quad (8.30)$$

For weak interactions, the average $\frac{1}{\hbar}\langle S_{\text{int}}[\phi^*,\phi]\rangle_0$ in the denominator of the above expectation value is small. To consistently evaluate the above expression up to first order in the interaction, we must expand the denominator, which yields

$$\left(1-\frac{1}{\hbar}\langle S_{\text{int}}[\phi^*,\phi]\rangle_0\right)^{-1} \simeq 1+\frac{1}{\hbar}\langle S_{\text{int}}[\phi^*,\phi]\rangle_0 + \frac{1}{\hbar^2}\langle S_{\text{int}}[\phi^*,\phi]\rangle_0^2 + \dots \quad (8.31)$$

Working out the multiplication of the numerator of (8.30) with (8.31), and collecting all the terms that are at most of first order in the interaction, we get

$$-\langle \phi_\alpha(\mathbf{x},\tau)\phi^*_{\alpha'}(\mathbf{x}',\tau')\rangle$$
$$= -\langle \phi_\alpha(\mathbf{x},\tau)\phi^*_{\alpha'}(\mathbf{x}',\tau')\rangle_0 + \frac{1}{\hbar}\langle \phi_\alpha(\mathbf{x},\tau)\phi^*_{\alpha'}(\mathbf{x}',\tau')S_{\text{int}}[\phi^*,\phi]\rangle_0$$
$$-\frac{1}{\hbar}\langle S_{\text{int}}[\phi^*,\phi]\rangle_0\langle \phi_\alpha(\mathbf{x},\tau)\phi^*_{\alpha'}(\mathbf{x}',\tau')\rangle_0. \quad (8.32)$$

This first-order approximation to the exact Green's function is shown diagrammatically in Fig. 8.3. It is important to note that the disconnected diagrams have exactly cancelled in the final result. This is a general feature that happens for any expectation value up to any order in the interaction, so the disconnected diagrams that occur in the numerator are always exactly cancelled by the denominator. It is left as an exercise to explicitly show this for the interacting Green's function up to second order.

8.3 Dyson's Equation

From the calculation of the previous section we find that in the presence of interactions between the atoms, the interacting Green's function can be systematically written as a sum of corrections to the noninteracting Green's function. These corrections are represented by connected diagrams that contain one incoming and one outgoing line. It is useful to introduce some nomenclature to be able to better describe the structure of the series of diagrams that contribute to the interacting Green's function. An amputated diagram is a diagram that arises when we remove the two external lines of a connected diagram contributing to the interacting Green's function. Mathematically, this means that we remove the incoming and outgoing noninteracting Green's functions in the corresponding expression for the Feynman diagram. The selfenergy diagrams are then precisely those amputated diagrams that cannot be broken up into two disconnected parts by removing one noninteracting line. Such diagrams are also said to be one-particle irreducible. Examples of selfenergy diagrams are shown in Fig. 8.4.

From the above definitions, it follows that the interacting Green's function satisfies the exact Dyson equation shown diagrammatically in Fig. 8.5 [39]. This equation is readily understood in terms of diagrams. The selfenergy part consists by definition of all the one-particle irreducible Feynman diagrams, whereas the recursive solution of the Dyson equation automatically generates the rest of the possible diagrams, i.e. all the one-particle reducible Feynman diagrams. To study the Dyson equation more concretely, we assume in first instance that the interacting Green's function is in diagonal form. This can always be arranged for in a homogeneous system, which is translationally invariant, such that the Green's function only depends on the difference in the coordinates. Therefore, it can be expanded exactly in terms of plane waves as

$$G_{\alpha,\alpha'}(\mathbf{x}-\mathbf{x}';\tau-\tau') = \sum_{\mathbf{k},n} G_{\alpha,\alpha'}(\mathbf{k},\mathrm{i}\omega_n)\frac{\mathrm{e}^{\mathrm{i}\mathbf{k}(\mathbf{x}-\mathbf{x}')}}{V}\frac{\mathrm{e}^{-\mathrm{i}\omega_n(\tau-\tau')}}{\hbar\beta} . \tag{8.33}$$

For an inhomogeneous system, the above argument does not hold. However, if the interaction energies are small compared to the energy splittings between the eigenstates $\chi_{\mathbf{n}}(\mathbf{x})$ in the external potential, then it is a good approximation to consider these eigenstates to be unaffected by the interaction effects. This is called the weak-coupling limit. Then, the interacting propagator is also to a good approximation diagonal, such that we have

Fig. 8.4 One-particle irreducible selfenergy diagrams that contribute to the interacting Green's function.

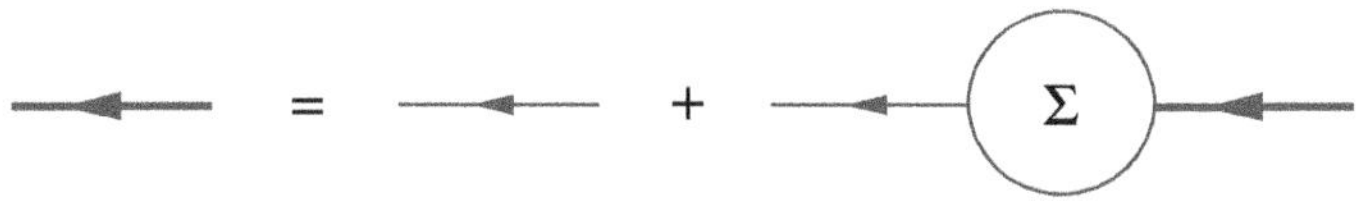

Fig. 8.5 Exact Dyson equation for the interacting Green's function. The thick line corresponds to the interacting Green's function, whereas the thin line corresponds to the noninteracting Green's function.

$$G_{\alpha,\alpha'}(\mathbf{x},\tau;\mathbf{x}',\tau') = \sum_{\mathbf{n},n} G_{\alpha,\alpha'}(\mathbf{n},\mathrm{i}\omega_n)\chi_{\mathbf{n}}(\mathbf{x})\chi^*_{\mathbf{n}}(\mathbf{x}')\frac{\mathrm{e}^{-\mathrm{i}\omega_n(\tau-\tau')}}{\hbar\beta}\,, \tag{8.34}$$

where it turns out that for realistic trapped atomic gases the weak-coupling limit is almost always realized in the normal phase of the quantum gas.

As a result, we find that the Dyson equation becomes

$$\begin{aligned} G_{\alpha,\alpha'}(\mathbf{n},\mathrm{i}\omega_n) &= G_{0;\alpha,\alpha'}(\mathbf{n},\mathrm{i}\omega_n) \\ &\quad + \sum_{\alpha'',\alpha'''} G_{0;\alpha,\alpha''}(\mathbf{n},\mathrm{i}\omega_n)\Sigma_{\alpha'',\alpha'''}(\mathbf{n},\mathrm{i}\omega_n)G_{\alpha''',\alpha'}(\mathbf{n},\mathrm{i}\omega_n)\,, \end{aligned} \tag{8.35}$$

which follows from Fig. 8.5 and the diagonal form of the corresponding Green's functions and selfenergies, such that the matrix multiplication implied by the right diagram simply amounts to the product of the diagonal elements. Since we have $G_{0;\alpha,\alpha'}(\mathbf{n},\mathrm{i}\omega_n) = G_{0;\alpha}(\mathbf{n},\mathrm{i}\omega_n)\delta_{\alpha,\alpha'}$ and since the interaction potential is spin-independent, we may easily convince ourselves that perturbation theory yields an interacting Green's function that is also diagonal in spin space $G_{\alpha,\alpha'}(\mathbf{n},\mathrm{i}\omega_n) = G_{\alpha}(\mathbf{n},\mathrm{i}\omega_n)\delta_{\alpha,\alpha'}$. Indeed, following the approach of the previous subsection, we can expand the interacting Green's function up to any order in the interaction, after which we apply Wick's theorem to factorize the resulting expectation values into noninteracting Green's functions. However, this procedure leads to zero in the case of unequal spin indices $-\langle\phi_\alpha\phi^*_{\alpha'}\rangle$, because each generated term has now in total an odd number of spin indices α and α', as follows directly from the form of the interaction given by (8.24). As a result, there must be in each generated term at least one noninteracting Green's function with unequal spin indices, which is not possible.

Having established that the interacting and the noninteracting Green's function are both also diagonal in spin space, we can multiply (8.35) from the right with G^{-1} and from the left with G_0^{-1}, where both the inverse matrices are of course also diagonal, leading to

$$\begin{aligned} \frac{1}{G_\alpha(\mathbf{n},\mathrm{i}\omega_n)} &= \frac{1}{G_{0;\alpha}(\mathbf{n},\mathrm{i}\omega_n)} - \Sigma_\alpha(\mathbf{n},\mathrm{i}\omega_n) \\ &= -\frac{1}{\hbar}(-\mathrm{i}\hbar\omega_n + \varepsilon_{\mathbf{n},\alpha} - \mu) - \Sigma_\alpha(\mathbf{n},\mathrm{i}\omega_n)\,, \end{aligned} \tag{8.36}$$

which shows that then also the selfenergy is diagonal in spin space. Hence, we arrive at

$$G_\alpha(\mathbf{n},\mathrm{i}\omega_n) = \frac{-\hbar}{-\mathrm{i}\hbar\omega_n + \varepsilon_{\mathbf{n},\alpha} + \hbar\Sigma_\alpha(\mathbf{n},\mathrm{i}\omega_n) - \mu}\,, \tag{8.37}$$

revealing that interaction effects have shifted the poles in the Green's function.

Example 8.3. The two simplest selfenergy diagrams that we can think of are shown in Fig. 8.6. We now calculate them in the weak-coupling limit, when they can be considered diagonal, i.e. $\Sigma_\alpha(\mathbf{n}, i\omega_n)$. In real space and imaginary time, the first self-energy diagram of Fig. 8.6 leads to

$$\begin{aligned}&\Sigma_{\alpha,\alpha'}(\mathbf{x},\tau;\mathbf{x}',\tau') \\ &= \mp\delta_{\alpha,\alpha'}\delta(\mathbf{x}-\mathbf{x}')\delta(\tau-\tau')\sum_{\alpha''}\int \mathrm{d}\mathbf{x}''\,\frac{V(\mathbf{x}-\mathbf{x}'')}{\hbar}G_{0;\alpha'',\alpha''}(\mathbf{x}'',\tau;\mathbf{x}''\tau^+)\,.\end{aligned} \tag{8.38}$$

We can expand this selfenergy in terms of the single-particle eigenstates and the Matsubara modes, giving

$$\begin{aligned}\Sigma_\alpha(\mathbf{n},\mathrm{i}\omega_n) &= \int_0^{\hbar\beta}\mathrm{d}\tau\,\mathrm{d}\tau'\int \mathrm{d}\mathbf{x}\,\mathrm{d}\mathbf{x}'\,\Sigma_\alpha(\mathbf{x},\tau;\mathbf{x}',\tau')\chi_\mathbf{n}(\mathbf{x})\chi^*_\mathbf{n}(\mathbf{x}')\frac{\mathrm{e}^{-\mathrm{i}\omega_n(\tau-\tau')}}{\hbar\beta} \\ &= \mp\frac{1}{\hbar^2\beta}\sum_{\mathbf{n}',n',\alpha'}V_{\mathbf{n},\mathbf{n}',\mathbf{n},\mathbf{n}'}G_{0;\alpha'}(\mathbf{n}';\mathrm{i}\omega_{n'}),\end{aligned} \tag{8.39}$$

where we used the expansion of the noninteracting Green's function from (7.51) and the expression for the matrix elements of the two-body interaction from (6.22). The above expression can be further evaluated, namely

$$\begin{aligned}\Sigma_\alpha(\mathbf{n},\mathrm{i}\omega_n) &= \mp\lim_{\eta\downarrow 0}\frac{1}{\hbar^2\beta}\sum_{\mathbf{n}',n',\alpha'}V_{\mathbf{n},\mathbf{n}',\mathbf{n},\mathbf{n}'}\frac{-\hbar\mathrm{e}^{\mathrm{i}\omega_{n'}\eta}}{-\mathrm{i}\hbar\omega_{n'}+\varepsilon_{\mathbf{n}',\alpha'}-\mu} \\ &= \frac{1}{\hbar}\sum_{\mathbf{n}',\alpha'}V_{\mathbf{n},\mathbf{n}',\mathbf{n},\mathbf{n}'}\frac{1}{\mathrm{e}^{\beta(\varepsilon_{\mathbf{n}',\alpha'}-\mu)}\mp 1},\end{aligned} \tag{8.40}$$

where we used the result for the sum over Matsubara frequencies of (7.31). Note that the sign of the convergence factor η eventually comes from the correct equal time limiting procedure as explained for (8.27). In the same way, we can calculate the second Feynman diagram of Fig. 8.6, which in real space and imaginary time is given by

$$\Sigma_{\alpha,\alpha'}(\mathbf{x},\tau;\mathbf{x}',\tau') = -\delta(\tau-\tau')\frac{V(\mathbf{x}-\mathbf{x}')}{\hbar}G_{0;\alpha,\alpha'}(\mathbf{x},\tau;\mathbf{x}',\tau^+), \tag{8.41}$$

which finally leads to

$$\begin{aligned}
\Sigma_\alpha(\mathbf{n},\mathrm{i}\omega_n) &= \int_0^{\hbar\beta} \mathrm{d}\tau\, \mathrm{d}\tau' \int \mathrm{d}\mathbf{x}\, \mathrm{d}\mathbf{x}'\, \Sigma_\alpha(\mathbf{x},\tau;\mathbf{x}',\tau')\chi_\mathbf{n}(\mathbf{x})\chi^*_\mathbf{n}(\mathbf{x}')\frac{\mathrm{e}^{-\mathrm{i}\omega_n(\tau-\tau')}}{\hbar\beta} \\
&= -\frac{1}{\hbar^2\beta}\sum_{\mathbf{n}',n'} V_{\mathbf{n}',\mathbf{n},\mathbf{n},\mathbf{n}'} G_{0;\alpha}(\mathbf{n}';\mathrm{i}\omega_{n'}) \\
&= \pm\frac{1}{\hbar}\sum_{\mathbf{n}'} V_{\mathbf{n}',\mathbf{n},\mathbf{n}'}\frac{1}{\mathrm{e}^{\beta(\varepsilon_{\mathbf{n}',\alpha}-\mu)}\mp 1}.
\end{aligned} \tag{8.42}$$

8.4 Hartree-Fock Approximation

In our first-order calculation of Sect. 8.2 we found two different corrections to the noninteracting Green's function, which are shown in Fig. 8.3. The corresponding selfenergy diagrams are then simply obtained by removing the external lines, which gives the diagrammatic result shown in Fig. 8.6. This is the most simple approximation to the exact selfenergy that we can think of. In the weak-coupling limit or for a homogeneous system, when the selfenergy can be obtained in diagonal form, the Feynman diagrams of Fig. 8.6 yield

$$\begin{aligned}
\hbar\Sigma_\alpha(\mathbf{n},\mathrm{i}\omega_n) &= \sum_{\mathbf{n}',\alpha'} V_{\mathbf{n},\mathbf{n}';\mathbf{n},\mathbf{n}'}\frac{1}{\mathrm{e}^{\beta(\varepsilon_{\mathbf{n}',\alpha'}-\mu)}\mp 1} \pm \sum_{\mathbf{n}'} V_{\mathbf{n}',\mathbf{n};\mathbf{n},\mathbf{n}'}\frac{1}{\mathrm{e}^{\beta(\varepsilon_{\mathbf{n}',\alpha}-\mu)}\mp 1} \\
&= \sum_{\mathbf{n}',\alpha'} (V_{\mathbf{n},\mathbf{n}';\mathbf{n},\mathbf{n}'} \pm V_{\mathbf{n}',\mathbf{n};\mathbf{n},\mathbf{n}'}\delta_{\alpha,\alpha'})\frac{1}{\mathrm{e}^{\beta(\varepsilon_{\mathbf{n}',\alpha'}-\mu)}\mp 1},
\end{aligned} \tag{8.43}$$

as was explained in Example 8.3. The first term on the right-hand side of equation (8.43), which corresponds to the middle Feynman diagram in Fig. 8.6, is known as the direct or Hartree contribution, whereas the second term, which corresponds to the right Feynman diagram, is known as the exchange or Fock contribution to the selfenergy. From (8.43), we conclude that the matrix elements of the interaction only enter in the combination

$$V_{\mathbf{n},\mathbf{n}';\mathbf{n},\mathbf{n}'} \pm V_{\mathbf{n}',\mathbf{n};\mathbf{n},\mathbf{n}'}\delta_{\alpha,\alpha'}, \tag{8.44}$$

which in the fermionic case is a reflection of the Pauli principle, since it forces the effective interaction between two fermionic atoms in the same state to vanish.

To obtain a nonperturbative and fully self-consistent Hartree-Fock theory, we should not use the noninteracting propagators in the expression for the selfenergy, but rather precisely those propagators G^{HF} that follow from the Dyson equation, which is then given by $(G^{\mathrm{HF}})^{-1} = G_0^{-1} - \Sigma^{\mathrm{HF}}$. The diagramatic representation of the resulting Hartree-Fock approximation to the exact selfenergy is shown in Fig. 8.7. In the weak-coupling limit, it leads to the new dispersion relation

$$\hbar\omega_{\mathbf{n},\alpha} = \varepsilon_{\mathbf{n},\alpha} + \hbar\Sigma^{\mathrm{HF}}_\alpha(\mathbf{n},0) - \mu \tag{8.45}$$

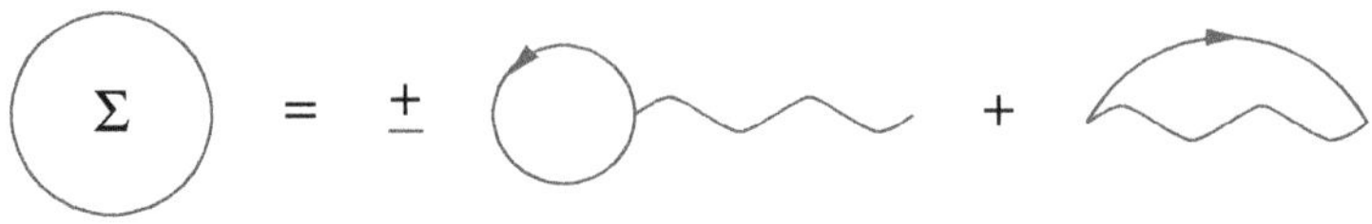

Fig. 8.6 Selfenergy up to first order in the interaction. The middle diagram is known as the direct or Hartree contribution, while the right diagram is known as the exchange or Fock contribution.

for the 'dressed' particles or quasiparticles of the gas. The selfenergy is essentially still given by (8.43), where we only have to replace on the right-hand side $\varepsilon_{\mathbf{n}',\alpha'}$ by

$$\varepsilon'_{\mathbf{n}',\alpha'} \equiv \varepsilon_{\mathbf{n}',\alpha'} + \hbar\Sigma^{\mathrm{HF}}_{\alpha'}(\mathbf{n}',0), \tag{8.46}$$

because we have replaced the noninteracting propagator by the propagator in the Hartree-Fock approximation. In this manner, we have thus obtained an approximation to the interacting Green's function that is nonperturbative in the interaction and effectively sums an infinite number of Feynman diagrams. Although we have assumed weak coupling so far, the Hartree-Fock approximation can also be used in the strong-coupling limit. In that case, it diagrammatically still corresponds to the solution of the Dyson equation with a selfenergy as shown in Fig. 8.7. However, we are then no longer allowed to assume that the exact Green's function is diagonal in the eigenstates $\chi_{\mathbf{n}}(\mathbf{x})$ of the external trapping potential.

8.5 Variational Approach

To gain more insight in the Hartree-Fock approximation, we are going to derive the Hartree-Fock selfenergy also by means of a variational calculation in the zero-temperature limit. We consider here only the fermionic case, while the treatment of the bosonic case is left to Exercise 11.1. In the variational calculation, we assume that the ground state of the system is given by a single Slater determinant of one-particle states $\chi'_{\mathbf{n},\alpha}$ with energies $\varepsilon'_{\mathbf{n},\alpha}$, which we both want to determine variationally. Since at zero temperature the Fermi distribution becomes the stepfunction $\theta(\mu - \varepsilon'_{\mathbf{n},\alpha})$, we have that the many-body ground state $|0;N\rangle$ is the fully antisym-

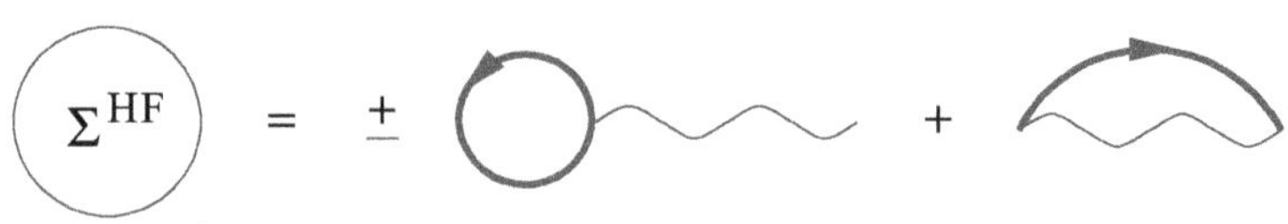

Fig. 8.7 Hartree-Fock approximation for the selfenergy. The thick line represents the interacting Green's function in the selfconsistent Hartree-Fock approximation. The middle diagram is the direct or Hartree contribution, while the right diagram is the exchange or Fock contribution.

metrized product of all single-particle states $\chi'_{\mathbf{n},\alpha}$ that have an energy $\varepsilon'_{\mathbf{n},\alpha}$ that is lower than the chemical potential μ. In the language of second quantization, this can be written as

$$|0;N\rangle = \left(\prod_{\mathbf{n},\alpha}{}' \hat{\psi}^{\dagger}_{\mathbf{n},\alpha}\right)|0\rangle, \tag{8.47}$$

where Π' denotes the restricted product over all states with quantum numbers $\mathbf{n}$ and α that satisfy $\varepsilon'_{\mathbf{n},\alpha} < \mu$.

Next, we want to calculate the expectation value of the energy $\langle 0;N|\hat{H}|0;N\rangle$ for this ground state, where the Hamiltonian $\hat{H}$ is given by

$$\begin{aligned}\hat{H} &= \sum_{\alpha}\int d\mathbf{x}\, \hat{\psi}^{\dagger}_{\alpha}(\mathbf{x})\left\{-\frac{\hbar^2\nabla^2}{2m} + V^{\text{ex}}(\mathbf{x}) + \varepsilon_{\alpha}\right\}\hat{\psi}_{\alpha}(\mathbf{x}) \\ &+ \frac{1}{2}\sum_{\alpha,\alpha'}\int d\mathbf{x}\int d\mathbf{x}'\, \hat{\psi}^{\dagger}_{\alpha}(\mathbf{x})\hat{\psi}^{\dagger}_{\alpha'}(\mathbf{x}')V(\mathbf{x}-\mathbf{x}')\hat{\psi}_{\alpha'}(\mathbf{x}')\hat{\psi}_{\alpha}(\mathbf{x}) \,. \end{aligned} \tag{8.48}$$

To do this, we expand the field operators in terms of the unknown states $\chi'_{\mathbf{n},\alpha}(\mathbf{x})$, i.e. $\hat{\psi}_{\alpha}(\mathbf{x}) = \sum_{\mathbf{n}} \hat{\psi}_{\mathbf{n},\alpha}\chi'_{\mathbf{n},\alpha}(\mathbf{x})$ and $\hat{\psi}^{\dagger}_{\alpha}(\mathbf{x}) = \sum_{\mathbf{n}} \hat{\psi}^{\dagger}_{\mathbf{n},\alpha}\chi'^{*}_{\mathbf{n},\alpha}(\mathbf{x})$. These states should not be confused with the solutions to the single-particle Schrödinger equation from (6.3). If we substitute these expansions into the Hamiltonian, we obtain

$$\begin{aligned}\hat{H} &= \sum_{\alpha}\sum_{\mathbf{n},\mathbf{n}'} \hat{\psi}^{\dagger}_{\mathbf{n},\alpha}E_{\alpha;\mathbf{n},\mathbf{n}'}\hat{\psi}_{\mathbf{n}',\alpha} \\ &+ \frac{1}{2}\sum_{\alpha,\alpha'}\sum_{\mathbf{n},\mathbf{n}'}\sum_{\mathbf{m},\mathbf{m}'} \hat{\psi}^{\dagger}_{\mathbf{n},\alpha}\hat{\psi}^{\dagger}_{\mathbf{n}',\alpha'}V_{\mathbf{n},\mathbf{n}';\mathbf{m},\mathbf{m}'}\hat{\psi}_{\mathbf{m}',\alpha'}\hat{\psi}_{\mathbf{m},\alpha}, \end{aligned} \tag{8.49}$$

where the quadratic coefficients are determined by

$$E_{\alpha;\mathbf{n},\mathbf{n}'} \equiv \int d\mathbf{x}\, \chi'^{*}_{\mathbf{n},\alpha}(\mathbf{x})\left\{-\frac{\hbar^2\nabla^2}{2m} + V^{\text{ex}}(\mathbf{x}) + \varepsilon_{\alpha}\right\}\chi'_{\mathbf{n}',\alpha}(\mathbf{x}), \tag{8.50}$$

and the fourth-order coefficients by

$$V_{\mathbf{n},\mathbf{n}';\mathbf{m},\mathbf{m}'} \equiv \int d\mathbf{x}\int d\mathbf{x}'\, \chi'^{*}_{\mathbf{n},\alpha}(\mathbf{x})\chi'^{*}_{\mathbf{n}',\alpha'}(\mathbf{x}')V(\mathbf{x}-\mathbf{x}')\chi'_{\mathbf{m}',\alpha'}(\mathbf{x}')\chi'_{\mathbf{m},\alpha}(\mathbf{x}). \tag{8.51}$$

The expectation value for the total energy thus becomes

$$\begin{aligned}\langle 0;N|\hat{H}|0;N\rangle &= \sum_{\alpha}\sum_{\mathbf{n},\mathbf{n}'} E_{\alpha;\mathbf{n},\mathbf{n}'}\langle 0;N|\hat{\psi}^{\dagger}_{\mathbf{n},\alpha}\hat{\psi}_{\mathbf{n}',\alpha}|0;N\rangle \\ &+ \frac{1}{2}\sum_{\alpha,\alpha'}\sum_{\mathbf{n},\mathbf{n}'}\sum_{\mathbf{m},\mathbf{m}'} V_{\mathbf{n},\mathbf{n}',\mathbf{m},\mathbf{m}'}\langle 0;N|\hat{\psi}^{\dagger}_{\mathbf{n},\alpha}\hat{\psi}^{\dagger}_{\mathbf{n}',\alpha'}\hat{\psi}_{\mathbf{m}',\alpha'}\hat{\psi}_{\mathbf{m},\alpha}|0;N\rangle. \end{aligned} \tag{8.52}$$

Considering the first term on the right-hand side, we observe that due to the orthonormality of the Fock-states $\hat{\psi}_{\mathbf{n},\alpha}|0;N\rangle$ and $\hat{\psi}_{\mathbf{n}',\alpha}|0;N\rangle$, the quantum numbers $\mathbf{n}$

and $\mathbf{n}'$ have to be equal to have a nonzero overlap. For the interaction term, the situation is a bit more complicated. There, we see that we have the freedom to choose which pairs of states are equal. We have two possibilities. The first choice is to have

$$\mathbf{n} = \mathbf{m} \quad \text{and} \quad \mathbf{n}' = \mathbf{m}', \tag{8.53}$$

while the second option is

$$\mathbf{n} = \mathbf{m}' \quad \text{and} \quad \mathbf{n}' = \mathbf{m}. \tag{8.54}$$

In the second case, we also have that $\alpha = \alpha'$. As a result, we find for the total energy

$$\langle 0;N|\hat{H}|0;N\rangle = \sum_{\mathbf{n},\alpha}{}' E_{\alpha;\mathbf{n},\mathbf{n}} + \frac{1}{2}\sum_{\mathbf{n},\alpha}{}' \sum_{\mathbf{n}',\alpha'}{}' \left(V_{\mathbf{n},\mathbf{n}';\mathbf{n},\mathbf{n}'} - V_{\mathbf{n},\mathbf{n}';\mathbf{n}',\mathbf{n}}\delta_{\alpha,\alpha'}\right), \tag{8.55}$$

where the restricted sum $\sum'$ is over all states with an energy $\varepsilon'_{\mathbf{n},\alpha}$ less than μ.

Next, we want to determine the unknown single-particle wavefunctions $\chi'_{\mathbf{n},\alpha}(\mathbf{x})$ that minimize the total energy under the condition that the wavefunctions remain properly normalized, that is

$$\int d\mathbf{x}\ |\chi'_{\mathbf{n},\alpha}(\mathbf{x})|^2 = 1. \tag{8.56}$$

The standard procedure to find the minimum of a function with an additional constraint is to introduce a Lagrange multiplier for the constraint. Then, the product of the constraint and the Lagrange multiplier is added to the original function, which is consequently minimized as a whole with respect to both the original variables and the Lagrange multiplier. In this case, we add the following term to the expectation value $\langle 0;N|\hat{H}|0;N\rangle$ of the total energy

$$\sum_{\mathbf{n},\alpha}{}' \varepsilon'_{\mathbf{n},\alpha}\left(1 - \int d\mathbf{x}\ \chi'^*_{\mathbf{n},\alpha}(\mathbf{x})\chi'_{\mathbf{n},\alpha}(\mathbf{x})\right),$$

after which we minimize the resulting expression by taking the functional derivative with respect to $\chi'^*_{\mathbf{n},\alpha}(\mathbf{x})$, that is

$$\frac{\delta}{\delta\chi'^*_{\mathbf{n},\alpha}(\mathbf{x})}\left\{\langle 0;N|\hat{H}|0;N\rangle + \sum_{\mathbf{n}',\alpha'}{}' \varepsilon'_{\mathbf{n}',\alpha'}\left(1 - \int d\mathbf{x}'\ \chi'^*_{\mathbf{n}',\alpha'}(\mathbf{x}')\chi'_{\mathbf{n}',\alpha'}(\mathbf{x}')\right)\right\} = 0. \tag{8.57}$$

From this, we find with the use of (8.50) and (8.51) a Schrödinger-like equation that determines the one-particle states and energies by

$$\left\{-\frac{\hbar^2\nabla^2}{2m}+V^{\text{ex}}(\mathbf{x})+\varepsilon_\alpha\right\}\chi'_{\mathbf{n},\alpha}(\mathbf{x})$$
$$+\sum_{\mathbf{n}',\alpha'}{}'\int d\mathbf{x}'\, V(\mathbf{x}-\mathbf{x}')\chi'^*_{\mathbf{n}',\alpha'}(\mathbf{x}')\chi'_{\mathbf{n}',\alpha'}(\mathbf{x}')\chi'_{\mathbf{n},\alpha}(\mathbf{x})$$
$$-\sum_{\mathbf{n}'}{}'\int d\mathbf{x}'\, V(\mathbf{x}-\mathbf{x}')\chi'^*_{\mathbf{n}',\alpha}(\mathbf{x}')\chi'_{\mathbf{n}',\alpha}(\mathbf{x})\chi'_{\mathbf{n},\alpha}(\mathbf{x}')=\varepsilon'_{\mathbf{n},\alpha}\chi'_{\mathbf{n},\alpha}(\mathbf{x}). \quad (8.58)$$

These equations take a particularly convenient form, if we introduce the single-particle density matrix $n_\alpha(\mathbf{x},\mathbf{x}')$ according to

$$n_\alpha(\mathbf{x},\mathbf{x}')=\sum_{\mathbf{n}}{}'\chi'^*_{\mathbf{n},\alpha}(\mathbf{x})\chi'_{\mathbf{n},\alpha}(\mathbf{x}'), \quad (8.59)$$

which for $\mathbf{x}=\mathbf{x}'$ becomes equal to the particle density in spin state $|\alpha\rangle$, i.e. $n_\alpha(\mathbf{x})$. The total density of particles is therefore given by

$$n(\mathbf{x})=\sum_\alpha n_\alpha(\mathbf{x}).$$

With the help of these definitions, (8.58) finally becomes

$$\left\{-\frac{\hbar^2\nabla^2}{2m}+V^{\text{ex}}(\mathbf{x})+\varepsilon_\alpha+\int d\mathbf{x}'\, V(\mathbf{x}-\mathbf{x}')n(\mathbf{x}')\right\}\chi'_{\mathbf{n},\alpha}(\mathbf{x})$$
$$-\int d\mathbf{x}'\, V(\mathbf{x}-\mathbf{x}')n_\alpha(\mathbf{x},\mathbf{x}')\chi'_{\mathbf{n},\alpha}(\mathbf{x}')=\varepsilon'_{\mathbf{n},\alpha}\chi'_{\mathbf{n},\alpha}(\mathbf{x}). \quad (8.60)$$

In the weak-coupling limit the eigenstates can be considered unaffected, i.e. $\chi'_{\mathbf{n},\alpha}(\mathbf{x}) = \chi_{\mathbf{n},\alpha}(\mathbf{x})$, so that we can calculate them from the one-particle Schrödinger equation and substitute them in the above equation to obtain the expression for the quasiparticle energy $\varepsilon'_{\mathbf{n},\alpha}$ in the Hartree-Fock approximation. However, with the use of (8.60), we can also deal with the strong-coupling limit. Starting from the noninteracting wavefunctions, we can then numerically calculate the strong-coupling wavefunctions $\chi'_{\mathbf{n},\alpha}(\mathbf{x})$ in an iterative way. To do this, we calculate the density matrices at each step of the iteration with the wavefunctions found in the previous step.

8.6 Hubbard-Stratonovich Transformation

In the previous sections, we showed how to do Hartree-Fock theory by using either diagrammatic perturbation theory or a zero-temperature variational method. In this section, we show how Hartree-Fock theory can also be derived with the use of the Hubbard-Stratonovich transformation, which we already briefly encountered for path integrals in Sect. 5.8. Here, we generalize the transformation to the case of quantum field theory. The advantages of the Hubbard-Stratonovich transformation over the other two techniques are the following. First, it is a technique that is easy to deal with in calculations. Second, it corresponds to an exact transformation and thus leads to nonperturbative results. Third, it is by no means restricted to zero temperature, and fourth, it is a very versatile technique, as also explained in Sect. 5.8. Since it is very useful for a wide range of physical problems, we use the Hubbard-Stratonovich transformation many times throughout the rest of this book.

8.6.1 Hartree Theory

Before we treat the full Hartree-Fock theory, we first consider the Hartree and Fock contributions separately. We start with the Hartree contribution and consider the action for an interacting gas of spinless atoms, given by

$$S[\phi^*,\phi] = \int_0^{\hbar\beta} d\tau \int d\mathbf{x}\, \phi^*(\mathbf{x},\tau) \left\{ \hbar\frac{\partial}{\partial\tau} - \frac{\hbar^2\nabla^2}{2m} + V^{\mathrm{ex}}(\mathbf{x}) - \mu \right\} \phi(\mathbf{x},\tau) \tag{8.61}$$
$$+\frac{1}{2}\int_0^{\hbar\beta} d\tau \int d\mathbf{x} \int d\mathbf{x}'\, \phi^*(\mathbf{x},\tau)\phi(\mathbf{x},\tau)V(\mathbf{x}-\mathbf{x}')\phi^*(\mathbf{x}',\tau)\phi(\mathbf{x}',\tau)\,.$$

Next, we consider the following identity for the functional-integral over the real field $\kappa(\mathbf{x},\tau)$

$$1 = \int d[\kappa] \exp\left\{ \frac{1}{2\hbar} \left(\kappa - V\phi^*\phi | V^{-1} | \kappa - V\phi^*\phi \right) \right\}, \tag{8.62}$$

where the integration measure now contains the factor $\exp\{\mathrm{Tr}[\log(-V^{-1}/\hbar)]/2\}$, which is thus seen to cancel the result coming from the Gaussian functional integral. This last procedure is mainly done for notational convenience, where we also note that the absorbed term merely amounts to a numerical prefactor, which is therefore often of little physical importance. As explained in Sect. 7.2.3, we note that in the above equation the short-hand notation in the exponent actually means

$$\frac{1}{2\hbar}\int_0^{\hbar\beta} d\tau\, d\tau' \int d\mathbf{x}\, d\mathbf{x}' \left(\kappa(\mathbf{x},\tau) - \int d\mathbf{x}''\, \phi^*(\mathbf{x}'',\tau)\phi(\mathbf{x}'',\tau)V(\mathbf{x}''-\mathbf{x}) \right)$$
$$\times V^{-1}(\mathbf{x}-\mathbf{x}')\delta(\tau-\tau') \left(\kappa(\mathbf{x}',\tau') - \int d\mathbf{x}''\, V(\mathbf{x}'-\mathbf{x}'')\phi^*(\mathbf{x}'',\tau')\phi(\mathbf{x}'',\tau') \right),$$

where we also explicitly wrote out the matrix structure in imaginary time.

Inserting the identity of (8.62) in the partition function conveniently cancels the fourth-order interaction term in the original action. However, it also introduces an extra path integral over the field κ, as we explicitly see below. This generalizes the Hubbard-Stratonovich transformation introduced in Sect. 5.8 to the present case of quantum field theory. To also be able to calculate the exact atomic Green's function $G(\mathbf{x},\tau;\mathbf{x}',\tau')$ after the Hubbard-Stratonovich transformation, we add the current terms $-\hbar\,(J|\phi) - \hbar\,(\phi|J)$ to the action, leading finally to the partition function

$$
\begin{aligned}
Z[J,J^*] &= \int \mathrm{d}[\phi^*]\mathrm{d}[\phi] \int \mathrm{d}[\kappa] \exp\left\{-\frac{1}{\hbar}S[\phi^*,\phi] + (J|\phi) + (\phi|J)\right\} \\
&\quad \times \exp\left\{\frac{1}{2\hbar}\left(\kappa - V\phi^*\phi|V^{-1}|\kappa - V\phi^*\phi\right)\right\} \\
&= \int \mathrm{d}[\phi^*]\mathrm{d}[\phi] \int \mathrm{d}[\kappa] \exp\left\{\frac{1}{2\hbar}\left(\kappa|V^{-1}|\kappa\right) + (J|\phi) + (\phi|J)\right\} \\
&\quad \times \exp\left\{\left(\phi|(G_0^{-1} - \Sigma)|\phi\right)\right\},
\end{aligned}
\tag{8.63}
$$

where the Hartree-like selfenergy is given by

$$
\hbar\Sigma(\mathbf{x},\tau;\mathbf{x}',\tau';\kappa) = \delta(\tau-\tau')\delta(\mathbf{x}-\mathbf{x}')\kappa(\mathbf{x},\tau). \tag{8.64}
$$

Since the resulting functional integral has become quadratic in the atomic fields, we can integrate them out exactly. In this manner, we obtain an action only for the κ field, which we call the effective action $S^{\text{eff}}[\kappa]$. As a result, we find

$$
\begin{aligned}
Z[J,J^*] &= \int \mathrm{d}[\kappa] \exp\left\{\frac{1}{2\hbar}\left(\kappa|V^{-1}|\kappa\right) \mp \mathrm{Tr}\left[\log\left(-G^{-1}\right)\right] - (J|G|J)\right\} \\
&\equiv \int \mathrm{d}[\kappa] \exp\left\{-\frac{1}{\hbar}S^{\text{eff}}[\kappa] - (J|G|J)\right\},
\end{aligned}
\tag{8.65}
$$

where the inverse Green's function $G^{-1}(\mathbf{x},\tau;\mathbf{x}',\tau';\kappa)$ satisfies the equation

$$
\begin{aligned}
G^{-1}(\mathbf{x},\tau;\mathbf{x}',\tau';\kappa) &= G_0^{-1}(\mathbf{x},\tau;\mathbf{x}',\tau') - \Sigma(\mathbf{x},\tau;\mathbf{x}',\tau';\kappa) \\
&= -\frac{1}{\hbar}\left\{\hbar\frac{\partial}{\partial\tau} - \frac{\hbar^2\boldsymbol{\nabla}^2}{2m} - \mu + V^{\text{ex}}(\mathbf{x}) + \kappa(\mathbf{x},\tau)\right\} \\
&\quad \times \delta(\tau-\tau')\delta(\mathbf{x}-\mathbf{x}'),
\end{aligned}
\tag{8.66}
$$

which follows from (7.48) for the noninteracting inverse Green's function and (8.64) for the selfenergy. By inverting the above equation, we obtain

$$
\left\{\hbar\frac{\partial}{\partial\tau} - \frac{\hbar^2\boldsymbol{\nabla}^2}{2m} - \mu + V^{\text{ex}}(\mathbf{x}) + \kappa(\mathbf{x},\tau)\right\} G(\mathbf{x},\tau;\mathbf{x}',\tau';\kappa) = -\hbar\delta(\tau-\tau')\delta(\mathbf{x}-\mathbf{x}'),
\tag{8.67}
$$

where this Green's function is actually not equal to the exact atomic Green's function, but rather is related to it. To see this, we consider the definition of the exact atomic Green's function $G(\mathbf{x},\tau;\mathbf{x}',\tau')$, which is given by

$$G(\mathbf{x},\tau;\mathbf{x}',\tau') = -\int \mathrm{d}[\phi^*]\mathrm{d}[\phi]\ \phi(\mathbf{x},\tau)\phi^*(\mathbf{x}',\tau')\exp\left\{-\frac{1}{\hbar}S[\phi^*,\phi]\right\}$$
$$= \frac{\mp 1}{Z[0,0]}\frac{\delta^2}{\delta J^*(\mathbf{x},\tau)\delta J(\mathbf{x}',\tau')}Z[J,J^*]\bigg|_{J=0}, \tag{8.68}$$

Using (8.65), we find that

$$G(\mathbf{x},\tau;\mathbf{x}',\tau') = \frac{\int \mathrm{d}[\kappa]\ G(\mathbf{x},\tau;\mathbf{x}',\tau';\kappa)\ \exp\{-S^{\mathrm{eff}}[\kappa]/\hbar\}}{\int \mathrm{d}[\kappa]\ \exp\{-S^{\mathrm{eff}}[\kappa]/\hbar\}}, \tag{8.69}$$

where we note that the factor that is absorbed in the measure of (8.62) cancels in the above equation. In this way, we can actually express all higher-order correlation functions of the product $\phi^*(\mathbf{x},\tau)\phi(\mathbf{x},\tau)$ in terms of the effective action $S^{\mathrm{eff}}[\kappa]$ by considering all higher-order derivatives with respect to the currents.

However, in the present case it gives more insight to perform the calculation of the exact atomic correlation functions slightly differently. Instead of adding the current terms $-\hbar(J|\phi)-\hbar(\phi|J)$, we add the following source term to the atomic action $S[\phi^*,\phi]$

$$-\hbar\int_0^{\hbar\beta}\mathrm{d}\tau\int \mathrm{d}\mathbf{x}\ \phi^*(\mathbf{x},\tau)\phi(\mathbf{x},\tau)I(\mathbf{x},\tau) = -\hbar(\phi^*\phi|I), \tag{8.70}$$

where we take I to be real, because $\phi^*\phi$ is also real. The generating functional is then given by

$$Z[I] = \int \mathrm{d}[\phi^*]\mathrm{d}[\phi]\exp\left\{-\frac{1}{\hbar}S[\phi^*\phi] + (\phi^*\phi|I)\right\}. \tag{8.71}$$

Next, we perform a Hubbard-Stratonovich transformation that simultaneously decouples the interaction between the fermions and also removes the above source term from the fermionic part of the action. We therefore insert the following identity

$$1 = \int \mathrm{d}[\kappa]\exp\left\{\frac{1}{2\hbar}\left(\kappa - V\phi^*\phi + \hbar I|V^{-1}|\kappa - V\phi^*\phi + \hbar I\right)\right\} \tag{8.72}$$

into the integrand of $Z[I]$, where the measure is the same as in (8.62). As a result, the generating functional $Z[I]$ is now given by

$$Z[I] = \int \mathrm{d}[\phi^*]\mathrm{d}[\phi]\exp\{(\phi|G_0^{-1}|\phi)\}$$
$$\times\int \mathrm{d}[\kappa]\exp\left\{\frac{1}{2}\left\{\hbar(I|V^{-1}|I) + (\kappa|V^{-1}|I) + (I|V^{-1}|\kappa)\right\}\right\}$$
$$\times\exp\left\{\frac{1}{2\hbar}\left\{(\kappa|V^{-1}|\kappa) - (\kappa|\phi^*\phi) - (\phi^*\phi|\kappa)\right\}\right\}, \tag{8.73}$$

where we can integrate out the atomic fields and calculate the effective action $S^{\text{eff}}[\kappa;I]$, for which we then find

$$Z[I] = \int d[\kappa] \exp\left\{ \frac{1}{2\hbar} (\kappa|V^{-1}|\kappa) \mp \text{Tr}[\log(-G^{-1})] + \frac{1}{2} \left\{ \hbar (I|V^{-1}|I) + (\kappa|V^{-1}|I) + (I|V^{-1}|\kappa) \right\} \right\}. \tag{8.74}$$

By definition, we have from (8.71) that

$$\langle \phi^*(\mathbf{x},\tau)\phi(\mathbf{x},\tau) \rangle = \frac{1}{Z[I]} \frac{\delta Z[I]}{\delta I(\mathbf{x},\tau)} \bigg|_{I=0}, \tag{8.75}$$

where we can calculate the right-hand side with the use of (8.74), such that

$$\int d\mathbf{x}'\, V^{-1}(\mathbf{x}-\mathbf{x}') \langle \kappa(\mathbf{x}',\tau) \rangle = \langle \phi^*(\mathbf{x},\tau)\phi(\mathbf{x},\tau) \rangle, \tag{8.76}$$

which can be rewritten as

$$\langle \kappa(\mathbf{x},\tau) \rangle = \int d\mathbf{x}'\, V(\mathbf{x}-\mathbf{x}') \langle \phi^*(\mathbf{x}',\tau)\phi(\mathbf{x}',\tau) \rangle. \tag{8.77}$$

Note that the above relation does not come as a complete surprise, because if we look at (8.72) for zero I we see that the minimum for the Gaussian integral over κ is given by $V\phi^*\phi$. Since (8.72) is consequently substituted in the path integral over ϕ^* and ϕ, this minimum is actually fluctuating, which then on average leads to $\langle \kappa \rangle = V\langle \phi^*\phi \rangle$ as given by (8.77). We can also take the second-order functional derivative with respect to I in (8.71) and (8.74), leading to

$$\int d\mathbf{x}''\, d\mathbf{x}'''\, V(\mathbf{x}-\mathbf{x}'')V(\mathbf{x}'-\mathbf{x}''') \langle \phi^*(\mathbf{x}'',\tau)\phi(\mathbf{x}'',\tau)\phi^*(\mathbf{x}''',\tau')\phi(\mathbf{x}''',\tau') \rangle = \langle \kappa(\mathbf{x},\tau)\kappa(\mathbf{x}',\tau') \rangle + \hbar V(\mathbf{x}-\mathbf{x}')\delta(\tau-\tau') \,. \tag{8.78}$$

This last equation shows that the operator $\hat{\kappa}(\mathbf{x},\tau)$ associated with the Hubbard-Stratonovich field $\kappa(\mathbf{x},\tau)$ is not identical to the operator

$$\int d\mathbf{x}'\, V(\mathbf{x}-\mathbf{x}') \hat{\psi}^\dagger(\mathbf{x}',\tau)\hat{\psi}(\mathbf{x}',\tau) \,.$$

Looking at (8.72), we indeed see that κ is not strictly equal to $V\phi^*\phi$, but actually is free to fluctuate around this value according to a Gaussian distribution. As a result, the Hubbard-Stratonovich transformation does not have a clear analogue in the operator formalism. We consider this a reason to prefer functional methods over operator methods, since this transformation is a very powerful technique in practice.

To summarize the results of this subsection, we recall that we have performed a Hubbard-Stratonovich transformation to a new collective field κ, which is an exact transformation and allows us to remove or decouple the fourth-order interaction term. The downside of the procedure is that we have now also introduced an additional path integral over this collective field κ. Since the action becomes quadratic in the atomic fields after the transformation, we can integrate these fields out exactly, leading to (8.65). The resulting path integral over κ can no longer be performed exactly, because the logarithm contains terms up to any order in κ. To still be able to extract physical results, we are thus forced to make approximations. To this end, we first note that we have also obtained an expression for the average of κ, which is given by (8.77). In the mean-field approximation, we simply approximate the field κ by its average value $\langle\kappa\rangle$, which is given by the minimum of the effective action $S^{\text{eff}}[\kappa]$. This actually means that we replace the whole path integral over κ by only its maximum contribution, which is given by the minimum of the effective action. Therefore, the mean-field approximation is also called the saddle-point approximation or the stationary-phase approximation, since no fluctuations in the collective field are considered. Note that by substituting in (8.64) for the selfenergy the average value of κ as given by (8.77), we obtain the familiar expression for the Hartree approximation to the selfenergy, because the right-hand side of (8.77) corresponds to the middle Feynman diagram of Fig. 8.7. Thus, although the Hubbard-Stratonovich to the field κ is in itself exact, mean-field theory in κ actually corresponds to the Hartree approximation. Then, if we want to improve on Hartree theory, we could try to take the fluctuations around the mean-field into account. In Sect. 8.7, when we discuss the explicit example of the Jellium model, the concepts sketched above are worked out more concretely.

8.6.2 Fock Theory

In this section, we show how the Hubbard-Stratonovich transformation can also be used to obtain the Fock theory. The approach that we use here is completely analogous to the one that we used in the previous section, when we discussed the Hartree theory. The starting point is also the action of (8.61), which describes a quantum gas of interacting atoms. To deal with the fourth-order term, we perform again a Hubbard-Stratonovich transformation, but this time to a field $\lambda(\mathbf{x},\mathbf{x}',\tau)$ that on average is proportional to the one-particle density matrix of the atomic quantum gas, that is

$$\langle\lambda(\mathbf{x},\mathbf{x}',\tau)\rangle = V(\mathbf{x}-\mathbf{x}')\langle\phi^*(\mathbf{x}',\tau)\phi(\mathbf{x},\tau)\rangle. \tag{8.79}$$

As for Hartree theory, the Hubbard-Stratonovich transformation is achieved by inserting a Gaussian integral over the field λ that decouples the interaction. However, in order to introduce a collective field λ that satisfies (8.79), we have to perform a somewhat more involved transformation. We must insert into the partition function the following identity

$$
\begin{aligned}
1 = \int \mathrm{d}[\lambda] \exp\Bigg\{ &\pm \frac{1}{2\hbar} \int_0^{\hbar\beta} \mathrm{d}\tau \int_0^{\hbar\beta} \mathrm{d}\tau' \int \mathrm{d}\mathbf{x} \int \mathrm{d}\mathbf{x}' \int \mathrm{d}\mathbf{x}'' \int \mathrm{d}\mathbf{x}''' \\
&\times \left(\lambda(\mathbf{x},\mathbf{x}',\tau) - V(\mathbf{x}-\mathbf{x}')\phi^*(\mathbf{x}',\tau)\phi(\mathbf{x},\tau)\right) \\
&\times \frac{1}{V(\mathbf{x}''-\mathbf{x}''')} \delta(\mathbf{x}-\mathbf{x}''')\delta(\mathbf{x}'-\mathbf{x}'')\delta(\tau-\tau') \\
&\times \left(\lambda(\mathbf{x}'',\mathbf{x}''',\tau') - V(\mathbf{x}''-\mathbf{x}''')\phi^*(\mathbf{x}''',\tau')\phi(\mathbf{x}'',\tau')\right)\Bigg\}, \qquad (8.80)
\end{aligned}
$$

where the present notation $1/V(\mathbf{x}-\mathbf{x}')$ should not be confused with $V^{-1}(\mathbf{x}-\mathbf{x}')$, since now we have that $V(\mathbf{x}-\mathbf{x}')/V(\mathbf{x}-\mathbf{x}') = 1$, whereas the inverse is defined through (7.44). As explained before, in the above notation the measure conveniently contains a prefactor that precisely cancels the result that comes from the Gaussian path integral over λ. To facilitate our calculations, we introduce again a short-hand notation, such that the above expression is represented by

$$
1 = \int \mathrm{d}[\lambda] \exp\left\{ \pm \frac{1}{2\hbar} \left(\lambda - V\phi^*\phi||V^{-1}||\lambda - V\phi^*\phi\right) \right\}, \qquad (8.81)
$$

where we note that in the present structure the notation $V\phi^*\phi$ denotes the product of the interaction with the atomic fields at different positions, whereas in the previous subsection a similar notation implied that the atomic fields were at the same position over which was consequently also integrated.

For the same reasons as in the case of the Hartree theory, we then add the current terms $-\hbar(\phi|J) - \hbar(J|\phi)$ to the action, such that we obtain

$$
\begin{aligned}
Z[J,J^*] = \int \mathrm{d}[\phi^*]\mathrm{d}[\phi] \int \mathrm{d}[\lambda] &\exp\left\{ \pm \frac{1}{2} (\lambda||V^{-1}||\lambda) + (\phi|J) + (J|\phi) \right\} \\
&\times \exp\left\{ (\phi|(G_0^{-1} - \Sigma)|\phi) \right\}, \qquad (8.82)
\end{aligned}
$$

where the Fock-like selfenergy is given by

$$
\hbar\Sigma(\mathbf{x},\tau;\mathbf{x}',\tau';\lambda) = \pm\lambda(\mathbf{x},\mathbf{x}',\tau)\delta(\tau-\tau'). \qquad (8.83)
$$

After integrating out the fermions, we obtain the effective action $S^{\text{eff}}[\lambda]$ for the λ field, namely

$$
\begin{aligned}
Z[J,J^*] &= \int \mathrm{d}[\lambda] \, \exp\left\{ \pm \frac{1}{2\hbar} (\lambda||V^{-1}||\lambda) - (J|G|J) \mp \mathrm{Tr}[\log(-G^{-1})] \right\} \\
&\equiv \int \mathrm{d}[\lambda] \, \exp\left\{ -\frac{1}{\hbar} S^{\text{eff}}[\lambda] - (J|G|J) \right\}, \qquad (8.84)
\end{aligned}
$$

where we have that

$$
G^{-1}(\mathbf{x},\tau;\mathbf{x}',\tau';\lambda) = G_0^{-1}(\mathbf{x},\tau;\mathbf{x}',\tau') - \Sigma(\mathbf{x},\tau;\mathbf{x}',\tau';\lambda). \qquad (8.85)
$$

The exact atomic Green's function $G(\mathbf{x},\tau;\mathbf{x}',\tau')$ can be expressed in terms of the effective action and the above Green's function $G(\mathbf{x},\tau;\mathbf{x}',\tau';\lambda)$. Analogously to (8.69) for Hartree theory, we obtain for Fock theory

$$-\langle\phi(\mathbf{x},\tau)\phi^*(\mathbf{x}',\tau')\rangle = \frac{\int \mathrm{d}[\lambda]\; G(\mathbf{x},\tau;\mathbf{x}',\tau';\lambda)\mathrm{e}^{-S^{\mathrm{eff}}[\lambda]/\hbar}}{\int \mathrm{d}[\lambda]\; \mathrm{e}^{-S^{\mathrm{eff}}[\lambda]/\hbar}}, \tag{8.86}$$

where we could also express the exact atomic correlation function in terms of λ by considering the source term

$$-\hbar\int_0^{\hbar\beta}\mathrm{d}\tau\int\mathrm{d}\mathbf{x}\int\mathrm{d}\mathbf{x}'\;\phi^*(\mathbf{x},\tau)\phi(\mathbf{x}',\tau)I(\mathbf{x}',\mathbf{x},\tau).$$

It is left as an exercise to show that, in complete analogy with (8.77) for Hartree theory, this last approach indeed leads to (8.79). Performing a mean-field theory in the collective field $\langle\lambda(\mathbf{x},\mathbf{x}',\tau)\rangle$ then leads to the Fock approximation to the selfenergy, because the right-hand side of (8.79) corresponds to the right diagram of Fig. 8.7.

8.6.3 Hartree-Fock Theory for an Atomic Fermi Gas

Having discussed both the Hartree and the Fock theory separately, we combine them into the Hartree-Fock theory using the Hubbard-Stratonovich transformation. We consider for simplicity a fermionic mixture with an equal number of atoms in two hyperfine states and start by splitting our spin-independent interaction $V(\mathbf{x}-\mathbf{x}')$ into two spin-dependent parts such that one part contributes only to the Hartree diagram and the other part only to the Fock diagram. Denoting a spin-dependent interaction by $V_{\alpha',\beta';\alpha,\beta}(\mathbf{x}-\mathbf{x}') = \langle\alpha',\beta'|\hat{V}|\alpha,\beta\rangle$, we thus want

$$V(\mathbf{x}-\mathbf{x}')\delta_{\alpha,\alpha'}\delta_{\beta,\beta'} = V^{\mathrm{H}}_{\alpha',\beta';\alpha,\beta}(\mathbf{x}-\mathbf{x}') + V^{\mathrm{F}}_{\alpha',\beta';\alpha,\beta}(\mathbf{x}-\mathbf{x}'), \tag{8.87}$$

with

$$\sum_\beta V^{\mathrm{H}}_{\beta,\alpha;\alpha,\beta}(\mathbf{x}-\mathbf{x}') = \sum_\beta V^{\mathrm{F}}_{\alpha,\beta;\alpha,\beta}(\mathbf{x}-\mathbf{x}') = 0. \tag{8.88}$$

Using operators in spin space, it is left as an exercise to show that a possible solution to these equations is

$$\hat{V}^{\mathrm{H}} = \frac{2}{3}(2-\hat{P}_{12})V(\mathbf{x}-\mathbf{x}') \tag{8.89}$$

and

$$\hat{V}^{\mathrm{F}} = \frac{1}{3}(2\hat{P}_{12}-1)V(\mathbf{x}-\mathbf{x}')\,, \tag{8.90}$$

where $\langle\alpha',\beta'|\hat{P}_{12}|\alpha,\beta\rangle = (1+\boldsymbol{\sigma}_{\beta',\beta}\cdot\boldsymbol{\sigma}_{\alpha',\alpha})/2$ are the matrix elements of the spin-exchange operator, such that $\hat{P}_{12}|\alpha,\beta\rangle = |\beta,\alpha\rangle$, and $\boldsymbol{\sigma}$ is a vector of Pauli matrices. Note that we are describing the atoms as having effectively a spin one half. As a result, we now have

$$S_{\text{int}}[\phi^*,\phi] = \frac{1}{2}\sum_{\alpha,\alpha';\beta,\beta'}\int_0^{\hbar\beta}\mathrm{d}\tau\int\mathrm{d}\mathbf{x}\int\mathrm{d}\mathbf{x}'$$
$$\times\Big\{\phi^*_{\alpha'}(\mathbf{x},\tau)\phi_\alpha(\mathbf{x},\tau)V^{\text{H}}_{\alpha',\beta';\alpha,\beta}(\mathbf{x}-\mathbf{x}')\phi^*_{\beta'}(\mathbf{x}',\tau)\phi_\beta(\mathbf{x}',\tau)$$
$$-\,\phi^*_{\alpha'}(\mathbf{x},\tau)\phi_\beta(\mathbf{x}',\tau)V^{\text{F}}_{\alpha',\beta';\alpha,\beta}(\mathbf{x}-\mathbf{x}')\phi^*_{\beta'}(\mathbf{x}',\tau)\phi_\alpha(\mathbf{x},\tau)\Big\}\,, \tag{8.91}$$

which we write as

$$S_{\text{int}}[\phi^*,\phi] \equiv \frac{1}{2}(\phi^*\phi|V^{\text{H}}|\phi^*\phi) - \frac{1}{2}(\phi^*\phi||V^{\text{F}}||\phi^*\phi)\,. \tag{8.92}$$

Next, we apply a Hubbard-Stratonovich transformation to both the Hartree and the Fock parts of the interaction. We start with the Hartree part, where we generalize the discussion of Sect. 8.6.1 to the spin-dependent case. Then, we note that $\mathrm{e}^{-S^{\text{H}}_{\text{int}}[\phi^*,\phi]}$ can be written as a functional integral over the four real fields $\kappa_{\alpha,\alpha'}$, which we may parameterize as $\kappa_{\alpha,\alpha'}(\mathbf{x},\tau) \equiv \kappa_0(\mathbf{x},\tau)\delta_{\alpha,\alpha'} + \boldsymbol{\kappa}(\mathbf{x},\tau)\cdot\boldsymbol{\sigma}_{\alpha,\alpha'}$. The Hubbard-Stratonovich transformation for the Hartree term is based on the identity

$$\exp\left\{-\frac{1}{2\hbar}(\phi^*\phi|V^{\text{H}}|\phi^*\phi)\right\}$$
$$=\int\mathrm{d}[\kappa]\,\exp\left\{\frac{1}{2\hbar}(\kappa|V^{\text{H}^{-1}}|\kappa)-\frac{1}{2\hbar}(\kappa|\phi^*\phi)-\frac{1}{2\hbar}(\phi^*\phi|\kappa)\right\}\,, \tag{8.93}$$

where the inner products now also contain sums over spin and where the measure incorporates the inverse determinant of the inverse interaction. If we ignore the Fock part for a moment, we can substitute this equality in the integrand of the partition function to obtain $Z=\int\mathrm{d}[\kappa]\mathrm{d}[\phi^*]\mathrm{d}[\phi]\mathrm{e}^{-S[\kappa,\phi^*,\phi]/\hbar}$ with the action

$$S[\kappa,\phi^*,\phi] = -\frac{1}{2}(\kappa|V^{\text{H}^{-1}}|\kappa)+\sum_{\alpha,\alpha'}\int_0^{\hbar\beta}\mathrm{d}\tau\int\mathrm{d}\mathbf{x}\;\phi^*_\alpha(\mathbf{x},\tau) \tag{8.94}$$
$$\times\left\{\left(\hbar\frac{\partial}{\partial\tau}-\frac{\hbar^2\boldsymbol{\nabla}^2}{2m}+V^{\text{ex}}(\mathbf{x})+\varepsilon_\alpha-\mu\right)\delta_{\alpha,\alpha'}+\kappa_{\alpha,\alpha'}(\mathbf{x},\tau)\right\}\phi_{\alpha'}(\mathbf{x},\tau)\,.$$

We see that in this manner the action for the fermions has become quadratic with a selfenergy $\hbar\Sigma_{\alpha,\alpha'}(\mathbf{x},\tau;\mathbf{x}',\tau') = \kappa_{\alpha,\alpha'}(\mathbf{x},\tau)\delta(\mathbf{x}-\mathbf{x}')\delta(\tau-\tau')$, such that we can integrate out the fermion fields to obtain $Z=\int\mathrm{d}[\kappa]\mathrm{e}^{-S^{\text{eff}}[\kappa]/\hbar}$ and

$$S^{\text{eff}}[\kappa] = -\frac{1}{2}(\kappa|V^{\text{H}^{-1}}|\kappa)-\hbar\text{Tr}[\log(-G^{-1})]\,, \tag{8.95}$$

where $G^{-1}=G_0^{-1}-\Sigma$ depends on $\kappa_{\alpha,\alpha'}(\mathbf{x},\tau)$. Up to now, we have not made any approximations and have performed an exact rewriting of the partition function. However, the resulting effective action $S^{\text{eff}}[\kappa]$ contains all powers of the fields $\kappa_{\alpha,\alpha'}(\mathbf{x},\tau)$

and is thus rather complicated. To proceed, we therefore need to make an approximation.

We start by noting that the contribution to the partition function is largest for configurations that minimize the action $S^{\rm eff}[\kappa]$. To make use of this observation, we expand the action around its minimum, i.e. we put

$$\kappa_{\alpha,\alpha'}(\mathbf{x},\tau) = \langle\kappa_{\alpha,\alpha'}(\mathbf{x})\rangle + \kappa'_{\alpha,\alpha'}(\mathbf{x},\tau), \tag{8.96}$$

which is a shift in the integration variables. To this end, we require that

$$\left.\frac{\delta S^{\rm eff}[\kappa]}{\delta\kappa_{\alpha,\alpha'}(\mathbf{x},\tau)}\right|_{\kappa=\langle\kappa\rangle} = 0,$$

and if we neglect the fluctuations, we obtain $Z \simeq \mathrm{e}^{-S^{\rm eff}[\langle\kappa\rangle]/\hbar}$. This turns out to be the Hartree approximation, as also discussed in Sect. 8.6.1, but which we show in more detail now. Anticipating this result, we introduce in basis-independent notation

$$G^{-1} = G_0^{-1} - \langle\kappa\rangle/\hbar - \kappa'/\hbar \equiv G^{\rm H^{-1}} - \kappa'/\hbar = G^{\rm H^{-1}}(1 - G^{\rm H}\kappa'/\hbar)\,, \tag{8.97}$$

which we substitute in (8.95), such that we obtain for the terms linear in the fluctuations

$$-\hbar\mathrm{Tr}[-G^{\rm H}\kappa'/\hbar] - (\kappa'|V^{\rm H^{-1}}|\langle\kappa\rangle)\,.$$

If $\langle\kappa_{\alpha,\alpha'}(\mathbf{x})\rangle$ is indeed a minimum of the action $S^{\rm eff}[\kappa]$, these linear terms have to vanish, which implies that

$$\langle\kappa_{\alpha',\alpha}(\mathbf{x})\rangle = \sum_{\beta,\beta'}\int \mathrm{d}\mathbf{x}\; V^{\rm H}_{\alpha',\beta';\alpha,\beta}(\mathbf{x}-\mathbf{x}')G^{\rm H}_{\beta,\beta'}(\mathbf{x}',\tau;\mathbf{x}',\tau^+)\,. \tag{8.98}$$

As promised, this is precisely the most general expression for the Hartree contribution to the selfenergy in Fig. 8.7. Considering as before a selfenergy that is diagonal in spin space, $\langle\kappa_{\alpha',\alpha}(\mathbf{x})\rangle = \kappa_\alpha(\mathbf{x})\delta_{\alpha',\alpha}$, we find that the Hartree approximation to the one-particle propagator obeys $G^{\rm H}_{\beta,\beta'}(\mathbf{x}',\tau;\mathbf{x}',\tau^+) = n(\mathbf{x}')\delta_{\beta,\beta'}/2$, with $n(\mathbf{x}')$ the total average atomic density in the Hartree approximation. As a result, we obtain the usual expression for the Hartree selfenergy

$$\kappa_\alpha(\mathbf{x}) = \sum_\beta\int \mathrm{d}\mathbf{x}'\; V^{\rm H}_{\alpha,\beta;\alpha,\beta}(\mathbf{x}-\mathbf{x}')\frac{n(\mathbf{x}')}{2} = \int \mathrm{d}\mathbf{x}'\; V(\mathbf{x}-\mathbf{x}')n(\mathbf{x}')\,, \tag{8.99}$$

where we also used (8.88).

Next, we want to include the Fock part of the interaction and treat this contribution also by a Hubbard-Stratonovich transformation. This requires introducing four real fields that depend on two spatial coordinates, which are denoted by $\lambda_{\alpha,\alpha'}(\mathbf{x},\mathbf{x}',\tau)$. We then use

$$\exp\left\{\frac{1}{2\hbar}(\phi^*\phi||V^{\mathrm{F}}||\phi^*\phi)\right\} \tag{8.100}$$
$$= \int d[\lambda]\, \exp\left\{-\frac{1}{2\hbar}(\lambda||V^{\mathrm{F}^{-1}}||\lambda)+\frac{1}{2\hbar}(\lambda||\phi^*\phi)+\frac{1}{2\hbar}(\phi^*\phi||\lambda)\right\} ,$$

where the inner product $(\phi^*\phi||\lambda)$ is given by

$$(\phi^*\phi||\lambda) = \sum_{\alpha,\alpha'}\int_0^{\hbar\beta} d\tau \int d\mathbf{x}\, d\mathbf{x}'\phi^*_\alpha(\mathbf{x},\tau)\phi_{\alpha'}(\mathbf{x}',\tau)\lambda_{\alpha,\alpha'}(\mathbf{x},\mathbf{x}',\tau). \tag{8.101}$$

This leads to the total selfenergy

$$\hbar\Sigma_{\alpha,\alpha'}(\mathbf{x},\tau;\mathbf{x}',\tau') = [\kappa_{\alpha,\alpha'}(\mathbf{x},\tau)\delta(\mathbf{x}-\mathbf{x}') - \lambda_{\alpha,\alpha'}(\mathbf{x},\mathbf{x}',\tau)]\delta(\tau-\tau'), \tag{8.102}$$

which after integration over the fermion fields gives rise to the effective action

$$S^{\mathrm{eff}}[\kappa,\lambda] = -\frac{1}{2}(\kappa|V^{\mathrm{H}^{-1}}|\kappa)+\frac{1}{2}(\lambda||V^{\mathrm{F}^{-1}}||\lambda) - \hbar\mathrm{Tr}[\log(-G^{-1})] . \tag{8.103}$$

Expanding the above action around the expectation value $\langle\lambda\rangle$ and demanding that the terms linear in the fluctuation are zero, then indeed leads to the expected expression for the Fock selfenergy

$$\langle\lambda_{\alpha',\alpha}(\mathbf{x},\mathbf{x}')\rangle = \sum_{\beta,\beta'} V^{\mathrm{F}}_{\beta',\alpha';\alpha,\beta}(\mathbf{x}-\mathbf{x}')G^{\mathrm{HF}}_{\beta',\beta}(\mathbf{x},\tau;\mathbf{x}',\tau^+) . \tag{8.104}$$

In the case of a diagonal selfenergy in spin space $\langle\lambda_{\alpha',\alpha}(\mathbf{x},\mathbf{x}')\rangle = \lambda_\alpha(\mathbf{x},\mathbf{x}')\delta_{\alpha',\alpha}$, we have that also the Green's function $G^{\mathrm{HF}}_{\beta,\beta'}(\mathbf{x},\tau;\mathbf{x}',\tau^+)$ is diagonal in spin space, such that

$$\lambda_\alpha(\mathbf{x},\mathbf{x}') = V(\mathbf{x}-\mathbf{x}')n_\alpha(\mathbf{x}',\mathbf{x}) \tag{8.105}$$

with $n_\alpha(\mathbf{x}',\mathbf{x}) = G^{\mathrm{HF}}_{\alpha,\alpha}(\mathbf{x},\tau;\mathbf{x}',\tau^+) = \langle\hat{\psi}^\dagger_\alpha(\mathbf{x}',\tau^+)\hat{\psi}_\alpha(\mathbf{x},\tau)\rangle$.

To perform a true Hartree-Fock calculation in practice, we need to be able to actually determine the Green's function $G^{\mathrm{HF}}_{\alpha,\alpha'}(\mathbf{x},\tau;\mathbf{x}',\tau')$ in terms of the parameters that specify the system, such as the strength of the trapping potential and the chemical potential, for instance. The easiest way to do so is by realizing that G^{HF} is the Green's function of the operator describing the fermionic piece of the action $S[\kappa,\lambda,\phi^*,\phi]$ obtained after the Hubbard-Stratonovich transformations. If we diagonalize this operator by solving the eigenvalue problem

$$\left\{-\frac{\hbar^2\nabla^2}{2m}+V^{\mathrm{ex}}(\mathbf{x})+\kappa_\alpha(\mathbf{x})-\varepsilon'_{\mathbf{n}}\right\}\chi'_{\mathbf{n}}(\mathbf{x}) - \int d\mathbf{x}'\,\lambda_\alpha(\mathbf{x},\mathbf{x}')\chi'_{\mathbf{n}}(\mathbf{x}') = 0\,, \tag{8.106}$$

the desired one-particle propagator acquires the ideal-gas form

$$G^{\rm HF}_{\alpha,\alpha'}(\mathbf{x},\tau;\mathbf{x}',\tau') = \delta_{\alpha,\alpha'}\sum_{\mathbf{n},n}\frac{-\hbar}{-i\hbar\omega_n+\varepsilon'_{\mathbf{n},\alpha}-\mu}\chi'_{\mathbf{n}}(\mathbf{x})\chi'^{*}_{\mathbf{n}}(\mathbf{x}')\frac{e^{-i\omega_n(\tau-\tau')}}{\hbar\beta}, \quad (8.107)$$

with new one-particle energies $\varepsilon'_{\mathbf{n},\alpha}=\varepsilon'_{\mathbf{n}}+\varepsilon_\alpha$ and eigenstates $\chi'_{\mathbf{n}}(\mathbf{x})$ that incorporate the average effect of the interactions of an atom with all the other atoms in the gas. Note that the eigenvalue problem in (8.106) is the same as the coupled equations we found by the variational calculation in (8.58), as can be seen from (8.99) and (8.105).

In the case that the eigenstates are not affected by these mean-field effects, we also recover the weak-coupling results of Sect. 8.4. Although we have thus precisely reproduced our diagrammatic result, there are two important advantages in using the Hubbard-Stratonovich transformation. First, it is in principle exact, and allows us to also calculate corrections to the Hartree-Fock approximation. For example, if we expand $S^{\rm eff}[\kappa,\lambda]$ up to quadratic order in κ' and λ', and neglect all higher orders, we find the generalized random-phase approximation or GRPA. The latter approach actually gives us also the opportunity to study the density fluctuations and therefore the collective excitations of the gas. Second, the Hubbard-Stratonovich transformation is by no means restricted to only Hartree-Fock theory. As it turns out, it actually allows for a beautiful way to describe phase transitions, as we show in the following chapters.

8.7 The Jellium Model

To give a concrete physical application of the techniques developed in this chapter, we consider the Hartree theory for an electron gas in the presence of a homogeneous positively charged background, where the average density of the electron gas $n_{\rm e}$ and the background are the same. This corresponds to what is called the jellium model for electrons in a metal. In particular, we are going to derive the dispersion for the density fluctuations in the jellium model with the use of the field-theoretical methods discussed in the previous section. It will be insightful to first discuss the model in the familiar setting of classical mechanics, before turning to the more abstract formulation of quantum field theory. In classical mechanics, the equation of motion for the total density of electrons $n(\mathbf{x},t)$ is obtained from the continuity equation

$$\frac{\partial n(\mathbf{x},t)}{\partial t}+\nabla\cdot\mathbf{J}(\mathbf{x},t)=0, \quad (8.108)$$

and Newton's law for the current density of particles $\mathbf{J}(\mathbf{x},t)$

$$m\frac{\partial \mathbf{J}(\mathbf{x},t)}{\partial t}+\nabla p(\mathbf{x},t)=-en(\mathbf{x},t)\mathbf{E}(\mathbf{x},t), \quad (8.109)$$

where m is the electron mass, $p(\mathbf{x},t)$ is the local pressure of the electron gas, $-e$ is the electron charge and $\mathbf{E}(\mathbf{x},t)$ is the electric field, which is determined by Gauss' law

$$\nabla\cdot\mathbf{E}(\mathbf{x},t) = -e(n(\mathbf{x},t)-n_\mathrm{e})/\varepsilon_0. \tag{8.110}$$

We rewrite (8.108) and (8.109) such that they both contain the term $m\nabla\cdot\partial\mathbf{J}(\mathbf{x},t)/\partial t$, after which subtraction of the two equations yields

$$m\frac{\partial^2 n(\mathbf{x},t)}{\partial t^2} - \nabla^2 p(\mathbf{x},t) = e\nabla\cdot(n(\mathbf{x},t)\mathbf{E}(\mathbf{x},t)). \tag{8.111}$$

In the case of small density fluctuations $n'(\mathbf{x},t) = n(\mathbf{x},t) - n_\mathrm{e}$, we may linearize (8.108) and (8.109) around the average value n_e to obtain a single equation for the fluctuations. Since the electric field $\mathbf{E}(\mathbf{x},t)$ is proportional to $n'(\mathbf{x},t)$, we have to first order in the fluctuations

$$\nabla\cdot(n(\mathbf{x},t)\mathbf{E}(\mathbf{x},t)) = n_\mathrm{e}\nabla\cdot\mathbf{E}(\mathbf{x},t).$$

As a result, we find

$$m\frac{\partial^2 n'(\mathbf{x},t)}{\partial t^2} - \nabla^2 p(\mathbf{x},t) = -\frac{e^2 n_\mathrm{e}}{\varepsilon_0}n'(\mathbf{x},t). \tag{8.112}$$

Linearizing $p(n)$ by means of

$$p(n(\mathbf{x},t)) = p(n_\mathrm{e}) + \left(\left.\frac{\partial p}{\partial n}\right|_{n=n_\mathrm{e}}\right)n'(\mathbf{x},t), \tag{8.113}$$

we find that the density fluctuations obey the wave equation

$$\left\{m\frac{\partial^2}{\partial t^2} - \left(\left.\frac{\partial p}{\partial n}\right|_{n=n_\mathrm{e}}\right)\nabla^2\right\}n'(\mathbf{x},t) = -\frac{e^2 n_\mathrm{e}}{\varepsilon_0}n'(\mathbf{x},t). \tag{8.114}$$

The solutions to this equation are travelling waves and we are interested in the dispersion relation for these waves. As a solution, we insert

$$n'(\mathbf{x},t) = n_0'\exp\{\mathrm{i}\mathbf{k}\cdot\mathbf{x} - \mathrm{i}\omega(\mathbf{k})t\} \tag{8.115}$$

into (8.114), which gives

$$\left\{-\omega^2(\mathbf{k}) + \frac{\mathbf{k}^2}{m}\left(\left.\frac{\partial p}{\partial n}\right|_{n=n_\mathrm{e}}\right)\right\}n_0' = -\frac{e^2 n_\mathrm{e}}{m\varepsilon_0}n_0'. \tag{8.116}$$

For this equation to be valid, the frequency $\omega(\mathbf{k})$ of the density fluctuations has to satisfy the dispersion relation

$$\omega(\mathbf{k}) = \sqrt{\frac{1}{m}\left(\frac{\partial p}{\partial n}\bigg|_{n=n_e}\right)\mathbf{k}^2 + \omega_p^2}\,, \tag{8.117}$$

with the plasma frequency given by $\omega_p = (e^2 n_e/m\varepsilon_0)^{1/2}$.

8.7.1 Field-Theory Approach

We are now going to reproduce these results in two steps with the use of field-theoretical methods. The action for the electron fields ϕ_α is given by

$$\begin{aligned} S[\phi^*,\phi] = & \\ & \sum_{\alpha=\uparrow,\downarrow}\int_0^{\hbar\beta} d\tau \int d\mathbf{x}\, \phi^*_\alpha(\mathbf{x},\tau)\left\{\hbar\frac{\partial}{\partial\tau} - \frac{\hbar^2\nabla^2}{2m} + V^{\text{ex}}(\mathbf{x}) - \mu\right\}\phi_\alpha(\mathbf{x},\tau) \\ & + \frac{1}{2}\sum_{\alpha,\alpha'}\int_0^{\hbar\beta} d\tau \int d\mathbf{x}\int d\mathbf{x}'\, \phi^*_\alpha(\mathbf{x},\tau)\phi^*_{\alpha'}(\mathbf{x}',\tau)V(\mathbf{x}-\mathbf{x}')\phi_{\alpha'}(\mathbf{x}',\tau)\phi_\alpha(\mathbf{x},\tau)\,, \end{aligned} \tag{8.118}$$

where the interaction potential is given by the Coulomb potential

$$V(\mathbf{x}-\mathbf{x}') = \frac{e^2}{4\pi\varepsilon_0}\frac{1}{|\mathbf{x}-\mathbf{x}'|}\,, \tag{8.119}$$

while the external potential $V^{\text{ex}}(\mathbf{x})$ is added to take into account the positively charged background in which the electrons move, such that

$$V^{\text{ex}}(\mathbf{x}) = -n_e \int d\mathbf{x}'\, V(\mathbf{x}-\mathbf{x}'). \tag{8.120}$$

Analogous to the discussion of Sects. 8.6.1 and 8.6.3, we may introduce a collective field $\kappa(\mathbf{x},\tau)$ by means of a Hubbard-Stratonovich transformation, such that on average

$$\langle\kappa(\mathbf{x},\tau)\rangle = \sum_\alpha \int d\mathbf{x}'\, V(\mathbf{x}-\mathbf{x}')\langle\phi^*_\alpha(\mathbf{x}',\tau)\phi_\alpha(\mathbf{x}',\tau)\rangle, \tag{8.121}$$

where it is left as an exercise to explicitly perform the transformation that leads to the above equation. The transformation then results in the following partition sum

$$Z = \int d[\kappa]\exp\left\{-\frac{1}{\hbar}S^{\text{eff}}[\kappa]\right\} \tag{8.122}$$

with the effective action given by

$$S^{\text{eff}}[\kappa] = -\frac{1}{2}(\kappa|V^{-1}|\kappa) - \hbar\text{Tr}[\log(-G^{-1})]. \tag{8.123}$$

The above inverse Green's function satisfies

$$G^{-1}_{\alpha,\alpha'}(\mathbf{x},\tau;\mathbf{x}',\tau') = G^{-1}_{0;\alpha,\alpha'}(\mathbf{x},\tau;\mathbf{x}',\tau') - \Sigma_{\alpha,\alpha'}(\mathbf{x},\tau;\mathbf{x}',\tau'), \tag{8.124}$$

with

$$\hbar\Sigma_{\alpha,\alpha'}(\mathbf{x},\tau;\mathbf{x}',\tau') = \kappa(\mathbf{x},\tau)\delta(\mathbf{x}-\mathbf{x}')\delta(\tau-\tau')\delta_{\alpha,\alpha'}. \tag{8.125}$$

8.7.2 Effective Action

As mentioned in Sect. 8.6.3, the largest contribution to the partition function is given by the configuration that minimizes the action $S^{\text{eff}}[\kappa]$. We put $\kappa(\mathbf{x},\tau) = \langle\kappa(\mathbf{x})\rangle + \kappa'(\mathbf{x},\tau)$, where $\langle\kappa(\mathbf{x})\rangle$ minimizes the effective action $S^{\text{eff}}[\kappa]$. As a result, we have in the basis-independent formulation that

$$G^{-1} = G_0^{-1} - \langle\kappa\rangle/\hbar - \kappa'/\hbar \equiv G^{\mathrm{H}^{-1}} - \Sigma' = G^{\mathrm{H}^{-1}}(1 - G^{\mathrm{H}}\Sigma')\,, \tag{8.126}$$

with $G^{\mathrm{H}^{-1}} = G_0^{-1} - \langle\kappa\rangle/\hbar$ and $\hbar\Sigma' = \kappa'$. Substituting this in (8.123) and using the series expansion of the logarithm, we find to second order in the fluctuations that

$$\log\left(1 - G^{\mathrm{H}}\Sigma'\right) = -G^{\mathrm{H}}\Sigma' - \frac{1}{2}\left(G^{\mathrm{H}}\Sigma' G^{\mathrm{H}}\Sigma'\right) + \dots\,. \tag{8.127}$$

As before, the terms that are linear in the fluctuations should be zero

$$-\hbar\mathrm{Tr}[-G^{\mathrm{H}}\Sigma'] - (\kappa'|V^{\mathrm{H}^{-1}}|\langle\kappa\rangle) = 0, \tag{8.128}$$

such that $\langle\kappa(\mathbf{x})\rangle$ is indeed the minimum of $S^{\text{eff}}[\kappa]$. This actually results in (8.121) for the Hartree contribution to the selfenergy.

Next, we expand the effective action $S^{\text{eff}}[\kappa]$ in (8.123) up to quadratic terms in the fluctuations κ'. Doing so, we find

$$S^{\text{eff}}[\kappa'] = -\frac{1}{2}\left(\kappa'|V^{-1}|\kappa'\right) + \frac{\hbar}{2}\mathrm{Tr}\left[G^{\mathrm{H}}\Sigma' G^{\mathrm{H}}\Sigma'\right] \equiv -\frac{\hbar}{2}\left(\kappa'|G^{-1}_{\kappa'}|\kappa'\right), \tag{8.129}$$

where we introduced the inverse Green's function $G^{-1}_{\kappa'}$ for the κ'-field, which is by definition the quadratic part of $S^{\text{eff}}[\kappa']$. It is diagrammatically represented by Fig. 8.8, as we soon find. Writing out the trace from (8.129) explicitly, we find

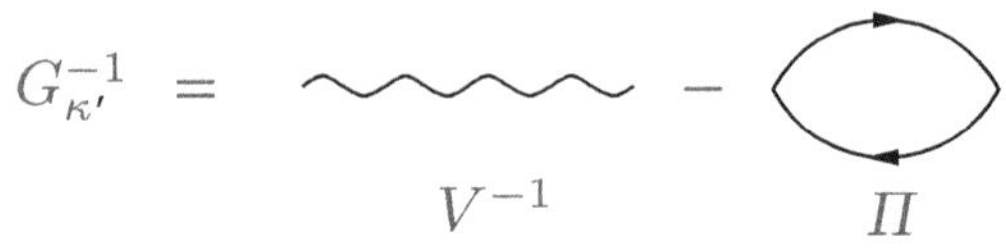

Fig. 8.8 Diagramatic representation of the (inverse) Green's function of the density fluctuations, which describes the screening of the Coulomb interaction. The diagram on the right-hand side is also called the bubble diagram.

$$
\begin{aligned}
\mathrm{Tr}\left[G^{\mathrm{H}}\Sigma' G^{\mathrm{H}}\Sigma'\right] &= \sum_{\alpha,\alpha'}\sum_{\alpha'',\alpha'''}\int_0^{\hbar\beta} \mathrm{d}\tau\,\mathrm{d}\tau'\,\mathrm{d}\tau''\,\mathrm{d}\tau''' \int \mathrm{d}\mathbf{x}\,\mathrm{d}\mathbf{x}'\,\mathrm{d}\mathbf{x}''\,\mathrm{d}\mathbf{x}''' \\
&\quad \times G^{\mathrm{H}}_{\alpha,\alpha'}(\mathbf{x},\tau;\mathbf{x}',\tau')\Sigma'_{\alpha',\alpha''}(\mathbf{x}',\tau';\mathbf{x}'',\tau'') \\
&\quad \times G^{\mathrm{H}}_{\alpha'',\alpha'''}(\mathbf{x}'',\tau'';\mathbf{x}''',\tau''')\Sigma'_{\alpha''',\alpha}(\mathbf{x}''',\tau''';\mathbf{x},\tau) \\
&= \sum_{\alpha}\int_0^{\hbar\beta}\mathrm{d}\tau\int\mathrm{d}\mathbf{x}\int_0^{\hbar\beta}\mathrm{d}\tau'\int\mathrm{d}\mathbf{x}' \\
&\quad \times G^{\mathrm{H}}_{\alpha,\alpha}(\mathbf{x},\tau,\mathbf{x}',\tau')\frac{\kappa'(\mathbf{x}',\tau')}{\hbar}G^{\mathrm{H}}_{\alpha,\alpha}(\mathbf{x}',\tau',\mathbf{x},\tau)\frac{\kappa'(\mathbf{x},\tau)}{\hbar}, \qquad (8.130)
\end{aligned}
$$

where in the last step we used $\hbar\Sigma'_{\alpha,\alpha'}(\mathbf{x},\tau;\mathbf{x}',\tau') = \kappa'(\mathbf{x},\tau)\delta(\mathbf{x}-\mathbf{x}')\delta(\tau-\tau')\delta_{\alpha,\alpha'}$.
In Fourier space, the above quadratic effective action becomes

$$
\begin{aligned}
&S^{\mathrm{eff}}[\kappa'] \qquad (8.131)\\
&= -\frac{1}{2}\sum_{\mathbf{k},n}\kappa'^{*}_{\mathbf{k},n}\left\{V^{-1}(\mathbf{k}) - \frac{1}{\hbar^2\beta V}\sum_{\alpha,\mathbf{k}',n'} G^{\mathrm{H}}_{\alpha}(\mathbf{k}+\mathbf{k}',\mathrm{i}\omega_n+\mathrm{i}\omega_{n'})G^{\mathrm{H}}_{\alpha}(\mathbf{k}',\mathrm{i}\omega_{n'})\right\}\kappa'_{\mathbf{k},n},
\end{aligned}
$$

where we used that $\kappa'_{-\mathbf{k},-n} = \kappa'^{*}_{\mathbf{k},n}$ because $\kappa'(\mathbf{x},\tau)$ is real. To explicitly Fourier transform the Coulomb potential, it is convenient to consider first

$$
V(\mathbf{x}-\mathbf{x}') = \frac{e^2}{4\pi\varepsilon_0}\frac{\mathrm{e}^{-\eta|\mathbf{x}-\mathbf{x}'|}}{|\mathbf{x}-\mathbf{x}'|} \qquad (8.132)
$$

and take the limit $\eta \to 0$ at the end of the calculation. Introducing the relative coor-

where the limit $\eta \to 0$ thus yields $V(\mathbf{k}) = e^2/\varepsilon_0\mathbf{k}^2$, such that $V^{-1}(\mathbf{k}) = \varepsilon_0\mathbf{k}^2/e^2$. For the other term in the effective action of (8.131) we need the Fourier transform of the Green's function $G^{\mathrm{H}}_{\alpha,\alpha'}(\mathbf{k},i\omega_n)$, which is given by

$$G^{\mathrm{H}}_{\alpha,\alpha'}(\mathbf{k},\mathrm{i}\omega_n) = \frac{-\hbar}{-\mathrm{i}\hbar\omega_n + \varepsilon_{\mathbf{k}} - \mu}\delta_{\alpha,\alpha'}, \tag{8.134}$$

where we note that in the above Green's function the expectation value $\langle\kappa\rangle$ has cancelled against the contribution coming from the positively charged background given by (8.120). This follows from comparing (8.126) and (8.121) with (8.120) and remembering that the density of the jellium n_e is equal to the total electron density. Using (8.134), we then calculate the product of Green's functions in (8.131). First, we introduce

$$\hbar\Pi(\mathbf{k},\mathrm{i}\omega_n) \equiv \frac{1}{\hbar\beta}\sum_{\alpha}\sum_{n'}\int\frac{\mathrm{d}\mathbf{k}'}{(2\pi)^3}\, G^{\mathrm{H}}_{\alpha}(\mathbf{k}+\mathbf{k}',\mathrm{i}\omega_n+\mathrm{i}\omega_{n'})G^{\mathrm{H}}_{\alpha}(\mathbf{k}',\mathrm{i}\omega_{n'}), \tag{8.135}$$

where we note that ω_n and $\omega_{n'}$ are bosonic and fermionic Matsubara frequencies, respectively. The diagrammatic representation of $\Pi(\mathbf{k},\mathrm{i}\omega_n)$ is shown in Fig. 8.8, where we remark that it is commonly referred to as the bubble diagram. To further evaluate the expression for the bubble diagram we perform the Matsubara sum in (8.135) by first splitting the fraction, namely

$$\begin{aligned}&\sum_{n'}\int\frac{\mathrm{d}\mathbf{k}'}{(2\pi)^3}\, G^{\mathrm{H}}_{\alpha}(\mathbf{k}',\mathrm{i}\omega_{n'})G^{\mathrm{H}}_{\alpha}(\mathbf{k}'+\mathbf{k},\mathrm{i}\omega_n+\mathrm{i}\omega_{n'}) \\ &= \sum_{n'}\int\frac{\mathrm{d}\mathbf{k}'}{(2\pi)^3}\,\frac{-\hbar}{-\mathrm{i}\hbar\omega_n+\varepsilon_{\mathbf{k}+\mathbf{k}'}-\varepsilon_{\mathbf{k}'}}\left\{G^{\mathrm{H}}_{\alpha}(\mathbf{k}',\mathrm{i}\omega_{n'}) - G^{\mathrm{H}}_{\alpha}(\mathbf{k}'+\mathbf{k},\mathrm{i}\omega_n+\mathrm{i}\omega_{n'})\right\}.\end{aligned} \tag{8.136}$$

Now, the sum over Matsubara frequencies can be performed with the use of the techniques developed in Sect. 7.2.2, giving

$$\Pi(\mathbf{k},\mathrm{i}\omega_n) = 2\int\frac{\mathrm{d}\mathbf{k}'}{(2\pi)^3}\,\frac{N_{\mathrm{FD}}(\varepsilon_{\mathbf{k}+\mathbf{k}'}) - N_{\mathrm{FD}}(\varepsilon_{\mathbf{k}'})}{\varepsilon_{\mathbf{k}+\mathbf{k}'}-\varepsilon_{\mathbf{k}'}-\mathrm{i}\hbar\omega_n}, \tag{8.137}$$

where $N_{\mathrm{FD}}(\varepsilon)$ is the Fermi distribution and the factor of 2 follows from evaluating the sum over the spin variable α. This implies that

$$G^{-1}_{\kappa'}(\mathbf{k},\mathrm{i}\omega_n) = \frac{1}{\hbar}\left\{\frac{1}{V(\mathbf{k})} - \Pi(\mathbf{k},\mathrm{i}\omega_n)\right\}. \tag{8.138}$$

By inverting this expression, we obtain the Green's function for the density fluctuations, i.e. for the κ' field

$$G_{\kappa'}(\mathbf{k},\mathrm{i}\omega_n) = \frac{\hbar V(\mathbf{k})}{1 - V(\mathbf{k})\Pi(\mathbf{k},\mathrm{i}\omega_n)}. \tag{8.139}$$

8.7.3 Dispersion and Screened Coulomb Interaction

Next, we want to examine the poles of $G_{\kappa'}(\mathbf{k},\omega)$, which should in this case give us the dispersion for the density fluctuations. We are primarily interested in the low-energy physics, which means that we make an expansion around $\mathbf{k}=\mathbf{0}$ to work in the long-wavelength limit. To first order, we obtain

$$N_{\rm FD}(\varepsilon_{\mathbf{k}+\mathbf{k}'}) \simeq \tag{8.140}$$
$$N_{\rm FD}(\varepsilon_{\mathbf{k}'}) + \frac{\partial N_{\rm FD}(\varepsilon_{\mathbf{k}'})}{\partial \mathbf{k}'}\cdot\mathbf{k} = N_{\rm FD}(\varepsilon_{\mathbf{k}'}) + \frac{\hbar^2}{m}\frac{\partial N_{\rm FD}(\varepsilon)}{\partial\varepsilon}\bigg|_{\varepsilon=\varepsilon_{\mathbf{k}'}}\mathbf{k}'\cdot\mathbf{k}+\ldots .$$

Substituting this into (8.137), we obtain for $k\to 0$

$$\Pi(\mathbf{k},\omega) \simeq 2\int\frac{\mathrm{d}\mathbf{k}'}{(2\pi)^3}\frac{\hbar^2\mathbf{k}\cdot\mathbf{k}'}{\hbar^2\mathbf{k}\cdot\mathbf{k}'-m\hbar\omega}\frac{\partial N_{\rm FD}(\varepsilon)}{\partial\varepsilon}\bigg|_{\varepsilon=\varepsilon_{\mathbf{k}'}} . \tag{8.141}$$

In the limit when $k/\omega\to 0$, we can expand the denominator of the above equation, giving

$$\Pi(\mathbf{k},\omega) \simeq -2\int\frac{\mathrm{d}\mathbf{k}'}{(2\pi)^3}\left(1+\frac{\hbar^2}{m}\frac{\mathbf{k}\cdot\mathbf{k}'}{\hbar\omega}\right)\frac{\hbar^2}{m}\frac{\mathbf{k}\cdot\mathbf{k}'}{\hbar\omega}\frac{\partial N_{\rm FD}(\varepsilon)}{\partial\varepsilon}\bigg|_{\varepsilon=\varepsilon_{\mathbf{k}'}} . \tag{8.142}$$

First, note that the integral linear in $\mathbf{k}\cdot\mathbf{k}'$, vanishes because the integrand is antisymmetric. Second, since $\partial N_{\rm FD}(\varepsilon)/\partial\varepsilon$ is strongly peaked at low temperatures around the chemical potential, we have that

$$\partial N_{\rm FD}(\varepsilon)/\partial\varepsilon \simeq -\delta(\mu-\varepsilon). \tag{8.143}$$

As a result, we find that

$$\begin{aligned}\Pi(\mathbf{k},\omega) &= -\frac{2k^2}{(2\pi)^2}\left(\frac{\hbar}{m\omega}\right)^2\int k'^4\mathrm{d}k'\int_0^\pi \mathrm{d}\vartheta\,\sin(\vartheta)\,\cos^2(\vartheta)\frac{\partial N_{\rm FD}(\varepsilon)}{\partial\varepsilon}\bigg|_{\varepsilon=\varepsilon_{\mathbf{k}'}}\\ &\simeq \frac{2\sqrt{2}}{3\pi^2}\frac{k^2}{(\hbar\omega)^2}\frac{\sqrt{m}}{\hbar}\mu^{3/2}.\end{aligned} \tag{8.144}$$

Using that the chemical potential is equal to the Fermi energy for the low temperatures in the present case, we have that $\mu\simeq\varepsilon_{\rm F}=(\hbar^2/2m)(3\pi^2 n_{\rm e})^{2/3}$, giving

$$\Pi(\mathbf{k},\omega) = \frac{n_{\rm e}\mathbf{k}^2}{m\omega^2} . \tag{8.145}$$

As a result, for long wavelengths we indeed have a pole in the Green's function $G_{\kappa'}(\mathbf{k},\omega)$ at the plasma frequency $\omega=\omega_{\rm p}=(e^2 n_{\rm e}/m\varepsilon_0)^{1/2}$, as follows from combining (8.133), (8.138) with (8.144).

Example 8.4. Note that if the particles are neutral with a repulsive interaction that can be considered constant $V_0 > 0$ for low momenta, we find that (8.138) now becomes

$$G_{\kappa'}(\mathbf{k},\omega) = \hbar \left\{ \frac{1}{V_0} - \frac{n_e \mathbf{k}^2}{m\omega^2} \right\}^{-1} . \tag{8.146}$$

The dispersion that follows from this is $\omega = ck$, which is the familiar dispersion for phonons and

$$c = \sqrt{\frac{n_e V_0}{m}} \tag{8.147}$$

is the speed of sound in the gas. Note that we actually need strong interactions for this result to be valid, because we have considered the limit $k/\omega \to 0$ in the evaluation of $\Pi(\mathbf{k},\omega)$.

The diagrammatic result from Fig. 8.8, can be interpreted as follows. The coupling between the density fluctuations, which is described by $G_{\kappa'}^{-1}$, is not simply given by the Coulomb potential because the presence of the medium also plays a role. Indeed, the Coulomb potential is screened by the presence of particle-hole excitations, which are described by the bubble diagram in Fig. 8.8. To discuss this screening more explicitly we calculate the static screened interaction $V^{\mathrm{sc}}(\mathbf{k}) \equiv G_{\kappa'}(\mathbf{k},0)/\hbar$ between the electrons, for which the long-wavelength behavior follows from setting $\omega = 0$ in (8.141), such that

$$\Pi(\mathbf{k},0) \simeq 2 \int \frac{d\mathbf{k}'}{(2\pi)^3} \left. \frac{\partial N_{\mathrm{FD}}(\varepsilon)}{\partial \varepsilon} \right|_{\varepsilon=\varepsilon_{\mathbf{k}'}} \simeq -\frac{\sqrt{2}}{\pi^2} \left(\frac{m}{\hbar^2} \right)^{3/2} \sqrt{\varepsilon_{\mathrm{F}}}. \tag{8.148}$$

As a result, the screened interaction is in this limit determined by

$$G_{\kappa'}(\mathbf{k},0) = \hbar \left\{ \frac{\varepsilon_0 \mathbf{k}^2}{e^2} + \frac{\sqrt{2}}{\pi^2} \left(\frac{m}{\hbar^2} \right)^{3/2} \sqrt{\varepsilon_{\mathrm{F}}} \right\}^{-1} . \tag{8.149}$$

Using also that $\varepsilon_{\mathrm{F}} = (\hbar^2/2m)(3\pi^2 n_e)^{2/3}$, this Green's function can also be written as

$$G_{\kappa'}(\mathbf{k},0) = \hbar \frac{e^2}{\varepsilon_0} \left\{ \mathbf{k}^2 + \frac{e^2}{\varepsilon_0} \frac{3}{2} \frac{n_e}{\varepsilon_{\mathrm{F}}} \right\}^{-1} . \tag{8.150}$$

To find the screened interaction in real space $V^{\mathrm{sc}}(\mathbf{x}-\mathbf{x}')$, we perform the Fourier transform

$$V^{\mathrm{sc}}(\mathbf{x}-\mathbf{x}') = \int \frac{d\mathbf{k}}{(2\pi)^3} \frac{1}{\hbar} G_{\kappa'}(\mathbf{k},0) e^{i\mathbf{k}\cdot(\mathbf{x}-\mathbf{x}')}, \tag{8.151}$$

which we have encountered before, such that from Exercise 2.5 we obtain that

$$V^{\mathrm{sc}}(\mathbf{x}-\mathbf{x}') = \frac{e^2}{4\pi\varepsilon_0}\frac{1}{|\mathbf{x}-\mathbf{x}'|}\mathrm{e}^{-|\mathbf{x}-\mathbf{x}'|/\lambda_{\mathrm{TF}}}, \tag{8.152}$$

where the Thomas-Fermi screening length λ_{TF} of the electron gas is at low temperatures given by

$$\lambda_{\mathrm{TF}} = \left(\frac{2\varepsilon_0\varepsilon_{\mathrm{F}}}{3n_{\mathrm{e}}e^2}\right)^{1/2}. \tag{8.153}$$

8.8 Problems

Exercise 8.1. Considering a homogeneous noninteracting spinless Bose gas at zero temperature, show that the Lehmann representation of (8.11) leads to the familiar expression of the noninteraction Green's function from (8.12). What does the ground state look like? What is the value of the chemical potential?

Exercise 8.2. By making use of Exercise 7.2, calculate

$$\langle \hat{\psi}_\alpha(\mathbf{x},\tau)\hat{\psi}^\dagger_\alpha(\mathbf{x},\tau)\rangle = -G_{\alpha,\alpha}(\mathbf{x},\tau^+;\mathbf{x},\tau) \tag{8.154}$$

in the same way as (8.2). Where does the difference with (8.2) come from?

Exercise 8.3. Show with the use of Feynman diagrams that up to second order in the interaction the interacting Green's function contains no disconnected diagrams.

Exercise 8.4. Calculate the Hartree-Fock selfenergy for spinless particles with a point-like interaction $V(\mathbf{x}-\mathbf{x}') = V_0\delta(\mathbf{x}-\mathbf{x}')$. Express your answer in terms of the average particle density $n(\mathbf{x})$. Do you understand why the selfenergy vanishes for the fermionic case?

Exercise 8.5. Hubbard-Stratonovich Transformation I
a) Show that the Hubbard-Stratonovich transformation to the collective $\lambda(\mathbf{x},\mathbf{x}',\tau)$ field of (8.80) gives rise to

$$\langle\lambda(\mathbf{x},\mathbf{x}',\tau)\rangle = V(\mathbf{x}-\mathbf{x}')\langle\phi^*(\mathbf{x}',\tau)\phi(\mathbf{x},\tau)\rangle. \tag{8.155}$$

b) Perform the Hubbard-Stratonovich transformation leading to (8.121)

Exercise 8.6. Spin Exchange
(a) Show that $\langle\alpha',\beta'|\hat{P}_{12}|\alpha,\beta\rangle = (1+\boldsymbol{\sigma}_{\beta',\beta}\cdot\boldsymbol{\sigma}_{\alpha',\alpha})/2$ are indeed the matrix elements of the spin-exchange operator, for which $\hat{P}_{12}|\alpha,\beta\rangle = |\beta,\alpha\rangle$.
Hint: use the spin raising and lowering operators $\hat{\sigma}_\pm = \hat{\sigma}_x \pm \mathrm{i}\hat{\sigma}_y$.
(b) Prove that (8.89) and (8.90) satisfy (8.87) and (8.88).

Exercise 8.7. Hubbard-Stratonovich Transformation II

In this exercise, we explicitly show with an example that the Hubbard-Stratonovich transformation can also be used to decouple quadratic terms in the action, whereas it is usually used to decouple quartic terms. Consider a noninteracting gas of N bosons in a box with volume V, such that the action in frequency-momentum representation is given by

$$S_0[\phi^*,\phi] = \sum_{\mathbf{k},n} \phi^*_{\mathbf{k},n} \left(-\mathrm{i}\hbar\omega_n + \varepsilon_{\mathbf{k}} - \mu\right) \phi_{\mathbf{k},n} \,, \tag{8.156}$$

where ω_n are the bosonic Matsubara frequencies and $\varepsilon_{\mathbf{k}} = \hbar^2\mathbf{k}^2/2m$ is the kinetic energy of a boson with momentum $\hbar\mathbf{k}$. We want to perform a Hubbard-Stratonovich transformation to the field $\Phi_{\mathbf{k},n}$ that decouples the kinetic energy term in the action, such that on average it obeys $\langle\Phi_{\mathbf{k},n}\rangle = \langle\phi_{\mathbf{k},n}\rangle$. To still be able to calculate the exact bosonic Green's function $G_0(\mathbf{k},\mathrm{i}\omega_n) = \langle\phi^*_{\mathbf{k},n}\phi_{\mathbf{k},n}\rangle_0$, even after the Hubbard-Stratonovich transformation and integrating out the bosonic field ϕ, we also add current terms to the action to obtain

$$S[\phi^*,\phi;J,J^*] = S_0[\phi^*,\phi] - \hbar\sum_{\mathbf{k},n} \left(\phi^*_{\mathbf{k},n}J_{\mathbf{k},n} + J^*_{\mathbf{k},n}\phi_{\mathbf{k},n}\right). \tag{8.157}$$

(a) Perform the desired Hubbard-Stratonovich transformation and integrate out the field ϕ to determine the effective action for the Φ field, $S[\Phi^*,\Phi;J,J^*]$, where in first instance you may omit terms that do not depend on Φ, Φ^*, J and J^*. Note that the Hubbard-Stratonovich transformation follows directly from requiring that the kinetic energy term $\phi^*_{\mathbf{k},n}\varepsilon_{\mathbf{k}}\phi_{\mathbf{k},n}$ is cancelled from the original action, and by requiring that $\langle\Phi_{\mathbf{k},n}\rangle = \langle\phi_{\mathbf{k},n}\rangle$. Show in particular that

$$\begin{aligned} &S[\Phi^*,\Phi;J,J^*] \\ &= -\hbar\sum_{\mathbf{k},n}\left\{\Phi^*_{\mathbf{k},n}G^{-1}_{\Phi}(\mathbf{k},\mathrm{i}\omega_n)\Phi_{\mathbf{k},n} + \Phi^*_{\mathbf{k},n}J_{\mathbf{k},n} + J^*_{\mathbf{k},n}\Phi_{\mathbf{k},n} + \hbar\frac{J_{\mathbf{k},n}J^*_{\mathbf{k},n}}{\varepsilon_{\mathbf{k}}}\right\}, \end{aligned} \tag{8.158}$$

where

$$-\hbar G^{-1}_{\Phi}(\mathbf{k},\mathrm{i}\omega_n) = -\varepsilon_{\mathbf{k}} - \frac{\varepsilon^2_{\mathbf{k}}}{-\mathrm{i}\hbar\omega_n - \mu}. \tag{8.159}$$

(b) Prove now from $S[\Phi^*,\Phi;J,J^*]$ that the exact bosonic Green's function G_0 equals

$$G_0(\mathbf{k},\mathrm{i}\omega_n) = \frac{-\hbar}{-\mathrm{i}\hbar\omega_n + \varepsilon_{\mathbf{k}} - \mu}\,, \tag{8.160}$$

as expected.

(c) Show also that the thermodynamic potential of the gas can be written as

$$\Omega = \frac{1}{\beta}\sum_{\mathbf{k},n}\left\{\log[\beta(-\mathrm{i}\hbar\omega_n - \mu)] + \log[-\hbar\beta G^{-1}_{\Phi}(\mathbf{k},\mathrm{i}\omega_n)]\right\}. \tag{8.161}$$

(d) Calculate from this result the total average density $n = N/V$ of the gas and show that it gives the same answer as obtained from the exact bosonic Green's function determined in question (b).

Additional Reading

- For a introductory text on Feynman diagrams in the many-body problem the reader is referred to
  R. D. Mattuck, *A guide to Feynman Diagrams in the Many-Body problem*, (Dover, New York, 1992).
- D. Pines, *The Many-Body Problem*, Addison-Wesley, New York, 1997.
- H. Kleinert, Forts. Phys. **26**, 565 (1978).

Chapter 9
Landau Theory of Phase Transitions

> *Of course, I found what had long been known. However close are physics and mathematics their connection can be most different. In particular, it is possible to make progress in theoretical physics using but a very modest mathematical apparatus, for instance, not exceeding the limits of what is taught at physical departments of universities. By contrast, in some cases theoretical physicists both use most complicated modern mathematics and develop it.*
> *–Vitaly L. Ginzburg*

A phase transition is the phenomenon that a many-body system may suddenly change its properties in a rather drastic way due to the change of an externally controllable variable. Familiar examples in everyday life are the transitions from gases to liquids or from liquids to solids, due to for example a change in the temperature or the pressure. Another example is the transition from a disordered to a magnetized state in a ferromagnetic material as a function of temperature or magnetic field. One property that all these transitions share is that the order of the system, described for example by the density or the magnetization, differs at each side of the transition. We consider as an example an Ising-like spin system at a low, but nonzero temperature, such that the ferromagnetic state with many spins pointing in the same direction corresponds to an absolute minimum of the free energy. Upon applying a magnetic field in the opposite direction, the equilibrium state may change to a state where most spins point in the opposite direction. As a result, the system is initially in a local minimum of the free energy, and it has to overcome a large energy barrier in order to reach the new equilibrium state. Still, eventually the system reaches the new equilibrium state due to the thermal activation of random spin flips in the system, such that the corresponding transition can be said to be driven by thermal fluctuations. Note that the magnetization makes a large jump by going from one equilibrium state to the other. Such a transition, when the parameter describing the order in the system is discontinuous, we call a first-order phase transition.

Phase transitions can also be continuous, which is the case when the order parameter changes from zero to a nonzero value in a continuous way. Continuous, or second-order, phase transitions can be very spectacular, because we will see that they give rise to a diverging correlation length and hence to behavior known as critical phenomena. The infinite correlation length implies that fluctuations extend over the whole many-body system, such that they are present at each length scale. As a result, the system looks similar at every length scale, i.e. it is scale invariant, which can be used to recursively describe the critical system at increasing wavelengths. This leads to the very powerful renormalization group method, which is able to go far beyond mean-field theory and which is the topic of Chap. 14. Finally we remark that phase transitions can also occur at zero temperature, and are then called quantum phase transitions because they are solely driven by quantum fluctuations. We will see a detailed example of a quantum phase transition in Chap. 16.

In this chapter, we discuss the Landau approach to phase transitions. As an introduction, we start with a mean-field treatment of the d-dimensional Ising model in order to discuss an explicit example of a phase transition. This allows us to introduce various important concepts, such as the order parameter and the critical temperature. Then, we formalize the theory of phase transition using the Landau free-energy functional, which is an effective functional for both the equilibrium value of the order parameter and the fluctuations around this equilibrium. Also, we show how the Landau free-energy functional is directly related to the effective action for the order parameter that can be obtained by using the Hubbard-Stratonovich transformation. Finally, we discuss the effect of fluctuations and the phenomenon of spontaneous symmetry breaking. In this chapter, we mostly consider the ferromagnetic transition in the Ising model as a concrete example, whereas in the following chapters we encounter more exotic phase transitions, such as the normal-superfluid transition in an ultracold quantum gas.

9.1 Ising Model in d Dimensions

In more than one dimension, the Ising model is one of the simplest microscopic models that gives rise to a second-order phase transition at a nonzero temperature. The model describes a system of spins $\hat{S}_{z,i}$ on lattice sites $\mathbf{x}_i$, that can take on values $m_i = \pm\hbar/2$, and which are coupled to their nearest neighbors with strength J. In one dimension, the lattice is a chain and the number of nearest neighbors is two, in two dimensions, the lattice is square and the number of nearest neighbors is four, and in three dimensions, the lattice is cubic and the number of nearest neighbors is six. The Hamiltonian of the d-dimensional Ising model is given by

$$\hat{H} = -\frac{2J}{\hbar^2}\sum_{\langle i,j\rangle}\hat{S}_{z,i}\cdot\hat{S}_{z,j}, \tag{9.1}$$

where $\langle i,j\rangle$ denotes the summation over all nearest-neighbors. From the Hamiltonian, we see that for positive J, the system lowers its energy by having adjacent spins with the same eigenvalue of the spin operator $\hat{S}_z$. On the other hand, for negative J the spins tend to be aligned in an antiparallel manner. From now on, we consider the case of positive J, which leads to ferromagnetic order. To be able to describe this, we introduce the magnetization $\langle M\rangle$, which is defined as the expectation value of the average spin, i.e.

$$\langle M \rangle = \frac{1}{N}\left\langle \sum_i \hat{S}_{z,i} \right\rangle$$
$$= \frac{1}{Z}\mathrm{Tr}\left[\left(\frac{1}{N}\sum_i \hat{S}_{z,i}\right)\exp\left\{\frac{2J}{\hbar^2 k_B T}\sum_{\langle i,j\rangle}\hat{S}_{z,i}\cdot\hat{S}_{z,j}\right\}\right], \tag{9.2}$$

where the trace is over all microstates of the d-dimensional Ising model, Z is the canonical partition function, and N the total number of spins. We have seen that the partition function for the one-dimensional Ising model, which we discussed in Chap. 4, can be solved exactly. It turns out that in two dimensions an exact solution is also possible, although it is much more difficult to obtain than for one dimension. In three dimensions, however, an exact solution is not known and we have to resort to approximate methods.

The standard approach is to start with a fluctuation expansion around a certain fixed magnetization M, i.e.

$$\hat{S}_{z,i} = M + \hat{S}'_{z,i}, \tag{9.3}$$

where this magnetization M has to be determined selfconsistently at the end of the calculation. After this expansion the fluctuation effects due to $\hat{S}'_{z,i}$ are usually assumed to be small, such that these effects are taken into account up to linear order, leading to a mean-field approximation. Substituting the fluctuation expansion into the Hamiltonian, we find up to first order in the fluctuations

$$-\frac{2J}{\hbar^2}\sum_{\langle i,j\rangle}\hat{S}_{z,i}\cdot\hat{S}_{z,j} = -\frac{2J}{\hbar^2}\sum_{\langle i,j\rangle}\left(M+\hat{S}'_{z,i}\right)\cdot\left(M+\hat{S}'_{z,j}\right)$$
$$= -\frac{2J}{\hbar^2}zNM^2 - \frac{4J}{\hbar^2}zM\sum_i \hat{S}'_{z,i} + \mathscr{O}((\hat{S}'_z)^2), \tag{9.4}$$

where $z = 2d$ denotes the number of nearest neighbors in d dimensions. Neglecting the higher-order terms in the fluctuations, and using (9.3) to express the above Hamiltonian in terms of $\hat{S}_{z,i}$ again, we find

$$-\frac{2J}{\hbar^2}\sum_{\langle i,j\rangle}\hat{S}_{z,i}\cdot\hat{S}_{z,j} \simeq \frac{2J}{\hbar^2}zNM^2 - \frac{4J}{\hbar^2}zM\sum_i \hat{S}_{z,i}. \tag{9.5}$$

Note that the second term in the above equation can be regarded as a Zeeman interaction caused by the effective magnetic field that is generated by the presence of a nonzero magnetization in the system. Using the Hamiltonian from (9.5), we find that

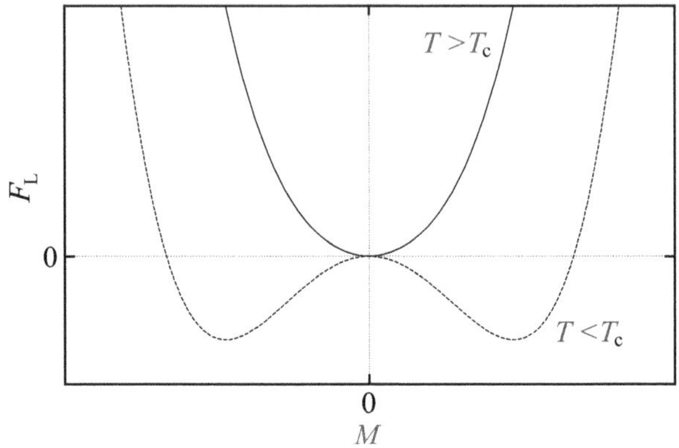

Fig. 9.1 Qualitative sketch of the Landau free energy for the Ising model as a function of the magnetization M for temperatures above and below the critical temperature T_c

$$\begin{aligned}
Z(M) &\simeq \mathrm{Tr}\left[\exp\left\{-\frac{2zJNM^2}{\hbar^2 k_B T} + \frac{4zJM}{\hbar^2 k_B T}\sum_i \hat{S}_{z,i}\right\}\right] \\
&= \exp\left\{-\frac{2zJNM^2}{\hbar^2 k_B T}\right\}\prod_i \sum_{m_i=\pm\hbar/2} \exp\left\{\frac{4zJMm_i}{\hbar^2 k_B T}\right\}. \\
&= \exp\left\{-\frac{2zJNM^2}{\hbar^2 k_B T} + N\log\left(\exp\left\{\frac{2zJM}{\hbar k_B T}\right\} + \exp\left\{-\frac{2zJM}{\hbar k_B T}\right\}\right)\right\} \\
&\equiv \mathrm{e}^{-F_L(M)/k_B T},
\end{aligned} \tag{9.6}$$

where in the second line of the above equation $m_i = \pm\hbar/2$ denotes the two possible values for the quantum number belonging to the spin operator $\hat{S}_{z,i}$. In the last line of the above equation, we introduced the Landau free energy

$$F_L(M) = \frac{2zJNM^2}{\hbar^2} - Nk_B T \log\left(\exp\left\{\frac{2zJM}{\hbar k_B T}\right\} + \exp\left\{\frac{-2zJM}{\hbar k_B T}\right\}\right), \tag{9.7}$$

where we still have to determine the equilibrium value of the magnetization $\langle M \rangle$, that is actually realized by the many-body system. This then gives us the actual free energy $F = F_L(\langle M \rangle)$ in the mean-field approximation. To this end, we should determine the minimum of $F_L(M)$ with respect to M, which we can do directly by differentiating the above equation.

However, to gain more insight into the ferromagnetic phase transition, it is more instructive to expand the Landau free energy in powers of M such that we obtain

$$\begin{aligned}
F_L(M) &= -k_B T N \log(2) \\
&\quad + \frac{1}{\hbar^2}\left(2JNz - \frac{2J^2Nz^2}{k_B T}\right) M^2 + \frac{1}{\hbar^4}\frac{4J^4Nz^4}{3(k_B T)^3} M^4 + \ldots,
\end{aligned} \tag{9.8}$$

where the first term in the above expansion is the entropy contribution to the free energy at zero magnetization. Since it is independent of M, it is not relevant to us in the following. The second term is then seen to be quadratic in M and its coefficient we designate with $\alpha(T)/2$, while the third term is quartic in M and its coefficient we designate with $\beta(T)/4$. If $\alpha(T)$ and $\beta(T)$ are both positive, then we have a minimum of $F_L(M)$ at $M = \langle M \rangle = 0$, which in this case is also a global minimum. We call the corresponding thermodynamic phase the paramagnetic or disordered phase, because there is no ferromagnetic order. However, if $\alpha(T)$ changes sign, then $F_L(M = 0)$ actually becomes a maximum and the minimum of the Landau free energy occurs at a nonzero value of the magnetization, as also illustrated by Fig. 9.1. Since we now have ferromagnetic ordering in the system, described by the nonzero expectation value $|\langle M \rangle| > 0$, a phase transition has occurred. It is a continuous phase transition, because the evolution to a nonzero order parameter has happened in a continuous manner. The transition is also commonly referred to as being of second order, for which the reason will become clear in Sect. 9.2. Note that the transition takes place at the critical temperature T_c, which is determined by the quadratic coefficient $\alpha(T)$ changing sign, i.e. by $\alpha(T_c) = 0$. As a result, we have that

$$k_B T_c = Jz, \tag{9.9}$$

for the critical temperature determined within mean-field theory. Moreover, close to the critical temperature, the coefficients for the Landau free energy

$$F_L(M) = -k_B T N \log(2) + \frac{1}{2} N \left\{ \alpha(T) M^2 + \frac{1}{2} \beta(T) M^4 + \dots \right\}, \tag{9.10}$$

are given by

$$\alpha(T) = \frac{4Jz}{\hbar^2} \left(1 - \frac{Jz}{k_B T} \right) \simeq \frac{4k_B}{\hbar^2} (T - T_c) \tag{9.11}$$

and

$$\beta(T) \simeq \frac{16}{3\hbar^4} k_B T_c. \tag{9.12}$$

Slightly below the critical temperature $T < T_c$, when the order parameter $\langle M \rangle$ is small, the corresponding minimum of the free energy is determined by the first two terms in the expansion of (9.10), such that the Landau free energy is minimized when

$$\langle M \rangle = \sqrt{\frac{|\alpha(T)|}{\beta(T)}} \simeq \hbar \sqrt{\frac{3|T - T_c|}{4T_c}}. \tag{9.13}$$

To obtain $\langle M \rangle$ also further below the critical temperature, we can determine it self-consistently. This is achieved by using (9.2) with the Hamiltonian from (9.5), such that

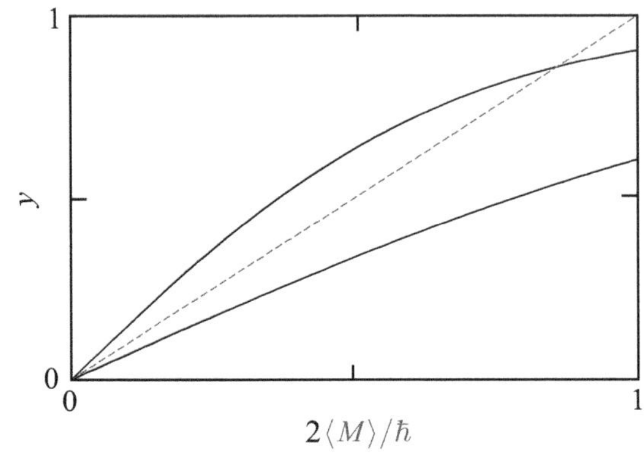

Fig. 9.2 Graphical solution to (9.16). The dashed line is given by $y = 2\langle M\rangle/\hbar$. The top curve corresponds to $y = \tanh(2zJ\langle M\rangle/\hbar k_B T)$ with $zJ/k_B T > 1$, while the bottom curve has $zJ/k_B T < 1$.

$$\langle M\rangle = \frac{1}{N}\left\langle \sum_i \hat{S}_{z,i} \right\rangle = \frac{1}{N}\sum_i \langle \hat{S}_{z,i}\rangle, \tag{9.14}$$

where we have in mean-field theory that

$$\begin{aligned}\langle \hat{S}_{z,j}\rangle &= \frac{1}{Z}\mathrm{Tr}\left[\hat{S}_{z,j}\,\exp\left\{-\frac{2zJN\langle M\rangle^2}{\hbar^2 k_B T} + \frac{4zJ\langle M\rangle}{\hbar^2 k_B T}\sum_i \hat{S}_{z,i}\right\}\right] \\ &= \sum_{m_j=\pm\hbar/2}\left\{m_j\,\exp\left\{\frac{4zJ\langle M\rangle}{\hbar^2 k_B T}m_j\right\}\right\} \Big/ \sum_{m_j=\pm\hbar/2}\exp\left\{\frac{4zJ\langle M\rangle}{\hbar^2 k_B T}m_j\right\},\end{aligned} \tag{9.15}$$

such that we finally obtain

$$\langle M\rangle = \frac{\hbar}{2}\tanh\left(\frac{2zJ\langle M\rangle}{\hbar k_B T}\right). \tag{9.16}$$

Note that this result also follows directly from solving

$$\left.\frac{dF_L(M)}{dM}\right|_{M=\langle M\rangle} = 0, \tag{9.17}$$

where $F_L(M)$ is given by (9.7). In Fig. 9.2, we show how the solution to (9.16) can be obtained graphically. We see that $\langle M\rangle = 0$ is the only solution, when $zJ/k_B T < 1$. However, another solution at nonzero $\langle M\rangle$ sets in when the slope on the right-hand side of (9.16) becomes larger than one at $\langle M\rangle = 0$. This leads again to the previously obtained condition

$$k_B T_c = zJ. \tag{9.18}$$

9.2 Landau Approach

The mean-field treatment of the Ising model gives us some important insights into the theory of phase transitions. For example, we have seen how the ferromagnetic phase and the paramagnetic phase are distinguished by the value of the magnetization $\langle M \rangle$. This is an example of an order parameter that discriminates between the two phases involved in the phase transition. Most phase transitions can be described with an appropriate order parameter, although a rare exception to this rule is found in the Kosterlitz-Thouless transition whose treatment is therefore postponed to Chap. 15. In the Landau approach to phase transitions, the central role is played by the Landau free-energy functional $F_{\text{L}}[m]$, which, as we show next, corresponds to an effective Hamiltonian that describes the system on a macroscopic scale after the microscopic fluctuations have been integrated out.

To make this statement more concrete, we apply the Landau approach to the particular case of the Ising model. The Landau free-energy functional can then be obtained from the Ising Hamiltonian of (9.1) in the following way. First, we define the continuous spin density

$$\hat{s}_z(\mathbf{x}) = \sum_i \delta(\mathbf{x} - \mathbf{x}_i)\hat{S}_{z,i}, \tag{9.19}$$

where the summation is over all lattice sites. Moreover, we introduce the following formal identity for the Dirac delta functional

$$\hat{1} = \int \mathrm{d}[m]\delta\left[m(\mathbf{x}) - \hat{s}_z(\mathbf{x})\right], \tag{9.20}$$

where the functional integral is over all possible functions $m(\mathbf{x})$. If the Ising lattice would consist of a single point, then (9.20) would correspond to the ordinary delta function $\delta(m - \hat{s}_z)$ integrated over the variable m. In general, the relation from (9.20) can be considered as an infinite dimensional integral over infinitely many delta functions, namely one for each spatial point $\mathbf{x}$. Inserting (9.20) into the partition function, we find

$$Z = \exp\{-\beta F\} = \mathrm{Tr}\left[\int \mathrm{d}[m]\ \delta\left[m(\mathbf{x}) - \hat{s}_z(\mathbf{x})\right]\exp\left\{-\beta\hat{H}\right\}\right], \tag{9.21}$$

after which we interchange the order of the summation and integration and perform the sum over all spin configurations that are consistent with the position-dependent magnetization $m(\mathbf{x})$ as set by the delta functional. This then defines the Landau free energy functional, such that it is given by

$$Z = \int \mathrm{d}[m]\exp\left\{-\frac{1}{k_B T}F_{\text{L}}[m]\right\}. \tag{9.22}$$

Although this rather formal derivation is exact, it might seem that it does not help us much, because we are often not able to perform the microscopic sum over spins exactly to obtain an explicit form for the Landau free energy. Progress can, however, be made by using phenomenological approaches to obtain the correct form of the Landau free energy functional, and by using appropriate approximations such as for example the mean-field approximation, which only takes into account the largest contribution to the partition function Z from (9.22). This maximal contribution comes from the equilibrium configuration $\langle m \rangle$ of the order parameter, which minimizes the Landau free energy and is given by

$$\left. \frac{\delta F_{\mathrm{L}}[m]}{\delta m(\mathbf{x})} \right|_{m=\langle m \rangle} = 0. \tag{9.23}$$

We thus have $Z \simeq \mathrm{e}^{-F_{\mathrm{L}}(\langle m \rangle)/k_{\mathrm{B}}T}$, which is sometimes also called the Landau approximation.

Looking in more detail at the Ising model, we start by noting that near the critical temperature for the second-order phase transition to the ferromagnetic phase, the order parameter $\langle m \rangle$ is small, such that the minimum can be found most easily by expanding the Landau free energy in powers of $m(\mathbf{x})$ and $\boldsymbol{\nabla} m(\mathbf{x})$. Moreover, we have that the symmetries that are present in the microscopic model continue to exist in the Landau free energy. Due to spin reversal symmetry, there are only even powers of $m(\mathbf{x})$ and $\boldsymbol{\nabla} m(\mathbf{x})$ in the expansion for the Landau free energy of the Ising model, such that we only keep

$$\begin{aligned} &F_{\mathrm{L}}[m] \\ &= \frac{1}{2} \int \mathrm{d}\mathbf{x} \, \left\{ \gamma(T)(\boldsymbol{\nabla} m(\mathbf{x}))^2 + \alpha(T) m(\mathbf{x})^2 + \frac{\beta(T)}{2} m(\mathbf{x})^4 + \frac{\delta(T)}{3} m(\mathbf{x})^6 + \ldots \right\}, \end{aligned} \tag{9.24}$$

where the expansion coefficients in general depend on temperature. Furthermore, it can be shown that $\gamma(T)$ is positive in the case of a positive J, which reflects the fact that it costs energy to have spins in the system that are not aligned due to a spatially varying magnetization. This implies that the minimum of the Landau free energy $\langle m \rangle$ should be independent of position, such that we only have to determine the scalar $\langle m \rangle$ that minimizes

$$F_{\mathrm{L}}[m] = V f_{\mathrm{L}}(m) = \frac{V}{2} \left\{ \alpha(T) m^2 + \frac{\beta(T)}{2} m^4 + \frac{\delta(T)}{3} m^6 + \ldots \right\} \tag{9.25}$$

with V the volume of the system and $f_{\mathrm{L}}(m)$ the Landau free-energy density.

In general, if a phase transition occurs, the behavior of the corresponding Landau free-energy density $f_{\mathrm{L}}(m)$ falls into two different categories. First, we consider the case for which the expansion coefficients of the free-energy density are positive for all terms that are of higher order than quadratic in m. Then, if the quadratic coefficient $\alpha(T)$ is also positive, we have a minimum at $m = \langle m \rangle = 0$. However, if $\alpha(T)$ becomes negative, the minimum shifts away in a continuous manner to a nonzero value of $\langle m \rangle$, resulting in a continuous phase transition. Note that the Landau free-energy density $f_{\mathrm{L}}(m)$ has a single minimum and that the order parameter has no

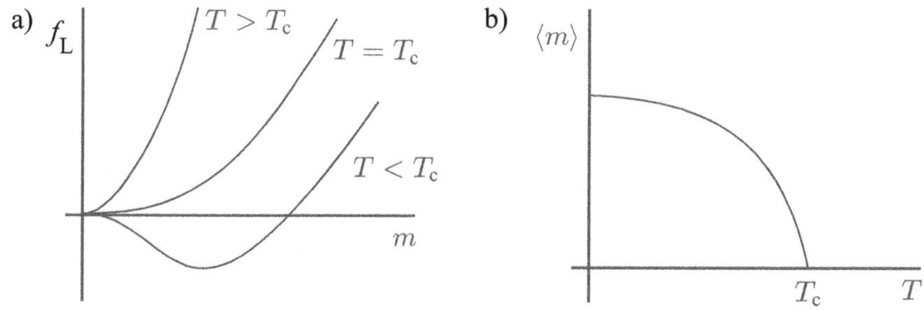

Fig. 9.3 Qualitative behavior of a) the Landau free energy density $f_L(m)$ and b) the order parameter $\langle m \rangle$ for a continuous phase transition.

discontinuity at the critical temperature T_c, as illustrated in Fig. 9.3. Close to the critical temperature, when the quadratic coefficient changes sign, we have

$$\alpha(T) \simeq \alpha_c k_B (T - T_c) \quad \text{and} \quad \beta(T) \simeq \beta_c, \tag{9.26}$$

as we also explicitly found in the mean-field treatment of the previous section. As a result, we find from (9.25) that the order parameter slightly below the critical temperature is given by

$$\langle m \rangle = \sqrt{-\frac{\alpha(T)}{\beta(T)}} \simeq \sqrt{\frac{\alpha_c k_B |T - T_c|}{\beta_c}}. \tag{9.27}$$

Note that the corresponding free-energy density f yields

$$f = f_L(\langle m \rangle) = -\frac{\alpha(T)^2}{2\beta_c} = -\frac{\alpha_c^2}{2\beta_c} k_B^2 (T - T_c)^2, \tag{9.28}$$

such that there is a discontinuity in the second derivative of the free-energy density with respect to the temperature. For example, we have from mean-field theory that in the disordered phase this second derivative is zero, whereas in the ferromagnetic phase it is $-(k_B \alpha_c)^2/\beta_c$. This kind of nonanalytic behavior is historically the reason why this transition is referred to as a second-order phase transition.

We may compare these results with the behavior of the free-energy density and the order parameter for the second class of phase transitions, namely the discontinuous or first-order phase transitions. The corresponding behavior is illustrated in Fig. 9.4. At temperatures very high compared to the critical temperature T_c, the system is fully disordered and the Landau free energy density $f_L(m)$ has a single minimum at $m = 0$ corresponding to an order parameter $\langle m \rangle$ equal to zero. However, by bringing the temperature closer to T_c the free energy may develop a second local minimum which is initially higher than the minimum at $m = 0$, such that the equilibrium value of $\langle m \rangle$ is still zero and no phase transition has occurred. Lowering the temperature even further, the value of the free energy in the second minimum decreases until,

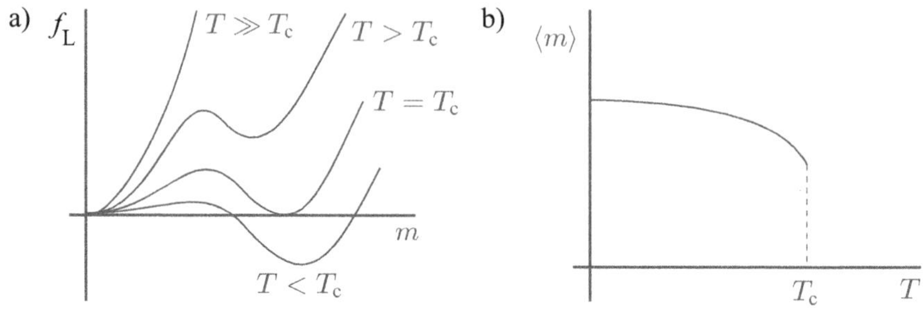

Fig. 9.4 Qualitative behavior of a) the Landau free energy density $f_L(m)$ and b) the order parameter $\langle m \rangle$ for a discontinuous or first-order phase transition.

precisely at the critical temperature T_c, it is equal to the free energy at $m = 0$. For temperatures below this critical value, the second minimum has actually become the global minimum of the free energy, which implies that $\langle m \rangle \neq 0$ and we are in the ordered phase. In this scenario, the order parameter has always a discontinuity at the critical temperature, such that it corresponds to a discontinuous, or first-order, phase transition.

To make the discussion of the first-order phase transitions more quantitative, we look again at a Landau free energy density of the form of (9.25), where we consider δ to be positive. Then, for minima of $f_L(m)$ both at a zero and a nonzero value of m, we need α to be positive and β to be negative. The first-order transition is determined by the condition that the two minima give rise to an equal free energy density, i.e

$$\alpha m^2 + \frac{\beta}{2} m^4 + \frac{\delta}{3} m^6 = 0, \tag{9.29}$$

while for a minimum also the first derivative needs to vanish, namely

$$\alpha m + \beta m^3 + \delta m^5 = 0. \tag{9.30}$$

Dividing out the trivial solution $m = 0$ and consequently eliminating m, we find a relation between α and β at the first-order transition, i.e. $\beta = -(16\alpha\delta/3)^{1/2}$, as is also illustrated in Fig. 9.5. A special point in this phase diagram occurs when both the second-order coefficient α as well as the fourth-order coefficient β are equal to zero. Then, the transition changes its nature from second-order to first-order and from Fig. 9.5, we see that three phases acquire the same free energy. This point is therefore also called a tricritical point. It is interesting to note that the behavior at the tricritical point is different from the behavior at the second-order critical points, although the order parameter still vanishes continuously. Indeed, if the coefficients $\alpha(T)$ and $\beta(T)$ vanish near the tricritical temperature T_{c3} as $(T - T_{c3})$, we have from from (9.30) that $\langle m \rangle$ vanishes as $(T - T_{c3})^{1/4}$, which is qualitatively different

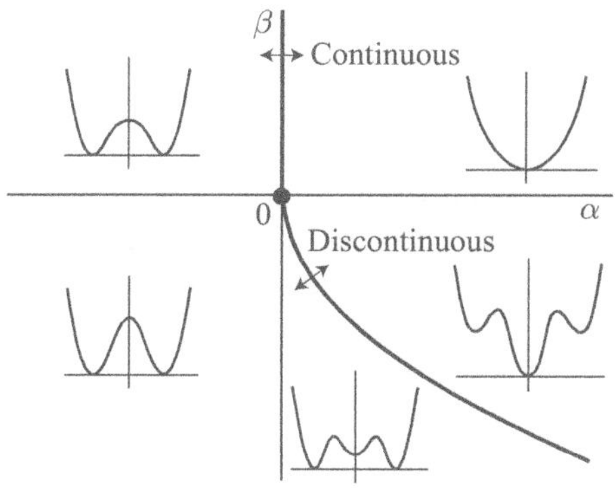

Fig. 9.5 Phase diagram for a system described by the Landau free energy density $f_L(m)$ from (9.25) as a function of the quadratic coefficient α and the quartic coefficient β, while the sextic coefficient δ is positive. Continuous, or second-order transitions occur along the positive β axis, where α changes sign, while discontinuous, or first-order transitions occur along the line $\beta = -(16\delta\alpha/3)^{1/2}$. These two lines are connected by a tricritial point. For each region in the phase diagram also a sketch of the Landau free energy as a function of m is given.

from (9.27). The power with which the order parameter vanishes is an example of a critical exponent.

9.3 Hubbard-Stratonovich Transformation

In the previous chapter, we used the Hubbard-Stratonovich transformation to arrive at the Hartree-Fock theory of an interacting quantum gas. As we discuss next, the transformation is also a very powerful tool for describing phase transitions. First, remember that for a homogeneous gas consisting of two spin species with a spin-independent interaction, we find with the use of perturbation theory up to any finite order that the interacting Green's function satisfies

$$G_{\alpha,\alpha'}(\mathbf{x},\tau;\mathbf{x},\tau^+) = \mp n\delta_{\alpha,\alpha'}/2, \tag{9.31}$$

where n is the total density. This follows from Wick's theorem and the form of the noninteracting Green's function $G_{0;\alpha,\alpha'}(\mathbf{x},\tau;\mathbf{x},\tau^+)$, which are discussed more extensively in Sect. 8.2. More generally speaking, the result of (9.31) is the consequence of both the translational invariance as well as the rotational symmetry in spin space of the action describing the many-body system.

However, we can imagine that a state with a constant particle density, as described by (9.31), is not always the energetically most favorable state for the interacting quantum gas. For example, in the case of strong repulsive long-range interactions, it is under certain conditions expected that at a (Wigner) crystalline phase is energetically more favorable, because it minimizes the interaction energy. We then have a spontaneous breaking of the translational symmetry, where we can introduce a corresponding order parameter to describe the associated phase transition in the system. Although spontaneous symmetry breaking cannot be described with perturbation theory, it can very conveniently be taken into account by using the Hubbard-Stratonovich transformation to bring the corresponding order parameter into the action.

Consider for example a Hubbard-Stratonovich transformation for an interacting Fermi gas to the collective field $\rho_{\alpha,\alpha'}(\mathbf{x},\tau)$ that is on average related to the single-particle density matrix by $\langle\rho_{\alpha,\alpha'}(\mathbf{x},\tau)\rangle = \langle\phi^*_{\alpha'}(\mathbf{x},\tau^+)\phi_\alpha(\mathbf{x},\tau)\rangle$. Indeed, the corresponding transformation would be achieved by inserting

$$1 = \int \mathrm{d}[\rho] \exp\left\{\frac{1}{2\hbar}(\rho - \phi^*\phi|V|\rho - \phi^*\phi)\right\}, \tag{9.32}$$

into the partition sum for the interacting Fermi gas, analogous to the discussion from Sect. 8.6. Since this results in a quadratic action for the atomic field $\phi(\mathbf{x},\tau)$, we can integrate this atomic field out exactly and obtain an effective action for the collective field $\rho_{\alpha,\alpha'}(\mathbf{x},\tau) = n(\mathbf{x},\tau)\delta_{\alpha,\alpha'}/2 + \mathbf{m}(\mathbf{x},\tau)\cdot\boldsymbol{\sigma}_{\alpha,\alpha'}$. It is not at all necessary that the minimum of the resulting effective action is given by the expectation value $\langle\rho_{\alpha,\alpha'}(\mathbf{x},\tau)\rangle = n\delta_{\alpha,\alpha'}$ as would be the conclusion from perturbation theory. Depending on for example the strength of the interaction and the temperature, the minimum of the effective action could perfectly well be given by a position-dependent expectation value

$$\langle\rho_{\alpha,\alpha'}(\mathbf{x},\tau)\rangle = G_{\alpha,\alpha'}(\mathbf{x},\tau;\mathbf{x},\tau^+) = n(\mathbf{x})\delta_{\alpha,\alpha'}/2, \tag{9.33}$$

which could describe a Wigner crystal or a charge-density wave, for which the particle density oscillates in space. If on the other hand, the Green's function is constant in space, but the minimum of the effective action for $\rho_{\alpha,\alpha'}(\mathbf{x},\tau)$ is given by

$$\langle\rho_{\alpha,\alpha'}(\mathbf{x},\tau)\rangle = G_{\alpha,\alpha'}(\mathbf{x},\tau;\mathbf{x},\tau^+) = n\delta_{\alpha,\alpha'}/2 + \mathbf{m}\cdot\boldsymbol{\sigma}_{\alpha,\alpha'}, \tag{9.34}$$

then the gas is in the ferromagnetic phase with a nonzero magnetization $\mathbf{m}$. Moreover, for a position-dependent magnetization $\mathbf{m}(\mathbf{x})$, we say that the system gives rise to a spin-density wave, such that we have

$$\mathbf{m}(\mathbf{x}) = \langle\mathbf{s}(\mathbf{x})\rangle = \langle\hat{\psi}^\dagger_\alpha(\mathbf{x},\tau)\boldsymbol{\sigma}_{\alpha,\alpha'}\hat{\psi}_{\alpha'}(\mathbf{x},\tau)\rangle/2, \tag{9.35}$$

where the vector $\boldsymbol{\sigma}$ consists of the three Pauli matrices. The possibility of a phase transition to a spin-density wave in an ultracold atomic gas is the topic of Exercise 9.2.

The main message of the above discussion is that with an appropriate Hubbard-Stratonovich transformation, the order parameter is introduced in an exact manner into the many-body theory. Often, it is then also possible to integrate out the atomic fields exactly, leading to an exact effective action for the collective field, which then in a natural way gives rise to the appropriate Landau theory of the phase transition. As an example, we note that after integrating out the atomic fields, the Hubbard-Stratonovich transformation of (9.32) would lead to

$$Z = \int d[n]\, d[\mathbf{m}]\, e^{-S^{\text{eff}}[n,\mathbf{m}]/\hbar} = \int d[\mathbf{m}]\, e^{-S^{\text{eff}}[\mathbf{m}]/\hbar}\,, \tag{9.36}$$

where in the second step we also integrated out the particle-density fluctuations. A minimum of the effective action at the expectation value $\langle \mathbf{m}(\mathbf{x},\tau)\rangle = \mathbf{m} \neq \mathbf{0}$, then signals a phase transition to a ferromagnetic phase with the magnetization $\mathbf{m}$ as the order parameter for this transition. For a space and time-independent magnetization $\mathbf{m}$, we compare (9.22) with (9.36), such that in mean-field theory the effective action $S^{\text{eff}}[\mathbf{m}]$ is then directly related to $\hbar F_{\text{L}}(\mathbf{m})/k_{\text{B}}T = \hbar\beta V f_{\text{L}}(\mathbf{m})$, where $f_{\text{L}}(\mathbf{m})$ is the Landau free-energy density introduced in the previous section. To go beyond mean-field theory, we can proceed by taking for instance quadratic fluctuations around the minimum into account when evaluating the functional integral in (9.22). This approach is discussed in the next section. Taking higher-order fluctuations into account turns out to be a difficult task which requires renormalization group methods to be discussed in Chap. 14.

9.4 Gaussian Fluctuations

In the Landau-Ginzburg theory of second-order phase transitions, the fluctuations around the order parameter are considered up to quadratic order. In the case of the Ising model, we start by writing

$$m(\mathbf{x}) = \langle m\rangle + m'(\mathbf{x}) \tag{9.37}$$

which we substitute into the functional integral from (9.22). Using also the expansion in (9.24), we obtain

$$\begin{aligned} Z &= \int d[m]\, \exp\left\{-\frac{1}{k_{\text{B}}T}F_{\text{L}}[m]\right\} \\ &= \int d[m']\exp\left\{-\frac{1}{2k_{\text{B}}T}\int d\mathbf{x}\left\{\gamma(\boldsymbol{\nabla} m'(\mathbf{x}))^2 + \alpha(T)\left(\langle m\rangle + m'(\mathbf{x})\right)^2\right.\right. \\ &\qquad\qquad \left.\left. + \frac{\beta}{2}\left(\langle m\rangle + m'(\mathbf{x})\right)^4\right\}\right\}. \end{aligned} \tag{9.38}$$

We may introduce the length scale $\xi(T)$ for the Ising model

$$\xi(T) = \begin{cases} \sqrt{\dfrac{\gamma}{|\alpha(T)|}} & T > T_{\text{c}} \\ \sqrt{\dfrac{\gamma}{2|\alpha(T)|}} & T < T_{\text{c}} \end{cases}\,, \tag{9.39}$$

which we call the correlation length for reasons that soon become clear. Note that close to T_{c}, the correlation length diverges as a function of temperature according to $\xi(T) \propto |T - T_{\text{c}}|^{-\nu}$, where $\nu = 1/2$, due to the fact that $|\alpha(T)|$ goes linearly to zero. The exponent ν with which the correlation length diverges is the second example of a critical exponent that we encounter. We come back to critical exponents in Chap. 14, when we treat critical phenomena more extensively. From (9.38) and (9.39), we obtain to second order in the fluctuations that

$$Z = \exp\left\{-\frac{1}{k_B T} F_L(\langle m\rangle)\right\} \times \int d[m'] \exp\left\{-\frac{\gamma}{2k_B T}\int d\mathbf{x}\left\{(\nabla m'(\mathbf{x}))^2 + \frac{1}{\xi(T)^2} m'(\mathbf{x})^2\right\}\right\}, \tag{9.40}$$

where we also used (9.27).

We further evaluate this functional integral by making use of the Fourier expansion

$$m'(\mathbf{x}) = \sum_{\mathbf{k}} m'_{\mathbf{k}} \frac{e^{i\mathbf{k}\cdot\mathbf{x}}}{\sqrt{V}}, \tag{9.41}$$

for which we have that $m'^*_{\mathbf{k}} = m'_{-\mathbf{k}}$, because the magnetization $m(\mathbf{x})$ is real. Substituting the Fourier expansion into the functional integral of (9.40) and performing the integration over position, we obtain the Gaussian integral

$$\int d[m'] \exp\left\{-\frac{\gamma}{2k_B T}\sum_{\mathbf{k}} m'^*_{\mathbf{k}} \left(\mathbf{k}^2 + \xi^{-1}(T)^2\right) m'_{\mathbf{k}}\right\},$$

which can be performed exactly. Note that in the above expression not all variables are independent, because $m'^*_{\mathbf{k}} = m'_{-\mathbf{k}}$. As a result, we should perform the functional integration only over half of the Fourier space in order to be able to use (2.67). Moreover, from the above expression, we can also read off the correlation function

$$\langle m'^*_{\mathbf{k}} m'_{\mathbf{k}}\rangle = \frac{k_B T}{\gamma} \frac{1}{\mathbf{k}^2 + \xi^{-1}(T)^2}. \tag{9.42}$$

By performing the inverse Fourier transform, we determine the correlation function for the fluctuations in real space

$$\begin{aligned}\langle m'(\mathbf{x}) m'(\mathbf{x}')\rangle &= \frac{k_B T}{\gamma}\int \frac{d\mathbf{k}}{(2\pi)^3} \frac{e^{i\mathbf{k}\cdot(\mathbf{x}-\mathbf{x}')}}{\mathbf{k}^2 + \xi^{-1}(T)^2} \\ &= \frac{k_B T}{4\pi\gamma} \frac{e^{-|\mathbf{x}-\mathbf{x}'|/\xi(T)}}{|\mathbf{x}-\mathbf{x}'|},\end{aligned} \tag{9.43}$$

using the result from Exercise 2.5. We thus find that the correlation length sets the length scale over which the correlations between the spins decay, thereby explaining its name.

In our treatment of the Gaussian approximation for the fluctuations, we have not answered yet the important question of the validity of this procedure. To obtain an estimation for the answer to this question, we start by noting that the divergence for small separations in (9.43) is actually an artefact of the approximation that we have used. In particular, we have performed a gradient expansion in (9.24), where we only took into account second-order gradients, leading to the $1/\mathbf{k}^2$ dependence for large momenta of the Green's function from (9.42). Taking higher-order gradients into account would lead to more rapid decay at higher momenta of the Green's function, which would consequently remove the divergence for small separations present in (9.43). As a result, we expect that the gradient expansion is valid for the physics at momenta smaller than $\hbar/\xi$ or equivalently at separations larger than ξ. Then, to determine whether or not the fluctuations become important, we compare the correlation function $\langle m'(\mathbf{x})m'(\mathbf{x}')\rangle$ at the typical separation of $|\mathbf{x}-\mathbf{x}'| \simeq \xi(T)$ with the square of the mean order parameter, i.e. $\langle m\rangle^2$. Fluctuations are considered not to be important as long as we satisfy

$$\langle m'(\mathbf{x})m'(\mathbf{x}')\rangle \ll \langle m\rangle^2, \tag{9.44}$$

for $|\mathbf{x}-\mathbf{x}'| \simeq \xi(T)$. This criterium is known as the Ginzburg criterium. Working out both sides of (9.44), the criterium can be written in the vicinity of the critical temperature as

$$\frac{k_B T_c}{4\pi\gamma}\xi(T)^{-1} \ll \langle m\rangle^2 = \frac{|\alpha(T)|}{\beta}. \tag{9.45}$$

Ignoring the factor of two difference in the definition of $\xi(T)$ for $T < T_c$ and $T > T_c$, we see that the Ginzburg criterium becomes

$$\xi(T) \ll \frac{4\pi\gamma^2}{\beta k_B T_c}. \tag{9.46}$$

Note that we have derived the above results for the specific case of the Ising model. In Chap. 14, we find with more sophisticated methods that the criterium is generally correct for quantum theories with a phase transition at $T_c \neq 0$. For ultracold atomic gases, the effects of fluctuations are then seen to be important only in a small temperature interval around the critical temperature, such that for many applications it is possible to neglect them. What usually cannot be neglected is the effect of the inhomogeneity of the gas due to the presence of an external potential. In the context of Landau theory, this would imply that the expansion coefficients $\alpha(T)$, $\beta(T)$ and $\gamma(T)$ become dependent on the spatial position in the trapping potential, which leads to the possibility of having different (homogeneous) thermodynamic phases as a function of position in the trap. For example, the bimodality of the middle velocity distribution in Fig. 4.4 was caused by such an inhomogeneity, because a condensate could only form in the center of the trap where the density was highest, while for the low densities in the outer regions of the cloud the gas was still normal.

9.5 Spontaneous Symmetry Breaking

Above the critical temperature, the minimum of the Landau free energy from (9.24) is given by $\langle m \rangle = 0$. Below the critical temperature we have that $\langle m \rangle$ is nonzero, which therefore results in $\langle m \rangle \neq \langle -m \rangle$, although the Landau free energy itself still satisfies reflection symmetry, $F_{\rm L}[m] = F_{\rm L}[-m]$. This is an example of a phenomenon called spontaneous symmetry breaking, where the phase of the system breaks a symmetry of the underlying microscopic Hamiltonian. For the case of a one-component real scalar order parameter $\langle m \rangle$, the Landau free energy $F_{\rm L}[m]$ has precisely two minima located at $\pm\langle m \rangle$, as shown in Fig. 9.1. In cooling the system below the critical temperature, the system arbitrarily has to choose one of the two minima. For the more general case, when the order parameter becomes an n-component vector $\mathbf{m}$ and the Landau free energy has an $O(n)$ symmetry, there is actually a continuum of minima that are connected by a rotation of the order parameter. In this case, the Landau free energy in (9.24) can be expanded as

$$F_{\rm L}[\mathbf{m}] = \frac{1}{2}\int \mathrm{d}\mathbf{x} \left\{ \gamma(\boldsymbol{\nabla}\mathbf{m}(\mathbf{x}))^2 + \alpha(T)\mathbf{m}(\mathbf{x})^2 + \frac{\beta}{2}\mathbf{m}(\mathbf{x})^4 + \ldots \right\}. \quad (9.47)$$

Then, analogous to the calculation for the single-component case, we may expand the order parameter as $\mathbf{m} = \langle \mathbf{m} \rangle + \mathbf{m}'(\mathbf{x})$ and calculate the correlation function

$$\langle m_i'(\mathbf{x}) m_j'(\mathbf{x}') \rangle \equiv -G_{ij}(\mathbf{x}-\mathbf{x}'). \quad (9.48)$$

When the temperature is above $T_{\rm c}$, the quadratic part of the Landau free energy simply becomes the sum of three separate Landau free energy contributions $F_{\rm L}[\mathbf{m}'(\mathbf{x})] = \sum_i F_{\rm L}[m_i'(\mathbf{x})]$, such that we can copy our calculation of the previous section to obtain

$$G_{ij}(\mathbf{x}-\mathbf{x}') = -\delta_{ij}\frac{k_{\rm B}T}{4\pi\gamma}\frac{\mathrm{e}^{-|\mathbf{x}-\mathbf{x}'|/\xi(T)}}{|\mathbf{x}-\mathbf{x}'|}. \quad (9.49)$$

The interesting case is when $T < T_{\rm c}$, as we show now. Writing down the fourth-order term explicitly and keeping only the quadratic terms in the fluctuations yields

$$\frac{\beta}{4}\sum_{i,j}\int \mathrm{d}\mathbf{x}\ \left\{2\langle m_i \rangle^2 m_j'^2(\mathbf{x}) + 4\langle m_i \rangle \langle m_j \rangle m_i'(\mathbf{x}) m_j'(\mathbf{x})\right\}.$$

Making again a Fourier expansion and combining the above terms with the other quadratic terms from (9.47), we find that the second-order contribution of the fluctuations to the partition function is given by the functional integral

$$\int \mathrm{d}[\mathbf{m}'] \exp\left\{ \frac{1}{2}\sum_{\mathbf{k}} \mathbf{m}_{\mathbf{k}}'^{*}\ \mathbf{G}^{-1}(\mathbf{k})\ \mathbf{m}_{\mathbf{k}}', \right\},$$

where the matrix elements of $\mathbf{G}^{-1}$ are given by

$$G_{ij}^{-1}(\mathbf{k}) = -\frac{1}{k_{\mathrm{B}}T}\delta_{i,j}\left(\gamma\mathbf{k}^2 + \alpha(T) + \beta\langle\mathbf{m}\rangle^2\right) - \frac{2\beta}{k_{\mathrm{B}}T}\langle m_i\rangle\langle m_j\rangle. \tag{9.50}$$

Introducing the unit vector $\mathbf{v}$ along the direction of the magnetization

$$\mathbf{v} = \frac{\langle\mathbf{m}\rangle}{|\langle\mathbf{m}\rangle|}, \tag{9.51}$$

the matrix element (9.50) becomes

$$G_{ij}^{-1}(\mathbf{k}) = -\frac{1}{k_{\mathrm{B}}T}\left(\gamma\mathbf{k}^2 - 2\alpha(T)\right)v_iv_j - \frac{\gamma\mathbf{k}^2}{k_{\mathrm{B}}T}\left(\delta_{i,j} - v_iv_j\right), \tag{9.52}$$

where we have used that $\langle\mathbf{m}\rangle^2 = -\alpha(T)/\beta$, and where we have explicitly separated the propagator for fluctuating modes that are parallel and orthogonal to $\mathbf{v}$. To see this, we can consider the three-dimensional case with the magnetization in the z direction. Then, the first term on the right-hand side of (9.52) describes G_{zz}^{-1}, whereas the second term describes G_{xx}^{-1} and G_{yy}^{-1}. Moreover, note that the matrix product of the tensors v_iv_j and $\delta_{i,j} - v_iv_j$ is zero as can be seen from

$$v_iv_j\left(\delta_{j,i'} - v_jv_{i'}\right) = v_iv_{i'} - v_iv_{i'} = 0, \tag{9.53}$$

where repeated indices are summed over, such that $v_iv_jv_jv_{i'} = v_iv_{i'}$ for $\mathbf{v}$ a unit vector. As a result, we also have that

$$\left(\delta_{i,j} - v_iv_j\right)\left(\delta_{j,i'} - v_jv_{i'}\right) = \left(\delta_{i,i'} - v_iv_{i'}\right). \tag{9.54}$$

With this decomposition of $\mathbf{G}^{-1}$ into longitudinal and transverse contributions, we find that the Green's function $\mathbf{G}$ is given by

$$G_{ij}(\mathbf{k}) = -\frac{k_{\mathrm{B}}T}{\gamma\mathbf{k}^2 - 2\alpha(T)}v_iv_j - \frac{k_{\mathrm{B}}T}{\gamma\mathbf{k}^2}\left(\delta_{i,j} - v_iv_j\right). \tag{9.55}$$

We thus have found that the correlation function of the longitudinal fluctuations behaves analogously to the scalar case, whereas the fluctuations $\mathbf{m}'$ that are perpendicular to the magnetization $\langle\mathbf{m}\rangle$ are described by the propagator

$$G_{ij}(\mathbf{k}) = \frac{-k_{\mathrm{B}}T}{\gamma\mathbf{k}^2}\left(\delta_{i,j} - v_iv_j\right). \tag{9.56}$$

In real-space, we have that this transverse part of the correlation function is given by

$$\begin{aligned}\langle m_i'(\mathbf{x})m_j'(\mathbf{x}')\rangle &= -\int\frac{\mathrm{d}\mathbf{k}}{(2\pi)^3}G_{ij}(\mathbf{k})\mathrm{e}^{\mathrm{i}\mathbf{k}\cdot(\mathbf{x}-\mathbf{x}')}\\ &= \frac{k_{\mathrm{B}}T}{4\pi\gamma|\mathbf{x}-\mathbf{x}'|}\left(\delta_{i,j} - v_iv_j\right).\end{aligned} \tag{9.57}$$

The last line of the previous equation shows that the correlation function of the transverse fluctuations no longer falls off exponentially, but only algebraically. This is a general feature of a phase transition that spontaneously breaks a continuous symmetry and is embodied in the Goldstone theorem. It states that with each spontaneously broken continuous symmetry there is an associated massless or Goldstone mode. The reason for calling the transverse Goldstone mode massless comes from high-energy physics, where the Green's functions or propagators of the elementary particles are typically of the form

$$G(\mathbf{k}) \propto \frac{1}{(\hbar c\mathbf{k})^2 + m^2c^4},$$

with c the speed of light and m the rest mass of the particle. In our case, we find that after the spontaneous symmetry breaking the system develops transverse modes that behave just like an elementary particle with zero mass.

For the propagator of the longitudinal modes, we have that

$$G_{ij}(\mathbf{k}) = -\frac{k_{\mathrm{B}}T}{\gamma\mathbf{k}^2 + \gamma/\xi(T)^2} v_i v_j, \tag{9.58}$$

such that we obtain

$$\begin{aligned} \langle m'_i(\mathbf{x}) m'_j(\mathbf{x}')\rangle &= -\int \frac{\mathrm{d}\mathbf{k}}{(2\pi)^3} G_{ij}(\mathbf{k}) \mathrm{e}^{\mathrm{i}\mathbf{k}\cdot(\mathbf{x}-\mathbf{x}')} \\ &= \frac{k_{\mathrm{B}}T}{4\pi\gamma} \frac{\mathrm{e}^{-|\mathbf{x}-\mathbf{x}'|/\xi(T)}}{|\mathbf{x}-\mathbf{x}'|} v_i v_j. \end{aligned} \tag{9.59}$$

Combining (9.57) and (9.59), we see that the total correlation function of the order parameter field is given by

$$\begin{aligned} \langle m_i(\mathbf{x}) m_j(\mathbf{x}')\rangle &= \langle\mathbf{m}\rangle^2 v_i v_j + \frac{k_{\mathrm{B}}T}{4\pi\gamma} \frac{\mathrm{e}^{-|\mathbf{x}-\mathbf{x}'|/\xi(T)}}{|\mathbf{x}-\mathbf{x}'|} v_i v_j + \frac{k_B T}{4\pi\gamma|\mathbf{x}-\mathbf{x}'|}(\delta_{i,j} - v_i v_j) \\ &= \langle\mathbf{m}\rangle^2 \left\{ v_i v_j + \frac{k_{\mathrm{B}}T\beta\xi(T)^2}{2\pi\gamma^2} \frac{\mathrm{e}^{-|\mathbf{x}-\mathbf{x}'|/\xi(T)}}{|\mathbf{x}-\mathbf{x}'|} v_i v_j \right. \\ &\qquad \left. + \frac{k_{\mathrm{B}}T\beta\xi(T)^2}{2\pi\gamma^2|\mathbf{x}-\mathbf{x}'|}(\delta_{i,j} - v_i v_j) \right\}, \end{aligned} \tag{9.60}$$

from which we see that at very large separations, the correlation function approaches the constant $\langle\mathbf{m}\rangle^2 v_i v_j$. This property is also known as long-range order. In Chap. 13 we discuss several more aspects of symmetries and symmetry breaking, including conservation laws and the Ward identities.

9.6 Problems

Exercise 9.1. Calculate the correlation function $\langle m'(\mathbf{x})m'(\mathbf{x}')\rangle$ for the case when the order parameter $m(\mathbf{x})$ in the Landau free energy from (9.38) is complex valued.

Exercise 9.2. Spin-Density Wave
Consider an interacting gas of N spin-1/2 fermions on a one-dimensional line with length L. The action for this gas at temperature $T = 1/k_{\mathrm{B}}\beta$ and chemical potential μ is given by

$$S[\phi^*,\phi] = \sum_{\alpha} \int_0^{\hbar\beta} \mathrm{d}\tau \int \mathrm{d}x\, \phi^*_{\alpha}(x,\tau) \left\{ \hbar\frac{\partial}{\partial\tau} - \frac{\hbar^2}{2m}\frac{\partial^2}{\partial x^2} - \mu \right\} \phi_{\alpha}(x,\tau) \tag{9.61}$$
$$+\frac{1}{2}\sum_{\alpha\neq\alpha'} \int_0^{\hbar\beta} \mathrm{d}\tau \int \mathrm{d}x \int \mathrm{d}x'\, \phi^*_{\alpha}(x,\tau)\phi^*_{\alpha'}(x',\tau)V(x-x')\phi_{\alpha'}(x',\tau)\phi_{\alpha}(x,\tau)\,,$$

where m is the mass of the fermions and $V(x-x')$ the interaction potential. Even though the above action is translationally invariant, the gas can in principle undergo a phase transition to a phase in which the spin density oscillates with a wavelength given by $2\pi/Q$. This phase is called a spin-density wave. To explore this possibility, we can imagine that we have performed a Hubbard-Stratonovich transformation to the field $\kappa_{\alpha}(x,\tau)$, which on average is related to the spin densities of the gas by $\int \mathrm{d}x'\, V(x-x')n_{\alpha}(x')$.
(a) Argue on physical grounds, i.e. without explicit calculations, that in the case of a spin density wave the average of the auxiliary field $\kappa_{\alpha}(x,\tau)$ has the form

$$\langle \kappa_{\alpha}(x,\tau)\rangle = V_0\frac{n}{2} + \frac{\alpha}{2}\left(\Delta \mathrm{e}^{\mathrm{i}Qx} + \Delta^* \mathrm{e}^{-\mathrm{i}Qx}\right), \tag{9.62}$$

with $V(k) = \int \mathrm{d}x\, V(x)\mathrm{e}^{-ikx}$ the Fourier transform of the interaction potential, $n = N/L$ the total average density of the gas, and $\alpha = \pm 1$. The complex number Δ is the order parameter of the spin-density wave.
(b) Using the above expression for the mean field, give the momentum-space action for the fermions in mean-field theory. Use $\phi_{\alpha}(x,\tau) = \sum_k \phi_{k,\alpha}(\tau)\mathrm{e}^{\mathrm{i}kx}/\sqrt{L}$ and the notation $\varepsilon_k = \hbar^2k^2/2m$. Note that the spin-density wave can transfer a wavevector $\pm Q$ to the fermions.
(c) Assuming that the spin-density wave can only exist at temperatures far below the Fermi temperature, argue that the most likely value of Q is equal to twice the Fermi wavevector, i.e. $2k_{\mathrm{F}}$.
(d) In an excellent approximation we can, therefore, consider only the coupling between left-moving states with $k<0$ and right-moving states with $k>0$. Determine in this approximation the dispersion $\hbar\omega_k$ of the single-particle excitations in the presence of a spin-density wave.
(e) Finally, show that in general the mean-field critical temperature is determined by the condition

$$\frac{V(Q)}{L}\sum_{p}\frac{N_{\mathrm{FD}}(\varepsilon_{Q+p})-N_{\mathrm{FD}}(\varepsilon_{p})}{\varepsilon_{Q+p}-\varepsilon_{p}}=-1\ , \tag{9.63}$$

To this end, take the following steps: 1) Perform the desired Hubbard-Stratonovich transformation to the collective κ_α fields, 2) integrate out the atomic fields to obtain the partition sum and the effective action in terms of the collective fields, 3) neglect fluctuations and consider the effective action up to quadratic order in $\langle\kappa_\alpha(x,\tau)\rangle \equiv \kappa_\alpha(x)$, 4) transform to momentum space and note that you obtain a 2×2 matrix structure in spin-space that can be diagonalized by introducing new variables, namely the density modes $\rho_k \equiv (\kappa_{k,+}+\kappa_{k,-})/\sqrt{2}$ and the magnetization modes $m_k \equiv (\kappa_{k,+}-\kappa_{k,-})/\sqrt{2}$, 5) the condition for the critical temperature of the second-order phase transition to the spin-density wave is then ultimately obtained by demanding that the quadratic coefficient for the magnetization mode m_Q changes sign.

Additional Reading

- An alternative introduction to the Landau theory of phase transitions can be found in J. W. Negele and H. Orland, *Quantum Many-Particle Systems*, (Westview Press, Boulder, 1998).
- See also the introductory chapters in E. Fradkin, *Field Theories of Condensed Matter Systems*, (Addison-Wesley, Redwood City, 1991).

Chapter 10
Atomic Physics

Joy in looking and comprehending is Nature's most beautiful gift.
–Albert Einstein

Ultracold quantum gases are like man-made universes that allow us, as we will see, to study interesting quantum many-body phenomena in detail. Although it turns out that these quantum gases are extremely well suited for manipulation by experimentalists, there are some limitations. A typical trapped alkali gas consists of about $10^5 - 10^9$ atoms and has, for realistic trap parameters, a central density of $n \simeq 10^{12} - 10^{15}$ cm^{-3}. This is many orders of magnitude less dense than air, which has a typical density of about 10^{19} cm^{-3}. Nevertheless, the gas can be cooled down to such low temperatures that it reaches the quantum degenerate regime where the thermal de Broglie wavelength $\Lambda = (2\pi\hbar^2/mk_{\mathrm{B}}T)^{1/2}$ is on the same order as the average interatomic distance $n^{-1/3}$. For the densities quoted this means that the temperature has to be as low as $1 - 100$ nK. This makes ultracold gases the coldest objects in the universe. The physics of how to cool dilute alkali gases to quantum degeneracy is extensively described elsewhere [40, 41] and we do not cover this subject here. A crucial ingredient, however, is that the gas is trapped in an external potential to keep the gas away from physical walls that can never be cooled to such low temperatures. The traps used in practice can almost always be well approximated by an anisotropic harmonic oscillator potential with frequencies of about $1 - 100$ Hz. The associated energy level splittings $\hbar\omega_i$ are then typically much smaller than the thermal energy $k_{\mathrm{B}}T$ or the chemical potential μ. However, this is not always the case, and it turns out to be also possible to create such steep potentials in certain directions, that in these directions only the quantum state with the lowest energy can be occupied by the atoms. As a result, the atomic motion is essentially frozen out in these directions, which allows for the creation of effectively one-dimensional and two-dimensional quantum gases.

In this chapter, we look at the relevant atomic physics that is necessary to understand the microscopic origin of the relevant physical parameters for interacting atomic quantum gases. In the following chapters, we then combine this knowledge with the quantum field theory formalism developed in the previous chapters to describe realistic ultracold atomic many-body systems. We start with discussing the fine and hyperfine structure of the atomic energy levels, as well as the Zeeman effect, because these are crucial for the ability to trap atoms by applying external magnetic or electric fields. The electric fields that are generated by a laser can also be used to create periodic trapping potentials, called optical lattices, which are discussed in Chap. 16. Next, we treat two-body scattering of atoms, which is the dominant interaction mechanism in ultracold atomic gases. We first discuss two-body scattering in vacuum, and then generalize the treatment to include also the presence of a medium. Finally, we end this chapter with a detailed discussion of the different physical regimes that we can explore with ultracold atomic gases, depending on the atomic density, the temperature and the interaction strength.

10.1 Atomic Structure

The understanding of the atomic structure resulted primarily from studying the absorption and emission of light by atoms. Roughly speaking, it was found that an atom is composed of a nucleus and a surrounding cloud of electrons which interact with each other and with the nucleus through the Coulomb interaction. Additional effects, such as the coupling of the electron spins with their orbital angular momentum and the weaker coupling of the nuclear spin with the electron spins, give rise to a more detailed and more complex picture of the atom. In fact, the atomic level structure is affected quantitatively by even more exotic physical effects, such as relativistic corrections to the kinetic energy of the electrons and the Lamb shift arising from the quantization of the electromagnetic field. These effects are, however, neglected in the following.

To make the above sketched picture a bit more concrete, we consider the for us relevant case of an alkali atom, which can be considered as a hydrogen-like atom with an inner core of charge e and a single outer electron with charge $-e$ and mass m_{e}. The dominant interaction is the Coulomb interaction between the electron and the core, which depends in a good approximation only on the relative coordinate r between the two. The Hamiltonian is then given by

$$\hat{H} = \frac{\hat{\mathbf{p}}^2}{2m_{\mathrm{e}}} - \frac{e^2}{4\pi\varepsilon_0 \hat{r}} , \tag{10.1}$$

where ε_0 is the permittivity of free space. Moreover, we have considered the core to be fixed in space because it is much heavier than the electron. If desired, this approximation can be easily relaxed by interpreting m_{e} as the reduced electron mass. Most textbooks on quantum mechanics show how the resulting time-independent Schrödinger equation can be solved. The spherical symmetry of the Coulomb potential allows for a factorization of the wavefunction $\psi_{n\ell m}(\mathbf{r}) = R_{n\ell}(r)Y_{\ell m}(\vartheta, \varphi)$ into a radial and an angular part respectively. The angular part $Y_{\ell m}(\vartheta, \varphi)$ is given by the spherical harmonics from (3.60), while the radial solution can is of the form $R_{n\ell}(r) = f_{n\ell}(r)\mathrm{e}^{-r/(n+1)a_0}$ with $n = 0, 1, 2, \ldots$. Here, the function $f_{n\ell}(r)$ is a polynomial of order n and $a_0 = 4\pi\varepsilon_0\hbar^2/m_{\mathrm{e}}e^2$ is called the (first) Bohr radius. In this chapter, particularly relevant are the s ($\ell = 0$) and p ($\ell = 1$) orbitals with lowest energy, for which we have $f_{00}(r) = 2/a_0^{3/2}$ and $f_{11}(r) = r/2\sqrt{6}a_0^{5/2}$ respectively. To see that these s and p orbitals have indeed the lowest energy, we remember that the energy eigenvalues of the hydrogenic atom are given by

$$E_n = -\frac{1}{2} m_e \alpha^2 c^2 \frac{1}{(n+1)^2}, \tag{10.2}$$

where $\alpha = e^2/4\pi\varepsilon_0\hbar c \simeq 1/137$ is the fine-structure constant. The angular momentum ℓ can take on integer values between 0 and n and the magnetic quantum number m runs from $-\ell$ to ℓ. As a result, the above energy levels are degenerate in both ℓ and m. These degeneracies are lifted when we take into account the spin degrees of freedom and the associated spin-dependent corrections to the Hamiltonian, which are typically small. However, understanding the spin structure of the alkali atoms is crucial in order to explain the trapping of ultracold atoms. Moreover, the Zeeman effect and the hyperfine structure of the atoms are the fundamental ingredients for the precise tuning of the interatomic interaction strength. The corresponding mechanism, called a Feshbach resonance, is explained in Chap. 17.

10.1.1 Fine Structure

For the Coulomb Hamiltonian from (10.1), both the electronic orbital angular momentum and the electronic spin angular momentum are separately conserved, because both quantities commute with the Hamiltonian. In reality, however, this is not the case, because the electronic orbital angular momentum operator $\hat{\mathbf{L}}$ and the electronic spin angular momentum operator $\hat{\mathbf{S}}$ are coupled by the spin-orbit coupling Hamiltonian

$$\hat{H}_{so} = \frac{\alpha_{so}}{\hbar^2} \hat{\mathbf{L}} \cdot \hat{\mathbf{S}}, \tag{10.3}$$

where the constant α_{so} determines the strength of the coupling. Because of this additional term, only the total electronic angular momentum $\hat{\mathbf{J}} = \hat{\mathbf{L}} + \hat{\mathbf{S}}$ commutes with the Hamiltonian and is conserved, where we note that also $\hat{\mathbf{L}}^2$ and $\hat{\mathbf{S}}^2$ still give rise to good quantum numbers. The splitting of the electronic states due to the spin-orbit coupling is known as the fine-structure splitting of the atom. For alkali atoms, the fine-structure eigenstates can be written more precisely as $|n; (\ell s) j m_j\rangle$, with $s = 1/2$ the electron spin, j the total electronic angular momentum, and m_j its projection on the quantization axis. To find the fine-structure splitting explicitly, we note that the square of the total angular momentum operator is given by

$$\hat{\mathbf{J}}^2 = \hat{\mathbf{L}}^2 + \hat{\mathbf{S}}^2 + 2\hat{\mathbf{L}} \cdot \hat{\mathbf{S}}. \tag{10.4}$$

Using this observation, we can determine the fine structure energy shift of an atom in state $|n; (\ell s) j m_j\rangle$, which is given by

$$\begin{aligned}\Delta E_j &= \langle n;(\ell s)jm_j|\hat{H}_{\text{so}}|n;(\ell s)jm_j\rangle \\ &= \frac{\alpha_{\text{so}}}{2}\left(j(j+1)-\ell(\ell+1)-s(s+1)\right).\end{aligned} \tag{10.5}$$

For our purposes, the fine structure splitting of a p orbital is most important. In that case, we have $\ell = 1$, and we find $\Delta E_{3/2} = \alpha_{\text{so}}/2$ and $\Delta E_{1/2} = -\alpha_{\text{so}}$. We examine the consequences of this fine-structure splitting further in Chap. 16 when we discuss the interaction of atoms with light, which is important for understanding optical lattices.

10.1.2 Hyperfine Structure

Now, we recall that in the ground state of the alkali atoms all electrons in the core occupy closed shells, whereas the outer valence electron is in a s orbital. Since the total orbital angular momentum is thus zero for ground-state alkali atoms, i.e. $\ell = 0$ and $j = s$, they do not have a fine-structure splitting due to the spin-orbit coupling. However, a splitting of this ground state does nevertheless occur, when we also include the coupling of the nuclear spin with the electron spin due to the magnetic dipole-dipole interaction. This coupling is described by the hyperfine interaction

$$\hat{H}_{\text{hf}} = \frac{\alpha_{\text{hf}}}{\hbar^2}\hat{\mathbf{I}}\cdot\hat{\mathbf{S}}, \tag{10.6}$$

where α_{hf} yields the strength of the coupling and $\hat{\mathbf{I}}$ is the total nuclear spin. As a result, only the total angular momentum $\hat{\mathbf{F}} = \hat{\mathbf{I}} + \hat{\mathbf{J}}$ is conserved and the resulting level splitting is known as the hyperfine structure. Recalling that we are considering the special case for which $\mathbf{L} = 0$, we have that the hyperfine spin state is completely determined by the total spin quantum number f, its projection on the quantization axis m_f, and the nuclear and electronic spin quantum numbers i and s, respectively. As their names suggest, the fine structure leads to a small splitting of the eigenstates for the Coulomb problem, whereas the hyperfine structure leads to an even smaller splitting of the fine-structure energy levels.

For the s-orbital ground state of an alkali atom, the hyperfine eigenstates can thus be written as $|n\ell;(is)fm_f\rangle = |00;(is)fm_f\rangle$. To find the hyperfine structure, we note that

$$\hat{\mathbf{F}}^2 = \hat{\mathbf{I}}^2 + \hat{\mathbf{S}}^2 + 2\hat{\mathbf{I}}\cdot\hat{\mathbf{S}}, \tag{10.7}$$

where we have used that $\hat{\mathbf{J}} = \hat{\mathbf{S}}$ for the orbital ground state of the alkali atoms. As a result, the hyperfine energy shift of an atom in state $|00;(is)fm_f\rangle$ is given by

$$\begin{aligned}\Delta E_f &= \langle 00;(is)fm_f|\hat{H}_{\text{hf}}|00;(is)fm_f\rangle \\ &= \frac{\alpha_{\text{hf}}}{2}\left(f(f+1)-i(i+1)-s(s+1)\right).\end{aligned} \tag{10.8}$$

Since for alkali atoms $s = 1/2$, the allowed values for f are $i \pm 1/2$ in the orbital ground state, where from now on we simply write $|fm_f\rangle$ to specify the hyperfine states of interest. The manifold $f = i + 1/2$ contains the states with $m_f = -(i + 1/2), \ldots, (i + 1/2)$ and the manifold with total hyperfine spin $f = i - 1/2$ contains the states with $m_f = -(i-1/2), \ldots, (i-1/2)$. Note that in the absence of an external magnetic field, the states within each of these two manifolds are degenerate. The energy difference between the two manifolds $f = i \pm 1/2$ is known as the hyperfine splitting $\hbar\omega_{\text{hf}}$. From (10.8), it follows that

$$\hbar\omega_{\text{hf}} = \alpha_{\text{hf}}(i + \frac{1}{2}). \tag{10.9}$$

Note that for atomic hydrogen the hyperfine splitting of 1420 MHz corresponds precisely to the famous 21-cm emission line, whose detailed study across the sky has led to most of what is known about the distribution of cold gases in our galaxy, including the mapping of the nearby spiral arms.

10.2 Zeeman Effect

Of particular experimental relevance is the case of an atom in the presence of an applied magnetic field. The coupling of both the nuclear and electron spin magnetic moments to the external magnetic field causes an energy shift of the hyperfine states known as the Zeeman shift. As a result, the degeneracy in m_f of the hyperfine states is lifted. Put differently, by controlling the applied magnetic field it is possible to control the energies of the atoms, and this is widely used for many interesting applications and experiments with ultracold gases. To make this statement more concrete, we consider an alkali atom in a magnetic field of strength B directed along the quantization axis. Considering again only the s-orbital groundstate, the relevant Hamiltonian for the hyperfine structure becomes

$$\hat{H} = \hat{H}_{\text{hf}} - \gamma_{\text{N}} B \hat{I}_z + \gamma B \hat{S}_z, \tag{10.10}$$

where we have taken into account both the Zeeman coupling of the magnetic field to the electronic spin with coupling strength γ and to the nuclear spin with coupling strength γ_{N}. In practice, we have that $\gamma_{\text{N}} \ll \gamma$, which comes about because the Zeeman coupling is inversely proportional to the mass, and the nucleus is much heavier than an electron. By diagonalizing the Hamiltonian of (10.10), we then obtain the magnetic field dependence of the atomic energy levels.

First, we solve the Hamiltonian for the simplest nontrivial case when $i = 1/2$, which corresponds to atomic hydrogen. It is convenient to rewrite the hyperfine interaction by using

$$\hat{\mathbf{I}} \cdot \hat{\mathbf{S}} = \hat{I}_x \hat{S}_x + \hat{I}_y \hat{S}_y + \hat{I}_z \hat{S}_z = \hat{I}_z \hat{S}_z + \frac{1}{2}\left(\hat{I}_+ \hat{S}_- + \hat{I}_- \hat{S}_+\right), \tag{10.11}$$

where in the last line we introduced the familiar raising and lowering operators $\hat{I}_\pm = \hat{I}_x \pm i\hat{I}_y$ and $\hat{S}_\pm = \hat{S}_x \pm i\hat{S}_y$, respectively. They have the property that for $i = 1/2$ we have

$$\hat{I}_\pm |im_i\rangle = \hbar |im_i \pm 1\rangle, \tag{10.12}$$

where a similar equation holds for $\hat{S}_\pm$. In the basis of the states $|im_i, sm_s\rangle \equiv |m_i, m_s\rangle$, the matrix elements of the Hamiltonian from (10.10) are given by

$$\begin{aligned}\langle m_i, m_s|\hat{H}|m_i', m_s'\rangle &= (\alpha_{\rm hf} m_i m_s - \hbar\gamma_{\rm N} m_i B + \hbar\gamma m_s B)\, \delta_{m_i,m_i'}\delta_{m_s,m_s'} \\ &\quad + \frac{\alpha_{\rm hf}}{2}\delta_{m_i,m_i'+1}\delta_{m_s,m_s'-1} + \frac{\alpha_{\rm hf}}{2}\delta_{m_i,m_i'-1}\delta_{m_s,m_s'+1}. \end{aligned} \tag{10.13}$$

From this, we see that the two states $|-1/2,-1/2\rangle$ and $|1/2,1/2\rangle$ are uncoupled and have energies $\alpha_{\rm hf}/4 \mp \hbar(\gamma - \gamma_{\rm N})B/2$. These states thus show a linear Zeeman effect with a magnetic moment equal to $\hbar(\gamma - \gamma_{\rm N})m_f/2$. The other two states mix, but the associated Hamiltonian is easily diagonalized by writing down its matrix representation. Explicitly, we find

$$\frac{1}{2}\begin{pmatrix} -\frac{\alpha_{\rm hf}}{2} + \hbar\gamma_{\rm N} B + \hbar\gamma B & \alpha_{\rm hf} \\ \alpha_{\rm hf} & -\frac{\alpha_{\rm hf}}{2} - \hbar\gamma_{\rm N} B - \hbar\gamma B \end{pmatrix},$$

whose eigenvalues are given by

$$E = -\frac{\alpha_{\rm hf}}{4} \pm \frac{\alpha_{\rm hf}}{2}\sqrt{1 + \left(\frac{\hbar(\gamma_{\rm N} + \gamma)B}{\alpha_{\rm hf}}\right)^2}. \tag{10.14}$$

For small magnetic fields, these states show a quadratic Zeeman effect, which is caused by the fact that $m_f = 0$ and the magnetic moment vanishes. Note that by making the magnetic field position dependent, also the energy of the atoms in the various hyperfine states become position dependent. Put differently, the magnetic field acts as a potential energy for the atoms, which has been used to confine ultracold atomic gases in space. Note that this potential energy then also depends on the specific hyperfine state, which can be used to trap atoms in a selected number of quantum states.

The above calculation can be generalized to more involved cases, where the nucleus has a higher spin. For example, the alkali fermion ^{6}Li has a nuclear spin $i = 1$, such that for its ground state, we can have either $f = 3/2$ or $f = 1/2$, which are split by the hyperfine splitting discussed in the previous section. For a weak magnetic field, the Zeeman interaction can be considered as a perturbation to the hyperfine interaction and the Zeeman splitting is therefore determined by the hyperfine states $|fm_f\rangle$. The states with $f = 3/2$ split up into four levels, whereas the states with $f = 1/2$ split up into two, where initially the splittings go proportional to Bm_f. At large values of the magnetic field, we should not use the hyperfine states $|fm_f\rangle$ anymore, because now the hyperfine interaction has become a perturbation to the eigenstates of the Zeeman interaction, given by $|im_i, sm_s\rangle$. Since $\gamma \gg \gamma_N$, the Zee-

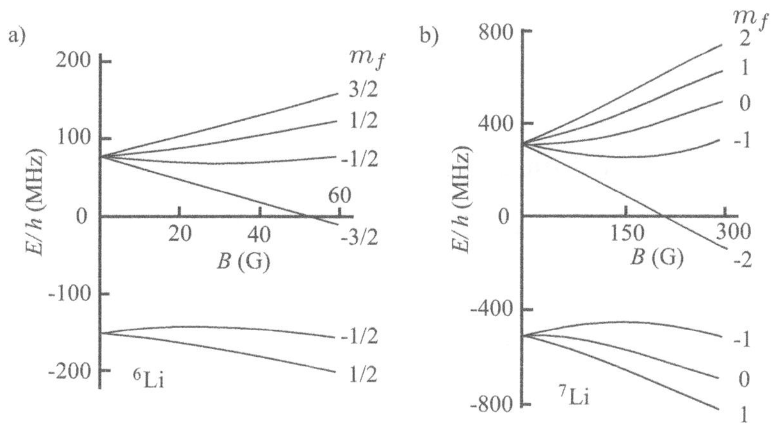

Fig. 10.1 Energy level diagram of the s-orbital ground state for a) fermionic ^{6}Li, which has a nuclear spin $i = 1$, and b) bosonic ^{7}Li, which has a nuclear spin $i = 3/2$. For both isoptopes, the hyperfine structure is shown, as well as the Zeeman splitting due to the applied magnetic field B.

man splitting is dominated by m_s for large magnetic field strengths. In between the weak and strong-field limit, some of the eigenstates then have to bend in energy as a function of magnetic-field strength, as observed in Fig. 10.1.

The boson ^{7}Li has nuclear spin $i = 3/2$, such that either $f = 2$ or $f = 1$, both separated initially by the hyperfine splitting. As a function of magnetic field strength, the states with $f = 2$ split into five levels, whereas the states with $f = 1$ into three, as is seen in Fig. 10.1. Many important experiments have been performed with atomic lithium in these ground states, where nowadays fermionic ^{6}Li is a particular favorite of the experimentalists. This is because it gives rise to an experimentally easily accessible Feshbach resonance with which the atomic interactions are very accurately tuned.

10.3 Two-body Scattering in Vacuum

Knowing only the properties of a single atom is of course not sufficient to describe an interacting quantum gas of atoms. At the end of Chap. 3, we gave an argument that for dilute gases the two-body interactions are most important. Here, we consider in more detail the scattering properties of two atoms in vacuum and show that for an ultracold gas of atoms the two-body interactions are characterized by a single parameter. This parameter is called the s-wave scattering length, and can be measured experimentally. Furthermore, we also show how to take into account the many-body corrections due to the presence of a medium. In first instance, we assume that the atoms in the gas are prepared in an initial state that does not change during the scattering process. We relax this assumption in Chap. 17 when we discuss Feshbach resonances.

10.3.1 Two-Body Transition Matrix

In Sect. 3.11 we encountered the Lippmann-Schwinger equation, where it was used to determine the general form of the scattering wavefunction, which was seen to consist of an incoming plane wave and an outgoing spherical wave at large separations between the colliding particles. Now, instead of dealing with the scattering states $|\psi_{\mathbf{p}}^{(+)}\rangle$ explicitly, we solve the scattering problem in a more convenient way [42]. To this end we introduce the operator $\hat{T}^{2B}$, which is called the two-body transition or T matrix, and is defined by

$$\hat{V}|\psi_{\mathbf{p}}^{(+)}\rangle \equiv \hat{T}^{2B}|\mathbf{p}\rangle. \tag{10.15}$$

The matrix elements of the two-body T matrix are directly related to the scattering amplitudes $f(\mathbf{p}',\mathbf{p})$ that were introduced in (3.124). Indeed, substituting (10.15) into (3.124), we find

$$f(\mathbf{p}',\mathbf{p}) = -\frac{1}{4\pi}(2\pi\hbar)^3\frac{m}{\hbar^2}\langle\mathbf{p}'|\hat{T}^{2B}|\mathbf{p}\rangle, \tag{10.16}$$

where the plane waves are normalized as $\langle\mathbf{x}|\mathbf{p}\rangle = e^{i\mathbf{p}\cdot\mathbf{x}/\hbar}/(2\pi\hbar)^{3/2}$, which is conventional in the treatment of two-body scattering. However, it turns out that for the many-body generalization of two-body scattering, which is of interest for the rest of this book, it is actually more convenient to introduce the wavevector $\mathbf{k} \equiv \mathbf{p}/\hbar$ and use from now on the normalization $\langle\mathbf{x}|\mathbf{k}\rangle = e^{i\mathbf{k}\cdot\mathbf{x}}$ for the plane waves. Doing so, (10.16) becomes

$$f(\mathbf{k}',\mathbf{k}) = -\frac{1}{4\pi}\frac{m}{\hbar^2}\langle\mathbf{k}'|\hat{T}^{2B}|\mathbf{k}\rangle, \tag{10.17}$$

while the Lippmann-Schwinger equation (3.118) is given by

$$\hat{T}^{2B}|\mathbf{k}\rangle = \hat{V}|\mathbf{k}\rangle + \hat{V}\frac{1}{E-\hat{H}_0+i0}\hat{T}^{2B}|\mathbf{k}\rangle, \tag{10.18}$$

which has to be satisfied for any incoming plane wave $|\mathbf{k}\rangle$. Therefore, we find for $\hat{T}^{2B}$ the operator equation

$$\hat{T}^{2B} = \hat{V} + \hat{V}\frac{1}{E-\hat{H}_0+i0}\hat{T}^{2B}, \tag{10.19}$$

where $(E-\hat{H}_0+i0)^{-1}$ is an operator that gives rise to a noninteracting propagator similar to (7.50). To see this, we first note that by definition

$$\left(E - \hat{H}_0 + \mathrm{i}0\right) \frac{1}{E - \hat{H}_0 + \mathrm{i}0} = \hat{1}. \tag{10.20}$$

Multiplying with $\langle \mathbf{x}|$ on the left and $|\mathbf{x}'\rangle$ on the right and inserting a completeness relation gives

$$\begin{aligned} \delta(\mathbf{x}-\mathbf{x}') &= \int \mathrm{d}\mathbf{x}'' \langle \mathbf{x}|E - \hat{H}_0 + \mathrm{i}0|\mathbf{x}''\rangle \langle \mathbf{x}''|\frac{1}{E - \hat{H}_0 + \mathrm{i}0}|\mathbf{x}'\rangle \\ &\equiv \left\{E - \frac{\hbar^2 \boldsymbol{\nabla}^2}{m}\right\} G_0(\mathbf{x},\mathbf{x}'), \end{aligned} \tag{10.21}$$

where we also used the position representation of $\hat{H}_0$.

The Lippmann-Schwinger equation (10.19) for the two-body T matrix can be solved iteratively, resulting in the Born series, which is given by

$$\hat{T}^{2\mathrm{B}}(z) = \hat{V} + \hat{V}\hat{G}_0(z)\hat{V} + \hat{V}\hat{G}_0(z)\hat{V}\hat{G}_0(z)\hat{V} + \cdots , \tag{10.22}$$

where the operator

$$\hat{G}_0(z) = \frac{1}{z - \hat{H}_0} \tag{10.23}$$

corresponds to the noninteracting propagator of the atoms at a (complex) energy z. A common approximation for the two-body T matrix is to take only the first term of the Born series into account, which is called the Born approximation. From (10.22) we see that the T matrix physically describes the outcome of a collision process, in which the particles interact quantum mechanically an arbitrary number of times, such that the T matrix sums over all elementary interaction processes that take place during a collision. We note that the formal solution of the operator equation for the transition matrix is given by

$$\hat{T}^{2\mathrm{B}}(z) = \hat{V} + \hat{V}\frac{1}{z - \hat{H}}\hat{V} , \tag{10.24}$$

where $\hat{H} = \hat{H}_0 + \hat{V}$, whose complete set of eigenstates we denote by $|\psi_\alpha\rangle$. Inserting this set, we find

$$\hat{T}^{2\mathrm{B}}(z) = \hat{V} + \sum_\alpha \hat{V}\frac{|\psi_\alpha\rangle\langle\psi_\alpha|}{z - \varepsilon_\alpha}\hat{V} , \tag{10.25}$$

where the summation over α is discrete for possible bound states of the interaction potential with energies $\varepsilon_\alpha < 0$, while the summation becomes an integration for the continuum of scattering states with energies $\varepsilon_\alpha > 0$. Explicitly, we thus have that

$$\hat{T}^{2\mathrm{B}}(z) = \hat{V} + \sum_\kappa \hat{V}\frac{|\psi_\kappa\rangle\langle\psi_\kappa|}{z - \varepsilon_\kappa}\hat{V} + \int \frac{\mathrm{d}\mathbf{k}}{(2\pi)^3}\hat{V}\frac{|\psi_{\mathbf{k}}^{(+)}\rangle\langle\psi_{\mathbf{k}}^{(+)}|}{z - 2\varepsilon_{\mathbf{k}}}\hat{V}, \tag{10.26}$$

which shows that the two-body T matrix has poles in the complex-energy plane that corresponding to the bound states of the interaction potential, while it has also a branch cut on the positive real axis due to the continuum of scattering states.

10.3.2 Partial-Wave Expansion

For spherically symmetric interaction potentials, the potential depends only on the distance between the atoms. As a result, the elastic scattering amplitude is determined by the magnitude of the incoming momentum k and a single angle ϑ with the outgoing momentum. To further evaluate the scattering wavefunction that we derived in Sect. 3.11, namely

$$\psi_{\mathbf{k}}^{(+)}(\mathbf{r}) = \mathrm{e}^{\mathrm{i}\mathbf{k}\cdot\mathbf{r}} + f(\mathbf{k}',\mathbf{k})\frac{\mathrm{e}^{\mathrm{i}kr}}{r}, \tag{10.27}$$

we decompose its scattering amplitudes $f(\mathbf{k}',\mathbf{k})$ with the method of partial waves. We have that

$$f(\mathbf{k}',\mathbf{k}) = \sum_{\ell=0}^{\infty}(2\ell+1)f_\ell(k)P_\ell(\cos\vartheta), \tag{10.28}$$

where $P_\ell(x)$ are the Legendre polynomials. To see the meaning of the partial-wave amplitudes $f_\ell(k)$ more clearly, we first note that the plane-wave part of the scattering wavefunction can be written as a sum of incoming and outgoing spherical waves. To this end, we use the identity

$$\mathrm{e}^{\mathrm{i}\mathbf{k}\cdot\mathbf{r}} = \sum_{\ell}(2\ell+1)\mathrm{i}^\ell j_\ell(kr)P_\ell(\cos\vartheta), \tag{10.29}$$

where, for large separation r, we are allowed to use the asymptotic behavior of the spherical Bessel functions $j_\ell(kr)$. We obtain

$$\mathrm{e}^{\mathrm{i}\mathbf{k}\cdot\mathbf{r}} \simeq \sum_{\ell}(2\ell+1)P_\ell(\cos\vartheta)\left(\frac{\mathrm{e}^{\mathrm{i}kr} - \mathrm{e}^{-\mathrm{i}(kr-\ell\pi)}}{2\mathrm{i}kr}\right), \tag{10.30}$$

which is indeed a superposition of outgoing and incoming spherical waves. Combining (10.27), (10.28) and (10.30), we see that the presence of an interaction potential changes the coefficient of the outgoing spherical wave according to

$$\frac{\mathrm{e}^{\mathrm{i}kr}}{r} \to \frac{(1+2\mathrm{i}kf_\ell(k))\mathrm{e}^{\mathrm{i}kr}}{r},$$

where, due to the conservation of probability flux, the magnitude of the coefficient $(1+2\mathrm{i}kf_\ell(k))$ has to be equal to one. This can explicitly be incorporated by writing the coefficient as an exponential

$$1 + 2ikf_\ell(k) \equiv e^{2i\delta_\ell(k)}, \tag{10.31}$$

where we have defined the phase shifts $\delta_\ell(k)$. The conclusion is that, at large distances, the change in the wavefunction due to the collision process is solely given by a shift in the phase of every outgoing partial wave.

It is remarkable that ultracold gases can be cooled to the point where only one partial wave is dominant and all the others can be neglected. Roughly speaking, at low enough temperatures the atoms cannot overcome the centrifugal barrier for nonzero angular momentum scattering, such that the collision process is fully determined by the part with zero angular momentum. To quantify this statement somewhat, we estimate the typical angular momentum $\hbar\ell$ of two scattering atoms as the range of the interaction R times their relative momentum, which is determined by the thermal de Broglie wavelength Λ. We find that typically

$$\hbar\ell \simeq \hbar\frac{R}{\Lambda}, \tag{10.32}$$

where, in practice, we have for ultracold atomic gases that $R/\Lambda \ll 1$. As a result, the dominant phase shift is given by the partial wave with zero angular momentum, $\ell = 0$, which we use to define the s-wave scattering length a as

$$a = -\lim_{k\downarrow 0}\frac{\delta_0(k)}{k}. \tag{10.33}$$

With the use of (10.31), we then find that

$$f(\mathbf{k},\mathbf{k}') \simeq f_0(k) = \frac{1}{k\cot\delta_0(k) - ik}, \tag{10.34}$$

where we note that this expression is spherically symmetric, i.e. independent of the zenith angle ϑ. We may evaluate the above expression further by using the following expansion

$$k\cot(\delta_0(k)) = -\frac{1}{a} + \frac{1}{2}r_{\text{eff}}k^2 + \cdots, \tag{10.35}$$

where the quadratic coefficient r_{eff} is called the effective range. Using (10.17) and (10.35), we obtain

$$\langle\mathbf{k}'|\hat{T}^{2B}|\mathbf{k}\rangle = T^{2B}(k) = -\frac{4\pi\hbar^2}{m}\frac{1}{k\cot\delta_0(k) - ik} \simeq \frac{4\pi\hbar^2}{m}\frac{a}{1 - ar_{\text{eff}}k^2 + iak}, \tag{10.36}$$

which we can express explicitly in terms of the small positive energies E as

$$T^{2B}(E + i0) \simeq \frac{4\pi a\hbar^2}{m}\frac{1}{1 + ia\sqrt{mE/\hbar^2} - ar_{\text{eff}}mE/2\hbar^2}. \tag{10.37}$$

By analytic continuation, we generalize the above result to

$$T^{2B}(z) \simeq \frac{4\pi a\hbar^2}{m} \frac{1}{1 - a\sqrt{-mz/\hbar^2} - ar_{\rm eff}mz/2\hbar^2}, \tag{10.38}$$

where for large and positive s-wave scattering length, the two-body T matrix is seen to have a pole at the negative energy $E_{\rm m} = -\hbar^2/ma^2$. Remembering the general form of the T matrix from (10.26), we see that this pole signals the presence of a two-body bound state with a small binding energy $E_{\rm m}$. To understand the important general results of this section better, we consider the following explicit example.

10.3.3 Scattering from a Square-Well Potential

We illustrate the physical meaning of the s-wave scattering length by calculating it for the simplified case that the interparticle potential is a square well [42]. The advantages of treating this model is that it can be solved exactly and that its properties are very common to scattering processes in general. We take the interaction potential of the form

$$V(r) = \begin{cases} V_0 \text{ if } r < R; \\ 0 \quad \text{if } r > R, \end{cases} \tag{10.39}$$

with $R > 0$. Note that this is a spherically symmetric potential and that the radial wavefunction $R(r) = u(r)/r$ for $\ell = 0$ is determined by the radial Schrödinger equation

$$\left\{ \frac{\mathrm{d}^2}{\mathrm{d}r^2} - \frac{mV(r)}{\hbar^2} + k^2 \right\} u(r) = 0. \tag{10.40}$$

The general solution to (10.40) is given by

$$\begin{aligned} u^<(r) &= A\mathrm{e}^{\mathrm{i}k^< r} + B\mathrm{e}^{-\mathrm{i}k^< r}, \qquad && \text{for } r < R, \\ u^>(r) &= C\mathrm{e}^{\mathrm{i}kr} + D\mathrm{e}^{-\mathrm{i}kr}, && \text{for } r > R, \end{aligned} \tag{10.41}$$

with $k^< = \sqrt{k^2 - mV_0/\hbar^2}$. Since the radial wavefunction $R(r)$ has to obey the Schrödinger equation at the origin, we demand that $u^<(r)$ vanishes at this point. This leads to the boundary condition $B = -A$. By comparing the explicit form of $u^>(r)$ with the s-wave component of the general scattering wavefunction for $r \to \infty$, we find that

$$\mathrm{e}^{2\mathrm{i}\delta_0(k)} = -\frac{C}{D}. \tag{10.42}$$

We determine the s-wave phase shift by demanding that the wavefunctions for $r < R$ and $r > R$ join smoothly. This leads to the equations

$$A\left(e^{ik^<R} - e^{-ik^<R}\right) = -e^{2i\delta_0(k)}e^{ikR} + e^{-ikR},$$
$$A\left(k^<e^{ik^<R} + k^<e^{-ik^<R}\right) = -e^{2i\delta_0(k)}ke^{ikR} - ke^{-ikR}, \qquad (10.43)$$

where we have chosen the normalization such that $D = 1$. Multiplication of the above equations with $e^{-i\delta_0(k)}$ and dividing the result leads to

$$k\tan(k^<R) = k^<\tan(\delta_0(k) + kR), \qquad (10.44)$$

from which it follows that

$$\delta_0(k) = -kR + \tan^{-1}\left[\frac{k}{k^<}\tan(k^<R)\right]. \qquad (10.45)$$

Note that for a repulsive hard-core potential, we have that $V_0 \to \infty$ and therefore, with the use of the definition in (10.33), that the scattering length $a = R$. This gives rise to a physical picture for the s-wave scattering length in the case that a is positive. Namely, for low energies and long wavelengths, we have that the details of a short-ranged potential are not probed and therefore unimportant. We are then allowed to model any short-ranged interaction potential as an effective hard-core potential with radius $R = a$, because they both lead to exactly the same scattering wavefunctions at large distances. As a result, the scattering length is sometimes also called the effective hard-core radius of the potential. We remark that for a fully repulsive potential the scattering length is always positive. However, for a (partially) attractive potential, the scattering length can be both negative and positive, corresponding to attractive and repulsive effective interactions, respectively.

This can be made explicit by calculating the scattering length for a square well with $V_0 < 0$. From (10.33) and (10.45), we find that the scattering length is given by

$$a = R\left(1 - \frac{\tan\gamma}{\gamma}\right), \qquad (10.46)$$

with $\gamma = R\sqrt{m|V_0|/\hbar^2}$ a dimensionless constant. The effective range, which the second-order coefficient in the expansion from (10.35), is in this example given by

$$r_{\text{eff}} = R\left[1 + \frac{3\tan\gamma - \gamma(3+\gamma^2)}{3\gamma(\gamma - \tan\gamma)^2}\right]. \qquad (10.47)$$

In Fig. 10.2, the scattering length is shown as a function of γ by the solid lines. We indeed find that the scattering length can be both negative and positive, while it becomes equal to zero when $\gamma = \tan\gamma$. In the same figure, the effective range is shown by the dashed line. Note that the effective range diverges when the scattering length becomes equal to zero. This comes about because the expansion in (10.35) is ill defined for $a = 0$. Indeed for $|a| \ll R$, we find that $\delta_0(k)$ behaves as $-ka + (kR)^3/6$ at small energies. When $\gamma = (n + 1/2)\pi$ with n a positive integer, the scattering length diverges and changes sign. This behavior is called a potential

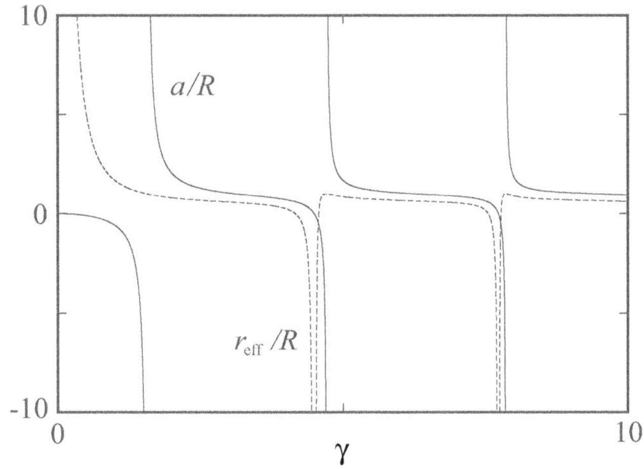

Fig. 10.2 Scattering length a (solid line) and effective range r_{eff} (dashed line) for an attractive square well in units of the radius of the potential R as a function of the dimensionless parameter $\gamma = R\sqrt{m|V_0|/\hbar^2}$. Figure adapted from [32].

or shape resonance and in fact occurs each time the potential is just deep enough to support a new bound state. We find that for large and positive scattering length, the square well has a bound state with an energy just below the continuum threshold. It turns out that there is an important relationship between the energy of this bound state and the scattering length.

To find this relation, we have to determine the bound-state energy by solving the Schödinger equation for negative energy $V_0 < E < 0$. This leads to the solutions

$$\begin{aligned} u^<(r) &= A\left(e^{ik^<r} - e^{-ik^<r}\right), && \text{for } r < R, \\ u^>(r) &= Be^{-\kappa r}, && \text{for } r > R, \end{aligned} \tag{10.48}$$

with $k^< = \sqrt{m(E-V_0)/\hbar^2}$ and $\kappa = \sqrt{m|E|/\hbar^2}$. Demanding that these solutions join smoothly at $r = R$, we find the equation for the bound-state energy

$$\sqrt{\frac{m}{\hbar^2}|E_m|} = -\sqrt{\frac{m}{\hbar^2}(E_m - V_0)}\cot\left(\sqrt{\frac{m}{\hbar^2}(E_m - V_0)}R\right) . \tag{10.49}$$

It is possible to show that for values of γ given by $(n-1/2)\pi < \gamma < (n+1/2)\pi$, this equation has n solutions for $V_0 < E_m < 0$. For small binding energy $|E_m| \ll |V_0|$, we have from the equation for the bound-state energy that

$$\sqrt{\frac{m}{\hbar^2}|E_m|} \simeq -\frac{\gamma}{R}\cot(\gamma) \simeq \frac{1}{a}, \tag{10.50}$$

where we also made use of (10.46) and the fact that γ has to be close to the resonant values $(n+1/2)\pi$ in this case. This leads to the desired relation between the energy of the molecular state and the scattering length, given by

$$E_{\rm m} = -\frac{\hbar^2}{ma^2} \, . \tag{10.51}$$

This relation turns out to be general, such that it does not depend on the specific details of the interaction potential, as can also be seen from our derivation of the same result using the two-body T matrix of (1.34). We conclude that any potential with a large positive scattering length has a bound state just below the continuum threshold with an energy given by (10.51).

10.4 Two-Body Scattering in a Medium

In the previous sections, we considered two-body scattering in vacuum using the T matrix approach. In this section, we generalize this approach to the case of an interacting many-body system consisting of bosonic atoms, where we approximate the interaction potential by a point interaction

$$V(\mathbf{x}-\mathbf{x}') \simeq V_0\delta(\mathbf{x}-\mathbf{x}'), \tag{10.52}$$

which is valid when the interactions are short ranged and when the interacting particles have a large de Broglie wavelength, i.e. a small relative momentum. As a result, the action from (8.61) becomes

$$\begin{aligned} S[\phi^*,\phi] &= \int_0^{\hbar\beta} \mathrm{d}\tau \int \mathrm{d}\mathbf{x}\, \phi^*(\mathbf{x},\tau)\left\{\hbar\frac{\partial}{\partial\tau} - \frac{\hbar^2\nabla^2}{2m} - \mu\right\}\phi(\mathbf{x},\tau) \\ &\quad + \frac{V_0}{2}\int_0^{\hbar\beta} \mathrm{d}\tau \int \mathrm{d}\mathbf{x}\, \phi^*(\mathbf{x},\tau)\phi^*(\mathbf{x},\tau)\phi(\mathbf{x},\tau)\phi(\mathbf{x},\tau)\,, \\ &= \sum_{\mathbf{k},n} \phi^*_{\mathbf{k},n}(-\mathrm{i}\hbar\omega_n + \varepsilon_{\mathbf{k}} - \mu)\phi^*_{\mathbf{k},n} \\ &\quad + \frac{V_0}{2\hbar\beta V} \sum_{\substack{\mathbf{k},\mathbf{k}',\mathbf{K} \\ n,n',m}} \phi^*_{\mathbf{K}/2-\mathbf{k}',m-n'}\phi^*_{\mathbf{K}/2+\mathbf{k}',n'}\phi_{\mathbf{K}/2+\mathbf{k},n}\phi_{\mathbf{K}/2-\mathbf{k},m-n}\,, \end{aligned} \tag{10.53}$$

where we used the expansion from (7.22) and we see that the contact interaction is constant in momentum space.

To investigate the properties of this point interaction a bit further, we consider its Lippmann-Schwinger equation at zero energy $z=0$. Multiplying both sides in (10.18) by $\langle\mathbf{k}'|$ and inserting a completeness relation in the second term on the right-hand side, we obtain

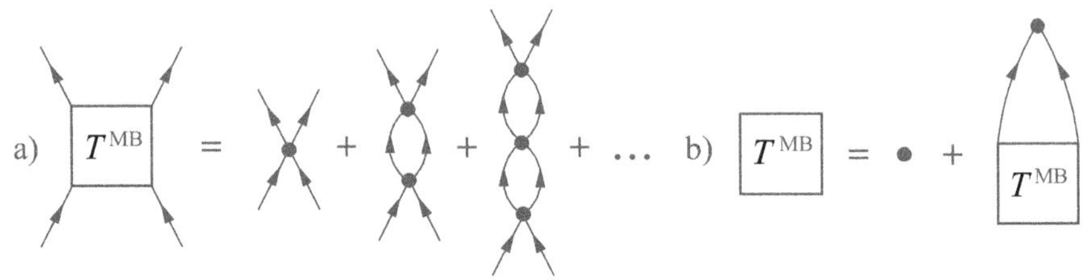

Fig. 10.3 Many-body T matrix approximation to the exact effective interatomic interaction, where the dot represents the point interaction potential. a) Ladder diagrams contributing to the many-body T matrix, leading to b) the many-body Lippmann-Schwinger or Bethe-Salpeter equation.

$$\frac{1}{T^{2B}(0)} = \frac{m}{4\pi\hbar^2 a} = \frac{1}{V_0} + \int \frac{d\mathbf{k}}{(2\pi)^3} \frac{1}{2\varepsilon_{\mathbf{k}}}, \tag{10.54}$$

where we also used (10.36). The second term on the right-hand side of (10.54) is seen to contain an ultraviolet divergence, and this directly results from the unphysical behavior of a point interaction at high momenta. A true atomic interaction potential would decay as $1/\mathbf{k}^2$, which would lift the ultraviolet divergence in the Lippmann-Schwinger equation. Although this explicitly shows that we should be careful with the high-momentum behavior of a point interaction, we note that this is usually not a problem because at ultralow temperatures we are primarily interested in the long-wavelength physics. Furthermore, there are various ways to deal with the encountered ultraviolet divergence by introducing a cutoff or using an improved potential. However, there is an even simpler way, as we see explicitly soon.

Next, we show how we can obtain an equation that generalizes the two-body Lippmann-Schwinger equation, such that it also takes into account many-body corrections due to the medium in which the atoms scatter. To this end, we use the diagrammatic perturbation theory for the many-body system, which was developed in Chap. 8. We remember that in Sect. 8.4, a nonperturbative approximation to the exact irreducible two-point vertex G^{-1} was obtained by summing over certain classes of diagrams, which led to Hartree-Fock theory. We now try to apply a similar approach to the exact irreducible four-point vertex, which is also known as the exact effective interaction. The approach amounts to summing over a certain class of one-particle irreducible Feynman diagrams, which are called the ladder diagrams. Diagrammatically the procedure is summarized in Fig. 10.3, and it goes under the name of the (many-body) T matrix approximation. The reason for summing over this class of diagrams is that in the zero-density or vacuum limit, the approach reduces to the two-body Lippmann-Schwinger equation, as we show next. As a result, the procedure is seen to contain the relevant two-body physics exactly. In Chap. 13, we discuss one-particle irreducible actions in a more general context.

Example 10.1. It is instructive to calculate a single ladder diagram, which remains when we take away the four outer legs and the two interaction dots from the middle

Feynman diagram of Fig. 10.3a. Note the difference in the direction of the arrows compared to the bubble diagram from Fig. 8.8. The ladder diagram is given in real space by

$$-\hbar\Xi(\mathbf{x},\tau;\mathbf{x}',\tau') = G_0(\mathbf{x},\tau;\mathbf{x}',\tau')G_0(\mathbf{x},\tau;\mathbf{x}',\tau') \tag{10.55}$$

which we may Fourier transform as

$$\begin{aligned}
\Xi(\mathbf{K},\mathrm{i}\Omega_n) &= \int_0^{\hbar\beta} \mathrm{d}\tau\,\mathrm{d}\tau' \int \mathrm{d}\mathbf{x}\,\mathrm{d}\mathbf{x}'\;\Xi(\mathbf{x},\tau;\mathbf{x}',\tau')\frac{\mathrm{e}^{-\mathrm{i}\mathbf{K}(\mathbf{x}-\mathbf{x}')}}{V}\frac{\mathrm{e}^{\mathrm{i}\Omega_n(\tau-\tau')}}{\hbar\beta} \\
&= \frac{-1}{\hbar^2\beta V}\sum_{\mathbf{k}',n',} G_0(\mathbf{k}',\mathrm{i}\omega_{n'})G_0(\mathbf{K}-\mathbf{k}',\mathrm{i}\Omega_n-\mathrm{i}\omega_{n'}) \\
&= \frac{1}{\hbar\beta V}\sum_{n',\mathbf{k}'} \frac{G_0(\mathbf{k}',\mathrm{i}\omega_{n'})+G_0(\mathbf{K}-\mathbf{k}',\mathrm{i}\Omega_n-\mathrm{i}\omega_{n'})}{-\mathrm{i}\hbar\Omega_n+\varepsilon_{\mathbf{K}-\mathbf{k}'}+\varepsilon_{\mathbf{k}'}-2\mu} \\
&= \frac{1}{V}\sum_{\mathbf{k}'} \frac{1+N_{\mathrm{BE}}(\varepsilon_{\mathbf{K}-\mathbf{k}'})+N_{\mathrm{BE}}(\varepsilon_{\mathbf{k}'})}{\mathrm{i}\hbar\Omega_n-\varepsilon_{\mathbf{K}-\mathbf{k}'}-\varepsilon_{\mathbf{k}'}+2\mu},
\end{aligned} \tag{10.56}$$

where $N_{\mathrm{BE}}(\varepsilon_{\mathbf{k}}) = 1/\{\mathrm{e}^{\beta(\varepsilon_{\mathbf{k}}-\mu)}-1\}$ is the Bose-Einstein distribution. To obtain (10.56), we used (7.31), (7.54) and the result from problem 7.2.

Noting that the interaction is independent of momentum and frequency, while the ladder diagram only depends on the center-of mass momentum **K** and frequency Ω_n, we have that the T-matrix equation from Fig. 10.3b is given in momentum space by

$$T^{\mathrm{MB}}(\mathbf{K},\mathrm{i}\Omega_n) = V_0 + \frac{V_0}{V}\sum_{\mathbf{k}'}\frac{1+N_{\mathrm{BE}}(\varepsilon_{\mathbf{K}-\mathbf{k}'})+N_{\mathrm{BE}}(\varepsilon_{\mathbf{k}'})}{\mathrm{i}\hbar\Omega_n-\varepsilon_{\mathbf{K}-\mathbf{k}'}-\varepsilon_{\mathbf{k}'}+2\mu}T^{\mathrm{MB}}(\mathbf{K},\mathrm{i}\Omega_n)\,, \tag{10.57}$$

where we used the result from Example 10.1. (10.57) is then solved by

$$\frac{1}{T^{\mathrm{MB}}(\mathbf{K},\mathrm{i}\Omega_n)} = \frac{1}{V_0} - \frac{1}{V}\sum_{\mathbf{k}'}\frac{1+N_{\mathrm{BE}}(\varepsilon_{\mathbf{K}-\mathbf{k}'})+N_{\mathrm{BE}}(\varepsilon_{\mathbf{k}'})}{\mathrm{i}\hbar\Omega_n-\varepsilon_{\mathbf{K}-\mathbf{k}'}-\varepsilon_{\mathbf{k}'}+2\mu}\,. \tag{10.58}$$

At zero center-of-mass momentum **K** and frequency Ω_n, we take the two-body limit as $N_{\mathrm{BE}}(\varepsilon_{\mathbf{k}}) \to 0$ and $\mu \to 0$, so that we retrieve the two-body Lippmann-Schwinger equation (10.54). Moreover, we have that the momentum sum on the right-hand side of (10.58) has an ultraviolet divergence, and that it is exactly the same unphysical divergence that we encountered in (10.54) from using the point-interaction approximation. The argument for using this approximation was that the thermal de Broglie wavelength Λ of the atoms is for the ultralow temperatures of interest always much larger than the typical range of the interaction. We can see that this argument is not fully correct because, if we calculate corrections in perturbation theory, we have to deal with momentum sums which are not always restricted to momenta of order $\hbar/\Lambda$. Then, to deal with the resulting divergence most conveniently, we note that the same ultraviolet divergence is present in (10.54) and (10.58). Therefore, by subtracting the two equations, the divergence precisely cancels and we find that the many-body T matrix obeys

$$\frac{1}{T^{\rm MB}(\mathbf{K},{\rm i}\Omega_n)} = \frac{m}{4\pi a\hbar^2} - \frac{1}{V}\sum_{\mathbf{k}'}\left\{\frac{1+N_{\rm BE}(\varepsilon_{\mathbf{K}-\mathbf{k}'})+N_{\rm BE}(\varepsilon_{\mathbf{k}'})}{{\rm i}\hbar\Omega_n-\varepsilon_{\mathbf{K}-\mathbf{k}'}-\varepsilon_{\mathbf{k}'}+2\mu}+\frac{1}{2\varepsilon_{\mathbf{k}'}}\right\}. \quad (10.59)$$

Note that this renormalization procedure is rather elegant, because we do not have to introduce an arbitrary cutoff by hand. Also note that we have eliminated V_0 in terms of a, where the latter is the experimentally more relevant quantity because it can be measured very precisely. As a result, we often use this convenient procedure when calculating ladder diagrams with a point interaction.

10.5 Physical Regimes

We are now in the position to quantify the possible physical regimes that can be explored with ultracold atomic quantum gases. We start by considering the homogeneous ideal gas, which only gives rise to two physical length scales, namely the thermal de Broglie wavelength Λ of the particles and the average interparticle distance $n^{-1/3}$. Equivalently, this implies that the only two energy scales are the thermal energy $k_{\rm B}T$ and the kinetic energy $(\hbar^2/2m)(2\pi n^{1/3})^2$, where for a Fermi gas, the latter energy scale is proportional to the Fermi energy. From these two length or energy scales only one dimensionless parameter can be formed, which is usually taken to be $n\Lambda^3$. This leads to two different physical regimes for the ideal gas. In the nondegenerate regime, $n\Lambda^3 \ll 1$, the occupation numbers of all the one-particle states are small compared to one and the statistical properties are the same for a Bose gas and a Fermi gas. To be more precise, the atomic gas behaves as a classical ideal gas, whose occupation numbers are given by the Maxwell-Boltzmann distribution of (4.41). In what is called the degenerate regime, $n\Lambda^3 \gg 1$, the fundamental difference in quantum statistics between bosons and fermions becomes clearly visible, since the occupation numbers follow the Bose-Einstein and Fermi-Dirac distributions from (4.40) and (4.41) respectively. While the ideal Bose gas condenses in the degenerate regime, the ideal Fermi gas gives rise to an incompressible Fermi surface.

For an interacting gas, two new length scales enter the many-body problem, where we first consider the range R of the interaction. The typical value of this range is for alkali gases around $100a_0$, where the Bohr radius is given by $a_0 \simeq 0.529\cdot 10^{-10}$ m. There are now two more dimensionless parameters to consider, namely nR^3 and R/Λ. Under realistic conditions, the densities and the temperatures of ultracold atomic quantum gases are always such that $nR^3 \ll 1$ and $R/\Lambda \ll 1$. These conditions imply that atoms only interact with each other if they have a zero relative angular momentum ℓ, i.e. via s-wave interactions. This is because the highest relative kinetic energy with which the atoms can scatter of each other is either the

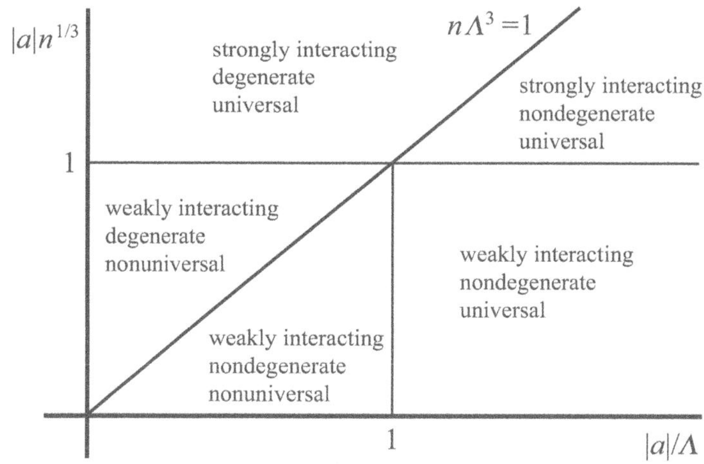

Fig. 10.4 Physical regimes that can be explored with ultracold atomic gases as a function of the density n, the thermal de Broglie wavelength Λ and the s wave scattering length a

thermal energy $k_B T$ or the kinetic energy $(\hbar^2/2m)(2\pi n^{1/3})^2$. However, under the above conditions, both energies are not high enough to overcome the centrifugal barrier $\ell(\ell+1)\hbar^2/mR^2$. This means that two atoms for which $\ell \neq 0$, do not experience the interatomic potential at distances $r < R$. As a result, the scattering of atoms is always a quantum-mechanical process that cannot be described classically, such that in this sense the ultracold atomic gases are always quantum gases even in the nondegenerate regime where their statistical properties are purely classical.

On the basis of the above arguments, it appears that the presence of an atomic interaction potential does not lead to any new physical regimes to explore. However, it turns out that the situation is more interesting. In the previous sections, we have seen that the scattering wavefunction of two atoms can be affected by the interaction at separations that are much bigger than the range of the interaction R if the s-wave scattering length a is much bigger than R. This shows that the relevant physical parameters of the gas are $n|a|^3$ and $|a|/\Lambda$. In Sect. 10.3.3 we found that the magnitude of the scattering length $|a|$ is usually on the same order as the range of the interaction R, but can become much bigger than R if the scattering process shows a resonance. This happens when there is a bound state in the interaction potential close to the continuum of scattering states. Unfortunately, the interatomic potentials are set by nature and their depths cannot be manipulated easily by external electromagnetic fields. Nevertheless, when we discuss Feshbach resonances in Chap. 17 we explain how it is still possible to experimentally shift a bound state through the scattering continuum, such that we can essentially tune the scattering length at will. The important point for now is that by using these Feshbach resonances it is possible to explore the full range of both the parameters $n|a|^3$ and $|a|/\Lambda$.

What, then, is the physical significance of these two parameters? If $n|a|^3 \ll 1$, then it is very unlikely that there is a third atom present when two atoms are interacting with each other. As a result, only two-body scattering processes take place and the gas is said to be weakly interacting. If $n|a|^3 \gg 1$, more-body scattering processes can take place and the gas is said to be strongly interacting. If we neglect the effective range r_{eff} of the potential, the two-body T matrix from (10.38) simply becomes

$$T^{2\text{B}}(z) = \frac{4\pi\hbar^2}{m}\left[\frac{1}{1/a - \sqrt{-mz/\hbar^2}}\right] . \tag{10.60}$$

In realistic equilibrium calculations, the typical energy at which the T matrix has to be evaluated is given by the maximum of $k_{\text{B}}T$ and $(\hbar^2/2m)(2\pi n^{1/3})^2$. In the regime where either $n|a|^3 \gg 1$ or $|a|/\Lambda \gg 1$, the T matrix is thus well approximated by

$$T^{2\text{B}}(z) = -\frac{4\pi\hbar^2}{m}\frac{1}{\sqrt{-mz/\hbar^2}} . \tag{10.61}$$

When the two-body T matrix becomes independent of the scattering length we are said to be in the unitarity limit, which is formally given by the limit $|a| \to \infty$. In this regime the properties of the gas become universal, meaning that they do not depend on the scattering length anymore, but only on $n\Lambda^3$. This particular feature is similar to the ideal-gas case, but now the gas is actually strongly interacting. We have also seen explicitly in Sect. 10.3.3 that the effective range r_{eff} is typically of order R near a resonance. Therefore, the corrections to universality are expected to be small in ultracold atomic gases, since the conditions $nR^3 \ll 1$ and $R/\Lambda \ll 1$ are always well satisfied. A summary of these results is shown in Fig. 10.4.

10.6 Problems

Exercise 10.1. Repeat the calculation of Sect. 10.2 for the case of a nuclear spin $I = 3/2$, which corresponds to the bosonic isotope ^{7}Li

Exercise 10.2. Calculate, in a similar manner as for the Bose case, the many-body T matrix for two fermions in different hyperfine states $|\uparrow,\downarrow\rangle$ interacting with the contact interaction $V_0\delta(\mathbf{x}-\mathbf{x}')$. Note that only fermions with opposite spin interact, as will be explained in Sect. 12.2, so that the interacting action is given by (12.2). Note that, similar to Example 10.1, you have to calculate the ladder diagram $G_{0,\uparrow}G_{0,\downarrow}$ which ultimately gives rise to

$$\Xi(\mathbf{K}, \mathrm{i}\Omega_n) = \frac{1}{V}\sum_{\mathbf{k}'}\frac{1 - N_{\text{FD}}(\varepsilon_{\mathbf{K}-\mathbf{k}'}) - N_{\text{FD}}(\varepsilon_{\mathbf{k}'})}{\mathrm{i}\hbar\Omega_n - \varepsilon_{\mathbf{K}-\mathbf{k}'} - \varepsilon_{\mathbf{k}'} + 2\mu} . \tag{10.62}$$

Express your result for the T matrix in terms of the scattering length a as was done in (10.59).

Additional Reading

- More extensive discussions on elementary atomic structure and atomic physics can be found, for example, in G. K. Woodgate, *Elementary Atomic Structure*, (Oxford University Press, Oxford, 1989), and also in
- B. H. Bransden and C. J. Joachain, *Physics of Atoms and Molecules*, (Longman, New York, 1983).
- More detailed discussions of two-body scattering can be found in W. Glöckle, *The Quantum Mechanical Few-Body Problem*, (Springer Verlag, Berlin, 1983), and also in
- J. J. Sakurai, *Modern Quantum Mechanics*, (Addison-Wesley, Reading, 1994), or in
- B. H. Bransden and C. J. Joachain, *Quantum Mechanics*, (Prentice Hall, New York, 2000).
- Relevant atomic physics for ultracold Bose gases is discussed in the introductory chapters of C. J. Pethick and H. Smith, *Bose-Einstein Condensation in Dilute Gases*, (Cambridge University, Cambridge, 2002).

Chapter 11
Bose-Einstein Condensation

Superfluidity and lasing were two of my favorite topics in physics, but each was surrounded by a vast thicket of lore and literature.
– Eric A. Cornell

States of matter, such as the familiar gas, liquid and solid phases, are characterized by certain specific correlations between particles. For instance, the solid phase is characterized by the existence of a periodicity in the atomic density $n(\mathbf{x}) = \langle \hat{\psi}^\dagger(\mathbf{x})\hat{\psi}(\mathbf{x})\rangle$, such that the Fourier transform of $n(\mathbf{x})$ signals the periodic lattice structure of the solid. This kind of order is called diagonal long-range order, because the periodic structure that extends over the whole size of the solid shows itself in the diagonal elements of the one-particle density matrix $n(\mathbf{x},\mathbf{x}') = \langle \hat{\psi}^\dagger(\mathbf{x})\hat{\psi}(\mathbf{x}')\rangle$. As we soon see, in the state of matter that is known as a Bose-Einstein condensate the long-range order is actually off-diagonal in the one-particle density matrix, which makes the Bose-Einstein condensed gas behave very differently from the other phases of matter that we have encountered so far. In particular, the intrinsic quantum-mechanical nature of this many-body state results in intriguing properties such as the possibility for the gas to flow without friction, i.e. superfluidity.

The subjects of this chapter are the Bose-Einstein condensation (BEC) and the superfluidity of ultracold atomic Bose gases, as first observed in gases of rubidium [10], lithium [11], sodium [12] and hydrogen [43]. The experimental realization of BEC for rubidium is shown in Fig. 4.4. We can describe Bose-Einstein condensed gases elegantly using statistical field theory, allowing for a treatment of interaction effects, which leads to a quantitative comparison with experiments. We start with a discussion of the order parameter for Bose-Einstein condensation, after which we give a criterion due to Landau, telling us when a quasiparticle dispersion gives rise to superfluid flow. Using the functional form of the Bogoliubov theory for Bose-Einstein condensation, we then derive the quasiparticle dispersion for excitations above the ground state, showing that it satisfies the Landau criterion. We also obtain the celebrated Gross-Pitaevskii equation for the condensate wavefunction, the Bogliubov-de Gennes equation that describes the inhomogeneous case, the Popov theory that takes fluctuations into account, and the collective modes using a hydrodynamic-like approach. Finally, we briefly discuss what happens when we try to bring the Bose-Einstein condensed gas into rotation, and what happens when the condensate has effectively attractive interactions, such that it is metastable.

11.1 Definitions for a Bose-Einstein Condensate

As discussed in Sect. 4.3.2, Bose-Einstein condensation corresponds to a macroscopic occupation of the same single-particle quantum state, where macroscopic means that the density $\langle N_{\mathbf{0}} \rangle / V$ does not go to zero in the thermodynamic limit. As a result, we expect that a Bose-Einstein condensed gas develops strong correlations over the size of the system. Formally, this can be made explicit by considering the one-particle density matrix [44, 45, 46], which is given by

$$n(\mathbf{x},\mathbf{x}') = \langle \hat{\psi}^{\dagger}(\mathbf{x}) \hat{\psi}(\mathbf{x}') \rangle. \tag{11.1}$$

This density matrix tells us whether or not the presence of a particle at position $\mathbf{x}'$ is correlated with the presence of a particle at position $\mathbf{x}$. Typically, as for instance in an ordinary gas or liquid, this kind of correlation decays exponentially fast with increasing separation $|\mathbf{x}-\mathbf{x}'|$, where the length scale associated with the exponential decay is called the correlation length. Since the above density matrix, whose elements are labelled by the continuous variables $\mathbf{x}$ and $\mathbf{x}'$, is a Hermitian matrix, it can be diagonalized with real eigenvalues. This means that there exists a set of single-particle eigenfunctions $\chi'_{\mathbf{n}}(\mathbf{x})$ of $n(\mathbf{x},\mathbf{x}')$ with eigenvalues $\lambda_{\mathbf{n}}$, that satisfy

$$\int d\mathbf{x}''\, n(\mathbf{x},\mathbf{x}'') \chi'^{*}_{\mathbf{n}}(\mathbf{x}'') = \lambda_{\mathbf{n}} \chi'^{*}_{\mathbf{n}}(\mathbf{x}). \tag{11.2}$$

Multiplying this last equation on both sides with $\chi'_{\mathbf{n}}(\mathbf{x}')$ and summing over $\mathbf{n}$, we obtain

$$\sum_{\mathbf{n}} \int d\mathbf{x}''\, n(\mathbf{x},\mathbf{x}'') \chi'_{\mathbf{n}}(\mathbf{x}') \chi'^{*}_{\mathbf{n}}(\mathbf{x}'') = n(\mathbf{x},\mathbf{x}') = \sum_{\mathbf{n}} \lambda_{\mathbf{n}} \chi'^{*}_{\mathbf{n}}(\mathbf{x}) \chi'_{\mathbf{n}}(\mathbf{x}'), \tag{11.3}$$

where we also used the completeness of the eigenfunctions $\chi'_{\mathbf{n}}(\mathbf{x})$.

The system is now called Bose-Einstein condensed if one of the eigenvalues $\lambda_{\mathbf{0}}$ is on the order of the number of particles $\langle N \rangle$ and all the other eigenvalues are of order one. If this occurs, then $\lambda_{\mathbf{0}} = \langle N_{\mathbf{0}} \rangle$ is the number of condensate particles and the density matrix is also said to have off-diagonal long-range order [47]. To get more insight in the above order parameter, we consider the noninteracting homogeneous Bose gas. Since the system is now translationally invariant, the single-particle density matrix depends only on the difference in the coordinates $\mathbf{x}-\mathbf{x}'$. As a result, this density matrix is given by

$$n(\mathbf{x},\mathbf{x}') = \langle \hat{\psi}^{\dagger}(\mathbf{x}) \hat{\psi}(\mathbf{x}') \rangle = \frac{1}{V} \sum_{\mathbf{k}} e^{-i\mathbf{k}\cdot(\mathbf{x}-\mathbf{x}')} \langle \hat{\psi}^{\dagger}_{\mathbf{k}} \hat{\psi}_{\mathbf{k}} \rangle, \tag{11.4}$$

and the eigenstates of the density matrix are just plane waves. This follows from

$$\int d\mathbf{x}'' \frac{1}{V} \sum_{\mathbf{k}'} \langle N_{\mathbf{k}'} \rangle e^{-i\mathbf{k}'\cdot(\mathbf{x}-\mathbf{x}'')} \frac{e^{-i\mathbf{k}\cdot\mathbf{x}''}}{\sqrt{V}} = \langle N_{\mathbf{k}} \rangle \frac{e^{-i\mathbf{k}\cdot\mathbf{x}}}{\sqrt{V}}, \tag{11.5}$$

where the corresponding eigenvalues $\lambda_{\mathbf{k}}$ are thus simply the average occupation numbers $\langle N_{\mathbf{k}}\rangle = \langle \hat{\psi}^{\dagger}_{\mathbf{k}} \hat{\psi}_{\mathbf{k}}\rangle$, as given by the Bose-Einstein distribution. For the ideal Bose gas, we found in Sect. 4.3.2 that the largest eigenvalue is given by $\langle N_{\mathbf{0}}\rangle = N_{\mathrm{BE}}(\varepsilon_{\mathbf{0}})$. This becomes on the order of $\langle N\rangle$ below the critical temperature T_{c}, at which the chemical potential reaches $\varepsilon_{\mathbf{0}} \to 0$ in the thermodynamic limit. To see more clearly that this corresponds to off-diagonal long-range order, we consider the density matrix near T_{c} when the chemical potential is slightly below or equal to zero. Separating explicitly the zero-momentum contribution and writing the sum over momenta as an integral, this expression becomes

$$\begin{aligned} n(\mathbf{x},\mathbf{x}') &= \frac{\langle \hat{\psi}^{\dagger}_{\mathbf{0}} \hat{\psi}_{\mathbf{0}}\rangle}{V} + \int \frac{\mathrm{d}\mathbf{k}}{(2\pi)^3} \mathrm{e}^{-\mathrm{i}\mathbf{k}\cdot(\mathbf{x}-\mathbf{x}')} \frac{1}{\mathrm{e}^{\beta(\varepsilon_{\mathbf{k}}-\mu)}-1} \\ &\simeq \frac{\langle \hat{\psi}^{\dagger}_{\mathbf{0}} \hat{\psi}_{\mathbf{0}}\rangle}{V} + \frac{1}{\beta}\int \frac{\mathrm{d}\mathbf{k}}{(2\pi)^3} \frac{\mathrm{e}^{-\mathrm{i}\mathbf{k}\cdot(\mathbf{x}-\mathbf{x}')}}{\varepsilon_{\mathbf{k}}-\mu} \\ &\simeq \frac{\langle \hat{\psi}^{\dagger}_{\mathbf{0}} \hat{\psi}_{\mathbf{0}}\rangle}{V} + \frac{mk_{\mathrm{B}}T}{2\pi|\mathbf{x}-\mathbf{x}'|}\mathrm{e}^{-\sqrt{2m|\mu|}|\mathbf{x}-\mathbf{x}'|/\hbar}, \end{aligned} \tag{11.6}$$

where in the first step we substituted the Bose distribution from (4.40). Since we are particularly interested in the limit $|\mathbf{x}-\mathbf{x}'| \to \infty$, we focus in the second step on the low-momentum or long-wavelength modes for which $\beta(\varepsilon_{\mathbf{k}}-\mu)$ is small. As the chemical potential becomes zero and the occupation of the ground state becomes macroscopic, we have that $\langle \hat{\psi}^{\dagger}_{\mathbf{0}} \hat{\psi}_{\mathbf{0}}\rangle/V = \langle N_{\mathbf{0}}\rangle/V$ is nonzero in the thermodynamic limit. Then, we find from (11.6) that the first term, which describes physically the constant nonzero density of Bose-Einstein condensed atoms, does not vanish at large separations $|\mathbf{x}-\mathbf{x}'|$. This is the off-diagonal long-range order in the single-particle density matrix.

Next, consider the zero-temperature limit of the homogeneous Bose gas with a weak repulsive interaction. The case of an attractive interaction actually only allows for a metastable condensate, whose discussion we leave for Sect. 11.9. Then, the ground state wavefunction of the interacting Bose gas is in general a complicated linear superposition of states $|N_{\mathbf{0}},\ldots\rangle$, where the occupation number of the zero-momentum state $N_{\mathbf{0}}$ fluctuates around the average value $\langle N_{\mathbf{0}}\rangle$ but is always a substantial fraction of the total number of atoms $\langle N\rangle \gg 1$. As a result, we have from (6.14) and (6.15) that within the subspace of these relevant states both $\hat{\psi}^{\dagger}_{\mathbf{0}} \hat{\psi}_{\mathbf{0}}|N_{\mathbf{0}},\ldots\rangle$ and $\hat{\psi}_{\mathbf{0}} \hat{\psi}^{\dagger}_{\mathbf{0}}|N_{\mathbf{0}},\ldots\rangle$ are very well approximated by $\langle N_{\mathbf{0}}\rangle|N_{\mathbf{0}},\ldots\rangle$. This seems to suggest that we can neglect the fact that the operators $\hat{\psi}_{\mathbf{0}}$ and $\hat{\psi}^{\dagger}_{\mathbf{0}}$ do not commute and replace them by the ordinary complex numbers $\sqrt{\langle N_{\mathbf{0}}\rangle}\mathrm{e}^{\mathrm{i}\theta}$ and $\sqrt{\langle N_{\mathbf{0}}\rangle}\mathrm{e}^{-\mathrm{i}\theta}$, which is called the Bogoliubov substitution. This implies that the operators obtain a nonzero expectation value in the Bose-Einstein condensed phase, i.e. $\langle \hat{\psi}_{\mathbf{0}}\rangle = \sqrt{\langle N_{\mathbf{0}}\rangle}\mathrm{e}^{\mathrm{i}\theta} = \sqrt{\lambda_{\mathbf{0}}}\mathrm{e}^{\mathrm{i}\theta}$, or $\langle \hat{\psi}(\mathbf{x})\rangle = \langle \hat{\psi}_{\mathbf{0}}\rangle \chi_{\mathbf{0}}(\mathbf{x}) = \sqrt{\langle n_{\mathbf{0}}\rangle}\mathrm{e}^{\mathrm{i}\theta}$ with $\langle n_{\mathbf{0}}\rangle = \langle N_{\mathbf{0}}\rangle/V$, which shows that we may use $\langle \hat{\psi}(\mathbf{x})\rangle$ as the order parameter for the transition.

Now, why would such a replacement be a useful thing to do? An important reason is that we cannot solve the interacting quantum gas exactly, and we have to make approximations that take the dominant physics into account. As we see next, the above substitution allows us to reduce the fourth-order interaction term to a number of quadratic terms, which take the dominant interaction of the noncondensed particles with the condensate into account. We can then deal analytically with the resulting Gaussian action, which is the essence of the Bogoliubov theory, giving rise to the famous Bogoliubov dispersion for the quasiparticles above the condensate. Moreover, the Bogoliubov replacement is easily generalized to the inhomogeneous interacting Bose gas at nonzero temperatures, for which it yields $\langle\hat{\psi}(\mathbf{x})\rangle = \sqrt{\lambda_0}\chi'_0(\mathbf{x}) = \sqrt{\langle N_0\rangle}\chi'_0(\mathbf{x})$. In general, we may thus define $\langle\hat{\psi}(\mathbf{x})\rangle \equiv \phi_0(\mathbf{x})$ as the order parameter, where its normalization is given by $\int d\mathbf{x}|\phi_0(\mathbf{x})|^2 = \langle N_0\rangle$ such that $\phi_0(\mathbf{x})$ is also known as the macroscopic wavefunction of the Bose-Einstein condensate.

Having introduced the definition of the order parameter, we can immediately apply Landau theory that was developed in Chap. 9. In Sect. 9.5, we discussed a Landau free energy with an $O(n)$ symmetry for a real n-component order parameter $\mathbf{m}$, such that this free energy had a continuum of minima connected by a rotation of the order parameter. The phase transition to a specific phase where $\langle\mathbf{m}\rangle \neq \mathbf{0}$ was then seen to spontaneously break the continuous $O(n)$ symmetry and give rise to Goldstone modes. We will see that Bose-Einstein condensation actually gives rise to a phase that spontaneously breaks a $U(1)$ symmetry of the microscopic action. The effects of this symmetry and the corresponding symmetry breaking are discussed in more detail in Chap. 13.

11.2 Superfluidity

Closely related to the phenomenon of Bose-Einstein condensation is the phenomenon of superfluidity, which is the ability of the Bose-Einstein condensed gas to flow without experiencing friction. In this section we present a physical argument due to Landau, showing that a gas with a quasiparticle dispersion that is linear in momentum gives rise to a critical flow velocity below which it behaves as a superfluid. Later this chapter, when we have set up the quantum field theory to describe Bose-Einstein condensation, we show that the dispersion of the condensed gas is indeed linear in momentum. However, we hasten to remark that the correspondence between superfluidity and Bose-Einstein condensation is not one to one. To illustrate this we note that in Chap. 15, where we consider low-dimensional systems, the two-dimensional Bose gas is seen to give rise to superfluidity without actually being Bose-Einstein condensed.

11.2.1 Landau Criterion

To study the physical meaning of superfluidity, we consider a body of mass M with no internal degrees of freedom that is moving with a momentum $\mathbf{p} = M\mathbf{v}$ through an atomic gas that resides in its ground state. This body can then experience friction by creating excitations in the gas, which must occur in such a manner that both momentum and energy are conserved in the process. For a process creating an excitation with momentum $\mathbf{p}'$, the condition of energy conservation yields

$$\Delta E = \frac{(\mathbf{p}-\mathbf{p}')^2}{2M} + \hbar\omega_{\mathbf{p}'} - \frac{\mathbf{p}^2}{2M} = \frac{-\mathbf{p}\cdot\mathbf{p}'}{M} + \frac{\mathbf{p}'^2}{2M} + \hbar\omega_{\mathbf{p}'} = 0, \tag{11.7}$$

where $\hbar\omega_{\mathbf{p}'}$ is the energy of the excitation, which by definition is positive. Since $\mathbf{p}'^2/2M$ is also positive, we have that $-\mathbf{p}\cdot\mathbf{p}'/M$ should be negative in order to obey energy conservation. We can rewrite (11.7) in a slightly more illuminating way, namely

$$v\cos(\vartheta) = \frac{\hbar\omega_{\mathbf{p}'}}{p'} + \frac{p'}{2M}, \tag{11.8}$$

where ϑ is the angle between the two momenta and we also used $v = p/M$. For convenience, we consider the mass M to be very large such that the second term in the right-hand side can be neglected. We then immediately find the desired Landau criterion, which states that there is a critical velocity v_c of the body below which it cannot create excitations in the atomic gas, where the critical velocity is given by the minimum of $\hbar\omega_{\mathbf{p}'}/p'$. Considering, for example, sound waves with $\hbar\omega_{\mathbf{p}'} = cp'$ and c the speed of sound, we find that it is not possible to satisfy (11.8) for velocities v smaller then the speed of sound c. As a result, the body fails to cause excitations in the gas, because this would violate energy conservation. Since the object now simply moves through the gas as if the latter were not there, the gas does not exert any friction on the body, and the gas is said to be superfluid. Note that for velocities larger than the critical velocity, it is always possible to find angles ϑ and momenta p' such that (11.8) is satisfied. Then, the body can dissipate its energy by causing excitations in the gas and therefore the superfluid behavior of the gas breaks down.

Although the above argument turns out to be not fully rigorous, superfluid flow has been observed for Bose-Einstein condensates, as well as critical velocities beyond which the superfluidity breaks down. Note that when the excitation spectrum is gapped, for example when $\hbar\omega_{\mathbf{p}'} = \Delta + (p'-p_0)^2/2m^*$ with the energy gap Δ a positive nonzero energy, then the minimum $\hbar\omega_{\mathbf{p}'}/p'$ is also nonzero, yielding again a critical velocity below which the gas is superfluid. In the next chapter we find that this kind of dispersion occurs in Bose-Einstein condensates of Cooper pairs, which occur in atomic Fermi gases with an attractive interaction. The superfluidity of these atomic Fermi gases is analogous to the conventional superconductivity in metals, where the electrons can form Cooper pairs due to a mutual attractive interaction generated by lattice vibrations, i.e. by phonons.

11.2.2 Superfluid Density

We can extend the discussion of the previous paragraph a bit further to determine the superfluid density. To this end, we look at a partly Bose-Einstein condensed ideal Bose gas of volume V that enters a tube with a velocity $-\mathbf{v}$, or, equivalently, we look at a stationary Bose gas in a tube whose walls are moving with the velocity $\mathbf{v}$. As explained in the previous paragraph, below a critical velocity the superfluid part of the gas with density n_s is not affected by the motion of the tube, whereas the normal part with density n_n eventually equilibrates at the tube's velocity $\mathbf{v}$ due to friction with the walls. As a result, the total momentum of the gas in the rest frame for the condensate is after equilibration given by

$$\langle \mathbf{P} \rangle = V n_n m \mathbf{v} = \sum_{\mathbf{k}} \hbar \mathbf{k} \frac{1}{e^{\beta(E_{\mathbf{k}}-\mu)} - 1}, \tag{11.9}$$

where we have for $E_{\mathbf{k}}$

$$E_{\mathbf{k}} = \frac{1}{2m}(\hbar \mathbf{k} - m\mathbf{v})^2. \tag{11.10}$$

The latter is the kinetic energy of a particle in the frame moving with the tube, where the normal part is in equilibrium. In the limit $\mathbf{v} \to 0$, we may look at the linear response of the system by expanding the obtained expressions up to first order in $\mathbf{v}$, giving for the kinetic energy

$$E_{\mathbf{k}} \simeq \frac{\hbar^2 \mathbf{k}^2}{2m} - \hbar \mathbf{k} \cdot \mathbf{v}. \tag{11.11}$$

Substitution of the above result in (11.9) then yields to first order in $\mathbf{v}$

$$\langle \mathbf{P} \rangle = \sum_{\mathbf{k}} \beta \hbar^2 \mathbf{k} \frac{e^{\beta(\varepsilon_{\mathbf{k}}-\mu)}}{(e^{\beta(\varepsilon_{\mathbf{k}}-\mu)} - 1)^2} \mathbf{k} \cdot \mathbf{v}. \tag{11.12}$$

Converting the sum over $\mathbf{k}$ into an integral, we see that the integral vanishes if the integrand is proportional to $k_i k_j$ with $i \neq j$, because then the integrand is odd. For k_i^2 we have that the integration over angles leads to $4\pi k^2/3$, such that the combination of (11.9) with (11.12) yields

$$n_n = \int \frac{d\mathbf{k}}{(2\pi)^3} \frac{\beta}{3} \frac{\hbar^2 \mathbf{k}^2}{m} \frac{e^{\beta(\varepsilon_{\mathbf{k}}-\mu)}}{(e^{\beta(\varepsilon_{\mathbf{k}}-\mu)} - 1)^2}. \tag{11.13}$$

Now that we have obtained an expression for the normal part of the gas, the superfluid density is given by the difference between the total density n and the normal density n_n. We thus have for the ideal Bose gas that

$$n_s = n_0 + \int \frac{d\mathbf{k}}{(2\pi)^3} \frac{1}{(e^{\beta(\varepsilon_{\mathbf{k}}-\mu)} - 1)} - \int \frac{d\mathbf{k}}{(2\pi)^3} \frac{2}{3} \beta \varepsilon_{\mathbf{k}} \frac{e^{\beta(\varepsilon_{\mathbf{k}}-\mu)}}{(e^{\beta(\varepsilon_{\mathbf{k}}-\mu)} - 1)^2}. \tag{11.14}$$

The last two integrals on the right-hand side turn out to be equal to each other. To show this, we call the first integral I_1 and the second I_2, and we note that

$$I_2 = -\frac{2}{3}\left\{\beta\frac{\partial I_1}{\partial \beta} - \mu\frac{\partial I_1}{\partial \mu}\right\}, \tag{11.15}$$

where in Sect. 4.3.2.1, we showed that $I_1 = g_{3/2}(e^{\beta\mu})/\Lambda^3$. From the expression for the thermal de Broglie wavelength $\Lambda = (2\pi\hbar^2\beta/m)^{1/2}$, it immediately follows that the right-hand side of the above equation is indeed equal to I_1. As a result, (11.14) shows that for the ideal Bose gas, the superfluid density n_s is equal to the condensate density n_0.

However, in obtaining this result we used explicitly the ideal gas dispersion $\varepsilon_{\mathbf{k}} - \mu$ for the noncondensed atoms. This changes in the presence of interactions, as we show explicitly in the next section. If we denote the dispersion in the interacting case by $\hbar\omega_{\mathbf{k}}$, we find in the same manner as before that the normal density of the gas is given by

$$n_n = \int \frac{d\mathbf{k}}{(2\pi)^3}\frac{\beta}{3}\frac{\hbar^2\mathbf{k}^2}{m}\frac{e^{\beta\hbar\omega_{\mathbf{k}}}}{(e^{\beta\hbar\omega_{\mathbf{k}}}-1)^2} = \frac{1}{3}\int \frac{d\mathbf{k}}{(2\pi)^3}\frac{\hbar^2\mathbf{k}^2}{m}\left(-\frac{dN_{\rm BE}(\hbar\omega_{\mathbf{k}})}{d\hbar\omega_{\mathbf{k}}}\right). \tag{11.16}$$

Then, in general, the normal density does not cancel against the noncondensed density as in (11.14), such that the condensate density deviates from the superfluid density $n_s = n - n_n$. An extreme example of this is the strongly-interacting superfluid helium-4. The condensate fraction amounts at low temperatures only to about 10% of the total liquid, while the superfluid fraction is essentially 100%.

11.3 Field-Theory Approach

To treat Bose-Einstein condensation within the functional-integral formalism, we start from the following action for an ultracold gas of bosonic atoms

$$\begin{aligned} S[\phi^*,\phi] = &\int_0^{\hbar\beta} d\tau \int d\mathbf{x}\, \phi^*(\mathbf{x},\tau)\left\{\hbar\frac{\partial}{\partial\tau} - \frac{\hbar^2\nabla^2}{2m} + V^{\rm ex}(\mathbf{x}) - \mu\right\}\phi(\mathbf{x},\tau) \\ &+ \frac{1}{2}\int_0^{\hbar\beta} d\tau \int d\mathbf{x}\, T^{\rm 2B}\phi^*(\mathbf{x},\tau)\phi^*(\mathbf{x},\tau)\phi(\mathbf{x},\tau)\phi(\mathbf{x},\tau)\,, \end{aligned} \tag{11.17}$$

where we have replaced the true interatomic potential $V(\mathbf{x}-\mathbf{x}')$ by the pseudopotential of the form $T^{\rm 2B}\delta(\mathbf{x}-\mathbf{x}')$. The two-body T matrix was introduced in Sect. 10.3.1, and we have $T^{\rm 2B} \equiv T^{\rm 2B}(0) = 4\pi\hbar^2 a/m$ in terms of the interatomic scattering length a. The use of the pseudopotential is allowed when the thermal de Broglie wavelength $\Lambda = (2\pi\hbar^2/mk_BT)^{1/2}$ of the atoms is much larger than the range of the interatomic interaction, such that the atoms only probe the long-wavelength, or low-momentum behavior of the true interaction. In Sect. 10.3.1 we found that for low momenta the true interatomic interaction is fully characterized by the s-wave scattering length a. Since the Fourier transform of the above pseudopotential is equal to $4\pi\hbar^2 a/m$, this pseudopotential is seen to yield the exact result for the scattering amplitude of the true interatomic potential already in the Born approximation. Put differently, the first term of the Born series from (10.22) now effectively incorporates all higher order terms of the true interaction. This also means that we should be careful with using the pseudopotential in higher orders of perturbation theory since this could lead to double-counting problems. We will see some examples of this shortly. For that reason, the pseudopotential considered here is quite different from the contact potential $V_0\delta(\mathbf{x}-\mathbf{x}')$ considered in Sect. 10.4. The latter is related to the exact two-body scattering properties via the Lippmann-Schwinger equation, rather then immediately in the Born approximation. In the next chapter, we find that the latter contact potential is more convenient to study fermionic condensates. Finally, we remark that an ab-initio calculation for the s-wave scattering length a is quite difficult in practice because it depends sensitively on the details of the true interatomic potential, which is often not precisely known. However, the value of the s-wave scattering length can be measured experimentally, and the result of the experiment is then consequently used as an input parameter for our many-body theory.

As explained in Sect. 11.1, the order parameter for Bose-Einstein condensation is given in field-theory language by $\langle\phi(\mathbf{x},\tau)\rangle \equiv \langle\phi_0(\mathbf{x})\rangle$. Using this information, we note that for time-independent fields, the action of (11.17) leads to a Landau 'free energy' for the order parameter, given by

$$F_{\mathrm{L}}[\phi^*,\phi] = \int \mathrm{d}\mathbf{x} \left\{ \frac{\hbar^2}{2m}|\nabla\phi(\mathbf{x})|^2 + (V^{\mathrm{ex}}(\mathbf{x})-\mu)|\phi(\mathbf{x})|^2 + \frac{T^{2\mathrm{B}}}{2}|\phi(\mathbf{x})|^4 \right\}, \quad (11.18)$$

where, although we are now explicitly working grand-canonically, we still call the above functional a 'free energy' to establish a direct link with the discussion of Landau theory in Chap. 9. There, it was also explained how the critical temperature for a second-order phase transition is determined by the change of sign in the quadratic coefficient of the Landau free energy. To determine the critical temperature for Bose-Einstein condensation to lowest (zeroth) order in the interaction, we expand the field $\phi(\mathbf{x})$ into the single-particle eigenstates for the external potential as given by (6.3). Then, considering the part proportional to the single-particle ground state $\phi_0\chi_0(\mathbf{x})$ and neglecting the interaction terms, we find that the corresponding quadratic part of the Landau free energy becomes $(\varepsilon_0-\mu)|\phi_0|^2$. Upon lowering the temperature, the critical temperature is reached when $\mu(T_c)=\varepsilon_0$, such that the quadratic coefficient vanishes. This result for the critical temperature makes sense, because it is precisely the condition that we encountered in Sect. 4.3.2 when we discussed Bose-Einstein condensation for an ideal Bose gas. Note that we must have that $T^{2\mathrm{B}} > 0$ for the Landau theory to be stable. This case with repulsive interactions we discuss first, since it is most relevant for experiments with ultracold alkali gases. The case of attractive interactions with a negative scattering length is sometimes also realized, but is postponed until Sect. 11.9.

11.3.1 Bogoliubov Theory and the Gross-Pitaevskii Equation

To be able to determine the corrections to the critical temperature for the ideal Bose gas, we start by performing the following fluctuation expansion

$$\phi(\mathbf{x},\tau) = \phi_0(\mathbf{x}) + \phi'(\mathbf{x},\tau), \tag{11.19}$$

which is the field-theory equivalent of the Bogoliubov substitution discussed in Sect. 11.1. To consistently define the fluctuations $\phi'(\mathbf{x},\tau)$ in this manner, we have to require

$$\int \mathrm{d}\mathbf{x}\; \phi_0^*(\mathbf{x})\phi'(\mathbf{x},\tau) + \int \mathrm{d}\mathbf{x}\; \phi_0(\mathbf{x})\phi'^*(\mathbf{x},\tau) = 0\,, \tag{11.20}$$

such that the fluctuations $\phi'(\mathbf{x},\tau)$ contain all configurations that are orthogonal to the condensate $\phi_0(\mathbf{x}) \equiv \langle\phi(\mathbf{x},\tau)\rangle$. This condition can be understood better once we have obtained the Bogoliubov theory of Bose-Einstein condensation, so we come back to this condition in Sect. 11.5. Note that (11.20) even allows for fluctuations that only multiply $\phi_0(\mathbf{x})$ by a global phase factor $\mathrm{e}^{i\theta(\tau)}$. These kind of fluctuations turn out to give rise to the phenomenon of phase diffusion, which is discussed in more detail in Sect. 13.5. There, we also find that these subtle fluctuations do not influence the thermodynamics, so in this chapter we consider the fluctuations to occupy only states with $\mathbf{n} \neq \mathbf{0}$.

After substituting (11.19) into the action of (11.17), we find that

$$S[\phi'^*,\phi'] = \hbar\beta F_{\mathrm{L}}[\phi_0^*,\phi_0] + S_{\mathrm{Bog}}[\phi'^*,\phi'] + S_{\mathrm{int}}[\phi'^*,\phi']\,, \tag{11.21}$$

where the linear and quadratic terms are given by

$$\begin{aligned}
&S_{\mathrm{Bog}}[\phi'^*,\phi'] \\
&= \int_0^{\hbar\beta}\mathrm{d}\tau \int \mathrm{d}\mathbf{x}\; \phi'^*(\mathbf{x},\tau)\left\{-\frac{\hbar^2\boldsymbol{\nabla}^2}{2m} + V^{\mathrm{ex}}(\mathbf{x}) - \mu + T^{2\mathrm{B}}|\phi_0(\mathbf{x})|^2\right\}\phi_0(\mathbf{x}) \\
&+ \int_0^{\hbar\beta}\mathrm{d}\tau \int \mathrm{d}\mathbf{x}\; \phi'(\mathbf{x},\tau)\left\{-\frac{\hbar^2\boldsymbol{\nabla}^2}{2m} + V^{\mathrm{ex}}(\mathbf{x}) - \mu + T^{2\mathrm{B}}|\phi_0(\mathbf{x})|^2\right\}\phi_0^*(\mathbf{x}) \\
&+ \int_0^{\hbar\beta}\mathrm{d}\tau \int \mathrm{d}\mathbf{x}\; \phi'^*(\mathbf{x},\tau) \\
&\qquad\qquad \times\left\{\hbar\frac{\partial}{\partial\tau} - \frac{\hbar^2\boldsymbol{\nabla}^2}{2m} + V^{\mathrm{ex}}(\mathbf{x}) - \mu + 2T^{2\mathrm{B}}|\phi_0(\mathbf{x})|^2\right\}\phi'(\mathbf{x},\tau) \\
&+ \frac{1}{2}\int_0^{\hbar\beta}\mathrm{d}\tau \int \mathrm{d}\mathbf{x}\; T^{2\mathrm{B}}(\phi_0(\mathbf{x}))^2\phi'^*(\mathbf{x},\tau)\phi'^*(\mathbf{x},\tau) \\
&+ \frac{1}{2}\int_0^{\hbar\beta}\mathrm{d}\tau \int \mathrm{d}\mathbf{x}\; T^{2\mathrm{B}}(\phi_0^*(\mathbf{x}))^2\phi'(\mathbf{x},\tau)\phi'(\mathbf{x},\tau)\,,
\end{aligned} \tag{11.22}$$

and the cubic and quartic terms by

$$S_{\rm int}[\phi'^*,\phi'] = \int_0^{\hbar\beta} d\tau \int d\mathbf{x}\, T^{2B} \phi_0(\mathbf{x})\phi'^*(\mathbf{x},\tau)\phi'^*(\mathbf{x},\tau)\phi'(\mathbf{x},\tau) \qquad (11.23)$$
$$+\int_0^{\hbar\beta} d\tau \int d\mathbf{x}\, T^{2B} \phi_0^*(\mathbf{x})\phi'^*(\mathbf{x},\tau)\phi'(\mathbf{x},\tau)\phi'(\mathbf{x},\tau)$$
$$+\frac{1}{2}\int_0^{\hbar\beta} d\tau \int d\mathbf{x}\, T^{2B} \phi'^*(\mathbf{x},\tau)\phi'^*(\mathbf{x},\tau)\phi'(\mathbf{x},\tau)\phi'(\mathbf{x},\tau)\,.$$

Note that we used partial integrations in (11.22) to bring all the terms that are linear in the fluctuations into the same form.

In the Bogoliubov approximation, we neglect all the terms in the action that are higher than second-order in the fluctuations [48]. This thus means that we simply ignore $S_{\rm int}[\phi'^*,\phi']$, which consists of the third and fourth-order terms. Moreover, since we want $\phi_0(\mathbf{x})$ to describe the Bose-Einstein condensate, i.e. $\phi_0(\mathbf{x}) = \langle\phi(\mathbf{x},\tau)\rangle$ and $\langle\phi'(\mathbf{x},\tau)\rangle = 0$, we need to make sure that in the action the linear terms in the fluctuations ϕ' and ϕ'^* vanish such that $\phi_0(\mathbf{x})$ really minimizes the action. In the Bogoliubov approximation, we then require that the linear terms in ϕ' and ϕ'^* should be zero in the action $S_0[\phi'^*,\phi']$, which is the case when

$$\left\{-\frac{\hbar^2\nabla^2}{2m} + V^{\rm ex}(\mathbf{x}) + T^{2B}|\phi_0(\mathbf{x})|^2\right\}\phi_0(\mathbf{x}) = \mu\phi_0(\mathbf{x})\,. \qquad (11.24)$$

In the context of trapped atomic gases, this nonlinear differential equation is better known as the Gross-Pitaevskii equation [49, 50]. Note that it can also be obtained by minimizing the Landau 'free-energy' $F_{\rm L}[\phi^*,\phi]$ of (11.18) with respect to $\phi^*(\mathbf{x})$. Satisfying the Gross-Pitaevskii equation makes sure that our fluctuation expansion has been performed correctly around the minimum of the action, while at the same time it also determines the macroscopic wavefunction of the condensate $\phi_0(\mathbf{x})$. As mentioned previously, the reason for calling $\phi_0(\mathbf{x})$ a macroscopic wavefunction follows from the fact that the total density of the gas obeys

$$n(\mathbf{x}) = \langle\phi(\mathbf{x},\tau)\phi^*(\mathbf{x},\tau^+)\rangle = |\phi_0(\mathbf{x})|^2 + \langle\phi'(\mathbf{x},\tau)\phi'^*(\mathbf{x},\tau^+)\rangle\,, \qquad (11.25)$$

where the total number of condensate atoms equals $\langle N_0\rangle = \int d\mathbf{x}\, |\phi_0(\mathbf{x})|^2$, which is macroscopic in the case of Bose-Einstein condensation. From (11.25), it thus follows that $\langle N_0\rangle$ is typically smaller than the total number of atoms $\langle N\rangle$ due to the presence of fluctuations. Note that the average $\langle\phi'(\mathbf{x},\tau)\phi'^*(\mathbf{x},\tau^+)\rangle$ not only describes the depletion of the condensate due to the thermal occupation of excited states, which happens already for the ideal Bose gas, but also the depletion due to interaction effects. As we see later, due to the presence of interactions the average $\langle\phi'(\mathbf{x},\tau)\phi'^*(\mathbf{x},\tau^+)\rangle$ is nonzero even at zero temperature. This depletion clearly cannot be caused by thermal fluctuations, and is thus entirely caused by quantum fluctuations. In Chap. 16, we find that these quantum fluctuations can even drive phase transitions at zero temperature, just like thermal fluctuations drive phase transitions at nonzero temperature. These kinds of phase transitions are therefore appropriately called quantum phase transitions.

11.3.2 Dyson Equation

As follows from (11.22), the fluctuation effects are described in Bogoliubov theory by a quadratic action of the form

$$S_{\text{Bog}}[\phi'^*,\phi'] = \tag{11.26}$$
$$-\frac{\hbar}{2}\int_0^{\hbar\beta} d\tau\, d\tau' \int d\mathbf{x}\, d\mathbf{x}' \, [\phi'^*(\mathbf{x},\tau),\phi'(\mathbf{x},\tau)]\cdot \mathbf{G}^{-1}(\mathbf{x},\tau;\mathbf{x}',\tau')\cdot \begin{bmatrix} \phi'(\mathbf{x}',\tau') \\ \phi'^*(\mathbf{x}',\tau') \end{bmatrix},$$

where we have assumed that we have solved the Gross-Pitaevskii equation, such that the linear terms are zero. The Green's function $\mathbf{G}$ has now an additional 2×2 matrix structure, because not only the normal average $\langle\phi'(\mathbf{x},\tau)\phi'^*(\mathbf{x}',\tau')\rangle$, but also the anomalous average $\langle\phi'(\mathbf{x},\tau)\phi'(\mathbf{x}',\tau')\rangle$, is unequal to zero. This additional 2×2 matrix structure is also called Nambu space. We then have that

$$-\mathbf{G}(\mathbf{x},\tau;\mathbf{x}',\tau') = \left\langle \begin{bmatrix} \phi'(\mathbf{x},\tau) \\ \phi'^*(\mathbf{x},\tau) \end{bmatrix} \cdot [\phi'^*(\mathbf{x}',\tau'),\phi'(\mathbf{x}',\tau')] \right\rangle, \tag{11.27}$$

where it follows from (11.22) that in the Bogoliubov approximation we find

$$\mathbf{G}^{-1}(\mathbf{x},\tau;\mathbf{x}',\tau') = \mathbf{G}_0^{-1}(\mathbf{x},\tau;\mathbf{x}',\tau') \tag{11.28}$$
$$-\frac{1}{\hbar}\begin{bmatrix} 2T^{2\text{B}}|\phi_0(\mathbf{x})|^2 & T^{2\text{B}}(\phi_0(\mathbf{x}))^2 \\ T^{2\text{B}}(\phi_0^*(\mathbf{x}))^2 & 2T^{2\text{B}}|\phi_0(\mathbf{x})|^2 \end{bmatrix}\delta(\mathbf{x}-\mathbf{x}')\delta(\tau-\tau'),$$

with the noninteracting Green's function matrix $\mathbf{G}_0$ defined by

$$\mathbf{G}_0^{-1}(\mathbf{x},\tau;\mathbf{x}',\tau') = \begin{bmatrix} G_0^{-1}(\mathbf{x},\tau;\mathbf{x}',\tau') & 0 \\ 0 & G_0^{-1}(\mathbf{x}',\tau';\mathbf{x},\tau) \end{bmatrix}, \tag{11.29}$$

where, as before, we have

$$G_0^{-1}(\mathbf{x},\tau;\mathbf{x}',\tau') = -\frac{1}{\hbar}\left\{\hbar\frac{\partial}{\partial\tau} - \frac{\hbar^2\nabla^2}{2m} + V^{\text{ex}}(\mathbf{x}) - \mu\right\}\delta(\mathbf{x}-\mathbf{x}')\delta(\tau-\tau'). \tag{11.30}$$

Note that (11.28) is actually the lowest-order result for the Green's function of the fluctuations in the Bose-Einstein condensed phase. This is because we have neglected $S_{\text{int}}[\phi'^*,\phi']$, whose perturbative treatment would then lead to all higher-order corrections. However, it turns out that the exact Dyson equation can always be written in the form

$$\begin{bmatrix} G_{11} & G_{12} \\ G_{21} & G_{22} \end{bmatrix}^{-1} = \begin{bmatrix} G_0^{-1} & 0 \\ 0 & G_0^{-1} \end{bmatrix} - \begin{bmatrix} \Sigma_{11} & \Sigma_{12} \\ \Sigma_{21} & \Sigma_{22} \end{bmatrix}, \tag{11.31}$$

where the Dyson equations for G_{11} and G_{21} are diagrammatically given by Fig. 11.1. As mentioned before, the off-diagonal elements are called anomalous, because they are not present in the normal phase of the gas. The selfenergy in the Bogoliubov

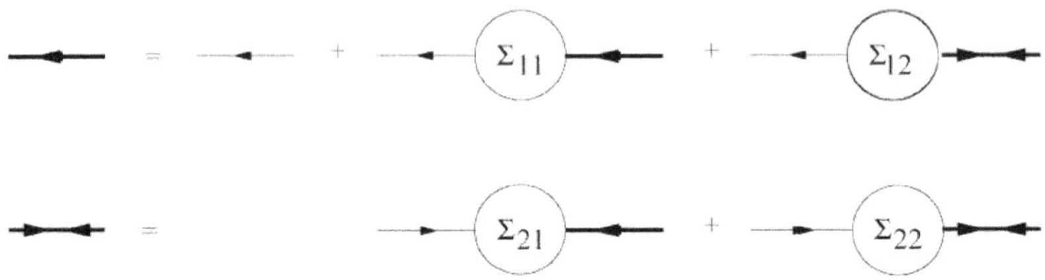

Fig. 11.1 Exact Dyson equation for the interacting a) normal and b) anomalous Green's functions. The thin lines represent G_0, whereas the thick lines represent G_{11} (one arrowhead) and G_{12} (two arrowheads).

approximation is then given by

$$\hbar\Sigma(\mathbf{x},\tau;\mathbf{x}',\tau') = \begin{bmatrix} 2T^{2B}|\phi_0(\mathbf{x})|^2 & T^{2B}(\phi_0(\mathbf{x}))^2 \\ T^{2B}(\phi_0^*(\mathbf{x}))^2 & 2T^{2B}|\phi_0(\mathbf{x})|^2 \end{bmatrix} \delta(\mathbf{x}-\mathbf{x}')\delta(\tau-\tau') . \quad (11.32)$$

Since we have completely neglected the cubic and quartic terms in the fluctuations to obtain the above selfenergy, the Bogoliubov approach is only expected to be valid if the fluctuations are sufficiently small, which also implies that the depletion of the condensate should be small. As a result, it can not be applied to liquid helium-4, for which the condensate fraction is only on the order of 10%. However, it is expected to be valid for a weakly-interacting dilute atomic Bose gas at ultralow temperatures, where the condensate fraction is nearly 100%. Under these conditions, the Bogoliubov theory not only predicts the condensate density profile $n_0(\mathbf{x}) = |\phi_0(\mathbf{x})|^2$, but also the quasiparticle dispersion and the collective modes of the condensate. In the next paragraph we study the quasiparticle excitation spectrum, and in Sect. 11.9.2 we discuss the collective modes.

11.3.3 Quasiparticle Dispersion

In order to simplify our discussion, we turn to the homogeneous Bose gas in a box with volume $V = L^3$ and periodic boundary conditions. The single-particle wavefunctions are then equal to $\chi_{\mathbf{k}}(\mathbf{x}) = e^{i\mathbf{k}\cdot\mathbf{x}}/\sqrt{V}$, and are characterized by the wavevector $\mathbf{k} = 2\pi\mathbf{n}/L$ with $n_i = 0, \pm 1, \ldots$. The corresponding single-particle energies are given by $\varepsilon_{\mathbf{k}} = \hbar^2\mathbf{k}^2/2m$. For the homogenous case, the Gross-Pitaevskii equation is easily solved by considering the constant solution $\phi_0(\mathbf{x}) = \phi_0$, which leads to $T^{2B}|\phi_0|^2 = \mu$. Moreover, we can consider the Fourier transform of the Green's function matrix from (11.28), which is given by

$$-\hbar\mathbf{G}^{-1}(\mathbf{k},\mathrm{i}\omega_n) = -\hbar\int_0^{\hbar\beta} \mathrm{d}\tau\,\mathrm{d}\tau' \int \mathrm{d}\mathbf{x}\,\mathrm{d}\mathbf{x}'\,\mathbf{G}^{-1}(\mathbf{x},\tau;\mathbf{x}',\tau')\frac{\mathrm{e}^{-\mathrm{i}\mathbf{k}(\mathbf{x}-\mathbf{x}')}}{V}\frac{\mathrm{e}^{\mathrm{i}\omega_n(\tau-\tau')}}{\hbar\beta}$$
$$= \begin{bmatrix} -\mathrm{i}\hbar\omega_n + \varepsilon_{\mathbf{k}} + T^{2\mathrm{B}}|\phi_{\mathbf{0}}|^2 & T^{2\mathrm{B}}\phi_{\mathbf{0}}^2 \\ T^{2\mathrm{B}}\phi_{\mathbf{0}}^{*2} & \mathrm{i}\hbar\omega_n + \varepsilon_{\mathbf{k}} + T^{2\mathrm{B}}|\phi_{\mathbf{0}}|^2 \end{bmatrix}, \tag{11.33}$$

where we used (7.55) and eliminated the chemical potential μ with the use of the homogeneous Gross-Pitaevskii equation. Note that the different signs in front of the Matsubara frequencies come from the reversed order of the imaginary-time arguments in (11.29). We can invert (11.33) to obtain the Green's function $\mathbf{G}^{-1}(\mathbf{k},\mathrm{i}\omega_n)$, given by

$$\mathbf{G}(\mathbf{k},\mathrm{i}\omega_n) \tag{11.34}$$
$$= \frac{-\hbar}{\hbar^2\omega_n^2 + \varepsilon_{\mathbf{k}}^2 + 2T^{2\mathrm{B}}|\phi_{\mathbf{0}}|^2\varepsilon_{\mathbf{k}}} \begin{bmatrix} \mathrm{i}\hbar\omega_n + \varepsilon_{\mathbf{k}} + T^{2\mathrm{B}}|\phi_{\mathbf{0}}|^2 & -T^{2\mathrm{B}}\phi_{\mathbf{0}}^2 \\ -T^{2\mathrm{B}}\phi_{\mathbf{0}}^{*2} & -\mathrm{i}\hbar\omega_n + \varepsilon_{\mathbf{k}} + T^{2\mathrm{B}}|\phi_{\mathbf{0}}|^2 \end{bmatrix}.$$

There are poles in the Green's function $G(\mathbf{k},\omega)$, when

$$\hbar\omega = \hbar\omega_{\mathbf{k}} \equiv \sqrt{\varepsilon_{\mathbf{k}}^2 + 2T^{2\mathrm{B}}|\phi_{\mathbf{0}}|^2\varepsilon_{\mathbf{k}}} = \sqrt{\varepsilon_{\mathbf{k}}^2 + 2T^{2\mathrm{B}}n_{\mathbf{0}}\varepsilon_{\mathbf{k}}}\,, \tag{11.35}$$

where it was explained in Sect. 8.1 that the poles of $G(\mathbf{k},\omega)$ correspond to the elementary, or single-particle, excitations of the system. As a result, (11.35) gives us the famous Bogoliubov dispersion for the quasiparticle excitations in the presence of a Bose-Einstein condensate. In Sect. 8.1.3, it was mentioned that collective excitations are described by poles in the correlation function of the form $\langle\phi^*\phi^*\phi\phi\rangle$. Since this two-particle correlation function contains also the contribution $|\phi_{\mathbf{0}}|^2\langle\phi'^*\phi'\rangle$, the poles in the Green's function of (11.34) are also poles of the two-particle correlation function, and we have that the Bogoliubov quasiparticle excitations can also be interpreted as collective excitations. In the following, we use both names for the excitations associated with the dispersion from (11.35). Note that for small momenta, or large wavelengths, the dispersion of (11.35) is of the phonon-like form $\omega_{\mathbf{k}} = ck$ with $c = (T^{2\mathrm{B}}n_{\mathbf{0}}/m)^{1/2}$, which according to the Landau criterion indeed gives rise to superfluid flow.

To finish the calculation, we still wish to obtain the condensate density $n_{\mathbf{0}} = |\phi_{\mathbf{0}}|^2$ in term of the total density n of the atomic Bose gas. The expression for the total density is given by (11.25), such that

$$n = n_{\mathbf{0}} + n' = |\phi_{\mathbf{0}}|^2 - G_{11}(\mathbf{x},\tau;\mathbf{x},\tau^+)\,, \tag{11.36}$$

where n' is the density of noncondensed particles. This density is equal to

$$
\begin{aligned}
n' &= \lim_{\eta \downarrow 0} \frac{\hbar}{V\hbar\beta} \sum_{\mathbf{k}\neq\mathbf{0},n} e^{i\omega_n\eta} \frac{i\hbar\omega_n + \varepsilon_{\mathbf{k}} + T^{2B} n_0}{(\hbar\omega_n)^2 + (\hbar\omega_{\mathbf{k}})^2} \\
&= \lim_{\eta \downarrow 0} \frac{1}{V\hbar\beta} \sum_{\mathbf{k}\neq\mathbf{0},n} \left\{ \frac{-e^{i\omega_n\eta}}{i\omega_n - \omega_{\mathbf{k}}} + \frac{\varepsilon_{\mathbf{k}} + T^{2B} n_0 - \hbar\omega_{\mathbf{k}}}{2\hbar\omega_{\mathbf{k}}} \left(\frac{-e^{i\omega_n\eta}}{i\omega_n - \omega_{\mathbf{k}}} + \frac{e^{i\omega_n\eta}}{i\omega_n + \omega_{\mathbf{k}}} \right) \right\} \\
&= \frac{1}{V} \sum_{\mathbf{k}\neq\mathbf{0}} \left(\frac{\varepsilon_{\mathbf{k}} + T^{2B} n_0}{\hbar\omega_{\mathbf{k}}} \frac{1}{e^{\beta\hbar\omega_{\mathbf{k}}} - 1} + \frac{\varepsilon_{\mathbf{k}} + T^{2B} n_0 - \hbar\omega_{\mathbf{k}}}{2\hbar\omega_{\mathbf{k}}} \right), \qquad (11.37)
\end{aligned}
$$

where in the last step we used the results of the Matsubara summation from (7.31) and Exercise 7.2. For a given density and temperature, the last two equations thus fully determine the condensate density. Note that (11.37) explicitly shows that the condensate is indeed depleted both by thermal effects, described by the first term in the last line, as well as quantum effects, described by the second term, which does not depend explicitly on temperature.

11.4 Thermodynamic Potential for Bosons

Next, we calculate the thermodynamic potential Ω for the homogeneous condensed Bose gas in the Bogoliubov approximation. From Chap. 4, we know that various thermodynamic quantities of interest can be calculated from this thermodynamic potential. To carry out the calculation we have to remember a subtlety explained in Example 7.2 from Chap. 7, on which we elaborate further in the following example.

Example 11.1. In Example 7.2, we found how the equal-time commutation relation between the creation and annihilation operators is encoded in the functional formalism by using the correct time ordering. Considering the same partition sum as in (7.38), we can introduce the following inverse Green's function

$$
G^{-1}(\tau,\tau') = -\frac{1}{\hbar}\left\{\hbar\frac{\partial}{\partial\tau} + \varepsilon - \mu\right\}\delta(\tau-\tau'), \qquad (11.38)
$$

such that we have for the partition sum from (7.38)

$$
\begin{aligned}
Z &= \int d[\phi^*]d[\phi]\exp\left\{\int_0^{\hbar\beta} d\tau\, d\tau' \phi^*(\tau^+)G^{-1}(\tau,\tau')\phi(\tau')\right\} \qquad (11.39)\\
&= \int d[\phi^*]d[\phi]\exp\left\{\int_0^{\hbar\beta} d\tau\, d\tau' \left[\phi^*(\tau^+),\phi(\tau)\right] \frac{\mathbf{G}^{-1}(\tau,\tau')}{2} \begin{bmatrix}\phi(\tau')\\ \phi^*(\tau'^+)\end{bmatrix}\right\} \\
&= e^{\beta(\varepsilon-\mu)/2}\int d[\phi^*]d[\phi]\exp\left\{\int_0^{\hbar\beta} d\tau\, d\tau' \left[\phi^*(\tau),\phi(\tau)\right] \frac{\mathbf{G}^{-1}(\tau,\tau')}{2} \begin{bmatrix}\phi(\tau'^+)\\ \phi^*(\tau'^+)\end{bmatrix}\right\}
\end{aligned}
$$

with the corresponding Green's function matrix

$$\mathbf{G}^{-1}(\tau,\tau') = \begin{bmatrix} G^{-1}(\tau,\tau') & 0 \\ 0 & G^{-1}(\tau',\tau) \end{bmatrix}. \tag{11.40}$$

In the last step of (11.39), we had to add a factor $e^{\beta(\varepsilon-\mu)/2}$ to compensate for the change of limits, as follows from Example 7.2. The reason for wanting the elements of the Nambu-space vectors to be at equal time is that in the following, we often perform operations on the vectors as a whole, such as Fourier transforming and rotating in Nambu space. Also note that in the last line of (11.39) we could even omit the imaginary-time limiting procedure, because in this symmetric form there is no discontinuity at equal time.

Using (11.18), (11.21), (11.22), and (11.33), we find for the total action in the Boguliobov approximation

$$S_{\mathrm{Bog}}[\phi^*,\phi] = -\frac{\hbar\beta}{2}T^{2\mathrm{B}}n_{\mathbf{0}}N_{\mathbf{0}} - \frac{\hbar\beta}{2}\sum_{\mathbf{k}\neq 0}\left(\varepsilon_{\mathbf{k}} + T^{2\mathrm{B}}n_{\mathbf{0}}\right) + \frac{1}{2}\sum_{\mathbf{k}\neq 0,n}\left[\phi'^*_{\mathbf{k},n},\phi'_{-\mathbf{k},-n}\right]$$
$$\cdot \begin{bmatrix} -i\hbar\omega_n + \varepsilon_{\mathbf{k}} + T^{2\mathrm{B}}n_{\mathbf{0}} & T^{2\mathrm{B}}n_{\mathbf{0}} \\ T^{2\mathrm{B}}n_{\mathbf{0}} & i\hbar\omega_n + \varepsilon_{\mathbf{k}} + T^{2\mathrm{B}}n_{\mathbf{0}} \end{bmatrix} \cdot \begin{bmatrix} \phi'_{\mathbf{k},n} \\ \phi'^*_{-\mathbf{k},-n} \end{bmatrix}, \tag{11.41}$$

where the second term on the right-hand side is explained by the discussion from Example 11.1. Note that the linear terms have been cancelled by satisfying the homogeneous Gross-Pitaevskii equation $\mu = T^{2\mathrm{B}}n_{\mathbf{0}}$, which was also used to eliminate the chemical potential in the above equation. Moreover, the expression for the chemical potential can also be understood physically, because the energy needed to add a particle to the condensate is in the presence of interactions given by the Hartree contribution $T^{2\mathrm{B}}n_{\mathbf{0}}$.

The matrix in (11.41) can be diagonalized by a Bogoliubov transformation. To this end, we introduce new fields $\psi^*_{\mathbf{k},n}$ and $\psi_{\mathbf{k},n}$ via the transformation

$$\begin{bmatrix} \psi_{\mathbf{k},n} \\ \psi^*_{-\mathbf{k},-n} \end{bmatrix} = \begin{bmatrix} u_{\mathbf{k}} & v_{\mathbf{k}} \\ v^*_{\mathbf{k}} & u^*_{\mathbf{k}} \end{bmatrix} \cdot \begin{bmatrix} \phi'_{\mathbf{k},n} \\ \phi'^*_{-\mathbf{k},-n} \end{bmatrix}. \tag{11.42}$$

Since the part of the matrix depending on the Matsubara frequency $i\omega_n$ is already diagonal, it is most convenient that it remains invariant under this transformation. This is achieved by demanding that the coefficients obey

$$|u_{\mathbf{k}}|^2 - |v_{\mathbf{k}}|^2 = 1. \tag{11.43}$$

In the operator formalism, this condition ensures that the operators associated with the fields $\psi^*_{\mathbf{k},n}$ and $\psi_{\mathbf{k},n}$ still obey the standard commutation relations for bosonic creation and annihilation operators. Using this normalization, we find that the inverse transformation is given by

$$\begin{bmatrix} \phi_{\mathbf{k},n} \\ \phi^*_{-\mathbf{k},-n} \end{bmatrix} = \begin{bmatrix} u^*_{\mathbf{k}} & -v_{\mathbf{k}} \\ -v^*_{\mathbf{k}} & u_{\mathbf{k}} \end{bmatrix} \cdot \begin{bmatrix} \psi_{\mathbf{k},n} \\ \psi^*_{-\mathbf{k},-n} \end{bmatrix}. \tag{11.44}$$

If we now substitute the transformation from (11.44) into (11.41) and demand that the resulting quadratic part of the action reduces to the diagonal form

$$\sum_{\mathbf{k}\neq 0,n} (-i\hbar\omega_n + \hbar\omega_{\mathbf{k}})\, \psi^*_{\mathbf{k},n}\psi_{\mathbf{k},n},$$

we obtain the following equations for $u_{\mathbf{k}}$, $v_{\mathbf{k}}$ and $\hbar\omega_{\mathbf{k}}$

$$\begin{aligned}
\left(u_{\mathbf{k}}^2 + v_{\mathbf{k}}^2\right) T^{2B} n_{\mathbf{0}} - 2u_{\mathbf{k}}v_{\mathbf{k}}(\varepsilon_{\mathbf{k}} + T^{2B}n_{\mathbf{0}}) &= 0, \\
\left(|u_{\mathbf{k}}|^2 + |v_{\mathbf{k}}|^2\right)(\varepsilon_{\mathbf{k}} + T^{2B}n_{\mathbf{0}}) - (u^*_{\mathbf{k}}v_{\mathbf{k}} + u_{\mathbf{k}}v^*_{\mathbf{k}})T^{2B}n_{\mathbf{0}} &= \hbar\omega_{\mathbf{k}}.
\end{aligned} \tag{11.45}$$

Using the normalization from (11.43), we find that these equations are solved by

$$\begin{aligned}
\hbar\omega_{\mathbf{k}} &= \sqrt{\varepsilon_{\mathbf{k}}^2 + 2T^{2B}n_{\mathbf{0}}\varepsilon_{\mathbf{k}}}, \\
|v_{\mathbf{k}}|^2 &= |u_{\mathbf{k}}|^2 - 1 = \frac{1}{2}\left(\frac{\varepsilon_{\mathbf{k}} + T^{2B}n_{\mathbf{0}}}{\hbar\omega_{\mathbf{k}}} - 1\right).
\end{aligned} \tag{11.46}$$

As a result, the action from (11.41) becomes

$$\begin{aligned}
S[\psi^*,\psi] = &-\frac{\hbar\beta}{2}T^{2B}n_{\mathbf{0}}N_{\mathbf{0}} + \frac{\hbar\beta}{2}\sum_{\mathbf{k}\neq 0}\left[\hbar\omega_{\mathbf{k}} - \left(\varepsilon_{\mathbf{k}} + T^{2B}n_{\mathbf{0}}\right)\right] \\
&+ \sum_{\mathbf{k}\neq 0,n} (-i\hbar\omega_n + \hbar\omega_{\mathbf{k}})\, \psi^*_{\mathbf{k},n}\psi_{\mathbf{k},n},
\end{aligned} \tag{11.47}$$

where the additional $\hbar\omega_{\mathbf{k}}$ in the second term on the right-hand side again comes from interchanging the order of the fields and the corresponding time-limiting procedure, which in the operator formalism corresponds to commuting the corresponding creation and annihilation operator. In terms of the transformed fields the action is diagonal, and we can thus perform the corresponding Matsubara sum, which then gives the ideal gas result from (7.30).

However, note that the second term on the right-hand side of (11.47) actually does not converge at high momenta. This is most easily seen by expanding the summant in powers of $T^{2B}n_{\mathbf{0}}/\varepsilon_{\mathbf{k}}$ and converting the sum into an integral, which gives

$$\frac{1}{V}\sum_{\mathbf{k}\neq 0}\left[\hbar\omega_{\mathbf{k}} - \left(\varepsilon_{\mathbf{k}} + T^{2B}n_{\mathbf{0}}\right)\right] = \int \frac{d\mathbf{k}}{(2\pi)^3}\left\{-\frac{(T^{2B}n_{\mathbf{0}})^2}{2\varepsilon_{\mathbf{k}}} + \ldots\right\}. \tag{11.48}$$

We find that after integration over the angles the integrand becomes a nonzero constant for large k, meaning that we encounter an ultraviolet divergence. This divergence is an artifact from the use of the pseudopotential $T^{2B}\delta(\mathbf{x}-\mathbf{x}')$, whose Fourier transform is a constant for all momenta. A realistic interatomic potential always falls off as $1/\mathbf{k}^2$ at high momenta, which would result in a finite integral. Moreover, there is another issue with the use of the pseudopotential. Note that both the first term on the right-hand side of (11.47) and on the right-hand side of (11.48) are proportional to n_0^2. If we would have used a true interatomic potential, then the term from (11.47) would have corresponded to the first-order term in the Born series from (10.22), whereas the term from (11.48) would have corresponded to the second-order term in the Born series. However, in Sect. 11.3 it was explained how the pseudopotential conveniently incorporates in the first term of the Born series the information of all the higher-order terms. As a result, we have a double-counting problem in the contributions to the thermodynamic potential that are proportional to n_0^2. This can be avoided by cancelling the first term on the right-hand side of (11.48). Then, the momentum sum becomes finite and the thermodynamic potential is finally given by

$$\begin{aligned}\frac{\Omega}{V} &= -\frac{1}{\beta}\log(Z) = -\frac{1}{2}n_0^2 T^{2B} + \frac{1}{2V}\sum_{\mathbf{k}\neq 0}\left(\hbar\omega_{\mathbf{k}} - \varepsilon_{\mathbf{k}} - n_0 T^{2B} + \frac{(T^{2B}n_0)^2}{2\varepsilon_{\mathbf{k}}}\right)\\ &\quad + \frac{1}{\beta V}\sum_{\mathbf{k}\neq 0}\log\left(1 - e^{-\beta\hbar\omega_{\mathbf{k}}}\right)\\ &= -\frac{1}{2}n_0^2 T^{2B}\left(1 - \frac{128}{15}\sqrt{\frac{n_0 a^3}{\pi}}\right) + \frac{1}{\beta V}\sum_{\mathbf{k}\neq 0}\log\left(1 - e^{-\beta\hbar\omega_{\mathbf{k}}}\right), \end{aligned} \tag{11.49}$$

where in the last step we converted the sum over momenta of the ground-state contribution into an integral, which is analytically solvable. The correction to the ground-state energy proportional to $\sqrt{n_0 a^3}$ is also called the Lee-Huang-Yang correction. It is usually the smallest term in the thermodynamic potential. Note that from (4.35) we find that the thermodynamic potential immediately also gives us the pressure of the Bose-Einstein condensed gas. Neglecting the small Lee-Huang-Yang correction, the pressure of the condensate is $p = n_0^2 T^{2B}/2$. From the discussion in Sect. 8.7, we recall that the speed of sound for the propagation of density fluctuations equals $c = \sqrt{(\mathrm{d}p/\mathrm{d}n)/m} = \sqrt{n_0 T^{2B}/m}$, which agrees with the result $\hbar\omega_{\mathbf{k}} = \hbar c k$ for the long-wavelength limit of the Bogoliubov dispersion. This shows explicitly that the quasiparticle and collective excitations are identical in a Bose-Einstein condensed gas.

11.5 Bogoliubov-de Gennes Equation

The generalization of the field-theory approach presented in the previous section to the inhomogeneous case is straightforward. First, we have to solve the inhomogeneous Gross-Pitaevskii equation at a fixed chemical potential, such that our fluctuation expansion is again correctly performed around the true minimum of the action. Given the condensate wavefunction, we can then calculate the quasiparticle dispersion by finding the poles of $\mathbf{G}$ or, equivalently but more conveniently, the zero's of $\mathrm{Det}[\mathbf{G}^{-1}]$. This quasiparticle dispersion then also describes collective modes, as mentioned in the previous section. The zeroes of $\mathrm{Det}[\mathbf{G}^{-1}]$ are found by solving the eigenvalue equation

$$\int d\mathbf{x}' \mathbf{G}^{-1}(\mathbf{x},\mathbf{x}',\omega) \cdot \begin{bmatrix} u_{\mathbf{n}}(\mathbf{x}') \\ v_{\mathbf{n}}(\mathbf{x}') \end{bmatrix} = 0, \tag{11.50}$$

where $u_{\mathbf{n}}(\mathbf{x})$ and $v_{\mathbf{n}}(\mathbf{x})$ are the nontrivial eigenvectors of the inverse Green's function matrix with eigenvalue zero. We thus find that the poles of the Green's function matrix are located at $\hbar\omega = \hbar\omega_{\mathbf{n}}$, where $\hbar\omega_{\mathbf{n}}$ is found from

$$\begin{bmatrix} \hat{K} + 2T^{2B}|\phi_0(\mathbf{x})|^2 & T^{2B}(\phi_0(\mathbf{x}))^2 \\ T^{2B}(\phi_0^*(\mathbf{x}))^2 & \hat{K} + 2T^{2B}|\phi_0(\mathbf{x})|^2 \end{bmatrix} \cdot \begin{bmatrix} u_{\mathbf{n}}(\mathbf{x}) \\ v_{\mathbf{n}}(\mathbf{x}) \end{bmatrix} = \hbar\omega_{\mathbf{n}} \begin{bmatrix} 1 & 0 \\ 0 & -1 \end{bmatrix} \cdot \begin{bmatrix} u_{\mathbf{n}}(\mathbf{x}) \\ v_{\mathbf{n}}(\mathbf{x}) \end{bmatrix}, \tag{11.51}$$

where we used (11.28) and (11.50), and introduced the operator

$$\hat{K} = -\hbar^2\nabla^2/2m + V^{\text{ex}}(\mathbf{x}) - \mu. \tag{11.52}$$

(11.51) is the Bogoliubov-de Gennes equation that has been applied with great success to the collective modes of a Bose-Einstein condensed rubidium and sodium gas [51, 52], where the collective modes are discussed in more detail in the next section from a somewhat different perspective. Note that a special solution with $\hbar\omega_{\mathbf{0}} = 0$ is given by $[u_{\mathbf{0}}(\mathbf{x}), v_{\mathbf{0}}(\mathbf{x})] = [\phi_0(\mathbf{x}), -\phi_0^*(\mathbf{x})]$, which physically describes the dynamics of the global phase of the condensate [53]. Due to the $U(1)$ symmetry of the action from (11.17), i.e. its invariance under the transformation $\phi(\mathbf{x},\tau) \to e^{i\theta}\phi(\mathbf{x},\tau)$ and $\phi^*(\mathbf{x},\tau) \to e^{-i\theta}\phi^*(\mathbf{x},\tau)$ that changes the global phase, this solution essentially does not influence the thermodynamic properties of the macroscopic quantum gas and is therefore usually safely neglected. Nevertheless, this zero mode has from a fundamental point of view some interesting consequences which we study in more detail in Sect. 13.5. By noting that the left-hand side of the Bogoliubov-de Gennes equation involves a Hermitian operator, it follows that the solutions obey

$$\int d\mathbf{x}\, \phi_0^*(\mathbf{x}) u_{\mathbf{n}}(\mathbf{x}) + \int d\mathbf{x}\, \phi_0(\mathbf{x}) v_{\mathbf{n}}(\mathbf{x}) = 0\,, \tag{11.53}$$

corresponding to the orthogonality condition imposed in (11.20). Moreover, it is also possible to show that the solutions with $\hbar\omega_{\mathbf{n}} > 0$ can always be properly normalized as [54]

$$\int d\mathbf{x}\, (|u_{\mathbf{n}}(\mathbf{x})|^2 - |v_{\mathbf{n}}(\mathbf{x})|^2) = 1\,, \tag{11.54}$$

such that the quasiparticle creation and annihilation operators also have bosonic commutation relations in the inhomogeneous case. Knowing the eigenstates of $\mathbf{G}^{-1}$, we can perform the inversion and determine the density profile of the noncondensed atoms for the inhomogeneous case. Ultimately, we find

$$n'(\mathbf{x}) = \sum_{\mathbf{n}\neq\mathbf{0}} \left\{ (|u_{\mathbf{n}}(\mathbf{x})|^2 + |v_{\mathbf{n}}(\mathbf{x})|^2) \frac{1}{e^{\beta\hbar\omega_{\mathbf{n}}} - 1} + |v_{\mathbf{n}}(\mathbf{x})|^2 \right\}, \tag{11.55}$$

which, together with (11.54), are to be directly compared with (11.37) and (11.46) for the homogeneous case.

11.6 Popov Theory

At temperatures near absolute zero, we have to a good approximation that $n'(\mathbf{x}) = 0$ and the Bogoliubov theory is valid. However, at nonzero temperatures we thermally excite particles and $n'(\mathbf{x})$ becomes nonzero, so the resulting fluctuation effects are not negligible anymore. In Chap. 8, we used three different methods to derive the celebrated Hartree-Fock theory for the interacting quantum gas at nonzero temperature. We now apply this theory to treat the effect of the noncondensed part of the gas in the Hartree-Fock approximation. First note that for a point interaction the Hartree and Fock approximations are equal, which is diagrammatically readily understood because for a point interaction the direct and the exchange diagram from Fig. 8.7 are exactly the same. We saw how to perform the Hartree and the Fock theory simultaneously for fermions in Sect. 8.6.3, which amounted to a self-consistent mean-field theory in the collective κ and λ fields that were both introduced by a Hubbard-Stratonovich transformation. For the present point interaction, this would simply yield $\langle \kappa(\mathbf{x},\tau)\rangle = T^{2\mathrm{B}}n(\mathbf{x})$ and $\langle \lambda((\mathbf{x},\mathbf{x}',\tau))\rangle = T^{2\mathrm{B}}n(\mathbf{x},\mathbf{x}')\delta(\mathbf{x}-\mathbf{x}') = T^{2\mathrm{B}}n(\mathbf{x})\delta(\mathbf{x}-\mathbf{x}')$, leading for the bosonic case to the following effect on the interacting part of the action

$$
\begin{aligned}
&\frac{T^{2\mathrm{B}}}{2}\int_0^{\hbar\beta} \mathrm{d}\tau \int \mathrm{d}\mathbf{x}\, \phi^*(\mathbf{x},\tau)\phi(\mathbf{x},\tau)\phi^*(\mathbf{x},\tau)\phi(\mathbf{x},\tau) \rightarrow \\
&2T^{2\mathrm{B}}\int_0^{\hbar\beta} \mathrm{d}\tau \int \mathrm{d}\mathbf{x}\, n(\mathbf{x})\phi^*(\mathbf{x},\tau)\phi(\mathbf{x},\tau) - T^{2\mathrm{B}}\int_0^{\hbar\beta} d\tau \int d\mathbf{x}\, n^2(\mathbf{x}). \quad (11.56)
\end{aligned}
$$

This shows that the prescription is to reduce the fourth-order interaction term to the quadratic level by taking all four possible normal averages $\langle \phi^*\phi\rangle$, and subtract a zeroth-order term to avoid double-counting problems in the partition sum.

Applying this procedure to the interacting action for the fluctuations $S_{\mathrm{int}}[\phi'^*,\phi']$ from (11.23), we find that the fields that are of third and fourth order in the fluctuations, give rise to the following terms that are linear and quadratic in the fluctuations

$$
\begin{aligned}
S_{\mathrm{int}}^{\mathrm{HF}}[\phi'^*,\phi'] = &\ 2\int_0^{\hbar\beta} \mathrm{d}\tau \int \mathrm{d}\mathbf{x}\, T^{2\mathrm{B}} n'(\mathbf{x})\phi'^*(\mathbf{x},\tau)\phi_{\mathbf{0}}(\mathbf{x}) \\
&+2\int_0^{\hbar\beta} \mathrm{d}\tau \int \mathrm{d}\mathbf{x}\, T^{2\mathrm{B}} n'(\mathbf{x})\phi'(\mathbf{x},\tau)\phi_{\mathbf{0}}^*(\mathbf{x}) \\
&+2\int_0^{\hbar\beta} \mathrm{d}\tau \int \mathrm{d}\mathbf{x}\, T^{2\mathrm{B}} n'(\mathbf{x})\phi'^*(\mathbf{x},\tau)\phi'(\mathbf{x},\tau)\,. \quad (11.57)
\end{aligned}
$$

The new Gross-Pitaevskii equation thus directly follows from the new action in the fluctuations when the above terms are added to $S_0[\phi'^*,\phi']$ from (11.22). In Fig.

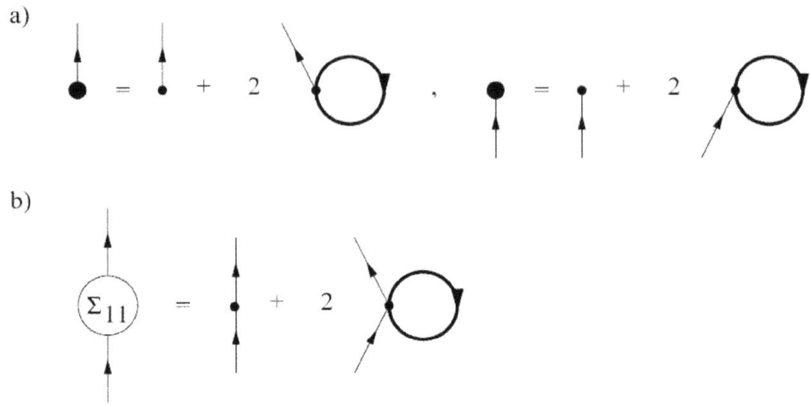

Fig. 11.2 Hartree-Fock corrections to a) the linear and b) the quadratic interaction terms of the Bogoliubov theory. The black dot denotes the point interaction with strength T^{2B}. The thick line represents the interacting Green's function, yielding the density of noncondensed particles. The open circles represent a source for noncondensed particles. We note that condensed particles are not explicitly shown in the diagrams, since they only give rise to a multiplicative factor. This represents the Popov theory.

11.2 we show how this can be represented diagrammatically, indicating that we are indeed performing a Hartree-Fock theory for the Bose-Einstein condensed case. Performing the same analysis as in Sect. 11.3, we then conclude that the Gross-Pitaevskii equation is modified to

$$\left\{ -\frac{\hbar^2 \nabla^2}{2m} + V^{\text{ex}}(\mathbf{x}) + 2T^{2B} n'(\mathbf{x}) + T^{2B} |\phi_0(\mathbf{x})|^2 \right\} \phi_0(\mathbf{x}) = \mu \phi_0(\mathbf{x}) , \tag{11.58}$$

while the diagonal or normal selfenergies from (11.32) are changed into

$$2T^{2B} |\phi_0(\mathbf{x})|^2 + 2T^{2B} n'(\mathbf{x}) = 2T^{2B} n(\mathbf{x}).$$

As a result, the Bogoliubov-de Gennes equation for the elementary excitations is now given by

$$\begin{bmatrix} \hat{K} + 2T^{2B} n(\mathbf{x}) & T^{2B} (\phi_0(\mathbf{x}))^2 \\ T^{2B} (\phi_0^*(\mathbf{x}))^2 & \hat{K} + 2T^{2B} n(\mathbf{x}) \end{bmatrix} \cdot \begin{bmatrix} u_{\mathbf{n}}(\mathbf{x}) \\ v_{\mathbf{n}}(\mathbf{x}) \end{bmatrix} = \hbar \omega_{\mathbf{n}} \begin{bmatrix} 1 & 0 \\ 0 & -1 \end{bmatrix} \cdot \begin{bmatrix} u_{\mathbf{n}}(\mathbf{x}) \\ v_{\mathbf{n}}(\mathbf{x}) \end{bmatrix} . \tag{11.59}$$

These last two equations in combination with (11.55) are known as the Popov theory [55], which has been studied extensively in the context of Bose-Einstein condensation in ultracold atomic Bose gases. It has been successfully applied to the equilibrium density profile of the condensed gas below the critical temperature [56]. The theory has also been used to determine the collective mode frequencies of the gas at nonzero temperatures, however, with much less success [57, 58]. The reason for the failure of the Popov theory in this case is that the Bogoliubov-de Gennes equation, (11.59), describes physically only the motion of the condensate in the presence of a static noncondensed cloud and not the dynamics of the noncondensed cloud itself. The correct description of the condensate's collective modes just below the critical temperature then turns out to be a difficult nonequilibrium problem. However, at very low temperatures, when Bogoliubov theory applies, the collective modes of a condensate can be studied conveniently using a hydrodynamic-like approach. This is the topic of the next section.

11.7 Hydrodynamic-Like Approach

As we have seen several times before, the equilibrium quantum field theory in terms of imaginary time τ and Matsubara frequencies ω_n can also be used to obtain information on the dynamical properties of the system, achieved via the substitution $\omega_n \rightarrow -\mathrm{i}\omega$ or, equivalently, $\tau \rightarrow \mathrm{i}t$. Indeed, the imaginary-time evolution operator $\mathrm{e}^{-\hat{H}\tau/\hbar}$, which we use to incorporate the appropriate statistical averaging into our many-body theory, then becomes the familiar real-time evolution operator $\mathrm{e}^{-\mathrm{i}\hat{H}t/\hbar}$ which is known to describe the dynamics of a quantum system. However, we should sometimes be careful with this procedure, since the example of Popov theory for Bose-Einstein condensation tells us that we may not always end up with the correct physics needed for the full description of a particular experiment. A possible problem with the substitution $\omega_n \rightarrow -\mathrm{i}\omega$ is that we never perform both the full statistical averaging and the complete dynamics at the same time. For some situations, it is important to take both the statistical averaging and the dynamics accurately into account, meaning that we need to use a truly nonequilibrium statistical field theory. This can be achieved by using the Keldysh formalism, which is, however, beyond the scope of this book [59]. Still, there are many interesting situations where we do not need the Keldysh formalism in order to still correctly describe the dynamics. This is, for example, the case when we only have to deal with the time evolution of mean-field quantities, such as the collective modes of the condensate at very low temperatures, for which the Bogoliubov theory is valid.

11.7.1 Time-Dependent Gross-Pitaevskii Equation

The time evolution of the Bose-Einstein condensed gas can be derived in various ways. We have seen that for time-independent fields $\phi(\mathbf{x})$ the action from (11.17) takes the form of a Landau free energy. The time-independent Gross-Pitaevskii equation is then conveniently obtained by minimizing the Landau free energy with respect to $\phi^*(\mathbf{x})$. Analogously, by considering the action from (11.17) in real time and by minimizing it with respect to the field $\phi^*(\mathbf{x},t)$, we obtain the time-dependent Gross-Pitaevskii equation, given by

$$\mathrm{i}\hbar\frac{\partial\phi_0(\mathbf{x},t)}{\partial t} = \left\{-\frac{\hbar^2\nabla^2}{2m} + V^{\mathrm{ex}}(\mathbf{x}) - \mu + T^{2\mathrm{B}}|\phi_0(\mathbf{x},t)|^2\right\}\phi_0(\mathbf{x},t). \quad (11.60)$$

This result can also be obtained with operator methods in the language of second-quantization. Then, the appropriate second-quantized Hamiltonian $\hat{H}$ corresponding to the action used above is given by (6.29). The imaginary-time evolution of the field operator $\hat{\psi}(\mathbf{x},\tau)$ is determined by the Heisenberg equation of motion in (6.43). Similar to how we treated Bose condensation in the functional formalism, we can then also expand the field operator as a sum of its expectation value and fluctuations, i.e. use the Bogoliubov shift $\hat{\psi}(\mathbf{x},\tau) = \phi_0(\mathbf{x},\tau) + \hat{\psi}'(\mathbf{x},\tau)$. Performing the substitution $\tau \to it$, then also leads to the time-dependent Gross-Pitaevskii equation given by (11.60).

Noting the analogy between the time-dependent Gross-Pitaesvkii equation and the time-dependent Schrödinger equation, we can obtain a continuity equation for the condensate wavefunction in the usual way as for a single-particle wavefunction. This then leads to the familiar expression

$$\frac{\partial}{\partial t}|\phi_0(\mathbf{x},t)|^2 + \nabla\cdot\mathbf{J}(\mathbf{x},t) = 0, \quad (11.61)$$

where the current density is given by

$$\mathbf{J}(\mathbf{x},t) = \frac{\hbar}{2mi}\left\{\phi_0^*(\mathbf{x},t)\nabla\phi_0(\mathbf{x},t) - \phi_0(\mathbf{x},t)\nabla\phi_0^*(\mathbf{x},t)\right\}. \quad (11.62)$$

The above current density is seen to describe the current of condensed particles, such that we may introduce the concept of the superfluid velocity $\mathbf{v}_s(\mathbf{x},t)$ via

$$\mathbf{J}(\mathbf{x},t) = n_0(\mathbf{x},t)\mathbf{v}_\mathrm{s}(\mathbf{x},t). \quad (11.63)$$

Since we can always represent the complex-valued condensate wavefunction in terms of its real amplitude and phase, we have that $\phi_0(\mathbf{x},t) = \mathrm{e}^{\mathrm{i}\theta(\mathbf{x},t)}\sqrt{n_0(\mathbf{x},t)}$. Substituting this expression in (11.62), then yields for the superfluid velocity

$$\mathbf{v}_\mathrm{s}(\mathbf{x},t) = \frac{\hbar}{m}\nabla\theta(\mathbf{x},t). \quad (11.64)$$

An intriguing property that follows immediately from this result is that the flow of the superfluid is irrotational, i.e.

$$\nabla\times\mathbf{v}_\mathrm{s} = 0, \quad (11.65)$$

showing that the superfluid cannot support rigid-body rotation. The behavior of the condensate in the presence of rotation is the topic of the next section.

An important and experimentally relevant case for which we can solve the Gross-Pitaevskii equation is when the kinetic energy is small with respect to the interaction energy, which is also called the Thomas-Fermi limit [60, 61]. In this limit, which physically boils down to $N_0 a/l \gg 1$ with l the effectively isotropic harmonic oscillator length, we are allowed to neglect the kinetic-energy term. The time-independent Gross-Pitaevskii equation simply becomes

$$\left\{V^{\mathrm{ex}}(\mathbf{x})+T^{2\mathrm{B}}|\phi_{\mathbf{0}}(\mathbf{x})|^2\right\}\phi_{\mathbf{0}}(\mathbf{x})=\mu\phi_{\mathbf{0}}(\mathbf{x}). \tag{11.66}$$

The solution to this equation is immediately obtained as [62]

$$|\phi_{\mathbf{0}}(\mathbf{x})|^2=\frac{\mu-V^{\mathrm{ex}}(\mathbf{x})}{T^{2\mathrm{B}}}, \tag{11.67}$$

which for the case of a harmonic external trapping potential shows that the shape of the condensate is thus described by an upside-down parabolic density profile. Of course, the density should be taken equal to zero whenever the right-hand side of (11.67) becomes negative. In particular, for an isotropic harmonic trapping potential with frequency ω,

$$V^{\mathrm{ex}}(\mathbf{x})=\frac{1}{2}m\omega^2\mathbf{x}^2, \tag{11.68}$$

the condensate wavefunction has to vanish for $x>R_{\mathrm{TF}}$, where the Thomas-Fermi radius R_{TF} is given by

$$R_{\mathrm{TF}}=\sqrt{\frac{2\mu}{m\omega^2}}. \tag{11.69}$$

Note that, because the interaction energy is proportional to the density, the central assumption for the Thomas-Fermi approach is no longer satisfied near the edge of the gas cloud where the condensate density goes to zero. Here, the Thomas-Fermi approximation breaks down and the kinetic energy operator should be included again in order to describe the behavior near the edges realistically. However, this region is usually rather small, and the Thomas-Fermi approximation is quite accurate for by far the largest part of the condensate density profile.

11.7.2 Collective Modes

To study the possible collective excitations around the equilibrium Thomas-Fermi profile, we derive the hydrodynamic-like equations for the condensate [63]. We start by substituting $\phi_{\mathbf{0}}(\mathbf{x},t)=\mathrm{e}^{\mathrm{i}\theta(\mathbf{x},t)}\sqrt{n_{\mathbf{0}}(\mathbf{x},t)}$ into (11.60), after which we perform the corresponding derivatives and divide out $\mathrm{e}^{\mathrm{i}\theta(\mathbf{x},t)}$. Then, the real part of the resulting equation is seen to satisfy

$$\hbar\frac{\partial\theta}{\partial t}=\frac{\hbar^2}{2m\sqrt{n_{\mathbf{0}}}}\nabla^2\sqrt{n_{\mathbf{0}}}-\frac{1}{2}m\mathbf{v}_{\mathrm{s}}^2-V^{\mathrm{ext}}+\mu-T^{2\mathrm{B}}n_{\mathbf{0}}, \tag{11.70}$$

where we also used (11.64) and omitted the explicit dependence on the temporal and spatial coordinates for notational convenience. As in the Thomas-Fermi approximation, the first kinetic term on the right-hand side is considered to be negligible compared to the last three terms. As a result, the dynamics of the condensate is governed by the continuity equation and the equation for the superfluid velocity, given by

$$\frac{\partial n_{\mathbf{0}}}{\partial t} = -\boldsymbol{\nabla} \cdot (n_{\mathbf{0}} \mathbf{v}_{\mathrm{s}}) \tag{11.71}$$

$$m\frac{\partial \mathbf{v}_{\mathrm{s}}}{\partial t} = -\boldsymbol{\nabla}\left(\frac{1}{2}m\mathbf{v}_{\mathrm{s}}^2 + V^{\mathrm{ext}} + T^{2\mathrm{B}} n_{\mathbf{0}} - \mu\right), \tag{11.72}$$

where the last equation was found by combining (11.64) and (11.70). The second equation is of the same form as the Euler equation from fluid mechanics that describes the flow of a fluid without viscosity, which thus shows that the condensate indeed behaves as a superfluid. In order to find analytic solutions to these hydrodynamic-like equations, we proceed by expanding the condensate density $n_{\mathbf{0}}(\mathbf{x},\mathbf{t}) = n_{\mathbf{0}}(\mathbf{x}) + \delta n_{\mathbf{0}}(\mathbf{x},\mathbf{t})$ around the equilibrium profile $n_{\mathbf{0}}(\mathbf{x})$ given by the Thomas-Fermi result from (11.67). Both the fluctuations around the equilibrium profile as well as the superfluid velocity are then assumed to be small, such that we may consider the hydrodynamic equations up to linear order in these two quantities, giving

$$\begin{aligned} m\frac{\partial^2 \delta n_{\mathbf{0}}}{\partial t^2} - T^{2\mathrm{B}} \boldsymbol{\nabla} \cdot (n_{\mathbf{0}} \boldsymbol{\nabla} \delta n_{\mathbf{0}}) &= m\frac{\partial^2 \delta n_{\mathbf{0}}}{\partial t^2} - T^{2\mathrm{B}} \boldsymbol{\nabla} n_{\mathbf{0}} \cdot \boldsymbol{\nabla} \delta n_{\mathbf{0}} - T^{2\mathrm{B}} n_{\mathbf{0}} \boldsymbol{\nabla}^2 \delta n_{\mathbf{0}} \\ &= 0. \end{aligned} \tag{11.73}$$

Note that in the absence of a trap, we have that $\boldsymbol{\nabla} n_{\mathbf{0}}(\mathbf{x}) = 0$ and the above equation reduces to a wave equation with a speed of sound equal to $\sqrt{T^{2\mathrm{B}} n_{\mathbf{0}}/m}$ as expected.

We may try to solve this equation also in the inhomogeneous case for solutions that oscillate periodic in time $\delta n_{\mathbf{0}}(\mathbf{x},t) = \delta n_{\mathbf{0}}(\mathbf{x})\mathrm{e}^{\mathrm{i}\omega t}$. For the experimentally relevant case of a cylindrically symmetric trap $V^{\mathrm{ext}}(\mathbf{x}) = m(\omega_\rho^2 \rho^2 + \omega_z^2 z^2)/2$, we then obtain

$$\omega^2 \delta n_{\mathbf{0}} = \left\{ \omega_\rho^2 \rho \frac{\partial}{\partial \rho} + \omega_z^2 z \frac{\partial}{\partial z} \right\} \delta n_{\mathbf{0}} - \frac{1}{2}(\mu - \omega_\rho^2 \rho^2 - \omega_z^2 z^2) \boldsymbol{\nabla}^2 \delta n_{\mathbf{0}}. \tag{11.74}$$

We only study the solutions to the above equation that are of the simple form

$$\delta n_{\mathbf{0}}(\rho, \varphi, z) \propto \rho^\ell \mathrm{e}^{\mathrm{i}\ell\varphi} \propto r^\ell Y_{\ell,\ell}, \tag{11.75}$$

where $Y_{\ell,\ell}$ are the spherical harmonics from (3.60), such that the Laplacian on the right-hand side of (11.74) gives zero. Solutions with angular momentum $\ell = 0$ are called monopole or breathing modes, with $\ell = 1$ are called dipole modes, with $\ell = 2$ quadrupole modes, and so on. The solutions of (11.75) have the property that their oscillation frequencies are related to the radial frequency of the trap via $\omega = \sqrt{\ell}\omega_\rho$.

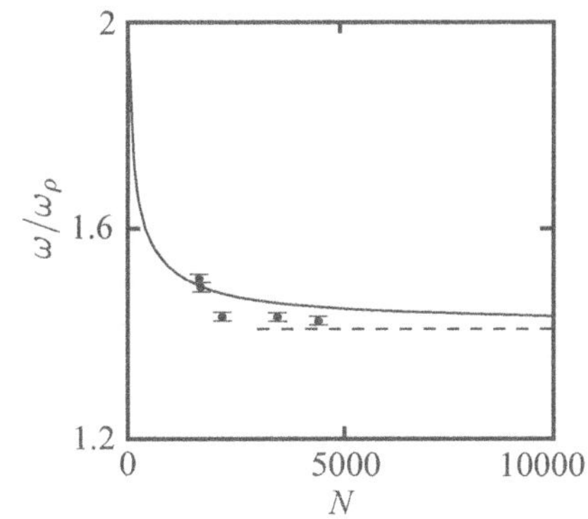

Fig. 11.3 Experimental observation of collective quadrupole oscillations. The dots are measurements of the oscillation frequency ω in units of the radial trap depth ω_ρ performed by Jin et al. as a function of the number of ^{87}Rb atoms N [64]. The dashed line at $\sqrt{2}$ is the analytic result in the hydrodynamic limit [63], while the solid line results from a Bogoliubov-de Gennes approach [52].

The prediction for the frequencies of the collective modes [63], and in particular for the quadrupole mode with $\ell = 2$, i.e. $\omega = \sqrt{2}\omega_\rho$, has been beautifully confirmed by experiments [64, 65, 66, 67], where the results from Jin et al. are shown in Fig. 11.3. Note that the important ingredients in obtaining the collective-mode frequencies are the time-dependent Gross-Pitaesvkii equation and the presence of interactions that are strong enough to validate the Thomas-Fermi approximation. As a result, we would have obtained very different quantitative behavior for a non-interacting condensate. The quadrupole frequency would then be $2\omega_\rho$, as we will see in Sect. 11.9.2. This means that the agreement with experiments is an important verification of the many-body theory for Bose-Einstein condensation in the presence of interactions.

11.8 Rotating Bose-Einstein Condensates

In the previous section, we found that the condensate wavefunction $\phi_0(\mathbf{x}) = e^{i\theta(\mathbf{x})}\sqrt{n_0(\mathbf{x})}$ gives rise to a superfluid velocity, given by

$$\mathbf{v}_s = \frac{\hbar}{m}\boldsymbol{\nabla}\theta(\mathbf{x}). \tag{11.76}$$

An important property that follows from this expression is that the flow of the superfluid is irrotational, i.e.

$$\boldsymbol{\nabla} \times \mathbf{v}_s = 0. \tag{11.77}$$

This last equation determines the system's response to an externally applied torque and shows that the superfluid cannot support rigid-body rotation. However, as we show next, the condensate can store angular momentum by forming singularities in the density around which the phase winds by a integer multiple of 2π, i.e. by forming vortices. This multiple of 2π comes about from demanding that the condensate wavefunction is single valued, such that we have for any closed contour

$$\Delta\theta = \oint \nabla\theta(\mathbf{x})\cdot \mathrm{d}\mathbf{l} = 2\pi j \tag{11.78}$$

with j an integer, which is also called the Onsager-Feynman quantization condition.

To study vortices, we derive the Gross-Pitaevski equation for a Bose-Einstein condensate subject to rotation. To this end, it is convenient to consider the Gross-Pitaevskii equation in a frame that is rotating around the z axis with frequency Ω. The coordinates x' and y' of the rotating frame are related to the coordinates in the lab frame by

$$\begin{aligned} x' &= \cos(\Omega t)x + \sin(\Omega t)y, \\ y' &= \cos(\Omega t)y - \sin(\Omega t)x. \end{aligned} \tag{11.79}$$

Next, we perform the coordinate transformation to the rotating frame and express the condensate wavefunction $\phi(x,y,z,t) = \phi'(x',y',z,t)$ in terms of the rotating coordinates. Then, we have for the temporal derivative that

$$\begin{aligned} \mathrm{i}\hbar\frac{\partial}{\partial t}\phi(x,y,z,t) &= \mathrm{i}\hbar\frac{\partial}{\partial t}\phi'(x'(x,y,t),y'(x,y,t),z,t) \\ &= -\mathrm{i}\hbar\Omega\left(x'\frac{\partial}{\partial y'} - y'\frac{\partial}{\partial x'}\right)\phi'(x',y',z,t) + \mathrm{i}\hbar\frac{\partial}{\partial t}\phi'(x',y',z,t), \end{aligned} \tag{11.80}$$

which is readily shown from (11.79) and the chain rule for differentiation. Furthermore, we have that the spatial derivatives satisfy

$$\left(\frac{\partial^2}{\partial x^2} + \frac{\partial^2}{\partial y^2} + \frac{\partial^2}{\partial z^2}\right) = \left(\frac{\partial^2}{\partial x'^2} + \frac{\partial^2}{\partial y'^2} + \frac{\partial^2}{\partial z^2}\right). \tag{11.81}$$

Considering a trapping potential $V(x'(x,y,t),y'(x,y,t),z)$ that is rotating in the lab frame, but stationary in the rotating frame, we find from (11.60), (11.80) and (11.81) that the Gross-Pitaevskii equation in the rotating frame becomes

$$\mathrm{i}\hbar\frac{\partial\phi(\mathbf{x},t)}{\partial t} = \left\{-\frac{\hbar^2\nabla^2}{2m} - \Omega\hat{L}_z + V^{\mathrm{ex}}(\mathbf{x}) - \mu + T^{2\mathrm{B}}|\phi(\mathbf{x},t)|^2\right\}\phi(\mathbf{x},t), \tag{11.82}$$

where all the primes are omitted for notational convenience. The angular momentum operator $\hat{L}_z$ in the rotating frame is then given by

$$\hat{L}_z = -\mathrm{i}\hbar\left(x\frac{\partial}{\partial y} - y\frac{\partial}{\partial x}\right). \tag{11.83}$$

It is interesting to note that the Gross-Pitaevskii equation, (11.82), can be rewritten in the form [68]

$$i\hbar\frac{\partial\phi(\mathbf{x},t)}{\partial t} = \tag{11.84}$$
$$\left\{\frac{(\hat{\mathbf{p}}-m\Omega\mathbf{e}_z\times\mathbf{x})^2}{2m}+V^{\mathrm{ex}}(\mathbf{x})-\frac{1}{2}m\Omega^2(x^2+y^2)-\mu+T^{2\mathrm{B}}|\phi(\mathbf{x},t)|^2\right\}\phi(\mathbf{x},t),$$

which in the absence of interactions is similar to the Schrödinger equation for a particle with charge e in a homogeneous magnetic field, described by the vector potential $\mathbf{A}(\mathbf{x}) = cm\Omega\mathbf{e}_z\times\mathbf{x}/e$.

To understand the structure of possible solutions to (11.82) it is instructive to look at the noninteracting case, such that Bose-Einstein condensation occurs for $\mu = \varepsilon_0$ in the absence of rotation. Furthermore, we consider a spherically symmetric harmonic trap, whose eigenstates we studied in Sect. 3.6. Note that the wavefunctions $\psi_{n\ell m}(\mathbf{x})$ from (3.65) are also eigenstates of the $\hat{L}_z$ operator, such that we find

$$\left\{-\frac{\hbar^2\nabla^2}{2m}-\Omega\hat{L}_z+V^{\mathrm{ex}}(\mathbf{x})\right\}\psi_{n\ell m}(\mathbf{x}) = E_{n\ell m}\psi_{n\ell m}(\mathbf{x}), \tag{11.85}$$

with eigenenergies $E_{n\ell m} = (2n+\ell+3/2)\hbar\omega - m\hbar\Omega$. Here m is the quantum number belonging to $\hat{L}_z$, which is not to be confused with the mass of the atoms. As a result, we find that for slow rotation $\Omega < \omega$ the ground state of the rotating system is given by $\phi(\mathbf{x}) = \sqrt{N_0}\psi_{000}(\mathbf{x})$ of the harmonic trap with the chemical potential given by $\mu = 3\hbar\omega/2$. However, at the critical rotation $\Omega_c = \omega$ the ground state changes to

$$\phi(r,\vartheta,\varphi) = \sqrt{N_0}\psi_{011}(r,\vartheta,\varphi) = \sqrt{\frac{N_0}{l^3\pi^{3/2}}}\frac{r}{l}\mathrm{e}^{-r^2/2l^2}\sin(\vartheta)\mathrm{e}^{\mathrm{i}\varphi}, \tag{11.86}$$

which corresponds to a vortex with angular momentum $\ell = 1$, and the chemical potential is then given by $\mu = 5\hbar\omega/2 - \hbar\Omega$. The superfluid velocity field of this vortex solution is seen to yield azimuthal flow, namely

$$\mathbf{v}_s = \frac{\hbar}{m}\nabla\theta(r,\vartheta,\varphi) = \frac{\hbar}{m}\nabla\varphi = \frac{\hbar}{mr\sin(\vartheta)}\mathbf{e}_\varphi, \tag{11.87}$$

where $l = \sqrt{\hbar/m\omega}$ is the harmonic oscillator length and $\mathbf{e}_\varphi$ the unit vector in the azimuthal direction. Note that the condensate density $|\phi(r,\vartheta,\varphi)|^2$ vanishes in the core of the vortex, which is directed along the z axis, where the superfluid velocity diverges. The energy level structure, the density profile and the velocity field of the vortex solution from (11.86) are shown in Fig. 11.4.

It turns out that in the presence of interactions the above discussion remains qualitatively the same. However, the size of the vortex is no longer given by the harmonic oscillator length l, but rather by the coherence or healing length $\xi = 1/\sqrt{16\pi n_0 a}$, which is much smaller than l in the Thomas-Fermi limit. Beautiful experiments, performed by rotating Bose-Einstein condensates, have indeed shown the presence of vortices beyond a critical rotation speed [69]. Since applying angular momentum to a normal viscous gas would simply give rise to rigid-body rotation, the onset of a vortex is seen to be the smoking gun for the presence of the superfluid state.

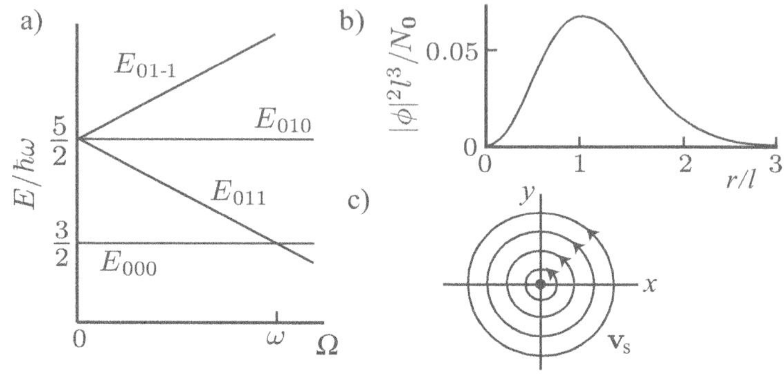

Fig. 11.4 a) Energy levels $E_{n\ell m}$ for a harmonically-trapped rotating condensate in the absence of interactions. When the rotation frequency Ω becomes equal to the trap frequency ω, the ground state becomes a vortex state, for which we show b) the azimuthally symmetric density profile $|\phi(r)|^2 = N_0|\psi_{011}(r,\pi/2,\varphi)|^2$ in the xy plane and c) the corresponding superfluid velocity field.

In fact, whole lattices of vortices have been observed experimentally by rotating Bose-Einstein condensates faster and faster [70], where the lattice structure can be understood from the repulsive long-range interaction between like vortices. This interaction is studied in Chap. 15 in the context of low-dimensional quantum gases.

11.9 Attractive Interactions

Up to now, we have considered an atomic Bose gas with repulsive interactions, i.e. a positive scattering length. However, experiments with attractive atomic gases have also been performed [11, 71]. In this section, we briefly discuss the very different physics that occurs in this case. We first discuss the homogeneous situation. From Bogoliubov theory, we know that small density fluctuations in a homogeneous Bose-Einstein condensate evolve according to

$$\delta n_{\mathbf{0}}(\mathbf{x},t) = \frac{1}{V}\sum_{\mathbf{k}} \delta N_{\mathbf{0}}(\mathbf{k}) e^{i\mathbf{k}\cdot\mathbf{x} - i\omega_{\mathbf{k}}t} , \tag{11.88}$$

where $\hbar\omega_{\mathbf{k}} = \sqrt{\varepsilon_{\mathbf{k}}^2 + 2T^{2B} n_{\mathbf{0}} \varepsilon_{\mathbf{k}}}$ is the Bogoliubov dispersion. If the scattering length is positive, we have that $T^{2B} > 0$, and the Bogoliubov dispersion is real for all momenta. As a result, the magnitude of the density fluctuations $|\delta n_{\mathbf{0}}(\mathbf{x},t)|$ does not grow and the Bose-Einstein condensate is stable. However, if the scattering length is negative, the two-body T-matrix is also negative, and the Bogoliubov dispersion is imaginary for momenta obeying $\varepsilon_{\mathbf{k}} < 2|T^{2B}|n_{\mathbf{0}}$. This implies that the oscillating behavior of $\delta n_{\mathbf{0}}(\mathbf{x},t)$ becomes exponential behavior, so the magnitude of long-wavelength density fluctuations will grow exponentially in time and the Bose-

Einstein condensate is therefore now unstable. Another way to physically understand this instability is by considering the pressure of the condensate $p = T^{2B}n_0^2/2$. For repulsive interactions, the pressure increases if the density increases, which opposes a further increase of the density. For attractive interactions, however, a small increase in the density leads to a reduction of the pressure in the gas which results in a further increase of the density.

The conclusion thus appears to be that a Bose-Einstein condensate with attractive interactions is always unstable and cannot be realized in an experiment. The situation is actually more interesting. We have just seen that only density fluctuations with wavenumbers obeying $k\xi < 1$ are unstable, where $\xi = 1/\sqrt{16\pi n_0|a|}$. Therefore, if we can confine the Bose-Einstein condensate in a box with hard walls of size $L \equiv V^{1/3} < \pi\xi$, it actually becomes stable, because the unstable long-wavelength fluctuations are no longer possible due to the small size of the confining box. Since the density of the Bose-Einstein condensate is given by $n_0 = N_0/L^3$, the above stability condition is satisfied if the number of atoms in the condensate is sufficiently small, i.e.

$$N_0 < \frac{\pi}{16}\frac{L}{|a|}. \tag{11.89}$$

If the number of atoms is larger than this maximum, the Bose-Einstein condensate will collapse. Note that the physics here is somewhat analogous to the physics of white dwarfs and neutron stars which, due to the gravitational attraction, also become unstable and collapse if the total mass of the compact object becomes too large. In actual experiments ultracold atoms are not trapped in a box, but in an harmonic oscillator potential. This does not affect our qualitative discussion, but quantitatively some differences occur, as we now show explicitly.

11.9.1 Effective Action

As just explained, a condensed Bose gas with effectively attractive interactions, i.e. with a negative scattering length a, has the tendency to collapse to a high-density state [72] such that the most important question in this context is if a condensate can exist sufficiently long to be experimentally observed. To address this question, we neglect the variation of the noncondensate density over the size of the condensate [73, 74] and we use the functional-integral formalism to describe the quantum dynamics of the collapse. We start from the grand-canonical partition function

$$Z(\mu) = \int \mathrm{d}[\phi^*]\mathrm{d}[\phi]\, \exp\left\{-\frac{1}{\hbar}S[\phi^*,\phi]\right\}, \tag{11.90}$$

with the Euclidian action

$$S[\phi^*,\phi] = \int_0^{\hbar\beta} \mathrm{d}\tau \int \mathrm{d}\mathbf{x}\, \phi^*(\mathbf{x},\tau) \left\{ \hbar\frac{\partial}{\partial\tau} - \frac{\hbar^2\nabla^2}{2m} + V^{\mathrm{ex}}(\mathbf{x}) - \mu + \frac{2\pi a\hbar^2}{m}|\phi(\mathbf{x},\tau)|^2 \right\} \phi(\mathbf{x},\tau) . \quad (11.91)$$

Next, we introduce explicitly the density fluctuations $\rho(\mathbf{x},\tau)$ and the phase fluctuations $\theta(\mathbf{x},\tau)$ of the condensate by performing the canonical variable transformation [55]

$$\phi(\mathbf{x},\tau) = \sqrt{\rho(\mathbf{x},\tau)}\mathrm{e}^{\mathrm{i}\theta(\mathbf{x},\tau)}$$

in the functional integral for the partition function. As a result, we find

$$Z(\mu) = \int \mathrm{d}[\rho]\mathrm{d}[\theta]\, \exp\left\{ -\frac{1}{\hbar}S[\rho,\theta;\mu] \right\} , \quad (11.92)$$

with

$$S[\rho,\theta;\mu] = \int_0^{\hbar\beta} \mathrm{d}\tau \int \mathrm{d}\mathbf{x} \left\{ \mathrm{i}\hbar\rho(\mathbf{x},\tau)\frac{\partial\theta(\mathbf{x},\tau)}{\partial\tau} + \frac{\hbar^2\rho(\mathbf{x},\tau)}{2m}(\nabla\theta(\mathbf{x},\tau))^2 \right. \quad (11.93)$$
$$\left. + \frac{\hbar^2}{8m\rho(\mathbf{x},\tau)}(\nabla\rho(\mathbf{x},\tau))^2 + (V^{\mathrm{ex}}(\mathbf{x}) - \mu)\rho(\mathbf{x},\tau) + \frac{2\pi a\hbar^2}{m}\rho^2(\mathbf{x},\tau) \right\} .$$

We notice that this action is quadratic in the phase fluctuations, such that the field $\theta(\mathbf{x},\tau)$ can be integrated out exactly.

Compared to ordinary Gaussian integrals there is, however, one slight complication, which is associated with the fact that $\theta(\mathbf{x},\tau)$ is a phase variable. This implies that the periodicity of the original field $\phi(\mathbf{x},\tau)$ only constraints the phase field $\theta(\mathbf{x},\tau)$ to be periodic up to a multiple of 2π. To evaluate the grand-canonical partition function in (11.92) we must therefore first integrate over all fields $\theta(\mathbf{x},\tau)$ that obey the boundary condition $\theta(\mathbf{x},\hbar\beta) = \theta(\mathbf{x},0) + 2\pi j$ and subsequently sum over all possible integers j. Because $2\pi j$ is independent of $\mathbf{x}$, it affects only the zero-momentum part of $\theta(\mathbf{x},\tau)$. Therefore, we start with evaluating the sum

$$\sum_j \int^{\theta_{\mathbf{0}}(\hbar\beta)=\theta_{\mathbf{0}}(0)+2\pi j} \mathrm{d}[\theta_{\mathbf{0}}]\, \exp\left\{ -\mathrm{i}\int_0^{\hbar\beta} \mathrm{d}\tau\, N_{\mathbf{0}}(\tau)\frac{\partial\theta_{\mathbf{0}}(\tau)}{\partial\tau} \right\} ,$$

where $N_{\mathbf{0}}(\tau) \equiv \int \mathrm{d}\mathbf{x}\, \rho(\mathbf{x},\tau)$. After performing a partial integration in the exponent, we may carry out the path integration over $\theta_{\mathbf{0}}(\tau)$ to obtain

$$\sum_j \mathrm{e}^{2\pi\mathrm{i}N_{\mathbf{0}}j}\delta\left[\frac{\partial N_{\mathbf{0}}(\tau)}{\partial\tau}\right] .$$

As expected, the integration over the global phase of the condensate leads to the constraint of a constant number of condensate particles, i.e. $N_{\mathbf{0}}(\tau) = N_{\mathbf{0}}$. Moreover, we have $\sum_j e^{2\pi i N_0 j} = \sum_j \delta(N_0 - j)$, which restricts the number of condensate particles to an integer. As a result we find that the integration over the zero-momentum part of $\rho(\mathbf{x},\tau)$ is only a sum over the number of condensate particles, such that

$$Z(\mu) = \sum_{N_0} e^{\beta\mu N_0} Z_{N_0} . \tag{11.94}$$

Here, we introduced the canonical partition function of the condensate, which is apparently equal to the functional integral

$$Z_{N_0} = \int d[\rho]d[\theta] \exp\left\{-\frac{1}{\hbar}S[\rho,\theta;0]\right\} \tag{11.95}$$

over all the nonzero momentum components of the density and phase fields.

Now the integration over the nonzero momentum components of the phase field $\theta(\mathbf{x},\tau)$ is readily performed, because it involves an ordinary Gaussian integral. Introducing the Green's function for the phase fluctuations $G(\mathbf{x},\mathbf{x}';\rho)$ by

$$\frac{\hbar}{m}\left((\boldsymbol{\nabla}\rho)\cdot\boldsymbol{\nabla}+\rho\boldsymbol{\nabla}^2\right)G(\mathbf{x},\mathbf{x}';\rho) = \delta(\mathbf{x}-\mathbf{x}') , \tag{11.96}$$

we obtain the effective action for the density field

$$\begin{aligned} S^{\text{eff}}[\rho] = & \int_0^{\hbar\beta} d\tau \int d\mathbf{x} \int d\mathbf{x}' \left\{-\frac{\hbar}{2}\frac{\partial\rho(\mathbf{x},\tau)}{\partial\tau}G(\mathbf{x},\mathbf{x}';\rho)\frac{\partial\rho(\mathbf{x}',\tau)}{\partial\tau}\right\} \\ & + \int_0^{\hbar\beta} d\tau \int d\mathbf{x} \left\{\frac{\hbar^2}{8m\rho(\mathbf{x},\tau)}(\boldsymbol{\nabla}\rho(\mathbf{x},\tau))^2 + V^{\text{ex}}(\mathbf{x})\rho(\mathbf{x},\tau)\right. \\ & \left. + \frac{2\pi a\hbar^2}{m}\rho^2(\mathbf{x},\tau)\right\} + \frac{\hbar}{2}\text{Tr}[\log(-G^{-1})] , \end{aligned} \tag{11.97}$$

where it turns out that it is allowed for our purposes to neglect the last term of the effective action. Being an action for the collective density fluctuations of the condensate, $S^{\text{eff}}[\rho]$ also describes the collective modes of the condensate. This is important for our purposes because we wish to describe how an attractive condensate collapses, and this is determined by the mode that becomes unstable first. Moreover, this mode also yields the probability for the collapse, because the energy barrier is the smallest in the corresponding direction of configuration space.

11.9.2 Breathing Mode

We start our discussion of the collective modes with the ideal Bose gas, i.e. $a = 0$, and consider the time-dependent Gaussian profile

$$\rho(\mathbf{x};q(\tau)) = N_\mathbf{0}\left(\frac{1}{\pi q^2(\tau)}\right)^{3/2} \exp\left\{-\frac{\mathbf{x}^2}{q^2(\tau)}\right\}, \tag{11.98}$$

which we expect to describe a collisionless mode of the condensate, where collisionless refers to the fact that we are considering the noninteracting limit. This is expected, because we can make a density fluctuation by taking one particle from the condensate and putting it in one of the excited states of the external potential. The corresponding density fluctuation then obeys in real time

$$\delta\rho(\mathbf{x},t) \propto \mathrm{e}^{-\mathrm{i}(\varepsilon_\mathbf{n}-\varepsilon_\mathbf{0})t/\hbar}\chi^*_\mathbf{n}(\mathbf{x})\chi_\mathbf{0}(\mathbf{x}) \ . \tag{11.99}$$

For the experimentally relevant case of an isotropic harmonic oscillator of strength $\hbar\omega$ it is convenient to use the two angular-momentum quantum numbers ℓ and m, and the quantum number n that counts the number of nodes in the radial wavefunction $R_{n\ell}(r)$. The density fluctuation then becomes

$$\delta\rho(\mathbf{x},t) \propto \mathrm{e}^{-\mathrm{i}(2n+\ell)\omega t}R_{n\ell}(r)Y^*_{\ell m}(\hat{\mathbf{x}})\frac{\mathrm{e}^{-r^2/2l^2}}{(\pi l^2)^{3/4}} \ , \tag{11.100}$$

with $r = |\mathbf{x}|$ and with

$$\varepsilon_{n\ell m} - \varepsilon_{000} = (2n+\ell)\hbar\omega \tag{11.101}$$

the excitation energy and $l = (\hbar/m\omega)^{1/2}$ the size of the condensate wavefunction. Expanding the Gaussian profile in (11.98) around the ground-state density profile, which boils down to substituting $q(\tau) = l + \delta q(\tau)$, we find that

$$\delta\rho(\mathbf{x},\tau) = -\sqrt{6}N_\mathbf{0}\frac{\delta q(\tau)}{l}R_{10}(r)Y^*_{00}(\hat{\mathbf{x}})\frac{\mathrm{e}^{-r^2/2l^2}}{(\pi l^2)^{3/4}}, \tag{11.102}$$

which has a similar form to (11.100) for the case that $(n\ell m) = (100)$.

To show that the Gaussian profile from (11.98) indeed describes a breathing mode with a frequency equal to 2ω, as expected from (11.100), we evaluate the effective action $S^{\mathrm{eff}}[\rho]$ and in particular the Green's function $G(\mathbf{x},\mathbf{x}';\rho)$ for the Gaussian density profile. Substituting (11.98) into (11.96) leads to

$$G(\mathbf{x},\mathbf{x}';\rho) = G(\mathbf{x},\mathbf{x}';q)/\rho(\mathbf{x}';q), \tag{11.103}$$

where

$$\frac{\hbar}{m}\left\{-\frac{2}{q^2}\mathbf{x}\cdot\boldsymbol{\nabla} + \boldsymbol{\nabla}^2\right\}G(\mathbf{x},\mathbf{x}';q) = \delta(\mathbf{x}-\mathbf{x}') \ . \tag{11.104}$$

To solve the latter equation, we first consider the eigenvalue problem

$$\left\{\boldsymbol{\nabla}^2 - \frac{2r}{q^2}\frac{\partial}{\partial r}\right\}\xi(\mathbf{x}) = \lambda\xi(\mathbf{x}) \ , \tag{11.105}$$

which upon substitution of

$$\xi_{n\ell m}(\mathbf{x}) = \xi_{n\ell}(r)\frac{\mathrm{e}^{r^2/2q^2}}{r}Y_{\ell m}(\hat{\mathbf{x}}) \tag{11.106}$$

yields the radial Schrödinger equation for an isotropic harmonic oscillator with frequency $\omega_q = \hbar/mq^2$, i.e.

$$\begin{aligned} &\lambda_{n\ell}\xi_{n\ell}(r) = \\ &-\frac{2m}{\hbar^2}\left\{-\frac{\hbar^2}{2m}\frac{\partial^2}{\partial r^2} + \frac{1}{2}m\omega_q^2 r^2 + \frac{\hbar^2\ell(\ell+1)}{2mr^2} - \frac{3}{2}\hbar\omega_q\right\}\xi_{n\ell}(r). \end{aligned} \tag{11.107}$$

The desired eigenfunctions are therefore

$$\xi_{n\ell m}(\mathbf{x};q) = \psi_{n\ell m}(\mathbf{x})\mathrm{e}^{r^2/2q^2}, \tag{11.108}$$

where $\psi_{n\ell m}(\mathbf{x})$ are the harmonic oscillator states with energies $(2n+\ell+3/2)\hbar\omega_q$, so that

$$\lambda_{n\ell}(q) = -2(2n+\ell)/q^2. \tag{11.109}$$

Introducing also

$$\bar{\xi}_{n\ell m}(\mathbf{x};q) \equiv \psi^*_{n\ell m}(\mathbf{x})\mathrm{e}^{-r^2/2q^2}, \tag{11.110}$$

we find that the Green's function $G(\mathbf{x},\mathbf{x}';q)$ is given by

$$G(\mathbf{x},\mathbf{x}';q) = \sum_{n\ell m}{}' \xi_{n\ell m}(\mathbf{x};q)\frac{m}{\hbar\lambda_{n\ell}(q)}\bar{\xi}_{n\ell m}(\mathbf{x}';q), \tag{11.111}$$

where the primed sum means that $(n\ell m) = (000)$ is excluded. This is because the corresponding eigenfunction $\xi_{000}(\mathbf{x};q)$ is a constant and does not contribute to $G(\mathbf{x},\mathbf{x}';\rho)$, which is defined as the Green's function for all phase fluctuations with nonvanishing momenta.

Putting everything together, we ultimately find that the dynamics of the collective variable $q(\tau)$ is determined by the action

$$\begin{aligned} S^{\mathrm{eff}}[q] &= \int_0^{\hbar\beta}\mathrm{d}\tau\left\{\frac{3mN_0}{4}\left(\frac{\mathrm{d}q}{\mathrm{d}\tau}\right)^2 + N_0\left(\frac{3\hbar^2}{4mq^2} + \frac{3}{4}m\omega^2 q^2\right)\right\} \\ &\equiv \int_0^{\hbar\beta}\mathrm{d}\tau\left\{\frac{1}{2}m^*\left(\frac{\mathrm{d}q}{\mathrm{d}\tau}\right)^2 + V(q)\right\}, \end{aligned} \tag{11.112}$$

which is equivalent to the action of a particle with effective mass $m^* = 3mN_0/2$ in a potential

$$V(q) = N_0 \left(\frac{3\hbar^2}{4mq^2} + \frac{3}{4} m\omega^2 q^2 \right) . \tag{11.113}$$

This potential has a minimum for $q = l$, around which it is expanded as

$$V(q) \simeq \frac{3}{2} N_0 \hbar\omega + \frac{1}{2} m^* (2\omega)^2 (\delta q)^2 , \tag{11.114}$$

so that the Gaussian profile indeed describes a breathing mode with frequency 2ω around an equilibrium density profile that is given by

$$\rho(\mathbf{x}; l) = N_0 |\psi_{000}(\mathbf{x})|^2 . \tag{11.115}$$

11.9.3 Metastability of the Condensate

The next question is how the interactions affect this result. It turns out that for the Gaussian density profile of (11.98), the action $S^{\text{eff}}[q]$ is again that of a particle with effective mass $m^* = 3mN_0/2$, but now in the potential [75]

$$V(q) = N_0 \left(\frac{3\hbar^2}{4mq^2} + \frac{3}{4} m\omega^2 q^2 - \frac{N_0}{\sqrt{2\pi}} \frac{\hbar^2 |a|}{mq^3} \right) . \tag{11.116}$$

The most important feature of this potential is that it is unbounded from below, because $V(q) \to -\infty$, if $q \downarrow 0$. Hence, the condensate always has the tendency to collapse to the high-density state $\lim_{q\downarrow 0} \rho(\mathbf{x}; q) = N_0 \delta(\mathbf{x})$. However, if the number of condensate particles is sufficiently small, or more precisely if [76]

$$N_0 < \frac{2\sqrt{2\pi}}{5^{5/4}} \frac{l}{|a|} \simeq 0.68 \frac{l}{|a|} , \tag{11.117}$$

then the condensate has to overcome a macroscopic energy barrier before it can collapse. Under these conditions, the condensate is really metastable and can in principle be observed experimentally. The most important question is how metastable the condensate actually is. Within the Gaussian approximation this question can be readily answered, because the dynamics of the condensate is then equivalent to the dynamics of a particle in an unstable potential. The corresponding tunneling rate can thus be evaluated using the WKB expression for the penetration through a (macroscopic) energy barrier, while the rate of decay due to thermal fluctuations is given by [77]

$$\Gamma = \frac{\omega}{2\pi} \mathrm{e}^{-\{V(q_{\max}) - V(q_{\min})\}/k_{\mathrm{B}}T} , \tag{11.118}$$

where $q_{\min}$ and $q_{\max}$ denote the position of the metastable minimum and the unstable maximum of $V(q)$ respectively. For realistic conditions of experiments with atomic ^{7}Li [78], it turns out that the thermal rate is dominant over the tunneling rate, and that a small condensate can indeed live long enough to be observed.

11.10 Problems

Exercise 11.1. Gross-Pitavskii Equation
Using a variational approach, we may show that the Gross-Pitaevskii equation is equivalent to the Hartree approximation. To this end, we assume that the ground state wavefunction of the system is a product of one-particle states $\chi'_{\mathbf{0}}$ with an energy $\varepsilon'_{\mathbf{0}}$, which we both want to determine variationally. We thus consider the many-body groundstate $|0;N\rangle$, where all atoms are in the same state $\chi'_{\mathbf{0}}$ with energy $\varepsilon'_{\mathbf{0}}$.
a) Calculate the average energy $\langle 0;N|\hat{H}|0;N\rangle$ for this ground state, where $\hat{H}$ is the Hamiltonian from (6.29).
b) Minimize the average energy by introducing the Lagrange multiplier $\varepsilon'_{\mathbf{0}}$ that takes into account the fact that the state $\chi'_{\mathbf{0}}$ should be properly normalized. In this manner, you arrive at a Schrödinger-like equation that determines the one-particle wavefunction and energy.
c) By introducing $\phi_{\mathbf{0}} = \sqrt{N_{\mathbf{0}}}\chi'_{\mathbf{0}}$ and using $V(\mathbf{x}-\mathbf{x}') = V_{\mathbf{0}}\delta(\mathbf{x}-\mathbf{x}')$, show that you obtain the Gross-Pitaevskii equation. Do you understand why $\varepsilon'_{\mathbf{0}}$ should be interpreted as the chemical potential?

Exercise 11.2. Show that the exact normal and anomalous Green's functions for a homogeneous Bose-Einstein condensed gas obey $G_{22}(\mathbf{k},\mathrm{i}\omega_n) = G_{11}(-\mathbf{k},-\mathrm{i}\omega_n)$ and $G_{12}(\mathbf{k},\mathrm{i}\omega_n) = G^*_{21}(-\mathbf{k},-\mathrm{i}\omega_n)$. What do these relations imply for the exact normal and anomalous selfenergies?

Exercise 11.3. Determine the dispersion relation $\hbar\omega_{\mathbf{k}}$ for the collective excitations of the homogeneous Bose gas with Popov theory. Is it gapless?

Exercise 11.4. Charged Bose Gas
Consider a charged gas of $\langle N\rangle$ spinless bosons in an external trapping potential $V^{\mathrm{ex}}(\mathbf{x})$. The action for this gas at temperature $T = 1/k_{\mathrm{B}}\beta$ and chemical potential μ is given by

$$S[\phi^*,\phi] = \int_0^{\hbar\beta} \mathrm{d}\tau \int \mathrm{d}\mathbf{x}\, \phi^*(\mathbf{x},\tau)\left\{\hbar\frac{\partial}{\partial\tau} - \frac{\hbar^2\boldsymbol{\nabla}^2}{2m} + V^{\mathrm{ex}}(\mathbf{x}) - \mu\right\}\phi(\mathbf{x},\tau)$$
$$+\frac{1}{2}\int \mathrm{d}\tau \int \mathrm{d}\mathbf{x} \int \mathrm{d}\mathbf{x}'\, \phi^*(\mathbf{x},\tau)\phi^*(\mathbf{x}',\tau)V(\mathbf{x}-\mathbf{x}')\phi(\mathbf{x}',\tau)\phi(\mathbf{x},\tau)\,, \tag{11.119}$$

where m is the mass of the bosons and $V(\mathbf{x}-\mathbf{x}') = e^2/4\pi\varepsilon_0|\mathbf{x}-\mathbf{x}'|$ the Coulomb potential between the particles. Note that this is a long-range potential that cannot be approximated by a delta function.
(a) We first consider the gas at sufficiently low temperatures that we can apply the Bogoliubov approximation to the gas. Put $\phi(\mathbf{x},\tau) = \phi_0(\mathbf{x}) + \phi'(\mathbf{x},\tau)$, where $\phi_0(\mathbf{x}) \equiv \langle\phi(\mathbf{x},\tau)\rangle$ is the condensate wavefunction. Determine from the above action the Gross-Pitaevskii equation for the condensate wavefunction.
(b) At higher temperatures, we have to use the Popov approximation. Determine the Gross-Pitaevskii equation for the condensate wave function in this case. Express the

answer in terms of the single-particle density matrix $n'(\mathbf{x},\mathbf{x}') \equiv \langle \phi'^*(\mathbf{x},\tau)\phi'(\mathbf{x}',\tau)\rangle$ of the fluctuations, i.e. of the thermal particles that are not in the condensate.
Now we consider the homogeneous case, for which the gas is in a box with volume V. In this case, we can make use of Fourier transformation. In particular, we have for the Coulomb potential that $V(\mathbf{k}) = e^2/\varepsilon_0\mathbf{k}^2$.
(c) Determine now the Bogoliubov dispersion $\hbar\omega_{\mathbf{k}}$ of the excitations of the Bose-Einstein condensed gas, by considering the Gaussian (quadratic) fluctuations around $\phi_0(\mathbf{x})$. Sketch the behaviour of this dispersion as a function of momentum $\hbar\mathbf{k}$.
(d) Can you physically understand your result for $\mathbf{k} = \mathbf{0}$?

Exercise 11.5. Supersymmetric Bose-Fermi Mixture
Consider an interacting mixture of spinless fermions and bosons. The fermions are described by the field operator $\hat{\psi}(\mathbf{x})$ and the bosons by the field operator $\hat{\phi}(\mathbf{x})$. The Hamiltonian for this system is given by

$$\hat{H} = \int d\mathbf{x}\, \hat{\phi}^\dagger(\mathbf{x})\left\{-\frac{\hbar^2}{2m}\frac{\partial^2}{\partial\mathbf{x}^2} - \mu\right\}\hat{\phi}(\mathbf{x}) + \int d\mathbf{x}\, \hat{\psi}^\dagger(\mathbf{x})\left\{-\frac{\hbar^2}{2m}\frac{\partial^2}{\partial\mathbf{x}^2} - \mu\right\}\hat{\psi}(\mathbf{x})$$
$$+\frac{U}{2}\int d\mathbf{x}\left\{\hat{\phi}^\dagger(\mathbf{x})\hat{\phi}(\mathbf{x}) + \hat{\psi}^\dagger(\mathbf{x})\hat{\psi}(\mathbf{x})\right\}^2,$$

where m is the mass of both the fermions and the bosons and U gives the interaction strength. Consider first the noninteracting case, i.e. $U = 0$.
a) Perform a Fourier transform of all the field operators and determine the dispersion relations $\varepsilon_{\mathbf{k}}^{\mathrm{B,F}}$ for the bosons and the fermions, respectively. If the dispersions of the bosons and fermions are the same, then the mixture is supersymmetric. Is that the case here?

Consider now the case with interactions, i.e. $U \neq 0$. Assume that $\mu > 0$, so we have a Bose-Einstein condensation of the bosons, which we are going to treat in the Bogoliubov approximation.
(b) Give the action corresponding to the above Hamiltonian and determine the expression for $\langle\phi\rangle$ in the Bogoliubov approximation, which for the present case means that only terms up to quadratic order in both the bosonic fluctuation fields and the fermionic fields are considered.
(c) Determine also the Green's functions $G^{\mathrm{B,F}}$ of the bosons and fermions in the Bogoliubov approximation.
(d) Using the result obtained in the previous question, derive the dispersion relations of the bosons and the fermions. Are the dispersions gapped? Is the system still supersymmetric?
(e) Reconsider question (c), but now treat the fermions in the Hartree-Fock approximation. How are the results for $\langle\phi\rangle$ and the dispersions changed?

Exercise 11.6. Condensate in a Harmonic Trap
For very low temperatures, the Bose-Einstein condensate wavefunction $\phi_{\mathbf{0}}(\mathbf{x})$ is the solution of the Gross-Pitaevskii equation

$$\left\{-\frac{\hbar^2\nabla^2}{2m}+V^{\text{ex}}(\mathbf{x})+T^{2\text{B}}|\phi_\mathbf{0}(\mathbf{x})|^2\right\}\phi_\mathbf{0}(\mathbf{x})=\mu\phi_\mathbf{0}(\mathbf{x}),\tag{11.120}$$

where the condensate wavefunction $\phi_\mathbf{0}(\mathbf{x})$ is normalized to the number of condensate atoms $N_\mathbf{0}$. This is a nonlinear equation, which is in general hard to solve analytically. However, there are certain limits, where analytic expressions can be obtained. We want to solve this equation for the case of the harmonic potential

$$V^{\text{ex}}(\mathbf{x})=\frac{1}{2}m\omega^2(x^2+y^2+z^2).\tag{11.121}$$

a) Suppose that the gas is weakly interacting, such that we can neglect the interaction term. Then the Gross-Pitaevskii equation reduces to the Schrödinger equation for a three-dimensional harmonic oscillator. Solve the Gross-Pitaevskii equation for this case and give the condensate wavefunction. What is the chemical potential μ? Can you understand your result? Does the condensate size depend on the number of condensate atoms? Give the physical reason for your answer.
b) For a weakly-interacting gas, the interaction term can be treated in perturbation theory. Calculate the first-order correction to the chemical potential. What is the physical interpretation of this correction term?
c) When the gas is strongly interacting, the kinetic term $-\hbar^2\nabla^2/2m$ can be neglected. Solve the Gross-Pitaevskii equation in this approximation and prove that the condensate density can be written as

$$n_\mathbf{0}(\mathbf{x})=\phi_\mathbf{0}^*(\mathbf{x})\phi_\mathbf{0}(\mathbf{x})=\begin{cases}\frac{15N_\mathbf{0}}{8\pi R^3}\left(1-\frac{r^2}{R^2}\right) & \text{if } r\le R;\\ 0 & \text{if } r>R.\end{cases}\tag{11.122}$$

This is called the Thomas-Fermi profile and R is called the Thomas-Fermi radius.
d) Express the Thomas-Fermi radius R and the chemical potential μ in terms of the number of condensate atoms $N_\mathbf{0}$ and the interaction strength $T^{2\text{B}}$. What do you expect for the dependence of R and μ as a function of $N_\mathbf{0}$?
c) Give an estimate for the magnitude of the kinetic term in the Gross-Pitaevskii equation within the Thomas-Fermi approximation. This term should be small compared to the interaction term. Use this to prove that the Thomas Fermi approximation is valid if the number of condensate atoms obeys

$$N_\mathbf{0}\gg\frac{\hbar^2 R}{mT^{2\text{B}}}.\tag{11.123}$$

Additional Reading

- M. Inguscio, S. Stringari, and C. E. Wieman, Eds., *Bose-Einstein Condensation in Atomic Gases, Proceedings of the International School of Physics Enrico Fermi, Course CXL,* (IOS, Amsterdam, 1999), and references therein.

- F. Dalfolvo, S. Giorgini, L. P. Pitaevskii, and S. Stringari, Rev. Mod. Phys. **71**, 463 (1999) and references therein.
- A. J. Leggett, Rev. Mod. Phys. **73**, 307 (2001) and references therein.
- C.J. Pethick and H. Smith, *Bose-Einstein Condensation in Dilute Gases*, (Cambridge University, Cambridge, 2002).

Chapter 12
Condensation of Fermionic Pairs

Our present understanding of superconductivity has arisen from a close interplay of theory and experiment. It would have been very difficult to have arrived at the theory by purely deductive reasoning from the basic equations of quantum mechanics. Even if someone had done so, no one would have believed that such remarkable properties would really occur in Nature.
– John Bardeen

In this chapter, we discuss the many-body theory for fermions with an attractive interaction. We focus in particular on the phase transition to the superfluid state. To this end, we use the functional formalism to elegantly incorporate the Bardeen-Cooper-Schrieffer order parameter into our theory by means of the Hubbard-Stratonovich transformation. The traditional approach used by Bardeen, Cooper, and Schrieffer, which is based on a variational wavefunction to minimize the second-quantized Hamiltonian for interacting fermions, has been left as Exercise 6.6. We determine the critical temperature, the gapped quasiparticle dispersion and the thermodynamic potential of the superfluid state with the use of mean-field theory. Since the interatomic interaction between fermionic atoms is precisely tunable with the use of a Feshbach resonance, which is discussed in more detail in Chap. 17, it turns out to be possible to study both experimentally and theoretically a crossover between a Bardeen-Cooper-Schrieffer (BCS) superfluid and a Bose-Einstein condensate (BEC) of diatomic molecules. This is the last topic of the chapter.

12.1 Introduction

Two identical fermions cannot occupy the same quantum state due to the Pauli principle. As a result, the ideal Fermi gas does not undergo Bose-Einstein condensation and form a superfluid state, unlike the ideal Bose gas. However, in the presence of an (effective) attractive interaction between the fermions, no matter how weak, the behavior of the many-body Fermi gas changes drastically. The physical reason for this change was first pointed out by Cooper, who showed that in the presence of a filled Fermi sea it is actually energetically favorable for two electrons that interact via an attractive phonon-mediated interaction to form a bound pair, also called a Cooper pair. As a result, the Fermi sea turns out to be unstable against the formation of a new many-body ground state consisting, loosely speaking, of a Bose-Einstein condensate of Cooper pairs. This many-body ground state was introduced by Bardeen, Cooper, and Schrieffer as

$$|\Psi_{\rm BCS}\rangle = \prod_{\mathbf{k}} \left\{ u_{\mathbf{k}} + v_{\mathbf{k}} \hat{\psi}^{\dagger}_{\mathbf{k},\uparrow} \hat{\psi}^{\dagger}_{-\mathbf{k},\downarrow} \right\} |0\rangle, \tag{12.1}$$

which shows explicitly that the Cooper pairs form between electrons with opposite momentum and spin. This ansatz formed the basis of their celebrated microscopic theory of ordinary superconducting metals, which has become known as Bardeen-Cooper-Schrieffer (BCS) theory. Almost fifty years after the discovery of superconductivity by Kamerlingh Onnes in 1911, BCS theory was finally able to microscopically account for the second-order phase transition to the superconducting phase and make quantitative statements about the critical temperature. Furthermore, the theory could explain the Meissner effect that an applied magnetic field is completely expelled from the interior of the superconductor. Another great success was the derivation of the observed energy gap in the single-particle excitation spectrum, and the associated exponentially suppressed specific heat below the critical temperature. Much later, in 1986, it was discovered that certain ceramic materials become superconducting at temperatures much higher than metals. Critical temperatures of above 100 K have now been observed. However, the properties of these materials are not described with ordinary BCS theory, and the microscopic mechanism behind this high-temperature superconductivity is arguably the biggest unsolved problem in condensed-matter physics.

In recent years, the equivalent of the BCS transition has also been observed in neutral ultracold atomic Fermi gases with an attractive s-wave interaction, leading to a fermionic superfluid state consisting of paired atoms. The big excitement in the study of atomic Fermi gases is caused by the impressive amount of experimental control that is achievable in manipulating these fundamental many-body quantum gases. Various important parameters, such as the interaction strength between the fermions, the experienced external potential and the number of fermions in each quantum state, are under full experimental control. Note that this is not the case in a typical condensed-matter system, such as an ordinary metal or a high-temperature superconductor, where these parameters are unchangeable properties of the material being used. As a result, fundamental questions about the properties of strongly-correlated quantum many-body systems are systematically addressable with unprecedented experimental detail. This provides accurate tests for quantum many-body theory.

12.2 Thouless Criterion

To study BCS theory we start from the action for interacting fermions in two different hyperfine states, which we label with a spin index $\alpha = \uparrow, \downarrow$. Since the hyperfine space is now only two dimensional, it is often also referred to as an effective or pseudo spin-$1/2$ space. We consider atomic fields $\phi_\alpha(\mathbf{x}, \tau)$ that interact only via s-wave interactions, because this is the dominant scattering mechanism at the low momenta or ultralow temperatures of interest, as explained in Sect. 10.3.1. It is important to note that s-wave scattering only occurs between two identical fermions if they are in different spin states. This is most easily understood by considering the two-particle scattering problem, where for identical fermions the total wavefunction describing the scattering process must be antisymmetric. However, the spatial part of a wavefunction with zero angular momentum is symmetric, such that in order to obtain an antisymmetric wavefunction we should have antisymmetry in the spin part of the wavefunction. With two fermions in the same spin state it is impossible to make an antisymmetric spin state, which implies that two fermionic particles in identical spin states cannot interact via s-wave scattering and obey the Pauli principle at the same time. In the case of two fermionic particles in different spin states it is of course possible to make an antisymmetric combination, and s-wave scattering is then consequently allowed. From the above discussion, it thus follows that the interacting fermionic action is given by

$$\begin{aligned} S[\phi^*,\phi] = &\sum_{\alpha=\uparrow,\downarrow} \int_0^{\hbar\beta} \mathrm{d}\tau \int \mathrm{d}\mathbf{x}\, \phi_\alpha^*(\mathbf{x},\tau) \left\{ \hbar\frac{\partial}{\partial\tau} - \frac{\hbar^2\nabla^2}{2m_\alpha} - \mu_\alpha \right\} \phi_\alpha(\mathbf{x},\tau) \\ &+ \int_0^{\hbar\beta} \mathrm{d}\tau \int \mathrm{d}\mathbf{x}\, V_0 \phi_\uparrow^*(\mathbf{x},\tau)\phi_\downarrow^*(\mathbf{x},\tau)\phi_\downarrow(\mathbf{x},\tau)\phi_\uparrow(\mathbf{x},\tau) , \end{aligned} \tag{12.2}$$

where μ_α is the chemical potential for particles with mass m_α in spin state $|\alpha\rangle$, while the short-ranged interaction $V_0\delta(\mathbf{x}-\mathbf{x}')$ was discussed in more detail in Sect. 10.4. In this chapter we discuss interacting particles with identical masses, so $m_\uparrow = m_\downarrow = m$, and we consider an equal amount of particles in each hyperfine state, which is achieved by considering equal chemical potentials $\mu_\uparrow = \mu_\downarrow = \mu$. In Sect. 14.3.1 we then discuss the interesting consequences of an imbalance in the spin populations, which turns out to have profound effects on the Cooper pairing that preferably takes place for equal spin densities. Moreover, the case of unequal masses is relevant for strongly-interacting quarks in neutron stars, which are also believed to undergo a transition to the superfluid paired state. However, the latter case is beyond the scope of this text. In the first instance we focus on the homogeneous case, while we later show how the effect of the external trapping potential is conveniently incorporated by using the local-density approximation which we encountered in Sect. 4.3.1.2.

As mentioned in the previous section, the ground state for the noninteracting Fermi gas, i.e. the filled Fermi sea, is unstable towards a nontrivial many-body ground state in the presence of an attractive interaction $V_0 < 0$. To discuss this instability more quantitatively, we consider the many-body T matrix, which was introduced in Sect. 10.4 as the many-body generalization of the two-body T matrix discussed in Sect. 10.3.1. There, it was shown that for large and positive s-wave scattering length a, the two-body T-matrix from (10.38) has a pole located at $z = -\hbar^2/ma^2$ which then corresponded precisely to the energy of a two-body bound state in the interaction potential. This observation may be generalized to the many-body case, such that the onset of a pole in the many-body T matrix corresponds to long-lived pairs occurring in the system. This is also known as the Thouless criterion. We apply the Thouless criterion to the above described system of effective spin-$1/2$ fermions interacting via s-wave scattering with a negative scattering length

a, such that the two-body interaction potential does not give rise to a bound state. We recall that the inverse many-body T-matrix is given by (10.59), which for the fermionic case yields

$$\frac{1}{T(\mathbf{K},\mathrm{i}\Omega_n)} = \frac{m}{4\pi a\hbar^2} - \frac{1}{V}\sum_{\mathbf{k}}\left\{\frac{1-N_{\mathrm{FD}}(\varepsilon_{\mathbf{K}/2+\mathbf{k}})-N_{\mathrm{FD}}(\varepsilon_{\mathbf{K}/2-\mathbf{k}})}{\mathrm{i}\hbar\Omega_n-\varepsilon_{\mathbf{K}/2+\mathbf{k}}-\varepsilon_{\mathbf{K}/2-\mathbf{k}}+2\mu}+\frac{1}{2\varepsilon_{\mathbf{k}}}\right\}, \quad (12.3)$$

where $\mathbf{K}$ is the center-of-mass momentum of the scattering particles, $\mathrm{i}\Omega_n$ the corresponding bosonic Matsubara frequencies, and $N_{\mathrm{FD}}(\varepsilon_{\mathbf{k}}) = 1/\{\exp(\beta(\varepsilon_{\mathbf{k}}-\mu))+1\}$ the Fermi-Dirac distribution. To study the possibility of a pole at zero frequency, we note that the right-hand side is maximal for $\mathbf{K} = \mathbf{0}$, such that the pole occurs first in the many-body T-matrix, when

$$\frac{1}{T(\mathbf{0},0)} = 0 = \frac{m}{4\pi a\hbar^2} + \frac{1}{V}\sum_{\mathbf{k}}\left\{\frac{1-2N_{\mathrm{FD}}(\varepsilon_{\mathbf{k}})}{2(\varepsilon_{\mathbf{k}}-\mu)} - \frac{1}{2\varepsilon_{\mathbf{k}}}\right\}, \quad (12.4)$$

which then, according to the Thouless criterion, corresponds to a new bound state that has become accessible to the interacting fermions. In Sect. 12.4, we find that the above equation indeed gives the critical temperature for the transition to the superfluid state. Note that the pairing of fermions and the occurrence of a pole in the many-body T matrix is truly a many-body effect induced by the presence of the Fermi sea, in the sense that the two-body limit of the T matrix does not yield a pole, since the considered two-body potential has no bound states.

12.3 Hubbard-Stratonovich Transformation

BCS theory can be physically interpreted as the Bose-Einstein condensation of Cooper pairs, meaning that the corresponding order parameter is proportional to the expectation value $\langle\phi_\downarrow(\mathbf{x},\tau)\phi_\uparrow(\mathbf{x},\tau)\rangle$ in analogy to the order parameter $\langle\phi(\mathbf{x},\tau)\rangle$ for Bose-Einstein condensation. For the transition to the BCS state we require that the two-body interaction potential is attractive, because otherwise the formation of pairs would not be energetically favorable. From now on, we therefore consider an attractive potential. Note that this does not necessarily mean that the corresponding scattering length a needs to be negative. In Sect. 10.3.1, we saw that attractive potentials can also give rise to a positive scattering length when there is a two-body bound state present in the potential. To introduce the BCS order parameter elegantly into our theory, we use the Hubbard-Stratonovich transformation, with which we introduce a complex pairing field $\Delta(\mathbf{x},\tau)$ that on average is related to the expectation value of $\phi_\downarrow(\mathbf{x},\tau)\phi_\uparrow(\mathbf{x},\tau)$ through

$$\langle\Delta(\mathbf{x},\tau)\rangle = V_{\mathbf{0}}\langle\phi_\downarrow(\mathbf{x},\tau)\phi_\uparrow(\mathbf{x},\tau)\rangle. \quad (12.5)$$

In analogy with the discussion of the Hubbard-Stratonovich transformation for the Hartree-Fock theory in Sect. 8.6, this is achieved by inserting the following identity into the integrand of the partition function

$$1 = \int \mathrm{d}[\Delta^*]\mathrm{d}[\Delta] \, \exp\left\{\frac{1}{\hbar}\left(\Delta - V_{\mathbf{0}}\phi_{\downarrow}\phi_{\uparrow}|V_{\mathbf{0}}^{-1}|\Delta - V_{\mathbf{0}}\phi_{\downarrow}\phi_{\uparrow}\right)\right\}, \tag{12.6}$$

where the measure contains conveniently the factor $\exp\{\mathrm{Tr}[\log(-V_{\mathbf{0}}^{-1}/\hbar)]\}$ that cancels the outcome of the functional integral and where the inner product in the exponent is a short hand for

$$\int_0^{\hbar\beta} \mathrm{d}\tau \int \mathrm{d}\mathbf{x} \left(\Delta^*(\mathbf{x},\tau) - \phi_{\uparrow}^*(\mathbf{x},\tau)\phi_{\downarrow}^*(\mathbf{x},\tau)V_{\mathbf{0}}\right) V_{\mathbf{0}}^{-1} \left(\Delta(\mathbf{x},\tau) - V_{\mathbf{0}}\phi_{\downarrow}(\mathbf{x},\tau)\phi_{\uparrow}(\mathbf{x},\tau)\right)$$

as explained in Sect. 7.2.3.

As a result, upon inserting the Hubbard-Stratonovich transformation into the partition sum, the fourth-order term in the fermionic fields coming from the transformation exactly cancels the interaction term from the original action $S[\phi^*,\phi]$ in (12.2). The resulting action $S[\Delta^*,\Delta,\phi^*,\phi]$ determining the partition function then depends only quadratically on the atomic fields $\phi_\alpha(\mathbf{x},\tau)$. Explicitly, it is given by

$$\begin{aligned} S[\Delta^*,\Delta,\phi^*,\phi] = &-\int_0^{\hbar\beta} \mathrm{d}\tau \int \mathrm{d}\mathbf{x} \, \frac{|\Delta(\mathbf{x},\tau)|^2}{V_{\mathbf{0}}} \\ &-\hbar \sum_{\alpha=\uparrow,\downarrow} \int_0^{\hbar\beta} \mathrm{d}\tau \int \mathrm{d}\mathbf{x} \int_0^{\hbar\beta} \mathrm{d}\tau' \int \mathrm{d}\mathbf{x}' \; \phi_\alpha^*(\mathbf{x},\tau) G_{0;\alpha}^{-1}(\mathbf{x},\tau;\mathbf{x}',\tau')\phi_\alpha(\mathbf{x}',\tau') \\ &+\int_0^{\hbar\beta} \mathrm{d}\tau \int \mathrm{d}\mathbf{x} \, \left\{\phi_{\uparrow}^*(\mathbf{x},\tau)\phi_{\downarrow}^*(\mathbf{x},\tau)\Delta(\mathbf{x},\tau) + \Delta^*(\mathbf{x},\tau)\phi_{\downarrow}(\mathbf{x},\tau)\phi_{\uparrow}(\mathbf{x},\tau)\right\}, \end{aligned} \tag{12.7}$$

where the noninteracting Green's function for atoms with spin index $\alpha = \uparrow,\downarrow$ is given by

$$G_{0;\alpha}^{-1}(\mathbf{x},\tau;\mathbf{x}',\tau') = -\frac{1}{\hbar}\left\{\hbar\frac{\partial}{\partial\tau} - \frac{\hbar^2\nabla^2}{2m} - \mu\right\}\delta(\mathbf{x}-\mathbf{x}')\delta(\tau-\tau'). \tag{12.8}$$

We can write the action more compactly as a matrix multiplication, namely

$$\begin{aligned} S[\Delta^*,\Delta,\phi^*,\phi] = &-\int_0^{\hbar\beta} \mathrm{d}\tau \int \mathrm{d}\mathbf{x} \, \frac{|\Delta(\mathbf{x},\tau)|^2}{V_{\mathbf{0}}} \\ &-\hbar \int_0^{\hbar\beta} \mathrm{d}\tau \int \mathrm{d}\mathbf{x} \int_0^{\hbar\beta} \mathrm{d}\tau' \int \mathrm{d}\mathbf{x}' \; \left[\phi_{\uparrow}^*(\mathbf{x},\tau), \phi_{\downarrow}(\mathbf{x},\tau)\right] \cdot \mathbf{G}^{-1} \cdot \begin{bmatrix} \phi_{\uparrow}(\mathbf{x}',\tau') \\ \phi_{\downarrow}^*(\mathbf{x}',\tau') \end{bmatrix}. \end{aligned} \tag{12.9}$$

Note that by interchanging the fermionic fields and the corresponding equal-time limiting procedure, we pick up a constant term as explained in Sect. 11.4. This is important for determining the thermodynamic potential, such that we discuss this point more carefully in Sect. 12.7. However, for the present discussion it does not play a role. Expressing the interacting Green's function matrix in terms of the noninteracting part and the selfenergy part, we have

$$\mathbf{G}^{-1}(\mathbf{x},\tau;\mathbf{x}',\tau') = \mathbf{G}_0^{-1}(\mathbf{x},\tau;\mathbf{x}',\tau') - \mathbf{\Sigma}(\mathbf{x},\tau;\mathbf{x}',\tau') \tag{12.10}$$

where the noninteracting Green's function matrix $\mathbf{G}_0$ yields

$$\mathbf{G}_0^{-1}(\mathbf{x},\tau;\mathbf{x}',\tau') = \begin{bmatrix} G_{0;\uparrow}^{-1}(\mathbf{x},\tau;\mathbf{x}',\tau') & 0 \\ 0 & -G_{0;\downarrow}^{-1}(\mathbf{x}',\tau';\mathbf{x},\tau) \end{bmatrix}. \tag{12.11}$$

Note the minus sign in the $G_{0;22}^{-1}$ component, which comes about from interchanging the fermionic Grassmann variables, such that it is absent in the bosonic counterpart from (11.29). The selfenergy is then given by

$$\mathbf{\Sigma}(\mathbf{x},\tau;\mathbf{x}',\tau') \equiv \begin{bmatrix} \Sigma_{11} & \Sigma_{12} \\ \Sigma_{21} & \Sigma_{22} \end{bmatrix} = \frac{1}{\hbar}\begin{bmatrix} 0 & \Delta(\mathbf{x},\tau) \\ \Delta^*(\mathbf{x},\tau) & 0 \end{bmatrix} \delta(\mathbf{x}-\mathbf{x}')\delta(\tau-\tau'). \tag{12.12}$$

12.4 Bardeen-Cooper-Schrieffer Theory

Since the action is quadratic in the fermion fields we can integrate them out exactly, such that we obtain the effective action for the pairing field, given by

$$S^{\text{eff}}[\Delta^*,\Delta] = -\int_0^{\hbar\beta} \mathrm{d}\tau \int \mathrm{d}\mathbf{x}\, \frac{|\Delta(\mathbf{x},\tau)|^2}{V_0} - \hbar\mathrm{Tr}[\log(-\mathbf{G}^{-1})]. \tag{12.13}$$

Next, we expand the effective action in powers of $\Delta(\mathbf{x},\tau)$ by using

$$\mathbf{G}^{-1} = \mathbf{G}_0^{-1} - \mathbf{\Sigma} = \mathbf{G}_0^{-1}(\mathbf{1} - \mathbf{G}_0\mathbf{\Sigma}), \tag{12.14}$$

where the selfenergy is given by (12.12). We use

$$-\hbar\mathrm{Tr}[\log(-\mathbf{G}^{-1})] = -\hbar\mathrm{Tr}[\log(-\mathbf{G}_0^{-1})] + \hbar\sum_{m=1}^{\infty}\frac{1}{m}\mathrm{Tr}[(\mathbf{G}_0\mathbf{\Sigma})^m]. \tag{12.15}$$

where the trace is to be taken over real space, imaginary time, and the 2×2 matrix structure of the Green's function and the selfenergy, which is also called Nambu space. Note that the first-order term, $m=1$, vanishes because it only gives rise to off-diagonal terms in the 2×2 matrix $\mathbf{G}_0\mathbf{\Sigma}$. The second-order term in the expansion of the logarithm yields

$$\frac{\hbar}{2}\mathrm{Tr}\left[(\mathbf{G}_0\mathbf{\Sigma})^2\right] = \frac{\hbar}{2}\int_0^{\hbar\beta} \mathrm{d}\tau\, \mathrm{d}\tau'\, \mathrm{d}\tau''\, \mathrm{d}\tau''' \int \mathrm{d}\mathbf{x}\, \mathrm{d}\mathbf{x}'\, \mathrm{d}\mathbf{x}''\, \mathrm{d}\mathbf{x}''' \tag{12.16}$$
$$\mathrm{tr}\left[\mathbf{G}_0(\mathbf{x},\tau;\mathbf{x}',\tau')\mathbf{\Sigma}(\mathbf{x}',\tau';\mathbf{x}'',\tau'')\mathbf{G}_0(\mathbf{x}'',\tau'';\mathbf{x}''',\tau''')\mathbf{\Sigma}(\mathbf{x}''',\tau''';\mathbf{x},\tau)\right],$$

where the reduced trace operation $\mathrm{tr}[\ldots]$ means that we still have to take the trace over the two-dimensional Nambu space, i.e. we have to sum the diagonal elements of the 2×2 matrix $\mathbf{G}_0\mathbf{\Sigma}\mathbf{G}_0\mathbf{\Sigma}$. Substituting the expression for the selfenergy from

(12.12) and evaluating the trace over Nambu space, we find for the second-order term

$$\frac{\hbar}{2}\mathrm{Tr}\left[(\mathbf{G}_0\boldsymbol{\Sigma})^2\right] \tag{12.17}$$
$$= \frac{1}{\hbar}\int_0^{\hbar\beta} \mathrm{d}\tau\,\mathrm{d}\tau' \int \mathrm{d}\mathbf{x}\,\mathrm{d}\mathbf{x}'\; G_{0;11}(\mathbf{x},\tau;\mathbf{x}',\tau')\Delta(\mathbf{x}',\tau')G_{0;22}(\mathbf{x}',\tau';\mathbf{x},\tau)\Delta^*(\mathbf{x},\tau),$$

where from (12.11) we have that $G_{0;22}(\mathbf{x}',\tau';\mathbf{x},\tau) = -G_{0;\downarrow}(\mathbf{x},\tau;\mathbf{x}',\tau')$. In the same way, we can also compute the higher-order terms in the expansion, where all the odd terms are seen to give zero while the even terms lead to higher-order generalizations of (12.17).

So far we have not made any approximations, since we have performed an exact rewriting of the partition function with the use of the Hubbard-Stratonovich transformation to the collective $\Delta(\mathbf{x},\tau)$ field. Considering the Fourier transform $\Delta(\mathbf{k},\mathrm{i}\omega_n)$ of $\Delta(\mathbf{x},\tau)$ we expect physically that the collective paring field with zero-momentum corresponds to the lowest energy state for the Cooper pairs, because the nonzero momentum components $\Delta(\mathbf{k},\mathrm{i}\omega_n)$ describe Cooper pairing with additional kinetic energy. This zero-momentum pairing field is constant in space. By performing a fluctuation expansion around the mean-field, i.e. $\Delta(\mathbf{x},\tau) = \langle\Delta(\mathbf{x},\tau)\rangle + \Delta'(\mathbf{x},\tau)$, we are in the position to set up a self-consistent mean-field theory for the introduced expectation value. We then show that by considering a space-independent expectation value we obtain the functional formulation of the traditional BCS theory, which we study for the rest of this chapter. Condensates of Cooper pairs with nonzero momentum have also been considered in the literature. Under more exotic circumstances, for example in the case of low dimensionality and a high population imbalance in the spin species, such condensates have been theoretically predicted to yield the groundstate of the quantum system. They are also known as the Fulde-Ferrel [79] and the Larkin-Ovchinnikov [80] phases. Presently there is much interest to observe such exotic superfluids experimentally in ultracold atomic Fermi gases, so far without success.

To continue our discussion of BCS theory, we note that for a space and time-independent collective field Δ the effective action from (12.13) becomes a 'free-energy' density of the form known from the Landau theory of second-order phase transitions discussed in Chap. 9. It is given by

$$f_L(|\Delta|) = \alpha(T)|\Delta|^2 + \beta(T)|\Delta|^4 + \dots\,, \tag{12.18}$$

where the Landau 'free-energy' density is related to the effective action via

$$S^{\mathrm{eff}}[\Delta^*,\Delta] = \hbar\beta V f_{\mathrm{L}}(|\Delta|), \tag{12.19}$$

so that the microscopic derivation of the effective action from (12.13) makes it possible to calculate the coefficients $\alpha(T)$ and $\beta(T)$ in terms of the inverse temperature β, the chemical potential μ and the interaction parameter $V_{\mathbf{0}}$. Comparing (12.10), (12.13) and (12.17) with (12.18) and (12.19), we find that the coefficient $\alpha(T)$ is given by

$$\begin{aligned}\alpha(T) &= -\frac{1}{V_0} - \frac{1}{\hbar^2\beta V}\int_0^{\hbar\beta} \mathrm{d}\tau\,\mathrm{d}\tau' \int \mathrm{d}\mathbf{x}\,\mathrm{d}\mathbf{x}'\, G_{0;\uparrow}(\mathbf{x},\tau;\mathbf{x}',\tau')G_{0;\downarrow}(\mathbf{x},\tau;\mathbf{x}',\tau') \\ &= -\frac{1}{V_0} - \frac{1}{\hbar^2\beta V}\sum_{n,n'}\sum_{\mathbf{k},\mathbf{k}'} G_{0;\uparrow}(\mathbf{k},\mathrm{i}\omega_n)G_{0;\downarrow}(\mathbf{k}',\mathrm{i}\omega_{n'})\delta_{\mathbf{k},-\mathbf{k}'}\delta_{n,-n'} \\ &= -\frac{1}{V_0} - \frac{1}{\hbar^2\beta V}\sum_{n}\sum_{\mathbf{k}} \frac{-\hbar}{-\mathrm{i}\hbar\omega_n+\varepsilon_{\mathbf{k}}-\mu}\,\frac{-\hbar}{\mathrm{i}\hbar\omega_n+\varepsilon_{\mathbf{k}}-\mu},\end{aligned} \tag{12.20}$$

where we substituted the Fourier expansion for the homogeneous noninteracting Green's functions, as discussed in Example 7.3, after which the integrals over position and imaginary time give rise to the Kronecker deltas $\delta_{\mathbf{k},-\mathbf{k}'}$ and $\delta_{n,-n'}$. Next, we split the fraction and perform the sum over Matsubara frequencies, giving

$$\begin{aligned}\alpha(T) &= -\frac{1}{V_0} - \frac{1}{\hbar^2\beta V}\sum_{\mathbf{k}}\sum_{n} \frac{-\hbar}{2(\varepsilon_{\mathbf{k}}-\mu)}\left\{\frac{-\hbar e^{\mathrm{i}\omega_n\eta}}{-\mathrm{i}\hbar\omega_n+\varepsilon_{\mathbf{k}}-\mu} + \frac{-\hbar e^{\mathrm{i}\omega_n\eta}}{\mathrm{i}\hbar\omega_n+\varepsilon_{\mathbf{k}}-\mu}\right\} \\ &= -\frac{1}{V_0} - \frac{1}{V}\sum_{\mathbf{k}} \frac{1}{2(\varepsilon_{\mathbf{k}}-\mu)}\left\{1 - \frac{2}{e^{\beta(\varepsilon_{\mathbf{k}}-\mu)}+1}\right\},\end{aligned} \tag{12.21}$$

where we used (7.31) and the result from Exercise 7.2. As explained in Sect. 10.4, the interaction parameter V_0 is related to the experimentally known s-wave scattering length a. Using (10.54), we finally arrive at [81]

$$\begin{aligned}\alpha(T) &= -\frac{m}{4\pi a\hbar^2} - \frac{1}{V}\sum_{\mathbf{k}}\left\{\frac{1}{2(\varepsilon_{\mathbf{k}}-\mu)}\left(1 - \frac{2}{e^{\beta(\varepsilon_{\mathbf{k}}-\mu)}+1}\right) - \frac{1}{2\varepsilon_{\mathbf{k}}}\right\} \\ &= -\frac{m}{4\pi a\hbar^2} - \frac{1}{V}\sum_{\mathbf{k}}\left[\frac{\tanh(\beta(\varepsilon_{\mathbf{k}}-\mu)/2)}{2(\varepsilon_{\mathbf{k}}-\mu)} - \frac{1}{2\varepsilon_{\mathbf{k}}}\right].\end{aligned} \tag{12.22}$$

12.5 Critical Temperature

As explained in Chap. 9, the second-order coefficient $\alpha(T)$ of the Landau free energy $f_{\mathrm{L}}(|\Delta|)$ determines the critical temperature for a second-order phase transition. It is this term that changes sign at the critical temperature, such that the minimum of the Landau free energy shifts away from zero, yielding a nonzero order parameter $\langle\Delta\rangle$. As a result, the critical temperature $k_{\mathrm{B}}T_{\mathrm{c}} \equiv 1/\beta_{\mathrm{c}}$ is determined by the condition $\alpha(k_{\mathrm{B}}T_{\mathrm{c}}) = 0$. As we now show, in the weakly-interacting limit when the critical temperature is low, we can obtain an analytic expression for the critical temperature from BCS theory. We start with converting the sum on the right-hand side of (12.22) into an integral such that it becomes

$$
\begin{aligned}
\int \frac{d\mathbf{k}}{(2\pi)^3} & \left\{ \frac{\tanh(\beta(\varepsilon_{\mathbf{k}}-\mu)/2)}{2(\varepsilon_{\mathbf{k}}-\mu)} - \frac{1}{2\varepsilon_{\mathbf{k}}} \right\} \\
&= \frac{2\sqrt{2}}{(2\pi)^2} \frac{m^{3/2}}{\hbar^3} \int_0^\infty d\varepsilon \sqrt{\varepsilon} \left\{ \frac{(\tanh(\beta(\varepsilon-\mu)/2))}{2(\varepsilon-\mu)} - \frac{1}{2\varepsilon} \right\} \\
&= \frac{mk_{\mathrm{F}}}{2\pi^2\hbar^2} \int_0^\infty \sqrt{x}\, dx \left\{ \frac{\tanh(y(x-1))}{2(x-1)} - \frac{1}{2x} \right\}. \qquad (12.23)
\end{aligned}
$$

Here we introduced $x = \varepsilon/\mu$ and $y = \beta\mu/2$, where the latter is a large number for the present case of interest. Furthermore, for a weakly-interacting Fermi gas near zero temperature, the chemical potential is well approximated by the Fermi energy $\mu \simeq \varepsilon_{\mathrm{F}} = \hbar^2 k_{\mathrm{F}}^2/2m$, since the two are equal in the noninteracting zero-temperature limit as explained in Sect. 4.3.3. Moreover, we have for large y that $\tanh(y(x-1)) \simeq 1 - 2\theta(1-x)$ with $\theta(u)$ the step function, such that $\theta(u) = 0$ for $u < 0$ and $\theta(u) = 1$ for $u > 0$.

To further evaluate the integral from (12.23), we first use that

$$
\frac{\sqrt{x}}{x-1} = \frac{1}{\sqrt{x}+1} + \frac{1}{x-1}, \qquad (12.24)
$$

such that we have two integrals to perform. The first integral is given by

$$
\begin{aligned}
\frac{mk_{\mathrm{F}}}{2\pi^2\hbar^2} & \int_0^{x_M} dx \left\{ \frac{\tanh(y(x-1))}{2(\sqrt{x}+1)} - \frac{1}{2\sqrt{x}} \right\} \\
&\simeq \frac{mk_{\mathrm{F}}}{4\pi^2\hbar^2} \left\{ \int_0^1 dx \left(\frac{-1}{\sqrt{x}+1} - \frac{1}{\sqrt{x}} \right) + \int_1^{x_M} dx \left(\frac{1}{\sqrt{x}+1} - \frac{1}{\sqrt{x}} \right) \right\} \\
&= \frac{mk_{\mathrm{F}}}{2\pi^2\hbar^2} \left\{ -2 - \log(4) + \log(1+\sqrt{x_M}) \right\}, \qquad (12.25)
\end{aligned}
$$

where we take the limit $x_M \to \infty$ at the end of the calculation, when we have also taken the other term from (12.23) and (12.24) into account. Then we find that the above divergent term is exactly cancelled, as is expected, because the original integral from (12.23) is finite. The other contribution from (12.23) and (12.24) is given by

$$
\frac{mk_{\mathrm{F}}}{2\pi^2\hbar^2} \int_0^{x_M} dx \frac{\tanh(y(x-1))}{2(x-1)} = \frac{mk_{\mathrm{F}}}{4\pi^2\hbar^2} \int_{-1}^{x_M-1} dx \frac{\tanh(yx)}{x}, \qquad (12.26)
$$

which we can further evaluate by using

$$\begin{aligned}\int_{-1}^{x_M-1} \mathrm{d}x \frac{\tanh(yx)}{x} &= \int_0^1 \mathrm{d}x \frac{\tanh(yx)}{x} + \int_0^{x_M-1} \mathrm{d}x \frac{\tanh(yx)}{x} \\ &= \log(x_M-1) - \int_0^1 \frac{y\log(x)\mathrm{d}x}{\cosh^2(yx)} - \int_0^{x_M-1} \frac{y\log(x)\mathrm{d}x}{\cosh^2(yx)} \\ &= \log(x_M-1) - \int_0^y \frac{\log(z/y)\mathrm{d}z}{\cosh^2(z)} - \int_0^\infty \frac{\log(z/y)\mathrm{d}z}{\cosh^2(z)} \\ &= \log(x_M-1) + 2\{\gamma - \log(\pi/4y)\}, \end{aligned} \tag{12.27}$$

where we have used partial integration and the result from the standard integral $\int_0^\infty \mathrm{d}z \, \log(z/y)/\cosh^2(z) = \log(\pi \mathrm{e}^{-\gamma}/4y)$ and $\gamma = 0.5772$ is Euler's constant. Note that in the last step we also used that for large y it is allowed to extend the range of the integral to infinity, since the integrand's tail is exponentially suppressed.

The critical temperature T_c then follows from the condition $\alpha(T_\mathrm{c}) = 0$, such that from (12.22), (12.25), (12.26) and (12.27) we obtain that

$$\begin{aligned} -\frac{m}{4\pi a\hbar^2} &= \frac{mk_\mathrm{F}}{2\pi^2\hbar^2}\Big(-2+\log 4 - \log(1+\sqrt{x_M}) + \log(\sqrt{x_M-1}) + \gamma - \log(\pi/4y)\Big) \\ &= \frac{mk_\mathrm{F}}{2\pi^2\hbar^2}\left(-2+\log 4 + \gamma - \log(\pi/2\beta_\mathrm{c}\varepsilon_\mathrm{F})\right), \end{aligned} \tag{12.28}$$

where we finally took the limit $x_M \longrightarrow \infty$. The above equation can be solved for T_c, giving in the case of a small negative scattering length a the analytic result

$$T_\mathrm{c} = \frac{8}{\pi}\frac{\varepsilon_\mathrm{F}}{k_\mathrm{B}} \mathrm{e}^{\gamma-2} \mathrm{e}^{-\pi/2k_\mathrm{F}|a|}. \tag{12.29}$$

Note that at various moments during the calculation we assumed that the temperature was low, such that $\beta\mu$ was a large number and the Fermi distribution could be approximated by a step function. From (12.29) we see that this working assumption is justified in the case of weak interactions $|a| \to 0$, because then the critical temperature becomes exponentially small. In the weakly-interacting limit we can also ignore selfenergy effects, such that the zero-temperature ideal gas result $\mu = \hbar^2(3\pi^2 n)^{2/3}/2m \equiv \varepsilon_\mathrm{F}$ is justified to a very good approximation, where n is the total number of particles. An effect that cannot be ignored, even in the weakly-interacting limit, is the screening of the interaction by the bubble diagram which we encountered in Sect. 8.7. It turns out that this effect reduces the critical temperature by another factor of 2.2, which is also called the Gor'kov correction [82, 83]. In Sect. 14.3.1 we find that screening effects naturally enter the treatment of the interacting Fermi gas with the use of renormalization-group techniques, such that we then come back to the Gor'kov correction. The powerful renormalization-group approach can also be applied to the strongly-interacting regime, when the critical temperature becomes on the order of ε_F, and when selfenergy effects are important.

It is left as an exercise to calculate the coefficient of the fourth-order term $\beta(T)$ in the expansion for the Landau free energy in the weakly-interacting limit. It is given by

$$\beta(T) = \frac{\mathscr{D}(\varepsilon_F)}{V} \frac{7\zeta(3)}{16(\pi k_B T_c)^2}, \tag{12.30}$$

where $\mathscr{D}(\varepsilon_F) = mk_F V/(2\pi^2\hbar^2)$ is the density of states for a single spin state at the Fermi-energy $\varepsilon_F = \hbar^2 k_F^2/2m$. As a result, $\beta(T)$ is always positive, validating our discussion of a second-order phase transition.

12.6 Gap Equation

To perform a fully selfconsistent BCS theory, we can perform a treatment analogous to the treatment for Bose-Einstein condensation of Sect. 11.3.1, where we start with the fluctuation expansion $\Delta(\mathbf{x},\tau) = \Delta + \Delta'(\mathbf{x},\tau)$ and demand that the terms linear in the fluctuations are zero to obtain a selfconsistent equation for Δ. If the linear terms in the fluctuations are zero then we have performed the fluctuation expansion around the minimum of the action, giving rise to the correct equilibrium expectation value. To this end, it is useful to slightly reformulate the Green's function from (12.10) as

$$\mathbf{G}^{-1}(\mathbf{x},\tau;\mathbf{x}',\tau') = \mathbf{G}_\Delta^{-1}(\mathbf{x},\tau;\mathbf{x}',\tau') - \mathbf{\Sigma}_\Delta(\mathbf{x},\tau;\mathbf{x}',\tau'), \tag{12.31}$$

where the BCS propagator is defined through

$$\mathbf{G}_\Delta^{-1}(\mathbf{x},\tau;\mathbf{x}',\tau') = \tag{12.32}$$
$$\begin{bmatrix} G_{0;\uparrow}^{-1}(\mathbf{x},\tau;\mathbf{x}',\tau') & -\Delta\delta(\mathbf{x}-\mathbf{x}')\delta(\tau-\tau')/\hbar \\ -\Delta^*\delta(\mathbf{x}-\mathbf{x}')\delta(\tau-\tau')/\hbar & -G_{0;\downarrow}^{-1}(\mathbf{x}',\tau';\mathbf{x},\tau) \end{bmatrix},$$

whereas the new selfenergy matrix now only contains the fluctuations of the pairing field

$$\mathbf{\Sigma}_\Delta(\mathbf{x},\tau;\mathbf{x}',\tau') = \frac{1}{\hbar}\begin{bmatrix} 0 & \Delta'(\mathbf{x},\tau) \\ \Delta'^*(\mathbf{x},\tau) & 0 \end{bmatrix}\delta(\mathbf{x}-\mathbf{x}')\delta(\tau-\tau')\,. \tag{12.33}$$

Following the same lines as the discussion from Sect. 12.4, we find from the action of (12.13) that the linear terms in the fluctuations are given by

$$S[\Delta'^*,\Delta'] = \hbar\mathrm{Tr}[\mathbf{G}_\Delta\mathbf{\Sigma}_\Delta] - \int_0^{\hbar\beta} \mathrm{d}\tau \int \mathrm{d}\mathbf{x} \frac{\Delta'(\mathbf{x},\tau)\Delta^* + \Delta'^*(\mathbf{x},\tau)\Delta}{V_0}. \tag{12.34}$$

The condition for the terms linear in $\Delta'^*(\mathbf{x},\tau)$ to vanish, yields therefore

$$\int_0^{\hbar\beta} \mathrm{d}\tau \int \mathrm{d}\mathbf{x} \left\{ -\frac{\Delta}{V_0} + G_{\Delta;12}(\mathbf{x},\tau;\mathbf{x},\tau) \right\} \Delta'^*(\mathbf{x},\tau) = 0 \tag{12.35}$$

with a similar equation for $\Delta'(\mathbf{x},\tau)$. From (12.35), we thus find the following self-consistency equation for the BCS order parameter $\Delta \equiv \langle\Delta(\mathbf{x},\tau)\rangle$

$$\Delta = V_0 G_{\Delta;12}(\mathbf{x},\tau;\mathbf{x},\tau) = V_0 \langle \phi_\downarrow(\mathbf{x},\tau)\phi_\uparrow(\mathbf{x},\tau)\rangle, \tag{12.36}$$

which we can also write as

$$\Delta = \frac{V_0}{\hbar\beta V}\sum_{\mathbf{k},n} G_{\Delta;12}(\mathbf{k},\mathrm{i}\omega_n), \tag{12.37}$$

where we still have to determine $G_{\Delta;12}(\mathbf{k},\mathrm{i}\omega_n)$. This equation is also known as the BCS gap equation, for reasons which becomes clear soon. The off-diagonal element $G_{\Delta;12}(\mathbf{k},\mathrm{i}\omega_n)$ can be obtained by inverting the inverse BCS propagator in Fourier space, given by

$$-\hbar\mathbf{G}_\Delta^{-1}(\mathbf{k},\mathrm{i}\omega_n) = \begin{bmatrix} -\mathrm{i}\hbar\omega_n + \varepsilon_\mathbf{k} - \mu & \Delta \\ \Delta^* & -(\mathrm{i}\hbar\omega_n + \varepsilon_\mathbf{k} - \mu) \end{bmatrix}, \tag{12.38}$$

which gives rise to

$$\mathbf{G}_\Delta(\mathbf{k},\mathrm{i}\omega_n) = \frac{-\hbar}{(\hbar\omega_n)^2 + (\hbar\omega_\mathbf{k})^2}\begin{bmatrix} \mathrm{i}\hbar\omega_n + \varepsilon_\mathbf{k} - \mu & \Delta \\ \Delta^* & \mathrm{i}\hbar\omega_n - (\varepsilon_\mathbf{k} - \mu) \end{bmatrix}, \tag{12.39}$$

where we also introduced the notation $\hbar\omega_\mathbf{k} = \sqrt{(\varepsilon_\mathbf{k} - \mu)^2 + |\Delta|^2}$.

As a result, we find that there are poles in the Green's function $\mathbf{G}_\Delta(\mathbf{k},\omega)$, when $\hbar\omega = \pm\hbar\omega_\mathbf{k}$. As we see in the next section, these (eigen)values describe the time dependence of the Grassmann fields ψ_α and ψ_α^* respectively, corresponding to the Bogoliubov quasiparticles of the superfluid state. Therefore, the dispersion for the Bogoliubov quasiparticles is given by

$$\hbar\omega = \hbar\omega_\mathbf{k} = \sqrt{(\varepsilon_\mathbf{k} - \mu)^2 + |\Delta|^2}, \tag{12.40}$$

which describes the fermionic single-particle excitations in the presence of a Bose-Einstein condensate of Cooper pairs. Note that the minimum of the excitation spectrum, which is located at $\varepsilon_\mathbf{k} = \mu$, is given by $|\Delta|$ such that the excitation spectrum is gapped. As a result, it costs a nonzero amount of energy to make an elementary excitation. The physical interpretation of this result is that the minimum amount of energy needed to break up a Cooper pair is equal to 2Δ. Note that in Sect. 11.2.1, we saw that the application of the Landau criterion to a gapped dispersion indeed leads to a nonzero critical velocity below which the quantum gas behaves as a superfluid.

Now that we have obtained an expression for $G_{\Delta;12}(\mathbf{k},\mathrm{i}\omega_n)$, we can substitute this result into (12.37) to obtain

$$\begin{aligned} -\frac{1}{V_0} &= \frac{1}{\hbar\beta V}\sum_{\mathbf{k},n}\frac{\hbar}{(\hbar\omega_n)^2 + (\hbar\omega_\mathbf{k})^2} \\ &= \frac{1}{2\hbar^2\beta V}\sum_{\mathbf{k},n}\frac{1}{\omega_\mathbf{k}}\left\{\frac{\hbar}{-\mathrm{i}\hbar\omega_n + \hbar\omega_\mathbf{k}} + \frac{\hbar}{\mathrm{i}\hbar\omega_n + \hbar\omega_\mathbf{k}}\right\} = \frac{1}{V}\sum_\mathbf{k}\frac{1 - 2N_\mathrm{FD}(\hbar\omega_\mathbf{k})}{2\hbar\omega_\mathbf{k}}, \end{aligned} \tag{12.41}$$

with $N_{\rm FD}(\hbar\omega_{\mathbf{k}}) = 1/\{\exp(\beta\hbar\omega_{\mathbf{k}})+1\}$ the Fermi-Dirac distribution for the quasi-particles. Note that in the last step it does not matter which convergence factor we use to perform the Matsubara sum, because either choice gives the same result. This is because $G_{\Delta;12}(\mathbf{x},\tau;\mathbf{x},\tau')$, unlike $G_{\Delta;11}(\mathbf{x},\tau;\mathbf{x},\tau')$, does not contain a discontinuity at equal times, which simply reflects the fact that $\hat{\psi}_\alpha(\mathbf{x},\tau)$ and $\hat{\psi}_{-\alpha}(\mathbf{x},\tau')$ anticommute at equal times whereas $\hat{\psi}_\alpha(\mathbf{x},\tau)$ and $\hat{\psi}^\dagger_\alpha(\mathbf{x},\tau')$ do not. To relate the interaction potential to the scattering length, which is the experimentally known quantity, we use (10.54) to finally obtain

$$-\frac{1}{T^{\rm 2B}} = \frac{1}{V}\sum_{\mathbf{k}}\left\{\frac{1-2N_{\rm FD}(\hbar\omega_{\mathbf{k}})}{2\hbar\omega_{\mathbf{k}}} - \frac{1}{2\varepsilon_{\mathbf{k}}}\right\}. \tag{12.42}$$

This is indeed an equation for the gap or BCS order parameter $|\Delta|$, and represents the gap equation in its most practical form. Note that at the BCS transition temperature, we have that $\Delta = 0$ and we recover the equation for $T_{\rm c}$. Given μ, a and T, we can solve the gap equation for the order parameter $|\Delta|$.

It is instructive to study (12.42) first in the zero-temperature limit, such that $N_{\rm FD}(\hbar\omega_{\mathbf{k}}) = 0$. Then, we obtain for the gap equation

$$\begin{aligned}\frac{m}{4\pi|a|\hbar^2} &= \frac{1}{V}\sum_{\mathbf{k}}\left\{\frac{1}{2\sqrt{(\varepsilon_{\mathbf{k}}-\mu)^2+|\Delta|^2}} - \frac{1}{2\varepsilon_{\mathbf{k}}}\right\}\\ &= \frac{2\sqrt{2}}{(2\pi)^2}\frac{m^{3/2}}{\hbar^3}\int_0^\infty d\varepsilon\sqrt{\varepsilon}\left\{\frac{1}{2\sqrt{(\varepsilon-\mu)^2+|\Delta|^2}} - \frac{1}{2\varepsilon}\right\},\end{aligned} \tag{12.43}$$

which we may write as

$$\frac{\pi}{k_{\rm F}|a|} = \int_0^\infty dx\,\sqrt{x}\left\{\frac{1}{\sqrt{(x-1)^2+|\Delta/\varepsilon_{\rm F}|^2}} - \frac{1}{x}\right\}, \tag{12.44}$$

where $x = \varepsilon/\varepsilon_{\rm F}$ is the dimensionless integration variable. Note that we have used $\mu \simeq \varepsilon_{\rm F}$ again, which is the result for the noninteracting gas and thus only valid in the weakly-interacting limit, when $|\Delta|$ is small as we see next. The integral on the right-hand side can be performed as follows

$$\begin{aligned}&\int_0^\infty dx\,\sqrt{x}\left[\frac{1}{\sqrt{(x-1)^2+|\Delta/\varepsilon_{\rm F}|^2}} - \frac{1}{x}\right] \\ &\simeq \int_0^\infty dx\,(\sqrt{x}-1)\left[\frac{1}{\sqrt{(x-1)^2}} - \frac{1}{x}\right] + \int_0^\infty dx\left[\frac{1}{\sqrt{(x-1)^2+|\Delta/\varepsilon_{\rm F}|^2}} - \frac{1}{x}\right]\\ &= 5\log 2 - 4 - \log\left(-1+\sqrt{1+|\Delta/\varepsilon_{\rm F}|^2}\right) \simeq 6\log 2 - 4 - 2\log(|\Delta/\varepsilon_{\rm F}|),\end{aligned} \tag{12.45}$$

for small $|\Delta/\varepsilon_{\rm F}|$. Substituting this into (12.44), we finally obtain

$$|\Delta| = \frac{8}{e^2} \varepsilon_F e^{-\pi/2k_F|a|}. \tag{12.46}$$

Note that the critical temperature is thus related to the zero-temperature value of the gap by the convenient relation $k_B T_c = (e^\gamma/\pi)|\Delta|$, which is a universal result of BCS theory.

12.7 Thermodynamic Potential for Fermions

To complete our study of the thermodynamic properties of the BCS superfluid, we calculate the mean-field thermodynamic potential. For a homogeneous system, the mean-field action resulting from (12.9), which is quadratic in the atomic fields, can be written in momentum space as

$$\begin{aligned} S[\phi^*, \phi] = & -\hbar\beta V \frac{|\Delta|^2}{V_0} + \hbar\beta \sum_{\mathbf{k}} (\varepsilon_{\mathbf{k}} - \mu) \\ & -\hbar \sum_{\mathbf{k},n} \left[\phi^*_{\mathbf{k},n,\uparrow}, \phi_{-\mathbf{k},-n,\downarrow}\right] \cdot \mathbf{G}_\Delta^{-1}(\mathbf{k}, i\omega_n) \cdot \begin{bmatrix} \phi_{\mathbf{k},n,\uparrow} \\ \phi^*_{-\mathbf{k},-n,\downarrow} \end{bmatrix}, \end{aligned} \tag{12.47}$$

where the sum $\sum_{\mathbf{k}}(\varepsilon_{\mathbf{k}} - \mu)$ comes from interchanging the fermionic spin-down fields and the corresponding equal-time limiting procedure in order to write the above action in matrix form. This was more extensively explained in Sect. 11.4. As we show now, it is possible to diagonalize the above action by making a Bogoliubov transformation of the atomic fields, very similar to the transformation we performed for the Bose-Einstein condensed phase of the Bose gas. We introduce the fields $\psi_{\mathbf{k},n,\alpha}$, with $\alpha = \uparrow, \downarrow$, which are related to the fields $\phi_{\mathbf{k},n,\alpha}$ by means of the unitary transformation

$$\begin{bmatrix} \psi_{\mathbf{k},n,\uparrow} \\ \psi^*_{-\mathbf{k},-n,\downarrow} \end{bmatrix} = \begin{bmatrix} u_{\mathbf{k}} & -v_{\mathbf{k}} \\ v^*_{\mathbf{k}} & u^*_{\mathbf{k}} \end{bmatrix} \cdot \begin{bmatrix} \phi_{\mathbf{k},n,\uparrow} \\ \phi^*_{-\mathbf{k},-n,\downarrow} \end{bmatrix}. \tag{12.48}$$

Note the minus sign that is absent in the bosonic case. This transformation is indeed unitary if

$$|u_{\mathbf{k}}|^2 + |v_{\mathbf{k}}|^2 = 1, \tag{12.49}$$

which, in the operator formalism, ensures that the operators associated with the fields $\psi^\dagger_{\mathbf{k},n,\alpha}$ and $\psi_{\mathbf{k},n,\alpha}$ still obey the anticommutation relations for fermionic creation and annihilation operators. Using this normalization, we find that the inverse transformation is given by

$$\begin{bmatrix} \phi_{\mathbf{k},n,\uparrow} \\ \phi^*_{-\mathbf{k},-n,\downarrow} \end{bmatrix} = \begin{bmatrix} u^*_{\mathbf{k}} & v_{\mathbf{k}} \\ -v^*_{\mathbf{k}} & u_{\mathbf{k}} \end{bmatrix} \cdot \begin{bmatrix} \psi_{\mathbf{k},n,\uparrow} \\ \psi^*_{-\mathbf{k},-n,\downarrow} \end{bmatrix}. \tag{12.50}$$

Inserting this Bogoliubov transformation into (12.47) and demanding that the resulting expression is diagonal in terms of the new fields, we have the condition that the coefficient of $\psi^*_{\mathbf{k},n,\alpha}\psi^*_{-\mathbf{k},-n,-\alpha}$ has to be equal to zero, which gives

$$2(\varepsilon_{\mathbf{k}}-\mu)u_{\mathbf{k}}v_{\mathbf{k}}-v_{\mathbf{k}}^2\Delta^*+u_{\mathbf{k}}^2\Delta=0. \tag{12.51}$$

For the same reason, the coefficient of $\psi_{\mathbf{k},n,\alpha}\psi_{-\mathbf{k},-n,-\alpha}$ has to be zero, which leads to the complex conjugate of the above equation. We have that that (12.49) and (12.51) are solved by

$$|u_{\mathbf{k}}|^2=1-|v_{\mathbf{k}}|^2=\frac{1}{2}\left(1+\frac{\varepsilon_{\mathbf{k}}-\mu}{\hbar\omega_{\mathbf{k}}}\right), \tag{12.52}$$

where $\hbar\omega_{\mathbf{k}}$ is the quasiparticle dispersion from (12.40). Moreover, it is useful to note that

$$|u_{\mathbf{k}}|^2-|v_{\mathbf{k}}|^2=2|u_{\mathbf{k}}|^2-1=\frac{\varepsilon_{\mathbf{k}}-\mu}{\hbar\omega_{\mathbf{k}}}=-\frac{\mathrm{d}\hbar\omega_{\mathbf{k}}}{\mathrm{d}\mu}. \tag{12.53}$$

In terms of the Bogoliubov fields, the action in (12.47) thus becomes

$$\begin{aligned} S[\psi^*,\psi] &= -\hbar\beta V\frac{|\Delta|^2}{V_0}+\hbar\beta\sum_{\mathbf{k}}(-\hbar\omega_{\mathbf{k}}+\varepsilon_{\mathbf{k}}-\mu) \\ &\quad+\sum_{\alpha,\mathbf{k},n}(-\mathrm{i}\hbar\omega_n+\hbar\omega_{\mathbf{k}})\,\psi^*_{\mathbf{k},n,\alpha}\psi_{\mathbf{k},n,\alpha}, \end{aligned} \tag{12.54}$$

where the additional term $\hbar\omega_{\mathbf{k}}$ in the first sum again comes about because we have interchanged the Bogoliubov fields $\psi^*_\downarrow$ and $\psi_\downarrow$ together with the corresponding equal-time limiting procedure. In terms of the Bogoliubov fields, the effective action is diagonal and it is therefore straightforward to determine the thermodynamic potential Ω. Eliminating also the interaction parameter for the two-body scattering matrix, we ultimately find

$$\frac{\Omega}{V}=-\frac{|\Delta|^2}{T^{2\mathrm{B}}}+\frac{1}{V}\sum_{\mathbf{k}}\left\{-\hbar\omega_{\mathbf{k}}+\varepsilon_{\mathbf{k}}-\mu+\frac{|\Delta|^2}{2\varepsilon_{\mathbf{k}}}\right\} \tag{12.55}$$

$$-\frac{2}{\beta V}\sum_{\mathbf{k}}\log\left(1+\mathrm{e}^{-\beta\hbar\omega_{\mathbf{k}}}\right). \tag{12.56}$$

The thermodynamic potential can now be used to calculate the density using $\langle N\rangle = -\mathrm{d}\Omega/\mathrm{d}\mu$. Working out the derivative and using (12.53), we obtain

$$n=\frac{2}{V}\sum_{\mathbf{k}}\left\{\left(|u_{\mathbf{k}}|^2-|v_{\mathbf{k}}|^2\right)N_{\mathrm{FD}}(\hbar\omega_{\mathbf{k}})+|v_{\mathbf{k}}|^2\right\}, \tag{12.57}$$

which can be compared with (11.37), (11.46) and (11.55) for the Bose case.

12.8 The BEC-BCS Crossover

Consider an ultracold Fermi gas at zero temperature under the influence of a weak attractive interacting potential that does not support a two-body bound state, and which gives rise to a small negative scattering length. Then we can use BCS theory to calculate the BCS order parameter Δ, which is exponentially small as shown in (12.46). In analogy with the macroscopic wave function for the Bose-Einstein condensed Bose gas $\phi_0(\mathbf{x}) = \langle\phi(\mathbf{x},\tau)\rangle$, we can define the macroscopic Cooper-pair wave function $\phi_0(\mathbf{x}-\mathbf{x}')$ as

$$\begin{aligned}\phi_0(\mathbf{x}-\mathbf{x}') &= \langle\phi_\downarrow(\mathbf{x},\tau)\phi_\uparrow(\mathbf{x}',\tau)\rangle = G_{\Delta;12}(\mathbf{x}',\tau;\mathbf{x},\tau)\\ &= \frac{1}{V\hbar\beta}\sum_{\mathbf{k},n} G_{\Delta;12}(\mathbf{k},\mathrm{i}\omega_n)\mathrm{e}^{-\mathrm{i}\mathbf{k}(\mathbf{x}-\mathbf{x}')} = -\frac{1}{V}\sum_{\mathbf{k}}\frac{\Delta}{2\hbar\omega_{\mathbf{k}}}\mathrm{e}^{-\mathrm{i}\mathbf{k}(\mathbf{x}-\mathbf{x}')}\\ &= -\frac{1}{V}\sum_{\mathbf{k}} u_{\mathbf{k}}v_{\mathbf{k}}\mathrm{e}^{-\mathrm{i}\mathbf{k}(\mathbf{x}-\mathbf{x}')},\end{aligned} \tag{12.58}$$

where we used (12.36), (12.39) and (12.52). In the weakly-interacting case, the Fourier transform $u_{\mathbf{k}}v_{\mathbf{k}}$ of the Cooper-pair wavefunction is sharply peaked around the Fermi momentum k_{F} with a width that is on the order of $m|\Delta|/\hbar^2 k_{\mathrm{F}}$. This means that the spatial extent of the Cooper-pair wavefunction is on the order of $\hbar^2 k_{\mathrm{F}}/m|\Delta|$, which can thus be interpreted as the average size of a Cooper pair. As a result this average size of the Cooper pairs is typically very large in the weakly-interacting limit, much larger than the interatomic distance between fermions, showing the exotic nature of the pairs in the BCS limit.

Now suppose that we can manipulate the interatomic interaction potential, such that we can make the attraction stronger, leading to a larger gap and a smaller size of the Cooper pairs. If we keep increasing the strength of the attractive interacting potential, it will at some point give rise to a two-body bound state in the potential. From Sect. 10.3.1, we know that then the scattering length diverges. By making the interaction even more attractive the molecular bound state becomes more deeply bound, such that its size becomes smaller. In this limit, the zero-temperature ground state consists of a Bose-Einstein condensate of tightly-bound diatomic molecules, whose size is much smaller than the average interparticle distance. This is called the BEC limit. We note that the two limits are at first sight physically very different, because for example the bosonic molecules in the BEC limit are already stable at the two-body level, whereas the stability of the pairs in the BCS limit is truly a many-body effect caused by the presence of a Fermi sea. In Fig. 12.1, we show the relative two-body wavefunctions $\chi_0(\mathbf{r})$ for the condensed pairs both in the BCS regime and in the BEC regime, where the latter leads to

$$\chi_0(\mathbf{r}) = -\frac{\mathrm{e}^{-r/a}}{r\sqrt{2\pi a}}, \tag{12.59}$$

as follows from (10.48) and (10.51). The question now arises how this evolution between the extreme BCS limit of loosely bound Cooper pairs and the extreme BEC limit of tightly confined bosonic molecules actually takes place.

12.8.1 Theoretical Results

As first realized by Eagles [24] and later by Leggett [25], the evolution from the BEC to the BCS limit is smooth, meaning that there is no symmetry change or other reason for nonanalytic thermodynamic behavior as the Bose-Einstein condensed pairs become more tightly bound. As a result this evolution is known as the BEC-BCS crossover. This also implies that the wavefunction from (12.1) and the corresponding BCS theory can be used to qualitatively describe the whole crossover. Before we do this we remark that the BEC-BCS crossover can actually be studied experimentally with the use of ultracold atomic Fermi gases, since here the change in the interaction potential can be mimicked with the use of a Feshbach resonance, which allows for the control of the scattering length by applying an external magnetic field. This important tool for manipulating ultracold atomic quantum gases is discussed extensively in Chap. 17.

At zero temperature and at a fixed particle density $n = 2n_\alpha \equiv k_F^3/3\pi^2$, we can study the BEC-BCS crossover as a function of the dimensionless parameter $1/k_Fa$ by solving for each value of $1/k_Fa$ both the equation for the density

$$n = \frac{1}{V}\sum_{\mathbf{k}}\left\{1 - \frac{\varepsilon_{\mathbf{k}} - \mu}{\hbar\omega_{\mathbf{k}}}\right\}, \tag{12.60}$$

and the equation for the gap parameter

$$-\frac{m}{4\pi\hbar^2 a} = \frac{1}{V}\sum_{\mathbf{k}}\left\{\frac{1}{2\hbar\omega_{\mathbf{k}}} - \frac{1}{2\varepsilon_{\mathbf{k}}}\right\}, \tag{12.61}$$

which follow from (12.42) and (12.57) by taking the zero-temperature limit. By converting the above sums into integrals and numerically solving the two equations to find both the chemical potential μ and the BCS order parameter Δ, we obtain the results shown in Fig. 12.1. In the extreme BCS limit, $1/k_Fa \to -\infty$, we retrieve the analytic exponential behavior of the gap from (12.46) together with $\mu = \varepsilon_F$, as expected for the weakly-interacting Fermi gas.

Upon increase of the interaction strength $|a|$ we enter the strongly-interacting regime $1/k_F|a| < 1$, where there is no longer a natural small parameter in the theory, such that perturbation theory, and in particular mean-field theory, is no longer expected to hold quantitatively. This is especially true for the unitarity limit, $1/k_F|a| \to 0$, when the scattering length diverges. It is interesting to note that in this limit, the scattering length no longer yields a physical length scale, and the thermodynamics cannot depend on a anymore. As a result, thermodynamic quantities only

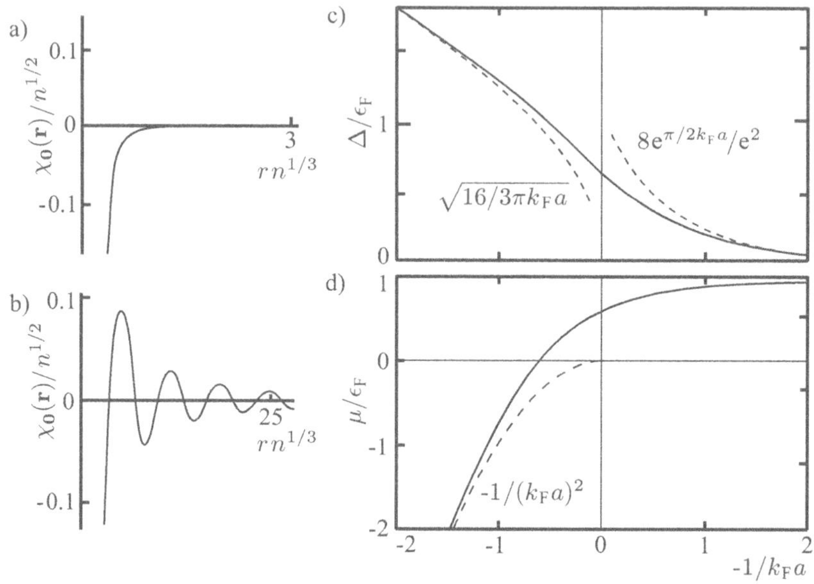

Fig. 12.1 The BEC-BCS crossover at zero temperature. a) Two-body condensed pair wavefunction $\chi_0(\mathbf{r})$ in the BEC regime, $1/k_F a = 3$, where the pairs are smaller than the interparticle distance $n^{-1/3}$. b) Two-body condensed pair wavefunction $\chi_0(\mathbf{r})$ in the BCS regime, $1/k_F a = -3$, where the pairs are larger than $n^{-1/3}$. c) The order parameter Δ and d) the chemical potential μ as a function of $-1/k_F a$ at a fixed particle density $n = k_F^3/3\pi^2$. The dashed lines show the analytic results in the BEC and BCS limit, respectively. Note that there is no nonanalytic behavior, indicating that mean-field BCS theory predicts a smooth crossover as a function of $1/k_F a$.

depend on the Fermi energy ε_F, such that we have for example $\mu = (1+\beta)\varepsilon_F$ and $\Delta = \xi\varepsilon_F$ where β an ξ are dimensionless constants which are not to be confused with the inverse temperature and the correlation length. This statement is called the universality hypothesis [84], and it has been tested by experiments and Monte-Carlo calculations [85, 86, 87]. Both have converged to $\beta \simeq -0.6$ and $\xi \simeq 0.5$, where the latter shows that the gap becomes on the order of the Fermi energy. We can also calculate these quantities within BCS theory, leading to $\beta = -0.4$ and $\xi = 0.7$, as can be read off from Fig. 12.1. As expected, BCS theory is quantitatively not fully correct at unitarity. However, even in the strongly-interacting regime it gives answers that are on the right order of magnitude.

Upon entering the BEC regime, i.e. $1/k_F a > 1$, a two-body bound state has entered the interaction potential with a molecular binding energy given by $E_m = -\hbar^2/ma^2$ as explained in Sect. 10.3.1. The chemical potential describing the condensation of diatomic molecules is expected to behave as $2\mu \simeq E_m$, which is indeed the behavior observed in Fig. 12.1d. Moreover, the gap behaves in this regime as $\Delta \simeq (16/3\pi k_F a)^{1/2}\varepsilon_F$. This can be understood by considering the two-body wavefunction from (12.59) and from the general expression for the gap as the integral of the (macroscopic) paired condensate wavefunction multiplied by the interaction, i.e.

$$\Delta = \int \mathrm{d}\mathbf{r}\, V(\mathbf{r}) \sqrt{\frac{N}{2V}} \chi_0(\mathbf{r}) = \int \mathrm{d}\mathbf{r}\, V(\mathbf{r}) \left(1 - \frac{a}{r}\right) \frac{\sqrt{n}}{a\sqrt{4\pi a}}$$
$$= \frac{4\pi a \hbar^2}{m} \frac{\sqrt{n}}{a\sqrt{4\pi a}} = \sqrt{\frac{16}{3\pi k_\mathrm{F} a}} \varepsilon_\mathrm{F}, \tag{12.62}$$

where in the first step we used that the number of pairs is $N/2$, while the center-of-mass part of the Cooper pair wavefunction is $1/\sqrt{V}$, because the pairs have no momentum. In the second step, we used that the interaction potential is short ranged, allowing us to expand the relative wavefunction for the small separations r of interest. In the third step we used the definition of the two-body T matrix from (10.15), which we can write as

$$\lim_{\mathbf{k}\to\mathbf{0}} \int \mathrm{d}\mathbf{r}\, V(\mathbf{r}) \psi_{\mathbf{k}}^{(+)}(\mathbf{r}) = \lim_{\mathbf{k}\to\mathbf{0}} \int \mathrm{d}\mathbf{r}\, T^{2\mathrm{B}}(\mathbf{r}) \mathrm{e}^{\mathrm{i}\mathbf{k}\mathbf{r}} = \frac{4\pi a \hbar^2}{m}, \tag{12.63}$$

where, for small $\mathbf{r}$ and $\mathbf{k}$, we have $\psi_{\mathbf{k}}^{(+)}(\mathbf{r}) = 1 - a/r$ as follows from (10.27), (10.34), and (10.35).

We may thus conclude that, at zero temperature, BCS theory leads to very reasonable results for the theoretical study of the BEC-BCS crossover. However, at nonzero temperatures, the results become less reliable. In particular, the study of the critical temperature as a function of $1/k_\mathrm{F}a$ leads to an incorrect behavior in the BEC limit. The reason for this is that the critical temperature obtained from BCS theory describes physically the temperature at which pairs break up into separate fermions. However, deep in the BEC limit when the molecules are tightly bound, the breakdown of superfluidity is caused not by the break-up of pairs but rather by the thermal occupation of the nonzero momentum states of the molecules and the associated thermal depletion of the Bose-Einstein condensate. Therefore the correct behavior of the critical temperature can only be obtained by incorporating also the effect of noncondensed pairs, the fluctuations of the pairing field $\Delta'(\mathbf{x},\tau)$, which are completely neglected in BCS theory. Taking fluctuations into account up to quadratic order around the normal state solution $\Delta = 0$ is called the Nozières-Schmitt-Rink approximation [88], which leads to an additional term Ω_NSR in the thermodynamic potential

$$\Omega_\mathrm{NSR} = -\frac{1}{\beta} \log\left(\int \mathrm{d}[\Delta'^*]\mathrm{d}[\Delta'] \mathrm{e}^{(\Delta'|G_{\Delta'}^{-1}|\Delta')} \right) = \frac{1}{\beta} \mathrm{Tr}[\log(-G_{\Delta'}^{-1})], \tag{12.64}$$

where we have from (12.13) and (12.17) that in real space $G_{\Delta'}^{-1}$ is given by

$$G_{\Delta'}^{-1} = \frac{1}{\hbar V_0} \delta(\mathbf{x}-\mathbf{x}')\delta(\tau-\tau') + \frac{1}{\hbar^2} G_{0;\uparrow}(\mathbf{x},\tau;\mathbf{x}',\tau') G_{0;\downarrow}(\mathbf{x},\tau;\mathbf{x}',\tau'), \tag{12.65}$$

which yields the following expression for Ω_NSR in momentum space

$$\Omega_{\rm NSR} = \frac{1}{\beta}\sum_{\mathbf{K},n}\log\left(-\frac{1}{\hbar T^{2\rm B}} - \frac{1}{\hbar V}\sum_{\mathbf{k}'}\left\{\frac{1-N_{\rm FD}(\varepsilon_{\mathbf{K}-\mathbf{k}'})-N_{\rm FD}(\varepsilon_{\mathbf{k}'})}{-i\hbar\Omega_n+\varepsilon_{\mathbf{K}-\mathbf{k}'}+\varepsilon_{\mathbf{k}'}-2\mu} - \frac{1}{2\varepsilon_{\mathbf{k}'}}\right\}\right), \tag{12.66}$$

where we used the result for the fermionic ladder diagram obtained in Exercise 10.2. The additional Nozières-Schmitt-Rink term in the thermodynamic potential then changes the number equation calculated from $n = \partial\Omega/\partial\mu$, which physically stems from taking also noncondensed pairs into account. This consequently modifies the ratio T_c/T_F, and gives a rather good description of the critical temperature throughout the whole crossover [81].

12.8.2 Comparison with Experiment

So far, we have discussed BCS theory and the BEC-BCS crossover for the homogeneous interacting Fermi gas. However, to perform an actual experiment, the atoms always need to be confined in space and be kept away from material walls. This can be done by applying an electric or magnetic field to create an external trapping potential. The easiest situation in which we can study the effects of an external potential is when the local-density approximation applies, which we already encountered in Sect. 4.3.1.2. In this approximation the external potential is simply absorbed in the chemical potential, such that we can locally perform the homogeneous theory, however, with a spatially varying chemical potential. Since we have already obtained the homogeneous theory, this procedure thus conveniently implies that we have also automatically obtained the inhomogeneous theory. Such an approach is only valid if the correlation length of the gas is smaller than the typical length scale associated with variations of the external potential. We found previously that the typical size of the correlated Cooper pairs is $\hbar^2 k_F/m|\Delta|$. In the strongly-interacting regime, when Δ is on the order of ε_F, we find for the correlation length $\xi \simeq 1/k_F$. In this case the local-density approximation is indeed well justified for realistic trapped gases. Another way to obtain the same condition is by thinking of the local-density approximation as a WKB approximation. The WKB approximation is valid when the de Broglie wavelength for a particle is small compared to the length scale over which the potential varies. For degenerate fermions, the typical de Broglie wavelength of the atoms is also given by $1/k_F$. Finally, if the local-density approximation fails, then we need to perform a truly inhomogeneous calculation. This can be done by using the Bogoliubov-de Gennes formalism from Sect. 11.5, which can be immediately generalized to the fermionic case.

As mentioned in the introductory chapter, many experiments have been performed recently on the BEC-BCS crossover using ultracold atomic gases [18, 19, 20, 21, 22, 23]. Here. we highlight the measurement of the density profiles for condensed ^{6}Li clouds as a function of the interaction strength by Bartenstein et al. [21]. The results are shown in Fig. 12.2. For ^{6}Li the scattering length a diverges at 834 G, where larger magnetic fields correspond to the BCS side, while lower magnetic fields correspond to the BEC side. We remark that all profiles in Fig. 12.2

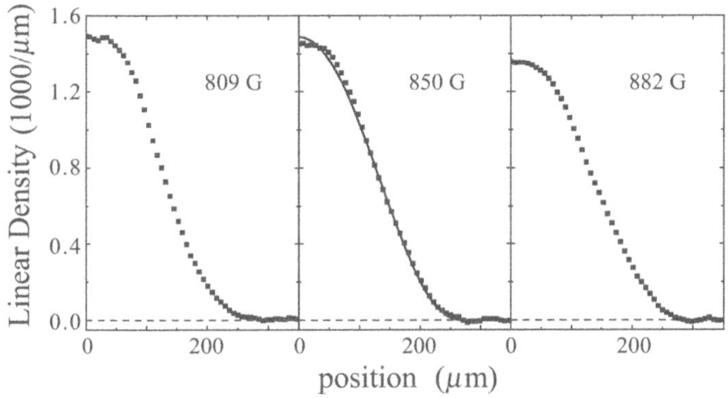

Fig. 12.2 Measured axial density profiles of atomic ^{6}Li clouds in the strongly-interacting regime. At 834 G the scattering length diverges, where for lower/higher magnetic field strengths the BEC/BCS side is approached. The evolution as a function of magnetic field is smooth, indicating a crossover, while at 850 G the predicted unitary density profile is shown by the solid line. Reprinted figure with permission from M. Bartenstein, A. Altmeyer, S. Riedl, S. Jochim, C. Chin, J. Hecker Denschlag, and R. Grimm, Phys. Rev. Lett. **92**, 203201 (2004). Copyright 2008 by the American Physical Society.

are actually in the strongly-interacting regime, for which $|k_F a| > 1$. Note that the three density profiles look both qualitatively and quantitatively very similar, which strongly suggests that there is a smooth evolution through the point where the scattering length diverges, a crossover rather than a phase transition. To study the inhomogeneous Fermi gas in the strongly-interacting regime we can use the universality hypothesis $\mu = (1+\beta)\varepsilon_{\rm F}$ in combination with the local-density approximation from Sect. 4.3.1.2. Determining the integrated density $n_z(z)$ along the z axis as in Sect. 4.3.3.1, we find in the unitarity limit

$$\begin{aligned} n_z(z) &\equiv \int_0^{2\pi} \mathrm{d}\varphi \int \mathrm{d}\rho \; n(\rho,z,\phi) = \int \rho \; \mathrm{d}\rho \; \frac{2}{3\pi}\left(\frac{2m\mu(\rho,z)}{(1+\beta)\hbar^2}\right)^{3/2} \\ &= \frac{2(1+\beta)}{15\pi m\omega_\rho^2}\left(\frac{m}{\hbar^2}\right)^{3/2}\left(\frac{2\mu - m\omega_z^2 z^2}{1+\beta}\right)^{5/2}, \end{aligned} \tag{12.67}$$

where we have an additional factor of 2 compared to (4.92) due to the presence of two spin species. Note that the strong attractions among the atoms act as an enhancement of the trapping potential and narrow the atomic cloud. The profile from (12.67) was compared with the measured axial profile in Fig. 12.2b, giving rise to the measured value $\beta \simeq -0.7 \pm 0.1$ in good agreement with the Monte-Carlo result of -0.6. Finally, we remark that the density profile from Fig. 4.5 was actually also measured in the superfluid state, but then in the BCS regime, for which $k_{\rm F}a < -1$ and the gap parameter is exponentially suppressed. As a result, the density profiles on the far BCS side are indistinguishable from the ideal gas profiles, making the superfluid state hard to detect. As mentioned in the introductory chapter, this problem can be overcome by either using Feshbach resonances to convert the BCS state into a BEC of molecules, or by using other means to probe superfluidity, such as rotating the gas to observe vortices [89].

12.9 Problems

Exercise 12.1. Prove (12.30). You can make use of the identity

$$\sum_n \frac{1}{|\hbar\omega_n|^3} = \frac{7\zeta(3)}{4}\frac{1}{(\pi k_B T)^3},$$

where ω_n are the fermionic Matsubara frequencies and $\zeta(3) \simeq 1.202$.

Exercise 12.2. Show by calculating the commutation relations of the Cooper pairs in second quantization that they behave very much like bosons, i.e. show that the commutator of the composite particle $b^\dagger_{\mathbf{k}} \equiv c^\dagger_{\mathbf{k},\uparrow}c^\dagger_{-\mathbf{k},\downarrow}$ is given by

$$[b_{\mathbf{k}}, b^\dagger_{\mathbf{k}'}]_- = \delta_{\mathbf{k},\mathbf{k}'}\left(1 - c^\dagger_{-\mathbf{k}',\downarrow}c_{-\mathbf{k},\downarrow} - c^\dagger_{\mathbf{k}',\uparrow}c_{\mathbf{k},\uparrow}\right). \tag{12.68}$$

Also calculate the commutators, $[b_{\mathbf{k}}, b_{\mathbf{k}'}]_-$ and $[b^\dagger_{\mathbf{k}}, b^\dagger_{\mathbf{k}'}]_-$. Note that in the functional integral approach, the order parameter introduced through the Hubbard-Stratonovich transformation is exactly a bosonic field.

Exercise 12.3. In principle, a BCS-like transition to a paired state could also occur in a Bose gas with attractive interactions, i.e. $V_{\mathbf{0}} < 0$. Show that the dispersion relation of the single-particle excitations is

$$\hbar\omega_{\mathbf{k}} = \sqrt{(\varepsilon_{\mathbf{k}} - \mu)^2 - |\Delta|^2}$$

in that case. What happens if $\mu = |\Delta|$?

Exercise 12.4. Imbalance in Spin Populations.
Recently, experimentalists have been able to study two-component Fermi gases as a function of spin polarization, which is defined as $P \equiv (N_\uparrow - N_\downarrow)/(N_\uparrow + N_\downarrow)$ where N_α is the number of particles in spin-state $|\alpha\rangle$. So for $P > 0$, we have more spin-up particles than spin-down. We can describe a system with a population imbalance by taking different chemical potentials μ_α for the two spin species, leading to the following action

$$S[\phi^*_\alpha, \phi_\alpha] = \int_0^{\hbar\beta} d\tau \int d\mathbf{x} \sum_{\alpha=\uparrow,\downarrow} \phi^*_\alpha(\mathbf{x},\tau)\left\{\hbar\partial_\tau - \frac{\hbar^2\nabla^2}{2m} - \mu_\alpha\right\}\phi_\alpha(\mathbf{x},\tau)$$
$$+ \int_0^{\hbar\beta} d\tau \int d\mathbf{x}\, V_{\mathbf{0}}\, \phi^*_\uparrow(\mathbf{x},\tau)\phi^*_\downarrow(\mathbf{x},\tau)\phi_\downarrow(\mathbf{x},\tau)\phi_\uparrow(\mathbf{x},\tau), \tag{12.69}$$

where we have to find a way to deal with the interaction term. An elegant way to do this, is by applying a Hubbard-Stratonovich transformation.

a) Perform a Hubbard-Stratonovich transformation to the fields Δ and Δ^*, such that Δ is on average given by

$$\langle \Delta(\mathbf{x},\tau) \rangle = V_{\mathbf{0}} \langle \phi_{\downarrow}(\mathbf{x},\tau)\phi_{\uparrow}(\mathbf{x},\tau) \rangle, \tag{12.70}$$

and show that the resulting action in terms of the fields Δ, Δ^*, ϕ^*_α, an ϕ_α can be written in the form

$$S[\Delta,\Delta^*,\phi^*_\alpha,\phi_\alpha] = -\int_0^{\hbar\beta} \mathrm{d}\tau \int \mathrm{d}\mathbf{x}\, \frac{|\Delta(\mathbf{x},\tau)|^2}{V_{\mathbf{0}}} - \hbar \int_0^{\hbar\beta} \mathrm{d}\tau\, \mathrm{d}\tau' \int \mathrm{d}\mathbf{x}\, \mathrm{d}\mathbf{x}' \times \left[\phi^*_\uparrow(\mathbf{x},\tau), \phi_\downarrow(\mathbf{x},\tau)\right] \cdot \mathbf{G}^{-1} \cdot \begin{bmatrix} \phi_\uparrow(\mathbf{x}',\tau') \\ \phi^*_\downarrow(\mathbf{x}',\tau') \end{bmatrix}. \tag{12.71}$$

Give the two-by-two matrix $\mathbf{G}^{-1}$ in terms of $G^{-1}_{0;\alpha}$, Δ and Δ^*.

If the polarization is not too large, then below a certain critical temperature, the gas is in a phase, where the field Δ has a nonzero expectation value $\langle \Delta(\mathbf{x},\tau) \rangle = \Delta$. In the following, we simply approximate the field $\Delta(\mathbf{x},\tau)$ by its average value Δ.
b) Transform the action from (12.71) to momentum space and frequency space. Show that the Green's function matrix has poles, when

$$\hbar\omega_{\mathbf{k}} = -h \pm \sqrt{(\varepsilon_{\mathbf{k}} - \mu)^2 + |\Delta|^2}, \tag{12.72}$$

where we introduced $\mu = (\mu_\uparrow + \mu_\downarrow)/2$ and $h = (\mu_\uparrow - \mu_\downarrow)/2$. Show that this means that the quasiparticle dispersions become

$$\hbar\omega_{\mathbf{k},\uparrow} = -h + \sqrt{(\varepsilon_{\mathbf{k}} - \mu)^2 + |\Delta|^2}. \tag{12.73}$$

and

$$\hbar\omega_{\mathbf{k},\downarrow} = h + \sqrt{(\varepsilon_{\mathbf{k}} - \mu)^2 + |\Delta|^2}. \tag{12.74}$$

It is interesting to note that for $h > |\Delta|$, the majority species has a gapless excitation spectrum. This exotic superfluid state is also called the superfluid Sarma phase.

Exercise 12.5. Josephson Junction
A Josephson junction is a layer of isolating material between two superconductors. Consider a circular shaped superconducting wire coupled by a Josephson junction as shown in Fig. 12.3, where a thick dot denotes the Josephson junction. Denote the radius of the circle by R and the angular coordinate along the wire by φ. The Josephson junction is located at $\varphi = 0$. In second quantization, the Hamiltonian of the system contains the usual kinetic and interaction terms for the Cooper pairs, but contains also a term describing effectively the tunnelling of Cooper pairs at the Josephson junction. The latter is given by

$$\hat{H}_\mathrm{T} = -\pi R t \left(\hat{\Delta}^\dagger(2\pi)\hat{\Delta}(0) + \hat{\Delta}^\dagger(0)\hat{\Delta}(2\pi)\right) , \tag{12.75}$$

where $\hat{\Delta}^\dagger(\varphi)$ and $\hat{\Delta}(\varphi)$ can be seen as creation and annihilation operators of Cooper pairs in the wire, respectively. In addition, t is the effective tunnelling amplitude for the Josephson junction.

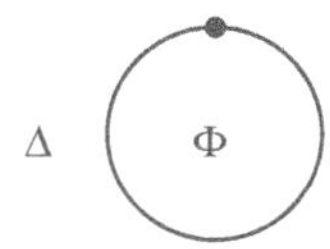

Fig. 12.3 A circular superconducting wire coupled by a Josephson junction. There is a flux Φ through the circle. Here, Δ is the Cooper pair wavefunction.

The action of the system can then be written as

$$S[\Delta^*,\Delta] = S_\mathrm{O}[\Delta^*,\Delta] + S_\mathrm{T}[\Delta^*,\Delta] , \tag{12.76}$$

where $S_\mathrm{O}[\Delta^*,\Delta]$ is the effective Ginzburg-Landau action for the collective field $\Delta(\varphi,t)$ with expectation value $\Delta_0 = \langle\Delta(\varphi,t)\rangle$, namely

$$S_\mathrm{O}[\Delta^*,\Delta] =$$
$$\frac{\mathscr{D}(\varepsilon_\mathrm{F})}{8\pi R}\int \mathrm{d}t \int_0^{2\pi} \mathrm{d}\varphi R \left\{ \frac{\hbar^2}{|\Delta_0|^2}\left|\frac{\partial\Delta}{\partial t}\right|^2 - \frac{\hbar^2 v_\mathrm{F}^2}{3|\Delta_0|^2}|\nabla\Delta|^2 + 2|\Delta|^2\left(1-\frac{|\Delta|^2}{2|\Delta_0|^2}\right)\right\} .$$

Here $\mathscr{D}(\varepsilon_\mathrm{F})$ is the density of states for one spin projection at the Fermi energy, and v_F is the Fermi velocity. In the presence of a vector potential A, the derivative $\nabla \equiv (1/R)\partial/\partial\varphi$ is replaced by the covariant derivative $\nabla - \mathrm{i}(q/\hbar)A$, with q the charge of the field Δ. Here, the vector potential A is given by $A = \Phi/2\pi R$ with Φ the magnetic flux through the circle.

a) What is the charge q of the field Δ in terms of the electron charge $-e$? Write down the action S_Φ for the superconducting wire in the presence of the flux Φ.

Assume now that the system is in equilibrium, i.e. that $|\Delta|$ takes its vacuum expectation value $|\Delta_0|$. Denote the phase of Δ by θ.

b) What is the effective action for the phase θ in the presence of the magnetic flux Φ?

Assume now that the phase θ is static, i.e. that it has no time dependence. Note that in general the phase is inhomogeneous, i.e. it depends on the angle φ.

c) How does $\theta(\varphi)$ depend on the flux Φ when the superconducting wire is in equilibrium and we neglect the effect of the tunnelling Hamiltonian?

d) Express the tunnelling energy in terms of the phases $\theta(0)$ and $\theta(2\pi)$. When is it minimal? Write down the total energy of the superconducting wire and the

Josephson junction. What are the values of the flux Φ for which the total energy is minimal?
The current flowing through the Josephson junction is given by

$$\hat{J} = \frac{2i\pi eRt}{\hbar}\left(\hat{\Delta}^\dagger(2\pi)\hat{\Delta}(0) - \hat{\Delta}^\dagger(0)\hat{\Delta}(2\pi)\right) . \tag{12.77}$$

e) In the mean-field approximation, what is the expectation value of the current J through the Josephson junction? What is the current J through the wire in terms of the flux Φ? Give the maximum amplitude of the current J_{max}. For what values of Φ is the amplitude of the current minimal? Give an interpretation of this result in view of your answer to question d).

Additional Reading

- J. Bardeen, L.N. Cooper, and J.R. Schrieffer, Phys. Rev., **108**, 1175 (1957).
- A.J. Leggett, Rev. Mod. Phys. **47**, 331 (1975).
- H. Kleinert, Forts. Phys. **26**, 565 (1978).
- J. R. Schrieffer, *Theory of Superconductivity*, (Perseus Books, Reading,1999).
- D. Vollhardt and P. Wölfle, *The Superfluid phases of Helium 3*, (Taylor & Francis, London, 1990).

Chapter 13
Symmetries and Symmetry Breaking

In physics, your solution should convince a reasonable person.
In math, you have to convince a person who's trying to make trouble. Ultimately, in physics, you're hoping to convince Nature. And I've found Nature to be pretty reasonable.
– Frank Wilczek

Symmetries play an important role in modern physics. For instance, the Standard Model of high-energy physics is to a large extent specified by its local symmetry group $U(1) \times SU(2) \times SU(3)$. Apart from the esthetic beauty of having a symmetric description of physical phenomena, the presence of a symmetry in the problem also often leads to useful practical simplifications. For example, the energy-level structure of a Hamiltonian, and in particular the degeneracies, can often be understood on the basis of the symmetry of the Hamiltonian without doing any calculations. Another use of symmetries was seen in Chap. 3, where the rotational symmetry of an isotropic harmonic potential allows for a separation of variables in the Schrödinger equation and an expansion of the wavefunction into spherical harmonics. This last example illustrates the fact that symmetries lead to conservation laws, in this case the conservation of angular momentum.

In condensed-matter physics symmetries also play a crucial role, because many phase transitions can be seen as a spontaneous breakdown of symmetry, i.e. the Hamiltonian has a certain symmetry which is not shared by a particular ground state of the many-body system. If this occurs, the application of the symmetry operation onto this ground state leads to another eigenstate of the system with the same energy. As a result, the ground state must be degenerate. This leads to the fundamental problem of how the system chooses a particular ground state when it goes through a phase transition. In atomic gases, the answer to this question is related to the interesting phenomena of phase diffusion that we discuss at the end of this chapter. In addition, we have seen in the previous chapters how phase transitions can be described by making use of a Landau free-energy functional. In the case of the Ising model, we explicitly showed that the Landau free energy is an effective Hamiltonian that comes about by integrating out the microscopic degrees of freedom. Moreover, starting from a microscopic action, we demonstrated how the Hubbard-Stratonovich transformation can be used to introduce the order parameter into the problem and to obtain the effective long-wavelength action of the system. In this chapter, we formalize the concept of effective actions and study some of their symmetry properties.

13.1 Effective Actions

One of the central quantities of interest in a quantum field theory is the generating functional $Z[J,J^*]$ of all the possible correlation functions of the system, which we encountered previously in (7.67) for the case of an ideal gas. To study the general case, we consider a field theory for an atomic gas that is described by the microscopic action $S[\psi^*,\psi]$, with $\psi(\mathbf{x},\tau)$ being the atomic field. For notational simplicity we neglect the possible spin degrees of freedom of the atoms, such that the generating functional in imaginary time is determined by

$$Z[J,J^*] = \int \mathrm{d}[\psi^*]\mathrm{d}[\psi] \exp\left\{-\frac{1}{\hbar}S[\psi^*,\psi] + S_J[\psi^*,\psi]\right\}, \tag{13.1}$$

where the source currents couple to the fields according to,

$$S_J[\psi^*,\psi] = \int_0^{\hbar\beta} \mathrm{d}\tau \int \mathrm{d}\mathbf{x}\,(\psi^*(\mathbf{x},\tau)J(\mathbf{x},\tau) + J^*(\mathbf{x},\tau)\psi(\mathbf{x},\tau))\,. \tag{13.2}$$

We see from (13.1) and (13.2) that by taking functional derivatives of $Z[J,J^*]$ with respect to the currents we can calculate all the correlation functions of the theory. However, instead of working with $Z[J,J^*]$, we usually prefer to work with the generating functional $W[J,J^*]$ of all the connected Green's functions. This is related to $Z[J,J^*]$ through

$$Z[J,J^*] = \exp\{W[J,J^*]\}. \tag{13.3}$$

The functional derivatives of $W[J,J^*]$ with respect to the currents then immediately yield the expectation values of the fields, i.e.

$$\begin{aligned}\frac{\delta W[J,J^*]}{\delta J(\mathbf{x},\tau)} &= \pm\langle\psi^*(\mathbf{x},\tau)\rangle \equiv \pm\phi^*(\mathbf{x},\tau)\\ \frac{\delta W[J,J^*]}{\delta J^*(\mathbf{x},\tau)} &= \langle\psi(\mathbf{x},\tau)\rangle \equiv \phi(\mathbf{x},\tau),\end{aligned} \tag{13.4}$$

where we established in Chap. 8 that expectation values only depend on connected Green's functions.

It is also possible to define a functional $\Gamma[\phi^*,\phi]$ that depends explicitly on the expectation values $\phi(\mathbf{x},\tau)$, which is related to $W[J,J^*]$ by means of a Legendre transformation, i.e.

$$\Gamma[\phi^*,\phi] = -W[J,J^*] + \int \mathrm{d}\tau \int \mathrm{d}\mathbf{x}\ (\phi^*(\mathbf{x},\tau)J(\mathbf{x},\tau) + J^*(\mathbf{x},\tau)\phi(\mathbf{x},\tau))\,. \tag{13.5}$$

From $\Gamma[\phi^*,\phi]$, we introduce the exact effective action as

$$S^{\mathrm{eff}}[\phi^*,\phi] = \hbar\Gamma[\phi^*,\phi]. \tag{13.6}$$

To understand the meaning of this effective action, we note that $\Gamma[\phi^*,\phi]$ is the generating functional of all one-particle irreducible vertex functions, which can be seen as follows

$$\begin{aligned}\frac{\delta\Gamma[\phi^*,\phi]}{\delta\phi(\mathbf{x},\tau)} &= -\int_0^{\hbar\beta}\mathrm{d}\tau'\int\mathrm{d}\mathbf{x}'\left\{\frac{\delta J^*(\mathbf{x}',\tau')}{\delta\phi(\mathbf{x},\tau)}\frac{\delta W[J,J^*]}{\delta J^*(\mathbf{x}',\tau')}+\frac{\delta J(\mathbf{x}',\tau')}{\delta\phi(\mathbf{x},\tau)}\frac{\delta W[J,J^*]}{\delta J(\mathbf{x}',\tau')}\right\}\\ &+\int_0^{\hbar\beta}\mathrm{d}\tau'\int\mathrm{d}\mathbf{x}'\left\{\pm\phi^*(\mathbf{x},\tau)\frac{\delta J(\mathbf{x}',\tau')}{\delta\phi(\mathbf{x},\tau)}+\frac{\delta J^*(\mathbf{x}',\tau')}{\delta\phi(\mathbf{x},\tau)}\phi(\mathbf{x},\tau)\right\}\\ &\pm J^*(\mathbf{x},\tau)=\pm J^*(\mathbf{x},\tau),\end{aligned} \tag{13.7}$$

where we used (13.4) and (13.5). For the second-order derivative, we then have

$$\begin{aligned}\left.\frac{\delta^2\Gamma[\phi^*,\phi]}{\delta\phi(\mathbf{x}',\tau')\delta\phi^*(\mathbf{x},\tau)}\right|_{\phi^*=\phi=0} &= \left.\frac{\delta J^*(\mathbf{x}',\tau')}{\delta\phi^*(\mathbf{x},\tau)}\right|_{\phi^*=\phi=0} = \left.\left(\frac{\delta\phi^*(\mathbf{x},\tau)}{\delta J^*(\mathbf{x}',\tau')}\right)^{-1}\right|_{J^*=J=0}\\ &= \pm\left.\left(\frac{\delta^2 W[J,J^*]}{\delta J^*(\mathbf{x}',\tau')\delta J(\mathbf{x}',\tau')}\right)^{-1}\right|_{J^*=J=0}\\ &= -G^{-1}(\mathbf{x},\tau;\mathbf{x}',\tau')\\ &= -G_0^{-1}(\mathbf{x},\tau;\mathbf{x}',\tau')+\Sigma(\mathbf{x},\tau;\mathbf{x}',\tau'),\end{aligned} \tag{13.8}$$

where we used that differentiating $W[J,J^*]$ with respect to the currents generates the exact interacting Green's function. This proves our claim for the quadratic term of $\Gamma[\phi^*,\phi]$, because we have seen in Chap. 8 that the selfenergy by definition contains all one-particle irreducible diagrams. By further differentiation, the same can be proven for all higher-order terms containing the vertex functions $\Gamma^{(2n)}$ with $n \geq 3$, such that we have

$$\begin{aligned}S^{\mathrm{eff}}[\phi^*,\phi] &= -\int_0^{\hbar\beta}\mathrm{d}\tau\int\mathrm{d}\mathbf{x}\int_0^{\hbar\beta}\mathrm{d}\tau'\int\mathrm{d}\mathbf{x}'\ \phi^*(\mathbf{x},\tau)\hbar G^{-1}(\mathbf{x},\tau;\mathbf{x}',\tau')\phi(\mathbf{x}',\tau')\\ &+\frac{1}{2}\int_0^{\hbar\beta}\mathrm{d}\tau\int\mathrm{d}\mathbf{x}\int_0^{\hbar\beta}\mathrm{d}\tau'\int\mathrm{d}\mathbf{x}'\int_0^{\hbar\beta}\mathrm{d}\tau''\int\mathrm{d}\mathbf{x}''\int_0^{\hbar\beta}\mathrm{d}\tau'''\int\mathrm{d}\mathbf{x}'''\\ &\times\phi^*(\mathbf{x},\tau)\phi^*(\mathbf{x}',\tau')\hbar\Gamma^{(4)}(\mathbf{x},\tau;\mathbf{x}',\tau';\mathbf{x}'',\tau'';\mathbf{x}''',\tau''')\phi(\mathbf{x}'',\tau'')\phi(\mathbf{x}''',\tau''')\\ &+\ldots,\end{aligned} \tag{13.9}$$

where $G(\mathbf{x},\tau;\mathbf{x}',\tau')$ is the exact propagator of the atoms and where the four-point vertex $\Gamma^{(4)}(\mathbf{x},\tau;\mathbf{x}',\tau';\mathbf{x}'',\tau'';\mathbf{x}''',\tau''')$ is the exact effective interaction. The dots denote all $2n$-point one-particle irreducible vertices $\Gamma^{(2n)}$ with $n \geq 3$, which turn out to be less relevant for our purposes since they correspond to three and higher-body processes that are usually negligible for dilute atomic quantum gases.

Although the formal definition of the exact effective action is in terms of the generating functional of all one-particle irreducible vertex functions, we have used the term effective action rather loosely. Typically, we have called any action effective that arises from the microscopic action $S[\psi^*,\psi]$ after having integrated out microscopic degrees of freedom of $\psi(\mathbf{x},\tau)$. In practice it is often impossible to find the exact effective action, and we are forced to make some approximation. This is, for example, the case when we sum only over a certain class of Feynman diagrams in calculating the effective action, as in Hartree-Fock theory. However, we need to be careful when using such approximations, because they can have consequences for the symmetry properties of the action and the conservation laws of the system. As we show in the next sections, an action that is invariant under continuous symmetries satisfies conservation laws that correspond to these symmetries. Explicitly, we will demonstrate that an action $S[\psi^*, \psi]$ that is invariant under a $U(1)$ transformations of the field $\psi(\mathbf{x}, \tau)$ obeys particle number conservation. If we do not carefully construct the approximate effective action, we can easily violate such conservation laws. To be able to ensure that we always satisfy the appropriate conservation laws of the system, we then derive the Ward identities between the various exact vertex functions in Sect. 13.3.

Example 13.1. As an example of a system where we can calculate the exact effective action, we consider again the homogeneous ideal gas. Using (7.71) and performing a Fourier transform, we see that the generating functional $Z[J, J^*]$ is given by

$$Z[J,J^*] = Z[0,0]\exp\left\{-\sum_{\mathbf{k},n} J^*_{\mathbf{k},n} G(\mathbf{k},\mathrm{i}\omega_n) J_{\mathbf{k},n}\right\}, \tag{13.10}$$

where the Green's function is given by (7.57)

$$G(\mathbf{k},\mathrm{i}\omega_n) = \frac{-\hbar}{-\mathrm{i}\hbar\omega_n + \varepsilon_{\mathbf{k}} - \mu}. \tag{13.11}$$

The effective action given in (13.5) now becomes

$$\Gamma[\phi^*,\phi] = \sum_{\mathbf{k},n} \left\{ J^*_{\mathbf{k},n} G(\mathbf{k},\mathrm{i}\omega_n) J_{\mathbf{k},n} + \phi^*_{\mathbf{k},n} J_{\mathbf{k},n} + J^*_{\mathbf{k},n}\phi_{\mathbf{k},n} \right\}, \tag{13.12}$$

where $J_{\mathbf{k},n}$ and $J^*_{\mathbf{k},n}$ must be eliminated by means of

$$\pm\phi^*_{\mathbf{k},n} = \frac{\delta W[J,J^*]}{\delta J_{\mathbf{k},n}} = \mp J^*_{\mathbf{k},n} G(\mathbf{k},\mathrm{i}\omega_n) \quad , \quad \phi_{\mathbf{k},n} = \frac{\delta W[J,J^*]}{\delta J^*_{\mathbf{k},n}} = -G(\mathbf{k},\mathrm{i}\omega_n) J_{\mathbf{k},n}.$$

This gives the desired result

$$\Gamma[\phi^*,\phi] = -\sum_{\mathbf{k},n} \phi^*_{\mathbf{k},n} G^{-1}(\mathbf{k},\mathrm{i}\omega_n)\phi_{\mathbf{k},n}. \tag{13.13}$$

Because there are by definition no interactions, the exact effective action in this case only has a quadratic term.

13.2 Noether's Theorem

In this section, we show that symmetries of the action are related to conservation laws of the system. To start the discussion, we first consider an action $S[\psi^*, \psi]$ which has a global $U(1)$ symmetry, i.e. the action is invariant under the global phase transformation

$$\psi(\mathbf{x},\tau) \rightarrow \mathrm{e}^{\mathrm{i}\theta}\psi(\mathbf{x},\tau) \quad \text{and} \quad \psi^*(\mathbf{x},\tau) \rightarrow \mathrm{e}^{-\mathrm{i}\theta}\psi^*(\mathbf{x},\tau).$$

The action for the interacting quantum gases of the form of (8.22) satisfies such a global $U(1)$ symmetry. Let us look at the implications of this global symmetry by considering the more general case of a local $U(1)$ transformation, i.e. with a phase $\theta(\mathbf{x},\tau)$ that depends on space and imaginary time. Under such a local transformation, the action of (8.22) is not invariant and the transformed action

$$S[(1-\mathrm{i}\theta)\psi^*,(1+\mathrm{i}\theta)\psi]$$

can be expanded to first order in $\theta(\mathbf{x},\tau)$ as

$$S[\psi^*,\psi] \rightarrow S[\psi^*,\psi] + \hbar\int_0^{\hbar\beta} \mathrm{d}\tau \int \mathrm{d}\mathbf{x}\, J^\mu(\mathbf{x},\tau)\partial_\mu\theta(\mathbf{x},\tau), \qquad (13.14)$$

where the linear coefficient of the expansion is by definition called the current $J^\mu(\mathbf{x},\tau)$. Moreover, we have introduced here the relativistic four-vector notation $\partial_\mu \equiv \partial/\partial x^\mu$ with $x^0 = c\tau = ict$, where c is the speed of light, and the spatial components satisfy $x^j = x_j$. As a result, we thus have $x^\mu = (c\tau, \mathbf{x})$, and to facilitate the notation we use in this chapter the convention that repeated space-time indices are summed over. The convenience of this four-vector notation becomes clear shortly. Note that the expression of (13.14) is a direct consequence of the global $U(1)$ symmetry, and that in general more terms would be generated. Indeed, if we take the phase $\theta(\mathbf{x},\tau)$ constant in (13.14), then we find that the action is invariant under the transformation, as desired for a global $U(1)$ symmetry.

As we show now, the current $J^\mu(\mathbf{x},\tau)$ actually corresponds to a conserved quantity. By performing a partial integration, the induced infinitesimal change in the action by the local phase transformation can be written as

$$\Delta S = -\hbar\int_0^{\hbar\beta} \mathrm{d}\tau \int \mathrm{d}\mathbf{x}\, \theta(\mathbf{x},\tau)\partial_\mu J^\mu(\mathbf{x},\tau). \qquad (13.15)$$

We remember that under a general variation $\psi(\mathbf{x},\tau) + \delta\psi(\mathbf{x},\tau)$ the induced first-order change in the action vanishes if $\psi(\mathbf{x},\tau)$ obeys the Euler-Lagrange equation $\delta S[\psi^*,\psi]/\delta\psi(\mathbf{x},\tau) = 0$. Since the local phase transformation is just a special case of this general variation, we conclude that automatically we must have that

$$\partial_\mu J^\mu(\mathbf{x},\tau) = 0, \qquad (13.16)$$

if the atomic field $\psi(\mathbf{x},\tau)$ obeys the equation of motion. This result is a manifestation of Noether's theorem, which states that there is a conservation law associated with each symmetry of the system.

As an explicit example, we consider a system of free atoms whose action is given by

$$S[\psi^*,\psi] = \int_0^{\hbar\beta} d\tau \int d\mathbf{x}\, \psi^*(\mathbf{x},\tau) \left\{ \hbar\frac{\partial}{\partial\tau} - \frac{\hbar^2\nabla^2}{2m} + V^{\text{ex}}(\mathbf{x}) - \mu \right\} \psi(\mathbf{x},\tau). \tag{13.17}$$

Under the infinitesimal transformation

$$\psi(\mathbf{x},\tau) \to (1+i\theta(\mathbf{x},\tau))\psi(\mathbf{x},\tau) \quad \text{and} \quad \psi^*(\mathbf{x},\tau) \to (1-i\theta(\mathbf{x},\tau))\psi^*(\mathbf{x},\tau),$$

we obtain to first order in $\theta(\mathbf{x},\tau)$ the result

$$\begin{aligned} S[\psi^*,\psi] \to\ & S[\psi^*,\psi] + \int_0^{\hbar\beta} d\tau \int d\mathbf{x} \left\{ i\psi^*(\mathbf{x},\tau)\psi(\mathbf{x},\tau)\hbar\frac{\partial}{\partial\tau}\theta(\mathbf{x},\tau) \right. \\ & \left. -2i\psi^*(\mathbf{x},\tau)\frac{\hbar^2}{2m}\nabla\psi(\mathbf{x},\tau)\cdot\nabla\theta(\mathbf{x},\tau) - i\psi^*(\mathbf{x},\tau)\psi(\mathbf{x},\tau)\frac{\hbar^2\nabla^2}{2m}\theta(\mathbf{x},\tau) \right\}. \end{aligned} \tag{13.18}$$

Performing the partial integrations on the time-derivative and the gradient terms, we obtain

$$\begin{aligned} \Delta S = & -\hbar \int d\tau \int d\mathbf{x}\, \frac{\partial}{\partial\tau}\left(i\psi^*(\mathbf{x},\tau)\psi(\mathbf{x},\tau)\right)\theta(\mathbf{x},\tau) \\ & +\hbar \int d\tau \int d\mathbf{x}\, \nabla\cdot\left\{ i\frac{\hbar}{2m}\left(\psi^*(\mathbf{x},\tau)\nabla\psi(\mathbf{x},\tau) - \nabla\psi^*(\mathbf{x},\tau)\psi(\mathbf{x},\tau)\right)\right\}\theta(\mathbf{x},\tau). \end{aligned} \tag{13.19}$$

Comparing this result with (13.15), we see that the source current $J^\mu(\mathbf{x},\tau)$ is given by

$$\begin{aligned} J^0(\mathbf{x},\tau) &= ic\psi^*(\mathbf{x},\tau)\psi(\mathbf{x},\tau) \\ \mathbf{J}(\mathbf{x},\tau) &= -\frac{i\hbar}{2m}\left(\psi^*(\mathbf{x},\tau)\nabla\psi(\mathbf{x},\tau) - \nabla\psi^*(\mathbf{x},\tau)\psi(\mathbf{x},\tau)\right), \end{aligned} \tag{13.20}$$

which, according to Noether's theorem, gives rise to the conservation law

$$\partial_\mu J^\mu(\mathbf{x},\tau) = i\frac{\partial}{\partial\tau}n(\mathbf{x},\tau) + \nabla\cdot\boldsymbol{J}(\mathbf{x},\tau) = 0 \tag{13.21}$$

for the total atomic density $n(\mathbf{x},\tau) = \psi^*(\mathbf{x},\tau)\psi(\mathbf{x},\tau)$. Performing a Wick rotation to real time, the conservation law in (13.21) becomes

$$\frac{\partial}{\partial t}\psi^*(\mathbf{x},t)\psi(\mathbf{x},t) - \frac{i\hbar}{2m}\nabla\cdot\{\psi^*(\mathbf{x},t)\nabla\psi(\mathbf{x},t) - \nabla\psi^*(\mathbf{x},t)\psi(\mathbf{x},t)\} = 0, \tag{13.22}$$

which we recognize as the continuity equation, (8.108),

$$\frac{\partial}{\partial t} n(\mathbf{x},t) + \nabla \cdot \mathbf{J}(\mathbf{x},t) = 0 \tag{13.23}$$

for the conservation of the total number of atoms. It is left as an exercise to show that it holds if $\psi(\mathbf{x},t)$ obeys its equation of motion

$$\mathrm{i}\hbar \frac{\partial}{\partial t} \psi(\mathbf{x},t) = \left\{ -\frac{\hbar^2}{2m} \nabla^2 + V^{\mathrm{ex}}(\mathbf{x}) - \mu \right\} \psi(\mathbf{x},t). \tag{13.24}$$

The generalization of the discussion to more complicated symmetries is then straightforward.

13.3 Ward Identities

In this section, we generalize the system of atoms with a global $U(1)$ symmetry to systems that are also invariant under local $U(1)$ transformations by including a coupling to a gauge field. Perhaps the most well-known example of a locally symmetric $U(1)$ theory is quantum electrodynamics, or QED. This quantum theory for electromagnetism is based on the coupling between charged particles, which are described by the fields $\psi(\mathbf{x},\tau)$, and the quantized electromagnetic field, which is described in terms of the photon field $A_\mu(\mathbf{x},\tau)$ also called a gauge field. Using the minimal-coupling prescription, local $U(1)$ invariance is achieved by replacing the derivatives $\hbar\partial/\partial\tau$ and $\mathrm{i}\hbar\nabla$ in (13.17) by

$$\hbar \frac{\partial}{\partial \tau} \to \hbar \frac{\partial}{\partial \tau} + e\phi(\mathbf{x},\tau) \quad \text{and} \quad \mathrm{i}\hbar \frac{\partial}{\partial \mathbf{x}} \to \mathrm{i}\hbar \frac{\partial}{\partial \mathbf{x}} - \frac{e}{c}\mathbf{A}(\mathbf{x},\tau). \tag{13.25}$$

The first line of the previous equation tells us for instance that the interaction with the scalar electromagnetic potential is given by $e\phi(\mathbf{x},\tau)\psi^*(\mathbf{x},\tau)\psi(\mathbf{x},\tau)$, as desired. In four-vector notation, the minimal-coupling prescription for a particle with a positive charge e can be written as

$$\mathrm{i}\hbar \frac{\partial}{\partial x^\mu} \to \mathrm{i}\hbar \frac{\partial}{\partial x^\mu} - \frac{e}{c} A_\mu(\mathbf{x},\tau),$$

where $A_\mu(\mathbf{x},\tau) = (-\mathrm{i}\phi(\mathbf{x},\tau), \mathbf{A}(\mathbf{x},\tau))$. The action for nonrelativistic particles with charge e is then given by

$$\begin{aligned} S[\psi^*,\psi,A_\mu] = \int \mathrm{d}\tau \int \mathrm{d}\mathbf{x}\; \psi^*(\mathbf{x},\tau) \left\{ \hbar \frac{\partial}{\partial \tau} + \mathrm{i}eA_0(\mathbf{x},\tau) + \frac{1}{2m}\left(\mathrm{i}\hbar\nabla - \frac{e}{c}\mathbf{A}(\mathbf{x},\tau)\right)^2 \right. \\ \left. + V^{\mathrm{ex}}(\mathbf{x}) - \mu + \frac{1}{2}\int \mathrm{d}\mathbf{x}'\; V(\mathbf{x}-\mathbf{x}')|\psi(\mathbf{x}',\tau)|^2 \right\} \psi(\mathbf{x},\tau). \end{aligned} \tag{13.26}$$

In the theory of quantum electrodynamics there is also a free term $S_0[A_\mu]$ for the gauge field $A_\mu(\mathbf{x},\tau)$ in the above action, which we have not included here. The reason is that it is not our goal to consider charged systems in the presence of real fluctuating electromagnetic fields, as we only wish to explore consequences of a local $U(1)$ invariance of the theory by introducing an external gauge field. As a result, in the applications to ultracold atomic gases that we discuss later on, the term $S_0[A_\mu]$ will not be present. Note that the action from (13.26) is indeed invariant under local $U(1)$ transformations if the fields are transformed according to

$$\psi(\mathbf{x},\tau) \rightarrow \mathrm{e}^{\mathrm{i}\theta(\mathbf{x},\tau)}\psi(\mathbf{x},\tau) \quad \text{and} \quad A_\mu(\mathbf{x},\tau) \rightarrow A_\mu(\mathbf{x},\tau) - \frac{\hbar c}{e}\frac{\partial\theta(\mathbf{x},\tau)}{\partial x^\mu}. \tag{13.27}$$

Proceeding in the same spirit as we did for the derivation of Noether's theorem, we now show that invariance of the action under a local $U(1)$ symmetry actually leads to a relation between the selfenergy of the particles and the (three-point) vertex between the particles and the gauge fields. This relation is also known as a Ward identity.

Let us start the discussion by giving a sketchy example of the relationship between selfenergies, vertex corrections, and gauge invariance. Suppose we want to calculate corrections to the photon propagator in QED. Then, the first Feynman diagram that we should include is the bubble diagram shown in Fig. 13.1a. It describes an incoming photon that turns into a particle-hole (electron-positron) pair, which then annihilates again to form a photon. It turns out that after including this diagram into the photon propagator, we still obey gauge invariance. If we go further in our calculations by also including Fock-like selfenergy corrections to the particle propagators, as shown in Fig. 13.1b, then we actually break gauge invariance. This breakdown of the gauge invariance is only because we have taken into account a limited set of Feynman diagrams, whereas inclusion of all diagrams would lead to a gauge-invariant result. However, it turns out that at this level of perturbation theory we can already restore the gauge invariance by adding only one diagram, namely that shown in Fig. 13.1c, which can be viewed as a correction to the three-point vertex between the particles and the gauge field. How all this comes about is what we want to show next.

To generalize the above discussion, we derive the relation between the vertex corrections and the selfenergies that allows us to preserve gauge invariance at any level in perturbation theory. Let $S[\psi^*,\psi,A_\mu]$ be a locally $U(1)$ gauge-invariant action, such as the action from (13.26). By introducing the source currents $J(\mathbf{x},\tau)$ for the atomic fields and $J^\mu(\mathbf{x},\tau)$ for the gauge field, we obtain the generating functional function $Z[J,J^*,J^\mu]$, given by

$$Z[J,J^*,J^\mu] = \int \mathrm{d}[\psi^*]\mathrm{d}[\psi]\mathrm{d}[A_\mu]\exp\left\{-\frac{1}{\hbar}S[\psi^*,\psi,A_\mu]+S_J\right\} \tag{13.28}$$

where we have

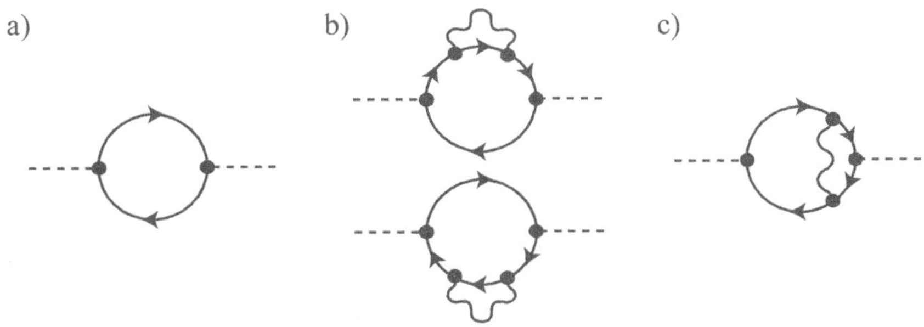

Fig. 13.1 Example of corrections to the noninteracting photon propagator, which is represented by the dashed line. a) Bubble correction to the propagator, where the noninteracting particle and hole propagators are represented by solid lines. b) Additional first-order Fock selfenergy corrections to the particle propagator, where the interparticle interaction is represented by a wiggly line. c) First-order vertex correction.

$$S_J = \int_0^{\hbar\beta} \mathrm{d}\tau \int \mathrm{d}\mathbf{x}\, \left(\psi^*(\mathbf{x},\tau)J(\mathbf{x},\tau) + J^*(\mathbf{x},\tau)\psi(\mathbf{x},\tau) + J^\mu(\mathbf{x},\tau)A_\mu(\mathbf{x},\tau)\right). \tag{13.29}$$

If we now apply an infinitesimal version of the gauge transformation from (13.27) the action itself is invariant, but the term S_J containing the source currents is not. As a result, we obtain to first order in the local phase $\theta(\mathbf{x},\tau)$ the following transformed partition function

$$\begin{aligned} Z'[J,J^*,J^\mu] &= \int \mathrm{d}[\psi^*]\mathrm{d}[\psi]\mathrm{d}[A_\mu] \exp\left\{-\frac{1}{\hbar}S[\psi^*,\psi,A_\mu] + S_J\right\} \\ &\times \left\{1 + \int_0^{\hbar\beta} \mathrm{d}\tau \int \mathrm{d}\mathbf{x} \left(-\mathrm{i}\theta(\mathbf{x},\tau)\psi^*(\mathbf{x},\tau)J(\mathbf{x},\tau) + \mathrm{i}\theta(\mathbf{x},\tau)J^*(\mathbf{x},\tau)\psi(\mathbf{x},\tau) \right.\right. \\ &\left.\left. -\frac{\hbar c}{e} J^\mu(\mathbf{x},\tau)\frac{\partial}{\partial x^\mu}\theta(\mathbf{x},\tau)\right)\right\}, \end{aligned} \tag{13.30}$$

where we have assumed that the integration measure remains invariant under the transformation. Since we have merely performed a transformation of integration variables, we must have that $Z = Z'$, which gives

$$\begin{aligned} &\int \mathrm{d}[\psi^*]\mathrm{d}[\psi]\mathrm{d}[A_\mu]\, \mathrm{e}^{-S[\psi^*,\psi,A_\mu]/\hbar + S_J} \int_0^{\hbar\beta} \mathrm{d}\tau \int \mathrm{d}\mathbf{x} \left\{-\mathrm{i}\theta(\mathbf{x},\tau)\psi^*(\mathbf{x},\tau)J(\mathbf{x},\tau) \right. \\ &\left. +\mathrm{i}\theta(\mathbf{x},\tau)J^*(\mathbf{x},\tau)\psi(\mathbf{x},\tau) - \frac{\hbar c}{e}J^\mu(\mathbf{x},\tau)\frac{\partial}{\partial x^\mu}\theta(\mathbf{x},\tau)\right\} = 0. \end{aligned} \tag{13.31}$$

Using the identities

$$\psi^*(\mathbf{x},\tau)\mathrm{e}^{S_J} = \pm\frac{\delta}{\delta J(\mathbf{x},\tau)}\mathrm{e}^{S_J} \quad \text{and} \quad \psi(\mathbf{x},\tau)\mathrm{e}^{S_J} = \frac{\delta}{\delta J^*(\mathbf{x},\tau)}\mathrm{e}^{S_J}, \tag{13.32}$$

we can write (13.31) more compactly as

$$\int_0^{\hbar\beta} d\tau \int d\mathbf{x} \left\{ -i\theta(\mathbf{x},\tau)J(\mathbf{x},\tau)\frac{\delta}{\delta J(\mathbf{x},\tau)} + i\theta(\mathbf{x},\tau)J^*(\mathbf{x},\tau)\frac{\delta}{\delta J^*(\mathbf{x},\tau)} + \frac{\hbar c}{e}\theta(\mathbf{x},\tau)\frac{\partial}{\partial x^\mu}J^\mu(\mathbf{x},\tau) \right\} Z[J,J^*,J^\mu] = 0, \tag{13.33}$$

where we also used partial integration. Because this equation is valid for all fields $\theta(\mathbf{x},\tau)$, we obtain the identity

$$\left\{ -iJ(\mathbf{x},\tau)\frac{\delta}{\delta J(\mathbf{x},\tau)} + iJ^*(\mathbf{x},\tau)\frac{\delta}{\delta J^*(\mathbf{x},\tau)} + \frac{\hbar c}{e}\frac{\partial}{\partial x^\mu}J^\mu(\mathbf{x},\tau) \right\} Z[J,J^*,J^\mu] = 0. \tag{13.34}$$

To proceed, we introduce as before the generating functional $W[J,J^*,J^\mu] = \log Z[J,J^*,J^\mu]$ of all connected Green's functions, whose derivatives with respect to the source currents are the expectation values of the atomic and electromagnetic fields, i.e.

$$\frac{\delta W[J,J^*,J^\mu]}{\delta J^\mu(\mathbf{x},\tau)} = A_\mu(\mathbf{x},\tau), \quad \frac{\delta W[J,J^*,J^\mu]}{\delta J^*(\mathbf{x},\tau)} \equiv \phi(\mathbf{x},\tau) = \langle \psi(\mathbf{x},\tau) \rangle,$$
$$\frac{\delta W[J,J^*,J^\mu]}{\delta J(\mathbf{x},\tau)} \equiv \pm\phi^*(\mathbf{x},\tau) = \pm\langle \psi^*(\mathbf{x},\tau) \rangle, \tag{13.35}$$

where we note that we use the same notation for the photon field and its expectation value, with which we do not mean to imply that the two are necessarily equal. Moreover, we define $\hbar\Gamma[\phi^*,\phi,A_\mu]$ as the effective action for the averaged fields $\phi^*(\mathbf{x},\tau)$, $\phi(\mathbf{x},\tau)$, and $A_\mu(\mathbf{x},\tau)$ via the Legendre transform

$$\Gamma[\phi^*,\phi,A_\mu] \equiv \int_0^{\hbar\beta} d\tau \int d\mathbf{x} \left(\phi^*(\mathbf{x},\tau)J(\mathbf{x},\tau) + J^*(\mathbf{x},\tau)\phi(\mathbf{x},\tau) + J^\mu(\mathbf{x},\tau)A_\mu(\mathbf{x},\tau) \right) - W[J,J^*,J^\mu]. \tag{13.36}$$

We can use these definitions to rewrite (13.34), namely we have

$$\left\{ -J(\mathbf{x},\tau)\frac{\delta}{\delta J(\mathbf{x},\tau)} + J^*(\mathbf{x},\tau)\frac{\delta}{\delta J^*(\mathbf{x},\tau)} \right\} W[J,J^*,J^\mu] = i\frac{\hbar c}{e}\frac{\partial}{\partial x^\mu}J^\mu(\mathbf{x},\tau), \tag{13.37}$$

where with the use of (13.35) and

$$\frac{\delta\Gamma[\phi^*,\phi,A_\mu]}{\delta\phi^*(\mathbf{x},\tau)} = J(\mathbf{x},\tau), \quad \frac{\delta\Gamma[\phi^*,\phi,A_\mu]}{\delta\phi(\mathbf{x},\tau)} = \pm J^*(\mathbf{x},\tau),$$
$$\frac{\delta\Gamma[\phi^*,\phi,A_\mu]}{\delta A_\mu(\mathbf{x},\tau)} = J^\mu(\mathbf{x},\tau). \tag{13.38}$$

we find that the exact effective action satisfies

$$\left\{ -\phi^*(\mathbf{x},\tau)\frac{\delta}{\delta\phi^*(\mathbf{x},\tau)} + \phi(\mathbf{x},\tau)\frac{\delta}{\delta\phi(\mathbf{x},\tau)} \right\}\Gamma[\phi^*,\phi,A_\mu] = \mathrm{i}\frac{\hbar c}{e}\frac{\partial}{\partial x^\mu}\frac{\delta\Gamma[\phi^*,\phi,A_\mu]}{\delta A_\mu(\mathbf{x},\tau)}. \tag{13.39}$$

This is a Ward identity, which the effective action satisfies exactly due to the underlying $U(1)$ symmetry of the microscopic action. If we approximate the exact effective action by using perturbation theory, this equation can be used to tell us whether or not we are still satisfying the exact gauge symmetry, or if we have artificially broken it with our approximation. Moreover, in the latter case, the Ward identity can also be used to find the Feynman diagrams that should be included to restore the gauge symmetry again. As a result, the Ward identities are not only important from a fundamental point of view but also have practical applications. They can be used to establish exact connections between experimentally relevant quantities, for example to correctly describe atom-light interactions, which is important for understanding spectroscopy experiments. We will see examples of this in the following sections.

13.3.1 Hugenholtz-Pines Theorem

We start with considering the case $A_\mu = 0$, which is of interest in the case of ultracold atomic gases, where we usually only have a global $U(1)$ symmetry. From (13.39), we then find that the expansion of the effective action $\Gamma[\phi^*,\phi]$ must be a functional of the form $\Gamma[\phi^*,\phi] = f(\phi^*\phi)$. This is also expected from the general arguments developed in the Landau theory of Chap. 9, where the effective Landau free energy functional always preserved the symmetry of the microscopic Hamiltonian, in this the case global $U(1)$ symmetry. A particularly useful application of these considerations is the derivation of the Hugenholtz-Pines theorem for a homogeneous interacting Bose gas, which can be written as

$$\mu = \hbar\Sigma_{11}(\mathbf{0},0) - \hbar\Sigma_{12}(\mathbf{0},0), \tag{13.40}$$

such that the chemical potential is seen to be exactly equal to the difference of the diagonal and off-diagonal selfenergies evaluated at zero momentum and zero frequency.

To show this, we first realize from the above discussion that the effective action for the Bose gas after integrating over all Fourier modes with momenta and frequencies unequal to zero is given by

$$S^{\mathrm{eff}}[\phi_{\mathbf{0}}^*,\phi_{\mathbf{0}}] = \hbar\beta V\sum_{n=1}^{\infty}\frac{\Gamma_0^{(2n)}}{n}|\phi_{\mathbf{0}}|^{2n}, \tag{13.41}$$

where $\Gamma_0^{(2n)}$ denote the exact $2n$-point vertex functions with all $2n-1$ external momentum and frequency arguments equal to zero. The free energy of the system acquires a minimum in a stationary point of the effective action, i.e. a solution of $\delta S^{\text{eff}}/\delta\phi_{\mathbf{0}}^{*}=0$. One possible solution, which corresponds to the symmetric phase, is always $\phi_{\mathbf{0}}=0$. However, at low-enough temperatures this stationary point actually becomes a maximum and a second solution is possible that obeys

$$\sum_{n=1}^{\infty}\Gamma_{0}^{(2n)}|\phi_{\mathbf{0}}|^{2n-2}=0. \tag{13.42}$$

Expanding the effective action around an arbitrary solution of (13.42) by means of $\phi_{\mathbf{0}}=\sqrt{n_{\mathbf{0}}}+\phi'_{\mathbf{0},0}/\sqrt{\hbar\beta V}$ gives for the quadratic part in the fluctuations

$$S^{\text{eff}}[\phi_{\mathbf{0}}^{*},\phi_{\mathbf{0}}]=S_{\text{n}}\phi'^{*}_{\mathbf{0},0}\phi'_{\mathbf{0},0}+S_{\text{a}}\left(\phi'_{\mathbf{0},0}\phi'_{\mathbf{0},0}+\phi'^{*}_{\mathbf{0},0}\phi'^{*}_{\mathbf{0},0}\right), \tag{13.43}$$

with

$$S_{\text{n}}=\sum_{n=1}^{\infty}n^{2}(n_{\mathbf{0}})^{n-1}\frac{\Gamma_{0}^{(2n)}}{n}, \tag{13.44}$$

$$S_{\text{a}}=\sum_{n=2}^{\infty}\frac{n(n-1)}{2}(n_{\mathbf{0}})^{n-1}\frac{\Gamma_{0}^{(2n)}}{n}. \tag{13.45}$$

From the definition of the normal and anomalous selfenergies from (11.26) and (11.31), on the other hand, we have that $S_{\text{n}}\equiv\hbar\Sigma_{11}(\mathbf{0},0)-\mu$ and $S_{\text{a}}\equiv\hbar\Sigma_{12}(\mathbf{0},0)/2$. Therefore, the Hugenholtz-Pines theorem is proven by

$$S_{\text{n}}-2S_{\text{a}}=\sum_{n=1}^{\infty}(n_{\mathbf{0}})^{n-1}\Gamma_{0}^{(2n)}=0 \tag{13.46}$$

as follows from the stationary condition (13.42) for the solution $\phi_{\mathbf{0}}=\sqrt{n_{\mathbf{0}}}$.

13.3.2 Bragg Scattering

To investigate the properties of ultracold gases we typically make use of some external probe to disturb the system, and its response then tells us something about the state it is in. In the following, we consider two examples of common experimental techniques that are used to make high-precision measurements. The effective action formalism and the Ward identities derived in this chapter then turn out to be very useful, and sometimes indispensable, for the correct theoretical analysis of such experiments. The first example is called Bragg spectroscopy, which in the field of ultracold atoms refers to the diffraction of an atomic cloud by a light grating. It can be seen as the matter-wave analogue of the same process known from optics. It has found many applications, such as coherently splitting a Bose-Einstein condensate into two momentum components [90], measuring the excitation spectrum of a trapped Bose-Einstein condensate [91], and measuring the light-shifted energy levels of an atom in an optical lattice [92]. In Chap. 16, we show how Bragg spectroscopy can also be used to probe the excitation spectrum of what is known as the Mott-insulator state.

In the traditional X-ray Bragg spectroscopy experiments, the momentum of an incoming electromagnetic wave is changed by diffraction from a periodic structure. Analogously, in experiments with ultracold gases, matter waves are diffracted from a periodic light grating. To this end, the setup from Fig. 16.10 is used, where two laser beams with a slight mismatch in wavevector and frequency create an intensity pattern that propagates in space. On the microscopic level, these two laser beams make excitations in the atoms, which allows for a controllable momentum and energy transfer. When an atom absorbs a photon from beam two, it is stimulated to emit a photon either back into beam two or into beam one, where the latter causes the atom to undergo a change of momentum $\hbar\mathbf{q} = \hbar\mathbf{k}_2 - \hbar\mathbf{k}_1$ and a change of energy $\hbar\omega = \hbar\omega_2 - \hbar\omega_1$. The magnitude of the momentum kick to the atom is then to a good approximation given by $\hbar q = 2\hbar k_{\text{ph}} \sin(\theta/2)$, where $\hbar k_{\text{ph}} = 2\pi\hbar/\lambda$ is the photon momentum in both lasers, λ is the wavelength of the laser light, and θ is the angle between the two laser beams. By varying the angle between the two laser beams any momentum between zero and $2\hbar k_{\text{ph}}$ can be transferred, while by varying the relative frequency between the beams the amount of energy that is transferred to the atoms can be controlled. Note that even though this kind of Bragg scattering is a two-photon or Raman process, it can also be thought of as an effective single-photon scattering process similar to Compton scattering.

For a theoretical treatment of Bragg spectroscopy, we describe the (effective) photons by gauge fields that are coupled to a gas of atoms, as given by the following effective action $\hbar\Gamma[\phi^*, \phi, A_\mu]$

$$
\begin{aligned}
\hbar\Gamma[\phi^*, \phi, A_\mu] = &- \int_0^{\hbar\beta} d\tau \int d\mathbf{x} \int_0^{\hbar\beta} d\tau' \int d\mathbf{x}' \phi^*(\mathbf{x},\tau)\hbar G^{-1}(\mathbf{x},\tau;\mathbf{x}',\tau')\phi(\mathbf{x}',\tau') \\
&+ \int_0^{\hbar\beta} d\tau \int d\mathbf{x} \int_0^{\hbar\beta} d\tau' \int d\mathbf{x}' \int_0^{\hbar\beta} d\tau'' \int d\mathbf{x}'' \\
&\times \gamma^\mu(\mathbf{x},\tau;\mathbf{x}',\tau;\mathbf{x}'',\tau'')\phi^*(\mathbf{x},\tau)A_\mu(\mathbf{x}',\tau')\phi(\mathbf{x}'',\tau'') \\
&+ \dots ,
\end{aligned}
\tag{13.47}
$$

where we introduced the exact effective atom-photon vertex $\gamma^\mu(\mathbf{x},\tau;\mathbf{x}',\tau';\mathbf{x}'',\tau'')$. Let us now assume that, with the use of diagrammatic perturbation theory, we have obtained an approximate expression for the atomic selfenergy $\hbar\Sigma(\mathbf{x},\tau;\mathbf{x}',\tau')$ by taking into account only certain classes of Feynman diagrams. In that case, (13.39) can be seen as a defining equation for the vertex $\gamma^\mu(\mathbf{x},\tau;\mathbf{x}',\tau';\mathbf{x}'',\tau'')$, which it has to satisfy in order to preserve the local $U(1)$ symmetry of the action at this particular level of perturbation theory. Differentiating this equation with respect to $\phi(\mathbf{x}',\tau')$ and $\phi^*(\mathbf{x}'',\tau'')$ and consequently putting all fields equal to zero, we obtain

$$\delta(\tau-\tau'')\delta(\mathbf{x}-\mathbf{x}'')\frac{\delta^2\Gamma[\phi^*,\phi,A_\mu]}{\delta\phi(\mathbf{x}',\tau')\delta\phi^*(\mathbf{x},\tau)} - \delta(\tau-\tau')\delta(\mathbf{x}-\mathbf{x}')\frac{\delta^2\Gamma[\phi^*,\phi,A_\mu]}{\delta\phi(\mathbf{x},\tau)\delta\phi^*(\mathbf{x}'',\tau'')}$$
$$= -\mathrm{i}\frac{\hbar c}{e}\frac{\partial}{\partial x^\mu}\frac{\delta^3\Gamma[\phi^*,\phi,A_\mu]}{\delta\phi(\mathbf{x}',\tau')\delta\phi^*(\mathbf{x}'',\tau'')\delta A_\mu(\mathbf{x},\tau)}\,. \tag{13.48}$$

This result can be further evaluated by using the expansion in (13.47), such that we find

$$\delta(\tau-\tau'')\delta(\mathbf{x}-\mathbf{x}'')\hbar G^{-1}(\mathbf{x},\tau;\mathbf{x}',\tau') - \delta(\tau-\tau')\delta(\mathbf{x}-\mathbf{x}')\hbar G^{-1}(\mathbf{x}'',\tau'';\mathbf{x},\tau)$$
$$= \mathrm{i}\frac{\hbar c}{e}\frac{\partial}{\partial x^\mu}\gamma^\mu(\mathbf{x}'',\tau'';\mathbf{x},\tau;\mathbf{x}',\tau'), \tag{13.49}$$

where the left-hand side is the difference between the exact atomic propagators and the right-hand side is the sum of derivatives of the atom-photon vertex.

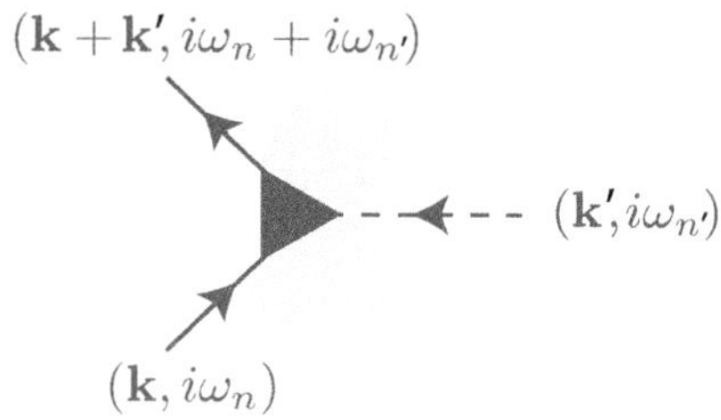

Fig. 13.2 An incoming photon with four-momentum k'_μ couples to an incoming particle with four-momentum k_μ, whose four-momentum is changed to $k_\mu + k'_\mu$

For a homogeneous system we can conveniently express this equation in momentum space, which yields

$$-\frac{c}{e}k'_\mu\gamma^\mu(\mathbf{k},\mathrm{i}\omega_n;\mathbf{k}',\mathrm{i}\omega_{n'}) = G^{-1}(\mathbf{k}+\mathbf{k}',\mathrm{i}\omega_n+\mathrm{i}\omega_{n'}) - G^{-1}(\mathbf{k},\mathrm{i}\omega_n), \tag{13.50}$$

where we have introduced the four-vector notation $k'_\mu = \left(-\omega_{n'}/c,\,\mathbf{k}'\right)$, and where the momentum and Matsubara frequency labels are depicted in Fig. 13.2. Using that the propagators can be written as the sum of a noninteracting part and a selfenergy contribution, we obtain

$$-\frac{c}{e}k'_\mu\gamma^\mu(\mathbf{k},\mathrm{i}\omega_n;\mathbf{k}',\mathrm{i}\omega_{n'}) = G_0^{-1}(\mathbf{k}+\mathbf{k}',\mathrm{i}\omega_n+\mathrm{i}\omega_{n'}) - G_0^{-1}(\mathbf{k},\mathrm{i}\omega_n)$$
$$+\Sigma(\mathbf{k},\mathrm{i}\omega_n) - \Sigma(\mathbf{k}+\mathbf{k}',\mathrm{i}\omega_n+\mathrm{i}\omega_{n'}). \tag{13.51}$$

The lowest-order contribution $\gamma_0^\mu(\mathbf{k},\mathrm{i}\omega_n;\mathbf{k}',\mathrm{i}\omega_{n'})$ to the vertex $\gamma^\mu(\mathbf{k},\mathrm{i}\omega_n;\mathbf{k}',\mathrm{i}\omega_{n'})$ can be calculated from the microscopic action in (13.26), and is given by

$$\gamma_0^0(\mathbf{k},\mathrm{i}\omega_n;\mathbf{k}',\mathrm{i}\omega_{n'}) = \mathrm{i}e,$$
$$\gamma_0^j(\mathbf{k},\mathrm{i}\omega_n;\mathbf{k}',\mathrm{i}\omega_{n'}) = \frac{e}{c}\frac{\hbar}{2m}(k'_j+2k_j). \tag{13.52}$$

The difference of the noninteracting Green's functions on the right-hand side of (13.51) is then seen to be exactly equal to the lowest-order contribution on the left-hand side, which formally proves that the lowest-order Feynman diagram in Fig. 13.1a leads to a gauge-invariant or conserving approximation. We make use of this result by writing the full vertex as

$$\gamma^{\mu}(\mathbf{k},\mathrm{i}\omega_n;\mathbf{k}',\mathrm{i}\omega_{n'}) = \gamma_0^{\mu}(\mathbf{k},\mathrm{i}\omega_n;\mathbf{k}',\mathrm{i}\omega_{n'}) + \Delta\gamma^{\mu}(\mathbf{k},\mathrm{i}\omega_n;\mathbf{k}',\mathrm{i}\omega_{n'}). \qquad (13.53)$$

We obtain the following identity for the vertex correction $\Delta\gamma^{\mu}$:

$$\frac{c}{e}k'_{\mu}\Delta\gamma^{\mu}(\mathbf{k},\mathrm{i}\omega_n;\mathbf{k}',\mathrm{i}\omega_{n'}) = \Sigma(\mathbf{k}+\mathbf{k}',\mathrm{i}\omega_n+\mathrm{i}\omega_{n'}) - \Sigma(\mathbf{k},\mathrm{i}\omega_n). \qquad (13.54)$$

A direct application of this equation to a physical system is postponed until Sect. 16.7, where we theoretically investigate the possibility of probing the excitation spectrum of the Mott-insulator phase using Bragg spectroscopy.

13.4 RF Spectroscopy

A second well-established tool to probe ultracold atomic gases is radio-frequency (RF) spectroscopy, which, unlike Bragg spectroscopy, makes explicit use of the internal hyperfine structure of the atoms by inducing coherent transfer between different internal states. The possibility of coherently transferring atoms from one hyperfine state to another by means of the RF photon field is described by adding an interaction term S_{RF} to the microscopic action, namely

$$S_{\mathrm{RF}}[\psi^*,\psi,A_{\mu}] = \sum_{\alpha,\beta}\int_0^{\hbar\beta}\mathrm{d}\tau\int\mathrm{d}\mathbf{x}\;\gamma^{\mu}_{0;\alpha\beta}\psi^*_{\alpha}(\mathbf{x},\tau)A_{\mu}(\mathbf{x},\tau)\psi_{\beta}(\mathbf{x},\tau), \qquad (13.55)$$

where $2\gamma^{\mu}_{0;\alpha\beta}/\hbar = 2(\gamma^{\mu}_{0;\beta\alpha})^*/\hbar$ is the (single-photon) Rabi frequency. It is defined more precisely in (16.11), and determines the strength of the coherent coupling between the RF photon field and the atoms in the internal states $|\alpha\rangle$ and $|\beta\rangle$. The wavelength of the RF field is usually much larger than the size of the atomic cloud, such that we are able to take the momentum of the RF photons equal to zero. A more detailed account of the origin of this interaction term is left for Chap. 16, where we consider explicitly the interaction between atoms and radiation. Interestingly, the high accuracy of RF spectroscopy is actually used to provide the standard of time, while it has also been applied to precisely determine the size and temperature of ultracold atomic gas clouds [93, 94]. In addition, RF pulses can be used to break up tightly-bound molecules or loosely-bound Cooper pairs, allowing for an experimental study of the binding energy or excitation gap of fermionic pairs [95, 96] throughout the BEC-BCS crossover.

In the following, we consider an interacting Fermi mixture consisting of atoms in the two hyperfine states $|1\rangle$ and $|2\rangle$ with a temperature above the superfluid criti-

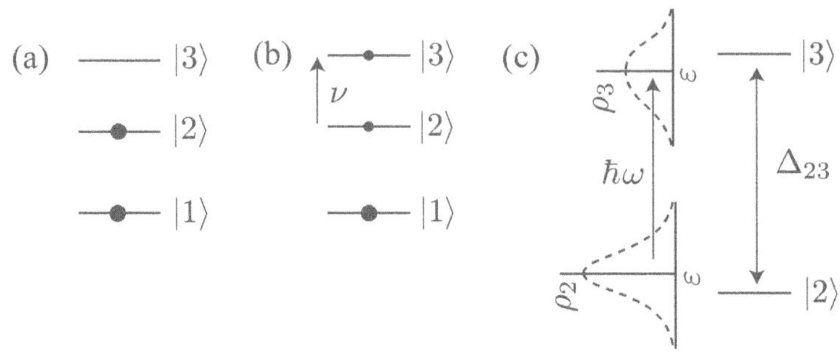

Fig. 13.3 Schematic of the mean-field shift experiment of Gupta et al. [97]. (a) First a mixture of atoms in state $|1\rangle$ and $|2\rangle$ is created. (b) A RF field with a variable frequency $\nu = \omega/2\pi$ drives transitions between state $|2\rangle$ and state $|3\rangle$. (c) An atom in state $|2\rangle$ that interacts with atoms in state $|1\rangle$ has a density of states $\rho_2(\varepsilon)$, while an atom in state $|3\rangle$ has a different density of states $\rho_3(\varepsilon)$ due to the different interaction potential with atoms in state $|1\rangle$. The noninteracting energy difference between states $|2\rangle$ and $|3\rangle$ is denoted by Δ_{23}.

cal temperature T_c. By means of a RF field with a tunable frequency $\hbar\omega$, a transfer of atoms between state $|2\rangle$ and a third state $|3\rangle$ is induced, where the resonance frequency for the transfer is sensitive to the interatomic interaction potentials. This comes about because the energy of atoms in state $|2\rangle$ is shifted due to interactions with atoms in state $|1\rangle$, where a simple Hartree estimate for the shift is given by $4\pi\hbar^2 a_{21} n_1/m$ with n_1 the density of particles in state $|1\rangle$ and a_{21} the scattering length characterizing the strength of the interactions. Similarly, atoms in state $|3\rangle$ are then estimated to experience an energy shift of $4\pi\hbar^2 a_{31} n_1/m$ due to interactions with state $|1\rangle$, so that the energy $\hbar\omega$ needed to induce transitions from $|2\rangle$ to $|3\rangle$ is now estimated as $\hbar\omega = \Delta_{23} + 4\pi\hbar^2(a_{31} - a_{21})n_1/m$, where Δ_{23} is the unperturbed difference in energy between state $|2\rangle$ and $|3\rangle$. More generally speaking, because the interaction potential between atoms in states $|1\rangle$ and $|3\rangle$ is different from that between atoms in states $|1\rangle$ and $|2\rangle$, the density of states in the hyperfine states $|2\rangle$ and $|3\rangle$ is also different. As a result, the resonant RF frequency may vary with interaction strength and is in general not equal to the noninteracting value Δ_{23}, where the difference is also called a mean-field shift or clock shift. This situation is sketched in Fig. 13.3.

We remark that in the absence of atoms in state $|1\rangle$ no mean-field shifts can occur, even if there are interactions between atoms in state $|2\rangle$ and $|3\rangle$. The physical reason is that the RF pulse coherently rotates the atomic spin, such that all atoms that initially are in state $|2\rangle$ are always in the same rotating spin state, which after the pulse is given by some coherent superposition of states $|2\rangle$ and $|3\rangle$. We have explained in Sect. 12.2 that the Pauli principle forbids s-wave interactions between atoms in the same spin state, explaining the absence of a mean-field shift in this particular case. Mathematically, this absence can also be shown rigorously by taking into account the proper vertex corrections [98]. However, here we do not wish to consider this problem in detail, so we simply ignore the presence of interactions between atoms in states $|2\rangle$ and $|3\rangle$.

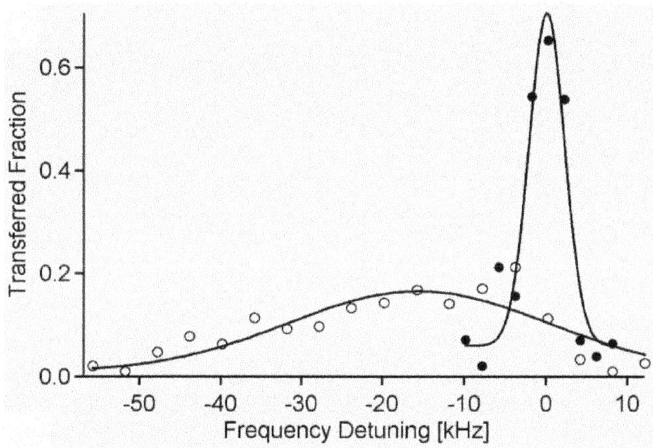

Fig. 13.4 The fraction of atoms transferred from hyperfine state $|2\rangle$ to hyperfine state $|3\rangle$ as a function of the RF frequency $\hbar\omega$ relative to the energy difference Δ_{23} between the two atomic states in vacuum. The difference $\hbar\omega - \Delta_{23}$ is also called the detuning. The solid circles represent data when atoms in state $|1\rangle$ are absent. As a result, the measured RF peak is narrow and unshifted. The open circles represent data when atoms in state $|1\rangle$ are present. Due to interaction effects, the peak is broadened and shifted. From S. Gupta, Z. Hadzibabic, M. W. Zwierlein, C. A. Stan, K. Dieckmann, C. H. Schunck, E. G. M. van Kempen, B. J. Verhaar, and W. Ketterle, Science **300**, 1723 (2003). Reprinted with permission from AAAS.

The sensitivity of the transition frequency between states $|2\rangle$ and $|3\rangle$ to the presence of atoms in state $|1\rangle$ can be probed with high precision using RF spectroscopy, as shown in Fig. 13.4 for a ^{6}Li cloud. The transferred fraction of atoms from state $|2\rangle$ to $|3\rangle$ is obtained as a function of the frequency of the RF field $\hbar\omega$ relative to Δ_{23}, where $\hbar\omega - \Delta_{23}$ is also called the detuning. In the absence of atoms in state $|1\rangle$ the experiment observes a strong peak around zero detuning, such that there is no mean-field shift, as mentioned above. In the presence of atoms in state $|1\rangle$, a clear broadening and a shift in the maximum of the RF spectrum is observed due to the presence of interactions. In Chap. 17, we explain how the interaction strength between the atoms can be tuned experimentally by applying an external magnetic field. In the case of ^{6}Li atoms, the dependence of the s-wave scattering lengths on the applied magnetic field has been determined by a combination of experimental and theoretical methods and is shown in Fig. 13.5. By recording the values of the shifted RF peak for many different values of the magnetic field, Gupta et al. [97] measured the mean-field shift of the spectrum from the weakly-interacting to the strongly-interacting regime. The results are shown in Fig. 13.6. The most surprising feature of these results is that the mean-field shift completely disappears above 650 Gauss, even though there are atoms in state $|1\rangle$ present.

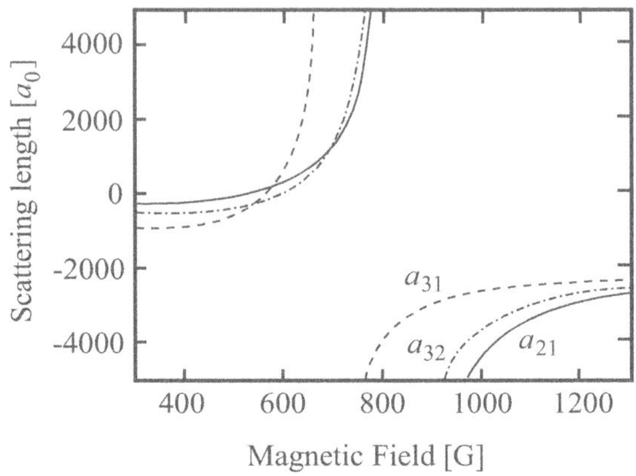

Fig. 13.5 The scattering lengths a_{31}, a_{32}, and a_{21} for the interactions between atoms in the three lowest hyperfine levels $|1\rangle$, $|2\rangle$, and $|3\rangle$ of ^{6}Li as a function of magnetic field.

A simple explanation for the vanishing of the interaction shifts may be based on the observation that beyond 650 G the scattering lengths a_{21} and a_{31} both become very large, as can be seen from Fig. 13.5. As a result, the corresponding two-body T matrices both become unitary limited, such that they do not depend on their scattering lengths anymore as explained in Sect. 10.5. This means that they actually describe identical scattering processes. Let us then consider for convenience the case when the two s-wave scattering lengths a_{21} and a_{31} are exactly equal. A rotation of the many-body wavefunction in spin space from state $|2\rangle$ to $|3\rangle$ now yields exactly the same interaction energy with state $|1\rangle$, such that the interaction shift is absent. If we ignore the energy difference Δ_{23}, the absence of interaction shifts can also be studied using the global $SU(2)$ symmetry of the interacting Hamiltonian under rotation in spin space. The Ward identity associated with this $SU(2)$ symmetry can be derived in the same manner as for the $U(1)$ case of Sect. 13.3. Since it is technically somewhat more involved, we present here only the for our purposes relevant final result, which reads

$$k'_\mu \Delta\gamma^\mu_{32}(\mathbf{k}, i\omega_n; \mathbf{k}', i\omega_{n'}) = \frac{\gamma_{0;32}}{c}\{\Sigma_3(\mathbf{k}+\mathbf{k}', i\omega_n + i\omega_{n'}) - \Sigma_2(\mathbf{k}, i\omega_n)\}. \quad (13.56)$$

In the case of zero momentum for the RF pulse, it becomes

$$-\omega_{n'}\Delta\gamma^0_{32}(\mathbf{k}, i\omega_n; \mathbf{0}, i\omega_{n'}) = \gamma_{0;32}\{\Sigma_3(\mathbf{k}, i\omega_n + i\omega_{n'}) - \Sigma_2(\mathbf{k}, i\omega_n)\}, \quad (13.57)$$

where in the following section we apply this fundamental identity to the RF experiments discussed here.

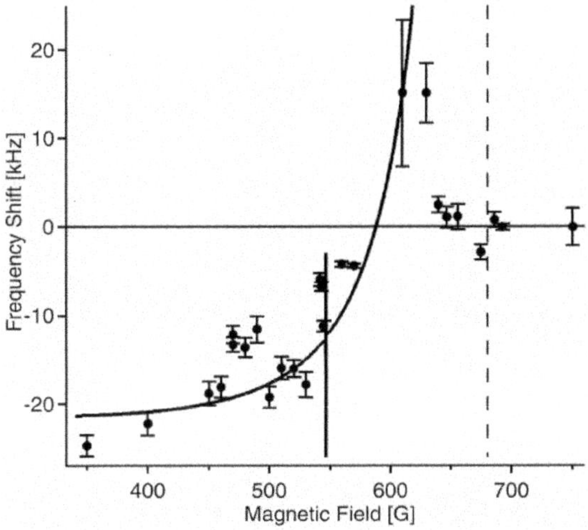

Fig. 13.6 Measurement of the clock shift in the RF spectrum for the transfer of atoms in state $|2\rangle$ to state $|3\rangle$ as a function of magnetic field. Varying the magnetic field causes a change in the interaction strengths a_{21} and a_{31}, so that also the corresponding clock shift changes. From S. Gupta, Z. Hadzibabic, M. W. Zwierlein, C. A. Stan, K. Dieckmann, C. H. Schunck, E. G. M. van Kempen, B. J. Verhaar, and W. Ketterle, Science **300**, 1723 (2003). Reprinted with permission from AAAS.

13.4.1 Second-Order Perturbation Theory

To make the connection between theory and experiment, we first have to identify the observable of the experiment. The RF field induces the transfer of atoms from state $|2\rangle$ to state $|3\rangle$, and we observe the total number of atoms in the final state after applying the RF pulse. This total number of atoms in state $|3\rangle$ depends on the rate at which the RF photons are absorbed by the atomic gas, i.e. it depends on the inverse lifetime of the photon. This is a convenient way to think about the measurement, because it shows that the experiment measures the imaginary part of the photon selfenergy $\Pi(\mathbf{0}, \omega + \mathrm{i}0)$, for which the Feynman diagram is shown in Fig. 13.7a. This Feynman diagram shows that we need to determine the exact atomic propagators, i.e. the exact atomic selfenergies, and the exact three-point vertex. Although we cannot obtain these quantities exactly, we can try to find a reliable approximation that builds upon the fact that we have an exact relation between the selfenergies and the three-point vertices at our disposal in the $SU(2)$-symmetric case, namely the Ward identity from (13.57).

In the next section we determine the selfenergy in the ladder or many-body T matrix approximation, which sums over all two-body scattering processes and is therefore a commonly used approximation. We then use the Ward identity to find the suitable vertex corrections that preserve the $SU(2)$ symmetry, such that we are not artificially breaking any conservation laws. As an example of how this works, we start by considering only the contributions up to second-order in the interaction,

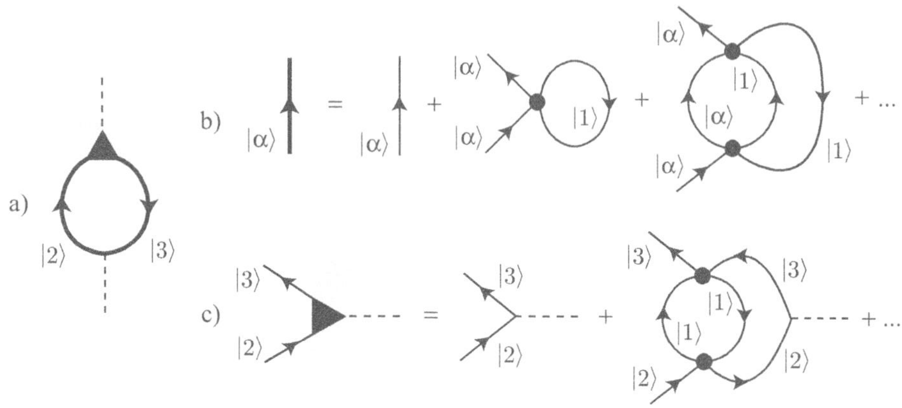

Fig. 13.7 a) Feynman diagram determining the photon selfenergy $\Pi(\mathbf{0},\omega+\mathrm{i}0)$, where the thick lines correspond to the exact atomic propagators and the black triangle denotes the exact atom-photon vertex. b) Feynman diagrams contributing to the exact atomic propagator in the ladder approximation up to second order. The interatomic interactions $V_{\mathbf{0};\alpha 1} = V_{\mathbf{0}}$ are denoted by a black dot. c) Vertex corrections up to second order in the ladder approximation.

where the corresponding Feynman diagrams for the atomic propagator are shown in Fig. 13.7b. The first-order selfenergy diagram is seen to be momentum and frequency independent, such that it cancels in (13.57). The second-order selfenergy contribution, as shown in Fig. 13.7b, is given by

$$\hbar\Sigma_\alpha(\mathbf{k},\mathrm{i}\omega_n+\mathrm{i}\omega_{n'}) = -\frac{V_{\mathbf{0}}^2}{\hbar^2}\frac{\hbar}{(\hbar\beta V)^2}\sum_{\mathbf{k}',m'}\sum_{\mathbf{k}'',m''} G_{0;1}(\mathbf{k}',\mathrm{i}\omega_{m'}) \\ \times G_{0;1}(\mathbf{k}+\mathbf{k}'-\mathbf{k}'',\mathrm{i}\omega_{m'}-\mathrm{i}\omega_{m''})G_{0;\alpha}(\mathbf{k}'',\mathrm{i}\omega_{m''}+\mathrm{i}\omega_n+\mathrm{i}\omega_{n'}). \quad (13.58)$$

Here, the subscript $\alpha = 2,3$ refers to the spin state of the atom and $V_{\mathbf{0}} \equiv V_{\mathbf{0};\alpha 1}$ is the atomic interaction between atoms in state $|\alpha\rangle$ and state $|1\rangle$, where we use a point interaction. The difference of the two selfenergies $\hbar\Sigma_3(\mathbf{k},\mathrm{i}\omega_n+\mathrm{i}\omega_{n'}) - \hbar\Sigma_2(\mathbf{k},\mathrm{i}\omega_n)$ is written as

$$-\frac{V_{\mathbf{0}}^2}{\hbar(\hbar\beta V)^2}\sum_{\mathbf{k}',m'}\sum_{\mathbf{k}'',m''} G_{0;1}(\mathbf{k}',\mathrm{i}\omega_{m'})G_{0;1}(\mathbf{k}+\mathbf{k}'-\mathbf{k}'',\mathrm{i}\omega_{m'}-\mathrm{i}\omega_{m''}) \\ \times\left\{G_{0;3}(\mathbf{k}'',\mathrm{i}\omega_{m''}+\mathrm{i}\omega_n+\mathrm{i}\omega_{n'}) - G_{0;2}(\mathbf{k}'',\mathrm{i}\omega_{m''}+\mathrm{i}\omega_n)\right\} \\ = \hbar\omega_{n'}\frac{\mathrm{i}V_{\mathbf{0}}^2}{(\hbar^2\beta V)^2}\sum_{\mathbf{k}',m'}\sum_{\mathbf{k}'',m''} G_{0;1}(\mathbf{k}',\mathrm{i}\omega_{m'})G_{0;1}(\mathbf{k}+\mathbf{k}'-\mathbf{k}'',\mathrm{i}\omega_{m'}-\mathrm{i}\omega_{m''}) \\ \times G_{0;3}(\mathbf{k}'',\mathrm{i}\omega_{m''}+\mathrm{i}\omega_n+\mathrm{i}\omega_{n'})G_{0;2}(\mathbf{k}'',\mathrm{i}\omega_{m''}+\mathrm{i}\omega_n). \quad (13.59)$$

The latter expression is exactly $-\hbar\omega_{n'}/\gamma_{0;32}$ times the vertex correction in second-order perturbation theory, which is the second diagram shown in Fig. 13.7c. As a

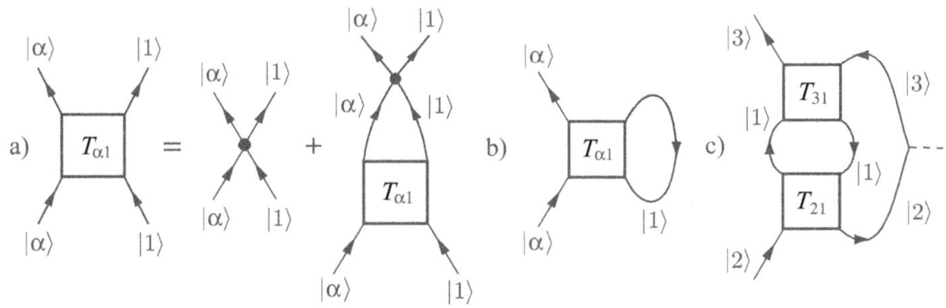

Fig. 13.8 The relevant Feynman diagrams for the calculation of the RF spectrum in the ladder or many-body T matrix approximation. a) The equation for the T matrix. b) Diagrammatic representation of the selfenergy. c) Diagrammatic representation of the vertex corrections.

result, up to second order in the interaction we still satisfy the Ward identity from (13.57).

13.4.2 Ladder Summations

We can then generalize the approach of the previous section by incorporating the complete ladder sum, such that the selfenergy is given by Fig. 13.8b. The spirit is precisely the same, only the notation becomes somewhat more cumbersome. Let us consider the jth-order contribution $\hbar\Sigma_{\alpha}^{(j)}(\mathbf{k}, i\omega_n + i\omega_{n'})$ to the selfenergy, for which we have

$$\begin{aligned}
&\hbar\Sigma_{\alpha}^{(j)}(\mathbf{k}, i\omega_n + i\omega_{n'}) = \\
&-\hbar\left(\frac{-V_0}{\hbar}\right)^j \left(\frac{1}{\hbar\beta V}\right)^j \sum_{\mathbf{k}_1, n_1, \ldots, \mathbf{k}_{j-1}, n_{j-1}} \sum_{\mathbf{k}', m'} \left\{ \prod_{p=1}^{j-1} G_{0;\alpha}(\mathbf{k}_p, i\omega_{n_p} + i\omega_n + i\omega_{n'}) \right. \\
&\qquad \left. \times G_{0;1}(\mathbf{k} + \mathbf{k}' - \mathbf{k}_p, i\omega_{m'} - i\omega_{n_p}) \right\} G_{0;1}(\mathbf{k}', i\omega_{m'}) \, . \qquad (13.60)
\end{aligned}$$

The Ward identity tells us that the appropriate vertex correction corresponding to the above j^{th}-order selfenergy contribution is proportional to $\hbar\Sigma_{3}^{(j)}(\mathbf{k}, i\omega_n + i\omega_{n'}) - \hbar\Sigma_{2}^{(j)}(\mathbf{k}, i\omega_n)$. To obtain a more insightful expression, we add and subtract from this difference the term

$$-\hbar\left(\frac{-V_{\mathbf{0}}}{\hbar^2\beta V}\right)^j \sum_{\mathbf{k}_1,n_1,\ldots,\mathbf{k}_{j-1},n_{j-1}} \sum_{\mathbf{k}',m'}$$
$$\times\left\{\prod_{p=1}^{m} G_{0;3}(\mathbf{k}_p,\mathrm{i}\omega_{n_p}+\mathrm{i}\omega_n+\mathrm{i}\omega_{n'})G_{0;1}(\mathbf{k}+\mathbf{k}'-\mathbf{k}_p,\mathrm{i}\omega_{m'}-\mathrm{i}\omega_{n_p})\right.$$
$$\left.\times\prod_{p=m+1}^{j-1} G_{0;2}(\mathbf{k}_p,\mathrm{i}\omega_{n_p}+\mathrm{i}\omega_n)G_{0;1}(\mathbf{k}+\mathbf{k}'-\mathbf{k}_p,\mathrm{i}\omega_{m'}-\mathrm{i}\omega_{n_p})\right\} G_{0;1}(\mathbf{k}',\mathrm{i}\omega_{m'})$$

with $m = 1,\ldots,j-2$. For two subsequent values of m, where $\hbar\Sigma_2^{(j)}$ corresponds to $m=0$ and $\hbar\Sigma_3^{(j)}$ to $m=j-1$, the obtained expression is rewritten as a common multiplier times the difference $G_{0;3}(\mathbf{k}_p,\mathrm{i}\omega_{n_p}+\mathrm{i}\omega_n+\mathrm{i}\omega_{n'})-G_{0;2}(\mathbf{k}_p,\mathrm{i}\omega_{n_p}+\mathrm{i}\omega_n)$, which is expressed as the product $-\mathrm{i}\omega_{n'}G_{0;3}(\mathbf{k}_p,\mathrm{i}\omega_{n_p}+\mathrm{i}\omega_n+\mathrm{i}\omega_{n'})G_{0;2}(\mathbf{k}_p,\mathrm{i}\omega_{n_p}+\mathrm{i}\omega_n)$. Diagramatically this means that, starting from a Feynman diagram contributing to the selfenergy $\hbar\Sigma_2$, we are cutting every one of the internal lines associated with an atom in state $|2\rangle$ that carries an energy $\mathrm{i}\omega_{n_p}+\mathrm{i}\omega_n$ and gluing it back together with a bare vertex that changes the spin state of the atom and adds an additional energy $\hbar\omega_{n'}$. As a result, the suitable vertex correction is given diagramatically in Fig. 13.8c, while its algebraic expression yields

$$\frac{\Delta\gamma^0_{32}(\mathbf{k},\mathrm{i}\omega_n;\mathbf{0},\mathrm{i}\omega_{n'})}{\gamma_{0;32}} = -\frac{1}{(\hbar^2\beta V)^2}\sum_{\mathbf{k}',m'}\sum_{\mathbf{k}'',m''}$$
$$\times G_{0;3}(\mathbf{k}'',\mathrm{i}\omega_{m''}+\mathrm{i}\omega_n+\mathrm{i}\omega_{n'})G_{0;2}(\mathbf{k}'',\mathrm{i}\omega_{m''}+\mathrm{i}\omega_n)$$
$$\times G_{0;1}(\mathbf{k}',\mathrm{i}\omega_{m'})G_{0;1}(\mathbf{k}+\mathbf{k}'-\mathbf{k}'',\mathrm{i}\omega_{m'}-\mathrm{i}\omega_{m''})$$
$$\times T^{\mathrm{MB}}_{31}\left(\mathbf{k}+\mathbf{k}',\mathrm{i}\omega_{m'}+\mathrm{i}\omega_n+\mathrm{i}\omega_{n'}\right)T^{\mathrm{MB}}_{21}\left(\mathbf{k}+\mathbf{k}',\mathrm{i}\omega_{m'}+\mathrm{i}\omega_n\right)\ . \tag{13.61}$$

13.4.3 Absence of Clock Shift

To summarize the result of the previous section, we now know explicitly what the correct vertex correction is when the selfenergy is given by a sum of ladder diagrams. If the gas is $SU(2)$ invariant, this correction ensures that the Ward identity is satisfied and that the corresponding conservation law is obeyed. The goal is now to use these results to determine the photon selfenergy and the desired photon scattering rate, which determines the RF spectrum. The photon selfenergy, which is also known as the polarization, was diagrammatically shown in Fig. 13.7a, and the corresponding algebraic expression is given by

$$\Pi(\mathbf{0},\mathrm{i}\omega_{n'}) = \tag{13.62}$$
$$\frac{|\gamma_{0;32}|^2}{\hbar^2\beta V}\sum_{\mathbf{k},n} G_3(\mathbf{k},\mathrm{i}\omega_n+\mathrm{i}\omega_{n'})G_2(\mathbf{k},\mathrm{i}\omega_n)\left(1+\frac{\Delta\gamma^0_{32}(\mathbf{k},\mathrm{i}\omega_n;\mathbf{0},\mathrm{i}\omega_{n'})}{\gamma_{0;32}}\right).$$

Next, we will show that in the nearly $SU(2)$ symmetric case it can be rewritten as

$$\Pi(\mathbf{0},\mathrm{i}\omega_{n'}) = \frac{|\gamma_{0;32}|^2}{\hbar^2\beta V}\sum_{\mathbf{k},n}\frac{-\hbar}{\mathrm{i}\hbar\omega_{n'}-\Delta_{32}}\{G_3(\mathbf{k},\mathrm{i}\omega_n+\mathrm{i}\omega_{n'}/2)-G_2(\mathbf{k},i\omega_n-\mathrm{i}\omega_{n'}/2)\}\ , \quad (13.63)$$

which implies that the single-photon absorption spectrum equals

$$I(\omega) = -\frac{2V}{\hbar}\mathrm{Im}\{\Pi(\mathbf{0},\omega+\mathrm{i}0)\} = \frac{2\pi}{\hbar}|\gamma_{0;32}|^2\delta(\hbar\omega-\Delta_{32})\{N_2-N_3\}\ , \quad (13.64)$$

as is also expected from a Fermi's Golden Rule calculation. Note that this equation tells us again that there is no mean-field shift in the nearly $SU(2)$ symmetric case, where we are not fully $SU(2)$ symmetric due to the energy difference Δ_{32} between spin state $|2\rangle$ and $|3\rangle$. The factor N_2-N_3 can be understood from the Pauli blocking that occurs when the state $|3\rangle$ is not empty at the beginning of the RF pulse. The net rate for transferring atoms from state $|2\rangle$ to state $|3\rangle$ is then proportional to $N_2(1-N_3)-N_3(1-N_2)=N_2-N_3$. As we see next, (13.63) is not only valid when the interactions are exactly $SU(2)$ symmetric, but also when $1/T^{2B}_{\alpha 1}$ is negligible compared to $\Xi_{\alpha 1}(\mathbf{K},\mathrm{i}\Omega_n)$. At low temperatures, when the balanced Fermi mixture is degenerate with a Fermi momentum k_F, the conditions $T^{2B}_{\alpha 1}\Xi_{\alpha 1}(\mathbf{K},\mathrm{i}\Omega_n)\gg 1$ can be translated into the conditions $k_F a_{\alpha 1}\gg 1$.

To summarize, from the many-body T matrix expression for the selfenergy, we can derive the corresponding vertex corrections that satisfy the Ward identity for the exactly $SU(2)$ symmetric case. This results in an absorption spectrum with no mean-field shift. Using these same vertex corrections also for strong, but not necessarily exactly $SU(2)$ symmetric interactions, we will find that equation (13.63) still holds. This means that then the mean-field shifts are still absent, in agreement with the observations in Fig. 13.6 for the strongly-interacting regime. The main thing left to do is to prove (13.63), which requires some algebra associated with summing ladder diagrams.

Proof. The first step in proving (13.63) is to substitute (13.61) into (13.62) and recast the result into the more symmetric form

$$\begin{aligned}\Pi(\mathbf{0},\mathrm{i}\omega_{n'}) &= \frac{|\gamma_{0;32}|^2}{\hbar^2\beta V}\sum_{\mathbf{k},n}\frac{-\hbar}{\mathrm{i}\hbar\omega_{n'}-\Delta_{32}}\{G_3(\mathbf{k},\mathrm{i}\omega_n+\mathrm{i}\omega_{n'}/2)-G_2(\mathbf{k},\mathrm{i}\omega_n-\mathrm{i}\omega_{n'}/2)\}\\ &\times\frac{\mathrm{i}\hbar\omega_{n'}-\Delta_{32}-\hbar\tilde{\Sigma}_3(\mathbf{k},\mathrm{i}\omega_n+\mathrm{i}\omega_{n'}/2)+\hbar\tilde{\Sigma}_2(\mathbf{k},\mathrm{i}\omega_n-\mathrm{i}\omega_{n'}/2)}{\mathrm{i}\hbar\omega_{n'}-\Delta_{32}-\hbar\Sigma_3(\mathbf{k},\mathrm{i}\omega_n+\mathrm{i}\omega_{n'}/2)+\hbar\Sigma_2(\mathbf{k},\mathrm{i}\omega_n-\mathrm{i}\omega_{n'}/2)}\ ,\end{aligned} \quad (13.65)$$

where we used

$$G_\alpha(\mathbf{k},\mathrm{i}\omega_n) = \frac{-\hbar}{-\mathrm{i}\hbar\omega_n+\varepsilon_{\mathbf{k},\alpha}-\mu+\hbar\Sigma_\alpha(\mathbf{k},\mathrm{i}\omega_n)}\ , \quad (13.66)$$

and where we have introduced the definitions

$$\hbar\tilde{\Sigma}_3(\mathbf{k}, \mathrm{i}\omega_n + \mathrm{i}\omega_{n'}/2) = -\frac{\hbar}{(\hbar^2\beta V)^2} \sum_{\mathbf{k}',m'} G_{0;1}(\mathbf{k}', \mathrm{i}\omega_{m'})$$
$$\times T_{31}^{\mathrm{MB}}\left(\mathbf{k}+\mathbf{k}', \mathrm{i}\omega_{m'} + \mathrm{i}\omega_n + \mathrm{i}\omega_{n'}/2\right) T_{21}^{\mathrm{MB}}(\mathbf{k}+\mathbf{k}', \mathrm{i}\omega_{m'} + \mathrm{i}\omega_n - \mathrm{i}\omega_{n'}/2)$$
$$\times \sum_{\mathbf{k}'',m''} G_{0;3}(\mathbf{k}'', \mathrm{i}\omega_{m''} + \mathrm{i}\omega_{n'}/2) G_{0;1}(\mathbf{k}+\mathbf{k}'-\mathbf{k}'', \mathrm{i}\omega_{m'} - \mathrm{i}\omega_{m''} + \mathrm{i}\omega_n), \quad (13.67)$$

as well as

$$\hbar\tilde{\Sigma}_2(\mathbf{k}, \mathrm{i}\omega_n - \mathrm{i}\omega_{n'}/2) = -\frac{\hbar}{(\hbar^2\beta V)^2} \sum_{\mathbf{k}',m'} G_{0;1}(\mathbf{k}', \mathrm{i}\omega_{m'})$$
$$\times T_{31}^{MB}\left(\mathbf{k}+\mathbf{k}', \mathrm{i}\omega_{m'} + \mathrm{i}\omega_n + \mathrm{i}\omega_{n'}/2\right) T_{21}^{\mathrm{MB}}(\mathbf{k}+\mathbf{k}', \mathrm{i}\omega_{m'} + \mathrm{i}\omega_n - \mathrm{i}\omega_{n'}/2)$$
$$\times \sum_{\mathbf{k}'',m''} G_{0;2}(\mathbf{k}'', i\omega_{m''} - \mathrm{i}\omega_{n'}/2) G_{0;1}(\mathbf{k}+\mathbf{k}'-\mathbf{k}'', \mathrm{i}\omega_{m'} - \mathrm{i}\omega_{m''} + \mathrm{i}\omega_n)\,. \quad (13.68)$$

Thus, we now have to show that the last factor on the right-hand side of (13.65) equals unity. To this end, we start with considering $\hbar\Sigma_3(\mathbf{k}, i\omega_n + i\omega_{n'}/2) - \hbar\Sigma_2(\mathbf{k}, i\omega_n - i\omega_{n'}/2)$, which in the ladder approximation is given by

$$\hbar\Sigma_3(\mathbf{k}, \mathrm{i}\omega_n + \mathrm{i}\omega_{n'}/2) - \hbar\Sigma_2(\mathbf{k}, \mathrm{i}\omega_n - \mathrm{i}\omega_{n'}/2)$$
$$= \frac{1}{\hbar\beta V} \sum_{\mathbf{k}',m'} \{T_{31}^{\mathrm{MB}}(\mathbf{k}+\mathbf{k}', \mathrm{i}\omega_{m'} + \mathrm{i}\omega_n + \mathrm{i}\omega_{n'}/2)$$
$$-T_{21}^{\mathrm{MB}}(\mathbf{k}+\mathbf{k}', \mathrm{i}\omega_{m'} + \mathrm{i}\omega_n - \mathrm{i}\omega_{n'}/2)\} G_{0;1}(\mathbf{k}', \mathrm{i}\omega_{m'})\,, \quad (13.69)$$

and where from Exercise (10.2) and Sect. 10.4, we recall that the many-body transition matrices can be expressed as $T_{\alpha 1}^{\mathrm{MB}}(\mathbf{K}, \mathrm{i}\Omega_n) = T_{\alpha 1}^{\mathrm{2B}}/(1 - T_{\alpha 1}^{\mathrm{2B}}\Xi_{\alpha 1}(\mathbf{K}, \mathrm{i}\Omega_n))$ with $T_{\alpha 1}^{\mathrm{2B}} = 4\pi a_{\alpha 1}\hbar^2/m$. Next, we rewrite (13.69) as

$$\hbar\Sigma_3(\mathbf{k}, \mathrm{i}\omega_n + \mathrm{i}\omega_{n'}/2) - \hbar\Sigma_2(\mathbf{k}, \mathrm{i}\omega_n - \mathrm{i}\omega_{n'}/2) = \frac{1}{\hbar\beta V} \sum_{\mathbf{k}',m'} G_{0;1}(\mathbf{k}', \mathrm{i}\omega_{m'}) \quad (13.70)$$
$$\times T_{31}^{\mathrm{MB}}(\mathbf{k}+\mathbf{k}', \mathrm{i}\omega_{m'} + \mathrm{i}\omega_n + \mathrm{i}\omega_{n'}/2) T_{21}^{\mathrm{MB}}(\mathbf{k}+\mathbf{k}', \mathrm{i}\omega_{m'} + \mathrm{i}\omega_n - \mathrm{i}\omega_{n'}/2)$$
$$\times \left\{ \frac{1}{T_{21}^{\mathrm{MB}}(\mathbf{k}+\mathbf{k}', \mathrm{i}\omega_{m'} + \mathrm{i}\omega_n - \mathrm{i}\omega_{n'}/2)} - \frac{1}{T_{31}^{\mathrm{MB}}(\mathbf{k}+\mathbf{k}', \mathrm{i}\omega_{m'} + \mathrm{i}\omega_n + \mathrm{i}\omega_{n'}/2)} \right\}.$$

We still have to prove that $\hbar\tilde{\Sigma}_3(\mathbf{k}, i\omega_n + \mathrm{i}\omega_{n'}/2) - \hbar\tilde{\Sigma}_2(\mathbf{k}, \mathrm{i}\omega_n - \mathrm{i}\omega_{n'}/2)$ is equal to $\hbar\Sigma_3(\mathbf{k}, \mathrm{i}\omega_n + \mathrm{i}\omega_{n'}/2) - \hbar\Sigma_2(\mathbf{k}, \mathrm{i}\omega_n - \mathrm{i}\omega_{n'}/2)$, where we start with exactly $SU(2)$ symmetric interactions, i.e. $V_{\mathbf{0};31} = V_{\mathbf{0};21}$. Consider to this end the last factor on the right-hand side of (13.67). The sum over Matsubara frequencies can be evaluated as

$$-\frac{1}{\hbar^2\beta V}\sum_{\mathbf{k}'',m''} G_{0;3}(\mathbf{k}'',\mathrm{i}\omega_{m''}+\mathrm{i}\omega_{n'}/2)G_{0;1}(\mathbf{k}+\mathbf{k}'-\mathbf{k}'',\mathrm{i}\omega_{m'}-\mathrm{i}\omega_{m''}+\mathrm{i}\omega_n)$$
$$=-\frac{1}{\hbar\beta V}\sum_{\mathbf{k}'',m''}\frac{G_{0;3}(\mathbf{k}'',\mathrm{i}\omega_{m''}+\mathrm{i}\omega_{n'}/2)+G_{0;1}(\mathbf{k}+\mathbf{k}'-\mathbf{k}'',i\omega_{m'}-\mathrm{i}\omega_{m''}+\mathrm{i}\omega_n)}{\mathrm{i}\hbar\omega_{m'}+\mathrm{i}\hbar\omega_n+\mathrm{i}\hbar\omega_{n'}/2-\varepsilon_{\mathbf{k}''}-\varepsilon_{\mathbf{k}+\mathbf{k}'-\mathbf{k}''}+2\mu-\Delta_{32}}$$
$$=\Xi_{31}(\mathbf{k}+\mathbf{k}',\mathrm{i}\omega_{m'}+\mathrm{i}\omega_n+\mathrm{i}\omega_{n'}/2)-\frac{1}{V}\sum_{\mathbf{k}''}\frac{1}{2\varepsilon_{\mathbf{k}''}}, \tag{13.71}$$

where we note that the second term in the last line of the above equation is divergent due to the use of the point interaction. Fortunately this divergence cancels, because in the same manner we find for the last factor in (13.68) the result

$$\Xi_{21}(\mathbf{k}+\mathbf{k}',\mathrm{i}\omega_{m'}+\mathrm{i}\omega_n+\mathrm{i}\omega_{n'}/2)-\frac{1}{V}\sum_{\mathbf{k}''}\frac{1}{2\varepsilon_{\mathbf{k}''}},$$

such that we have for the difference

$$\hbar\tilde{\Sigma}_3(\mathbf{k},\mathrm{i}\omega_n+\mathrm{i}\omega_{n'}/2)-\hbar\tilde{\Sigma}_2(\mathbf{k},\mathrm{i}\omega_n-\mathrm{i}\omega_{n'}/2)=\frac{1}{\hbar\beta V}\sum_{\mathbf{k}',m'}G_{0;1}(\mathbf{k}',\mathrm{i}\omega_{m'}) \tag{13.72}$$
$$\times T^{\mathrm{MB}}_{31}(\mathbf{k}+\mathbf{k}',\mathrm{i}\omega_{m'}+\mathrm{i}\omega_n+\mathrm{i}\omega_{n'}/2)T^{\mathrm{MB}}_{21}(\mathbf{k}+\mathbf{k}',\mathrm{i}\omega_{m'}+\mathrm{i}\omega_n-\mathrm{i}\omega_{n'}/2)$$
$$\times\left\{\Xi_{31}(\mathbf{k}+\mathbf{k}',\mathrm{i}\omega_{m'}+\mathrm{i}\omega_n+\mathrm{i}\omega_{n'}/2)-\Xi_{21}(\mathbf{k}+\mathbf{k}',\mathrm{i}\omega_{m'}+\mathrm{i}\omega_n-\mathrm{i}\omega_{n'}/2)\right\}.$$

This we may finally write as

$$\hbar\Sigma_3(\mathbf{k},\mathrm{i}\omega_n+\mathrm{i}\omega_{n'}/2)-\hbar\Sigma_2(\mathbf{k},\mathrm{i}\omega_n-\mathrm{i}\omega_{n'}/2)= \tag{13.73}$$
$$\hbar\tilde{\Sigma}_3(\mathbf{k},\mathrm{i}\omega_n+\mathrm{i}\omega_{n'}/2)-\hbar\tilde{\Sigma}_2(\mathbf{k},\mathrm{i}\omega_n-\mathrm{i}\omega_{n'}/2)$$
$$+\left\{\frac{1}{T^{2\mathrm{B}}_{21}}-\frac{1}{T^{2\mathrm{B}}_{31}}\right\}\frac{1}{\hbar\beta V}\sum_{\mathbf{k}',m'}G_{0;1}(\mathbf{k}',\mathrm{i}\omega_{m'})$$
$$\times T^{\mathrm{MB}}_{31}(\mathbf{k}+\mathbf{k}',\mathrm{i}\omega_{m'}+\mathrm{i}\omega_n+\mathrm{i}\omega_{n'}/2)T^{\mathrm{MB}}_{21}(\mathbf{k}+\mathbf{k}',\mathrm{i}\omega_{m'}+\mathrm{i}\omega_n-\mathrm{i}\omega_{n'}/2).$$

We have thus indeed proven that $\hbar\tilde{\Sigma}_3-\hbar\tilde{\Sigma}_2$ is equal to $\hbar\Sigma_3-\hbar\Sigma_2$ when $V_{\mathbf{0};31}=V_{\mathbf{0};21}$, because then $T^{2\mathrm{B}}_{21}=T^{2\mathrm{B}}_{31}$. Consequently, the last factor of (13.65) is equal to one, as required. Note that we come to exactly the same conclusion if $1/T^{2\mathrm{B}}_{\alpha1}$ is negligible compared to $\Xi_{\alpha1}(\mathbf{K},\mathrm{i}\Omega_n)$, such that $T^{\mathrm{MB}}_{\alpha1}(\mathbf{K},\mathrm{i}\Omega_n)\simeq-1/\Xi_{\alpha1}(\mathbf{K},\mathrm{i}\Omega_n)$.

Using the results from (13.69) and (13.72) we can also consider the weakly-interacting limit, for which $k_{\mathrm{F}}a_{\alpha1}\gg1$. In that case, we have $\hbar\tilde{\Sigma}_\alpha\simeq0$, while $\hbar\Sigma_\alpha=T^{2\mathrm{B}}_{\alpha1}n_1$. Hence, we then find

$$\Pi(\mathbf{0},\mathrm{i}\omega_{n'})=|\gamma_{0;32}|^2\frac{n_2-n_3}{\mathrm{i}\hbar\omega_{n'}-\Delta_{32}-(T^{2\mathrm{B}}_{31}-T^{2\mathrm{B}}_{21})n_1} \tag{13.74}$$

and a RF absorption spectrum of

$$I(\omega) = \frac{2\pi}{\hbar}|\gamma_{0;32}|^2\delta(\hbar\omega - \Delta_{32} - (T_{31}^{2B} - T_{21}^{2B})n_1)\{N_2 - N_3\} , \tag{13.75}$$

which shows that the transition line has a mean-field shift equal to $(T_{31}^{2B} - T_{21}^{2B})n_1 = 4\pi\hbar^2(a_{31} - a_{21})n_1/m$, as was mentioned in the beginning of Sect. 13.4 and was also visible in Fig. 13.6 at magnetic fields that corresponded to the weakly-interacting regime in Fig. 13.5.

13.4.4 Absence of Vertex Corrections

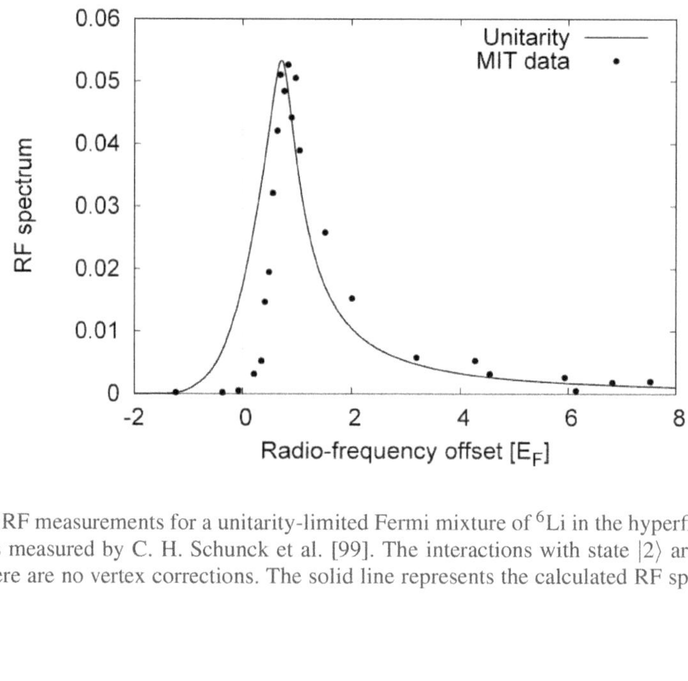

Fig. 13.9 RF measurements for a unitarity-limited Fermi mixture of ^{6}Li in the hyperfine states $|1\rangle$ and $|3\rangle$ as measured by C. H. Schunck et al. [99]. The interactions with state $|2\rangle$ are negligible, so that there are no vertex corrections. The solid line represents the calculated RF spectrum from (13.77).

The previous subsection revealed that a theoretical interpretation of the RF spectrum for interacting Fermi gases can become rather complicated if we insist on satisfying the fundamental Ward identity corresponding to $SU(2)$ symmetry. The full analysis is particularly difficult when both selfenergy effects and vertex corrections have to be taken accurately into account, which is the case when states $|2\rangle$ and $|3\rangle$ are strongly interacting with state $|1\rangle$. By changing the applied magnetic field other physical regimes can be explored, as can be seen explicitly from Fig. 13.5. In particular, by starting out from a mixture of atoms in states $|1\rangle$ and $|3\rangle$ and using RF pulses to induce transitions from $|3\rangle$ to $|2\rangle$ [99], it is possible to end up in a situation where the interactions of state $|2\rangle$ with both other states are negligible, while the interactions between state $|1\rangle$ and $|3\rangle$ are strong. This happens at around 650 Gauss according to Fig. 13.5. In that case, we have that the vertex corrections disappear and (13.62) becomes

$$\Pi(\mathbf{0}, i\omega_{n'}) = \frac{|\gamma_{0;32}|^2}{\hbar^2 \beta V} \sum_{\mathbf{k},n} G_{0;2}(\mathbf{k}, i\omega_n + i\omega_{n'}) G_3(\mathbf{k}, i\omega_n). \tag{13.76}$$

Performing again the Matsubara sum by a contour integration that takes into account the branch cut of $G_3(\mathbf{k}, i\omega_n)$ on the real axis leads ultimately to the result that the RF spectrum can be conveniently expressed in terms of the spectral function $\rho_3(\mathbf{k}, \omega)$ of the atoms in state $|3\rangle$ by means of

$$I(\omega) = \frac{2\pi}{\hbar} |\gamma_{0;32}|^2 \sum_{\mathbf{k}} \rho_3(\mathbf{k}, (\varepsilon_{\mathbf{k}} + \Delta_{23})/\hbar - \omega) N_{\text{FD}}(\varepsilon_{\mathbf{k}} + \Delta_{23} - \mu - \hbar\omega) , \tag{13.77}$$

where we have assumed that state $|2\rangle$ is empty, as is the case in experiment. The theoretical analysis of (13.77) has recently been carried out [100] and the resulting RF spectrum is shown in Fig. 13.9. The agreement with measurements performed by Schunck et al. is seen to be rather good, especially at higher frequencies. The discrepancy at low frequencies is probably due to the fact that the experiments are performed in the superfluid state, whereas we have considered the normal phase here.

13.5 Phase Diffusion

In this section, we discuss the striking phenomenon of phase diffusion [53], which is a direct consequence of the spontaneous $U(1)$ symmetry breaking in the superfluid phase and the finite size of a realistic condensate. In Chap. 11, we described a Bose-Einstein condensate as a coherent state which has a fixed phase and a nonzero expectation value of the atomic field operators. Having a fixed phase, however, implies that there are atom-number fluctuations in the condensate, so that if the number of atoms in the condensate is fixed, the phase of the condensate necessarily undergoes diffusion. This kind of behavior is common to phase transitions that spontaneously break a symmetry. For example, in the case of ferromagnetism, the direction of the magnetization turns out to be diffusing due to the conservation of the total spin angular momentum.

Before discussing the case of a trapped Bose gas, we first consider the same phenomenon for a neutral and homogeneous superconductor, which is technically slightly simpler. Using the approach of Chap. 12, it can be shown that at zero temperature the dynamics of the BCS gap parameter $\Delta(\mathbf{x}, t)$ is to a good approximation determined by a time-dependent Ginzburg-Landau theory [101, 102, 103, 104] with the action

$$S^{\text{eff}}[\Delta^*,\Delta] = \frac{\mathscr{D}(\varepsilon_{\text{F}})}{4V}\int \mathrm{d}t \int \mathrm{d}\mathbf{x} \left\{ \frac{\hbar^2}{|\Delta_0|^2}\left|\frac{\partial \Delta}{\partial t}\right|^2 \right.$$
$$\left. - \frac{\hbar^2 v_{\text{F}}^2}{3|\Delta_0|^2}|\nabla\Delta|^2 + 2|\Delta|^2\left(1-\frac{|\Delta|^2}{2|\Delta_0|^2}\right)\right\}, \qquad (13.78)$$

where $\mathscr{D}(\varepsilon_{\text{F}})$ is the density of states for one spin projection at the Fermi energy $\varepsilon_{\text{F}} = mv_{\text{F}}^2/2$ and Δ_0 is the equilibrium value of the order parameter. Writing the complex order parameter in terms of an amplitude and a phase, we observe that the amplitude fluctuations are gapped and can, therefore, be safely neglected at large length scales. The long-wavelength dynamics of the superconductor is then dominated by the phase fluctuations, for which we obtain the action

$$S^{\text{eff}}[\theta] = \frac{\mathscr{D}(\varepsilon_{\text{F}})\hbar^2}{4V}\int \mathrm{d}t \int \mathrm{d}\mathbf{x} \left\{ \left(\frac{\partial\theta}{\partial t}\right)^2 - \frac{v_{\text{F}}^2}{3}(\nabla\theta)^2 \right\}. \qquad (13.79)$$

This implies that the global phase $\theta_0(t) = \int \mathrm{d}\mathbf{x}\, \theta(\mathbf{x},t)/V$ of the superconductor has a dynamics that is governed by

$$S^{\text{eff}}[\theta_0] = \frac{\mathscr{D}(\varepsilon_{\text{F}})\hbar^2}{4}\int \mathrm{d}t \left(\frac{\mathrm{d}\theta_0}{\mathrm{d}t}\right)^2. \qquad (13.80)$$

Up to now, our discussion has been semiclassical. To consider quantum fluctuations, we have to quantize this theory by applying the usual rules of quantum mechanics. In the present case, the effective action in (13.80) describes the dynamics of the noninteracting degree of freedom $\theta_0(t)$. Just like the Feynman path integral for a free particle leads to the Schrödinger equation, we have that (13.80) leads to the wavefunction $\Psi(\theta_0;t)$, which obeys

$$\mathrm{i}\hbar\frac{\partial}{\partial t}\Psi(\theta_0;t) = -\frac{1}{\mathscr{D}(\varepsilon_{\text{F}})}\frac{\partial^2}{\partial\theta_0^2}\Psi(\theta_0;t), \qquad (13.81)$$

with a ‘diffusion’ constant $1/\mathscr{D}(\varepsilon_{\text{F}})$ that is proportional to $1/N$. In the thermodynamic limit $N \to \infty$, a state with a well-defined stationary phase is clearly a solution, which leads to a system with a spontaneously broken $U(1)$ symmetry. However, for a finite (and fixed) number of particles, the global phase cannot be well defined at all times and always has to ‘diffuse’ in accordance to the above Schrödinger equation. Also note that in the ground state the phase is actually fully undetermined, with $|\Psi(\theta_0;t)|^2 = 1/2\pi$.

Perhaps surprisingly, the same calculation is somewhat more complicated for a Bose gas, because the amplitude fluctuations of the order parameter cannot be neglected even at the largest length scales. However, taking these amplitude fluctuations into account, we nevertheless arrive at an action that is equivalent to (13.80) and hence again leads to the phenomenon of phase diffusion. This is what we show next. We start from the action $S[\rho,\theta;\mu]$ of (11.93), where the difference with Sect. 11.9 is that now we are not so much interested in the dynamics of the density, but rather in the dynamics of the phase. As a result, we now wish to integrate out the density field $\rho(\mathbf{x},\tau)$, which cannot be done exactly. Therefore, we consider here only the strong-coupling or Thomas-Fermi limit, which was treated by Lewenstein and You [53]. In that limit, we are allowed to neglect the gradient of the average density profile [62] and the action $S[\rho,\theta;\mu]$ is for the longest wavelengths well approximated by

$$S[\rho,\theta;\mu] = \int_0^{\hbar\beta} \mathrm{d}\tau \int \mathrm{d}\mathbf{x} \left\{ i\hbar\rho(\mathbf{x},\tau)\frac{\partial\theta(\mathbf{x},\tau)}{\partial\tau} \right. \\ \left. + V^{\mathrm{ex}}(\mathbf{x})\rho(\mathbf{x},\tau) - \mu\rho(\mathbf{x},\tau) + \frac{2\pi a\hbar^2}{m}\rho^2(\mathbf{x},\tau) \right\} . \quad (13.82)$$

In equilibrium, the average density profile of the condensate obeys

$$\langle\rho(\mathbf{x})\rangle = \frac{m}{4\pi a\hbar^2}\left(\mu - V^{\mathrm{ex}}(\mathbf{x})\right)\Theta(\mu - V^{\mathrm{ex}}(\mathbf{x})) . \quad (13.83)$$

Performing the shift $\rho(\mathbf{x},\tau) = \langle\rho(\mathbf{x})\rangle + \delta\rho(\mathbf{x},\tau)$, we find for the zero-momentum part of the action [105]

$$S[\delta N_\mathbf{0},\theta_\mathbf{0};\mu] = \hbar\beta E_\mathbf{0}(\mu) + \int_0^{\hbar\beta} \mathrm{d}\tau \left\{ i\hbar\delta N_\mathbf{0}\frac{\mathrm{d}\theta_\mathbf{0}}{\mathrm{d}\tau} + \frac{2\pi a\hbar^2}{mV_\mathbf{0}(\mu)}(\delta N_\mathbf{0})^2 \right\} , \quad (13.84)$$

where $E_\mathbf{0}(\mu)$ and $V_\mathbf{0}(\mu)$ correspond to the energy and the volume of the condensate in the Thomas-Fermi approximation [60, 61], while $\delta N_\mathbf{0}(\tau) = \int \mathrm{d}\mathbf{x}\ \delta\rho(\mathbf{x},\tau)$ represents the fluctuations in the total number of condensate particles.

Performing the integration over the number fluctuations $\delta N_\mathbf{0}(\tau)$ and Wick rotating to real times $\tau \to it$, we find that the effective action for the global phase of the condensate has precisely the same form as in (13.80), i.e.

$$S^{\mathrm{eff}}[\theta_\mathbf{0};\mu] = \frac{mV_\mathbf{0}(\mu)}{8\pi a}\int \mathrm{d}t \left(\frac{\mathrm{d}\theta_\mathbf{0}}{\mathrm{d}t}\right)^2 , \quad (13.85)$$

where the appropriate 'diffusion' constant is therefore equal to $2\pi a\hbar/mV_\mathbf{0}(\mu)$. This constant can be shown to equal $(1/2\hbar)\partial\mu/\partial N_\mathbf{0}$, if we make use of the the Thomas-Fermi approximation, such that the chemical potential obeys $\mu = m\omega^2 R_{\mathrm{TF}}^2/2$ and the radius of the condensate is given by [75]

$$R_{\mathrm{TF}} = (15a\hbar^2 N_\mathbf{0}/m^2\omega^2)^{1/5}.$$

Hence, the 'diffusion' constant is proportional to $1/N_\mathbf{0}^{3/5}$. For a large number of atoms, it thus takes a very long time for the condensate to diffuse. This calculation therefore explicitly shows why it is a good approximation to describe finite-sized condensates of ultracold atoms as symmetry-broken phases in the thermodynamic limit.

13.6 Problems

Exercise 13.1. Show that the continuity equation, (13.23), holds.

Exercise 13.2. Show explicitly by using the action (13.26) that the diagrams in Fig. 13.1 constitute all the diagrams to first order that contribute to the propagator of the gauge field.

Exercise 13.3. The effective action $S^{\text{eff}}[\theta_0;\mu]$ in principle also contains the topological term $i\hbar N_0(\mu)\int dt\, d\theta_0/dt$, with $N_0(\mu) = \int d\mathbf{x}\, \langle\rho(\mathbf{x})\rangle$ the average number of condensate atoms. Add this topological term to the effective action in (13.85) and derive the Schrödinger equation for the wave function $\Psi(\theta_0;t)$. What is the wave function of the ground state?

Additional Reading

- For a detailed account of effective actions and Ward identities, see J. Zinn-Justin, *Quantum Field Theory and Critical Phenomena*, (Oxford, New York, 1989).

Chapter 14
Renormalization Group Theory

I may not understand the microscopic phenomena at all, but I recognize that there is a microscopic level and I believe it should have certain general, overall properties especially as regards locality and symmetry: Those than serve to govern the most characteristic behavior on scales greater than atomic.
– Michael E. Fisher

In this chapter, we discuss the renormalization-group (RG) approach to quantum field theory. As we will see, renormalization group theory is not only a very powerful technique for studying strongly-interacting problems, but also gives a beautiful conceptual framework for understanding many-body physics in general. The latter comes about because in practice we are often interested in determining the physics of a many-body system at the macroscopic level, i.e. at long wavelengths or at low momenta. As a result we need to eliminate, or integrate out, the microscopic degrees of freedom with high momenta to arrive at an effective quantum field theory for the long-wavelength physics. The Wilsonian renormalization group approach is a very elegant procedure to arrive at this goal. The approach is a transformation that maps an action, characterized by a certain set of coupling constants, to a new action where the values of the coupling constants have changed. This is achieved by performing two steps. First, an integration over the high-momentum degrees of freedom is carried out, where the effect of this integration is absorbed in the coupling constants of the action that are now said to flow. Second, a rescaling of all momenta and fields is performed to bring the relevant momenta of the action back to their original domain. By repeating these two steps over and over again, it is possible to arrive at highly nonperturbative approximations to the exact effective action.

At a continuous phase transition the correlation length diverges, which implies that the critical fluctuations dominate at each length scale and that the system becomes scale invariant. This critical behavior is elegantly captured by the renormalization-group approach, where a critical system is described by a fixed point of the above two-step transformation. By studying the properties of these fixed points, it is possible to obtain accurate predictions for the critical exponents that characterize the nonanalytic behavior of various thermodynamic quantities near the critical point. In particular we find that the critical exponent ν, associated with the divergence of the correlation length, is in general not equal to $1/2$ due to critical fluctuations that go beyond the Landau theory of Chap. 9. It has recently been possible to beautifully confirm this theoretical prediction with the use of ultracold atomic gases. Moreover, the renormalization-group approach also explains universality, which is the observation that very different microscopic actions give rise to exactly the same critical exponents. It turns out that these different microscopic actions then flow to the same fixed point, which is to a large extent solely determined by the dimensionality and the symmetries of the underlying theory. As a result, critical phenomena can be categorized in classes of models that share the same critical behavior. In condensed-matter physics, many phase transitions of interest fall into the XY universality class, such as the transition to the superfluid state in interacting atomic Bose gases and liquid ^{4}He, and the transition to the superconducting state in an interacting Fermi gas. Therefore, we mainly focus on this universality class in the following.

14.1 The Renormalization-Group Transformation

As we have seen in Chap. 9, the order parameter is the central concept in the Landau theory of phase transitions. In the case of a continuous phase transition, two phases are separated by a critical point, and the physics of the system near the critical point are collectively known as critical phenomena. So far, we have made use of effective actions within the language of quantum field theory to describe these critical phenomena. Explicitly, we have seen that we can derive these effective actions from the underlying microscopic action by summing over certain classes of diagrams or by making use of a Hubbard-Stratonovich transformation. Generally speaking, to derive an effective description of the system at the low energy scales of interest, we have to integrate out physical processes that are associated with higher energy scales. In the following we develop another powerful tool to achieve this goal, namely the renormalization group theory of Wilson [106, 107].

We start our discussion of the Wilsonian renormalization group by considering the Landau free energy $F_{\mathrm{L}}[\phi^*,\phi] = F_0[\phi^*,\phi] + F_{\mathrm{int}}[\phi^*,\phi]$ with

$$F_0[\phi^*,\phi] = \int \mathrm{d}\mathbf{x} \left\{ \frac{\hbar^2}{2m} |\boldsymbol{\nabla}\phi(\mathbf{x})|^2 - \mu|\phi(\mathbf{x})|^2 \right\} = \sum_{\mathbf{k}} (\varepsilon_{\mathbf{k}} - \mu)\phi^*_{\mathbf{k}}\phi_{\mathbf{k}} \quad (14.1)$$

$$F_{\mathrm{int}}[\phi^*,\phi] = \int \mathrm{d}\mathbf{x}\, \frac{V_0}{2}|\phi(\mathbf{x})|^4 = \frac{V_0}{2V} \sum_{\mathbf{K},\mathbf{k},\mathbf{k}'} \phi^*_{\mathbf{K}-\mathbf{k}}\phi^*_{\mathbf{k}}\phi_{\mathbf{K}-\mathbf{k}'}\phi_{\mathbf{k}'}, \quad (14.2)$$

where, as we see soon, $F_{\mathrm{L}}[\phi^*,\phi]$ is a free-energy functional that describes phase transitions that belong to the XY universality class, which includes the superfluid-normal transition in liquid helium-4, in the interacting Bose gas and in the interacting Fermi gas. The above free energy only incorporates explicitly the classical fluctuations, which do not depend on imaginary time and do not incorporate the quantum statistics of the interacting particles. Since the interacting Bose and Fermi gases share many critical properties, we may expect that these common classical fluctuations are most important near the critical point. We show more rigorously that this is really the case in Sect. 14.2. Initially, we focus on the superfluid transition for bosons, whereas the fermionic case is left for Sect. 14.3.1. Moreover, to facilitate the discussion, we introduce a cutoff Λ for the wavevectors $\mathbf{k}$ of the order parameter field $\phi(\mathbf{x})$ such that we have

$$\phi(\mathbf{x}) = \frac{1}{\sqrt{V}} \sum_{\mathbf{k}<\Lambda} \phi_{\mathbf{k}} e^{i\mathbf{k}\cdot\mathbf{x}}, \tag{14.3}$$

where we find later that the universal critical properties actually do not depend on this cutoff. Moreover, to keep the discussion as general as possible, we consider the Landau free energy in an arbitrary number of dimensions d. The partition function of the theory is as usual given by

$$Z = \int d[\phi^*]d[\phi] e^{-\beta F_L[\phi^*,\phi]}. \tag{14.4}$$

If we want to calculate this partition function with the use of perturbation theory, as developed in Sect. 8.2, then it turns out that we run into severe problems near the critical point. This is readily seen for the condensation of the ideal Bose gas, where $\mu \to 0$, such that the noninteracting Green's function behaves as $G_0(\mathbf{k}) \propto 1/\mathbf{k}^2$. This leads to low-momentum, or infrared, divergencies in various Feynman diagrams and, as a result, the perturbative expansion breaks down.

To overcome this serious problem, we pursue the following renormalization-group procedure. First, we split the order parameter into two parts, i.e.

$$\phi(\mathbf{x}) = \phi_<(\mathbf{x}) + \phi_>(\mathbf{x}), \tag{14.5}$$

with the lesser and greater fields defined by

$$\phi_<(\mathbf{x}) = \sum_{\mathbf{k}<\Lambda/s} \phi_{\mathbf{k}} \frac{e^{i\mathbf{k}\cdot x}}{\sqrt{V}}, \quad \text{and} \quad \phi_>(\mathbf{x}) = \sum_{\Lambda/s<\mathbf{k}<\Lambda} \phi_{\mathbf{k}} \frac{e^{i\mathbf{k}\cdot x}}{\sqrt{V}} \tag{14.6}$$

with $s > 1$. Then, we integrate in the partition function over the high-momentum part $\phi_>(\mathbf{x})$, so that we obtain in the normal phase

$$\begin{aligned} Z &= \int d[\phi_<^*]d[\phi_<] \int d[\phi_>^*]d[\phi_>] e^{-\beta F_L[\phi_<^*+\phi_>^*,\phi_<+\phi_>]} \\ &= \int d[\phi_<^*]d[\phi_<] e^{-\beta F_0[\phi_<^*,\phi_<]} \int d[\phi_>^*]d[\phi_>] e^{-\beta F_0[\phi_>^*,\phi_>]} e^{-\beta F_{\text{int}}[\phi_<^*,\phi_<,\phi_>^*,\phi_>]} \\ &= \int d[\phi_<^*]d[\phi_<] e^{-\beta F_0[\phi_<^*,\phi_<]} Z_{0;>} \left\langle e^{-\beta F_{\text{int}}[\phi_<^*,\phi_<,\phi_>^*,\phi_>]} \right\rangle_{0;>} \\ &\equiv \int d[\phi_<^*]d[\phi_<] e^{-\beta F'[\phi_<^*,\phi_<;s]}, \end{aligned} \tag{14.7}$$

where we defined the effective free energy $F'[\phi_<^*, \phi_<; s]$ as the outcome of the integration over the high-momentum part $\phi_>(\mathbf{x})$. Moreover, we introduced the average

$$\left\langle e^{A[\phi^*_>,\phi_>]} \right\rangle_{0;>} \equiv \frac{1}{Z_{0;>}} \int d[\phi^*_>] d[\phi_>] e^{-\beta F_0[\phi^*_>,\phi_>]} e^{A[\phi^*_>,\phi_>]}$$
$$= e^{\langle A[\phi^*_>,\phi_>] \rangle_{0;>} + \left(\langle A^2[\phi^*_>,\phi_>] \rangle_{0;>} - \langle A[\phi^*_>,\phi_>] \rangle^2_{0;>} \right)/2 + \dots} , \qquad (14.8)$$

where $Z_{0;>}$ is the noninteracting partition function of the greater field and formally defined by the condition $\langle 1 \rangle_{0;>} = 1$. The second step in the above equation is also known as the cumulant expansion, which is left as an exercise to show up to quadratic order in the exponent. The integration over the high-momentum degrees of freedom is the most important step of the renormalization group, and we perform it in more detail soon.

The next step is calculationally simpler. We perform a scaling transformation $\mathbf{k} \to \mathbf{k}/s$, which brings the cutoff Λ/s back to Λ. As a result, the free energy after the renormalization group transformation is defined on the same momentum interval as before. We still have the freedom to scale the integration variables $\phi^*_<$ and $\phi_<$ in a convenient way, to which we come back in a moment. At the end of this second scaling step the partition function is again of the same form as in (14.4), but now with a new free energy $F[\phi^*,\phi;s]$. We may iterate the above two-step procedure over and over again, where it is convenient to parameterize the result of the jth iteration by the flow parameter $l \equiv \log s^j$. All calculated quantities that depend on the parameter l, in particular the free energy $F[\phi^*,\phi;l]$, are then said to flow. Note that for the flow of the free energy, we have that $F[\phi^*,\phi;0]$ is equal to $F_{\text{L}}[\phi^*,\phi]$, whereas we have calculated the exact free energy if $l \to \infty$, so $F[\phi^*,\phi;\infty] \equiv F$.

There are two important reasons for performing the above two steps. The first step turns out to solve our previously mentioned infrared divergences, because we are always integrating over momentum shells $\Lambda/s < \mathbf{k} < \Lambda$ that do not contain the origin. The second step is convenient, because it leads to an important relation between critical points and fixed points of the renormalization group. To see this we consider the correlation length ξ, which gives the distance over which the order-parameter correlations $\langle \phi^*(\mathbf{x}')\phi(\mathbf{x}) \rangle$ decay. Note that this correlation length is a property of the system, so calculating it from an exact effective free-energy functional should give the same result as calculating it from the corresponding microscopic functional. In particular, this implies that the correlation length does not change by only integrating out high-momentum modes. However, it does change due to the scaling transformation, with which we enhance all momenta and decrease all lengths in the system. This implies that the flowing correlation length $\xi(l)$ is related to the actual correlation length ξ according to

$$\xi(l) = e^{-l}\xi , \qquad (14.9)$$

such that for a nonzero and finite correlation length, we have that $\xi(l)$ goes exponentially to zero. However, as we will see, the renormalization-group transformation also gives rise to fixed points, for which the free energy $F[\phi^*,\phi;l] = F^*[\phi^*,\phi]$ is independent of l. This in particular means that $\xi(l) = \xi^*$, which can only be reconciled with (14.9) if either $\xi = 0$ or $\xi = \infty$. The latter case is of particular interest, since then we are exactly at a critical point. Our conclusion is, therefore, that we can study the critical properties of our system by studying the fixed points of the renormalization-group transformations.

14.1.1 Scaling

Let us focus first in more detail on the second step of the renormalization-group transformation, namely the scaling step, by making the approximation that integration over $\phi_>(\mathbf{x})$ has no effect on the effective free energy $F'[\phi^*_<, \phi_<; s]$ for the low-momentum degrees of freedom. This is exact for the noninteracting case, for which $F_{\text{int}}[\phi^*, \phi] = 0$, so the lesser and greater fields are uncoupled and the integration over fast degrees of freedom only gives rise to a constant shift in the free energy of the slow degrees of freedom. After integrating out the high-momentum part $\phi_>(\mathbf{x})$, we now simply have that

$$F'[\phi^*_<, \phi_<; s] = F_{\text{L}}[\phi^*_<, \phi_<], \tag{14.10}$$

after which we perform the scaling transformation $\mathbf{x} \longrightarrow s\mathbf{x}$ to obtain

$$F[\phi^*_<, \phi_<; s] = \int d\mathbf{x} \left\{ s^{d-2} \frac{\hbar^2}{2m} |\nabla \phi_<|^2 - s^d \mu |\phi_<|^2 + s^d \frac{V_0}{2} |\phi_<|^4 \right\}, \tag{14.11}$$

where for notational convenience we omit the coordinates on which the fields depend. Next, we may also scale the fields as $\phi_< \longrightarrow \phi / s^{(d-2)/2}$, which leads to

$$F[\phi^*, \phi; l] = \int d\mathbf{x} \left\{ \frac{\hbar^2}{2m} |\nabla \phi|^2 - \mu e^{2l} |\phi|^2 + \frac{V_0}{2} e^{(4-d)l} |\phi|^4 \right\}, \tag{14.12}$$

where we also substituted $l = \log(s)$. The reason for choosing this scaling of the fields is that for $\mu = V_0 = 0$ we know that we are at a critical point, namely the one that describes the superfluid transition of the ideal Bose gas. As a result, we also expect to be at a fixed point of the renormalization group, as explained in the previous section. This is indeed clearly seen from (14.12). By defining also $\mu(l) = \mu e^{2l}$ and $V_0(l) = V_0 e^{(4-d)l}$, the renormalization-group equations for $\mu(l)$ and $V_0(l)$ are given by

$$\frac{d\mu(l)}{dl} = 2\mu(l), \tag{14.13}$$

$$\frac{dV_0(l)}{dl} = (4-d)V_0(l). \tag{14.14}$$

What do these equations tell us? First of all, we see that if $d > 4$, then $V_0(l)$ goes exponentially to zero if $l \longrightarrow \infty$. In that case, V_0 is called an irrelevant variable, and we expect that the critical behavior of the system is just determined by the free energy

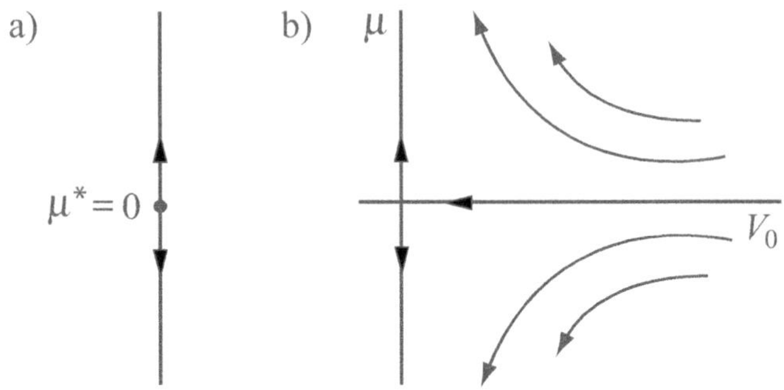

Fig. 14.1 a) Flow diagram for the running chemical potential $\mu(l)$, when $V_0 = 0$. b) Flow diagram for the running chemical potential $\mu(l)$ and the running interaction $V_0(l)$ in $d > 4$, where the interaction is irrelevant.

$$F[\phi^*, \phi; l] = \int \mathrm{d}\mathbf{x} \left\{ \frac{\hbar^2}{2m} |\nabla\phi|^2 - \mu(l)|\phi|^2 \right\}. \tag{14.15}$$

This free energy is fixed for $\mu(l) = \mu^* = 0$, which is also known as the Gaussian fixed point. The renormalization-group flow is now determined by $\mathrm{d}\mu(l)/\mathrm{d}l = 2\mu(l)$, which is graphically represented by Fig. 14.1a. If $\mu(l)$ is initially not exactly zero, then it increasingly deviates from the fixed point under the renormalization-group transformation, and for this reason it is called a relevant variable.

Next, we investigate what happens to the correlation length $\xi = \mathrm{e}^l \xi(l)$ as the chemical potential approaches its critical value $\mu^* = 0$. Since we know from Landau theory that for a free energy of the form of (14.15) we have that $\xi(l) \propto 1/\sqrt{|\mu(l)|}$, we find the behavior

$$\xi \propto \frac{\mathrm{e}^l}{\sqrt{|\mu(l)|}} = \frac{1}{\sqrt{|\mu - \mu^*|}}. \tag{14.16}$$

Introducing the deviation from criticality $\Delta\mu(0) = \mu - \mu^* = \mu$, this behavior can also be understood more formally from the observation that $\xi = \mathrm{e}^l \xi(\mu(l)) = \mathrm{e}^l \xi(\Delta\mu(0)\mathrm{e}^{2l})$ for any value of l. Therefore, we may evaluate the correlation length ξ at the specific value $l = \log(\Delta\mu_0/|\Delta\mu(0)|)/2$, where $\Delta\mu_0$ is an arbitrary energy scale larger than zero. This leads to

$$\xi = \sqrt{\frac{\Delta\mu_0}{|\Delta\mu(0)|}}\, \xi(\pm\Delta\mu_0) \propto \frac{1}{\sqrt{|\mu - \mu^*|}}, \tag{14.17}$$

such that on approach of the critical point, the correlation length diverges as $|\mu - \mu^*|^{-\nu}$ with a critical exponent $\nu = 1/2$. We find in the next section how this result changes, when we further include the effects of interactions. To end this section, we draw the general flow diagram for $d > 4$ in Fig. 14.1b, where we already remark that for $d < 4$ the flow turns out to be very different. Moreover, the case $d = 4$ is special because now V_0 does not flow in the present approximation, such that it is called a marginal variable.

14.1.2 Interactions

To study the critical properties for the case $d \leq 4$, we also need to perform the first step of the renormalization-group transformations. To first order in V_0, we find from (14.2), (14.7) and (14.8) the following correction to the free energy

$$\left\langle \frac{V_0}{2} \int d\mathbf{x} |\phi_<(\mathbf{x}) + \phi_>(\mathbf{x})|^4 \right\rangle_{0;>},$$

where the average is taken with respect to the Gaussian free energy $F_0[\phi_>^*, \phi_>^*]$ and only the high-momentum degrees of freedom are averaged over. The above average gives rise to 16 terms. One term has only lesser fields, namely $F_{\text{int}}[\phi_<^*, \phi_<^*]$, and one has only greater fields, which yields a constant shift for the slow degrees of freedom. Four terms have one greater field, which evaluates to zero after averaging over $F_0[\phi_>^*, \phi_>^*]$ as explained in Example 7.4 for the more general case of an action that also depends on imaginary time. The four terms with three greater fields also average to zero, after which there remain 6 terms with two greater fields. One term contains $\phi_>^* \phi_>^*$ and one $\phi_> \phi_>$, which both average to zero as explained in Example 7.4. Finally, there are four nonzero terms that give rise to

$$2V_0 \int d\mathbf{x} \phi_<^*(\mathbf{x}) \phi_<(\mathbf{x}) \left\langle \phi_>^*(\mathbf{x}) \phi_>(\mathbf{x}) \right\rangle_{0;>},$$

which we may further evaluate using the Fourier transform of (14.6), such that

$$\left\langle \phi_>^*(\mathbf{x}) \phi_>(\mathbf{x}) \right\rangle_{0;>} = \frac{1}{V} \sum_{\Lambda/s<\mathbf{k}<\Lambda} \left\langle \phi_{\mathbf{k}}^* \phi_{\mathbf{k}} \right\rangle_{0;>} = \frac{1}{V} \sum_{\Lambda/s<\mathbf{k}<\Lambda} \frac{1}{\beta(\varepsilon_{\mathbf{k}} - \mu)}, \tag{14.18}$$

where in the first step we used translational invariance, while in the second step we used (2.50) and (14.1) and (14.4).

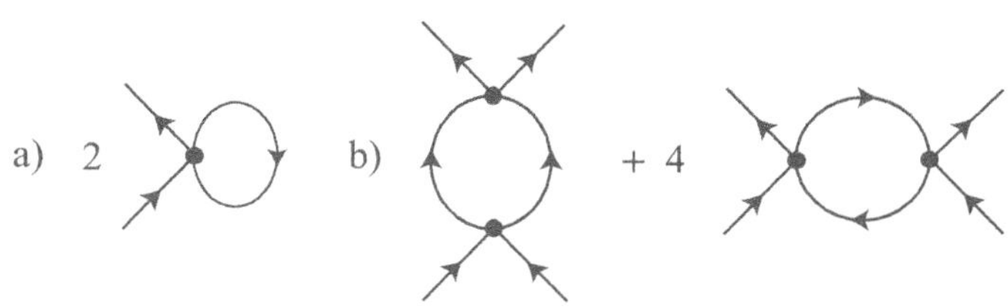

Fig. 14.2 Diagrammatic representation of the correction to a) the chemical potential $\mu(l)$ and b) the interaction $V_0(l)$

As a result, we have found to first order in V_0 that the integration over high-momentum shells gives a correction to the chemical potential μ, which is diagrammatically represented in Fig. 14.2a. Converting the sum into an integral, we find for the first step of the renormalization group

$$\mu' = \mu - 2V_0 \int_{\Lambda/s}^{\Lambda} \frac{\mathrm{d}\mathbf{k}}{(2\pi)^d} \frac{k_{\mathrm{B}}T}{\varepsilon_{\mathbf{k}} - \mu}, \tag{14.19}$$

which after rescaling the momenta and iterating j times, becomes

$$\mu_j = \mu_{j-1} - 2V_0 \int_{\Lambda/s}^{\Lambda} \frac{s^{-dj}\mathrm{d}\mathbf{k}}{(2\pi)^d} \frac{k_{\mathrm{B}}T}{s^{-2j}\varepsilon_{\mathbf{k}} - \mu_{j-1}}. \tag{14.20}$$

We take the continuum limit of this discrete result by renaming $s^j = \mathrm{e}^l$ and integrating out each time a momentum shell of infinitesimal width $\mathrm{d}\Lambda = \Lambda \mathrm{d}l$ and an area $2\pi^{d/2}\,\Lambda^{d-1}/\,\Gamma(d/2)$, where $2\pi^{d/2}/\Gamma(d/2)$ is the solid angle in d dimensions. Here, $\Gamma(z)$ is the Gamma function, and in three dimensions we recover the familiar solid angle of 4π. Substituting the above results in (14.20), we obtain

$$\mathrm{d}\mu = -2V_0 \frac{\Lambda^d}{(2\pi)^d} \frac{2\pi^{d/2}}{\Gamma(d/2)} \frac{k_{\mathrm{B}}T}{\varepsilon_\Lambda \mathrm{e}^{-2l} - \mu} \mathrm{e}^{-ld}\mathrm{d}l, \tag{14.21}$$

We may transform the above equation to a more convenient set of variables by using the same scaling as before, namely $\mu \to \mu \mathrm{e}^{-2l}$ and $V_0 \to V_0 \mathrm{e}^{-(4-d)l}$, so that we get

$$\frac{\mathrm{d}\mu}{\mathrm{d}l} = 2\mu - 2V_0 \frac{\Lambda^d}{(2\pi)^d} \frac{2\pi^{d/2}}{\Gamma(d/2)} \frac{k_{\mathrm{B}}T}{\varepsilon_\Lambda - \mu}, \tag{14.22}$$

where for notational convenience we omitted writing explicitly the l dependence of the flowing variables. Note that, by using the above scaling, we have that the renormalization of the physical chemical potential due to selfenergy effects is determined by $\mu \mathrm{e}^{-2l}$. Also note that if we are integrating out infinitesimal momentum shells we only have to incorporate the effect of one-loop Feynman diagrams, because each additional loop would introduce another factor of $\mathrm{d}\Lambda$. For the renormalization of V_0, we need to go to second order in the interaction to find the one-loop corrections. It is left as an exercise to show that this leads to the Feynman diagrams of Fig. 14.2b, which give rise to

$$\begin{aligned} \frac{\mathrm{d}V_0}{\mathrm{d}l} &= (4-d)V_0 - \frac{\Lambda^d}{(2\pi)^d} \frac{2\pi^{d/2}}{\Gamma(d/2)} V_0^2 \left[\frac{k_{\mathrm{B}}T}{(\varepsilon_\Lambda - \mu)^2} + 4\frac{k_{\mathrm{B}}T}{(\varepsilon_\Lambda - \mu)^2} \right] \\ &= (4-d)V_0 - 5\frac{\Lambda^d}{(2\pi)^d} \frac{2\pi^{d/2}}{\Gamma(d/2)} V_0^2 \frac{k_{\mathrm{B}}T}{(\varepsilon_\Lambda - \mu)^2} . \end{aligned} \tag{14.23}$$

We find that the second term on the right-hand side of (14.23) is negative, so that for $d \geq 4$ we have $V_0(l \to \infty) = 0$. As a result, the present treatment does not change the conclusions from the previous section for $d \geq 4$, so that we have the same flow

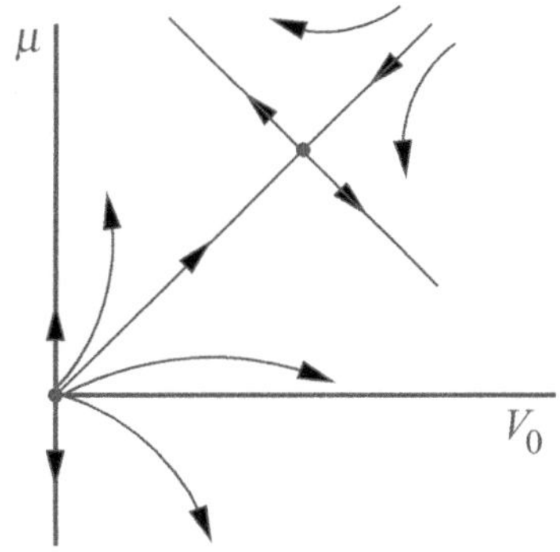

Fig. 14.3 Flow diagram of the running chemical potential μ and the running interaction V_0 for spatial dimensions $d < 4$, where the interaction is relevant and another nontrivial fixed point exists in the flow diagram.

diagram as shown in Fig. 14.1b, which only contains the Gaussian fixed point. Note that in the special case of $d = 4$, the decay of V_0 is actually not exponential, and the interaction is now called a marginally irrelevant variable. We thus come to the conclusion that the critical behavior in these dimensions is determined by the non-interacting theory, so that we find the critical exponent $\nu = 1/2$ for the divergence of the correlation length. However, for $d < 4$, we find the flow diagram shown in Fig. 14.3, which has a new, nontrivial fixed point determined by the equations

$$\begin{aligned} \mu &= V_0 \frac{\Lambda^d}{2^{d-1}} \frac{1}{\pi^{d/2}\Gamma(d/2)} \frac{k_B T}{\varepsilon_\Lambda - \mu}, \\ (4-d)V_0 &= 5V_0^2 \frac{\Lambda^d}{2^{d-1}} \frac{1}{\pi^{d/2}\Gamma(d/2)} \frac{k_B T}{(\varepsilon_\Lambda - \mu)^2}, \end{aligned} \tag{14.24}$$

which are solved by

$$\begin{aligned} \mu^* &= \frac{(4-d)}{(9-d)} \varepsilon_\Lambda, \\ V_0^* &= \frac{5(4-d)}{(9-d)^2} \frac{2^{d-1}\pi^{d/2}\Gamma(d/2)}{\Lambda^d} \frac{\varepsilon_\Lambda^2}{k_B T}. \end{aligned} \tag{14.25}$$

Having found the fixed point (μ^*, V_0^*), we can further investigate the behavior of the critical system by looking at small deviations $\Delta\mu$ and ΔV_0 around the fixed point and linearizing the renormalization-group equations (14.22) and (14.23) in these perturbations. The right-hand side of (14.22) and (14.23) are commonly known as the β functions $\beta_\mu(\mu, V_0)$ and $\beta_{V_0}(\mu, V_0)$ respectively, such that we have

$$\frac{\mathrm{d}\Delta\mu}{\mathrm{d}l} = \left.\frac{\partial \beta_\mu}{\partial \mu}\right|_{\mu^*, V_0^*} \Delta\mu + \left.\frac{\partial \beta_\mu}{\partial V_0}\right|_{\mu^*, V_0^*} \Delta V_0, \tag{14.26}$$

$$\frac{\mathrm{d}\Delta V_0}{\mathrm{d}l} = \left.\frac{\partial \beta_{V_0}}{\partial \mu}\right|_{\mu^*, V_0^*} \Delta\mu + \left.\frac{\partial \beta_{V_0}}{\partial V_0}\right|_{\mu^*, V_0^*} \Delta V_0, \tag{14.27}$$

where it is important to note that the derivatives of the β functions actually do not depend on the cutoff Λ. Moreover, we can diagonalize the above linear differential equations, so that the new coordinates $\Delta\mu'(l)$ and $\Delta V_0'(l)$ belonging to the eigenvalues λ_+ and λ_-, behave as $\Delta\mu'(l) = e^{\lambda_+ l}\Delta\mu'(0)$ and $\Delta V_0'(l) = e^{\lambda_- l}\Delta V'(0)$, where λ_+ is positive and λ_- is negative.

What can we learn from these results? As mentioned before, integrating out momentum shells does not change the correlation length. Therefore, we have that

$$\xi = e^{l}\xi(\mu(l), V_0(l)) , \tag{14.28}$$

which we may also express near the fixed point as

$$\xi = e^{l}\xi(\Delta\mu'(l), \Delta V_0'(l)) . \tag{14.29}$$

Since the eigenvalue λ_- turns out to be negative, we have for large values of l that

$$\xi \simeq e^{l}\xi(e^{\lambda_+ l}\Delta\mu'(0), 0) . \tag{14.30}$$

Taking in particular $l = \log(\Delta\mu_0'/|\Delta\mu'(0)|)/\lambda_+$ with $\Delta\mu_0'$ an arbitrary but nonzero energy scale, for example μ^*, we finally obtain the desired result

$$\xi \simeq \left(\frac{\Delta\mu_0'}{|\Delta\mu'(0)|}\right)^{1/\lambda_+} \xi(\pm\Delta\mu_0', 0) \equiv \left(\frac{\mu^*}{|\mu - \mu^*|}\right)^{1/\lambda_+} \xi_\pm , \tag{14.31}$$

where the different signs distinguish between the behavior on the two sides of the critical point. We see that when $|\Delta\mu'(0)|$ goes to zero, the correlation length diverges with the power $1/\lambda_+$, where if $d < 4$ the critical exponent ν is different from the mean-field value $1/2$. In the case of most interest, $d = 3$, we find $\lambda_+ = 1.878$ which implies that $\nu = 1/\lambda_+ = 0.532 > 1/2$. Even though the difference with the mean-field value may seem small, we note that this is a highly nontrivial result, because all theories we have encountered so far would have predicted $\nu = 1/2$. We also see explicitly that the critical exponent ν does not depend on the cutoff Λ or interaction strength, but only on the dimension d. This is an illustration of the important phenomenon of universality that we alluded to before. In fact, critical exponents in general depend essentially only on the dimensionality and the symmetry of the order parameter. As a result, many physical systems that are microscopically very different share the same critical properties and are therefore said to be in the same universality class.

Similar, but more involved, renormalization group calculations give that $\nu = 0.613$ [108]. At a fixed density, the chemical potential varies as a function of temperature, and our result for the divergence of the correlation length near the critical point can also be expressed as

$$\xi(T) = \left(\frac{T_c}{|T - T_c|}\right)^{\nu} \xi_\pm . \tag{14.32}$$

Recently, Donner et al. tested this fundamental relationship by measuring the critical exponent ν in an ultracold trapped atomic gas of rubidium-87, which was prepared slightly above the critical temperature [109]. Using radio-frequency fields, the spin of the trapped rubidium atoms was flipped at two different positions in the cloud, where the spin flip was from a trapped state to a state that was not trapped. In this way, atoms were extracted from the trapped cloud, after which they were free to expand and to interfere. The performed experiment can be seen as a modern version of a Young two-slit experiment. The interference pattern directly measured the correlation function, $\langle \hat{\psi}^{\dagger}(\mathbf{x})\hat{\psi}(\mathbf{x}')\rangle$, where $\mathbf{x}$ and $\mathbf{x}'$ denote the positions where the atoms were extracted from the cloud. By changing the distance between these positions, it was possible to determine the correlation length from the exponential decay at large separations. Performing the same experiment at different temperatures led to the results in Fig. 14.4, which beautifully confirms the theoretical prediction and in particular clearly shows that $\nu \neq 1/2$.

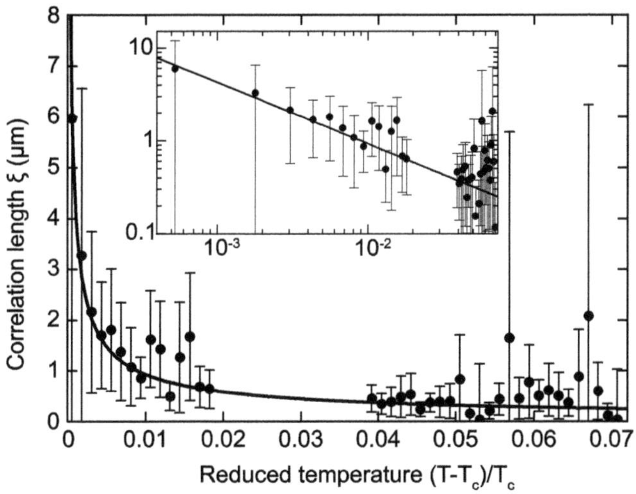

Fig. 14.4 Measurement of the diverging correlation length ξ as a function of temperature in a nearly critical gas of rubidium-87 atoms. The solid line is a fit using (14.32) and gives $\nu \simeq 0.67 \pm 0.13$. From T. Donner, S. Ritter, T. Bourdel, A Öttl, M. Köhl, and T. Esslinger, Science **315**, 1556 (2007). Reprinted with permission from AAAS.

To end this section, let us make a brief connection with the approach to renormalization group theory that is used in high-energy physics. In condensed-matter and statistical physics, we use the renormalization-group approach to regulate the infrared (long-wavelength) behavior of the system, whereas in high-energy physics it is used to regulate the ultra-violet (short-wavelength) behavior of the theory. The latter comes about because in high-energy physics the high-momentum degrees of freedom often lead to divergencies in the Feynman diagrams, which consequently have to be removed by an appropriate renormalization procedure in order to obtain predictive power. To make the translation between these two seemingly very different approaches, we may construct the following table for $|\phi|^4$ theory, where it should be noted that our classical $|\phi|^4$ theory in d dimensions corresponds to the imaginary-time action of a Lorentz invariant $|\phi|^4$ theory in $d-1$ spatial dimensions.

	Condensed-matter physics	High-energy physics
$d > 4$	Irrelevant theory	Nonrenormalizable theory
$d = 4$	Marginal theory	Renormalizable theory
$d < 4$	Relevant theory	Superrenormalizable theory

To explain this table, we first recall that a nonrenormalizable theory contains an infinite number of divergent Feynman diagrams that can only be absorbed by renormalizing an infinite number of coupling constants. An important example of such a theory is the quantum version of Einstein's theory of gravity. A renormalizable theory also contains an infinite number of divergent diagrams, but it requires only a renormalization of a finite number of coupling constants to absorb the infinities. The Standard Model is the ultimate example of a renormalizable theory. Finally, a superrenormalizable theory contains only a finite number of divergent diagrams. From these observations, we may infer that all quantum field theories of high-energy physics are thus actually effective theories, with all (usually unknown) high-energy degrees of freedom integrated out. The resulting theory is then finite at the long-wavelength scales of interest, where these wavelengths are actually still very short compared to the scales considered usually in condensed matter. Note that such an effective high-energy theory does not contain terms that are irrelevant in the long-wavelength limit, because their effect has been integrated out. Since we have seen that the $|\phi|^4$ term is not irrelevant only in $d \leq 4$, we can only obtain an effective renormalized $|\phi|^4$ theory in these dimensions.

14.2 Quantum Effects

Up to now we have considered only classical fluctuations, which we have mentioned to be most important close to the phase transition. To actually show this, we generalize our renormalization-group equations to also include quantum effects. To this end, we consider the partition function

$$Z = \int \mathrm{d}[\phi^*]\mathrm{d}[\phi] e^{-S_{\mathrm{L}}[\phi^*,\phi]/\hbar}, \qquad (14.33)$$

with $\phi(\mathbf{x}, \tau)$ the bosonic order parameter that is periodic in the imaginary-time interval $[0, \hbar\beta]$, and $S_{\mathrm{L}}[\phi^*, \phi]$ the Euclidean action of the order parameter field. This means that we now also consider the quantum dynamics of the order parameter, which was not incorporated in the classical case. More specifically, for our example of interacting bosonic alkali gases or liquid ^{4}He, we look at

$$S_{\mathrm{L}}[\phi^*,\phi] = \int_0^{\hbar\beta} \mathrm{d}\tau \int \mathrm{d}\mathbf{x} \left\{ \phi^*(\mathbf{x},\tau)\hbar\frac{\partial}{\partial\tau}\phi(\mathbf{x},\tau) + \frac{\hbar^2}{2m}|\nabla\phi(\mathbf{x},\tau)|^2 - \mu|\phi(\mathbf{x},\tau)|^2 \right.$$
$$\left. + \frac{V_0}{2}|\phi(\mathbf{x},\tau)|^4 \right\} \equiv S_0[\phi^*,\phi] + S_{\mathrm{int}}[\phi^*,\phi] \tag{14.34}$$

where S_0 is the quadratic part and S_{int} is the quartic part. We Fourier transform $\phi(\mathbf{x},\tau)$ as

$$\phi(\mathbf{x},\tau) = \frac{1}{\sqrt{\hbar\beta V}} \sum_n \sum_{\mathbf{k}<\Lambda} \phi_{\mathbf{k},n} \mathrm{e}^{\mathrm{i}(\mathbf{k}\cdot\mathbf{x}-\omega_n\tau)}, \tag{14.35}$$

with $\omega_n = 2\pi n/\hbar\beta$ the even Matsubara frequencies. We now set up a renormalization-group calculation in the same way as before, where we split the order parameter in lesser and greater fields as

$$\phi(\mathbf{x},\tau) = \phi_<(\mathbf{x},\tau) + \phi_>(\mathbf{x},\tau)$$

and integrate over $\phi_>(\mathbf{x},\tau)$, which contains only the momenta in the shell $\Lambda/s < \mathbf{k} < \Lambda$, but all Matsubara frequencies ω_n.

The next step is to perform a scaling transformation $\mathbf{k} \to \mathbf{k}/s$, $\omega \to \omega/s^z$, where z is called a dynamical critical exponent and is determined as follows. First, we make again the approximation that integrating over $\phi_>(\mathbf{x},\tau)$ has no effect, which is exact for the noninteracting theory, such that we have

$$S'[\phi^*_<,\phi_<;s] = S_{\mathrm{L}}[\phi^*_<,\phi_<]. \tag{14.36}$$

We perform the transformations $\mathbf{x} \to s\mathbf{x}$ and $\tau \to s^z\tau$, which give

$$S'[\phi^*_<,\phi_<;s] = \int_0^{\hbar\beta s^{-z}} \mathrm{d}\tau \int \mathrm{d}\mathbf{x} \left\{ s^d\phi^*_<\hbar\frac{\partial}{\partial\tau}\phi_< + s^{z+d-2}\frac{\hbar^2}{2m}|\nabla\phi_<|^2 \right.$$
$$\left. - s^{z+d}\mu|\phi_<|^2 + s^{z+d}\frac{V_0}{2}|\phi_<|^4 \right\}. \tag{14.37}$$

If we now take $z = 2$ and perform also $\phi_< \to \phi/s^{d/2}$, we obtain

$$S[\phi^*,\phi;l] = \int_0^{\hbar\beta \mathrm{e}^{-2l}} \mathrm{d}\tau \int \mathrm{d}\mathbf{x} \left\{ \phi^*\hbar\frac{\partial}{\partial\tau}\phi + \frac{\hbar^2}{2m}|\nabla\phi|^2 \right.$$
$$\left. - \mu\mathrm{e}^{2l}|\phi|^2 + \mathrm{e}^{(2-d)l}\frac{V_0}{2}|\phi|^4 \right\}, \tag{14.38}$$

where the reason for this scaling is again that it brings about explicitly the criticality of the ideal Bose gas when $\mu = V_0 = 0$. Transforming variables according to $\mu(l) = \mu\mathrm{e}^{2l}$, $V_0(l) = V_0\mathrm{e}^{(2-d)l}$, and $\beta(l) = \beta\mathrm{e}^{-2l}$ we now have three renormalization group equations, namely

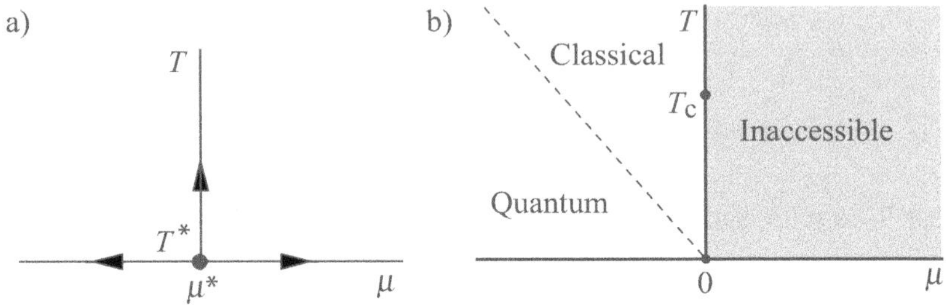

Fig. 14.5 a) Flow diagram and b) resulting phase diagram of the ideal Bose gas near its quantum critical point in $d > 2$.

$$\frac{\mathrm{d}\mu(l)}{\mathrm{d}l} = 2\mu(l), \tag{14.39}$$

$$\frac{\mathrm{d}V_0(l)}{\mathrm{d}l} = (2-d)V_0(l), \tag{14.40}$$

$$\frac{\mathrm{d}\beta(l)}{\mathrm{d}l} = -2\beta(l), \tag{14.41}$$

where the quantity $\beta(l)$ is the flowing upper boundary of the imaginary-time integration, and can be rewritten in terms of a running temperature $T(l) = \mathrm{e}^{2l}T$ as

$$\frac{\mathrm{d}T(l)}{\mathrm{d}l} = 2T(l). \tag{14.42}$$

We thus conclude that for $d > 2$, there is a fixed point for $\mu(l) = \mu^* = 0$ and $T(l) = T^* = 0$ where V_0 always renormalizes exponentially to zero and is thus irrelevant. This fixed point actually describes a quantum phase transition because it lies at zero temperature, where we have only quantum fluctuations. Its critical properties are determined by the action

$$S[\phi^*, \phi; l] = \int_0^{\hbar\beta(l)} \mathrm{d}\tau \int \mathrm{d}\mathbf{x} \left\{ \phi^* \hbar \frac{\partial}{\partial \tau} \phi + \frac{\hbar^2}{2m} |\nabla\phi|^2 - \mu(l)|\phi|^2 \right\}, \tag{14.43}$$

where the corresponding flow diagram is shown in Fig. 14.5a. Using similar arguments as in obtaining (14.17), we find that the correlation length ξ for the decay of the equal-time spatial correlations $\langle \phi^*(\mathbf{x}, \tau)\phi(\mathbf{x}', \tau) \rangle$ diverges as

$$\xi = \mathrm{e}^l \xi(l) \propto \sqrt{\frac{1}{|\mu - \mu^*|}}, \tag{14.44}$$

whereas the correlation time τ_c for the decay of the equal-position temporal correlations $\langle \phi^*(\mathbf{x}, \tau)\phi(\mathbf{x}, \tau') \rangle$ diverges as

$$\tau_c = e^{2l}\tau_c(l) \propto \frac{1}{|\mu - \mu^*|} \propto \frac{1}{|T - T^*|} \propto \xi^z \tag{14.45}$$

with $z = 2$. As a result, we find the phase diagram of the ideal Bose gas in Fig. 14.5b as a function of the chemical potential and temperature. The meaning of the classical and quantum regimes for this phase diagram were explained in Sect. 4.3. In our present language the crossover between quantum and classical behavior is determined by the condition $\tau_c \simeq \hbar\beta$. The phase diagram of Fig. 14.5b is a restricted version of the more general phase diagram shown in Fig. 14.6, which results when interactions are not taken strictly zero for $d > 2$. The effect of interactions is discussed next.

14.2.1 Interactions

To start our discussion of interaction effects, we explain the phase diagram of Fig. 14.6 more explicitly with the use of the Popov theory for Bose-Einstein condensation. Starting from a maximally condensed state at zero temperature, this theory predicts a phase transition to the normal state upon increasing the temperature, when $n_0(T_c) = 0$ and $\mu = 2T^{2B}n$ with n the atomic density. Since the critical temperature in Popov theory is still given by the ideal gas result, we obtain for the critical line

$$T_c = \frac{2\pi\hbar^2}{mk_B}\left(\frac{\mu}{2T^{2B}\zeta(3/2)}\right)^{2/3} . \tag{14.46}$$

Taking the limit of $T^{2B} \downarrow 0$, we recover the somewhat pathological phase diagram of the ideal Bose gas shown in Fig. 14.5, where the superfluid area of the phase diagram collapses into the single line $\mu = 0$. We now focus on the case that $T > 0$, where we have seen in Sect. 14.1.2 that there is a (classical) phase transition, for which we have argued that quantum effects are unimportant to obtain the critical properties. To actually show this, we derive the renormalization-group equation for the quantum theory in the same manner as for the classical theory in Sect. 14.1.2.

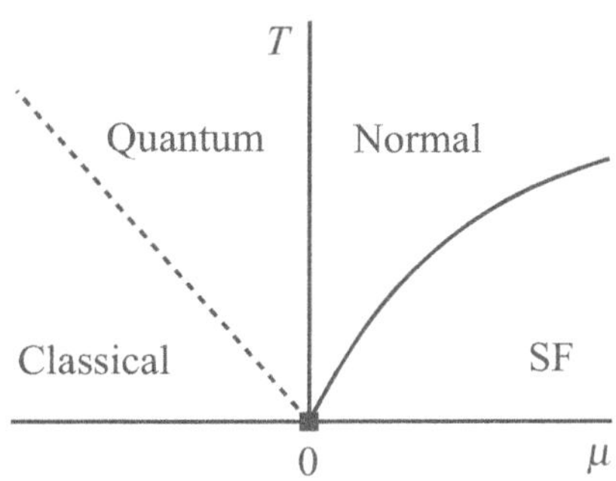

Fig. 14.6 Phase diagram around the quantum critical point of the weakly-interacting Bose gas. The solid line for positive values of the chemical potential distinguishes between the normal state and the superfluid state. The dashed diagonal line for negative chemical potential shows the crossover between the classical regime and the quantum regime.

This leads to exactly the same Feynman diagrams as before, only now we have to sum over all the Matsubara frequencies in each momentum shell. More explicitly, rather than the expression from (14.18), we now of course find

$$\begin{aligned}\langle \phi_{>}^{*}(\mathbf{x},\tau)\phi_{>}(\mathbf{x},\tau)\rangle_{0;>} &= \frac{1}{\hbar\beta V}\sum_{n}\sum_{\Lambda/s<\mathbf{k}<\Lambda}\left\langle \phi_{\mathbf{k},n}^{*}\phi_{\mathbf{k},n}\right\rangle_{0;>} \\ &= \frac{1}{\hbar\beta V}\sum_{n}\sum_{\Lambda/s<\mathbf{k}<\Lambda}\frac{\hbar}{-\mathrm{i}\hbar\omega_n+(\varepsilon_{\mathbf{k}}-\mu)} \\ &= \frac{1}{V}\sum_{\Lambda/s<\mathbf{k}<\Lambda}N_{\mathrm{BE}}(\varepsilon_{\mathbf{k}}-\mu),\end{aligned} \tag{14.47}$$

with $N_{\mathrm{BE}}(\mathbf{x}) = 1/(\mathrm{e}^{x}-1)$ the Bose-Einstein distribution function. In the limit of infinitesimal momentum-shells, we then obtain after rescaling the renormalization-group equations [110]

$$\frac{\mathrm{d}\beta}{\mathrm{d}l} = -2\beta, \tag{14.48}$$

$$\frac{\mathrm{d}\mu}{\mathrm{d}l} = 2\mu - 2V_0\frac{\Lambda^d}{(2\pi)^d}\frac{2\pi^{d/2}}{\Gamma(d/2)}N_{\mathrm{BE}}(\beta(\varepsilon_{\Lambda}-\mu)), \tag{14.49}$$

$$\begin{aligned}\frac{\mathrm{d}V_0}{\mathrm{d}l} &= -V_0 - V_0^2\frac{\Lambda^d}{(2\pi)^d}\frac{2\pi^{d/2}}{\Gamma(d/2)} \\ &\times\left[\frac{1+2N_{\mathrm{BE}}(\beta(\varepsilon_{\Lambda}-\mu))}{2(\varepsilon_{\Lambda}-\mu)} + 4\beta N_{\mathrm{BE}}(\beta(\varepsilon_{\Lambda}-\mu))\left(N_{\mathrm{BE}}(\beta(\varepsilon_{\Lambda}-\mu))+1\right)\right],\end{aligned} \tag{14.50}$$

where for notational convenience we omitted writing explicitly the l dependence of the flowing variables.

The above one-loop corrections for the case of infinitesimal momentum shells are particularly convenient to obtain with the use of the following procedure. The crucial observation here is that all one-loop corrections can be obtained by performing a Gaussian integral. Therefore, we use that in the Gaussian approximation for the greater field $\phi_{>}(\mathbf{x},\tau)$

$$\begin{aligned}Z &= \int \mathrm{d}[\phi_{<}^{*}]\mathrm{d}[\phi_{<}]\mathrm{e}^{-S_0[\phi_{<}^{*},\phi_{<}]/\hbar}\int \mathrm{d}[\phi_{>}^{*}]\mathrm{d}[\phi_{>}]\mathrm{e}^{-S_0[\phi_{>}^{*},\phi_{>}]/\hbar}\mathrm{e}^{-S_{\mathrm{int}}[\phi_{<}^{*},\phi_{<},\phi_{>}^{*},\phi_{>}]/\hbar} \\ &= \int \mathrm{d}[\phi_{<}^{*}]\mathrm{d}[\phi_{<}]\mathrm{e}^{-S_{\mathrm{L}}[\phi_{<}^{*},\phi_{<}]/\hbar}\int \mathrm{d}[\phi_{>}^{*}]\mathrm{d}[\phi_{>}]\exp\left\{\int_0^{\hbar\beta}\mathrm{d}\tau\,\mathrm{d}\tau'\int \mathrm{d}\mathbf{x}\,\mathrm{d}\mathbf{x}'\right. \\ &\left.\times\frac{1}{2}[\phi_{>}^{*}(\mathbf{x},\tau),\phi_{>}(\mathbf{x},\tau)]\cdot\mathbf{G}_{>}^{-1}(\mathbf{x},\tau;\mathbf{x}',\tau')\cdot\begin{bmatrix}\phi_{>}(\mathbf{x}',\tau')\\ \phi_{>}^{*}(\mathbf{x}',\tau')\end{bmatrix}\right\}, \\ &= \int \mathrm{d}[\phi_{<}^{*}]\mathrm{d}[\phi_{<}]\mathrm{e}^{-S_L[\phi_{<}^{*},\phi_{<}]}\exp\left\{-\mathrm{Tr}[\log(-\mathbf{G}_{>}^{-1})]/2\right\},\end{aligned} \tag{14.51}$$

where the trace is over Nambu space, all Matsubara frequencies and the high-momentum shell, while the inverse Green's function matrix $\mathbf{G}_{>}^{-1}$ is given by

$$\mathbf{G}_{>}^{-1}(\mathbf{x},\tau;\mathbf{x}',\tau') \equiv \begin{bmatrix} G_{0;>}^{-1}(\mathbf{x},\tau;\mathbf{x}',\tau') & 0 \\ 0 & G_{0;>}^{-1}(\mathbf{x}',\tau';\mathbf{x},\tau) \end{bmatrix}$$
$$-\frac{1}{\hbar}\begin{bmatrix} 2V_0|\phi_<(\mathbf{x})|^2 & V_0(\phi_<(\mathbf{x}))^2 \\ V_0(\phi_<^*(\mathbf{x}))^2 & 2V_0|\phi_<(\mathbf{x})|^2 \end{bmatrix}\delta(\mathbf{x}-\mathbf{x}')\delta(\tau-\tau')\,. \tag{14.52}$$

Note that in the second step of (14.51) we have neglected the linear terms in $\phi_>$ and $\phi_>^*$. This is allowed in the limit of infinitesimal momentum shells, because only the quadratic part leads to one-particle irreducible one-loop corrections. We can then expand the logarithm from (14.51) in terms of the interaction, just as we did in Sects. 8.7.2 and 12.4, so that to first order in the interaction we obtain the one-loop correction to the chemical potential. Expanding the logarithm to second order yields the one-loop corrections to the interaction, namely the ladder diagram and the bubble diagram from Fig. 14.2. This second-order calculation is analogous to those performed in Sects. 8.7.2 and 12.4, with the main difference that now the internal momenta are restricted to stay in the high-momentum shell. Moreover, for the present calculation we may set the external momenta equal to zero, because we are calculating the renormalization of the momentum-independent coupling V_0. In Sect. 14.3.1, we discuss the role of the external momentum in more detail. It turns out that the above procedure is also very convenient for extending the renormalization group to more difficult situations, such as to the superfluid phase [110].

Returning to the derived renormalization-group equations, we find that (14.48) is easily solved as $\beta(l) = \beta e^{-2l}$, which shows that for large l the temperature always flows to infinity for a nonzero initial temperature. This means that in the vicinity of the critical point the Bose distribution reduces for large values of l to

$$N(\beta(\varepsilon(\Lambda)-\mu)) \simeq \frac{1}{\beta(\varepsilon_\Lambda-\mu)} - \frac{1}{2} = \frac{k_B T e^{2l}}{\varepsilon_\Lambda - \mu} - \frac{1}{2}, \tag{14.53}$$

with which we almost reproduce the classical renormalization-group equations from (14.22) and (14.23). To obtain exactly the same equations we must use the appropriate classical scaling of V_0 again, i.e. $V_0(l) = V_0 e^l$, instead of the scaling $V_0(l) = V_0 e^{-l}$ appropriate for the quantum Gaussian fixed point. The reason why we reproduce the classical renormalization-group equations with the quantum theory near the critical point can be understood by considering the correlation time τ_c, which diverges as $1/|\mu-\mu^*|^{\nu z}$. For $T \neq 0$, we have that the time interval in the original functional integral is restricted to the finite interval $[0,\hbar\beta]$, so that near the critical point we are in the regime $\tau_c \gg \hbar\beta$. Then, the partition function is dominated by contributions from the fluctuations $\phi_{\mathbf{k},n}$ with zero Matsubara frequencies, where for these fluctuations we have that $S_L[\phi^*,\phi] = \hbar\beta F_L[\phi^*,\phi]$. This is the reason why near a classical critical point the quantum theory reduces to the classical theory.

14.2.2 Nonuniversal Quantities

Although we have introduced the renormalization-group equations for their use in studying critical phenomena, they are actually much more general and can be used at any temperature to include both quantum and thermal fluctuations beyond the Popov theory. In this manner it is also possible to determine nonuniversal quantities with the renormalization group, such as the shift in the critical temperature due to the interatomic interactions. Unfortunately, the nonuniversal quantities usually turn out to depend explicitly on the arbitrary high-momentum cutoff Λ that is used to start the renormalization-group flow. However, this problem can be solved by choosing the correct initial conditions for the renormalization-group equations, as we show next.

The initial conditions for the flow of the chemical potential and the temperature are simply equal to the actual physical chemical potential μ and temperature of interest to us. To determine the appropriate initial condition for the interaction $V_0(0)$ we consider the case of two atoms scattering in vacuum, so that $N(\beta(\varepsilon_\Lambda - \mu)) = 0$, $\mu = 0$, and the temperature does not play a role. Then, we obtain for the three-dimensional case $d = 3$ that

$$\frac{\mathrm{d}V_0}{\mathrm{d}l} = -V_0 - V_0^2 \frac{\Lambda^3}{2\pi^2} \frac{1}{2\varepsilon_\Lambda} , \tag{14.54}$$

which, after removing the trivial scaling by substituting $V_0 \to \mathrm{e}^{-l} V_0$, becomes

$$\frac{\mathrm{d}}{\mathrm{d}l} \frac{1}{V_0} = \frac{\Lambda^3}{4\pi^2} \frac{\mathrm{e}^{-l}}{\varepsilon_\Lambda} . \tag{14.55}$$

This equation is readily integrated to give

$$\frac{1}{V_0(\infty)} = \frac{1}{V_0(0)} + \frac{\Lambda^3}{4\pi^2} \frac{1}{\varepsilon_\Lambda} . \tag{14.56}$$

Finally, we make use of the fact that for two atoms we know that the exact effective interaction at low energies is given by the two-body T matrix. As a result, we use $V_0(\infty) = T^{2\mathrm{B}} = 4\pi\hbar^2 a/m$, so that the corresponding initial condition yields

$$V_0(0) = \frac{4\pi\hbar^2 a}{m} \frac{1}{1 - 2a\Lambda/\pi} . \tag{14.57}$$

The exact knowledge of the initial condition for the two-body interaction can consequently also be used for the many-body renormalization-group equations from the previous section. This turns out to eliminate the previously mentioned cutoff dependence, and gives us the possibility to determine with the renormalization group nonuniversal quantities that may be directly compared with experiments. A particularly interesting observable to determine is the shift in the critical temperature due to the interaction effects. It turns out that to study this subtle effect most accurately we need to go beyond the renormalization group equations from the previous section. This goes beyond the scope of this book, but is discussed at length in references [110, 111]. Next, we discuss another application of the possibility to study nonuniversal quantities with the renormalization group, determining the homogeneous phase diagram of a strongly-interacting imbalanced Fermi mixture.

14.3 Renormalization Group for Fermions

In Sect. 12.8 we considered an atomic Fermi gas in two different hyperfine states, which were populated by an equal number of particles. The mixture was at zero temperature and the scattering length between the particles with different spin could be tuned at will. This allowed us to study both theoretically and experimentally the crossover between a Bardeen-Cooper-Schrieffer (BCS) superfluid and a Bose-Einstein condensate (BEC) of diatomic molecules. In between these two extremes there is a region where the scattering length diverges, which is called the unitarity limit. In this strongly-interacting regime, there is no rigorous basis for perturbation theory because there is no natural small parameter. As a result, we found in Sect. 12.8 that the mean-field theory could not be trusted quantitatively, so that more sophisticated theoretical methods have to be invoked in order to get accurate results. In this section, we apply the Wilsonian renormalization-group method to the interacting atomic Fermi mixture, discussing both the weakly and the strongly-interacting case. Moreover, we consider both zero and nonzero temperatures, while we also look at the balanced and the imbalanced case, where the latter means that we have a different number of particles in each of the two hyperfine states.

The two-component Fermi mixture with an unequal number of particles in each spin state is actually a topic of great interest in atomic physics, condensed matter, nuclear matter, and astroparticle physics. Therefore, the landmark atomic-physics experiments with a trapped imbalanced mixture of ^{6}Li, performed at MIT by Zwierlein et al. [112] and at Rice University by Partridge et al. [113], have received a large amount of attention. It turned out that both experiments agree with a phase diagram for the trapped gas that has a tricritical point. This tricritical point separates the second-order superfluid-to-normal transitions from the first-order transitions that occur as a function of temperature and population imbalance [114, 115]. Moreover, the experiments at MIT turned out to be in agreement with predictions using the local-density approximation [116], which implies that the Fermi mixture can be seen as being locally homogeneous. As a result, the MIT group is in the unique position to also map out experimentally the homogeneous phase diagram by performing local measurements in the trap. Most recently, this important experiment was performed by Shin et al. [117], obtaining for the homogeneous tricritical point in the unitarity limit $P_{c3} = 0.20(5)$ and $T_{c3} = 0.07(2)\ T_{F,\uparrow}$, with P the local polarization given by $P = (n_\uparrow - n_\downarrow)/(n_\uparrow + n_\downarrow)$, n_α the density of atoms in spin state $|\alpha\rangle$, and $\varepsilon_{F,\alpha} = k_B T_{F,\alpha} = (6\pi^2 n_\alpha)^{2/3}\hbar^2/2m$ the Fermi energies with m the atomic mass. In this section, we use the Wilsonian renormalization-group approach to find

$P_{c3} = 0.24$ and $T_{c3} = 0.06\ T_{F,\uparrow}$ [118], in good agreement with the experiment by Shin *et al.* [117].

14.3.1 Renormalization-Group Equations

As we have seen, the central idea of Wilsonian renormalization is to subsequently integrate out degrees of freedom in shells at high momenta Λ of infinitesimal width $\mathrm{d}\Lambda$, and absorb the result of the integrations into various coupling constants, which are therefore said to flow. The first step is to calculate the Feynman diagrams renormalizing the coupling constants of interest, while keeping the integration over the internal momenta restricted to the considered high-momentum shell. Only one-loop diagrams contribute to the flow, because the thickness of the momentum shell is infinitesimal and each loop introduces a factor $\mathrm{d}\Lambda$. In order to obtain the exact partition sum, it is then needed to consider an infinite number of coupling constants. Although this is not possible in practice, the renormalization group is still able to distinguish between the relevance of the various coupling constants, such that a carefully selected set of them already leads to highly accurate results. If we wish to treat critical phenomena by looking at renormalization-group fixed points, it is useful to also perform the second step of the renormalization group, which is the rescaling of the momenta, frequencies, and fields. In this section, however, we use the renormalization group to calculate nonuniversal quantities such as the critical temperature for which rescaling is not particularly useful. As a result, we use the renormalization group mainly as a nonperturbative method to iteratively solve a many-body problem.

Consider the action of an interacting Fermi mixture of two different hyperfine states in momentum space, namely

$$S[\phi^*,\phi] = \sum_{\mathbf{k},n,\alpha} \phi^*_{\mathbf{k},n,\alpha}(-\mathrm{i}\hbar\omega_n + \varepsilon_{\mathbf{k}} - \mu_\alpha)\phi_{\mathbf{k},n,\alpha} + \frac{1}{\hbar\beta V} \sum_{\substack{\mathbf{k},\mathbf{k}',\mathbf{K}\\ n,n',m}} V_{\mathbf{K},m}\phi^*_{\mathbf{K}-\mathbf{k}',m-n',\uparrow}\phi^*_{\mathbf{k}',n',\downarrow}\phi_{\mathbf{K}-\mathbf{k},m-n,\downarrow}\phi_{\mathbf{k},n,\uparrow}\ , \tag{14.58}$$

where n and n' are odd, m is even, μ_α is the chemical potential for spin state $|\alpha\rangle$, $V_{\mathbf{K},m}$ is the interaction vertex, and $\alpha = \uparrow, \downarrow$. Note that by using two different chemical potentials we are in the position to also discuss the imbalanced Fermi gas. Moreover, we consider an interaction $V_{\mathbf{K},m}$ that in general depends on the center-of-mass frequency $\mathrm{i}\Omega_m$ and the center-of-mass momentum $\mathbf{K}$, for which the reason soon becomes clear.

In Fig. 14.7, we have drawn the by now familiar Feynman diagrams renormalizing μ_α and $V_{\mathbf{K},m}$. To start with a simple Wilsonian renormalization group, we take the interaction vertex to be frequency and momentum independent. If we then consider the three coupling constants μ_α and $V_{\mathbf{0},0}$, we obtain in a similar manner as for

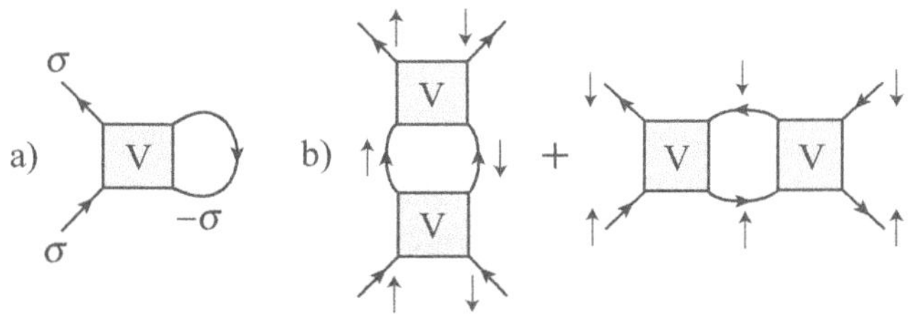

Fig. 14.7 Feynman diagrams renormalizing a) the chemical potentials and b) the interatomic interaction.

the Bose case that

$$\frac{\mathrm{d}V_{\mathbf{0},0}^{-1}}{\mathrm{d}\Lambda} = \frac{\Lambda^2}{2\pi^2}\left[\frac{1-N_\uparrow-N_\downarrow}{2(\varepsilon_\Lambda-\mu)} - \frac{N_\uparrow-N_\downarrow}{2h}\right], \tag{14.59}$$

$$\frac{\mathrm{d}\mu_\alpha}{\mathrm{d}\Lambda} = -\frac{\Lambda^2}{2\pi^2}\frac{N_{-\alpha}}{V_{\mathbf{0},0}^{-1}}, \tag{14.60}$$

where we have introduced $\mu=(\mu_\uparrow+\mu_\downarrow)/2$, $h=(\mu_\uparrow-\mu_\downarrow)/2$ and the Fermi distributions $N_\alpha = 1/\{\exp\{\beta(\varepsilon_\Lambda-\mu_\alpha)\}+1\}$. These expressions are readily obtained from the diagrams in Fig. 14.7 by setting all external frequencies and momenta equal to zero and by performing in each loop the full Matsubara sum over internal frequencies, while integrating the internal momenta over the infinitesimal shell $\mathrm{d}\Lambda$. The first term in (14.59) corresponds to the ladder diagram and describes the scattering between particles. The second term corresponds to the bubble diagram and describes screening of the interaction by particle-hole excitations. Also note that due to the coupling of the differential equations for μ_α and $V_{\mathbf{0},0}^{-1}$ we automatically generate an infinite number of Feynman diagrams, showing the nonperturbative nature of the renormalization group.

When the Fermi mixture becomes critical, the inverse many-body vertex $V_{\mathbf{0},0}^{-1}$ flows to zero according to the Thouless criterion discussed in Sect. 12.2. This unfortunately leads to anomalous behavior in (14.60), where the chemical potentials are seen to diverge. To solve this issue and calculate the critical properties realistically we need to go beyond our simple renormalization group, which can be achieved by taking also the frequency and momentum dependence of the interaction vertex into account. Although the two-body interaction in ultracold atomic gases is to a very good approximation constant in Fourier space, the renormalization-group transformation still generates momentum and frequency dependence of the interaction vertex due to many-body effects. The ladder and the bubble diagrams renormalizing the two-body interaction are both momentum and frequency dependent, where the ladder diagram was treated in more detail in Exercise 10.2. It depends only on the center-of-mass coordinates $\mathbf{K}$ and $\mathrm{i}\Omega_m$, so that its contribution to the renormalization of $V_{\mathbf{K},m}^{-1}$ is given by

$$\mathrm{d}\Xi(K^2, \mathrm{i}\Omega_m) = \int_{\mathrm{d}\Lambda} \frac{\mathrm{d}\mathbf{k}}{(2\pi)^3} \frac{1 - N_\uparrow(\varepsilon_\mathbf{k}) - N_\downarrow(\varepsilon_{\mathbf{K}-\mathbf{k}})}{\mathrm{i}\hbar\Omega_m - \varepsilon_\mathbf{k} - \varepsilon_{\mathbf{K}-\mathbf{k}} + 2\mu}, \tag{14.61}$$

where during integration both $\mathbf{k}$ and $\mathbf{K}-\mathbf{k}$ have to remain in the infinitesimal shell $\mathrm{d}\Lambda$. Since the ladder diagram is already present in the two-body limit, it is most important for two-body scattering properties. For this reason, the interaction vertex is mainly dependent on the center-of-mass coordinates, so we may neglect the dependence of the vertex on other frequencies and momenta. We will return to the validity of this approximation later. The way to treat the external frequency and momentum dependence in a Wilsonian renormalization group is to introduce new couplings by expanding the (inverse) interaction in the following way: $V^{-1}_{\mathbf{K},m} = V^{-1}_{\mathbf{0},0} - Z^{-1}_K K^2 + Z^{-1}_\Omega \mathrm{i}\hbar\Omega_m$. The flow equations for the additional coupling constants Z^{-1}_K and Z^{-1}_Ω are then obtained by

$$\mathrm{d}Z^{-1}_K = \left.\frac{\partial \mathrm{d}\Xi(K^2,\Omega)}{\partial K^2}\right|_{K=\Omega=0} \tag{14.62}$$

and

$$\mathrm{d}Z^{-1}_\Omega = -\left.\frac{\partial \mathrm{d}\Xi(K^2,\Omega)}{\partial \hbar\Omega}\right|_{K=\Omega=0}. \tag{14.63}$$

14.3.2 Extremely-Imbalanced Case

First, we apply the renormalization group to one spin-down particle in a Fermi sea of spin-up particles at zero temperature in the unitarity limit. The full equation of state for the normal state of a strongly-interacting Fermi mixture was obtained at zero temperature using Monte-Carlo techniques [116]. For large imbalances, the dominant feature in this equation of state is the selfenergy of the spin-down atoms in the sea of spin-up particles [116, 119]. We can also calculate this selfenergy with the renormalization group, where we consider the extreme imbalanced limit, which means that we have only one spin-down particle. The equations are now simplified, because $N_\downarrow$ can be set to zero and thus $\mu_\uparrow$ is not renormalized. Next, we have to incorporate the momentum and frequency dependence of the interaction in the one-loop Feynman diagram for the renormalization of $\mu_\downarrow$. In this particular case, the external frequency dependence of the ladder diagram can be taken into account exactly. It is namely possible to show with the use of contour integration that the one-loop Matsubara sum simply leads to the substitution $\mathrm{i}\hbar\Omega_m \to \varepsilon_\mathbf{K} - \mu_\uparrow$ in (14.61) [119]. The external momentum dependence is accounted for by the coupling Z^{-1}_K, giving

$$\frac{dV_{\mathbf{0},0}^{-1}}{d\Lambda} = \frac{\Lambda^2}{2\pi^2}\left[\frac{1-N_\uparrow}{2\varepsilon_\Lambda - \mu_\downarrow} - \frac{N_\uparrow}{2h}\right], \tag{14.64}$$

$$\frac{d\mu_\downarrow}{d\Lambda} = \frac{\Lambda^2}{2\pi^2}\frac{N_\uparrow}{-\Gamma_{\mathbf{0},0}^{-1} + Z_K^{-1}\Lambda^2}, \tag{14.65}$$

$$\frac{dZ_K^{-1}}{d\Lambda} = -\frac{\hbar^4\Lambda^4}{6\pi^2 m^2}\frac{1-N_\uparrow}{(2\varepsilon_\Lambda - \mu_\downarrow)^3}. \tag{14.66}$$

Note that these equations only have poles for positive values of $\mu_\downarrow$. Since this will not occur, we can use $\Lambda(l) = \Lambda_0 e^{-l}$ and $d\Lambda = -\Lambda_0 e^{-l} dl$ to integrate out all momentum shells, where we note that an additional minus sign is needed, because we are integrating from high to low momenta. We then obtain a system of three coupled ordinary differential equations in l which are easily solved numerically. In the unitarity limit, the initial condition from (14.57) becomes $V_{\mathbf{0},0}^{-1}(0) = -m\Lambda_0/2\pi^2\hbar^2$. The other initial conditions are $\mu_\downarrow(0) = \mu_\downarrow$ and $Z_K^{-1}(0) = 0$, because the interaction starts out as being momentum independent. Note that in this calculation $\mu_\downarrow(0) = \mu_\downarrow$ is initially negative and increases during the flow due to the strong attractive interactions. The quantum phase transition from a zero density to a nonzero density of spin-down particles occurs for the initial value $\mu_\downarrow$ that at the end of the flow precisely leads to $\mu_\downarrow(\infty) = 0$. This happens when $\mu_\downarrow = -0.598\mu_\uparrow$, which is therefore the selfenergy of a strongly-interacting spin-down particle in a sea of spin-up particles. It is in excellent agreement with the most recent Monte-Carlo result $\mu_\downarrow = -0.594\mu_\uparrow$ [120], although it is obtained with much less numerical effort. In particular, this result also implies that we agree with the prediction of a first-order quantum phase transition from the normal phase to the superfluid phase at a critical imbalance of $P = 0.39$, which was shown to follow from the Monte-Carlo calculations [116, 120].

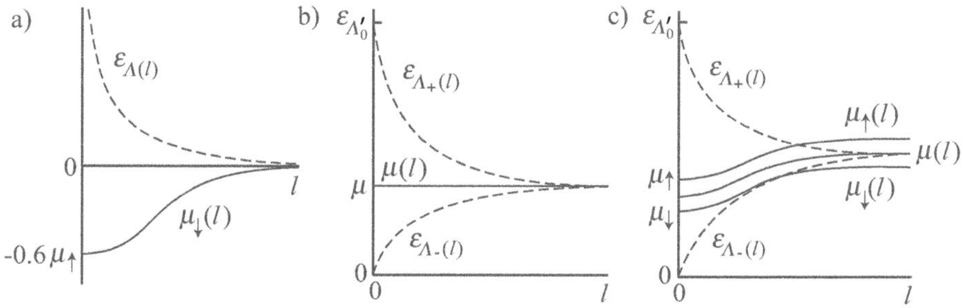

Fig. 14.8 a) Position of the momentum shells (dashed lines) and flow of the chemical potentials (solid lines) for a) the strongly-interacting extremely-imbalanced case, b) the weakly-interacting balanced case, and c) the strongly-interacting imbalanced case.

14.3.3 Homogeneous Phase Diagram

Next, we determine the critical properties of the strongly-interacting Fermi mixture at nonzero temperatures, and in particular the location of the tricritical point in the homogeneous phase diagram. Since it is not exact to make the substitution $i\hbar\Omega_m \to \varepsilon_{\mathbf{K}} - \mu_{-\alpha}$ at nonzero temperatures, we take the frequency dependence of the ladder diagram into account through the renormalization of the coupling Z_{Ω}^{-1}. While the flow of $V_{\mathbf{0},0}^{-1}$ is still given by (14.59), the expressions for the flow of μ_α and Z_{Ω}^{-1} become

$$\frac{\mathrm{d}\mu_\alpha}{\mathrm{d}\Lambda} = \frac{\Lambda^2}{2\pi^2} \frac{N_{-\alpha} + N_{\mathrm{B}}}{-V_{\mathbf{0},0}^{-1} + Z_K^{-1}\Lambda^2 - Z_{\Omega}^{-1}(\varepsilon_\Lambda - \mu_{-\alpha})}, \tag{14.67}$$

$$\frac{\mathrm{d}Z_{\Omega}^{-1}}{\mathrm{d}\Lambda} = \frac{\Lambda^2}{2\pi^2} \frac{1 - N_\uparrow - N_\downarrow}{4(\varepsilon_\Lambda - \mu)^2}, \tag{14.68}$$

with $N_{\mathrm{B}} = 1/\{\exp[\beta Z_{\Omega}(-V_{\mathbf{0},0}^{-1} + Z_K^{-1}\Lambda^2)] - 1\}$ coming from the bosonic frequency dependence of the interaction. The flow equation for Z_K^{-1} can be obtained analytically from (14.62), but is too cumbersome to write down explicitly. The initial conditions are the same as for the extremely imbalanced case with in addition $\mu_\uparrow(0) = \mu_\uparrow$ and $Z_{\Omega}^{-1}(0) = 0$. As mentioned before, the critical condition is that the inverse of the fully renormalized vertex $V_{\mathbf{0},0}^{-1}(\infty)$, which can be seen as the inverse many-body T matrix at zero external momentum and frequency, goes to zero. Physically, this implies that a (many-body) bound state is entering the system. From (14.67) we see that incorporating the coupling constants Z_K^{-1} and Z_{Ω}^{-1}, and thereby taking the dependence of the interaction on the center-of-mass momentum and frequency into account, is crucial to solve the previously mentioned problem of the diverging chemical potential.

We see that in the above renormalization group equations there is only a pole at the average Fermi level $\mu = (\mu_\uparrow + \mu_\downarrow)/2$. For fermions, the excitations of lowest energy lie near their Fermi energies, which is therefore the natural end point for a renormalization group flow [121]. A notorious problem for interacting fermions is that under renormalization the Fermi levels also flow to a priori unknown values, making the Wilsonian renormalization group difficult to perform in practice. Next, we show how to obtain renormalization-group equations that automatically flow to the final value of the renormalized average Fermi level. To this end, we integrate out all momentum shells with the following procedure. First, we start at a high momentum cutoff Λ_0 and flow to a momentum Λ_0' at roughly two times the average Fermi momentum, with the individual Fermi momenta given by $k_{\mathrm{F},\alpha} = \sqrt{2m\varepsilon_{\mathrm{F},\alpha}}/\hbar$. This integrates out the high-energy two-body physics, but hardly affects the chemical potentials. Then, we start integrating out the rest of the momentum shells symmetrically with respect to the flowing average Fermi level. This is achieved by using

$$\Lambda_+(l) = \left(\Lambda_0' - \sqrt{\frac{2m\mu}{\hbar^2}}\right) \mathrm{e}^{-l} + \sqrt{\frac{2m\mu(l)}{\hbar^2}} \tag{14.69}$$

and by

$$\Lambda_-(l) = -\sqrt{\frac{2m\mu}{\hbar^2}} \mathrm{e}^{-l} + \sqrt{\frac{2m\mu(l)}{\hbar^2}}. \tag{14.70}$$

Note that, as desired, $\Lambda_+(l)$ starts at Λ_0' and automatically flows from above to $\sqrt{2m\mu(\infty)}/\hbar$, whereas $\Lambda_-(l)$ starts at 0 and automatically flows from below to $\sqrt{2m\mu(\infty)}/\hbar$. By substituting $\Lambda_+(l), \Lambda_-(l)$ and their derivatives in (14.59), (14.62), (14.67) and (14.68) we obtain a set of coupled differential equations in l that can be solved numerically.

We first apply the above procedure to study the equal density case, i.e. $h = 0$, as a function of negative scattering length a. The scattering length enters the calculation through the initial condition of $V_{\mathbf{0},0}^{-1}$. To express our results in terms of the Fermi energy $\varepsilon_{\mathrm{F}} = \varepsilon_{\mathrm{F},\alpha}$, we calculate the densities of atoms with the flow equation $\mathrm{d}n_\alpha/\mathrm{d}\Lambda = \Lambda^2 N_\alpha/2\pi^2$. In the weak-coupling limit, $a \to 0^-$, the chemical potentials hardly renormalize, so that only (14.59) is important. The critical temperature becomes exponentially small, which allows us to integrate (14.59) exactly with the result $k_{\mathrm{B}}T_{\mathrm{c}} = 8\varepsilon_{\mathrm{F}}\mathrm{e}^{\gamma-3}\exp\{-\pi/2k_{\mathrm{F}}|a|\}/\pi$ and γ Euler's constant. Compared to the standard BCS result, we have an extra factor of $1/\mathrm{e}$ coming from the screening effect of the bubble diagram that is not present in BCS theory. It is to be compared with the Gor'kov correction, which is known to reduce the critical temperature by a factor of 2.2 in the weak-coupling BCS-limit [83]. The difference with our present result is that we have only allowed for a nonzero center-of-mass momentum, whereas to get precisely the Gor'kov correction we would also need to include the relative momentum. We see that due to our approximation of neglecting the relative momenta in the bubble diagram we are only off by 20%.

At larger values of $|a|$, the flow of the chemical potential becomes important and we obtain higher critical temperatures. In the unitarity limit, when a diverges, we obtain $T_{\mathrm{c}} = 0.13T_{\mathrm{F}}$ and $\mu(T_{\mathrm{c}}) = 0.55\varepsilon_{\mathrm{F}}$ in good agreement with the Monte-Carlo results $T_{\mathrm{c}} = 0.152(7)T_{\mathrm{F}}$ and $\mu(T_{\mathrm{c}}) = 0.493(14)\varepsilon_{\mathrm{F}}$ [87]. Note that, both in the weak and in the strong-coupling regime, our critical temperature seems to be only 20% too low at equal densities. However, upon increasing the imbalance, the effect of the bubble diagram becomes less pronounced and we expect to be even closer to the exact result. Keeping this in mind, we are in the unique position with our renormalization group approach to calculate the critical temperature as a function of polarization P and compare with the recent experiment of Shin et al.. The result is shown in Fig. 14.9. The inset of this figure shows the one-loop diagram determining the position of the tricritical point. If it changes sign, then the fourth-order coefficient in the Landau theory for the superfluid phase transition changes sign and the nature of the phase transition changes from second order to first order. This yields finally $P_{\mathrm{c3}} = 0.24$ and $T_{\mathrm{c3}} = 0.06\ T_{\mathrm{F},\uparrow}$ in good agreement with the experimental data. Our previous confirmation of the Monte-Carlo equation of state at $T = 0$ implies that we

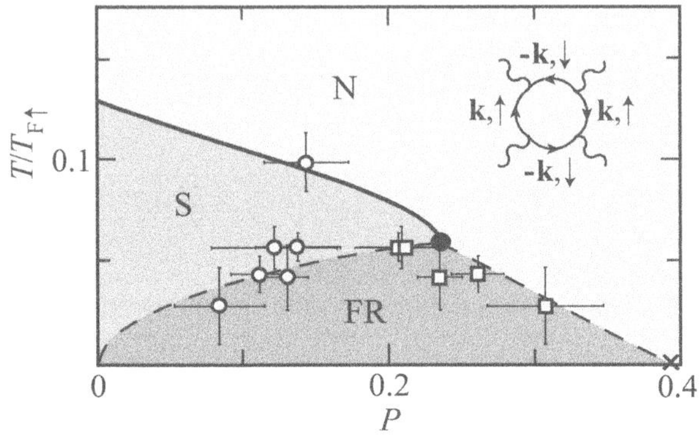

Fig. 14.9 The phase diagram of the homogeneous two-component Fermi mixture in the unitarity limit [118], consisting of the superfluid phase (S), the normal phase (N) and the forbidden region (FR). The solid black line is the result of the Wilsonian renormalization group calculations. The Monte-Carlo result of Lobo et al. [116], which is recovered by the renormalization group, is indicated by a cross. The open circles and squares are data along the phase boundaries from the experiment of Shin et al. [117]. The dashed lines are only guides to the eye. Also shown is the Feynman diagram determining the tricritical point.

also agree with the prediction of a quantum phase transition from the superfluid to the normal phase at a critical imbalance of $P_c = 0.39$ [116, 120]. To conclude, we would like to emphasize that the power of the Wilsonian renormalization group is rather impressive when we realize that so far no analytical theory has been able to yield a value for the tricritical point that fits on the scale of Fig. 14.9. Moreover, Monte-Carlo calculations, which are numerically very involved, have up to now been restricted to zero temperature or to the balanced case. However, our present calculations find good agreement with the experiments of Shin et al. in all limits.

14.4 Problems

Exercise 14.1. Show that the cumulant expansion from (14.8) is valid to second order in the interaction.

Exercise 14.2. Determine the trivial scaling of U_0 if the Landau free energy contains a term $\int d\mathbf{x}\, U_0|\phi|^6/3$. In what dimension is this a marginal variable?

Exercise 14.3. Derive the renormalization group equation for the interaction V_0, i.e. derive (14.23).

Additional Reading

- A review on renormalization group theory in statistical physics is given by K. G. Wilson and J. Kogut, *The renormalization group and the* ε*-expansion*, Phys. Rep. C **12**, 75 (1974) and M. E. Fisher, *Renormalization Group Theory: Its Basis and Formulation in Statistical Physics*, Rev. Mod. Phys. **70**, 653 (1998).
- A functional-integral formalism approach to the theory of critical phenomena is given by J. J. Binney, N. J. Dowrick, A. J. Fisher, and M. E. J. Newman, *The Theory of Critical Phenomena, an Introduction to the Renormalization Group*, (Clarendon Press, Oxford, 2001).
- D. J. Amit, *Field Theory, the Renormalization Group, and Critical Phenomena*, (World Scientific, Singapore, 1984).
- For a detailed account of quantum phase transitions in condensed-matter physics, see S. Sachdev, *Quantum Phase Transitions*, (Cambridge University Press, Cambridge, 2001).

Part III

Chapter 15
Low-Dimensional Systems

The idea is if you use those two shapes and try to colour the plane with them so the colours match, then the only way that you can do this is to produce a pattern which never repeats itself.
– Roger Penrose

In this chapter [122], we consider Bose-Einstein condensation in low-dimensional atomic quantum gases. The first low-dimensional Bose-Einstein condensates were created in the experiments of Görlitz et al. [123], where a magnetic trap was used. By lowering the mean-field interaction energy in a three-dimensional condensate below the energy splitting of the harmonic trap in one or two directions, the quantum states in these directions were not populated anymore. Since the dynamics in one or two directions was now frozen out, the experiment effectively realized a two-dimensional or a one-dimensional Bose-Einstein condensate respectively. Nowadays, low-dimensional Bose-Einstein condensates are more conveniently created in an optical trap [124], an optical lattice, or on an atom chip [125, 126].

The recent experimental and theoretical interest in these low-dimensional quantum gases stems from the fact that their physics is fundamentally different from the physics in three-dimensions quantum gases, primarily due to the enhanced importance of phase fluctuations [127, 128, 129, 130]. Because of these phase fluctuations, Bose-Einstein condensation cannot take place in a one-dimensional homogeneous Bose gas, while in a two-dimensional homogeneous Bose gas it only occurs at zero temperature. This is formalized in the Mermin-Wagner-Hohenberg theorem [131, 132], which we will prove by an argument ad absurdum. However, the one-dimensional Bose gas at zero temperature and the two-dimensional Bose gas below the famous Kosterlitz-Thouless temperature are still superfluid [133]. This shows that superfluidity does not require the presence of a Bose-Einstein condensate, for which the phase is truly coherent over a distance on the order of the system size, but requires only a "quasicondensate" [55], where the phase is coherent over a distance much less than the system size. This chapter aims at providing the theoretical background for a quantitative description of all these phenomena.

15.1 Modified Popov Theory

In the Popov theory for three-dimensional (partially) condensed Bose gases, which we discussed in Sect. 11.3, the phase fluctuations are taken into account up to second order around the mean field. In view of the above-mentioned importance of phase fluctuations in lower dimensions, this is in general insufficient and leads to infrared, i.e. low-momentum or long-wavelength, divergences. As we show next, it is possible to formulate a modified Popov theory that takes into account phase fluctuations up to all orders around the mean field. This results in a mean-field theory that is free of infrared divergences in every dimension. Moreover, both quantum and thermal depletion of the (quasi)condensate are taken into account and the theory is valid at temperatures where the condensate depletion, and therefore the thermal component in the gas, is not negligible.

15.1.1 Phase Fluctuations

In order to explain the infrared problems associated with the phase fluctuations of the condensate most clearly, we first treat a homogeneous Bose gas with $\langle N\rangle$ atoms in a box of volume V. Later, we generalize our discussion to the inhomogeneous case. The starting point is thus the grand-canonical action

$$\begin{aligned} S[\phi^*,\phi] = &\int_0^{\hbar\beta} d\tau \int d\mathbf{x}\, \phi^*(\mathbf{x},\tau)\left\{\hbar\frac{\partial}{\partial\tau} - \frac{\hbar^2}{2m}\nabla^2 - \mu\right\}\phi(\mathbf{x},\tau) \\ &+\frac{1}{2}\int_0^{\hbar\beta} d\tau \int d\mathbf{x}\, V_0|\phi(\mathbf{x},\tau)|^4 , \end{aligned} \tag{15.1}$$

where μ is the chemical potential and V_0 the atomic two-body interaction potential. The mass of the atoms is denoted by m. In the presence of a Bose-Einstein condensate, we use the Bogoliubov shift $\phi(\mathbf{x},\tau) = \sqrt{n_0} + \phi'(\mathbf{x},\tau)$, where n_0 is the condensate density and the field $\phi'(\mathbf{x},\tau)$ describes the fluctuations. Expressions for the density n and the chemical potential μ, known as the one-loop expressions, are then obtained in the following manner. After a quadratic approximation to the action in (15.1), we perform the Gaussian integral over the fluctuations to obtain the thermodynamic potential $\Omega(n_0,\mu)$. We then use the thermodynamic identity $\langle N\rangle = -\partial\Omega(n_0,\mu)/\partial\mu$ and the equilibrium condition for the condensate density $\partial\Omega(n_0,\mu)/\partial n_0 = 0$ to find, respectively,

$$n = n_0 + \frac{1}{V}\sum_{\mathbf{k}}\left\{\frac{\varepsilon_{\mathbf{k}} + n_0V_0 - \hbar\omega_{\mathbf{k}}}{2\hbar\omega_{\mathbf{k}}} + \frac{\varepsilon_{\mathbf{k}} + n_0V_0}{\hbar\omega_{\mathbf{k}}}N_{\mathrm{BE}}(\hbar\omega_{\mathbf{k}})\right\}, \tag{15.2}$$

$$\frac{\mu}{V_0} = n_0 + \frac{1}{V}\sum_{\mathbf{k}}\left\{\frac{2\varepsilon_{\mathbf{k}} + n_0V_0 - 2\hbar\omega_{\mathbf{k}}}{2\hbar\omega_{\mathbf{k}}} + \frac{2\varepsilon_{\mathbf{k}} + n_0V_0}{\hbar\omega_{\mathbf{k}}}N_{\mathrm{BE}}(\hbar\omega_{\mathbf{k}})\right\}, \tag{15.3}$$

where the zero-loop (Hartree) results $n = n_0$ and $\mu = V_0 n_0$ have been substituted into the expressions for the fluctuation corrections. Furthermore, the Bogoliubov dispersion $\hbar\omega_{\mathbf{k}} = (\varepsilon_{\mathbf{k}}^2 + 2n_0 V_0 \varepsilon_{\mathbf{k}})^{1/2}$, $N_{\mathrm{BE}}(x) = 1/(\mathrm{e}^{\beta x} - 1)$ is the Bose-Einstein distribution function, and $\beta = 1/k_{\mathrm{B}}T$ is the inverse thermal energy.

The momentum sums in (15.2) and (15.3) contain terms that are infrared divergent at all temperatures in one dimension and at any nonzero temperature in two dimensions. In fact, all the divergent terms are proportional to n_0, which allows for an ad absurdum proof of the Mermin-Wagner-Hohenberg theorem stating that a true condensate cannot exist at nonzero temperatures in one and two dimensions. Assuming the existence of a Bose-Einstein condensate, we arrived at the inconsistent result that the density of the gas is infinite, which invalidates our initial assumption. The physical reason for these divergences is that the above expressions have been derived by taking into account only quadratic fluctuations around the classical result n_0, i.e. by writing the atomic field as $\phi(\mathbf{x},\tau) = \sqrt{n_0} + \phi'(\mathbf{x},\tau)$ and neglecting terms in the action of third and fourth order in $\phi'(\mathbf{x},\tau)$. As a result, the phase fluctuations of the condensate give the quadratic contribution $n_0\langle\theta(\mathbf{x},\tau)\theta(\mathbf{x},\tau)\rangle$ to the right-hand side of (15.2) and (15.3), whereas an exact approach that sums up all the higher-order terms in the expansion would clearly give no contribution at all to these local quantities because

$$n_0\langle \mathrm{e}^{-\mathrm{i}\theta(\mathbf{x},\tau)}\mathrm{e}^{\mathrm{i}\theta(\mathbf{x},\tau)}\rangle = n_0\left(1 + \langle\theta(\mathbf{x},\tau)\theta(\mathbf{x},\tau)\rangle + \ldots\right) = n_0. \tag{15.4}$$

To correct for this, we thus need to subtract the quadratic contribution of the phase fluctuations, which from (15.2) and (15.3) is seen to be given by

$$n_0\langle\theta(\mathbf{x},\tau)\theta(\mathbf{x},\tau)\rangle = \frac{1}{V}\sum_{\mathbf{k}} \frac{n_0 V_0}{2\hbar\omega_{\mathbf{k}}}\left\{1 + 2N_{\mathrm{BE}}(\hbar\omega_{\mathbf{k}})\right\}, \tag{15.5}$$

which we show more rigorously later on. Note that all the infrared divergences that occur in the one and two-dimensional cases are removed by performing this subtraction.

After having removed the spurious contributions from the phase fluctuations of the condensate, the resulting expressions turn out to be ultraviolet divergent. These divergences are removed by the standard renormalization of the bare coupling constant V_0. Apart from a subtraction, this essentially amounts to replacing the bare two-body potential V_0 by the two-body T matrix evaluated at zero initial and final relative momenta and at the energy -2μ, which we denote from now on by $T^{2\mathrm{B}}(-2\mu)$. Generalizing (10.54) to nonzero energy, we have that

$$\frac{1}{T^{2\mathrm{B}}(-2\mu)} = \frac{1}{V_0} + \frac{1}{V}\sum_{\mathbf{k}} \frac{1}{2\varepsilon_{\mathbf{k}} + 2\mu}. \tag{15.6}$$

Note that the energy argument of the T matrix is -2μ, because this is precisely the energy it costs to excite two atoms from the condensate [134, 135]. After renormalization, the density and chemical potential are

$$n = n_0 + \frac{1}{V}\sum_{\mathbf{k}}\left\{\frac{\varepsilon_{\mathbf{k}} - \hbar\omega_{\mathbf{k}}}{2\hbar\omega_{\mathbf{k}}} + \frac{n_0 T^{2B}(-2\mu)}{2\varepsilon_{\mathbf{k}} + 2\mu} + \frac{\varepsilon_{\mathbf{k}}}{\hbar\omega_{\mathbf{k}}} N_{BE}(\hbar\omega_{\mathbf{k}})\right\}, \quad (15.7)$$

$$\mu = (2n - n_0)T^{2B}(-2\mu) = (2n' + n_0)T^{2B}(-2\mu), \quad (15.8)$$

where $n' = n - n_0$ represents the depletion of the condensate due to quantum and thermal fluctuations and the Bogoliubov quasiparticle dispersion now equals

$$\hbar\omega_{\mathbf{k}} = \sqrt{\varepsilon_{\mathbf{k}}^2 + 2n_0 T^{2B}(-2\mu)\varepsilon_{\mathbf{k}}}. \quad (15.9)$$

The most important feature of (15.7) and (15.8) is that they contain no infrared and ultraviolet divergences and therefore can be applied in any dimension and at all temperatures, even if no condensate exists.

Note that (15.5) is also ultraviolet divergent. The ultraviolet divergence is removed by the renormalization of the bare interaction V_0 and the final result is

$$\langle\theta(\mathbf{x},\tau)\theta(\mathbf{x},\tau)\rangle = \frac{T^{2B}(-2\mu)}{V}\sum_{\mathbf{k}}\left\{\frac{1 + 2N_{BE}(\hbar\omega_{\mathbf{k}})}{2\hbar\omega_{\mathbf{k}}} - \frac{1}{2\varepsilon_{\mathbf{k}} + 2\mu}\right\}. \quad (15.10)$$

We will return to the physics of this important expression in Sect. 15.1.3 below.

15.1.2 Many-Body T Matrix

In the previous section, we presented the modified Popov theory that takes the phase fluctuations into account exactly. The final results in (15.7), (15.8), and (15.10) involve the two-body T matrix, which takes into account successive two-body scattering processes in vacuum. However, it neglects the many-body effects of the surrounding gas. In order to take this into account as well, we must use the many-body T matrix instead of the two-body T matrix in (15.7), (15.8) and (15.10). Many-body effects have been shown to be appreciable in three dimensions only very close to the transition temperature [110], but turn out to be much more important in one and two dimensions [136]. Since the effects of the medium on the scattering properties of the atoms is only important at relatively high temperatures, we can apply a Hartree-Fock approximation to obtain for the many-body T matrix

$$T^{MB}(-2\mu) = \quad (15.11)$$
$$T^{2B}(-2\mu)\left\{1 + T^{2B}(-2\mu)\frac{1}{V}\sum_{\mathbf{k}}\frac{N_{BE}(\varepsilon_{\mathbf{k}} + n_0 T^{MB}(-2\mu))}{\varepsilon_{\mathbf{k}} + \mu}\right\}^{-1}.$$

The situation is in fact slightly more complicated, because now we actually need two coupling constants in the equation for the chemical potential, which is the homogeneous version of the Gross-Pitaevskii equation. When two atoms in the condensate collide at zero momentum, they both require an energy μ to be excited from the condensate, and thus their coupling is evaluated at -2μ. This is the coupling that multiplies n_0 in the Gross-Pitaevskii equation. On the other hand, the coupling that multiplies n' in the Gross-Pitaevskii equation involves one condensate atom and one atom in the thermal cloud, so that this coupling should now be evaluated at $-\mu$. The equation for the chemical potential thus becomes

$$\mu = 2n'T^{\rm MB}(-\mu) + n_0 T^{\rm MB}(-2\mu). \tag{15.12}$$

15.1.3 Long-Wavelength Physics

We have given physical arguments for how to identify and subtract the contribution to (15.2) and (15.3) from the phase fluctuations of the condensate. At this point, we would like to give a somewhat more rigorous field-theoretical argument. If we substitute $\phi(\mathbf{x},\tau) = \sqrt{n+\delta n(\mathbf{x},\tau)}\mathrm{e}^{\mathrm{i}\theta(\mathbf{x},\tau)}$ into (15.1), we obtain the action

$$\begin{aligned} S[\delta n,\theta] = \int_0^{\hbar\beta} \mathrm{d}\tau \int \mathrm{d}\mathbf{x} \Bigg\{ & \frac{1}{2}\hbar\frac{\partial \delta n}{\partial \tau} + \mathrm{i}\hbar(n+\delta n)\frac{\partial \theta}{\partial \tau} \\ & + \frac{\hbar^2}{2m} n (\boldsymbol{\nabla}\theta)^2 + \frac{1}{2}\delta n \left(-\frac{\hbar^2}{4mn}\boldsymbol{\nabla}^2 + V_0 \right) \delta n \Bigg\}. \end{aligned} \tag{15.13}$$

Here, n is the average total density of the gas and $\delta n(\mathbf{x},\tau)$ represents the fluctuations. At zero temperature, this action is exact in the long-wavelength limit, if $(\hbar^2\mathbf{k}^2/4mn + V_0)$ is replaced by $\chi_{nn}^{-1}(\mathbf{k})$, where $\chi_{nn}(\mathbf{k})$ is the exact static density-density correlation function.

By using the classical equation of motion to eliminate the phase $\theta(\mathbf{x},\tau)$, we obtain the following action for the density fluctuations $\delta n(\mathbf{x},\tau)$

$$S[\delta n] = \int_0^{\hbar\beta} \mathrm{d}\tau \int \mathrm{d}\mathbf{x} \left\{ -\frac{m}{n}\frac{\partial \delta n}{\partial \tau}\boldsymbol{\nabla}^{-2}\frac{\partial \delta n}{\partial \tau} + \frac{1}{2}\delta n \chi_{nn}^{-1}(-\mathrm{i}\boldsymbol{\nabla})\delta n \right\}. \tag{15.14}$$

The density fluctuations are therefore determined by

$$\langle \delta n(\mathbf{x},\tau)\delta n(\mathbf{x}',\tau)\rangle = \frac{1}{V}\sum_{\mathbf{k},n} \frac{n\varepsilon_{\mathbf{k}}}{\beta} \frac{1}{(\hbar\omega_n)^2 + (\hbar\omega_{\mathbf{k}})^2} \mathrm{e}^{\mathrm{i}\mathbf{k}\cdot(\mathbf{x}-\mathbf{x}')}, \tag{15.15}$$

where $\omega_n = 2\pi n/\hbar\beta$ are the even Matsubara frequencies and $\hbar\omega_{\mathbf{k}} = \sqrt{n\varepsilon_{\mathbf{k}}/\chi_{nn}(\mathbf{k})}$. Summing over these Matsubara frequencies, we obtain

$$\langle \delta n(\mathbf{x},\tau)\delta n(\mathbf{x}',\tau)\rangle = \frac{1}{V}\sum_{\mathbf{k}} n\varepsilon_{\mathbf{k}} \frac{1+2N_{\rm BE}(\hbar\omega_{\mathbf{k}})}{2\hbar\omega_{\mathbf{k}}} \mathrm{e}^{\mathrm{i}\mathbf{k}\cdot(\mathbf{x}-\mathbf{x}')}.$$

Similarly, by using the classical equation of motion for $\delta n(\mathbf{x},\tau)$, we obtain from (15.13) the following action for the phase fluctuations

$$S[\theta] = \int_0^{\hbar\beta} d\tau \int d\mathbf{x} \left\{ \hbar^2 \frac{\partial\theta}{\partial\tau} \chi_{nn}(-i\nabla) \frac{\partial\theta}{\partial\tau} + \frac{\hbar^2 n}{2m} (\nabla\theta)^2 \right\}. \tag{15.16}$$

From this action, it is straightforward to calculate the propagator for the field $\theta(\mathbf{x},\tau)$ and thereby the equal-time correlation function $\langle\theta(\mathbf{x},\tau)\theta(\mathbf{x}',\tau)\rangle$. The result is

$$\langle\theta(\mathbf{x},\tau)\theta(\mathbf{x}',\tau)\rangle = \frac{1}{V}\sum_{\mathbf{k}} \frac{1}{\chi_{nn}(\mathbf{k})} \frac{1+2N_{\mathrm{BE}}(\hbar\omega_{\mathbf{k}})}{2\hbar\omega_{\mathbf{k}}} e^{i\mathbf{k}\cdot(\mathbf{x}-\mathbf{x}')} . \tag{15.17}$$

Setting $\mathbf{x}' = \mathbf{x}$, we recover (15.10) in the long-wavelength limit if we use

$$\chi_{nn}(\mathbf{k}) \simeq 1/T^{\mathrm{MB}}(-2\mu) \tag{15.18}$$

for the static density-density correlation function in this limit. It is important to mention that (15.17) is often used for the short-wavelength part of the phase fluctuations as well [129, 130]. This is, however, problematic because it contains ultraviolet divergences due to the fact that the above procedure neglects interaction terms between density and phase fluctuations that are only irrelevant at large wavelengths. The appropriate short-wavelength behavior is given in (15.10).

15.2 Comparison with Popov Theory

We proceed to compare predictions based on (15.7), (15.8), and (15.10) with exact results in one dimension and results based on the Popov theory in two and three dimensions. We consider only the homogeneous case here and discuss the inhomogeneous Bose gas in Sect. 15.5.

15.2.1 One Dimension

To understand the physical meaning of the quantity n_0 in (15.7) and (15.8), i.e. whether it is the quasicondensate density or the true condensate density, we must determine the off-diagonal long-range behavior of the one-particle density matrix. Because this is a nonlocal property of the Bose gas, the phase fluctuations contribute and we find in the large-separation limit

$$\begin{aligned}\langle\phi^*(\mathbf{x},0)\phi(\mathbf{0},0)\rangle &\simeq n_0\langle e^{-i(\theta(\mathbf{x},0)-\theta(\mathbf{0},0))}\rangle \\ &= n_0 e^{-\frac{1}{2}\langle[\theta(\mathbf{x},0)-\theta(\mathbf{0},0)]^2\rangle} .\end{aligned} \tag{15.19}$$

Using (15.10), we obtain for the exponent in (15.19)

$$\langle[\theta(\mathbf{x},0)-\theta(\mathbf{0},0)]^2\rangle = \frac{T^{\mathrm{MB}}(-2\mu)}{V}\sum_{\mathbf{k}}\left\{\frac{1+2N_{\mathrm{BE}}(\hbar\omega_{\mathbf{k}})}{\hbar\omega_{\mathbf{k}}}-\frac{1}{\varepsilon_{\mathbf{k}}+\mu}\right\} \times\{1-\cos(\mathbf{k}\cdot\mathbf{x})\}\ . \tag{15.20}$$

Writing the sum over wavevectors $\mathbf{k}$ as an integral, the phase fluctuations at zero temperature can be written as

$$\langle[\theta(\mathbf{x},0)-\theta(\mathbf{0},0)]^2\rangle = \int_0^\infty \mathrm{d}k\frac{1-\cos(kx/\xi)}{2\pi n_0\xi}\left\{\frac{1}{k\sqrt{k^2+1}}-\frac{2}{2k^2+1}\right\}, \tag{15.21}$$

where $\xi=\hbar/[4mn_0T^{2\mathrm{B}}(-2\mu)]^{1/2}$ is the correlation length and we have also performed the substitution $k\xi\to k$ for convenience. Note that we have used that $T^{\mathrm{MB}}(-2\mu)=T^{2\mathrm{B}}(-2\mu)$ at zero temperature and that the chemical potential, as we show shortly, is to a good approximation equal to $n_0T^{2\mathrm{B}}(-2\mu)$. The integration can be performed analytically and the result is

$$\langle[\theta(\mathbf{x},0)-\theta(\mathbf{0},0)]^2\rangle = \frac{1}{2\pi n_0\xi}\left(\frac{\pi x}{2\xi}\,{}_1F_2(1/2;1,3/2;x^2/4\xi^2) - \frac{x^2}{2\xi^2}\,{}_2F_3(1,1;3/2,3/2;2x^2/4\xi^2)\right), \tag{15.22}$$

where ${}_iF_j(\alpha_1,\alpha_2,...\alpha_i;\beta_1,\beta_2,...\beta_j;x)$ are hypergeometric functions. In the limit $|\mathbf{x}|\to\infty$, (15.22) reduces to

$$\langle[\theta(\mathbf{x},0)-\theta(\mathbf{0},0)]^2\rangle \simeq \frac{1}{2\pi n_0\xi}\log(x/\xi)\ . \tag{15.23}$$

Using (15.23), we find that the one-particle density matrix behaves for $|\mathbf{x}|\to\infty$, as

$$\langle\phi^*(\mathbf{x},0)\phi(\mathbf{0},0)\rangle \simeq \frac{n_0}{(x/\xi)^{1/4\pi n_0\xi}}\ . \tag{15.24}$$

A few remarks are in order. First, the asymptotic behavior of the one-particle density matrix at zero temperature proves that the gas is not Bose-Einstein condensed and that n_0 should be identified with the quasicondensate density. Second, in the weakly-interacting limit $4\pi n\xi\gg1$ the depletion is small so, to first approximation, we can use $n_0\simeq n$ in the exponent $\eta=1/4\pi n_0\xi$. Indeed, from (15.7) and (15.8), we obtain the following expression for the fractional depletion of the quasicondensate

$$\frac{n-n_0}{n} = \frac{1}{4\pi n\xi}\left(\frac{\sqrt{2}}{4}\pi-1\right)\ . \tag{15.25}$$

We see that the expansion parameter is $1/4\pi n\xi$ and, therefore, that the depletion is very small. Keeping this in mind, (15.24) is in complete agreement with the exact result obtained by Haldane [137]. Note that our theory cannot describe the strongly-interacting case $4\pi n\xi \ll 1$, where the one-dimensional Bose gas behaves as a Tonks gas [138, 139].

Finally, our results show that at a nonzero temperature the phase fluctuations increase as $\langle[\theta(\mathbf{x},0)-\theta(\mathbf{0},0)]^2\rangle \propto |\mathbf{x}|$ for large distances, and thus that the off-diagonal one-particle density matrix vanishes exponentially. Hence, at nonzero temperatures, not even a quasicondensate exists and we can use the equation of state for the normal state to describe the gas, i.e.

$$n = \frac{1}{V}\sum_{\mathbf{k}} N_{\mathrm{BE}}(\varepsilon_{\mathbf{k}} + \hbar\Sigma - \mu)\ , \tag{15.26}$$

where the Hartree-Fock selfenergy satisfies

$$\hbar\Sigma = 2nT^{\mathrm{MB}}(-\hbar\Sigma)\ , \tag{15.27}$$

and the many-body T matrix obeys

$$T^{\mathrm{MB}}(-\hbar\Sigma) = T^{\mathrm{2B}}(-\hbar\Sigma)\left\{1 + T^{\mathrm{2B}}(-\hbar\Sigma)\frac{1}{V}\sum_{\mathbf{k}}\frac{N(\varepsilon_{\mathbf{k}}+\hbar\Sigma-\mu)}{\varepsilon_{\mathbf{k}}+\hbar\Sigma/2}\right\}^{-1}\ . \tag{15.28}$$

Note that the last three equations for the description of the normal phase of the Bose gas are again valid for an arbitrary number of dimensions.

15.2.2 Two Dimensions

By analogy with (15.21), we obtain for the phase fluctuations in two dimensions at zero temperature

$$\begin{aligned}&\langle[\theta(\mathbf{x},0)-\theta(\mathbf{0},0)]^2\rangle = \\ &\int_0^\infty \mathrm{d}k\frac{1-J_0(kx/\xi)}{\pi n_0\xi^2}\left\{\frac{1}{\sqrt{k^2+1}} - \frac{2k}{2k^2+1}\right\},\end{aligned} \tag{15.29}$$

where $J_n(z)$ is the Bessel function of the first kind. Therefore, we now find in the limit $|\mathbf{x}| \to \infty$ that

$$\langle[\theta(\mathbf{x},0)-\theta(\mathbf{0},0)]^2\rangle = \frac{\log 2}{2\pi}\frac{mT^{2B}(-2\mu)}{\hbar^2} \tag{15.30}$$

and

$$\langle \phi^*(\mathbf{x})\phi(0,0)\rangle = n_0 \exp\left\{-\frac{\log 2}{4\pi}\frac{mT^{2\text{B}}(-2\mu)}{\hbar^2}\right\} . \tag{15.31}$$

Clearly, the condensate density n_c of the gas is given by the right-hand side of this result and thus smaller than the quantity n_0. On the basis of (15.7) and (15.8) we have at zero temperature that

$$\frac{n-n_0}{n} = \frac{1}{4\pi}(1-\log 2)\frac{mT^{2\text{B}}(-2\mu)}{\hbar^2} , \tag{15.32}$$

where the chemical potential satisfies $\mu = nT^{2\text{B}}(-2\mu)$. Making use of the fact that for a weakly-interacting gas the argument of the exponent in (15.31) is small, the fractional depletion of the condensate in this case becomes

$$\frac{n-n_\text{c}}{n} = \frac{1}{4\pi}\frac{mT^{2\text{B}}(-2\mu)}{\hbar^2} , \tag{15.33}$$

which was first calculated by Schick [134] using the Popov approximation. However, at nonzero temperatures, no Bose-Einstein condensate exists because the correlation function behaves as

$$\langle \phi^*(\mathbf{x},0)\phi(0,0)\rangle \simeq \frac{n_0}{(x/\xi)^{1/n_0\Lambda^2}} , \tag{15.34}$$

where $\Lambda = \sqrt{2\pi\hbar^2/mk_\text{B}T}$ is the thermal de Broglie wavelength. Here, n_0 corresponds again to the quasicondensate density.

In a number of applications, we need to calculate many-body correlators. For instance, in order to calculate how a quasicondensate modifies the two-body relaxation constants of a spin-polarized two-dimensional Bose gas, we need to know

$$K^{(2)}(T) \equiv \langle \phi^*(\mathbf{x},\tau)\phi^*(\mathbf{x},\tau)\phi(\mathbf{x},\tau)\phi(\mathbf{x},\tau)\rangle/2n^2. \tag{15.35}$$

This correlator was considered in [140] using the many-body T matrix theory with an appropriate cutoff to remove the infrared divergences. An exact treatment of the phase fluctuations leads however directly to an infrared finite result as we show now. Using the same parametrization for the atomic fields as before, we obtain first of all

$$\begin{aligned}\langle \phi^*(\mathbf{x},\tau)\phi^*(\mathbf{x},\tau)\phi(\mathbf{x},\tau)\phi(\mathbf{x},\tau)\rangle &= n_0^2 + n_0\Big(\langle \phi'(\mathbf{x},\tau)\phi'(\mathbf{x},\tau)\rangle \\ &\quad +\langle \phi'^*(\mathbf{x},\tau)\phi'^*(\mathbf{x},\tau)\rangle + 4\langle \phi'^*(\mathbf{x},\tau)\phi'(\mathbf{x},\tau)\rangle\Big) \\ &\quad +2\langle \phi'^*(\mathbf{x},\tau)\phi'(\mathbf{x},\tau)\rangle^2 \\ &\quad +\langle \phi'(\mathbf{x},\tau)\phi'(\mathbf{x},\tau)\rangle\langle \phi'^*(\mathbf{x},\tau)\phi'^*(\mathbf{x},\tau)\rangle . \end{aligned} \tag{15.36}$$

The normal average is given by $\langle \phi'^*(\mathbf{x},\tau)\phi'(\mathbf{x},\tau)\rangle = n' + n_0\langle \theta(\mathbf{x},\tau)\theta(\mathbf{x},\tau)\rangle$ and the anomalous average obeys $\langle \phi'(\mathbf{x},\tau)\phi'(\mathbf{x},\tau)\rangle = -n_0\langle \theta(\mathbf{x},\tau)\theta(\mathbf{x},\tau)\rangle$, as we have seen. Using this, (15.36) can then be written as

$$\begin{aligned}\langle\phi^*(\mathbf{x},\tau)\phi^*(\mathbf{x},\tau)\phi(\mathbf{x},\tau)\phi(\mathbf{x},\tau)\rangle &= n_0^2(1+2\langle\theta(\mathbf{x},\tau)\theta(\mathbf{x},\tau)\rangle \\ &+3\langle\theta(\mathbf{x},\tau)\theta(\mathbf{x},\tau)\rangle^2) \\ &+4n_0(1+\langle\theta(\mathbf{x},\tau)\theta(\mathbf{x},\tau)\rangle)n'+2(n')^2 . \end{aligned} \quad (15.37)$$

Writing the correlator in this form, we explicitly see that the infrared divergences are due to spurious contributions from the phase fluctuations. Removing them, we obtain for the renormalized correlator

$$K_{\mathrm{R}}^{(2)}(T) = \frac{1}{2n^2}\left\{n_0^2+4n_0n'+2\left(n'\right)^2\right\} . \quad (15.38)$$

We would like to point out that critical fluctuations are not treated within our mean-field theory. This is of course essential in the study of the Kosterlitz-Thouless phase transition and we return to this issue in Sect. 15.4. Another important issue in ultracold atomic gases is the finite lifetime of the gas, which is often caused by the three-body recombination reaction $A+A+A \rightarrow A_2+A$. The decay of the atomic density n is then described by the rate equation $\mathrm{d}n/\mathrm{d}t = -3Ln^3$, because after the recombination event both the molecule and the atom have obtained too much energy to remain trapped. The three-body recombination rate constant L is an example of a physical observable where phase fluctuations are not important. We are therefore already in the position to determine the reduction of the three-body recombination rate constant due to the presence of a quasicondensate. This can be expressed as [140]

$$\frac{L^{\mathrm{N}}}{L(T)} \simeq \left\{\left(\frac{T^{2\mathrm{B}}(-2\mu)}{T^{2\mathrm{B}}(-2\hbar\Sigma)}\right)^6 K_{\mathrm{R}}^{(3)}(T)\right\}^{-1} , \quad (15.39)$$

where L^{N} is the recombination rate constant in the normal phase, which is essentially independent of temperature, and the selfenergy satisfies

$$\hbar\Sigma = 2nT^{2\mathrm{B}}(-\hbar\Sigma).$$

The renormalized three-body correlator

$$K_{\mathrm{R}}^{(3)}(T) = \frac{1}{6n^3}\left\{n_0^3+9n_0^2n'+18n_0(n')^2+6(n')^3\right\} \quad (15.40)$$

is obtained from the expression for the correlation function

$$\langle\phi^*(\mathbf{x},\tau)\phi^*(\mathbf{x},\tau)\phi^*(\mathbf{x},\tau)\phi(\mathbf{x},\tau)\phi(\mathbf{x},\tau)\phi(\mathbf{x},\tau)\rangle$$

by removing, as before, the spurious contributions from the phase fluctuations. Moreover, in two dimensions the two-body T matrix depends logarithmically on the chemical potential as

$$T^{2B}(-2\mu) = \frac{4\pi\hbar^2}{m} \frac{1}{\log(2\hbar^2/\mu m a^2)}, \tag{15.41}$$

where a is the two-dimensional s-wave scattering length. In the case of atomic hydrogen adsorbed on a superfluid helium film, the scattering length was found to be $a = 2.4a_0$ [141], where a_0 is the Bohr radius. However, there is some uncertainty in this number because the hydrogen wave function perpendicular to the helium surface is not known very accurately. In order to compare with experiment, we may therefore allow a to vary somewhat.

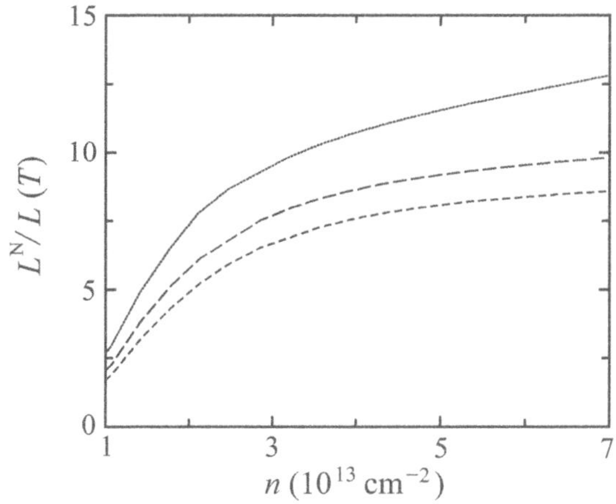

Fig. 15.1 Reduction of the three-body recombination rate as a function of the density for a temperature of $T = 190$ mK and three different values of the scattering length. The dotted line corresponds to $a = 2.4a_0$, the long-dashed line to $a = 1.2a_0$, and the dashed line to $a = 0.6a_0$, respectively.

In Fig. 15.1, we show the reduction of the three-body recombination rate as a function of the density at a fixed temperature $T = 190$ mK for three different values of a. As can clearly be seen, the reduction of the three-body recombination rate is very sensitive to the value of a. Our calculation shows that there is a large reduction of the recombination rate at high densities, which has also been observed experimentally by Safonov et al. [142]. However, a direct comparison between the results of our theory and the measurements of Safonov et al. cannot be made, since the density and temperature of the adsorbed hydrogen gas were not measured directly but instead inferred from the properties of the three-dimensional buffer gas. Because this procedure requires knowledge of the equation of state of the two-dimensional Bose gas absorbed on the superfluid helium film, the raw experimental data needs to be reanalyzed with the theory presented in this Chapter. We can, however, compare the density at which the recombination rate starts to deviate considerably from the result in the normal state. For the temperature of $T = 190$ mK, where most of the experimental data is taken, this is at a density of about 1.0×10^{13} cm^{-2}, which is in excellent agreement with experiment. In view of this and the above-mentioned problems we thus conclude that our results present a compelling theoretical explanation of the experimental findings.

15.2.3 Three Dimensions

The Popov theory has been very successful in describing the properties of dilute three-dimensional trapped Bose gases. Therefore, the exact treatment of the phase fluctuations in the three-dimensional case is expected to lead at most to small changes in the predictions compared to Popov theory. At zero temperature, the fractional depletion within the Popov theory was first calculated by Lee and Yang [143] and is given by

$$\frac{n - n_c}{n} = \frac{8}{3}\sqrt{\frac{na^3}{\pi}}\,, \tag{15.42}$$

where a is the s-wave scattering length and we have used

$$T^{2B}(-2\mu) = \frac{4\pi a \hbar^2}{m}\,. \tag{15.43}$$

The result that follows from (15.7) and (15.8) is

$$\frac{n - n_0}{n} = \left(\frac{32}{3} - 2\sqrt{2}\pi\right)\sqrt{\frac{na^3}{\pi}}\,. \tag{15.44}$$

However, we find again that the phase fluctuations give a nonzero contribution to the density matrix at large separations. More precisely, we find for $|\mathbf{x}| \to \infty$ that

$$\langle [\theta(\mathbf{x},0) - \theta(\mathbf{0},0)]^2 \rangle = (2 - \sqrt{2}\pi)\sqrt{\frac{na^3}{\pi}} \tag{15.45}$$

and thus

$$n_c = n_0 \exp\left\{-\left(1 - \frac{\sqrt{2}\pi}{2}\right)\sqrt{\frac{na^3}{\pi}}\right\}\,. \tag{15.46}$$

For a weakly-interacting Bose gas, the depletion of the condensate therefore exactly reduces to the result of Lee and Yang.

The critical temperature T_c is found by taking the limit $n_0 \to 0$ in (15.7) and (15.8). These expressions then reduce to the same expressions for the density and chemical potential as in the Popov theory. This implies that our critical temperature for Bose-Einstein condensation coincides with that obtained in Popov theory, i.e. the ideal gas result

$$T_c = \frac{2\pi\hbar^2}{mk_B}\left(\frac{n}{\zeta(3/2)}\right)^{2/3}, \tag{15.47}$$

where $\zeta(3/2) \simeq 2.612$.

15.3 Vortices in Two Dimensions

In the previous section we have seen that in two dimensions there is a superfluid to normal phase transition at a nonzero temperature, even though a Bose-Einstein condensate only exists in the gas at zero temperature. The transition between the superfluid and normal phases was first analyzed by Berezinskii, Kosterlitz, and Thouless and is of a topological nature, because it physically turns out to correspond to the unbinding of vortex pairs and the proliferation of unbound vortices. In Sect. 11.8, we briefly discussed the onset of a vortex solution in a three-dimensional condensate under rotation. To understand this physics better, we note that in a superfluid gas the kinetic energy due to a superfluid flow is given by

$$E_s = \frac{1}{2}mn_s \int d\mathbf{x}\, (\mathbf{v}_s(\mathbf{x}))^2 , \tag{15.48}$$

where n_s is the superfluid density. For a vortex in the origin, the circulating superfluid velocity obeys $\mathbf{v}_s(\mathbf{x}) = (\hbar/m|\mathbf{x}|^2)\,\mathbf{e}_z \times \mathbf{x}$ and the total energy of the vortex becomes

$$E_v = \frac{\pi\hbar^2}{m} n_s \log(R/\xi) . \tag{15.49}$$

Note that we have introduced the unit vector $\mathbf{e}_z$ perpendicular to the two-dimensional plane, and we have approximated the effect of the core of the vortex by simply using the correlation or healing length ξ as the lower bound on the radial integral.

The most important feature of the vortex energy is that it diverges logarithmically with the radius R of the system. This implies that for the large system sizes $R \gg \xi$ of interest to us, it is impossible to thermally excite vortices at low temperatures $T \ll \pi\hbar^2 n_s/mk_B$. At higher temperatures, however, we should not look at the energy of the vortex but at its free energy. The entropy of a vortex can be estimated as $2k_B \log(R/\xi)$, since, roughly speaking, the number of positions to place a vortex in the gas is $\pi R^2/\pi\xi^2$. We therefore find for the free energy

$$F_v = \left(\frac{\pi\hbar^2}{m} n_s - 2k_B T\right) \log(R/\xi) , \tag{15.50}$$

and see that for temperatures above the critical temperature

$$T_c = \frac{\pi\hbar^2 n_s}{2mk_B} \tag{15.51}$$

the free energy of a vortex is negative. Hence, the gas can lower its free energy by a proliferation of free vortices. This is the simplest argument for the Kosterlitz-Thouless transition based on noninteracting vortices, but it has nevertheless lead us to the exact critical condition $n_s\Lambda_c^2 = 4$ that we obtain from renormalization group theory in Sect. 15.4 below.

A more precise picture can be obtained by also considering the interactions between vortices. Because of the above-mentioned logarithmic divergence of the energy of a single vortex, it is at low temperature only possible to excite vortex-antivortex pairs, where for the antivortex the circulation of the superfluid flow is precisely opposite to that of the vortex. By adding the superfluid velocity profiles of a vortex at position $\mathbf{x}$ and of an antivortex at position $\mathbf{x}'$, and substituting this into (15.48), we find that the energy for such a pair is

$$V_v(\mathbf{x}-\mathbf{x}') = \frac{2\pi\hbar^2}{m} n_s \log(|\mathbf{x}-\mathbf{x}'|/\xi) . \tag{15.52}$$

A nice physical understanding of this result can be obtained by rotating all velocities in the velocity profile of the vortices by ninety degrees. Then, the velocity profiles become exactly equal to the electric-field configurations of a positive and negative point charge, and we know that two such point charges do indeed attract each other logarithmically in two dimensions.

Having obtained the attractive interaction potential between a vortex and an antivortex, we can now determine the average distance between two such vortices bound together into a pair. We have in first instance that

$$\langle r^2 \rangle = \frac{\displaystyle\int_\xi^\infty \mathrm{d}r\, r^3\, \mathrm{e}^{-\beta V_v(r)}}{\displaystyle\int_\xi^\infty \mathrm{d}r\, r\, \mathrm{e}^{-\beta V_v(r)}} , \tag{15.53}$$

which results in

$$\langle r^2 \rangle = \frac{n_s\Lambda^2 - 2}{n_s\Lambda^2 - 4} . \tag{15.54}$$

We thus again conclude that the unbinding of vortices occurs when $n_s\Lambda_c^2 = 4$, because then their average separation diverges. This famous Kosterlitz-Thouless transition and the associated unbinding of vortices has recently been observed in an atomic Bose gas by Hadzibabic et al. [144]. In this experiment, the proliferation of free vortices is directly imaged by letting two two-dimensional clouds expand and interfere with each other. The free vortices can then be counted individually by looking at the number of defects in the interference pattern [145, 146]. The result is shown in Fig. 15.2. In an optical lattice the same transition was also seen by V. Schweikhard et al. [147], using the usual absorption imaging of the vortex cores.

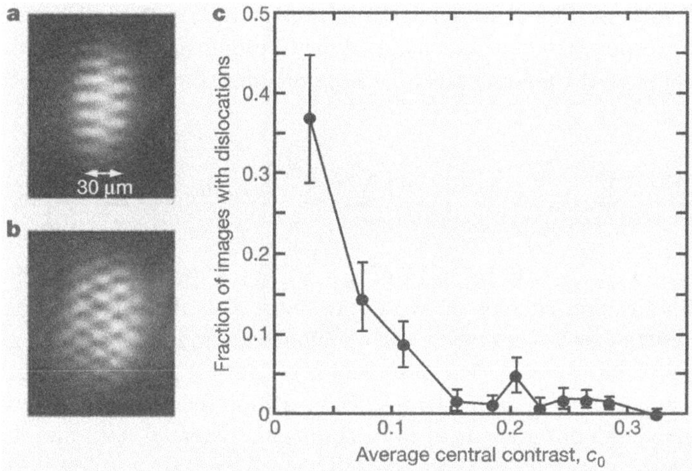

Fig. 15.2 Proliferation of free vortices as measured by Hadzibabic et al. [144]. a) Example of an interference pattern showing a sharp dislocation that is attributed to the presence of a free vortex in one of the interfering clouds. b) Interference pattern showing several dislocations. c) Fraction of images showing at least one dislocation, where the average central contrast c_0 is a measure for the degeneracy of the system. An average central contrast of $c_0 = 0.15$ corresponds to $n_0\Lambda^2 = 6 \pm 2$. For a lower contrast, i.e. a higher temperature, there are free vortices, whereas for a higher contrast, i.e. a lower temperature, there are no free vortices. Adapted by permission from Macmillan Publishers Ltd: Nature **441**, 1118 (2006), copyright (2006).

15.4 Kosterlitz-Thouless Phase Transition

In Sect. 15.2 we have compared our results using the modified many-body T matrix theory with established results in one, two, and three dimensions in the Popov approximation. Due to the mean-field nature of the modified many-body T matrix theory, the Kosterlitz-Thouless transition is absent. In this section, we improve upon this by explicitly including the effects of vortex pairs in the phase fluctuations. The idea is to use the modified many-body T matrix theory to determine the initial values of the superfluid density and the vortex fugacity, and to carry out a renormalization-group calculation to find the fully renormalized values of these quantities. In this manner, we can for example calculate the critical temperature T_c for the Kosterlitz-Thouless transition given the scattering length a and density n.

Let us for completeness first sketch the derivation of the renormalization group equations for the superfluid density and the vortex fugacity. Consider the velocity field of a vortex where the core is centered at the positions $\mathbf{x}_i$, which we for simplicity take to lie on a lattice with an area of the unit cell equal to Ω. By rotating the velocity field by ninety degrees, we have seen that we can map it onto the electric field of a point charge in two dimensions. Since the total energy in both systems is proportional to the square of the field integrated over space, there is complete analogy between a system of vortices and a two-dimensional Coulomb gas. This analogy is very useful and we will take advantage of it in the following. The total vorticity corresponds to the total charge of the Coulomb gas. For the analogous two-dimensional neutral Coulomb gas on a square lattice, the partition function can be written as

$$Z = \sum_{\mathbf{x}_1,\mathbf{x}_2,\ldots} \sum_{n_1,n_2,\ldots} \exp\left\{-\beta\left(\sum_{i\neq j} V(\mathbf{x}_i-\mathbf{x}_j)n_i n_j - E_c \sum_j n_j^2\right)\right\}, \tag{15.55}$$

where $V(\mathbf{x}_i - \mathbf{x}_j) = -(2\pi\hbar^2 n_s/m)\log(|\mathbf{x}_i - \mathbf{x}_j|/\xi)$ is the Coulomb interaction between two unit point charges in two dimensions, n_s is the superfluid density, and E_c is the energy associated with the spontaneous creation of a charge, i.e. the core energy of the vortices. The summation is over all possible configurations of charges n_i at positions $\mathbf{x}_i$ on the lattice. The partition function can be rewritten in a field-theoretic fashion in terms of the electrostatic potential $\phi(\mathbf{x})$ and the fugacity $y = e^{-\beta E_c}$ as

$$Z = \sum_{\mathbf{x}_1,\mathbf{x}_2,\ldots} \sum_{n_1,n_2,\ldots} \int d[\phi] e^{-\frac{1}{2}\int d\mathbf{x} K'(\nabla\phi(\mathbf{x}))^2} e^{-i\beta\Sigma_j n_j \phi(\mathbf{x}_j)} y^{\Sigma_j n_j^2}, \tag{15.56}$$

where $K' = (2\pi)^2 m/\hbar^2 k_B T n_s$. In the limit where $y \ll 1$, the charge density is very low, and thus only $n_j = 0, \pm 1$ contribute to the partition function. We can then write

$$\begin{aligned} Z &\simeq \int d[\phi]\, e^{-\frac{1}{2}\int d\mathbf{x} K'(\nabla\phi)^2} \prod_j \Big\{1 + y\exp(i\beta\phi(\mathbf{x}_j)) \\ &\qquad + y\exp(-i\beta\phi(\mathbf{x}_j)) + \ldots\Big\} \\ &\simeq \int d[\phi]\ e^{-\int d\mathbf{x}\left\{\frac{1}{2}K'(\nabla\phi)^2 - g\cos(\beta\phi)\right\}}, \end{aligned} \tag{15.57}$$

where $g = 2y/\Omega$. It is convenient to introduce a dimensionless dielectric constant K that is related to K' by $K = \beta^2/4\pi^2 K' = n_s \Lambda^2/2\pi$, where Λ is the thermal wavelength.

The renormalization group equations for K, which is thus proportional to the superfluid density and the fugacity y, can now be obtained by performing the usual momentum-shell integrations. For the Sine-Gordon model derived in (15.57), this results in

$$\frac{dK^{-1}(l)}{dl} = 4\pi^3 y^2(l) + O(y^3), \tag{15.58}$$

$$\frac{dy(l)}{dl} = \{2 - \pi K(l)\} y(l) + O(y^2). \tag{15.59}$$

The renormalization group equations to leading order in the variables $K(l)$ and $y(l)$ were first obtained by Kosterlitz [148], while the next-to-leading order terms were derived by Amit et al. [149]. The flow equations are not significantly changed by including the higher-order corrections and we do not include them in the following.

The renormalization group equations (15.58) and (15.59) can be solved analytically by separation of variables, and the solution is

$$y^2(l) - \frac{1}{2\pi^3}\left\{\frac{2}{K(l)} + \pi\log(K(l))\right\} = C\,, \tag{15.60}$$

where the integration constant C is determined by the initial conditions. For the critical trajectory it can be calculated by evaluating the left-hand side at the fixed point $(y(\infty),K(\infty)) = (0,2/\pi)$. In this manner, we find $C = [\log(\pi/2)-1]/2\pi^2 \simeq -0.0278$. In Fig. 15.3, we show the flow of the Kosterlitz renormalization group equations. There is a line of fixed points $y(\infty) = 0$ and $K(\infty) \geq 0$. The fixed point $(y(\infty),K(\infty)) = (0,2/\pi)$ corresponds to the critical condition for the Kosterlitz-Thouless transition, where the vortices start to unbind and superfluidity disappears. Physically this can be understood from the fact that below the transition the fugacity renormalizes to zero, which implies that at the largest length scales, single vortices cannot be created by thermal fluctuations. They are therefore forced to occur in pairs.

The initial conditions for the renormalization group equations are

$$K(0) = \frac{\hbar^2 n_0}{mk_{\mathrm{B}}T}\,, \tag{15.61}$$

$$y(0) = \mathrm{e}^{-\beta E_{\mathrm{c}}}\,, \tag{15.62}$$

where n_0 is the quasicondensate density and E_c is the core energy of a vortex. Both are obtained from the modified many-body T matrix theory considered previously. Writing the order parameter for a vortex configuration as $\psi_0(\mathbf{x}) = \sqrt{n_0}f(x/\xi)\mathrm{e}^{\mathrm{i}\vartheta}$,

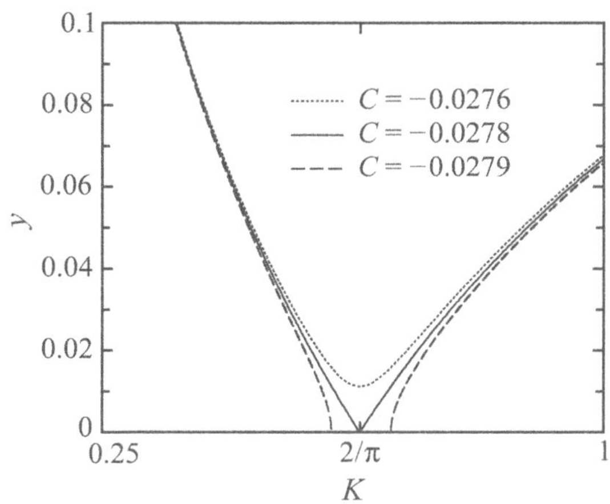

Fig. 15.3 Renormalization group flow for the coupling constants y and K. These curves are given by (15.60) for different values of C.

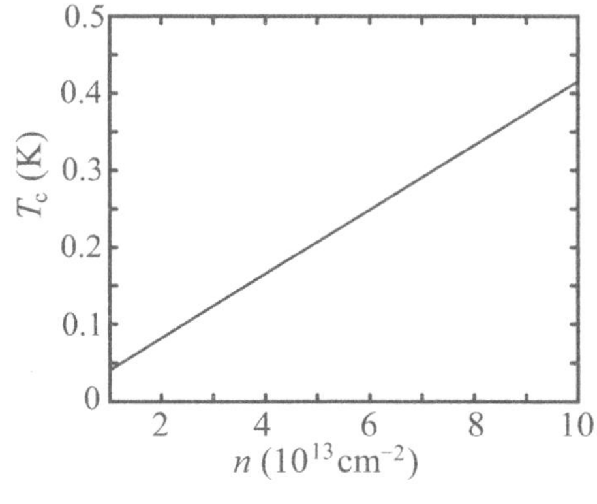

Fig. 15.4 The critical temperature for the Kosterlitz-Thouless transition as a function of the density for spin-polarized atomic hydrogen with $a = 2.4a_0$.

where ϑ is the azimuthal angle, the core energy of a vortex follows from the Gross-Pitaevskii energy functional. It reads

$$E_c = \frac{\hbar^2}{2m} n_0 \pi \int_0^\infty \mathrm{d}x\, x \left\{ \left(1 - f^2\right)^2 + 2\left(\frac{\mathrm{d}f}{\mathrm{d}x}\right)^2 \right\} . \tag{15.63}$$

The dimensionless integral was evaluated by Minnhagen and Nylén, and takes the value 1.56 [150]. Using the solution to the flow equations (15.60) and the initial conditions, we can calculate the temperature for the Kosterlitz-Thouless transition given the scattering length a and the density of the system. In the following, we consider again atomic hydrogen. In Fig. 15.4, we show the critical temperature as a function of density for $a = 2.4a_0$. We see that the critical temperature is essentially proportional to the density of the system. This can be seen in more detail in Fig. 15.5, where we plot $n\Lambda_c^2$ as a function of n. It is clear from this figure that $n\Lambda_c^2$ indeed changes only slightly over the density range considered.

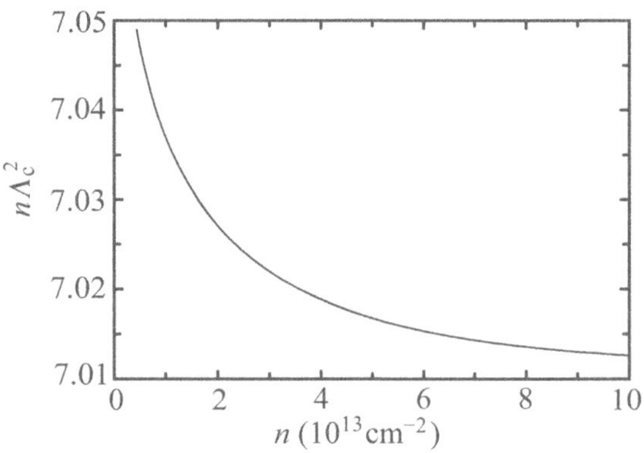

Fig. 15.5 The critical degeneracy parameter $n\Lambda_c^2$ as a function of the density for spin-polarized atomic hydrogen with $a = 2.4a_0$.

To understand the physics of the calculation better, we show in Fig. 15.6 the quasicondensate fraction n_0/n following from the many-body T matrix theory as a function of temperature for a total density $n = 1.25 \times 10^{13}\text{cm}^{-2}$. In addition, we show the superfluid density n_s as calculated from the renormalization-group procedure explained previously. The Kosterlitz-Thouless transition takes place when n_s lies on the line given by $n_s\Lambda^2 = 4$. Noticing that the left-hand side of (15.60) is a function of $n_0\Lambda^2$ only and solving the equation with respect to $n_0\Lambda^2$ using the value of C at the transition, we obtain the condition $n_0\Lambda^2 \simeq 6.65$ for the Kosterlitz-Thouless transition. It is therefore also seen in fig. 15.6 that the Kosterlitz-Thouless transition takes place when the line given by $n_0\Lambda^2 \simeq 6.65$ intersects with the curve for n_0.

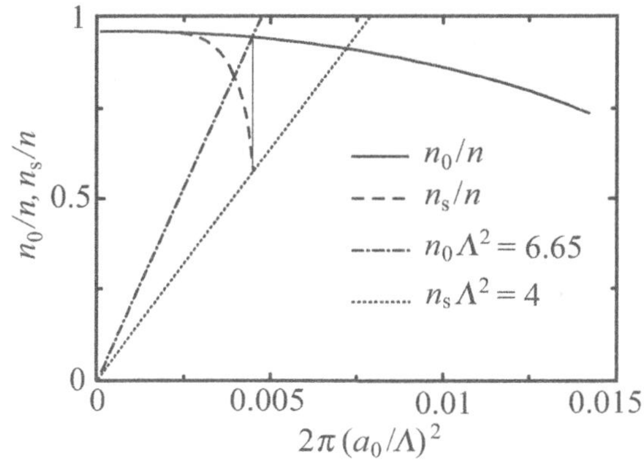

Fig. 15.6 Quasicondensate density n_0 (solid curve) and superfluid density n_s (dashed curve) as a function of temperature. Also plotted are the Kosterlitz-Thouless condition $n_s\Lambda^2 = 4$ (dotted line) and the condition $n_0\Lambda^2 = 6.65$ (dash-dotted line). The Kosterlitz-Thouless transition takes place when the dash-dotted line intersects the solid curve. At the intersection point, the dashed curve reaches the dotted line.

15.5 Trapped Bose Gases

In this section, we generalize the theory presented in Sects. 15.1 and 15.2 to inhomogeneous Bose gases. We also apply the results to a trapped one-dimensional Bose gas. We start by generalizing our previous expressions for the total density, (15.7), and the phase fluctuations, (15.10), to the inhomogeneous case. To do so we first consider the Gross-Pitaevskii equation

$$\left\{ -\frac{\hbar^2}{2m}\nabla^2 + V^{\mathrm{ex}}(\mathbf{x}) + T^{\mathrm{MB}}(-2\mu(\mathbf{x}))|\phi_0(\mathbf{x})|^2 \right.$$
$$\left. + 2T^{\mathrm{MB}}(-\mu(\mathbf{x}))n'(\mathbf{x}) \right\} \phi_0(\mathbf{x}) = \mu\phi_0(\mathbf{x}) , \qquad (15.64)$$

which generalizes (15.7) to trapped Bose-Einstein condensates. Here the local chemical potential equals $\mu(\mathbf{x}) = \mu - V^{\mathrm{ext}}(\mathbf{x})$. The noncondensed density $n'(\mathbf{x})$ is to be determined by solving the Bogoliubov-de Gennes equations

$$\hbar\omega_j u_j(\mathbf{x}) = \left\{ -\frac{\hbar^2}{2m}\nabla^2 + V^{\mathrm{HF}}(\mathbf{x}) - \mu \right\} u_j(\mathbf{x})$$
$$+ T^{\mathrm{MB}}(-2\mu(\mathbf{x}))n_0(\mathbf{x})v_j(\mathbf{x}) , \qquad (15.65)$$
$$-\hbar\omega_j v_j(\mathbf{x}) = \left\{ -\frac{\hbar^2}{2m}\nabla^2 + V^{\mathrm{HF}}(\mathbf{x}) - \mu \right\} v_j(\mathbf{x})$$
$$+ T^{\mathrm{MB}}(-2\mu(\mathbf{x}))n_0(\mathbf{x})u_j(\mathbf{x}) , \qquad (15.66)$$

where $n_0(\mathbf{x}) = |\phi_0(\mathbf{x})|^2$ and the Hartree-Fock potential $V^{\mathrm{HF}}(\mathbf{x})$ is given by

$$V^{\mathrm{HF}}(\mathbf{x}) = V^{\mathrm{ex}}(\mathbf{x}) + 2T^{\mathrm{MB}}(-\mu(\mathbf{x}))n'(\mathbf{x})$$
$$+ 2T^{\mathrm{MB}}(-2\mu(\mathbf{x}))n_0(\mathbf{x}) . \qquad (15.67)$$

The functions u_j and v_j are the usual Bogoliubov particle and hole amplitudes respectively, which are chosen to be real here. In some cases, for instance when ϕ_0 describes a vortex, we cannot choose these amplitudes real, however, our equations are readily generalized to incorporate this fact.

In terms of the Bogoliubov amplitudes, the expression for the total density in (15.7) reads

$$n(\mathbf{x}) = n_0(\mathbf{x}) + \sum_j \left\{ (u_j(\mathbf{x}) + v_j(\mathbf{x}))^2 N(\hbar\omega_j) \right.$$
$$\left. + v_j(\mathbf{x})(v_j(\mathbf{x}) + u_j(\mathbf{x})) + \frac{T^{\mathrm{MB}}(-2\mu(\mathbf{x}))n_0(\mathbf{x})}{2\varepsilon_j + 4\mu(\mathbf{x})}(\chi_j(\mathbf{x}))^2 \right\} . \qquad (15.68)$$

Here, χ_j is the large-j or high-energy limit of u_j which can be obtained by neglecting the interaction terms in (15.65), namely

$$\varepsilon_j \chi_j(\mathbf{x}) = \left\{ -\frac{\hbar^2}{2m}\nabla^2 + V^{\mathrm{ex}}(\mathbf{x}) - \mu \right\} \chi_j(\mathbf{x}) . \qquad (15.69)$$

In the large-j limit, we also have

$$v_j(\mathbf{x}) = -\frac{T^{\mathrm{MB}}(-2\mu(\mathbf{x}))n_0(\mathbf{x})}{2\varepsilon_j}\chi_j(\mathbf{x}) . \qquad (15.70)$$

It is clear that the expression of (15.68) for the total density is ultraviolet finite, because the second and third term cancel each other in the large-j limit.

Finally, the phase fluctuations in the trapped case are determined by the equal-time correlation function $\langle\theta(\mathbf{x},\tau)\theta(\mathbf{x}',\tau)\rangle$ which is given by

$$\begin{aligned}\langle\theta(\mathbf{x},\tau)\theta(\mathbf{x}',\tau)\rangle = -\sum_j \frac{1}{2\sqrt{n_0(\mathbf{x})n_0(\mathbf{x}')}}\Bigg\{ & u_j(\mathbf{x}')v_j(\mathbf{x})\left\{1+2N_{\mathrm{BE}}(\hbar\omega_j)\right\} \\ & + \frac{T^{\mathrm{MB}}(-2\mu(\mathbf{x}))n_0(\mathbf{x})}{2\varepsilon_j+4\mu(\mathbf{x})}\chi_j(\mathbf{x}')\chi_j(\mathbf{x}) + u_j(\mathbf{x})v_j(\mathbf{x}') \\ & \times\left\{1+2N_{\mathrm{BE}}(\hbar\omega_j)\right\} + \frac{T^{\mathrm{MB}}(-2\mu(\mathbf{x}'))n_0(\mathbf{x}')}{2\varepsilon_j+4\mu(\mathbf{x}')}\chi_j(\mathbf{x})\chi_j(\mathbf{x}')\Bigg\} .\end{aligned} \quad (15.71)$$

In particular the normalized form of the off-diagonal one-particle density matrix of (15.19) becomes for large distances $|\mathbf{x}-\mathbf{x}'|$ equal to

$$g^{(1)}(\mathbf{x},\mathbf{x}') = \exp\left\{-\langle[\theta(\mathbf{x},\tau)-\theta(\mathbf{x}',\tau)]^2\rangle/2\right\} . \quad (15.72)$$

15.5.1 Density Profiles

We are now ready to calculate the total density profile by solving (15.64) and (15.68) selfconsistently. In the rest of the Chapter, we restrict ourselves to one-dimensional harmonic traps with, therefore,

$$V^{\mathrm{ex}}(z) = \frac{1}{2}m\omega_z^2 z^2 . \quad (15.73)$$

For simplicity we use the local-density approximation, which allows us to calculate the densities directly using the many-body generalization of (15.7), and (15.12). In Fig. 15.7, the total density profile is shown at four different values of the temperature.

For the four different temperatures each of the four curves is composed of two parts. The first part near the center of the trap represents the superfluid part of the gas and contains the (quasi)condensate. The other part consists only of the noncondensed atoms. The small discontinuity between the two parts is caused by the use of two different equations of state for the superfluid and thermal phases of the gas. In the following we call the position of the discontinuity the temperature-dependent Thomas-Fermi radius of the (quasi)condensate. For distances below the discontinuity we use the above-mentioned equations, while for distances above the discontinuity, we simply use

$$n(z) = \int_{-\infty}^{\infty} \frac{\mathrm{d}k}{2\pi}\, N_{\mathrm{BE}}\left(\varepsilon_k + m\omega_z^2 z^2/2 + 2nT^{\mathrm{MB}}(-\hbar\Sigma(z)) - \mu\right) . \quad (15.74)$$

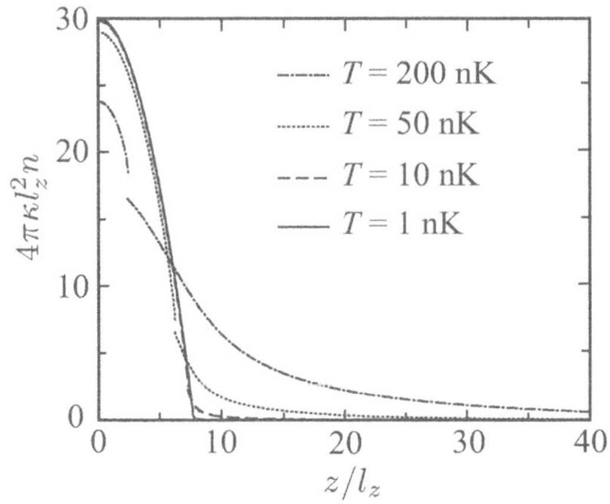

Fig. 15.7 Density profile of a trapped one-dimensional Bose gas at four different temperatures. The quantities l_z and κ are defined in the text.

For all these curves $\mu = 30\hbar\omega_z$. The remaining parameters used here are those of the experiment of Görlitz et al. [123]. In particular, we have used ^{23}Na in the trap with $\omega_z = 2\pi \times 3.5$ rad/sec, $l_z = \sqrt{\hbar/m\omega_z} \simeq 1.12 \times 10^{-5}$ m. The three-dimensional s-wave scattering length is $a \simeq 2.75$ nm which is related to the one-dimensional scattering length κ^{-1} defined by $T^{2B}(-2\mu) = 4\pi\kappa\hbar^2/m$. For harmonic confinement, we have $\kappa = a/2\pi l_\perp^2$, where $l_\perp$ is the harmonic oscillator length of the axially symmetric trap in the direction perpendicular to the z-axis. We have used $\omega_\perp = 2\pi \times 360$ rad/sec and $l_\perp = \sqrt{\hbar/m\omega_\perp} \simeq 1.10 \times 10^{-6}$ m.

As expected, the temperature-dependent Thomas-Fermi radius decreases with increasing temperature. At the temperature for which this radius vanishes the one-dimensional system reaches the crossover temperature for the formation of a (quasi)condensate. We have calculated this crossover temperature for different values of the scattering length at a constant value of the number of atoms, the latter being fixed by adjusting the chemical potential.

In Fig. 15.8 we show the result of this calculation, and plot the crossover temperature T_{QC} and the chemical potential against the scattering length. The inset in Fig. 15.8 shows that on a double logarithmic scale the temperature T_{QC} is clearly not a straight line, indicating that the relation between T_{QC} and κ is not a simple power law and may contain logarithmic dependence. It is shown in [151] that for $a = 0$, the transition temperature T_{QC} should satisfy $T_{QC} = N\hbar\omega_z/k_B \log(2N)$, where N is the number of atoms. In the case of Fig. 15.8 we have $N = 950$, which leads to $T_{QC} \simeq 164T_0$ for an ideal gas. Of course, this limit is not obtained in Fig. 15.8 because our calculation is based on a local-density approximation, which will always break down for sufficiently small values of κ. On the other hand, the curve for the chemical potential becomes almost a straight line on a double logarithmic scale. A calculation of the slope of this line shows that the slope starts at a value slightly larger than 2/3 at the lower end of the curve and saturates at this value

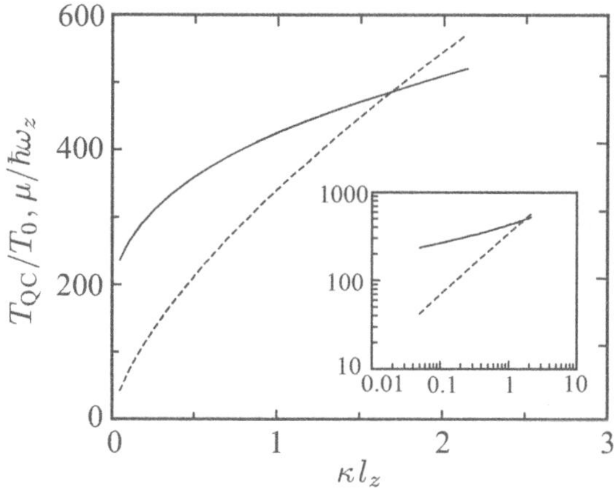

Fig. 15.8 The crossover temperature T_{QC} is shown with the solid curve, and the chemical potential at this temperature is shown with the dashed curve, both as a function of the coupling constant. The temperature is scaled to $T_0 = \hbar^2/mk_B l_z^2$. The inset shows the same curves on a double logarithmic scale.

near the upper end. The value 2/3 is what we expect, since in the Thomas-Fermi limit it is easy to show that $\mu = (3\pi/\sqrt{2})^{2/3}(N\kappa)^{2/3}\hbar\omega_z \simeq 3.5(N\kappa)^{2/3}\hbar\omega_z$. Calculating similar curves for different values of N we actually find numerically that $\mu \simeq 3.2(N\kappa)^{2/3}\hbar\omega_z$.

15.5.2 Phase Fluctuations

The aim of this section is to calculate the normalized off-diagonal density matrix given by (15.72). This function expresses the coherence in the system. It is calculated by solving the Bogoliubov-de Gennes equations in (15.65) and (15.66) using the density profile calculated in the previous subsection. Specifically from the (quasi)condensate density profile n_0, we determine a temperature-dependent Thomas-Fermi radius. This radius is then used to calculate the phase fluctuations at that specified temperature in the following manner.

We start by employing the following scaling: lengths are scaled to the trap length $l_z = (\hbar/m\omega_z)^{1/2}$, frequencies to ω_z, energies to $\hbar\omega_z$, and densities to $4\pi/l_z$. With this scaling, the Bogoliubov-de Gennes equations take the dimensionless form

$$\omega_j u_j = \left(-\frac{1}{2}\frac{d^2}{dz^2} + \frac{1}{2}z^2 - \mu + 2\kappa n\right) u_j - \kappa n_0 v_j \,, \tag{15.75}$$

$$-\omega_j v_j = \left(-\frac{1}{2}\frac{d^2}{dz^2} + \frac{1}{2}z^2 - \mu + 2\kappa n\right) v_j - \kappa n_0 u_j \,. \tag{15.76}$$

Using the same scaling, the Gross-Pitaevskii equation takes the form

$$\left\{-\frac{1}{2}\frac{\mathrm{d}^2}{\mathrm{d}z^2}+\frac{1}{2}z^2-\mu+\kappa(n_0+2n')\right\}\sqrt{n_0}=0\,. \tag{15.77}$$

Next, we define $F_j(z)=u_j(z)+v_j(z)$ and $G_j(z)=u_j(z)-v_j(z)$, and derive from (15.75) and (15.76) two equations for $F_j(z)$ and $G_j(z)$, namely

$$\frac{\mathrm{d}^4F}{\mathrm{d}z^4}-2(f+g)\frac{\mathrm{d}^2F}{\mathrm{d}z^2}-4\frac{\mathrm{d}g}{\mathrm{d}z}\frac{\mathrm{d}F}{\mathrm{d}z}-\left(4\omega_j^2+2\frac{\mathrm{d}^2g}{\mathrm{d}z^2}-4gf\right)F=0\,, \tag{15.78}$$

$$\frac{\mathrm{d}^4G}{\mathrm{d}z^4}-2(f+g)\frac{\mathrm{d}^2G}{\mathrm{d}z^2}-4\frac{\mathrm{d}f}{\mathrm{d}z}\frac{\mathrm{d}G}{\mathrm{d}z}-\left(4\omega_j^2+2\frac{\mathrm{d}^2f}{\mathrm{d}z^2}-4gf\right)G=0\,, \tag{15.79}$$

where the functions $f(z)$ and $g(z)$ are given by

$$f=\frac{1}{2}z^2+2\kappa n-\mu+\kappa n_0\,, \tag{15.80}$$

$$g=\frac{1}{2}z^2+2\kappa n-\mu-\kappa n_0\,. \tag{15.81}$$

For our purposes, we can use the Thomas-Fermi approximation which neglects the derivative term in (15.77). Hence

$$\left\{\frac{1}{2}z^2-\mu+\kappa(n_0+2n')\right\}\sqrt{n_0}=0\,. \tag{15.82}$$

In this limit, the functions $f(z)$ and $g(z)$ are given by $f(z)=2\kappa n_0(z)$ and $g(z)=0$. In the Thomas-Fermi approximation, we substitute these values for $f(z)$ and $g(z)$ into (15.78) and (15.79) and neglect the fourth-order derivative terms. These equations thus take the form

$$\kappa n_0\frac{\mathrm{d}^2F_j}{\mathrm{d}z^2}+\omega_j^2F_j=0\,, \tag{15.83}$$

$$\frac{\mathrm{d}^2(\kappa n_0G_j)}{\mathrm{d}z^2}+\omega_j^2G_j=0\,. \tag{15.84}$$

In reference [152], it was shown that $\sqrt{\kappa n_0(z)}G_j(z)$ corresponds to density fluctuations and $F_j(z)/\sqrt{\kappa n_0(z)}$ corresponds to phase fluctuations in the hydrodynamic-like approach [153]. We therefore define the function $h_j(z)$

$$h_j=\sqrt{\kappa n_0}G_j=F_j/\sqrt{\kappa n_0}\,. \tag{15.85}$$

Substituting this back in (15.83) and (15.84), both equations reduce to a single equation for $h_j(z)$, namely

$$\kappa n_0\frac{\mathrm{d}^2h_j}{\mathrm{d}z^2}+\kappa\frac{\mathrm{d}n_0}{\mathrm{d}z}\frac{\mathrm{d}h_j}{\mathrm{d}z}+\omega_j^2h_j=0\,. \tag{15.86}$$

This equation can finally be simplified using the Thomas-Fermi expression for $\kappa n_0(z)$ from (15.82), namely $\kappa n_0(z) \simeq \mu' - z^2/2$ where $\mu' = \mu - 2\kappa n'(0)$. Note that we have made the approximation that we take $n'(z)$ to be equal to its value at the center, namely $n'(0)$. This approximation is justified in view of the fact that the presence of the condensate repels noncondensate atoms from the center of the trap. This is also supported by a numerical solution of (15.7) and (15.12), where we find that $n'(z) \ll n_0(z)$, except at the Thomas-Fermi radius where they become of the same order. Moreover, the slope of $n'(z)$ is small for distances close to the center. Thus, the last equation becomes

$$(1-y^2)\frac{\mathrm{d}^2}{\mathrm{d}y^2}h_j(y) - 2y\frac{\mathrm{d}}{\mathrm{d}y}h_j(y) + 2\omega_j^2 h_j(y) = 0\,, \tag{15.87}$$

where $y = z/R_{\mathrm{TF}}(T)$ and $R_{\mathrm{TF}}(T) = \sqrt{2\mu'(T)}$ is the Thomas-Fermi radius.

In the following, we reinstate the units. Interestingly, (15.87) is the Legendre equation with the Legendre polynomials as solutions:

$$h_j(z) = P_j(z/R_{\mathrm{TF}}) = P_j(y)\,, \tag{15.88}$$

where the energy eigenvalues for $j = 0, 1, 2, \ldots$ are

$$\hbar\omega_j = \sqrt{\frac{j(j+1)}{2}}\hbar\omega_z\,. \tag{15.89}$$

The normalization condition for the Bogoliubov amplitudes is

$$\int_{-R_{TF}}^{R_{TF}} \mathrm{d}z\,\left\{|u_j|^2(z) - |v_j|^2(z)\right\} = 1\,, \tag{15.90}$$

which leads to

$$F_j(z) = \frac{1}{\sqrt{R_{\mathrm{TF}}}}\sqrt{\frac{(j+1/2)\mu'}{\hbar\omega_j}}\sqrt{1-y^2}P_j(y)\,, \tag{15.91}$$

$$G_j(z) = \frac{1}{\sqrt{R_{\mathrm{TF}}}}\sqrt{\frac{(j+1/2)\hbar\omega_j}{\mu'}}\frac{P_j(y)}{\sqrt{1-y^2}}\,. \tag{15.92}$$

These expressions are in agreement with those obtained in [154]. Consequently, we find

$$u_j(z) = \frac{1}{2}\left(A_j\sqrt{1-y^2} + \frac{B_j}{\sqrt{1-y^2}}\right)P_j(y)\,, \tag{15.93}$$

$$v_j(z) = \frac{1}{2}\left(A_j\sqrt{1-y^2} - \frac{B_j}{\sqrt{1-y^2}}\right)P_j(y)\,, \tag{15.94}$$

where

$$A_j = \frac{1}{\sqrt{R_{\text{TF}}}}\sqrt{\frac{(j+1/2)\mu'}{\hbar\omega_j}}, \tag{15.95}$$

$$B_j = \frac{1}{\sqrt{R_{\text{TF}}}}\sqrt{\frac{(j+1/2)\hbar\omega_j}{\mu'}}. \tag{15.96}$$

The expression for the phase fluctuations in (15.72) now reads, after neglect of the quantum contribution,

$$\begin{aligned}\langle[\theta(z,\tau)-\theta(z',\tau)]^2\rangle = \frac{4\pi\kappa l_z^4}{R_{\text{TF}}^2}\sum_j N(\hbar\omega_j)\Bigg\{A_j^2\big(P_j(y)-P_j(y')\big)^2 \\ -B_j^2\left(\frac{P_j(y)}{1-y^2}-\frac{P_j(y')}{1-y'^2}\right)^2\Bigg\}.\end{aligned} \tag{15.97}$$

It should be noted that the first term in this sum, $j=0$, does not diverge as one might think in first instance. It actually vanishes and the sum can start from $j=1$. Physically, this is a result of the fact that the global phase does not influence the phase fluctuations.

For the four values of temperature used in Fig. 15.7, we insert the corresponding $R_{\text{TF}}(T)$ in (15.97) to calculate the phase correlation function $g^{(1)}(0,z)$. In Fig. 15.9, we plot this quantity and we see that at sufficiently low temperatures the phase correlation function decreases only slightly over the condensate size. This indicates that a true condensate can exist at sufficiently low temperatures for interacting trapped one-dimensional Bose gases.

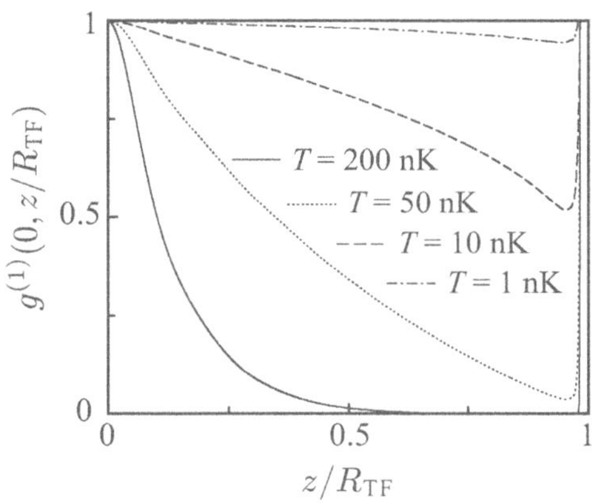

Fig. 15.9 Normalized first-order (phase) correlation function as a function of position for different temperatures.

15.5.3 Comparison with Exact Results

We next compare the above results to predictions based on a Langevin field equation for the order parameter of a trapped, one-dimensional condensate in contact with a three-dimensional Bose gas that acts as a "heat bath". Such a situation can be created experimentally in a magnetically trapped three-dimensional system by using a laser beam to provide an additional optical potential along two of the directions. The laser beam then needs to be focused such that the motion of the system freezes out along these directions. The gas in the potential "dimple" provided by the laser then becomes an effectively one-dimensional condensate, in contact with the three-dimensional thermal cloud in the magnetic trap which acts as its heat bath. The dynamics of the order parameter is governed in this case by [155, 156]

$$
\begin{aligned}
\mathrm{i}\hbar \frac{\partial \Phi(z,t)}{\partial t} = &\left\{ -\frac{\hbar^2 \nabla^2}{2m} + V^{\mathrm{ex}}(z) - \mu - \mathrm{i}R(z,t) \right. \\
&\left. + g|\Phi(z,t)|^2 \right\} \Phi(z,t) + \eta(z,t) \,, \qquad (15.98)
\end{aligned}
$$

where the external trapping potential in the weakly-confined direction $V^{\mathrm{ex}}(z)$ is given in (15.73) and μ is the effective chemical potential of the one-dimensional

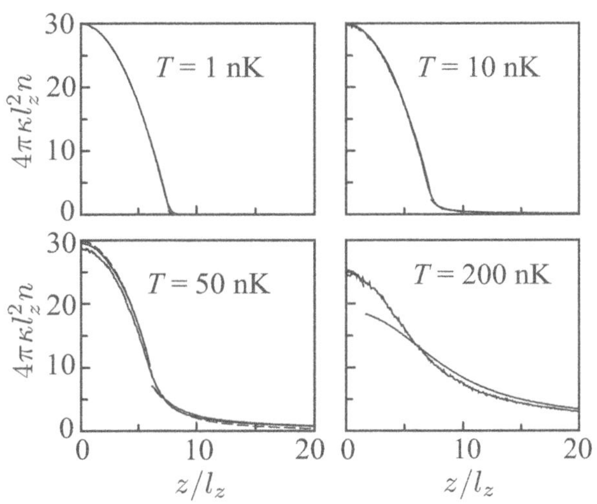

Fig. 15.10 Comparison of the mean-field density profiles (solid curves) to numerical solutions of the Langevin equation in (15.98) (noisy curves). All the above curves are calculated using the classical approximation of the Bose-Einstein distribution function. For the $T = 50$ nK case, we have also plotted the corresponding density profile calculated using the full Bose-Einstein distribution function (dashed curve) in order to show the difference between the classical and quantum mean-field approximations.

system. The one-dimensional coupling constant g is related to κ by $g = 4\pi\kappa\hbar^2/m$. Physically, the function $\mathrm{i}R(z,t)$ describes the pumping of the one-dimensional condensate from the surrounding thermal cloud, and $\eta(z,t)$ corresponds to the associated noise with Gaussian correlations. Both these quantities depend on the one-dimensional Keldysh selfenergy $\hbar\Sigma^{\mathrm{K}}(z)$, as discussed in detail in reference [156]. For our purposes, we only need that

$$\mathrm{i}R(z,t) = -\frac{\beta}{4}\hbar\Sigma^{\mathrm{K}}(z)\left(-\frac{\hbar^2\nabla^2}{2m} + V^{\mathrm{ex}}(z) - \mu + T^{\mathrm{2B}}|\Phi(z,t)|^2\right), \tag{15.99}$$

and

$$\langle\eta^*(z,t)\eta(z',t')\rangle = \frac{\mathrm{i}\hbar^2}{2}\Sigma^{\mathrm{K}}(z)\delta(z-z')\delta(t-t'), \tag{15.100}$$

where $\langle\ldots\rangle$ denotes averaging over the realizations of the noise $\eta(z,t)$. The numerical techniques employed are discussed in reference [156], where it also was shown that with the last two expressions the trapped gas relaxes to the correct equilibrium, as ensured by the fluctuation-dissipation theorem. To simplify the numerics, the noncondensed part in the dimple is here allowed to relax to the "classical" value $N(\varepsilon) = \{\beta(\varepsilon-\mu)\}^{-1}$, and the comparison to the previous mean-field predictions is therefore carried out by making the same approximation in the calculation of both $n_0(z)$ and $n'(z)$. The normalized first-order correlation function at equal time $g^{(1)}(0,z)$, corresponding to the previously computed phase correlation function, is calculated via numerical autocorrelation measurements i.e.

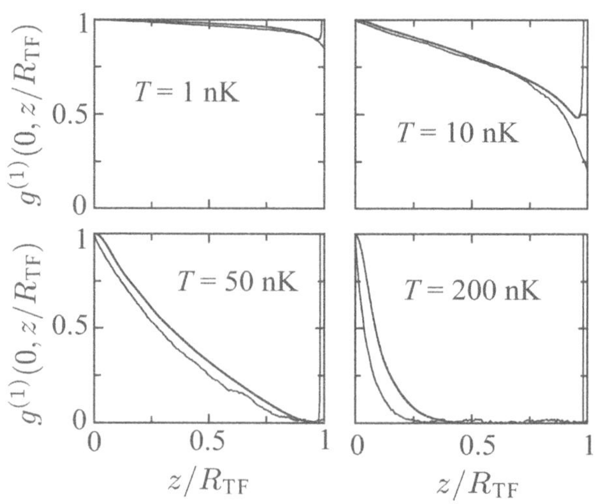

Fig. 15.11 Comparison of the normalized first-order (phase) correlation functions calculated using the present mean-field approach, given by the solid curves, and the numerical solution of the noisy Langevin equation in (15.98), shown by the noisy curves.

$$g^{(1)}(0,z,t) = \frac{\langle \Phi^*(0,t)\Phi(z,t)\rangle}{\sqrt{\langle|\Phi(0,t)|^2\rangle\langle|\Phi(z,t)|^2\rangle}}\,, \tag{15.101}$$

where the brackets again denote averaging over the different realizations of the noise. Of course, the time t must be sufficiently large so that the gas has relaxed to thermal equilibrium and $g^{(1)}(0,z,t)$ is independent of time.

In Figs. 15.10 and 15.11, we show the comparison of the many-body T matrix theory to the above Langevin calculations for the same temperatures used in Figs. 15.7 and 15.9. In Fig. 15.10, we compare the Langevin densities $\langle|\Phi(z,t)|^2\rangle$ to our classical mean-field density $n(z)$. This yields excellent agreement at low temperatures, except for a small region around the discontinuity in the mean-field theory, which can be understood from the fact that the local-density approximation always fails in a small region near the edge of the Thomas-Fermi radius. As expected, this region increases with increasing temperature. For $T = 50$ nK, Fig. 15.10 further shows the deviation of the "classical" prediction of our mean-field theory from the "quantum" one calculated previously in Sect. 15.5.1 and displayed in Fig. 15.7. Finally, Fig. 15.11 shows the corresponding phase correlation functions as a function of position. Here we find very good agreement in the entire temperature range. Note that the phase correlation functions are essentially indistinguishable for both classical and quantum treatments of the thermal cloud.

Finally it is worth mentioning again that, in obtaining our analytical expressions for the phase fluctuations and the density in Sects. 15.1 and 15.5, we have used the many-body T matrix for the interatomic interactions. As mentioned in Sect. 15.1.2, the many-body effects are important in one and two dimensions. To appreciate this importance, we recalculate the density profiles and phase fluctuations using the two-body T matrix. Thus for distances below R_{TF} the differences are due to (15.11),

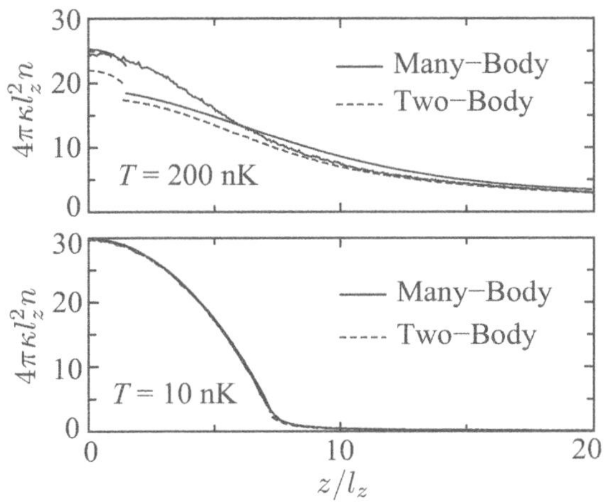

Fig. 15.12 Study of the many-body renormalization effects on the density profiles. The exact results are shown by the noisy curves.

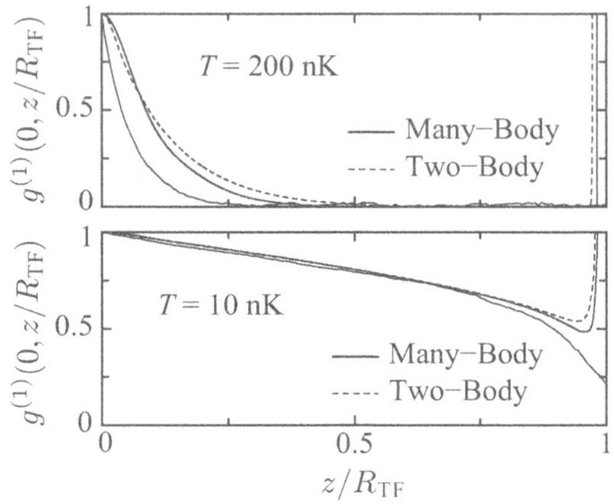

Fig. 15.13 Study of the many-body renormalization effects on the phase correlation function. The exact results are also shown with the noisy curves.

whereas for distances above R_{TF} they are a result of (15.28). In Figs. 15.12 and 15.13 it is clearly seen that the inclusion of many-body effects has led to a better agreement with the exact Langevin results. Moreover, the many-body corrections become more pronounced at higher temperatures. In Fig. 15.14, we show how the renormalized interatomic interaction strength $T^{\text{MB}}(-2\mu(z))$ depends on position. We notice that the effects of this renormalization becomes most significant near the edge of the condensate and for temperatures closer to the transition temperature, as expected from the results of references [110, 157].

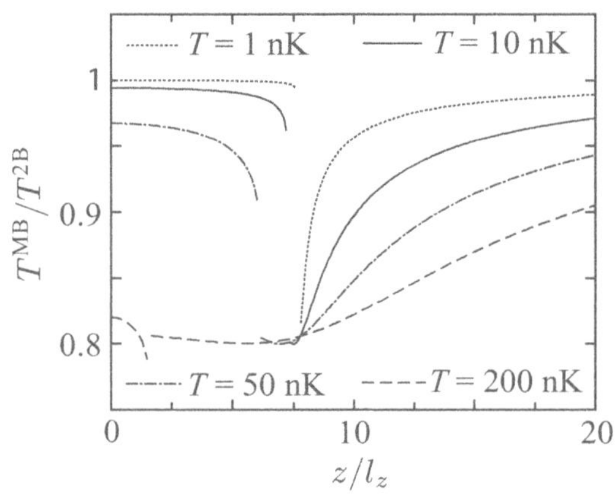

Fig. 15.14 The many-body transition matrix T^{MB} as a function of the distance from the center of the trap, for four different temperatures.

15.6 Problems

Exercise 15.1. Derive the one-loop results for the density and the chemical potential given in (15.2) and (15.3).

Exercise 15.2. Derive the logarithmic interaction potential between two vortices, two antivortices, and a vortex and an antivortex.

Chapter 16
Optical Lattices

The reductionist hypothesis does not by any means imply a "constructionist" one: The ability to reduce everything to simple fundamental laws does not imply the ability to start from those laws and reconstruct the universe.
–P. W. Anderson

In this chapter [158], we consider gases of ultracold atoms that are trapped in periodic potentials created by standing waves of laser light known as optical lattices. We start by considering in detail the atom-light interaction and derive from first principles the potential that an alkali atom experiences in an optical field. Then, we turn to many-body physics by showing that gases of ultracold atoms in a sufficiently deep optical lattice are described by the Hubbard models [159], which are very important in the fields of solid-state and condensed-matter physics. In particular, the high-temperature superconductors are often thought to be described by such a Hubbard model, so that a balanced Fermi mixture in an optical lattice might shed some new light in this unsolved problem of high-temperature superconductivity.

In this chapter we focus primarily on the Bose-Hubbard model, which applies to a Bose gas in an optical lattice. We show that this model contains a new quantum phase of matter called the Mott-insulator phase, as first discussed by Fisher et al. [160]. Moreover, the Bose gas is predicted to undergo a quantum phase transition from the superfluid state to the Mott-insulator state as a function of the potential depth of the optical lattice, i.e. as a function of the intensity of the lattice laser beams [159]. This quantum phase transition has recently been observed in a beautiful experiment by Greiner et al. [31], and has attracted much attention. It showed that ultracold atoms in an optical lattice can be used to simulate various lattice models of fundamental importance to condensed-matter physics, which are very difficult, if not impossible, to study in a controlled way in solid-state materials. At present, a large amount of effort from the community working on ultracold atoms is therefore directed on these exciting possiblities. Our recent proposal to create an ultracold superstring in the laboratory is motivated in the same spirit, namely with the aim of exploring high-energy physics problems with ultracold-atom experiments. In that proposal, a one-dimensional optical lattice plays a pivotal role, while the ultracold superstring is described by a supersymmetric version of the Bose-Hubbard model [161, 162].

16.1 Introduction

Before we discuss the interaction of atoms with light in the next section, we consider first the interference pattern caused by a set of crossed laser beams. A laser with wavelength λ emits photons with energy $2\pi\hbar c/\lambda$. Although the photons are quantum-mechanical entities, it is often possible to treat the laser beam and the corresponding electromagnetic field classically in the first instance. For a single plane wave travelling in the positive z direction, the electric field is given by

$$\mathbf{E}(\mathbf{x},t) = \boldsymbol{\varepsilon} E_0 \cos(kz-\omega t) \exp\left\{-\frac{x^2+y^2}{2w^2(z)}\right\}, \tag{16.1}$$

where we have $\omega = 2\pi c/\lambda$ for the frequency and $k = 2\pi/\lambda$ for the wavenumber. The polarization of the laser beam is described by $\boldsymbol{\varepsilon}$, which is a unit vector in the $x-y$ plane. It is a constant if the light is linearly polarized, or rotating around the z axis if the light is circularly polarized. The quantity $w(z)$ is called the waist of the Gaussian laser beam and defines its extension in the radial direction. The waist depends in general on the z coordinate, in particular when the laser beam is focussed, which is used experimentally to create an optical trap for atoms. However, for our purposes we can take the waist to be a constant, i.e. $w(z) = w$.

As we show more precisely in the next section, a single laser beam gives rise to a potential that is constant in the axial direction and Gaussian in the radial direction. This potential can be both repulsive and attractive depending on the frequency of the laser. In the latter case, atoms can be trapped in the laser beam and form a one-dimensional gas, as we considered in detail in Chap. 15. A single laser beam can also be used to rotate an atomic cloud, which has been used to create vortices in a Bose-Einstein condensate [69]. By superimposing two counter-propagating laser beams of the same wavelength and frequency, we can create a standing wave, which acts as a periodic potential for the atoms. This is known as an optical lattice. Assuming that the laser beams also have the same linear polarization, the electric field is given by

$$\mathbf{E}(\mathbf{x},t) = 2\boldsymbol{\varepsilon} E_0 \cos(\omega t)\cos(kz) \exp\left\{-\frac{x^2+y^2}{2w^2}\right\}. \tag{16.2}$$

By writing this as

$$\mathbf{E}(\mathbf{r},\mathbf{t}) \equiv \boldsymbol{\varepsilon} E(\mathbf{x}) \left(\frac{\mathrm{e}^{\mathrm{i}\omega t}+\mathrm{e}^{-\mathrm{i}\omega t}}{2}\right), \tag{16.3}$$

we obtain the quantized version of the laser field by simply replacing $\sqrt{\langle N_{\mathrm{ph}}\rangle}\mathrm{e}^{-\mathrm{i}\omega t}$ by the annihilation operator of a photon $\hat{a}$, where $\langle N_{\mathrm{ph}}\rangle = \langle \hat{a}^\dagger \hat{a}\rangle$ denotes the average number of photons present in the standing wave. Interestingly, we are here making use of the fact that a laser can be seen as a Bose-Einstein condensate of photons if the number of photons $\langle N_{\mathrm{ph}}\rangle$ is much larger than one, which is the appropriate limit for our purposes. As a result, the electric-field operator in the Schrödinger picture is given by

$$\hat{\mathbf{E}}(\mathbf{x}) = \boldsymbol{\varepsilon} E(\mathbf{x}) \left(\frac{\hat{a} + \hat{a}^\dagger}{2\sqrt{\langle N_{\text{ph}} \rangle}} \right). \tag{16.4}$$

In addition, the Hamiltonian for the photons is

$$\hat{H}_{\text{ph}} = \hbar\omega \hat{a}^\dagger \hat{a}. \tag{16.5}$$

It is straightforward to generalize this discussion to an electric field that is periodic in two or three directions. By having two counter-propagating laser beams in the z direction and two in the x direction, it is then possible to make a two-dimensional optical lattice in the $x-z$ plane such that atoms are confined in one-dimensional tubes along the y direction. If in addition two counter-propagating laser beams are placed along the y axis, we obtain a three-dimensional optical lattice.

16.2 Coupling between Atoms and Light

Ground-state alkali atoms couple to the electromagnetic field because of the quadratic Stark effect, i.e. via an induced electric dipole moment. The induced dipole moment is given by $\mathbf{d}(\mathbf{x}) = \alpha(\omega)\mathbf{E}(\mathbf{x})$, where $\alpha(\omega)$ is the polarizability of the atom, which in general depends on the frequency of the electric field, as we will see shortly. In principle, there is also a coupling to the magnetic dipole moment of the atom, but this is usually a negligible effect in the study of optical lattices. The resulting potential is then classically expected to be

$$V^{\text{ex}}(\mathbf{x}) = -\mathbf{d}(\mathbf{x}) \cdot \mathbf{E}(\mathbf{x}) = -\alpha(\omega) E^2(\mathbf{x}), \tag{16.6}$$

which can be either attractive or repulsive depending on the sign of the polarizability. In quantum mechanics, the above potential is obtained by taking into account laser-induced virtual excitations to an excited state. We first give the derivation for the simplest case of a two-level atom. After that, we take into account the fine structure of the alkali atoms, which means that we have to consider two excited states with a different total electronic angular momentum.

16.2.1 Two-Level Approximation

First, we consider a two-level atom whose ground state $|\text{g}\rangle$ has energy E_{g} and whose single excited state $|\text{e}\rangle$ has energy E_{e}. The interaction between the atom and the electric field is described by the Hamiltonian

$$\hat{H} = -\hat{\mathbf{d}} \cdot \hat{\mathbf{E}}(\mathbf{x}). \tag{16.7}$$

where $\hat{\mathbf{d}}$ is the quantum-mechanical dipole operator. It is given by

$$\hat{\mathbf{d}} = -e\sum_i \hat{\mathbf{r}}_i, \tag{16.8}$$

where $-e$ is the electron charge and $\hat{\mathbf{r}}_i$ are the position operators of the electrons relative to the nucleus of the atom. Usually, only the electrons in the outer shell are important, where the alkali atoms have only one electron in the outer shell. We denote the state of the atom and the electric field by $|\mathrm{g}, \langle N_{\mathrm{ph}}\rangle\rangle$ with $\langle N_{\mathrm{ph}}\rangle$ the average number of photons.

Next, we treat this interaction Hamiltonian in perturbation theory. Since for the atomic s-wave ground state the electronic orbital angular momentum is zero, we have that the first-order correction vanishes, i.e.

$$\langle \mathrm{g}, \langle N_{\mathrm{ph}}\rangle|\hat{H}|\mathrm{g}, \langle N_{\mathrm{ph}}\rangle\rangle = 0. \tag{16.9}$$

In the case of a p-wave excited state, the orbital angular momentum is one, which gives rise to a nonzero transition matrix element for the dipole operator. In second-order perturbation theory, this results in a potential for the atoms that is given by

$$V^{\mathrm{ex}}(\mathbf{x}) = \frac{|\langle \mathrm{g}|\hat{\mathbf{d}} \cdot \boldsymbol{\varepsilon}|\mathrm{e}\rangle|^2}{4}\left(\frac{1}{E_{\mathrm{g}} - E_{\mathrm{e}} + \hbar\omega} + \frac{1}{E_{\mathrm{g}} - E_{\mathrm{e}} - \hbar\omega}\right) E^2(\mathbf{x}). \tag{16.10}$$

Physically, the two-level approximation is based on the assumption that the energy of the photon $\hbar\omega$ is close to the energy difference $E_{\mathrm{e}} - E_{\mathrm{g}}$, so that this contribution dominates the expression for the energy shift in second-order perturbation theory, as given by 3.115. This then a posteriori justifies our neglect of the interaction of the light with the magnetic moment of the atom. In the next section, we will discuss what happens if there are two excited states with a small energy splitting due to the fine structure of the atom.

The first term on the right-hand side of (16.10) has the physical interpretation of the stimulated absorption of a photon, whereas the second term corresponds to the stimulated emission of a photon. The expectation value $\langle \mathrm{g}|\hat{\mathbf{d}} \cdot \boldsymbol{\varepsilon}|\mathrm{e}\rangle$ depends on the precise details of the atom and the polarization of the light, which is all conveniently lumped together into the Rabi frequency. The Rabi frequency for a single laser is defined as

$$\hbar\Omega = |\langle \mathrm{g}|\hat{\mathbf{d}} \cdot \boldsymbol{\varepsilon}|\mathrm{e}\rangle|E_0, \tag{16.11}$$

where E_0 is the maximum amplitude of the electric field as in (16.1). Moreover, it is customary to define the detuning from resonance δ as

$$\delta = \omega - (E_{\mathrm{e}} - E_{\mathrm{g}})/\hbar. \tag{16.12}$$

When the detuning is small, the first term in (16.10) is much bigger than the second and the latter can thus be neglected. This corresponds to the rotating-wave approximation and the potential is then given by

$$V^{\mathrm{ex}}(\mathbf{x}) = \frac{\Omega^2}{\delta}\cos^2(2\pi z/\lambda)\exp\left\{-\frac{x^2+y^2}{w^2}\right\}. \tag{16.13}$$

From this expression, we see that the potential can indeed be attractive or repulsive, depending on the sign of the detuning. When the detuning is negative, the atoms are attracted to the maxima of the laser intensity, where the potential energy of the atoms is minimal. In this case, the Gaussian profile of the laser beam automatically traps the atoms in the radial direction, and the laser light is also said to be red-detuned with respect to the atomic transition. When the detuning is negative, the atoms are attracted to the minima of laser intensity, where the potential energy of the atoms is also minimal. The laser light is then said to be blue-detuned with respect to the atomic transition. In this case, the Gaussian profile of the laser beam repels the atoms in the radial direction, so that an additional trapping mechanism has to be provided. This can be achieved by either a magnetic potential or by adding a second red-detuned laser beam.

To also take into account the finite lifetime of the excited state due to spontaneous emission of photons, we should add the imaginary part $-i\hbar\Gamma_e/2$ to the excited-state energy in (16.10). The real part of the resulting expression then corresponds to the experienced potential, which is typically only slightly modified if the detuning is not too close to resonance. The imaginary part can be written as

$$\hbar\Gamma^{\mathrm{eff}} = (\hbar\Omega)^2\hbar\Gamma_e\left(\frac{1}{(\hbar\delta)^2+(\hbar\Gamma_e/2)^2}+\frac{1}{(E_g-E_e-\hbar\omega)^2+(\hbar\Gamma_e/2)^2}\right), \tag{16.14}$$

where Γ^{eff} is the effective rate of photon emission. The inverse $1/\Gamma^{\mathrm{eff}}$ is the average time it takes to absorb one photon. This results in heating of the gas and eventually leads to the loss of atoms. The lifetime of the gas can therefore be estimated to be on the order of $1/\Gamma^{\mathrm{eff}}$. In the rotating-wave approximation and for a relatively large detuning, we have $\Gamma^{\mathrm{eff}} = (\Omega/\delta)^2\Gamma_e \ll \Gamma_e$, so that the lifetime in the optical lattice is much longer than the natural lifetime of the excited state.

16.2.2 Fine Structure

The two-level model that was introduced in the previous section explains how laser light gives rise to an external potential for atoms. For realistic alkali gases, however, this two-level model is often too simplistic. Then, it is important to include also the fine structure or even the hyperfine structure of the atoms to be sufficiently accurate. Next, we show how the fine structure can be taken into account, where we

ℓ, m_ℓ	s, m_s	j, m_j	$\langle \ell m_\ell; s m_s \| j m_j \rangle$
0,0	1/2,1/2	1/2,1/2	1
1,0	1/2,1/2	1/2,1/2	$1/\sqrt{3}$
1,0	1/2,1/2	3/2,1/2	$\sqrt{2/3}$
1,1	1/2,−1/2	1/2,1/2	$-\sqrt{2/3}$
1,1	1/2,−1/2	3/2,1/2	$1/\sqrt{3}$

Table 16.1 Relevant Clebsch-Gordan coefficients for the D_1 and D_2 optical transitions in alkali atoms.

assume that the laser is far enough detuned so that the hyperfine structure is not resolved and can be neglected. For alkali atoms, the ground state has a total electronic angular momentum of $j = 1/2$. The lowest-lying excited states have a total angular momentum $j = 1/2$, associated with the so called D_1 line, or a total angular momentum $j = 3/2$, which is associated with the D_2 line. The terminology D_1 and D_2 for the absorption lines is historical and originates from spectroscopy measurements.

In order to calculate the quadratic Stark effect again, we have to rewrite the total electronic angular momentum in terms of the electronic orbital angular momentum and the electron spin angular momentum using the Clebsch-Gordan coefficients for the addition of angular momentum. This means that we make the following basis transformation

$$|jm_j\rangle = \sum_{\ell,m_\ell} \sum_{s,m_s} |\ell m_\ell\rangle |sm_s\rangle \langle \ell m_\ell; sm_s | jm_j\rangle, \tag{16.15}$$

where the Clebsch-Gordan coefficients of relevance to us are given in Table 16.1. Neglecting lifetime effects, we find in the case of two counterpropagating laser beams with the same linear polarization that

$$V^{\text{ex}}(\mathbf{r}) = \frac{(\hbar\Omega)^2}{3} \cos^2(2\pi z/\lambda) \exp\left\{-\frac{x^2+y^2}{w^2}\right\} \tag{16.16}$$
$$\times \left(\frac{1}{E_{\text{g}} - E_{D_1} + \hbar\omega} + \frac{1}{E_{\text{g}} - E_{D_1} - \hbar\omega} + \frac{2}{E_{\text{g}} - E_{D_2} + \hbar\omega} + \frac{2}{E_{\text{g}} - E_{D_2} - \hbar\omega} \right).$$

The D_2 line is thus seen to contribute twice as much as the D_1 line, due to the difference in the Clebsch-Gordon coefficient of the transition. We note that when the detuning obeys $\hbar\delta \gg |E_{D_1} - E_{D_2}|$, the two different contributions add, such that we retrieve the previous two-level approximation of (16.10). This means that then the fine structure is not resolved and is safely neglected. In the same manner, the neglect of the hyperfine structure can also be justified.

16.3 Band Structure

The wavefunction of a free atom is a plane wave $e^{i\mathbf{k}\cdot\mathbf{x}}/\sqrt{V}$ and has an energy dispersion relation $\varepsilon_{\mathbf{k}} = \hbar^2\mathbf{k}^2/2m$. Interestingly, this result does not change too much for noninteracting atoms in a periodic optical-lattice potential. The relevant wavefunctions are now called Bloch waves and, as we will see, the dispersion develops a band structure. Indeed, the wavefunction can be written as a product of a plane wave and a function $u_{\mathbf{n},\mathbf{k}}(\mathbf{x})$ that is periodic with the lattice period, i.e.

$$\chi_{\mathbf{n},\mathbf{k}}(\mathbf{x}) = e^{i\mathbf{k}\cdot\mathbf{x}} u_{\mathbf{n},\mathbf{k}}(\mathbf{x}), \tag{16.17}$$

which is the well-known result from Bloch's theorem [163, 164]. Accordingly, the dispersion relation is no longer quadratic with the momentum but develops gaps at specific locations determined by the lattice structure. As a concrete example, for a weak one-dimensional optical lattice in the z-direction with lattice constant $\lambda/2$ the dispersion looks like the free-particle dispersion for small momenta. When the momentum approaches the boundary of the first Brillouin zone at $k_z = \pm 2\pi/\lambda$, with λ the wavelength of the laser light, the dispersion starts to deviate from the quadratic result. In fact, this happens around every value of k_z that is an odd multiple of $\pm 2\pi/\lambda$. If the momentum is increased past the boundary of the first Brillouin zone, the dispersion has a discontinuity and the difference in energy is the band gap. Equivalently, by folding the dispersion into the first Brillouin zone, the energy of an atom can be specified by a band index n_z and a momentum k_z that takes on values within the first Brillouin zone only.

Consider now an atom in an optical lattice with potential minima located at the lattice sites $\mathbf{x}_i$. It can be shown that for each band a set of Wannier functions $w_{\mathbf{n}}(\mathbf{x} - \mathbf{x}_i)$ exists, such that the exact Bloch wavefunctions can be written [163, 164] as

$$\chi_{\mathbf{n},\mathbf{k}}(\mathbf{x}) = \sum_i e^{i\mathbf{k}\cdot\mathbf{x}_i} w_{\mathbf{n}}(\mathbf{x} - \mathbf{x}_i). \tag{16.18}$$

The Wannier functions are orthogonal for different bands $\mathbf{n}$ as well as for different sites i. For deep optical lattices we can use the tight-binding limit, in which we approximate the lattice potential near each site $\mathbf{x}_i$ with a harmonic potential. Then, the exact Wannier functions $w_{\mathbf{n}}(\mathbf{x} - \mathbf{x}_i)$ are actually to a very good approximation given by the harmonic oscillator wavefunctions $\chi_{\mathbf{n}}(\mathbf{x} - \mathbf{x}_i)$.

16.4 Hubbard Models

We have seen in the previous sections that, by using standing waves of laser light, we can create a periodic potential for atoms. The imaginary-time action that describes a gas of atoms in such a periodic potential is given by

$$S[\psi^*,\psi] = \int_0^{\hbar\beta} d\tau \int d\mathbf{x}\, \psi^*(\mathbf{x},\tau)\left(\hbar\frac{\partial}{\partial\tau} - \frac{\hbar^2\nabla^2}{2m} + V^{\mathrm{ex}}(\mathbf{x})\right)\psi(\mathbf{x},\tau) \qquad (16.19)$$
$$+\frac{1}{2}\int_0^{\hbar\beta} d\tau \int d\mathbf{x} \int d\mathbf{x}'\, \psi^*(\mathbf{x},\tau)\psi^*(\mathbf{x}',\tau)V(\mathbf{x}-\mathbf{x}')\psi(\mathbf{x}',\tau)\psi(\mathbf{x},\tau),$$

where the (isotropic) optical-lattice potential is described by

$$V^{\mathrm{ex}}(\mathbf{x}) = \sum_j V_0 \cos^2(2\pi x_j/\lambda), \qquad (16.20)$$

with λ the wavelength of the laser light. Using the Wannier functions introduced in the previous section, we can expand the atomic fields as

$$\psi(\mathbf{x},\tau) = \sum_{\mathbf{n},i} a_{\mathbf{n},i}(\tau) w_{\mathbf{n}}(\mathbf{x}-\mathbf{x}_i). \qquad (16.21)$$

where the expansion coefficients $a^*_{\mathbf{n},i}(\tau)$ and $a_{\mathbf{n},i}(\tau)$ correspond to the creation and the annihilation operator respectively of an atom in the Wannier state $w_{\mathbf{n}}(\mathbf{x}-\mathbf{x}_i)$ at site i. In the tight-binding limit, the Wannier functions of the optical lattice are replaced by harmonic oscillator states on each site which, in Cartesian coordinates, depend on the three quantum numbers n_x, n_y and n_z. At sufficiently low temperatures and for sufficiently small interaction energies, the atoms only occupy the lowest $\mathbf{n}=\mathbf{0}$ state of the lattice. As a result, we find the following lattice action

$$S[a^*,a] = \int_0^{\hbar\beta} d\tau \left\{ \sum_{ij} a_i^*(\tau)\hbar\frac{\partial}{\partial\tau}a_j(\tau) \int d\mathbf{x}\, w_{\mathbf{0}}^*(\mathbf{x}-\mathbf{x}_i)w_{\mathbf{0}}(\mathbf{x}-\mathbf{x}_j) \right. \qquad (16.22)$$
$$+\sum_{ij} a_i^*(\tau)a_j(\tau)\int d\mathbf{x}\, w_{\mathbf{0}}^*(\mathbf{x}-\mathbf{x}_i)\left(-\frac{\hbar^2\nabla^2}{2m} + V^{\mathrm{ex}}(\mathbf{x}) - \mu\right) w_{\mathbf{0}}(\mathbf{x}-\mathbf{x}_j)$$
$$+\frac{1}{2}\sum_{ii'jj'} a_i^*(\tau)a_{i'}^*(\tau)a_j(\tau)a_{j'}(\tau)$$
$$\left. \times \int d\mathbf{x}\, d\mathbf{x}'\, w_{\mathbf{0}}^*(\mathbf{x}-\mathbf{x}_i)w_{\mathbf{0}}^*(\mathbf{x}'-\mathbf{x}_{i'})V(\mathbf{x}-\mathbf{x}')w_{\mathbf{0}}(\mathbf{x}'-\mathbf{x}_j)w_{\mathbf{0}}(\mathbf{x}-\mathbf{x}_{j'}) \right\},$$

where for notational convenience we have omitted the now redundant band index of the atomic fields. We can rewrite the above action more compactly as

$$S[a^*,a] = S_0[a^*,a] + S_{\mathrm{int}}[a^*,a], \qquad (16.23)$$

with the noninteracting part given by

$$S_0[a^*,a] = \qquad (16.24)$$
$$\int_0^{\hbar\beta} d\tau \left\{ \sum_i a_i^*(\tau)\left(\hbar\frac{\partial}{\partial\tau} + \varepsilon_i - \mu\right)a_i(\tau) - \sum_{i\neq j} a_i^*(\tau)t_{i,j}a_j(\tau) \right\},$$

Here, we have used the orthonormality of the Wannier functions in the same band, i.e. $\int \mathrm{d}\mathbf{x}\, w_{\mathbf{0}}^*(\mathbf{x}-\mathbf{x}_i)w_{\mathbf{0}}(\mathbf{x}-\mathbf{x}_j) = \delta_{i,j}$. We have also introduced the on-site energy

$$\varepsilon_i = \int \mathrm{d}\mathbf{x}\, w_{\mathbf{0}}^*(\mathbf{x}-\mathbf{x}_i)\left\{-\frac{\hbar^2\boldsymbol{\nabla}^2}{2m} + V^{\mathrm{ex}}(\mathbf{x})\right\} w_{\mathbf{0}}(\mathbf{x}-\mathbf{x}_i), \tag{16.25}$$

and the tunnelling or hopping amplitude between sites i and j

$$t_{i,j} = -\int \mathrm{d}\mathbf{x}\, w_{\mathbf{0}}^*(\mathbf{x}-\mathbf{x}_i)\left\{-\frac{\hbar^2\boldsymbol{\nabla}^2}{2m} + V^{\mathrm{ex}}(\mathbf{x})\right\} w_{\mathbf{0}}(\mathbf{x}-\mathbf{x}_j). \tag{16.26}$$

The double summation in the tunnelling or hopping term in (16.24) is over all combinations i,j for which $i \neq j$. For a deep lattice, the hopping energy $t_{i,j}$ will be exponentially suppressed for all sites that are not nearest neighbors. Therefore we restrict the summation to nearest neighbors only, where the corresponding summation is denoted by $\sum_{\langle i,j\rangle}$, while the nearest-neighbor hopping amplitude is denoted by t.

The interactions between the atoms are determined by the matrix elements

$$\int \mathrm{d}\mathbf{x} \int \mathrm{d}\mathbf{x}'\, w_{\mathbf{0}}^*(\mathbf{x}-\mathbf{x}_i)w_{\mathbf{0}}^*(\mathbf{x}'-\mathbf{x}_{i'})V(\mathbf{x}-\mathbf{x}')w_{\mathbf{0}}(\mathbf{x}'-\mathbf{x}_j)w_{\mathbf{0}}(\mathbf{x}-\mathbf{x}_{j'}),$$

which not only include on-site interactions but also interactions between atoms that are on remote sites. However, the latter interactions are typically exponentially suppressed, and for practical purposes it usually suffices to take only the on-site interactions into account. We then have

$$S_{\mathrm{int}}[a^*,a] = \int_0^{\hbar\beta} \mathrm{d}\tau\, \frac{U}{2}\sum_i a_i^*(\tau)a_i^*(\tau)a_i(\tau)a_i(\tau), \tag{16.27}$$

where we have defined the on-site interaction strength U as

$$U = \int \mathrm{d}\mathbf{x} \int \mathrm{d}\mathbf{x}'\, w_{\mathbf{0}}^*(\mathbf{x}-\mathbf{x}_i)w_{\mathbf{0}}^*(\mathbf{x}'-\mathbf{x}_i)V(\mathbf{x}-\mathbf{x}')w_{\mathbf{0}}(\mathbf{x}'-\mathbf{x}_i)w_{\mathbf{0}}(\mathbf{x}-\mathbf{x}_i). \tag{16.28}$$

For bosons, the Hamiltonian that corresponds to the action in (16.23) is known as the Bose-Hubbard model and is given by

$$\hat{H} = -t\sum_{\langle i,j\rangle} \hat{a}_i^\dagger \hat{a}_j + \sum_i(\varepsilon_i - \mu)\hat{a}_i^\dagger \hat{a}_i + \frac{U}{2}\sum_i \hat{a}_i^\dagger \hat{a}_i^\dagger \hat{a}_i \hat{a}_i. \tag{16.29}$$

For fermions, the on-site interaction disappears due to the Pauli principle if all the fermionic atoms are in the same hyperfine state. In that case, a mixture of two hyperfine states is more interesting, so that the relevant Hubbard Hamiltonian becomes

$$\hat{H} = -t_\alpha\sum_\alpha\sum_{\langle i,j\rangle} \hat{a}_{i,\alpha}^\dagger \hat{a}_{j,\alpha} + \sum_\alpha\sum_i(\varepsilon_{i,\alpha} - \mu_\alpha)\hat{a}_{i,\alpha}^\dagger \hat{a}_{i,\alpha} + U\sum_i \hat{a}_{i,\uparrow}^\dagger \hat{a}_{i,\downarrow}^\dagger \hat{a}_{i,\downarrow}\hat{a}_{i,\uparrow}. \tag{16.30}$$

where $\alpha = \uparrow, \downarrow$.

For deep lattices, we can work in the tight-binding limit, where the optical-lattice potential that we derived in Sect. 16.2 is approximated by its second-order expansion around a lattice site with coefficient $\hbar\omega$. To determine the hopping strength, we note that it is possible to solve the single-particle Schrödinger equation exactly in the tight-binding limit, where the final expression is given by [165]

$$t = \frac{4}{\sqrt{\pi}} E_{\mathrm{r}} \left(\frac{V_0}{E_{\mathrm{r}}} \right)^{3/4} \exp\left\{ -2 \left(\frac{V_0}{E_{\mathrm{r}}} \right)^{1/2} \right\}, \tag{16.31}$$

with $E_{\mathrm{r}} = 2\pi^2\hbar^2/m\lambda^2$ the recoil energy. The recoil energy gives the kinetic energy of an atom initially at rest after the absorption of a single photon, and is a commonly used energy scale in experiments with optical lattices. To determine the on-site interaction strength we make use of the pseudopotential approximation, such that we have

$$V(\mathbf{x} - \mathbf{x}') = \frac{4\pi a\hbar^2}{m} \delta(\mathbf{x} - \mathbf{x}'). \tag{16.32}$$

We substitute this expression into (16.27) and evaluate the integrals in the tight-binding limit, so that the Wannier functions are given by the harmonic-oscillator wavefunctions. This leads to

$$U = \frac{2\hbar\omega a}{l\sqrt{2\pi}}, \tag{16.33}$$

where λ is the wavelength of the laser light that is used to create the optical lattice, a is the s-wave scattering length, and $l = \sqrt{\hbar/m\omega} = (E_{\mathrm{r}}/V_0)^{1/4}\lambda/4\pi$ is the harmonic oscillator length. From this, we explicitly see that both the hopping strength and the interaction energy depend on the depth of the optical lattice potential and can therefore be easily tuned experimentally by varying the intensity of the laser beams.

16.5 Superfluid-Mott Insulator Transition

As we show in this section, a Bose-Einstein condensate in an optical lattice at zero temperature can undergo a quantum phase transition to a new quantum state of matter, namely via the superfluid-Mott insulator transition. The two competing ground states are a superfluid, with a well-defined global phase for the atoms, and a Mott insulator, which has the same integer occupation number at each lattice site such that the phase is fully undetermined. To gain a rough understanding of the physics, we show in Fig. 16.1 a one-dimensional optical lattice. Qualitatively we expect that, when there is an equal integer number of particles at each site i and $t \ll U$, then the strong repulsive interaction between the particles will make it energetically unfavorable for a particle to move from one site to another. As a result there is also

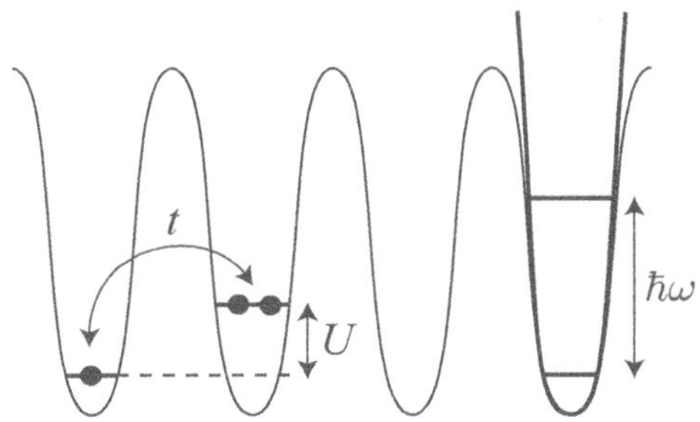

Fig. 16.1 In the Bose-Hubbard model, two bosons on the same site have an interaction energy U, while the probability to tunnel to a neighboring site is given by the hopping parameter t. A typical optical lattice is well approximated by a harmonic potential of strength $\hbar\omega$ for its lowest-lying quantum states.

a nonzero energy penalty to create an excitation, and the system is incompressible. In this situation the gas is said to be in the Mott insulator phase [166]. In the opposite limit the interactions are negligibly small, i.e. $t \gg U$, so that the atoms can hop freely around and minimize their energy by hybridizing over the optical lattice. The (nearly) ideal Bose gas then Bose-Einstein condenses in the lowest momentum state of the lowest Bloch band and becomes superfluid.

Suppose we are in the Mott phase at equal filling and that we add one particle to the system. Then this particle also minimizes its energy by delocalizing because its interaction energy is now the same on each site and it cannot occupy a free site. For this reason a gas which has, on average, a noninteger number of bosons at each site will also be in a superfluid phase at zero temperature. This qualitative picture has theoretically been investigated using Quantum Monte Carlo calculations [167] and several mean-field approaches [159, 168, 169]. Moreover, the superfluid-Mott insulator transition has also been observed experimentally in an pioneering experiment [31] and various additional experiments have investigated more detailed features of the transition [170, 171, 172, 173]. We come back to the experiments in Sect. 16.5.2 after we have more quantitatively discussed the phase diagram of the Bose-Hubbard model.

To do so, we use two different mean-field approaches with the aim of obtaining the phase diagram of an ultracold atomic Bose gas in an optical lattice. We determine the phase diagram using either perturbation theory in the interaction strength or perturbation theory in the hopping or kinetic energy. As we will see, the phase diagram indeed consists of various insulating phases and a superfluid phase. As our starting point, we consider the action of the Bose-Hubbard model, such that the grand-canonical partition function is given by

$$Z = \mathrm{Tr}\left[\mathrm{e}^{-\beta\hat{H}}\right] = \int \mathrm{d}[a^*]\mathrm{d}[a]\exp\left\{-\frac{1}{\hbar}S[a^*,a]\right\} \tag{16.34}$$

with the action

$$S[a^*,a] = \int_0^{\hbar\beta} \mathrm{d}\tau \left\{ \sum_i a_i^*(\tau) \left(\hbar \frac{\partial}{\partial \tau} - \mu \right) a_i(\tau) \right.$$
$$\left. - \sum_{\langle i,j \rangle} t_{ij} a_i^*(\tau) a_j(\tau) + \frac{U}{2} \sum_i a_i^*(\tau) a_i^*(\tau) a_i(\tau) a_i(\tau) \right\}. \quad (16.35)$$

In order to describe the zero-temperature phase transition from the superfluid to the Mott-insulating phase analytically, we need to make some appropriate mean-field approximation to the action in (16.34). In the following two subsections, we will follow two different approaches to bring out the physics and the problems involved more clearly.

16.5.1 Bogoliubov Approximation

In Chap. 11 we discussed the Bogoliubov theory to describe weakly-interacting atomic Bose gases. Here, we generalize the corresponding concepts to the presence of a lattice. For a Bose-Einstein condensed gas the average number of condensate atoms $\langle N_0 \rangle$ is a number much larger then one, and we have that $\langle a_i^*(\tau) \rangle = \langle a_i(\tau) \rangle = \sqrt{\langle N_0 \rangle / N_s}$, where N_s is the number of lattice sites and where we have chosen these expectation values to be real for simplicity. Therefore, we Fourier-transform the action of the Bose-Hubbard model by introducing the field amplitudes $a_{\mathbf{k},n}$ given by

$$a_i(\tau) = \frac{1}{\sqrt{\hbar\beta N_s}} \sum_{\mathbf{k},n} a_{\mathbf{k},n} e^{i(\mathbf{k}\cdot\mathbf{x}_i - \omega_n \tau)}, \quad (16.36)$$

where $\mathbf{x}_i$ is the coordinate of site i. The wavevector $\mathbf{k}$ runs only over the first Brillouin zone. For mathematical convenience we take only a finite volume V, so that the momenta $\hbar\mathbf{k}$ are discretized, which allows us to write sums instead of integrals in (16.36). Later we take the continuum limit $V \to \infty$. Using the fact that $\sum_i e^{-i(\mathbf{k}-\mathbf{k}')\cdot\mathbf{x}_i} = N_s \delta_{\mathbf{k},\mathbf{k}'}$ it is easily shown that the normalization factor $1/\sqrt{N_s}$ ensures that the average total number of atoms obeys $\langle N \rangle = \sum_i \langle a_i(\tau) a_i^*(\tau^+) \rangle = \sum_{\mathbf{k},n} \langle a_{\mathbf{k},n} a_{\mathbf{k},n}^* \rangle / \hbar\beta$, as desired. We also limit our description to cubic lattices with lattice distance $\lambda/2$, so $V = N_s(\lambda/2)^d$ with d the number of dimensions. Substituting explicitly

$$a_{\mathbf{0},0}^* \to \sqrt{\langle N_0 \rangle \hbar\beta} + a_{\mathbf{0},0}^* \quad \text{and} \quad a_{\mathbf{0},0} \to \sqrt{\langle N_0 \rangle \hbar\beta} + a_{\mathbf{0},0} \quad (16.37)$$

we find that up to quadratic order in the fluctuations the action is given by

$$
\begin{aligned}
S[a^*,a] = {} & \hbar\beta\left(-zt-\mu+\frac{1}{2}Un_0\right)\langle N_0\rangle + (-zt-\mu+Un_0)\sqrt{\langle N_0\rangle\hbar\beta}\,(a^*_{\mathbf{0},0}+a_{\mathbf{0},0}) \\
& + \sum_{\mathbf{k},n}(-\mathrm{i}\hbar\omega_n+\varepsilon_{\mathbf{k}}-\mu)\,a^*_{\mathbf{k},n}a_{\mathbf{k},n} \\
& + \frac{1}{2}Un_0\sum_{\mathbf{k},n}\left(a_{\mathbf{k},n}a_{-\mathbf{k},-n}+4a^*_{\mathbf{k},n}a_{\mathbf{k},n}+a^*_{-\mathbf{k},-n}a^*_{\mathbf{k},n}\right),
\end{aligned}
\tag{16.38}
$$

where for a cubic lattice the number of nearest neighbors is $z=2d$, the condensate fraction is defined by $n_0=\langle N_0\rangle/N_{\mathrm{s}}$, and the lattice dispersion reads

$$
\varepsilon_{\mathbf{k}} = -2t\sum_{j=1}^{d}\cos(k_j\lambda/2). \tag{16.39}
$$

Note that for small momenta this term can be written as

$$
\varepsilon_{\mathbf{k}} \simeq -zt+t\sum_{j=1}^{d}(k_j\lambda/2)^2 = -zt+\frac{t\lambda^2}{4}\mathbf{k}^2. \tag{16.40}
$$

Comparing this with the dispersion $\varepsilon_{\mathbf{k}}=\hbar^2\mathbf{k}^2/2m^*$ of a free particle in a homogeneous system, we see that the second term describes an atom moving with an effective mass $m^*=2\hbar^2/t\lambda^2$.

As before, the terms linear in the fluctuations must be put to equal zero, which implies that the chemical potential has to satisfy

$$
\mu = Un_0 - zt. \tag{16.41}
$$

This expression can be easily understood since the chemical potential is the energy needed to add one particle to the system. Adding one particle results in an energy increase due to the Hartree interaction with the n_0 particles already at each site, and an energy decrease due to the possible hopping to one of z nearest-neighbor sites. Substituting this result, the action thus becomes

$$
\begin{aligned}
S[a^*,a] = {} & -\frac{1}{2}\hbar\beta Un_0^2N_{\mathrm{s}} - \frac{1}{2}\hbar\beta\sum_{\mathbf{k}}(\bar{\varepsilon}_{\mathbf{k}}+Un_0) \\
& + \frac{1}{2}\sum_{\mathbf{k},n}\left[a^*_{\mathbf{k},n},a_{-\mathbf{k},-n}\right]\begin{bmatrix}\mathrm{i}\hbar\omega_n+\bar{\varepsilon}_{\mathbf{k}}+Un_0 & Un_0 \\ Un_0 & -\mathrm{i}\hbar\omega_n+\bar{\varepsilon}_{\mathbf{k}}+Un_0\end{bmatrix}\begin{bmatrix}a_{\mathbf{k},n} \\ a^*_{-\mathbf{k},-n}\end{bmatrix},
\end{aligned}
\tag{16.42}
$$

where the lattice dispersion with respect to the lower band edge is $\bar{\varepsilon}_{\mathbf{k}}=\varepsilon_{\mathbf{k}}+zt$ and the extra zeroth-order terms are again generated by the different time ordering of $a^*_{-\mathbf{k}}(\tau)$ and $a_{-\mathbf{k}}(\tau)$. The matrix in (16.42) can be diagonalized by the Bogoliubov transformation

$$
\begin{bmatrix}b_{\mathbf{k},n} \\ b^*_{-\mathbf{k},-n}\end{bmatrix} = \begin{bmatrix}u_{\mathbf{k}} & v_{\mathbf{k}} \\ v^*_{\mathbf{k}} & u^*_{\mathbf{k}}\end{bmatrix}\begin{bmatrix}a_{\mathbf{k},n} \\ a^*_{-\mathbf{k},-n}\end{bmatrix}, \tag{16.43}
$$

where we properly normalize the new quasiparticle fields by requiring that the coefficients of this transformation obey

$$|u_{\mathbf{k}}|^2 - |v_{\mathbf{k}}|^2 = 1. \tag{16.44}$$

If we now substitute (16.43) into (16.42) and demand that the result reduces to the diagonal action

$$\begin{aligned} S[b^*,b] = & -\frac{1}{2}\hbar\beta U n_0^2 N_s + \frac{1}{2}\hbar\beta\sum_{\mathbf{k}}(\hbar\omega_{\mathbf{k}} - \bar{\varepsilon}_{\mathbf{k}} - U n_0) \\ & + \sum_{\mathbf{k},n}(-\mathrm{i}\hbar\omega_n + \hbar\omega_{\mathbf{k}})\, b^*_{\mathbf{k},n} b_{\mathbf{k},n}, \end{aligned} \tag{16.45}$$

we find that $u_{\mathbf{k}}$ and $v_{\mathbf{k}}$ must be solutions of the following two equations

$$\begin{aligned} \left(u_{\mathbf{k}}^2 + v_{\mathbf{k}}^2\right) U n_0 - 2u_{\mathbf{k}} v_{\mathbf{k}}(\bar{\varepsilon}_{\mathbf{k}} + U n_0) &= 0, \\ \left(|u_{\mathbf{k}}|^2 + |v_{\mathbf{k}}|^2\right)(\bar{\varepsilon}_{\mathbf{k}} + U n_0) - (u^*_{\mathbf{k}} v_{\mathbf{k}} + u_{\mathbf{k}} v^*_{\mathbf{k}}) U n_0 &= \hbar\omega_{\mathbf{k}}. \end{aligned} \tag{16.46}$$

Using the normalization in (16.44), we then find the solution

$$\begin{aligned} \hbar\omega_{\mathbf{k}} &= \sqrt{\bar{\varepsilon}_{\mathbf{k}}^2 + 2U n_0 \bar{\varepsilon}_{\mathbf{k}}}, \\ |v_{\mathbf{k}}|^2 &= |u_{\mathbf{k}}|^2 - 1 = \frac{1}{2}\left(\frac{\bar{\varepsilon}_{\mathbf{k}} + U n_0}{\hbar\omega_{\mathbf{k}}} - 1\right). \end{aligned} \tag{16.47}$$

To also obtain the condensate fraction n_0, which until now has been arbitrary, we next need to calculate the total filling fraction $n = \langle N \rangle / N_s$ that follows from our action. The total filling fraction is given by

$$n = \frac{1}{\hbar\beta N_s}\sum_{\mathbf{k},n}\langle a^*_{\mathbf{k},n} a_{\mathbf{k},n}\rangle, \tag{16.48}$$

where as usual in the technical literature the appropriate convergence factor is left implicit. For a Bose-Einstein condensed gas, this filling fraction consists of two parts. These are the fraction associated with the macroscopic occupation of the one-particle ground state, i.e. the Bose-Einstein condensate, and the filling fraction due to the occupation of the energetically higher-lying one-particle states. In this case, the condensate filling fraction equals the parameter n_0 and the contribution of the noncondensate part is determined by the average over the quadratic fluctuations, which will be a function of n_0. Calculating the average over the quadratic fluctuations by means of (16.43) yields first of all

$$n = n_0 + \frac{1}{\hbar\beta N_s}\sum_{\mathbf{k}\neq 0}\left\{\left(|u_{\mathbf{k}}|^2 + |v_{\mathbf{k}}|^2\right)\frac{1}{e^{\beta\hbar\omega_{\mathbf{k}}} - 1} + |v_{\mathbf{k}}|^2\right\}. \tag{16.49}$$

If we then also use (16.47) we find that

$$n = n_0 + \frac{1}{N_s} \sum_{\mathbf{k} \neq 0} \left(\frac{\bar{\varepsilon}_\mathbf{k} + U n_0}{\hbar\omega_\mathbf{k}} \frac{1}{e^{\beta\hbar\omega_\mathbf{k}} - 1} + \frac{\bar{\varepsilon}_\mathbf{k} + U n_0 - \hbar\omega_\mathbf{k}}{2\hbar\omega_\mathbf{k}} \right). \quad (16.50)$$

In the zero-temperature limit $\beta \to \infty$, the first term in the summant is zero. Taking the continuum limit by using $\sum_\mathbf{k} \to V \prod_j \int_{-2\pi/\lambda}^{2\pi/\lambda} \mathrm{d}k_j/(2\pi)$, changing from momenta $\mathbf{k}$ to $\mathbf{q} = \mathbf{k}\lambda/4\pi$, and realizing that $N_s = V(2/\lambda)^d$, we arrive at the expression

$$n = n_0 + \frac{1}{2} \int_{-1/2}^{1/2} \mathrm{d}\mathbf{q} \left(\frac{\bar{\varepsilon}_\mathbf{q} + U n_0}{\hbar\omega_\mathbf{q}} - 1 \right), \quad (16.51)$$

with

$$\bar{\varepsilon}_\mathbf{q} = 2t \sum_{j=1}^{d} \{1 - \cos(2\pi q_j)\}. \quad (16.52)$$

We can now obtain the condensate density by solving (16.51) for n_0 for a fixed value of n. We expect that at integer n, for a fixed value of U/t there will be no superfluid solution and this will mark the phase transition to the insulating phase as predicted by [159, 167, 168, 169].

In Fig. 16.2a we plot the result of this calculation for a two-dimensional lattice. We see from this figure that there is only a marginal difference between the case that $n = 0.5$ and $n = 1.0$. In Fig. 16.2b we plotted the result for a three-dimensional lattice. In this case the difference between half filling and integer filling is somewhat larger, but there is clearly no critical value of U/t for which the condensate density goes to zero. These results lead to the suspicion that the phase transition to the

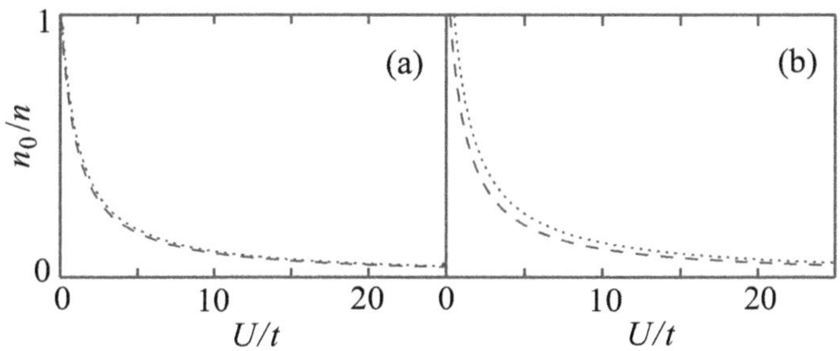

Fig. 16.2 The condensate fraction n_0/n (a) in a two-dimensional optical lattice and (b) a three-dimensional optical lattice, both as a function of U/t for $n = 0.5$ (dashed line) and $n = 1.0$ (dotted line).

insulating phase is not present in this approximation. To verify this, we investigate the limit of $U/t \to \infty$ in some detail.

When $U/t \to \infty$ we intuitively expect the system to become an insulator, because it effectively means that the hopping parameter goes to zero. We therefore expect that there are no superfluid solutions as $U/t \to \infty$. We can see that in this limit the integrand on the right-hand side of (16.51) behaves as $(Un_0/2\bar{\varepsilon}_{\mathbf{q}})^{1/2}$. One can also prove that $\bar{\varepsilon}_{\mathbf{q}} \leq 4\pi^2|\mathbf{q}|^2 t$. This means that

$$\int_{-1/2}^{1/2} d\mathbf{q} \frac{\bar{\varepsilon}_{\mathbf{q}} + Un_0}{\sqrt{\bar{\varepsilon}_{\mathbf{q}}^2 + 2U\bar{\varepsilon}_{\mathbf{q}} n_0}} \geq \frac{1}{2\pi}\sqrt{\frac{Un_0}{2t}} \int_{-1/2}^{1/2} \frac{d\mathbf{q}}{|\mathbf{q}|}. \tag{16.53}$$

The integral at the right-hand side of (16.53) can be done analytically in two dimensions and numerically in three dimensions. When we call the result of the integration in d dimensions I_d, we see that (16.51), for $U/t \to \infty$, reduces to

$$n \simeq n_0 + \frac{1}{4\pi}\sqrt{\frac{Un_0}{2t}} I_d - \frac{1}{2}, \tag{16.54}$$

where $I_2 = 2.22322$ and $I_3 = 2.38008$. This is a quadratic equation in $\sqrt{n_0}$ which always yields a positive solution for n_0 given by

$$n_0 = \left(\frac{1}{2}\sqrt{\frac{I_d{}^2}{16\pi^2}\frac{U}{2t} + 4n + 2} - \frac{I_d}{8\pi}\sqrt{\frac{U}{2t}} \right)^2. \tag{16.55}$$

We can correct for the error we made in (16.53) by using a higher value for I_d but, while this may change the value of n_0, it will still yield a positive solution. We see from (16.55) that only in the limit $U/t \to \infty$ we have that $n_0 = 0$, which leads us to the conclusion that the Bogoliubov approximation as described above does not predict the phase transition to the Mott-insulator phase in two and three dimensions. The reason for this is that the Bogoliubov approach only approximately treats the interactions. As a result, the Bogoliubov approach cannot describe large depletions of the condensate. We also see from (16.53) that in one dimension I_1 diverges. Substituting this into (16.54), we see that in one dimension there are no Bose-Einstein-condensed solutions, i.e. solutions with $n_0 \neq 0$. This is in accordance with the Mermin-Wagner-Hohenberg theorem [131, 132] which we discussed in Chap. 15. As the Bogoliubov approximation fails to predict the phase transition to the Mott-insulator phase, we now consider a different mean-field theory that treats the interactions exactly and approximates the kinetic energy of the atoms in the optical lattice.

16.5.2 Decoupling Approximation

In the next two sections we show how to arrive at a mean-field approach capable of describing the Mott-insulating phase. First we will use the Hamiltonian formalism, but in Sect. 16.46 we show how this mean-field theory can also be obtained by means of a Hubbard-Stratonovich transformation. The latter approach is actually more convenient because it will also allow us to analytically calculate the quasiparticle dispersions in the Mott insulator. We start with the Hamiltonian of the Bose-Hubbard model as given in (16.29). By analogy to the Bogoliubov approach, we introduce the superfluid order parameter

$$\psi = \sqrt{n_0} = \langle \hat{a}_i^\dagger \rangle = \langle \hat{a}_i \rangle, \tag{16.56}$$

where n_0 is the condensate fraction $\langle N_0 \rangle / N_s$ at site i. Note that we take the expectation values to be real, as before. We now, however, construct a consistent mean-field theory by making the substitution

$$\hat{a}_i^\dagger \hat{a}_j \simeq \langle \hat{a}_i^\dagger \rangle \hat{a}_j + \hat{a}_i^\dagger \langle a_j \rangle - \langle \hat{a}_i^\dagger \rangle \langle \hat{a}_j \rangle = \psi \left(\hat{a}_i^\dagger + \hat{a}_j \right) - \psi^2 \tag{16.57}$$

in the hopping term. Doing so, we obtain

$$\hat{H}^{\text{eff}} = zt\psi^2 N_s - zt \sum_i \psi \left(\hat{a}_i + \hat{a}_i^\dagger \right) + \frac{1}{2} U \sum_i \hat{a}_i^\dagger \hat{a}_i^\dagger \hat{a}_i \hat{a}_i - \mu \sum_i \hat{a}_i^\dagger \hat{a}_i. \tag{16.58}$$

This Hamiltonian is diagonal with respect to the site index i, and as a result we can use an effective on-site Hamiltonian. If we introduce $\bar{U} = U/zt$, $\bar{\mu} = \mu/zt$ and the number operator $\hat{n}_i = \hat{a}_i^\dagger \hat{a}_i$, we find $\hat{H}^{\text{eff}} \equiv zt \sum_i \hat{H}_i$ and

$$\hat{H}_i = \frac{1}{2} \bar{U} \hat{n}_i (\hat{n}_i - 1) - \bar{\mu} \hat{n}_i - \psi \left(\hat{a}_i^\dagger + \hat{a}_i \right) + \psi^2, \tag{16.59}$$

which is valid on each site i. We will therefore drop the subscript i in the following. Note that we scaled all the energies by a factor $1/zt$, making this Hamiltonian a dimensionless operator.

After writing (16.59) in matrix form with respect to an occupation number basis, we can solve the on-site problem numerically by explicitly diagonalizing the part of the matrix with occupation number below a certain maximum value [169]. Later we also follow this procedure, but we first determine the phase diagram analytically using second-order perturbation theory. As usual, we write the Hamiltonian as a sum of an exactly solvable part and a perturbation. Here we have $\hat{H} = \hat{H}^{(0)} + \psi \hat{V}$, where

$$\hat{H}^{(0)} = \frac{1}{2} \bar{U} \hat{n} (\hat{n} - 1) - \bar{\mu} \hat{n} + \psi^2, \tag{16.60}$$

and the perturbation is given by

$$\hat{V} = - \left(\hat{a}^\dagger + \hat{a} \right). \tag{16.61}$$

From the above Hamiltonian we see that in an occupation-number basis the odd powers of the expansion of the energy in ψ will always be zero. If we denote the unperturbed energy of the state with exactly n particles by $E_n^{(0)}$, we find that the unperturbed ground-state energy is given by

$$E_g^{(0)} = \left\{ E_n^{(0)} | n = 0, 1, 2, \ldots \right\}_{\min} .$$

Comparing $E_n^{(0)}$ and $E_{n+1}^{(0)}$ yields

$$E_g^{(0)} = \begin{cases} 0 & \text{if } \bar{\mu} < 0, \\ \frac{1}{2}\bar{U}g(g-1) - \bar{\mu}g & \text{if } \bar{U}(g-1) < \bar{\mu} < \bar{U}g. \end{cases} \tag{16.62}$$

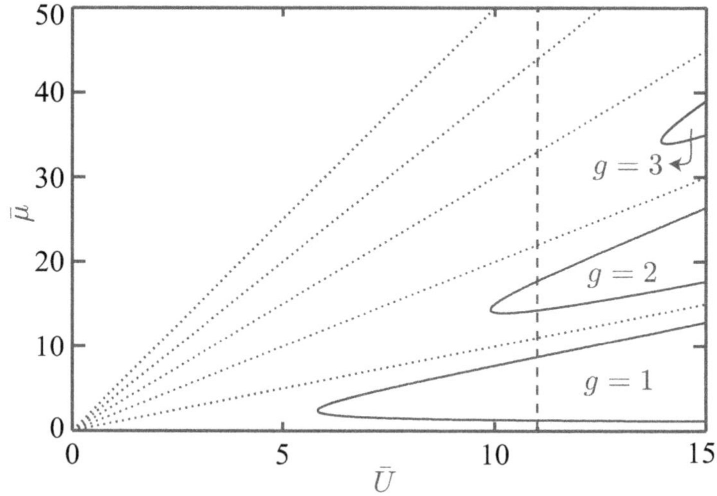

Fig. 16.3 Phase diagram of the Bose-Hubbard Hamiltonian as obtained from second-order perturbation theory (solid lines). The dotted lines indicate the zeroth-order phase diagram. Later on, Fig. 16.7 is taken along the dashed line in this figure.

Next, we calculate the second-order correction to the energy with the well-known expression

$$E_g^{(2)} = \psi^2 \sum_{n \neq g} \frac{|\langle n|V|g\rangle|^2}{E_g^{(0)} - E_n^{(0)}}, \tag{16.63}$$

where $|n\rangle$ denotes the unperturbed wavefunction with n particles, of which the state with $n = g$ particles is the ground state. Since the interaction V couples only to states with one more or one less atom than in the ground state, we find

$$E_g^{(2)} = \psi^2 \left[\frac{g}{\bar{U}(g-1) - \bar{\mu}} + \frac{g+1}{\bar{\mu} - \bar{U}g} \right]. \tag{16.64}$$

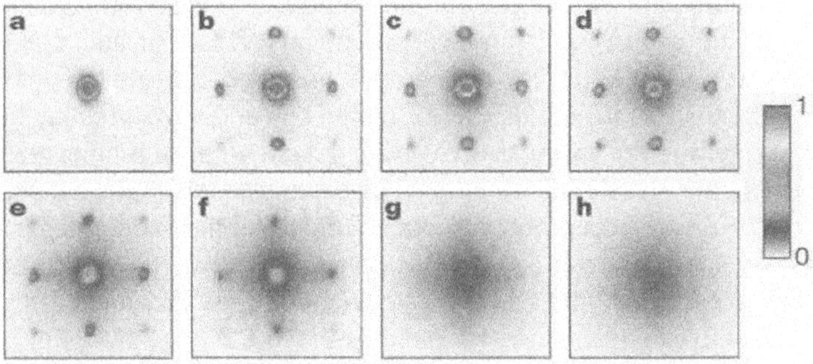

Fig. 16.4 Density distributions of ultracold atoms released from an optical lattice after time-of-flight expansion as measured by Greiner et al. [31]. The different pictures correspond to different lattice depths V_0, where the values of V_0 were: a, 0 E_r; b, 3 E_r; c, 7E_r; d, 10 E_r; e, 13 E_r; f, 14 E_r; g, 16 E_r; and h, 20 E_r. Reprinted by permission from Macmillan Publishers Ltd: Nature 415, 39 (2002), copyright (2002).

If we now follow the usual Landau procedure for second-order phase transitions by writing the ground-state energy as an expansion in ψ, i.e.

$$E_g(\psi) = a_0(g, \bar{U}, \bar{\mu}) + a_2(g, \bar{U}, \bar{\mu})\psi^2 + \mathcal{O}(\psi^4), \tag{16.65}$$

and minimize it as a function of the superfluid order parameter ψ, we find that $\langle\psi\rangle = 0$ when $a_2(g, \bar{U}, \bar{\mu}) \geq 0$ and that $\langle\psi\rangle \neq 0$ when $a_2(g, \bar{U}, \bar{\mu}) < 0$. This means that $a_2(g, \bar{U}, \bar{\mu}) = 0$ signifies the boundary between two phases. One of these has $\langle\psi\rangle \neq 0$ and corresponds to the superfluid phase. The other phase has $\langle\psi\rangle = 0$ and, as we will see shortly, corresponds to a Mott-insulator phase. The boundary is obtained by solving

$$a_2(g, \bar{U}, \bar{\mu}) = \frac{g}{\bar{U}(g-1) - \bar{\mu}} + \frac{g+1}{\bar{\mu} - \bar{U}g} + 1 = 0, \tag{16.66}$$

which yields

$$\bar{\mu}_\pm = \frac{1}{2}(\bar{U}(2g-1) - 1) \pm \frac{1}{2}\sqrt{\bar{U}^2 - 2\bar{U}(2g+1) + 1}, \tag{16.67}$$

where the subscript $\pm$ denotes the upper and lower halves of the Mott-insulating regions of phase space. In Fig. 16.3 we show a plot of (16.67) for $g = 1, 2, 3$. By equating $\bar{\mu}_+$ and $\bar{\mu}_-$ we can find the point of smallest $\bar{U}$ for each lobe. Denoting this critical value of $\bar{U}$ by $\bar{U}_c$ we have

$$\bar{U}_c = 2g + 1 + \sqrt{(2g+1)^2 - 1}. \tag{16.68}$$

which yields $\bar{U}_c \simeq 5.83$ for the $g = 1$ insulator.

Experimentally, the transition from the superfluid to the Mott insulator was first observed by Greiner et al. [31]. In this experiment a Bose-Einstein condensate was loaded into an optical lattice of variable potential depth. For each potential depth the phase coherence of the atoms was measured by suddenly switching of the lattice potential and letting the atoms evolve and interfere with each other for some time before imaging the density distribution. The experimental results are shown in Fig. 16.4. For an infinitely deep optical lattice which has the same integer number of atoms in each site, there can be no phase coherence between different sites and no interference is expected when the density distribution is measured after a time-of-flight expansion. This is the case for pictures *g* and *h* in Fig. 16.4. For a superfluid in a shallow optical lattice there is a strong phase coherence between the atoms, since the atoms are Bose-Einstein condensed, and as a result the density distribution after time-of-flight expansion shows a clear interference pattern. By changing the depth of the optical lattice, the experiment interpolates between these two limits. A detailed interpretation of the phase coherence of the atomic Mott insulator can be found in [174] and [175], which is complicated by the fact that the experiment is performed in a trap and the gas is therefore inhomogeneous. We come back to the consequences of the trap after we have also discussed the superfluid region of the phase diagram.

To find out more about the phase transition we need to perform also fourth-order perturbation theory to find the slope with which the particle density increases as a function of $\bar{\mu}$ in the superfluid state. Doing so we can write the ground-state energy as

$$E_g(\psi) = a_0(g,\bar{U},\bar{\mu}) + a_2(g,\bar{U},\bar{\mu})\psi^2 + a_4(g,\bar{U},\bar{\mu})\psi^4 + \mathcal{O}(\psi^6), \quad (16.69)$$

with

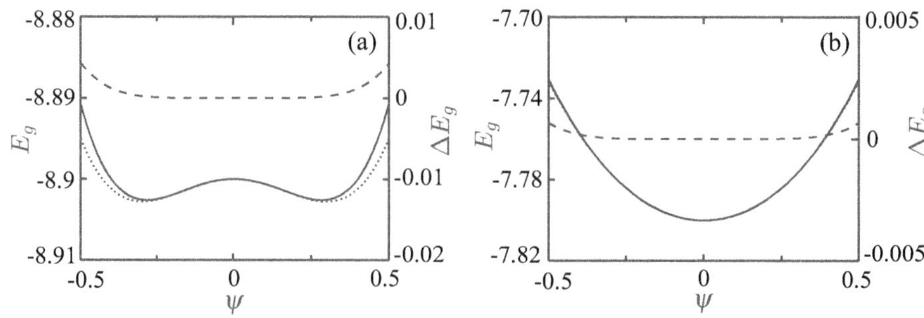

Fig. 16.5 Ground-state energy as a function of ψ for (a) $\bar{U} = 11$ and $\bar{\mu} = 8.9$ and for (b) $\bar{U} = 11$ and $\bar{\mu} = 7.8$. The solid line represents fourth-order perturbation theory whereas the dotted line represents a numerical diagonalization of the effective Hamiltonian. The dashed line is the difference between the two, plotted with the scale on the right.

$$a_4(g,\bar{U},\bar{\mu}) = \frac{g(g-1)}{(\bar{U}(g-1)-\bar{\mu})^2(\bar{U}(2g-3)-2\bar{\mu})} + \frac{(g+1)(g+2)}{(\bar{\mu}-\bar{U}g)^2(2\bar{\mu}-\bar{U}(2g+1))} - \left(\frac{g}{\bar{U}(g-1)-\bar{\mu}} + \frac{g+1}{\bar{\mu}-\bar{U}g}\right)\left(\frac{g}{(\bar{U}(g-1)-\bar{\mu})^2} + \frac{g+1}{(\bar{\mu}-\bar{U}g)^2}\right). \tag{16.70}$$

In Figs. 16.5a and b, we show plots of (16.69) together with the result of an exact numerical diagonalization of the effective Hamiltonian. As can be seen, the overlap is very good near the boundary given by (16.67). In Fig. 16.6, it can be seen that the numerical result exhibits a cusp when $\bar{U} = \bar{\mu}$, which is not predicted by (16.69). This is due to the fact that in this particular case we need to use first-order degenerate perturbation theory, because at $\bar{\mu} = n\bar{U}$ the states with $n-1$ and n particles per site form a doubly degenerate ground state. The resulting expression for the ground-state energy is now nonanalytic and given by

$$E_g(\psi)\big|_{\bar{\mu}=n\bar{U}} = -\frac{1}{2}\bar{U}n(n+1) + \psi^2 - |\psi|\sqrt{n+1}, \tag{16.71}$$

which is the solid line in Fig. 16.6. Note that the occurence of a cusp is analogous to the well-known Jahn-Teller effect in solid-state physics [176] and presents an interesting way in which a system can violate the general arguments of Landau theory.

We now continue by calculating the average number of particles per site in the grand-canonical ensemble by

$$n = -\frac{\partial\langle\hat{H}\rangle}{\partial\mu} = -\frac{\partial E_g(\langle\psi\rangle)}{\partial\bar{\mu}} = g - \frac{\partial}{\partial\bar{\mu}}\left(\frac{a_2(g,\bar{U},\bar{\mu})^2}{4a_4(g,\bar{U},\bar{\mu})}\right), \tag{16.72}$$

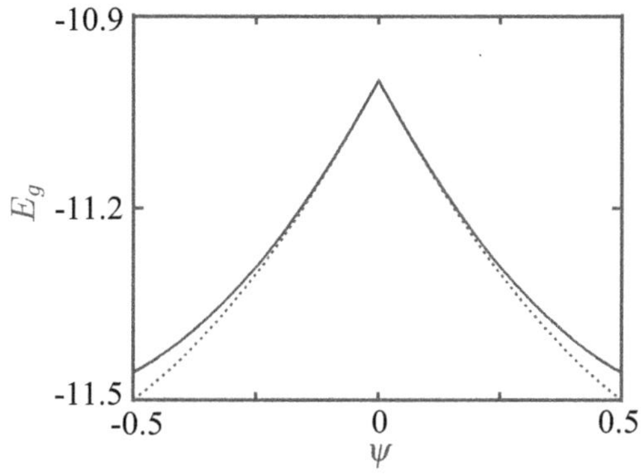

Fig. 16.6 Groundstate energy as a function of ψ for $\bar{U} = \bar{\mu} = 11$, as obtained from first-order perturbation theory (solid line) and from numerical diagonalization of the effective Hamiltonian (dotted line).

where

$$\langle\psi\rangle = \sqrt{\frac{-a_2(g,\bar{U},\bar{\mu})}{2a_4(g,\bar{U},\bar{\mu})}} \tag{16.73}$$

is the minimum of (16.69). Making use of the previous results, we can now plot the filling fraction as a function of $\bar{\mu}$ for a fixed value of $\bar{U}$. Between the edges $\bar{\mu}_{\pm}$, the filling fraction will remain constant because $\langle\psi\rangle = 0$ and the second term in the right-hand side of (16.72) does not contribute. Outside that region, the density will start to change with a nonzero slope. In Fig. 16.7 this is plotted for $\bar{U} = 11$. The solid line shows the result of the calculation described above and the dash-dotted line is a numerical result obtained by exact diagonalization. As can be seen, the analytical results are in good agreement with the numerical calculation.

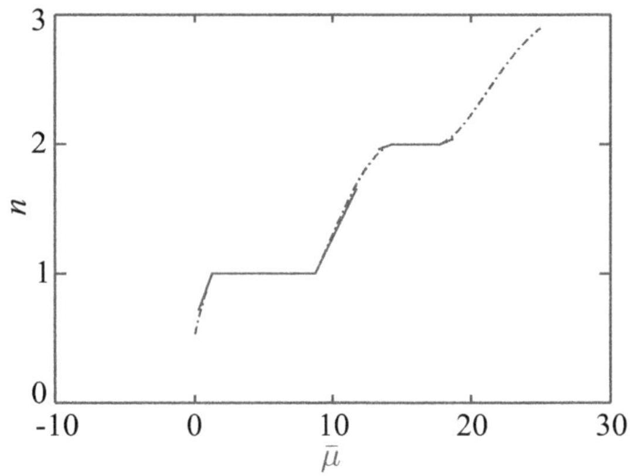

Fig. 16.7 The density as a function of chemical potential $\bar{\mu} = \mu/zt$ for an interaction strength of $\bar{U} = U/zt = 11$, i.e. along the dashed line in Fig. 16.3.

For realistic gases we also need to take into account the effect of an external trapping potential. In a first approximation, we can describe the effect of a slowly varying trapping potential by substituting for $\bar{\mu}$ in (16.72)

$$\bar{\mu}(\mathbf{x}) = \frac{\mu - V^{\text{ex}}(\mathbf{x})}{zt}, \tag{16.74}$$

where $V^{\text{ex}}(\mathbf{x})$ is the external trapping potential. Taking the trapping potential into account in this way is also known as the local-density approximation. Combining this with Fig. 16.7 yields the density profile for an optical lattice plus external trapping potential. For the case of a harmonic trapping potential, the density profile is now given by a wedding-cake profile. Moving outwards from the center we move through rings of Mott-insulating phases with decreasing density. Between two such rings there is a superfluid. This result had already been confirmed numerically in

[159]. Recent experiments have also directly investigated and confirmed the presence of this shell structure in the inhomogeneous trap [177, 178].

16.5.3 Hubbard-Stratonovich Transformation

An important property of the Mott-insulating phase is that the fluctuations in the average number of particles per site goes to zero at zero temperature. Since these fluctuations can be described as quasiparticle and quasihole excitations, we will study these now. We calculate the quasiparticle and quasihole dispersions using the functional-integral formalism. We consider again the action in (16.35). To decouple the hopping term, we perform a Hubbard-Stratonovich transformation to the action, which then becomes

$$\begin{aligned} S[a^*,a,\psi^*,\psi] &= S[a^*,a] \\ &+ \int_0^{\hbar\beta} \mathrm{d}\tau \sum_{i,j} (\psi_i^*(\tau) - a_i^*(\tau))\, t_{ij}\, (\psi_j(\tau) - a_j(\tau)) \end{aligned} \quad (16.75)$$

with $t_{ii} = 0$. Here, $\psi_i(\tau)$ is the complex order-parameter field. To obtain an effective action as a function of this field, we rewrite (16.75) as

$$\begin{aligned} S[a^*,a,\psi^*,\psi] &= \int_0^{\hbar\beta} \mathrm{d}\tau \sum_i \left\{ a_i^*(\tau) \left(\hbar \frac{\partial}{\partial \tau} - \mu \right) a_i(\tau) + \frac{U}{2} a_i^*(\tau) a_i^*(\tau) a_i(\tau) a_i(\tau) \right\} \\ &+ \int_0^{\hbar\beta} \mathrm{d}\tau \sum_{i,j} \left\{ -t_{ij} \left(a_i^*(\tau)\psi_j(\tau) + \psi_i^*(\tau) a_j(\tau) \right) + t_{ij}\psi_i^*(\tau)\psi_j(\tau) \right\}, \end{aligned} \quad (16.76)$$

and integrate over the original atomic field $a_i(\tau)$. If we denote by $S^{(0)}[a^*,a]$ the action for $t_{ij} = 0$, we have explicitely that

$$\begin{aligned} \exp\left\{ -\frac{1}{\hbar} S^{\text{eff}}[\psi^*,\psi] \right\} &\equiv \exp\left\{ -\frac{1}{\hbar} \int_0^{\hbar\beta} \mathrm{d}\tau \sum_{i,j} t_{ij} \psi_i^*(\tau)\psi_j(\tau) \right\} \\ &\times \int \mathrm{d}[a^*]\mathrm{d}[a] \exp\left\{ -\frac{1}{\hbar} S^{(0)}[a^*,a] \right\} \\ &\times \exp\left\{ -\frac{1}{\hbar} \int_0^{\hbar\beta} \mathrm{d}\tau \left(-\sum_{i,j} t_{ij} \left(a_i^*(\tau)\psi_j(\tau) + \psi_i^*(\tau) a_j(\tau) \right) \right) \right\}. \end{aligned} \quad (16.77)$$

We can now calculate $S^{\text{eff}}[\psi^*,\psi]$ perturbatively by performing a Taylor expansion of the exponent in the integrand of (16.77) and subsequently evaluating the various correlation functions of the field theory given by the action $S^{(0)}[a^*,a]$. The quadratic part of the effective action is given by

$$S^{(2)}[\psi^*,\psi] = -\frac{1}{2\hbar}\left\langle\left(\int_0^{\hbar\beta}\mathrm{d}\tau\sum_{i,j}t_{ij}\left(a_i^*(\tau)\psi_j(\tau)+\psi_i^*(\tau)a_j(\tau)\right)\right)^2\right\rangle_0 + \int_0^{\hbar\beta}\mathrm{d}\tau\sum_{i,j}t_{ij}\psi_i^*(\tau)\psi_j(\tau) \tag{16.78}$$

If we perform the multiplication in the first term in the right-hand side and use the information we have about the correlations in the unperturbed system, i.e.

$$\begin{aligned}\left\langle a_i^*(\tau)a_j^*(\tau)\right\rangle_0 &= \left\langle a_i(\tau)a_j(\tau)\right\rangle_0 = 0,\\ \left\langle a_i^*(\tau)a_j(\tau)\right\rangle_0 &= \left\langle a_i(\tau)a_j^*(\tau)\right\rangle_0 = \langle a_i(\tau)a_i^*(\tau)\rangle_0\,\delta_{i,j},\end{aligned} \tag{16.79}$$

we obtain

$$\begin{aligned}S^{(2)}[\psi^*,\psi] = \int_0^{\hbar\beta}\mathrm{d}\tau\Bigg\{&\sum_{i,j}t_{ij}\psi_i^*(\tau)\psi_j(\tau) \\ &-\frac{1}{\hbar}\int_0^{\hbar\beta}\mathrm{d}\tau'\sum_{i,j,i',j'}t_{ij}t_{i'j'}\psi_j^*(\tau)\left\langle a_i(\tau)a_{i'}^*(\tau')\right\rangle_0\psi_{j'}(\tau')\Bigg\}.\end{aligned} \tag{16.80}$$

Because we will only consider nearest-neighbor hopping, we also use

$$t_{ij} = t_{ji} = \begin{cases} t \text{ for nearest neighbours} \\ 0 \text{ otherwise.}\end{cases} \tag{16.81}$$

First, we treat the part of (16.80) that is linear in t_{ij}. We have

$$\sum_{i,j}t_{ij}\psi_i^*(\tau)\psi_j(\tau) = \sum_i t\,\psi_i^*(\tau)\psi_{i\pm\{1\}}(\tau), \tag{16.82}$$

where $\pm\{1\}$ denotes all possible jumps to nearest neighbors. In the case of one dimension this would simply be ± 1. If we call the lattice spacing $\lambda/2$ again and introduce Cartesian momentum components k_i with $i = 1,\cdots,d$, where d is again the number of dimensions, we find

$$\sum_{i,j}t_{ij}\psi_i^*(\tau)\psi_j(\tau) = \sum_{\mathbf{k}}2t\,\psi_{\mathbf{k}}(\tau)\psi_{\mathbf{k}}^*(\tau)\sum_{j=1}^{d}\cos\left(k_j\lambda/2\right). \tag{16.83}$$

Next we calculate the part that is quadratic in t_{ij}. We can treat this part by looking at double jumps. The expectation value of $\left\langle a_i(\tau)a_{i'}^*(\tau)\right\rangle_0$ is proportional to $\delta_{i,i'}$ and independent of the site i according to (16.79). This means that we find, with similar notation as before,

$$\begin{aligned}
&\sum_{j,i',j'} t_{ij}t_{i'j'}\psi_j^*(\tau)\langle a_i(\tau)a_{i'}^*(\tau')\rangle_0\,\psi_{j'}(\tau') \\
&= \langle a_i(\tau)a_i^*(\tau')\rangle_0 \sum_{j,j'} t_{ij}t_{ij'}\psi_j^*(\tau)\psi_{j'}(\tau') \\
&= t^2\langle a_i(\tau)a_i^*(\tau')\rangle_0 \sum_j \Big\{ z\psi_j^*(\tau)\psi_j(\tau') \\
&+ \psi_j^*(\tau)\psi_{j\pm\{2\}}(\tau') + \psi_j^*(\tau)\psi_{j\pm\{\sqrt{2}\}}(\tau') \Big\},
\end{aligned} \tag{16.84}$$

with z again the number of nearest neighbors. The first term in the summand is a jump in each direction, followed by a jump back. The second term indicates two jumps in the same direction and the third term is a jump in each direction followed by a jump in a perpendicular direction. Note that the third term is absent in one dimension. It can be shown that the complete double-jump term reduces to

$$\begin{aligned}
&\sum_{j,i',j'} t_{ij}t_{i'j'}\psi_j^*(\tau)\langle a_i(\tau)a_{i'}^*(\tau')\rangle_0\,\psi_{j'}(\tau') \\
&= \langle a_i(\tau)a_i^*(\tau')\rangle_0 \sum_{\mathbf{k}} \psi_{\mathbf{k}}^*(\tau)\psi_{\mathbf{k}}(\tau')\varepsilon_{\mathbf{k}}^2,
\end{aligned} \tag{16.85}$$

where we again used the lattice dispersion $\varepsilon_{\mathbf{k}} = -2t\sum_{j=1}^d \cos(k_j\lambda/2)$.

To translate the expectation value of the fields into the expectation value of operators, we introduce the imaginary time ordering operator T. As a result

$$\langle a_i(\tau)a_{i'}^*(\tau')\rangle_0 = \left\langle T\left[\hat{a}_i(\tau)\hat{a}_{i'}^\dagger(\tau')\right]\right\rangle_0. \tag{16.86}$$

The time ordering can be expressed by means of Heaviside functions as

$$\begin{aligned}
\left\langle T\left[\hat{a}_i(\tau)\hat{a}_{i'}^\dagger(\tau')\right]\right\rangle_0 &= \theta(\tau-\tau')\left\langle \hat{a}_i(\tau)\hat{a}_{i'}^\dagger(\tau')\right\rangle_0 \\
&\quad +\theta(\tau'-\tau)\left\langle \hat{a}_{i'}^\dagger(\tau')\hat{a}_i(\tau)\right\rangle_0.
\end{aligned} \tag{16.87}$$

If we use the unperturbed energies as given by (16.62), we thus find

$$\begin{aligned}
&\langle a_i(\tau)a_{i'}^*(\tau')\rangle_0 = \theta(\tau-\tau')(1+g)\exp\left\{-\left(E_{g+1}^{(0)}-E_g^{(0)}\right)(\tau-\tau')/\hbar\right\} \\
&+\theta(\tau'-\tau)g\exp\left\{\left(E_{g-1}^{(0)}-E_g^{(0)}\right)(\tau-\tau')/\hbar\right\}.
\end{aligned} \tag{16.88}$$

Because $E_g^{(0)}$ is by definition the grand-canonical energy of the ground state, we know that

$$\begin{aligned}
E_{g+1}^{(0)} - E_g^{(0)} &= -\mu + gU > 0, \\
E_g^{(0)} - E_{g-1}^{(0)} &= -\mu + (g-1)U < 0.
\end{aligned}$$

Note that we here still use the parameters μ and U instead of $\bar{\mu}$ and $\bar{U}$, because we have not yet divided out the factor zt. Combining the above with (16.80) we find

$$S^{(2)}[\psi^*,\psi] = -\sum_{\mathbf{k},n} |\psi_{\mathbf{k},n}|^2 \varepsilon_{\mathbf{k}} \left(1 + \frac{\varepsilon_{\mathbf{k}}}{\hbar}\int_{-\infty}^{0} \mathrm{d}\tau'(1+g)\exp\{(-\mathrm{i}\hbar\omega_n - \mu + gU)\tau'/\hbar\} \right.$$
$$\left. + \frac{\varepsilon_{\mathbf{k}}}{\hbar}\int_{0}^{\infty} \mathrm{d}\tau' g\exp\{-(\mathrm{i}\hbar\omega_n + \mu - (g-1)U)\tau'/\hbar\}\right). \quad (16.89)$$

Performing the τ' integration, we then obtain

$$S^{(2)}[\psi^*,\psi] = -\hbar\sum_{\mathbf{k},n} \psi^*_{\mathbf{k},n} G^{-1}(\mathbf{k},\mathrm{i}\omega_n)\psi_{\mathbf{k},n}, \quad (16.90)$$

where the inverse Green's function obeys

$$\hbar G^{-1}(\mathbf{k},\mathrm{i}\omega_n) =$$
$$\varepsilon_{\mathbf{k}} + \varepsilon_{\mathbf{k}}^2 \left(\frac{g+1}{-\mathrm{i}\hbar\omega_n - \mu + gU} + \frac{g}{\mathrm{i}\hbar\omega_n + \mu - (g-1)U}\right). \quad (16.91)$$

Note that this result is exact within our mean-field theory. If we want to make the connection with the Landau theory considered previously, we can identify the quadratic coefficient $a_2(g,U,\mu)$ in (16.69) with $-\hbar G^{-1}(\mathbf{0},0)/zt$. We come back to this in Sect. 16.6.2. The result in (16.91) contains all powers of the frequencies and momenta and no gradient expansion has been applied. This is important because the elementary excitations are gapped, as we will show in the next section. The energies of the quasiparticle and quasihole excitations $\hbar\omega_{\mathbf{k}}^{\mathrm{qp}}$ and $\hbar\omega_{\mathbf{k}}^{\mathrm{qh}}$, respectively, correspond to the zeros of $G^{-1}(\mathbf{k},\omega)$, where we have made the analytic continuation $\mathrm{i}\omega_n \rightarrow \omega$. This implies that

$$1 + \varepsilon_{\mathbf{k}} \left(\frac{g+1}{-\hbar\omega - \mu + gU} + \frac{g}{\hbar\omega + \mu - (g-1)U}\right) = 0. \quad (16.92)$$

Solving this equation for $\hbar\omega$ gives

$$\hbar\omega_{\mathbf{k}}^{\mathrm{qp,qh}} = -\mu + \frac{U}{2}(2g-1) + \frac{\varepsilon_{\mathbf{k}}}{2} \pm \frac{1}{2}\sqrt{\varepsilon_{\mathbf{k}}^2 + 2(2g+1)U\varepsilon_{\mathbf{k}} + U^2}. \quad (16.93)$$

In Fig. 16.8, we show for $\mathbf{k} = 0$ a plot of the above equations. The dotted lines indicate the asymptotes of (16.93), which are given by

$$\lim_{U\to\infty} \hbar\omega_{\mathbf{0}}^{\mathrm{qp}} = -\mu + gU + (g+1)\varepsilon_{\mathbf{0}} = E_{g+1}^{(0)} - E_g^{(0)} - (g+1)zt,$$
$$\lim_{U\to\infty} \hbar\omega_{\mathbf{0}}^{\mathrm{qh}} = -\mu + (g-1)U - g\varepsilon_{\mathbf{0}} = E_g^{(0)} - E_{g-1}^{(0)} + gzt, \quad (16.94)$$

with $E_{g+1}^{(0)} - E_g^{(0)}$ and $E_g^{(0)} - E_{g-1}^{(0)}$ given by (16.89). The difference between (16.94) and (16.89) is caused by the fact that (16.89) is calculated for $t = 0$. It can easily be

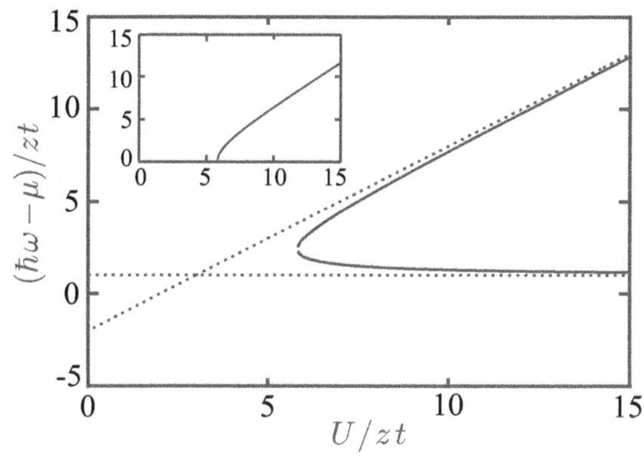

Fig. 16.8 The quasiparticle and the quasihole energy for $\mathbf{k} = 0$ in the $g = 1$ insulator lobe. The dotted lines are the asymptotes of the curves. The inset shows the resulting first-order approximation to the dispersion of the density fluctuations.

understood that for $t \neq 0$, the first-order correction is due to the hopping terms $ta_j^* a_i$, where site j is one of the nearest neighbors of site i. When we have g particles in all lattice sites and we add one particle to site i, we have $\langle a_j^* a_i \rangle = g+1$, so the effective hopping parameter for a particle is $(g+1)t$. However, when we remove a particle from site i, we have $\langle a_i^* a_j \rangle = g$, which represents a particle hopping to site i from one of its nearest neighbors. The effective hopping parameter for a hole is therefore only gt. In combination we see that, in the limit of $U \to \infty$, (16.93) indeed reduces to a physically intuitive result.

As shown above, the slopes of the asymptotes differ exactly by U, so in the limit of $U/zt \to \infty$ the gap for the creation of a quasiparticle-quasihole pair is equal to U. We can find a first approximation for the dispersion of the density fluctuations by subtracting the two solutions, which yields

$$\hbar\omega_{\mathbf{k}} = \hbar\omega_{\mathbf{k}}^{\mathrm{qp}} - \hbar\omega_{\mathbf{k}}^{\mathrm{qh}} = \sqrt{\varepsilon_{\mathbf{k}}^2 + 2(2g+1)U\varepsilon_{\mathbf{k}} + U^2}. \tag{16.95}$$

In the inset of Fig. 16.8 we show, again for $\mathbf{k} = 0$, a plot of the above equation as a function of $\bar{U} = U/zt$ for $g = 1$. We can see that there is a band gap, which proves that the Mott-insulator phase is indeed an insulator, and we also see that the band gap disappears as we approach the critical value $\bar{U}_c = U_c/zt \simeq 5.83$ that was found earlier. For smaller values of $\bar{U}$ we are in the superfluid phase, which according to the Hugenholtz-Pines theorem is expected to always have gapless density fluctuations.

As we have mentioned before, the superfluid-Mott insulator transition occurs, strictly speaking, only at zero temperature. For nonzero temperatures there is a 'classical' phase transition, i.e. a phase transition induced by thermal fluctuations, between a superfluid phase and a normal phase and there is only a crossover between the normal phase and the Mott insulator. It is important to mention here that a

Mott insulator is by definition incompressible. In principle there exists, therefore, no Mott insulator for any nonzero temperature where we always have a nonvanishing compressibility. Nevertheless, there is a region in the phase diagram where the compressibility is very close to zero and it is therefore justified to call this region a Mott insulator for all practical purposes. Qualitatively, this phase diagram is sketched in Fig. 16.9 for a fixed integer filling fraction. This figure shows how, if we start with

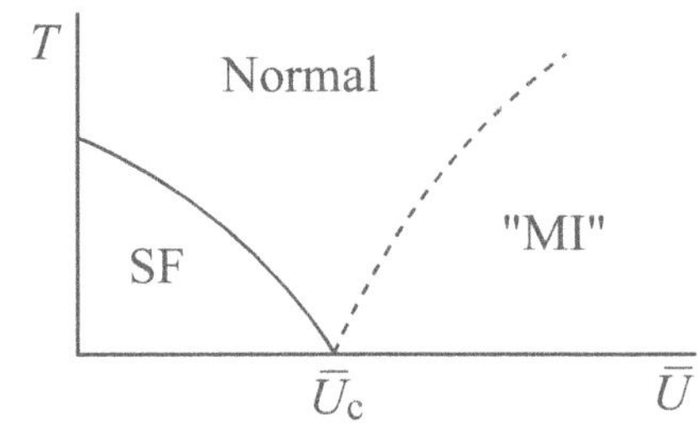

Fig. 16.9 Qualitative phase diagram for a fixed and integer filling fraction in terms of the temperature T and the dimensionless coupling constant $\overline{U} = U/zt$ (SF), normal and Mott insulating phases (MI). Only at $T = 0$ a true Mott insulator exists.

a superfluid for a small positive on-site interaction U and at a sufficiently small but nonzero temperature, we first encounter a phase transition to a normal phase as the interaction strength increases, and ultimately cross over to a Mott insulator for even higher values of the interaction strength.

16.6 Fluctuations

In this section, we study the effects of the Gaussian fluctuations and derive an identity between the atomic Green's function and the superfluid order-parameter Green's function in (16.91). We start again with the action of the Bose-Hubbard model introduced in (16.35)

$$S[a^*,a] = \int_0^{\hbar\beta} \mathrm{d}\tau \left\{ \sum_i a_i^*(\tau) \left(\hbar\frac{\partial}{\partial\tau} - \mu \right) a_i(\tau) - \sum_{i,j} t_{ij} a_i^*(\tau) a_j(\tau) \right.$$
$$\left. + \frac{U}{2} \sum_i a_i^*(\tau) a_i^*(\tau) a_i(\tau) a_i(\tau) \right\}. \qquad (16.96)$$

We are interested in calculating the $\langle a_i(\tau) a_{i'}^*(\tau') \rangle$ correlation function. Therefore we add currents $J_i(\tau)$, and $J_i^*(\tau)$ that couple to the $a_i^*(\tau)$ and $a_i(\tau)$ fields as

$$Z[J^*,J] = \int \mathrm{d}[a^*]\mathrm{d}[a] \exp\left\{ -\frac{1}{\hbar} S_0[a^*,a] + \frac{1}{\hbar} \int_0^{\hbar\beta} \mathrm{d}\tau \sum_{i,j} a_i^*(\tau) t_{ij} a_j(\tau) \right.$$
$$\left. + \int_0^{\hbar\beta} d\tau \sum_i \{ J_i^*(\tau) a_i(\tau) + a_i^*(\tau) J_i(\tau) \} \right\}. \qquad (16.97)$$

Here $S_0[a^*,a]$ denotes the action for $t_{ij}=0$. The most important step in the remainder of the calculation is to again perform a Hubbard-Stratonovich transformation by adding a complete square to the action. The latter can be written as

$$\int_0^{\hbar\beta} d\tau \sum_{i,j} \left(a_i^*(\tau) - \psi_i^*(\tau) + \hbar J_{i'}^*(\tau) t_{i'i}^{-1}\right) t_{ij} \left(a_j(\tau) - \psi_j(\tau) + \hbar t_{jj''}^{-1} J_{j''}(\tau)\right),$$

where the sums over i' and j' are left implicit for simplicity. Straightforward algebra yields

$$Z[J^*,J] = \int d[\psi^*]d[\psi] \exp\left\{ \sum_{\mathbf{k},n} \left(\psi_{\mathbf{k},n}^* G^{-1}(\mathbf{k}, i\omega_n) \psi_{\mathbf{k},n} \right.\right.$$
$$\left.\left. + J_{\mathbf{k},n}^* \psi_{\mathbf{k},n} + J_{\mathbf{k},n} \psi_{\mathbf{k},n}^* + \frac{\hbar}{\varepsilon_{\mathbf{k}}} J_{\mathbf{k},n}^* J_{\mathbf{k},n} \right) \right\}. \tag{16.98}$$

Differentiating twice with respect to the currents gives then the desired relation

$$\frac{1}{Z[J^*,J]} \frac{\delta^2}{\delta J_{\mathbf{k},n}^* \delta J_{\mathbf{k},n}} Z[J^*,J] \Bigg|_{J^*,J=0} = \langle a_{\mathbf{k},n} a_{\mathbf{k},n}^* \rangle = \langle \psi_{\mathbf{k},n} \psi_{\mathbf{k},n}^* \rangle + \frac{\hbar}{\varepsilon_{\mathbf{k}}}. \tag{16.99}$$

This is very useful indeed since the correlator $\langle \psi_{\mathbf{k},n} \psi_{\mathbf{k},n}^* \rangle = -G(\mathbf{k}, i\omega_n)$. At zero temperature the retarded Green's function can be written as

$$-\frac{1}{\hbar} G(\mathbf{k},\omega) = \frac{Z_{\mathbf{k}}}{-\hbar\omega + \hbar\omega_{\mathbf{k}}^{\mathrm{qp}}} + \frac{1 - Z_{\mathbf{k}}}{-\hbar\omega + \hbar\omega_{\mathbf{k}}^{\mathrm{qh}}} - \frac{1}{\varepsilon_{\mathbf{k}}}, \tag{16.100}$$

where the wavefunction renormalization factor is

$$Z_{\mathbf{k}} = \frac{U(2g+1) + \varepsilon_{\mathbf{k}} + \sqrt{\varepsilon_{\mathbf{k}}^2 + 2(2g+1)U\varepsilon_{\mathbf{k}} + U^2}}{2\sqrt{\varepsilon_{\mathbf{k}}^2 + 2(2g+1)U\varepsilon_{\mathbf{k}} + U^2}}, \tag{16.101}$$

and

$$\hbar\omega_{\mathbf{k}}^{\mathrm{qp,qh}} = -\mu + \frac{U}{2}(2g-1) + \frac{\varepsilon_{\mathbf{k}}}{2} \pm \frac{1}{2}\sqrt{\varepsilon_{\mathbf{k}}^2 + 2(2g+1)U\varepsilon_{\mathbf{k}} + U^2}. \tag{16.102}$$

Note that $Z_{\mathbf{k}}$ is always positive and in the limit where $U \to \infty$ we have that $Z_{\mathbf{k}} \to g+1$. The quasiparticle dispersion $\hbar\omega_{\mathbf{k}}^{\mathrm{qp}}$ is always greater than or equal to zero and $\hbar\omega_{\mathbf{k}}^{\mathrm{qh}}$ is always smaller than or equal to zero, in full agreement with our expectations based on the Lehmann representation discussed in Sect. 8.1.1.

16.6.1 Mott Insulator

In the Mott-insulator state, where $n_0 \equiv |\langle \psi_{\mathbf{0},0} \rangle|^2/\hbar\beta N_s = 0$, the thermodynamic potential is now easily calculated by integrating out the superfluid order parameter in Gaussian approximation. This gives

$$\Omega = \Omega_0 + \frac{1}{\beta}\mathrm{Tr}\left[\log\left(-\hbar\beta G^{-1}\right)\right], \tag{16.103}$$

where $\Omega_0 = N_s E_g^{(0)}$ at zero temperature. Note that in the above expression for the thermodynamic potential we have omitted the contribution from the Hubbard-Stratonovich transformation, which is due to the fact that the functional integral over the Hubbard-Stratonovich fields that we used to obtain (16.75) is not equal to one. However, this term does not depend on the chemical potential and in this section we are particularly interested in the density expression which follows from the thermodynamic potential by differentiating with respect to the chemical potential. Therefore, we are now allowed to neglect this point.

We differentiate minus the thermodynamic potential in (16.103) with respect to the chemical potential to find the average total number of atoms. Using the expression for $E_g^{(0)}$ from (16.62) we see that differentiating the zeroth-order term with respect to the chemical potential gives $-gN_s$, where g is the number of atoms per site in the Mott lobe that we consider. Due to the contribution of the fluctuation term we obtain in total

$$\begin{aligned} n &= -\frac{1}{N_s}\frac{\partial\Omega}{\partial\mu} = -\frac{1}{N_s}\frac{\partial\Omega_0}{\partial\mu} - \frac{1}{\beta N_s}\frac{\partial}{\partial\mu}\mathrm{Tr}\left[\log\left(-\hbar\beta G^{-1}\right)\right] \\ &= g - \frac{1}{\beta N_s}\sum_{\mathbf{k},n} G(\mathbf{k},\mathrm{i}\omega_n)\frac{\partial G^{-1}(\mathbf{k},\mathrm{i}\omega_n)}{\partial\mu}. \end{aligned} \tag{16.104}$$

To calculate the sum over the Matsubara frequencies in the above trace we first rewrite it as a complex integral, making use of the Bose distribution $N(z) = 1/(\mathrm{e}^{\hbar\beta z} - 1)$ that has poles at the even Matsubara frequencies $z = \mathrm{i}\omega_n$, where $\omega_n = 2n\pi/\hbar\beta$. The residue of the pole at $z = \mathrm{i}\omega_n$ of the function $\hbar\beta N(z)G(\mathbf{k},z)$ can be shown to be equal to $G(\mathbf{k},\mathrm{i}\omega_n)$, and as a result the sum over Matsubara frequencies in (16.104) is equal to the complex integral along the closed contour C that encloses the imaginary axis. This is also shown in Fig. 7.2.

From (16.100) we find that the poles of $G(\mathbf{k},z)$ are located at the quasiparticle $\hbar\omega_{\mathbf{k}}^{\mathrm{qp}}$ and quasihole $\hbar\omega_{\mathbf{k}}^{\mathrm{qh}}$ dispersions respectively. Using (16.91), we find that the derivative term $\partial G^{-1}(\mathbf{k},z)/\partial\mu$ has two second-order poles located at $z = (gU - \mu)/\hbar$ and $z = ((g-1)U - \mu)/\hbar$ respectively. Explicitly we thus have

$$
\begin{aligned}
&\sum_n G(\mathbf{k},\mathrm{i}\omega_n)\frac{\partial G^{-1}(\mathbf{k},\mathrm{i}\omega_n)}{\partial\mu}\\
&=\sum_n \varepsilon_{\mathbf{k}}^2\left(\frac{Z_{\mathbf{k}}}{-\mathrm{i}\hbar\omega_n+\hbar\omega_{\mathbf{k}}^{\mathrm{qp}}}+\frac{1-Z_{\mathbf{k}}}{-\mathrm{i}\hbar\omega_n+\hbar\omega_{\mathbf{k}}^{\mathrm{qh}}}-\frac{1}{\varepsilon_{\mathbf{k}}}\right)\\
&\times\left(\frac{g}{(-\mathrm{i}\hbar\omega_n-\mu+(g-1)U)^2}-\frac{g+1}{(-\mathrm{i}\hbar\omega_n-\mu+gU)^2}\right)\\
&=\frac{\hbar\beta\varepsilon_{\mathbf{k}}^2}{2\pi\mathrm{i}}\oint_C \mathrm{d}z\,N(z)\left(\frac{Z_{\mathbf{k}}}{-\hbar z+\hbar\omega_{\mathbf{k}}^{\mathrm{qp}}}+\frac{1-Z_{\mathbf{k}}}{-\hbar z+\hbar\omega_{\mathbf{k}}^{\mathrm{qh}}}-\frac{1}{\varepsilon_{\mathbf{k}}}\right)\\
&\times\left(\frac{g}{(-z-(\mu-(g-1)U)/\hbar)^2}-\frac{g+1}{(-z-(\mu-gU)/\hbar)^2}\right). \qquad (16.105)
\end{aligned}
$$

For the Mott-insulator state with filling fraction g, we have that $\overline{\mu}$ lies between $\overline{\mu}_{\pm}$, as given in (16.67). This implies that for the poles that originate from the derivative of G^{-1} we have that $\mu > (g-1)U$ and $\mu < gU$. At zero temperature only the quasihole pole is important, since only the quasihole dispersion is negative, and we find that the residue of that pole is proportional to

$$
\frac{(1-Z_{\mathbf{k}})\varepsilon_{\mathbf{k}}^2}{\hbar^2}\left(-\frac{g+1}{(-z-(\mu-gU)/\hbar)^2}+\frac{g}{(-z-(\mu-(g-1)U)/\hbar)^2}\right)_{z=\omega^{\mathrm{qh}}}=+1. \qquad (16.106)
$$

Similarly, the residue of the pole that is located at $z=-(\mu-gU)/\hbar$ is determined by

$$
\begin{aligned}
&-1= \qquad (16.107)\\
&\frac{-(g+1)\varepsilon_{\mathbf{k}}^2}{\hbar^2}\frac{\mathrm{d}}{\mathrm{d}z}\left\{N(z)\left(\frac{Z_{\mathbf{k}}}{-z+\hbar\omega_{\mathbf{k}}^{\mathrm{qp}}/\hbar}+\frac{1-Z_{\mathbf{k}}}{-z+\hbar\omega_{\mathbf{k}}^{\mathrm{qh}}/\hbar}+\frac{1}{\bar{\varepsilon}_{\mathbf{k}}/\hbar}\right)\right\}_{z=-(\mu-gU)/\hbar}.
\end{aligned}
$$

Adding these two contributions gives zero and as a result we have proven that the density in the Mott-insulator is, as expected, still exactly given by the integer g.

16.6.2 Superfluid Phase

In the superfluid phase the order parameter $|\psi_{\mathbf{0},0}|^2$ has a nonzero expectation value. We find this expectation value by calculating the minimum of the classical part of the action, i.e.

$$
-\hbar G^{-1}(\mathbf{0},0)|\psi_{\mathbf{0},0}|^2+\frac{zta_4(g,U,\mu)}{\hbar\beta N_{\mathrm{s}}}|\psi_{\mathbf{0},0}|^4.
$$

This minimum becomes nonzero when $-\hbar G^{-1}(\mathbf{0},0)$ becomes negative, and is then equal to

$$\frac{|\langle\psi_{\mathbf{0},0}\rangle|^2}{\hbar\beta N_s} = \frac{\hbar G^{-1}(\mathbf{0},0)}{2zta_4(g,U,\mu)} \equiv n_0 \tag{16.108}$$

As we have seen in Sect. 16.5.2, it is possible to calculate the coefficient $a_4(g,U,\mu)$ of the fourth-order term $|\psi_{\mathbf{0},0}|^4$. We approximate the prefactor of the fourth-order term, which in general depends on momenta and Matsubara frequencies, with the zero-momentum and zero-frequency value $zta_4(g,U,\mu) \equiv U^{\text{eff}}/2$ so that the approximate action up to fourth order becomes

$$\begin{aligned} S^{\text{eff}}[\psi^*,\psi] &= \hbar\beta\Omega_0 - \hbar\sum_{\mathbf{k},n}\psi^*_{\mathbf{k},n}G^{-1}(\mathbf{k},\mathrm{i}\omega_n)\psi_{\mathbf{k},n} \\ &+\frac{U^{\text{eff}}}{2}\frac{1}{\hbar\beta N_s}\sum_{\mathbf{k},\mathbf{k}',\mathbf{k}''}\sum_{n,n',n''}\psi^*_{\mathbf{k},n}\psi^*_{\mathbf{k}',n'}\psi_{\mathbf{k}'',n''}\psi_{\mathbf{k}+\mathbf{k}'-\mathbf{k}'',n+n'-n''}\,. \end{aligned} \tag{16.109}$$

We now write the order parameter as the sum of its expectation value plus fluctuations, i.e. $\psi_{\mathbf{0},0} \rightarrow \sqrt{n_0 N_s \hbar\beta} + \psi_{\mathbf{0},0}$ and a similar expression for $\psi^*_{\mathbf{0},0}$. If we put this into the action and only keep the terms up to second order, the contribution of the fourth-order term is given by

$$\frac{U^{\text{eff}}}{2}n_0\sum_{\mathbf{k},n}\left(\psi_{\mathbf{k},n}\psi_{-\mathbf{k},-n} + 4\psi^*_{\mathbf{k},n}\psi_{\mathbf{k},n} + \psi^*_{\mathbf{k},n}\psi^*_{-\mathbf{k},-n}\right).$$

There are also the contributions $-\hbar G^{-1}(\mathbf{0},0)n_0\hbar\beta N_s + U^{\text{eff}}n_0^2\hbar\beta N_s/2$, which, together with $\hbar\beta\Omega_0$, correspond to the on-site mean-field theory discussed in Sect. 16.5.2. To summarize, in the superfluid phase we can write the action from (16.109) to second order as

$$\begin{aligned} S^{\text{eff}}[\psi^*,\psi] &= \hbar\beta\Omega_0 - \hbar^2\beta G^{-1}(\mathbf{0},0)n_0N_s + \frac{1}{2}U^{\text{eff}}n_0^2\hbar\beta N_s \\ &-\frac{\hbar}{2}\sum_{\mathbf{k},n}\left[\psi^*_{\mathbf{k},n}\ \psi_{-\mathbf{k},-n}\right]\cdot\mathbf{G}^{-1}(\mathbf{k},\mathrm{i}\omega_n)\cdot\begin{bmatrix}\psi_{\mathbf{k},n}\\ \psi^*_{-\mathbf{k},-n}\end{bmatrix} \end{aligned} \tag{16.110}$$

with

$$\begin{aligned} &-\hbar\mathbf{G}^{-1}(\mathbf{k},\mathrm{i}\omega_n) \\ &= \begin{bmatrix} -\hbar G^{-1}(\mathbf{k},\mathrm{i}\omega_n) + 2U^{\text{eff}}n_0 & U^{\text{eff}}n_0 \\ U^{\text{eff}}n_0 & -\hbar G^{-1}(-\mathbf{k},-\mathrm{i}\omega_n) + 2U^{\text{eff}}n_0 \end{bmatrix}. \end{aligned} \tag{16.111}$$

Since (16.108) shows that the condensate fraction obeys $\hbar G^{-1}(\mathbf{0},0) = U^{\text{eff}}n_0$, we immediately have that $\text{Det}\left[\mathbf{G}^{-1}(\mathbf{0},0)\right] = 0$, which proves that the superfluid is gapless in contrast to the gapped Mott insulator. Moreover, the above result may be directly compared with the result in (16.42). The difference between these two Bogoliubov theories is that (16.110) is derived in the strong-coupling limit, whereas

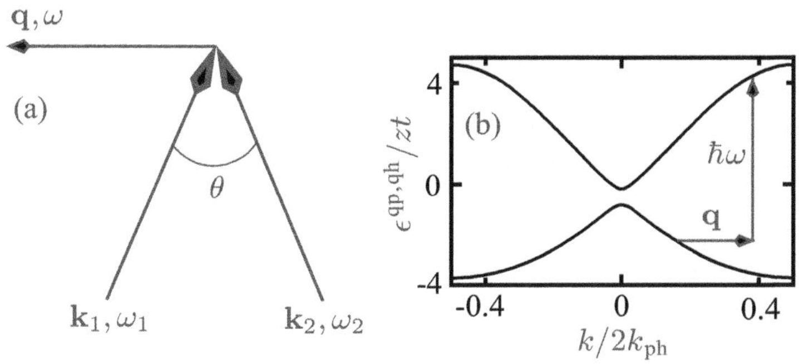

Fig. 16.10 (a) Setup for the proposed experiment. (b) Particle and hole dispersions in units of the tunneling parameter in a one dimensional lattice, for $U/zt = 6$. The horizontal arrow indicates absorption of momentum, the vertical arrow absorption of energy.

(16.42) applies to the weak-coupling situation. Although these should be smoothly connected to each other, at present it is not known how to formulate a mean-field theory that interpolates between these two extremes.

16.7 Bragg Spectroscopy

In this section, we calculate the excitation spectrum of the Mott-insulator state as probed by Bragg spectroscopy [179]. In particular, one can determine in this way the value of the particle-hole gap in the excitation spectrum and study the behavior of this gap as the system approaches the quantum critical point. Note that the excitation spectrum obtained from Bragg spectroscopy will not yield what is generally referred to as the Mott gap, because this gap is associated with single-particle excitations. The value of the particle-hole gap is an interesting quantity in the study of quantum critical phenomena, but it is also very important for the practical application of these systems in quantum information processing, since the gap determines how accurately one can experimentally realize the desired Mott ground state that can be used, for instance, as a memory for quantum bits. It is important to realize that the system of a Bose-Einstein condensate in an optical lattice is quite complicated, because in this case many-body effects and strong correlations have to be taken into account, as we will see.

In a Bragg spectroscopy experiment, two laser beams are used to make excitations in the system as explained in Sect. 13.3.2 and as shown in Fig. 16.10a. When an atom absorbs a photon from beam two and is stimulated to emit a photon into beam one, the atom undergoes a change of momentum $\hbar\mathbf{q} = \hbar\mathbf{k}_2 - \hbar\mathbf{k}_1$ and a change of energy $\hbar\omega = \hbar\omega_2 - \hbar\omega_1$. In principle various optical transitions could be used, but here we use the same transition that is employed to create the lattice potential. This means that the magnitude of the momentum is given by $\hbar q = 2\hbar k_{\rm ph}\sin(\theta/2)$, where to a good approximation $\hbar k_{\rm ph} = 2\pi\hbar/\lambda$ is the photon momentum of both the lasers, λ is equal to the wavelength of the lattice laser light and θ is the angle between the two laser beams. By varying the angle between the two laser beams any momentum between zero and $2\hbar k_{\rm ph}$ can be transferred, while by varying the relative detuning between the beams the amount of energy that is transferred to the system can be controlled.

Determining the scattering rate for a given momentum $\hbar\mathbf{q}$ and energy $\hbar\omega$ involves, roughly speaking, counting the number of ways in which the requirements of momentum and energy conservation can be met. To illustrate this process, we draw in Fig. 16.10b the quasiparticle and quasihole dispersions in the Mott insulator. The horizontal and vertical arrows in the figure indicate the transfer of momentum and energy respectively. Since energy is deposited in the system, this scattering rate can be measured in a trap loss experiment or by determining the increase in temperature of the atoms. To calculate the desired two-photon scattering rate theoretically we use Fermi's Golden Rule. In linear response, this can be expressed as

$$I(\mathbf{q},\omega) = -2\mathrm{Im}\left[\Pi(\mathbf{q},\omega)\right]/\hbar, \tag{16.112}$$

where $\Pi(\mathbf{q},\omega)$ is the polarizibility of the medium. The polarizibility can be written as

$$\Pi(\mathbf{q},\omega) = (\hbar\Omega/2)^2\chi(\mathbf{q},\omega), \tag{16.113}$$

with Ω the effective Rabi frequency for the two-photon process and χ the susceptibility. The retarded susceptibility is given by

$$\begin{aligned}\chi(\mathbf{q},\omega) &= -\frac{V}{\hbar}\int \mathrm{d}\mathbf{x}\int_0^\infty \mathrm{d}t\mathrm{e}^{-\mathrm{i}(\mathbf{q}\cdot\mathbf{x}-\omega t)} \\ &\quad\times\langle\left[\hat{a}^\dagger(\mathbf{x},t)\hat{a}(\mathbf{x},t),\hat{a}^\dagger(\mathbf{0},0)\hat{a}(\mathbf{0},0)\right]\rangle,\end{aligned} \tag{16.114}$$

with V the volume and $\hat{a}^\dagger(\mathbf{x},t)$ and $\hat{a}(\mathbf{x},t)$ creation and annihilation operators of the atoms. To calculate the correlation function in (16.114) we have to be careful to do it in such a way as to preserve particle-number conservation. As we have seen in Sect. 13.3 this can be achieved by using the relevant Ward identity. Physically this means that if the atom is dressed, we also have to dress the atom-photon coupling. Diagrammatically this is illustrated in Fig. 16.11.

We have seen that for particle-number conservation the relevant Ward identity for charged particles is given by (13.54)

$$\frac{c}{e}k'_\mu\Delta\gamma^\mu(\mathbf{k},\mathrm{i}\omega_n;\mathbf{k}',\mathrm{i}\omega_{n'}) = \Sigma(\mathbf{k},\mathrm{i}\omega_n) - \Sigma(\mathbf{k}+\mathbf{k}',\mathrm{i}\omega_{n+n'}). \tag{16.115}$$

According to (16.99) the atomic propagator in the Mott-insulator phase is given by

$$-\frac{1}{\hbar}G_{\rm a}(\mathbf{k},\omega) = \frac{Z_\mathbf{k}}{-\hbar\omega^+ + \hbar\omega_\mathbf{k}^{\rm qp}} + \frac{1-Z_\mathbf{k}}{-\hbar\omega^+ + \hbar\omega_\mathbf{k}^{\rm qh}}, \tag{16.116}$$

where the probabilities $Z_{\mathbf{k}}$ and $1-Z_{\mathbf{k}}$ account for the fact that an atomic excitation contains both quasiparticle and quasihole contributions. The notation $\hbar\omega^{+}$ is shorthand for $\hbar\omega + i0$. The dispersions for the quasiparticle and quasihole excitations are given by

$$\hbar\omega_{\mathbf{k}}^{\mathrm{qp,qh}} = -\mu + \frac{U}{2}(2g-1) + \frac{1}{2}\left(\varepsilon_{\mathbf{k}} \pm \hbar\omega_{\mathbf{k}}\right), \tag{16.117}$$

where g is the filling fraction of the lattice and $\varepsilon_{\mathbf{k}}$ corresponds to the lattice dispersion for the case of a regular square lattice. The momentum $\hbar\mathbf{k}$ is here and from now on always written in units of $2\hbar k_{\mathrm{ph}}$, which means that the first Brillouin zone runs from $k_j = -1/2$ to $1/2$. The energy $\hbar\omega_{\mathbf{k}}$ is given by

$$\hbar\omega_{\mathbf{k}} = \sqrt{U^2 + 2(2g+1)U\varepsilon_{\mathbf{k}} + \varepsilon_{\mathbf{k}}^2}$$

and the probability $Z_{\mathbf{k}}$ is given by

$$Z_{\mathbf{k}} = \frac{(2g+1)U + \varepsilon_{\mathbf{k}} + \hbar\omega_{\mathbf{k}}}{2\hbar\omega_{\mathbf{k}}}. \tag{16.118}$$

Using the Green's function in (16.116), we find in first approximation

$$\chi^0(\mathbf{q},\omega) = t(\mathbf{q})\left(\chi^0_+(\mathbf{q},\omega) - \chi^0_+(-\mathbf{q},-\omega)\right), \tag{16.119}$$

where $t(\mathbf{q})$ is a geometric factor that involves the appropriate overlap integral of the relevant Wannier functions [180], but that is unimportant for our purposes so we discuss it no further. Denoting integration over the first Brillouin zone as $\int_{1\mathrm{BZ}}$, the contribution due to the creation of a particle-hole pair is given by

$$\chi^0_+(\mathbf{q},\omega) = \frac{1}{2}\int_{1\mathrm{BZ}} d\mathbf{k}\, \frac{P(\mathbf{k},\mathbf{k}+\mathbf{q},\omega)}{-\hbar\omega^{+} + \hbar\omega^{\mathrm{qp}}_{\mathbf{k}+\mathbf{q}} - \hbar\omega^{\mathrm{qh}}_{\mathbf{k}}}, \tag{16.120}$$

and the time-reverse process can be written as $\chi^0_-(\mathbf{q},\omega) = \chi^0_+(-\mathbf{q},-\omega)$. This equation contains the probability

$$P(\mathbf{k},\mathbf{k}+\mathbf{q},\omega) = (1-Z_{\mathbf{k}})Z_{\mathbf{k}+\mathbf{q}} \tag{16.121}$$

for the creation of a hole with momentum $\mathbf{k}$ and a particle with momentum $\mathbf{k}+\mathbf{q}$, and an energy denominator that is associated with the energy cost $\hbar\omega^{\mathrm{qp}}_{\mathbf{k}+\mathbf{q}} - \hbar\omega^{\mathrm{qh}}_{\mathbf{k}}$ of that process. This can readily be verified by taking the imaginary part of the susceptibility, which is proportional to

$$\int_{1\mathrm{BZ}} d\mathbf{k}\, P(\mathbf{k},\mathbf{k}+\mathbf{q},\omega)\delta(\hbar\omega - \hbar\omega^{\mathrm{qp}}_{\mathbf{k}+\mathbf{q}} + \hbar\omega^{\mathrm{qh}}_{\mathbf{k}})$$

and can be understood as Fermi's Golden Rule. The actual computation of the above integral is too complicated to do analytically, so we have to resort to numerical methods. We achieve this by calculating the imaginary part of (16.120), which roughly corresponds to integrating over the surface in the Brillouin zone where the energy denominator vanishes. In practice, this amounts to numerically finding the poles of the expression and determining their residue. The real part is calculated from the imaginary part using a Kramers-Kronig relation [181].

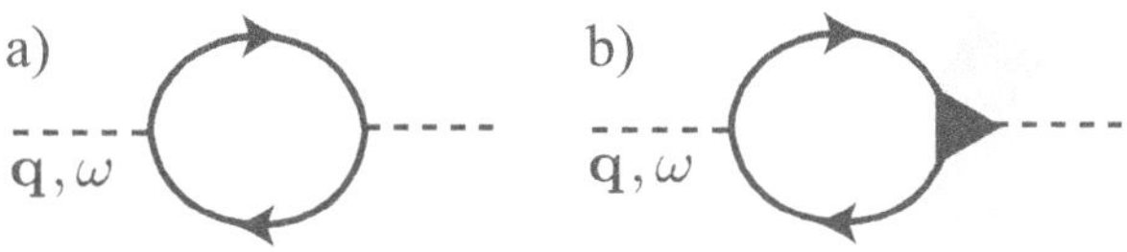

Fig. 16.11 Schematic representation of a) (16.120) and b) (16.123)

However, the results that one would obtain in this manner do not obey particle conservation. Physically, a Raman process with momentum $\mathbf{q}$ couples to a density fluctuation $\rho(\mathbf{q})$. For zero-momentum transfer, $\rho(\mathbf{0})$ corresponds to the total number of particles and fluctuations are impossible due to particle-number conservation. If we compute the imaginary part of (16.120) for $\mathbf{q} = 0$ we find a spectrum which is nonzero, which means that this approach is not sufficiently accurate. The problem is due to the fact that in (16.116) not the bare atomic propagator is used, but a dressed propagator which contains a large selfenergy correction given by

$$\hbar\Sigma(\mathbf{k},\omega) = 2gU + \frac{g(g+1)U^2}{-\hbar\omega - U - \mu}. \tag{16.122}$$

The first term on the right-hand side is the Hartree-Fock contribution, which is also present in a Bose-Einstein condensate. The second contribution is due to the correlations in the Mott insulator. Essentially this means that an atom moving through the Mott insulating background is dressed by all the other atoms.

Using the relevant Ward identity from (16.115), we can derive that the intuitive probability function given above has to be replaced by

$$P(\mathbf{k},\mathbf{k}+\mathbf{q},\omega) = \frac{2\hbar\omega - \hbar\omega^{\mathrm{qp}}_{\mathbf{k}+\mathbf{q}} + \hbar\omega^{\mathrm{qh}}_{\mathbf{k}}}{\hbar\omega_{\mathbf{k}+\mathbf{q}} + \hbar\omega_{\mathbf{k}}} \left(Z_{\mathbf{k}+\mathbf{q}} - Z_{\mathbf{k}}\right). \tag{16.123}$$

Note that the probability now vanishes when $\mathbf{q} \to 0$, so that particle conservation is indeed no longer violated. In fact, we can show that for small $\mathbf{q}$ and $\hbar\omega$ just above threshold $P \propto q^2/\Delta_0^2$, where Δ_0 is the gap for particle-hole excitations.

In Figs. 16.12 and 16.13 the result of a numerical integration is shown in two and three dimensions respectively. Both calculations have been carried out for a regular square lattice and the momentum $\mathbf{q}$ is chosen in a principal lattice direction. All energies in the following figures are given in units zt, where z is the coordination number of the lattice. The imaginary part of Fig. 16.12 clearly shows singularities around $\hbar\omega = U$. These singularities are due to the fact that there are saddle points

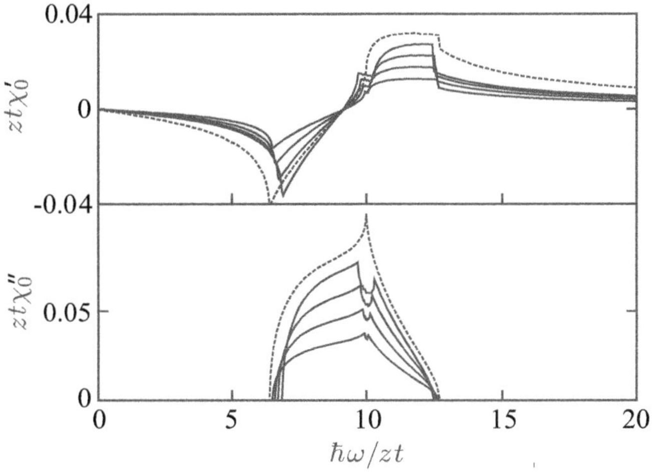

Fig. 16.12 Real and imaginary parts of the susceptibility for $U/zt = 10$ and $\mathbf{q} = 0.10, 0.14, 0.18, 0.20$ along a lattice direction, in two dimensions. The dotted line in the bottom figure is the result for $\mathbf{q} = 0.001$ multiplied by 250 to show the behaviour for small q.

in the dispersion and that a saddle point in the dispersion causes an integrable singularity in the density of states. These are called van Hove singularities [182]. It is interesting to see that the van Hove singularities split up as the momentum is increased, which is caused by the fact that the saddle-point energy in the direction of $\mathbf{q}$ and the saddle-point energy in the orthogonal direction(s) are shifted by different amounts. This is also visible in Fig. 16.13. However, it is less clear in this case, because the van Hove singularities are more smeared out in three dimensions. The opening of the threshold for the two-photon absorption in the three-dimensional case is far less steep than in the two-dimensional case. To investigate possible collective modes in this system, we determined higher-order corrections in the random-phase approximation (RPA). It can be shown that in RPA the susceptibility is given by

$$\chi(\mathbf{q}, \omega) = \frac{\chi^0(\mathbf{q}, \omega)}{1 - U\chi^0(\mathbf{q}, \omega)}. \tag{16.124}$$

This means that there is a resonance in the scattering rate when the real part of $\chi^0(\mathbf{q}, \omega)$ is equal to $1/U$. However, as can be seen from Figs. 16.12 and 16.13, the real parts in both cases are rather small compared to $1/U$ and in practice, including the RPA denominator does not qualitatively change our previous results.

In Fig. 16.14, we plot the imaginary part of $\chi^0(\mathbf{q}, \omega)$ for a range of values for the coupling constant U/zt and for a fixed momentum $\mathbf{q} = 0.10$. We see that the threshold behaviour becomes steeper as we approach the critical value of $U_c/zt \simeq 5.83$. We also see that there remains a nonzero gap when $U = U_c$. This is due to the fact that we are not considering a zero-momentum excitation, due to the reasons given above. In the inset of Fig. 16.14, we plot this gap Δ_q as a function of U/zt. For

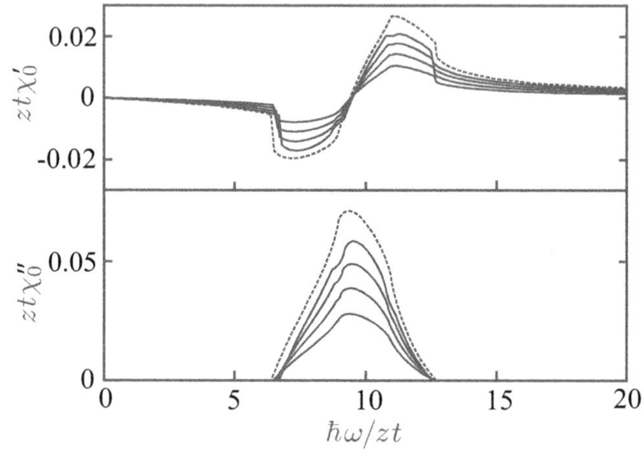

Fig. 16.13 Real and imaginary parts of the susceptibility for $U/zt = 10$ and $\mathbf{q} = 0.10, 0.14, 0.18, 0.20$ along a lattice direction, in three dimensions. The dotted line in the bottom figure is the result for $\mathbf{q} = 0.001$ multiplied by 250 to show the behaviour for small q.

large U the gap grows linearly with U, while for U close to U_c the gap closes more rapidly. In the case of $\mathbf{q} = \mathbf{0}$ the gap would in our mean-field approximation close as $\sqrt{U - U_c}$ when $U \downarrow U_c$, but for small nonzero $\mathbf{q}$ it closes as $\sqrt{U - U_c + \eta q^4}$, where the factor η is a positive function of U_c and t. In a recent experiment by Stöferle et al. [170] the authors use a setup where the laser beams are perfectly counterpropagating, which corresponds to a quasi-momentum transfer of zero. As we have argued above, there should be no scattering in that case and the signal can only be due to nonlinear response or to the fact that the system is inhomogeneous and of finite size. We have found that, by measuring the threshold behavior of the two-photon scattering rate at various quasi-momenta, it is possible to determine the gap by extrapolation. We have shown that for a theoretical description of Bragg spectroscopy on the Mott insulator it is absolutely essential to dress the photon-atom coupling, which is in a way unexpected, as the corrections are zero in the case of an harmonically trapped gas. As a result it turns out that although it is common to use the language of solid-state physics to describe these systems, the physics is qualitatively very different due to the many-body effects.

16.8 Problems

Exercise 16.1. **Particle Density**
In this problem, we first demonstrate for the noninteracting gas the equivalence of the calculation of the total particle density through the thermodynamic relation $N = -\partial\Omega/\partial\mu$ and through the use of source currents that couple to the atomic

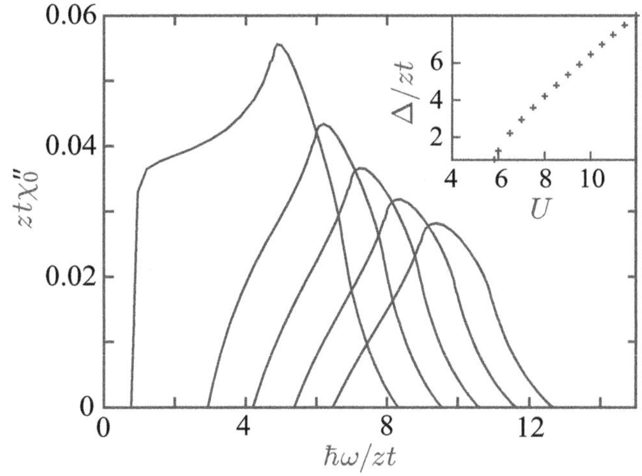

Fig. 16.14 Imaginary part of the susceptibility in a three dimensional lattice for $\mathbf{q} = 0.10$ along a lattice direction and $U/zt = 5.83, 7, 8, 9, 10$

fields. We consider a system of noninteracting bosons described by creation and annihilation fields $a_i^*(\tau)$ and $a_i(\tau)$ on a lattice.
a) Write down the expression for the generating functional $Z[J^*, J]$ for this system and decouple the hopping term by adding the following complete square to the action,

$$\sum_{ij}\left(a_i^*(\tau) - \psi_i^*(\tau) + \hbar\sum_{j'} t_{ij'}^{-1} J_{j'}^*(\tau)\right) t_{ij} \left(a_j(\tau) - \psi_j(\tau) + \hbar\sum_{j''} t_{ij''}^{-1} J_{j''}(\tau)\right).$$

Integrate out the atomic fields a^*, a and show that one arrives at the following expression for the generating functional,

$$Z[J^*, J] = \int d[\psi^*]d[\psi] \exp\left\{\sum_{\mathbf{k},n} \psi_{\mathbf{k},n}^* G^{-1}(\mathbf{k}, i\omega_n)\psi_{\mathbf{k},n} + J_{\mathbf{k},n}^*\psi_{\mathbf{k},n} + J_{\mathbf{k},n}\psi_{\mathbf{k},n}^* + \hbar\frac{J_{\mathbf{k},n}J_{\mathbf{k},n}^*}{\varepsilon_{\mathbf{k}}}\right\}, \tag{16.125}$$

where $\hbar G^{-1}(\mathbf{k}, i\omega_n) = \varepsilon_{\mathbf{k}} + \varepsilon_{\mathbf{k}}^2(-i\hbar\omega_n - \mu)^{-1}$.
b) The total density may be calculated from this expression by first calculating the correlator $\langle a_{\mathbf{k},n} a_{\mathbf{k},n}^* \rangle$ through functional differentiation with respect to the source-currents J, and then to sum over all momenta and Matsubara frequencies.
c) We can also calculate the density from the thermodynamic potential Ω, by using the relation $N = -\partial\Omega/\partial\mu$ where N is the total number of particles. Calculate the thermodynamic potential and demonstrate the equivalence of both methods.

d) Show that for an interacting system where the selfenergy contains the chemical potential, the above equivalence no longer holds.

Exercise 16.2. The Green's function of the atoms $G_{\mathrm{a}}(\mathbf{k}, \mathrm{i}\omega_n)$ is related to $G(\mathbf{k}, \mathrm{i}\omega_n)$ according to (16.99). Show that the Green's function of the atoms can also be written as

$$G_{\mathrm{a}}(\mathbf{k}, \mathrm{i}\omega_n) = \frac{-\hbar}{-i\hbar\omega_n + \varepsilon_{\mathbf{k}} - \mu + \hbar\Sigma(\mathbf{k}, i\omega_n)}, \tag{16.126}$$

where the selfenergy is given by

$$\hbar\Sigma(\mathbf{k}, \mathrm{i}\omega_n) = 2gU + \frac{g(g+1)U^2}{-\mathrm{i}\hbar\omega_n - \mu - U}. \tag{16.127}$$

Exercise 16.3. Estimate the temperature of an atomic gas that needs to be realized for the single-band approach to be valid.

Exercise 16.4. Derive the expressions for the hopping t and the on-site interaction U in (16.31) and (16.33).

Additional Reading

- For recent reviews on ultracold gases in optical lattices, see D. Jacksch, and P. Zoller, Ann. Phys. **315**, 52 (2005), and
- I. Bloch, J. B. Dalibard, and W. Zwerger, Rev. Mod. Phys. **80**, 885 (2008).
- For more details on atom-light interaction, see for instance H. J. Metcalf, and P. van der Straten, *Laser Cooling and Trapping*, (Springer-Verlag, New York, 1999), and
- R. Grimm, M. Weidemüller, and Y. B. Ovchinnikov, Adv. At. Mol. Opt. Phys. **42**, 95 (2000).
- For textbooks on solid-state physics, see J. M. Ziman, *Theory of Solids*, (Cambridge University Press, Cambridge, 1972), and
- N. W. Ashcroft and N. D. Mermin, *Solid State Physics*, (Saunders College Publishing, Orlando, 1976).
- For a textbook on quantum phase transitions, see S. Sachdev, *Quantum Phase Transitions*, (Cambridge University Press, New York, 2001).

Chapter 17
Feshbach Resonances

> *We're holding atoms in a laser beam, turn down the laser power and those atoms turn into molecules... If Nature had knocked on my door and said you have one free wish, wish for something you want in science, I wouldn't have been bold enough to ask for that!*
> – *Wolfgang Ketterle*

In Sect. 10.3.3 we discussed the possibility of a shape resonance in the scattering length of two atoms. Such a shape resonance occurs when the interatomic potential has a bound state that is very weakly bound and, therefore, has an energy that lies very close to the continuum threshold of the atoms. Because this bound state occurs in the same potential with which the atoms interact, it is difficult to experimentally control such a resonance, and thereby the scattering properties of atoms. To do so would require changing the actual shape of the interatomic potential, which is impossible by a static magnetic bias field. Put differently, for a shape resonance the magnetic moment of the weakly-bound molecule is exactly the same as the magnetic moment of the colliding atoms. As a result, an external magnetic field affects the energy of the bound state in exactly the same manner as the energy of the colliding atoms. One thus cannot change the energy difference between the bound state and the atomic continuum, which would drastically affect the outcome of the scattering process and therefore the effective interaction strength between the atoms. For completeness, we should mention that in principle time-dependent electric fields can be used to influence the scattering length of the atoms in this case, but the physics then actually turns out to be very similar to the physics of Feshbach resonances that we discuss next.

The crucial difference between a shape resonance and a Feshbach resonance is that in the latter case, the molecular state responsible for the resonance has a magnetic moment that is different from the magnetic moment of the colliding atoms [17, 183]. Therefore, the energy difference between the bound state and the atomic continuum can now be easily controlled experimentally by an external magnetic bias field. Theoretically this implies that the scattering process is not described by a single-channel Schrödinger equation, as in the case of a shape resonance, but by a multi-channel or matrix Schrödinger equation, because the scattering wavefunction has now a nonzero amplitude in a number of different spin states.

In this chapter [42], we discuss both the two-body and the many-body physics of ultracold atoms near a Feshbach resonance. To this end, we first present a simple two-channel model for a Feshbach resonance between two alkali atoms, and then show how this model can be captured by a quantum field theory. From this microscopic quantum field theory, we derive from first principles an effective quantum field theory of atoms and molecules that includes the two-body Feshbach physics exactly and can thus be used as a starting point to accurately discuss the many-body physics of the gas. To prove the latter explicitly, we discuss in detail the internal Josephson oscillations between an atomic and a molecular Bose-Einstein condensate. It is important to realize that during these oscillations the spin state changes, because the molecule has a different magnetic moment than the two colliding atoms. For completeness, we mention that in the limit of a broad resonance, it turns out that the amplitude in the molecular channel actually becomes very small, so that the two-channel model reduces effectively to a single-channel model [23, 184]. Then, the main effect of the Feshbach molecular state is to cause a magnetic-field dependent scattering length, which we can use as an input parameter for the single-channel model. Indeed, this was the approach in Sect. 12.8, where we discussed the BEC-BCS crossover for fermions. Note that in this single channel model we still have the possibility of forming tightly-bound states, which now have the same spin state as the two colliding atoms.

17.1 Example of a Feshbach Resonance

Up to now we always considered atom-atom interactions that did not change the internal states of the atoms. We consider now the more general, and more realistic, situation of atom-atom scattering whereby the atoms have various internal states that can change during the collision. In the case of alkali atoms, the simplest picture of such collisions is the following. The spin states that are coupled correspond, roughly speaking, to the eigenstates of the total spin operator of the valence electrons of the alkali atoms. Indeed, if we neglect hyperfine interactions the effective interaction potential between the atoms depends only on the state of the valence electrons of the colliding atoms. If these form a singlet the electrons are in principle allowed to be on top of each other. For a triplet this is forbidden. Hence, the singlet potential is generally much deeper than the triplet potential.

Of course, in reality every atom has also a nucleus with spin $\mathbf{I}$ that interacts with the spin $\mathbf{S}$ of a valence electron via the hyperfine interaction

$$\hat{H}_{\mathrm{hf}} = \frac{\alpha_{\mathrm{hf}}}{\hbar^2}\hat{\mathbf{I}}\cdot\hat{\mathbf{S}}, \tag{17.1}$$

with α_{hf} the hyperfine constant. This hyperfine interaction couples the singlet and triplet states. Moreover, in the presence of a magnetic field the different internal states of the atoms have a different Zeeman shift. In an experiment with magnetically-trapped gases, the energy difference between these states is therefore experimentally accessible. Putting these results together, we can write down the simplest two-channel Schödinger equation that models the above physics

$$\begin{bmatrix} -\frac{\hbar^2\nabla^2}{m} + V_{\mathrm{T}}(\mathbf{r}) - E & V_{\mathrm{hf}} \\ V_{\mathrm{hf}} & -\frac{\hbar^2\nabla^2}{m} + \Delta\mu B + V_{\mathrm{S}}(\mathbf{r}) - E \end{bmatrix} \begin{bmatrix} \psi_{\mathrm{T}}(\mathbf{r}) \\ \psi_{\mathrm{S}}(\mathbf{r}) \end{bmatrix} = 0 \,. \tag{17.2}$$

Here, $V_{\mathrm{T}}(\mathbf{r})$ and $V_{\mathrm{S}}(\mathbf{r})$ are the interaction potentials of atoms with internal state $|\mathrm{T}\rangle$ and $|\mathrm{S}\rangle$ respectively, and $\Delta\mu B$ is their difference in Zeeman energy due to the interaction with the magnetic field B, with $\Delta\mu$ the difference in magnetic moment. Finally, $V_{\mathrm{hf}} = \langle \mathrm{T}|\hat{H}_{\mathrm{hf}}|\mathrm{S}\rangle$ denotes the matrix element of the hyperfine coupling. In agreement with the above remarks, $|\mathrm{T}\rangle$ is referred to as the triplet channel, whereas $|\mathrm{S}\rangle$ is referred to as the singlet channel. The potentials $V_{\mathrm{T}}(\mathbf{r})$ and $V_{\mathrm{S}}(\mathbf{r})$ are the triplet and singlet interaction potentials respectively.

As a specific example, we use for both interaction potentials square-well potentials

$$V_{\mathrm{T,S}}(r) = \begin{cases} -V_{\mathrm{T,S}} & \text{if } r < R \\ 0 & \text{if } r > R \end{cases}, \tag{17.3}$$

where $V_{\mathrm{T,S}} > 0$. For convenience we have taken the range the same for both potentials. Furthermore, we assume that the potentials are such that $V_{\mathrm{T}} < V_{\mathrm{S}}$ and that V_{S} is just deep enough such that it contains exactly one bound state. Finally, we assume that $0 < V_{\mathrm{hf}} \ll V_{\mathrm{T}}, V_{\mathrm{S}}, \Delta\mu B$. The potentials are shown in Fig. 17.1.

To discuss the scattering properties of the atoms we have to diagonalize the Hamiltonian for $r > R$ in order to determine the incoming channels, which are superpositions of the triplet and singlet states $|\mathrm{T}\rangle$ and $|\mathrm{S}\rangle$. Since the kinetic energy operator is diagonal in the internal space of the atoms, we have to find the eigenvalues of the Hamiltonian

$$\mathbf{H}^{>} = \begin{bmatrix} 0 & V_{\mathrm{hf}} \\ V_{\mathrm{hf}} & \Delta\mu B \end{bmatrix} . \tag{17.4}$$

These are given by

$$\varepsilon_{\pm}^{>} = \frac{\Delta\mu B}{2} \pm \frac{1}{2}\sqrt{(\Delta\mu B)^2 + (2V_{\mathrm{hf}})^2}. \tag{17.5}$$

The Hamiltonian $\mathbf{H}^{>}$ is diagonalized by the matrix

$$\mathbf{Q}(\theta) = \begin{bmatrix} \cos\theta & \sin\theta \\ -\sin\theta & \cos\theta \end{bmatrix}, \tag{17.6}$$

according to

$$\mathbf{Q}(\theta^{>}) \cdot \mathbf{H}^{>} \cdot \mathbf{Q}^{-1}(\theta^{>}) = \begin{bmatrix} \varepsilon_{-}^{>} & 0 \\ 0 & \varepsilon_{+}^{>} \end{bmatrix}, \tag{17.7}$$

which determines $\tan 2\theta^{>} = -2V_{\mathrm{hf}}/\Delta\mu B$. We define now the hyperfine states $|\uparrow\uparrow\rangle$ and $|\downarrow\downarrow\rangle$, the precise reason for this particular choice of notation will become clear in the next section, according to

$$\begin{bmatrix} |\uparrow\uparrow\rangle \\ |\downarrow\downarrow\rangle \end{bmatrix} = \mathbf{Q}(\theta^{>}) \begin{bmatrix} |\mathrm{T}\rangle \\ |\mathrm{S}\rangle \end{bmatrix}, \tag{17.8}$$

which asymptotically represent the scattering channels. In this basis the Schrödinger equation for all $\mathbf{r}$ reads

$$\begin{bmatrix} -\frac{\hbar^2\nabla^2}{m} + V_{\uparrow\uparrow}(\mathbf{r}) - E & V_{\uparrow\downarrow}(\mathbf{r}) \\ V_{\uparrow\downarrow}(\mathbf{r}) & -\frac{\hbar^2\nabla^2}{m} + \varepsilon_{+}^{>} - \varepsilon_{-}^{>} + V_{\downarrow\downarrow}(\mathbf{r}) - E \end{bmatrix} \begin{bmatrix} \psi_{\uparrow\uparrow}(\mathbf{r}) \\ \psi_{\downarrow\downarrow}(\mathbf{r}) \end{bmatrix} = 0, \tag{17.9}$$

where the energy E is measured with respect to $\varepsilon_{-}^{>}$ and we have defined the potentials according to

$$\begin{bmatrix} V_{\uparrow\uparrow}(\mathbf{r}) & V_{\uparrow\downarrow}(\mathbf{r}) \\ V_{\uparrow\downarrow}(\mathbf{r}) & V_{\downarrow\downarrow}(\mathbf{r}) \end{bmatrix} = \mathbf{Q}(\theta^{>}) \cdot \begin{bmatrix} V_{\mathrm{T}}(\mathbf{r}) & 0 \\ 0 & V_{\mathrm{S}}(\mathbf{r}) \end{bmatrix} \cdot \mathbf{Q}^{-1}(\theta^{>}). \tag{17.10}$$

Since all these potentials vanish for $r > R$ we can study scattering of atoms in the states $|\uparrow\uparrow\rangle$ and $|\downarrow\downarrow\rangle$. Because the hyperfine interaction V_{hf} is small we have that $\varepsilon_{+}^{>} \simeq \Delta\mu B$ and $\varepsilon_{-}^{>} \simeq 0$. Moreover, for the experiments with magnetically-trapped gases we always have that $\Delta\mu B \gg k_{\mathrm{B}}T$ where k_{B} is Boltzmann's constant and T is the temperature. This means that in a realistic atomic gas, in which the states $|\uparrow\uparrow\rangle$ and $|\downarrow\downarrow\rangle$ are available, there are in equilibrium almost no atoms that scatter via the latter state. Because of this, the effects of the interactions of the atoms will be determined by the scattering amplitude in the state $|\uparrow\uparrow\rangle$. If two atoms scatter in this channel with energy $E \simeq k_{\mathrm{B}}T \ll \Delta\mu B$ they cannot come out in the other channel because of energy conservation. Therefore, the indices $\uparrow\uparrow$ refer to an open channel, whereas $\downarrow\downarrow$ is associated with a closed channel of the scattering process. The situation is further clarified in Fig. 17.1.

To calculate the s-wave scattering length in the open channel we have to solve the Schrödinger equation. In the region $r > R$ the solution is of the form

$$\begin{bmatrix} u_{\uparrow\uparrow}^{>}(r) \\ u_{\downarrow\downarrow}^{>}(r) \end{bmatrix} = \begin{bmatrix} C\mathrm{e}^{\mathrm{i}kr} + D\mathrm{e}^{-\mathrm{i}kr} \\ F\mathrm{e}^{-\kappa r} \end{bmatrix}, \tag{17.11}$$

where $\kappa = \sqrt{m(\varepsilon_{+}^{>} - \varepsilon_{-}^{>})/\hbar^2 - k^2}$ and, because we have used the same notation as in (10.41), the s-wave phase shift is again determined by (10.42). In the region $r < R$ the solutions are of the form

$$\begin{bmatrix} u_{\uparrow\uparrow}^{<}(r) \\ u_{\downarrow\downarrow}^{<}(r) \end{bmatrix} = \begin{bmatrix} A\left(\mathrm{e}^{\mathrm{i}k_{\uparrow\uparrow}^{<} r} - \mathrm{e}^{-\mathrm{i}k_{\uparrow\uparrow}^{<} r}\right) \\ B\left(\mathrm{e}^{\mathrm{i}k_{\downarrow\downarrow}^{<} r} - \mathrm{e}^{-\mathrm{i}k_{\downarrow\downarrow}^{<} r}\right) \end{bmatrix}, \tag{17.12}$$

where

$$k_{\uparrow\uparrow}^{<} = \sqrt{m(\varepsilon_{-}^{>} - \varepsilon_{-}^{<})/\hbar^2 + k^2}, \quad k_{\downarrow\downarrow}^{<} = \sqrt{m(\varepsilon_{-}^{>} - \varepsilon_{+}^{<})/\hbar^2 + k^2}, \tag{17.13}$$

and

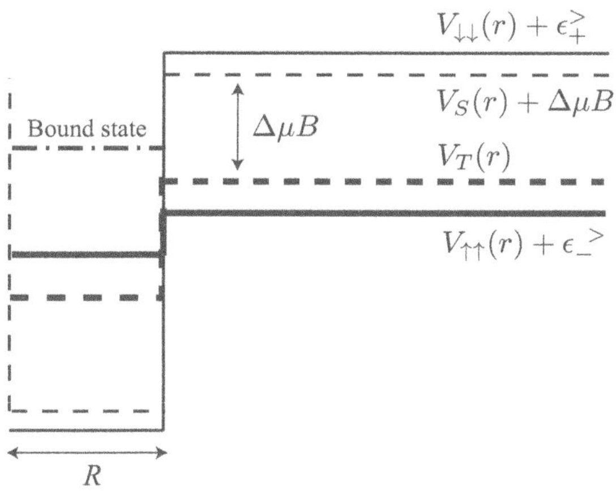

Fig. 17.1 Feshbach resonance in a two-channel system with square-well interaction potentials. The triplet potential $V_T(r)$ is indicated by the thick dashed line. The singlet potential that contains the bound state responsible for the Feshbach resonance is indicated by the thin dashed line. Due to the Zeeman interaction with the magnetic field, the energy difference between the singlet and triplet is equal to $\Delta\mu B$. The interactions in the open and closed hyperfine channels are indicated by $V_{\uparrow\uparrow}(r)$ and $V_{\downarrow\downarrow}(r)$ respectively.

$$\varepsilon_{\pm}^{<} = \frac{\Delta\mu B - V_{\rm T} - V_{\rm S}}{2} \mp \frac{1}{2}\sqrt{(V_{\rm S}-V_{\rm T}-\Delta\mu B)^2+(2V_{\rm hf})^2} \tag{17.14}$$

are the eigenvalues of the matrix

$$\mathbf{H}^{<} = \begin{bmatrix} -V_{\rm T} & V_{\rm hf} \\ V_{\rm hf} & \Delta\mu B - V_{\rm S} \end{bmatrix} . \tag{17.15}$$

In order to determine the phase shift we have to join the solution for $r<R$ and $r>R$ smoothly. This is done most easily by transforming to the singlet-triplet basis $\{|{\rm T}\rangle,|{\rm S}\rangle\}$ since this basis is independent of r. Demanding the solution to be continuously differentiable leads to the equations

$$\mathbf{Q}^{-1}(\theta^{<})\begin{bmatrix} u_{\uparrow\uparrow}^{<}(R) \\ u_{\downarrow\downarrow}^{<}(R)\end{bmatrix} = \mathbf{Q}^{-1}(\theta^{>})\begin{bmatrix} u_{\uparrow\uparrow}^{>}(R) \\ u_{\downarrow\downarrow}^{>}(R)\end{bmatrix} \quad \text{and}$$

$$\frac{\partial}{\partial r}\mathbf{Q}^{-1}(\theta^{<})\begin{bmatrix} u_{\uparrow\uparrow}^{<}(r) \\ u_{\downarrow\downarrow}^{<}(r)\end{bmatrix}\Bigg|_{r=R} = \frac{\partial}{\partial r}\mathbf{Q}^{-1}(\theta^{>})\begin{bmatrix} u_{\uparrow\uparrow}^{>}(r) \\ u_{\downarrow\downarrow}^{>}(r)\end{bmatrix}\Bigg|_{r=R}, \tag{17.16}$$

where $\tan 2\theta^{<} = 2V_{\rm hf}/(V_{\rm S}-V_{\rm T}-\Delta\mu B)$. These four equations determine the coefficients A,B,C,D and F up to a normalization factor, and therefore also the phase shift and the scattering length. Although it is possible to find an analytical expression for the scattering length as a function of the magnetic field, the resulting expression is rather formidable and is omitted here. The result for the scattering length is shown in Fig. 17.2, for $V_{\rm S} = 10\hbar^2/mR^2$, $V_{\rm T} = \hbar^2/mR^2$ and $V_{\rm hf} = 0.1\hbar^2/mR^2$, as a function of

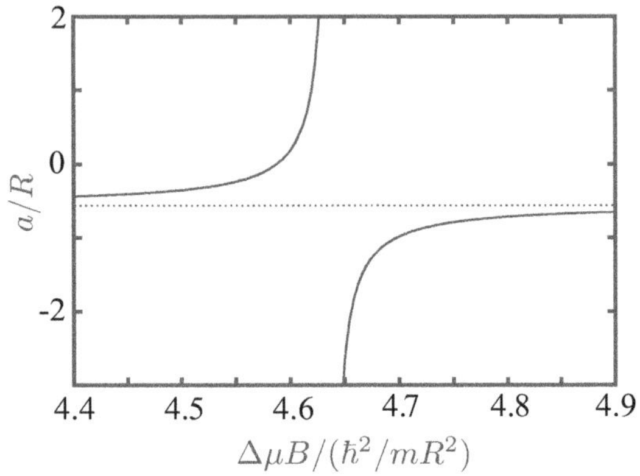

Fig. 17.2 Scattering length for two coupled square-well potentials as a function of $\Delta\mu B$. The depth of the triplet and singlet channel potentials is $V_T = \hbar^2/mR^2$ and $V_S = 10\hbar^2/mR^2$ respectively. The hyperfine coupling is $V_{hf} = 0.1\hbar^2/mR^2$. The dotted line shows the background scattering length a_{bg}.

$\Delta\mu B$. The resonant behavior is due to the bound state of the singlet potential $V_S(r)$. Indeed, solving the equation for the binding energy in (10.49) with $V_0 = -V_S$ we find that $|E_m| \simeq 4.62\hbar^2/mR^2$, which is approximately the position of the resonance in Fig. 17.2. The difference is due to the fact that the hyperfine interaction leads to a shift in the position of the resonance with respect to E_m.

The magnetic-field dependence of the scattering length near a Feshbach resonance is characterized experimentally by a width ΔB and position B_0 according to

$$a(B) = a_{bg}\left(1 - \frac{\Delta B}{B - B_0}\right). \tag{17.17}$$

This explicitly shows that the scattering length, and therefore the magnitude of the effective interatomic interaction, may be altered to any value by tuning the magnetic field. The first experimental proof of this fact is shown in Fig. 17.3, where the magnetic-field dependence of the scattering length is indeed seen to be given by (17.17). The off-resonant background scattering length is denoted by a_{bg} and is, in our example, approximately equal to the scattering length of the triplet potential $V_T(r)$. Using the expression for the scattering length of a square well in (10.46) for $\gamma = 1$, we find that $a_{bg} \simeq -0.56R$. Furthermore, we have for our example that the position of the resonance is given by $B_0 \simeq 4.64\hbar^2/m\Delta\mu R^2$ and that the width is equal to $\Delta B \simeq -0.05\hbar^2/m\Delta\mu R^2$.

Next we calculate the energy of the molecular state for the coupled-channel case, which is found by solving (17.9) for negative energy. In particular, we are interested in its dependence on the magnetic field. In the absence of the hyperfine coupling between the open and closed channel we simply have that $\varepsilon_m(B) = E_m + \Delta\mu B$. Here, E_m is the energy of the bound state responsible for the Feshbach resonance, which is

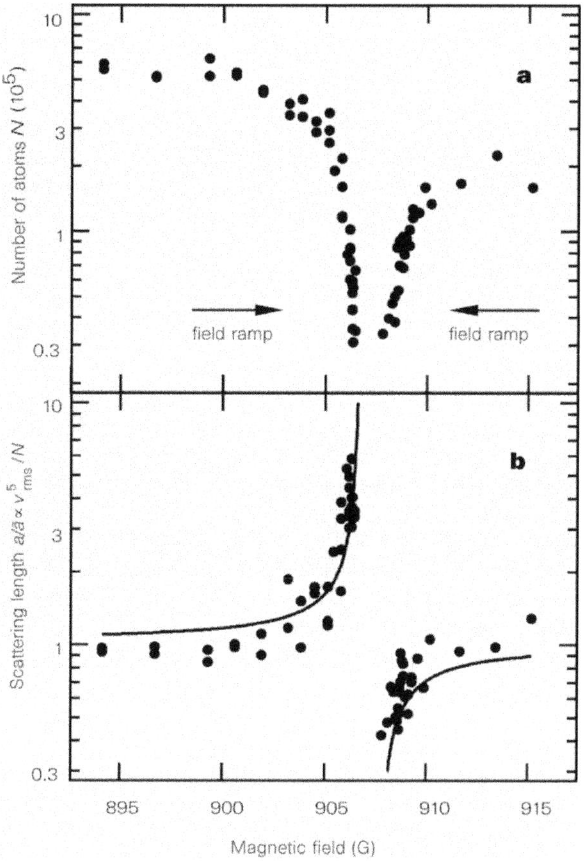

Fig. 17.3 The first experimental observation of a Feshbach resonance by Inouye et al. [185]. (a) The loss of atoms in a trap is greatly enhanced near a Feshbach resonance. (b) The characteristic behavior of the scattering length as a function of magnetic field, where the solid line is a fit to (17.17). Reprinted by permission from Macmillan Publishers Ltd: Nature **392**, 151 (1998), copyright (1998).

determined by solving the single-channel Schödinger equation for the singlet potential. This bound-state energy as a function of the magnetic field is shown in Fig. 17.4 by the dashed line. A nonzero hyperfine coupling drastically changes this result. For our example the bound-state energy is easily calculated. The result is shown by the solid line in Fig. 17.4 for the same parameters as before. Clearly, close to the resonance the dependence of the bound-state energy on the magnetic field is no longer linear, as the inset of Fig. 17.4 shows. Instead, it turns out to be quadratic. This was also seen in experiment, as shown in Fig. 17.5. Moreover, the magnetic field B_0 where the bound-state energy is equal to zero is shifted with respected to the case where $V_{hf} = 0$. It is at this shifted magnetic field that the resonance is observed experimentally. Moreover, for magnetic fields larger than B_0 there no longer exists

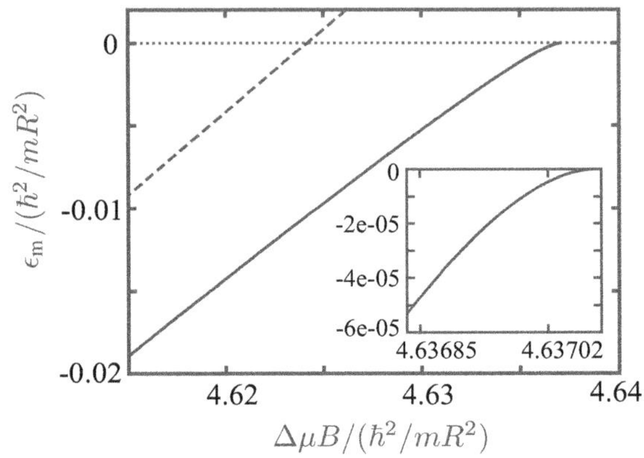

Fig. 17.4 Bound-state energy of the molecular state near a Feshbach resonance for two coupled square-well interaction potentials. The solid line and the inset show the result for $V_{\rm hf} = 0.1\hbar^2/mR^2$. The dashed line corresponds to $V_{\rm hf} = 0$. The other parameters are the same as in figure 17.2.

a bound state and the molecule now decays due to the hyperfine coupling into two free atoms, because its energy is above the two-atom continuum threshold.

Close to resonance the energy of the molecular state turns out to be related to the scattering length by

$$\varepsilon_{\rm m}(B) = -\frac{\hbar^2}{ma^2(B)}\,, \tag{17.18}$$

as in the single-channel case. As we will see in the next sections, the reason for this is that close to resonance the effective two-body T matrix again has a pole at the energy in (17.18). This important result will be proven analytically in Sect. 17.4.2. This is achieved by deriving the effective quantum field theory that offers a description of Feshbach-resonant interactions in terms of an atom-molecule action. For simplicity we consider from now on only bosonic atoms, because this requires only the introduction of two internal states. For fermions at least three internal states are needed for a s-wave Feshbach resonance, which makes the calculation unnecessary complicated. Nevertheless, the generalization to fermions is straightforward. To do so, we start from a microscopic atomic action that involves atoms with two internal states, i.e. we consider a situation with an open and a closed channel that are coupled by the exchange interaction. By performing a Hubbard-Stratonovich transformation we introduce a quantum field that incorporates into the theory the bound state in the closed channel responsible for the Feshbach resonance.

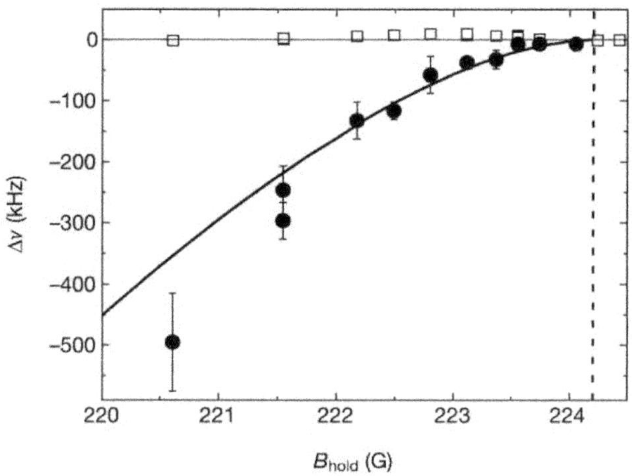

Fig. 17.5 Measurement of the binding energy of Feshbach molecules in atomic ^{40}K by Regal et al. [186]. The solid line shows the universal result $\varepsilon_{\mathrm{m}}(B) = -\hbar^2/ma^2(B)$. Reprinted by permission from Macmilan Publishers Ltd: Nature **427**, 47 (2003), copyright (2003).

17.2 Bare Atom-Molecule Theory

Without loss of generality, we can consider the simplest situation in which a Feshbach resonance arises, i.e. we consider a homogeneous gas of identical atoms in a box of volume V. These atoms have two internal states, denoted by $|\uparrow\rangle$ and $|\downarrow\rangle$, that are described by the fields $\phi_\uparrow(\mathbf{x},\tau)$ and $\phi_\downarrow(\mathbf{x},\tau)$ respectively. The atoms in these two states interact via the potentials $V_{\uparrow\uparrow}(\mathbf{x}-\mathbf{x}')$ and $V_{\downarrow\downarrow}(\mathbf{x}-\mathbf{x}')$ respectively. The state $|\downarrow\rangle$ has an energy $\Delta\mu B/2$ with respect to the state $|\uparrow\rangle$ due to the Zeeman interaction with the magnetic field B. The coupling between the two states, which from the point of view of atomic physics is due to the difference in singlet and triplet interactions, is denoted by $V_{\uparrow\downarrow}(\mathbf{x}-\mathbf{x}')$. Putting everything together we write the grand-canonical partition function for the gas as a functional integral given by

$$Z = \int \mathrm{d}[\phi_\uparrow^*]\mathrm{d}[\phi_\uparrow]\mathrm{d}[\phi_\downarrow^*]\mathrm{d}[\phi_\downarrow]\exp\left\{-\frac{1}{\hbar}S[\phi_\uparrow^*,\phi_\uparrow,\phi_\downarrow^*,\phi_\downarrow]\right\}. \tag{17.19}$$

Since we are dealing with bosons, the integration is over all fields that are periodic on the imaginary-time axis ranging from zero to $\hbar\beta$, with $\hbar$ Planck's constant and $\beta = 1/k_{\mathrm{B}}T$ the inverse thermal energy. The Euclidian action is given by

$$S[\phi_\uparrow^*,\phi_\uparrow,\phi_\downarrow^*,\phi_\downarrow] = \int_0^{\hbar\beta}\mathrm{d}\tau\left\{\int\mathrm{d}\mathbf{x}\left\{\phi_\uparrow^*(\mathbf{x},\tau)\hbar\frac{\partial}{\partial\tau}\phi_\uparrow(\mathbf{x},\tau)+\phi_\downarrow^*(\mathbf{x},\tau)\hbar\frac{\partial}{\partial\tau}\phi_\downarrow(\mathbf{x},\tau)\right\}\right.$$
$$\left. +H[\phi_\uparrow^*,\phi_\uparrow,\phi_\downarrow^*,\phi_\downarrow]\right\}, \tag{17.20}$$

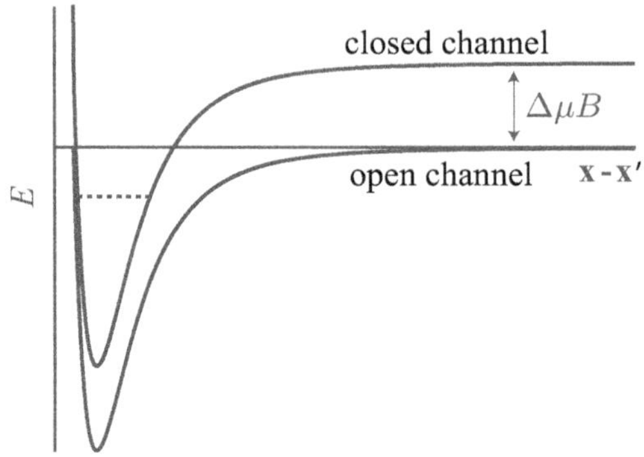

Fig. 17.6 Illustration of a Feshbach resonance. The upper potential curve corresponds to the closed-channel interaction potential $V_{\downarrow\downarrow}(\mathbf{x}-\mathbf{x}')$ that contains the bound state responsible for the Feshbach resonance, indicated by the dashed line. The lower potential curve corresponds to the open-channel interaction potential $V_{\uparrow\uparrow}(\mathbf{x}-\mathbf{x}')$.

with the grand-canonical Hamiltonian functional given by

$$\begin{aligned} H[\phi_\uparrow^*,\phi_\uparrow,\phi_\downarrow^*,\phi_\downarrow] = & \int \mathrm{d}\mathbf{x}\,\phi_\uparrow^*(\mathbf{x},\tau)\left\{-\frac{\hbar^2\nabla^2}{2m}-\mu\right. \\ & \left.+\frac{1}{2}\int \mathrm{d}\mathbf{x}'\;\phi_\uparrow^*(\mathbf{x}',\tau)V_{\uparrow\uparrow}(\mathbf{x}-\mathbf{x}')\phi_\uparrow(\mathbf{x}',\tau)\right\}\phi_\uparrow(\mathbf{x},\tau) \\ & +\int \mathrm{d}\mathbf{x}\,\phi_\downarrow^*(\mathbf{x},\tau)\left\{-\frac{\hbar^2\nabla^2}{2m}+\frac{\Delta\mu B}{2}-\mu\right. \\ & \left.+\frac{1}{2}\int \mathrm{d}\mathbf{x}'\phi_\downarrow^*(\mathbf{x}',\tau)V_{\downarrow\downarrow}(\mathbf{x}-\mathbf{x}')\phi_\downarrow(\mathbf{x}',\tau)\right\}\phi_\downarrow(\mathbf{x},\tau) \\ & +\frac{1}{2}\int \mathrm{d}\mathbf{x}\int \mathrm{d}\mathbf{x}'\phi_\uparrow^*(\mathbf{x},\tau)\phi_\uparrow^*(\mathbf{x}',\tau)V_{\uparrow\downarrow}(\mathbf{x}-\mathbf{x}')\phi_\downarrow(\mathbf{x}',\tau)\phi_\downarrow(\mathbf{x},\tau), \\ & +\frac{1}{2}\int \mathrm{d}\mathbf{x}\int \mathrm{d}\mathbf{x}'\phi_\downarrow^*(\mathbf{x},\tau)\phi_\downarrow^*(\mathbf{x}',\tau)V_{\uparrow\downarrow}(\mathbf{x}-\mathbf{x}')\phi_\uparrow(\mathbf{x}',\tau)\phi_\uparrow(\mathbf{x},\tau), \end{aligned} \tag{17.21}$$

where μ is the chemical potential of the atoms. Note that this Hamiltonian functional is the grand-canonical version of the Hamiltonian in (17.9). The indices $\uparrow$ and $\downarrow$ now refer again to single-particle states, and the two-particle hyperfine states are denoted by $|\uparrow\uparrow\rangle$ and $|\downarrow\downarrow\rangle$ respectively. The closed-channel potential is assumed again to contain the bound state responsible for the Feshbach resonance, as illustrated in Fig. 17.6. To introduce the molecular field that describes the center-of-mass motion of this bound state, we introduce the complex pairing field $\Delta(\mathbf{x},\mathbf{x}',\tau)$ and perform a Hubbard-Stratonovich transformation by inserting the identity

$$\exp\left\{-\frac{1}{2\hbar}\int_0^{\hbar\beta}\mathrm{d}\tau\int\mathrm{d}\mathbf{x}\int\mathrm{d}\mathbf{x}'\phi_\downarrow^*(\mathbf{x},\tau)\phi_\downarrow^*(\mathbf{x}',\tau)V_{\downarrow\downarrow}(\mathbf{x}-\mathbf{x}')\phi_\downarrow(\mathbf{x}',\tau)\phi_\downarrow(\mathbf{x},\tau)\right\}$$
$$\propto\int\mathrm{d}[\Delta^*]\mathrm{d}[\Delta]\exp\left\{-\frac{1}{2\hbar}\int_0^{\hbar\beta}\mathrm{d}\tau\int\mathrm{d}\mathbf{x}\int\mathrm{d}\mathbf{x}'\left(\Delta^*(\mathbf{x},\mathbf{x}',\tau)\phi_\downarrow(\mathbf{x}',\tau)\phi_\downarrow(\mathbf{x},\tau)\right.\right.$$
$$\left.\left.+\,\phi_\downarrow^*(\mathbf{x}',\tau)\phi_\downarrow^*(\mathbf{x},\tau)\Delta(\mathbf{x},\mathbf{x}',\tau)-\Delta^*(\mathbf{x},\mathbf{x}',\tau)V_{\downarrow\downarrow}^{-1}(\mathbf{x}-\mathbf{x}')\Delta(\mathbf{x},\mathbf{x}',\tau)\right)\right\}.$$

The functional integral over the field $\phi_\downarrow(\mathbf{x},\tau)$ has now become quadratic and we write this quadratic part as

$$-\frac{\hbar}{2}\int_0^{\hbar\beta}\mathrm{d}\tau\int\mathrm{d}\mathbf{x}\int_0^{\hbar\beta}\mathrm{d}\tau'\int\mathrm{d}\mathbf{x}'\left[\phi_\downarrow^*(\mathbf{x},\tau),\phi_\downarrow(\mathbf{x},\tau)\right]\cdot\mathbf{G}_{\downarrow\downarrow}^{-1}(\mathbf{x},\tau;\mathbf{x}',\tau')\cdot\begin{bmatrix}\phi_\downarrow(\mathbf{x}',\tau')\\ \phi_\downarrow^*(\mathbf{x}',\tau')\end{bmatrix},$$

where the Nambu-space Green's function for the closed channel obeys the Dyson equation

$$\mathbf{G}_{\downarrow\downarrow}^{-1}(\mathbf{x},\tau;\mathbf{x}',\tau')=\mathbf{G}_{0;\downarrow\downarrow}^{-1}(\mathbf{x},\tau;\mathbf{x}',\tau')-\mathbf{\Sigma}_{\downarrow\downarrow}(\mathbf{x},\tau;\mathbf{x}',\tau')\,. \tag{17.22}$$

The noninteracting Nambu-space Green's function is given by

$$\mathbf{G}_{0;\downarrow\downarrow}^{-1}(\mathbf{x},\tau;\mathbf{x}',\tau')=\begin{bmatrix}G_{0;\downarrow\downarrow}^{-1}(\mathbf{x},\tau;\mathbf{x}',\tau') & 0\\ 0 & G_{0;\downarrow\downarrow}^{-1}(\mathbf{x}',\tau';\mathbf{x},\tau)\end{bmatrix}, \tag{17.23}$$

where

$$\left\{\hbar\frac{\partial}{\partial\tau}-\frac{\hbar^2\nabla^2}{2m}+\frac{\Delta\mu B}{2}-\mu\right\}G_{0;\downarrow\downarrow}(\mathbf{x},\tau;\mathbf{x}',\tau')=-\hbar\delta(\tau-\tau')\delta(\mathbf{x}-\mathbf{x}'), \tag{17.24}$$

is the single-particle noninteracting Green's function. The selfenergy is purely off-diagonal in Nambu space and reads

$$\hbar\mathbf{\Sigma}_{\downarrow\downarrow}(\mathbf{x},\tau;\mathbf{x}',\tau')=\delta(\tau-\tau')\cdot\begin{bmatrix}0 & \sigma(\mathbf{x},\mathbf{x}',\tau)\\ \sigma^*(\mathbf{x},\mathbf{x}',\tau) & 0\end{bmatrix}, \tag{17.25}$$

where $\sigma(\mathbf{x},\mathbf{x}',\tau)\equiv\Delta(\mathbf{x},\mathbf{x}',\tau)+V_{\uparrow\downarrow}(\mathbf{x}-\mathbf{x}')\phi_\uparrow(\mathbf{x},\tau)\phi_\uparrow(\mathbf{x}',\tau)$. Note that a variation of the action with respect to the pairing field shows that

$$\langle\Delta(\mathbf{x},\mathbf{x}',\tau)\rangle=V_{\downarrow\downarrow}(\mathbf{x}-\mathbf{x}')\langle\phi_\downarrow(\mathbf{x})\phi_\downarrow(\mathbf{x}')\rangle\,, \tag{17.26}$$

which relates the auxiliary pairing field to the wavefunction of two atoms in the closed channel. Roughly speaking, to introduce the field that describes a pair of atoms in the closed-channel bound state we have to consider only contributions from this bound state to the pairing field. Close to resonance it is this contribution that dominates. Note that the average of the pairing field in (17.26) indeed shows that the pairing field is similar to the macroscopic wavefunction of the Cooper-pair Bose-Einstein condensate. However, in this case we are interested in the phase $\langle\Delta\rangle = 0$ and therefore need to consider also fluctuations.

Since the integration over the field $\phi_\downarrow(\mathbf{x},\tau)$ involves now a Gaussian integral, it is easily performed. This results in an effective action for the pairing field and the atomic fields that describes the open channel, given by

$$\begin{aligned}
S^{\text{eff}}[\phi_\uparrow^*,\phi_\uparrow,\Delta^*,\Delta] = &\int_0^{\hbar\beta} \mathrm{d}\tau \int \mathrm{d}\mathbf{x} \left\{ \phi_\uparrow^*(\mathbf{x},\tau)\hbar\frac{\partial}{\partial\tau}\phi_\uparrow(\mathbf{x},\tau) + \phi_\uparrow^*(\mathbf{x},\tau)\left\{ -\frac{\hbar^2\nabla^2}{2m} - \mu \right.\right. \\
&\left.\left. +\frac{1}{2}\int \mathrm{d}\mathbf{x}'\; \phi_\uparrow^*(\mathbf{x}',\tau)V_{\uparrow\uparrow}(\mathbf{x}-\mathbf{x}')\phi_\uparrow(\mathbf{x}',\tau) \right\} \phi_\uparrow(\mathbf{x},\tau) \right\} \\
&-\frac{1}{2}\int_0^{\hbar\beta} \mathrm{d}\tau \int \mathrm{d}\mathbf{x} \int \mathrm{d}\mathbf{x}' \Delta^*(\mathbf{x},\mathbf{x}',\tau)V_{\downarrow\downarrow}^{-1}(\mathbf{x}-\mathbf{x}')\Delta(\mathbf{x},\mathbf{x}',\tau) \\
&+\frac{\hbar}{2}\mathrm{Tr}\left[\log(-\mathbf{G}_{\downarrow\downarrow}^{-1})\right] . \qquad (17.27)
\end{aligned}$$

Because we are interested in the bare atom-molecule coupling we expand the effective action up to quadratic order in the field $\Delta(\mathbf{x},\mathbf{x}',\tau)$. Considering higher orders would lead to atom-molecule and molecule-molecule interaction terms that will be neglected here, since in our applications we always deal with a small density of molecules relative to the atomic density.

Hence, we expand the effective action by making use of

$$\mathrm{Tr}[\log(-\mathbf{G}_{\downarrow\downarrow}^{-1})] = \mathrm{Tr}[\log(-\mathbf{G}_{0;\downarrow\downarrow}^{-1})] - \sum_{m=1}^{\infty} \frac{1}{m}\mathrm{Tr}[(\mathbf{G}_{0;\downarrow\downarrow}\mathbf{\Sigma}_{\downarrow\downarrow})^m] . \qquad (17.28)$$

This gives for the part of the effective action that is quadratic in $\Delta^*(\mathbf{x},\mathbf{x}',\tau)$ and $\Delta(\mathbf{x},\mathbf{x}',\tau)$ the following result

$$\begin{aligned}
S[\Delta^*,\Delta] = &-\frac{1}{2}\int_0^{\hbar\beta} \mathrm{d}\tau \int \mathrm{d}\mathbf{x} \int \mathrm{d}\mathbf{x}' \int_0^{\hbar\beta} \mathrm{d}\tau' \int \mathrm{d}\mathbf{y} \int \mathrm{d}\mathbf{y}' \\
&\times \Delta^*(\mathbf{x},\mathbf{x}',\tau)\hbar G_\Delta^{-1}(\mathbf{x},\mathbf{x}',\tau;\mathbf{y},\mathbf{y}',\tau')\Delta(\mathbf{y},\mathbf{y}',\tau') , \qquad (17.29)
\end{aligned}$$

where the Green's function of the pairing field obeys the equation

$$\begin{aligned}
&G_\Delta(\mathbf{x},\mathbf{x}',\tau;\mathbf{y},\mathbf{y}',\tau') = \hbar V_{\downarrow\downarrow}(\mathbf{x}-\mathbf{x}')\delta(\mathbf{x}-\mathbf{y})\delta(\mathbf{x}'-\mathbf{y}')\delta(\tau-\tau') \\
&-\frac{1}{\hbar}\int_0^{\hbar\beta} \mathrm{d}\tau'' \int \mathrm{d}\mathbf{z} \int \mathrm{d}\mathbf{z}' V_{\downarrow\downarrow}(\mathbf{x}-\mathbf{x}')G_{0;\downarrow\downarrow}(\mathbf{x},\tau;\mathbf{z},\tau'')G_{0;\downarrow\downarrow}(\mathbf{x}',\tau;\mathbf{z}',\tau'') \\
&\times G_\Delta(\mathbf{z},\mathbf{z}',\tau'';\mathbf{y},\mathbf{y}',\tau') . \qquad (17.30)
\end{aligned}$$

From this equation we observe that the propagator of the pairing field is related to the many-body T matrix in the closed channel. More precisely, introducing the Fourier transform of the propagator to relative and center-of-mass momenta and Matsubara frequencies $\Omega_n = 2\pi n/\hbar\beta$, denoted by $G_\Delta(\mathbf{k},\mathbf{k}',\mathbf{K},\mathrm{i}\Omega_n)$, we have that

$$G_\Delta(\mathbf{k},\mathbf{k}',\mathbf{K},\mathrm{i}\Omega_n) = \hbar T^{\mathrm{MB}}_{\downarrow\downarrow}(\mathbf{k},\mathbf{k}',\mathbf{K},\mathrm{i}\hbar\Omega_n - \Delta\mu B + 2\mu)\ , \tag{17.31}$$

where the many-body T matrix in the closed channel obeys the equation

$$\begin{aligned} &T^{\mathrm{MB}}_{\downarrow\downarrow}(\mathbf{k},\mathbf{k}',\mathbf{K},z) = V_{\downarrow\downarrow}(\mathbf{k}-\mathbf{k}') \\ &+\frac{1}{V}\sum_{\mathbf{k}''} V_{\downarrow\downarrow}(\mathbf{k}-\mathbf{k}'')\frac{1+N_{\mathrm{BE}}\left(\varepsilon_{\mathbf{K}/2+\mathbf{k}''}-\mu+\frac{\Delta\mu B}{2}\right)+N_{\mathrm{BE}}\left(\varepsilon_{\mathbf{K}/2-\mathbf{k}''}-\mu+\frac{\Delta\mu B}{2}\right)}{z-\varepsilon_{\mathbf{K}/2+\mathbf{k}''}-\varepsilon_{\mathbf{K}/2-\mathbf{k}''}} \\ &\times T^{\mathrm{MB}}_{\downarrow\downarrow}(\mathbf{k}'',\mathbf{k}',\mathbf{K},z)\ . \end{aligned} \tag{17.32}$$

with $N_{\mathrm{BE}}(x) = (\mathrm{e}^{\beta x}-1)^{-1}$ the Bose distribution function. Here,

$$V_{\downarrow\downarrow}(\mathbf{k}) = \int \mathrm{d}\mathbf{x}\ V_{\downarrow\downarrow}(\mathbf{x})\mathrm{e}^{-\mathrm{i}\mathbf{k}\cdot\mathbf{x}} \tag{17.33}$$

denotes the Fourier transform of the atomic interaction potential. This equation describes the scattering of a pair of atoms from relative momentum $\mathbf{k}'$ to relative momentum $\mathbf{k}$ at energy z. Due to the fact that the scattering takes places in a medium the many-body T matrix also depends on the center-of-mass momentum $\mathbf{K}$, contrary to the two-body T matrix introduced in the previous section, which describes scattering in vacuum. The kinetic energy of a single atom is equal to $\varepsilon_{\mathbf{k}} = \hbar^2\mathbf{k}^2/2m$. The factor that involves the Bose-Einstein distribution function arises because the probability of a process where a boson scatters into a state that is already occupied by N_1 bosons is proportional to $1+N_1$. The reverse process is only proportional to N_1. This explains the factor

$$1+N_1+N_2 = (1+N_1)(1+N_2) - N_1N_2\ , \tag{17.34}$$

in the equation for the many-body T matrix [187].

The many-body T matrix is discussed in more detail in the next section when we calculate the renormalization of the interatomic interactions. For now we only need to realize that, for the conditions of interest to us, we are always in the situation where we are allowed to neglect the many-body effects in (17.32) because the Zeeman energy $\Delta\mu B/2$ strongly suppresses the Bose occupation numbers for atoms in the closed channel. This is certainly true for the experimental applications of interest because in the current experiments with magnetically-trapped ultracold gases the Zeeman splitting of the magnetic trap is much larger than the thermal energy. This reduces the many-body T-matrix equation to the Lippmann-Schwinger equation for the two-body T matrix in the closed channel $T^{\mathrm{2B}}_{\downarrow\downarrow}(\mathbf{k},\mathbf{k}',z-\varepsilon_{\mathbf{K}}/2)$, which, in its basis-independent operator formulation, reads

$$\hat{T}^{\mathrm{2B}}_{\downarrow\downarrow}(z) = \hat{V}_{\downarrow\downarrow} + \hat{V}_{\downarrow\downarrow}\frac{1}{z-\hat{H}_0}\hat{T}^{\mathrm{2B}}_{\downarrow\downarrow}(z)\ , \tag{17.35}$$

with $\hat{H}_0 = \hat{p}^2/m$. As we have seen in Sect. 10.3.1, this equation is formally solved by

$$\hat{T}^{2B}_{\downarrow\downarrow}(z) = \hat{V}_{\downarrow\downarrow} + \hat{V}_{\downarrow\downarrow}\frac{1}{z-\hat{H}_{\downarrow\downarrow}}\hat{V}_{\downarrow\downarrow} , \tag{17.36}$$

with $\hat{H}_{\downarrow\downarrow} = \hat{H}_0 + \hat{V}_{\downarrow\downarrow}$, where we also remember that the two-body T matrix has poles at the bound states of the closed-channel potential. We assume that we are close to resonance and hence that one of these bound states dominates. Therefore, we approximate the two-body T matrix by

$$\hat{T}^{2B}_{\downarrow\downarrow}(z) \simeq \hat{V}_{\downarrow\downarrow}\frac{|\chi_{\mathrm{m}}\rangle\langle\chi_{\mathrm{m}}|}{z-E_{\mathrm{m}}}\hat{V}_{\downarrow\downarrow} , \tag{17.37}$$

where the properly normalized and symmetrized bound-state wavefunction $\chi_{\mathrm{m}}(\mathbf{x}) \equiv \langle\mathbf{x}|\chi_{\mathrm{m}}\rangle$ obeys the Schrödinger equation

$$\left\{-\frac{\hbar^2\nabla^2}{m} + V_{\downarrow\downarrow}(\mathbf{x})\right\}\chi_{\mathrm{m}}(\mathbf{x}) = E_{\mathrm{m}}\chi_{\mathrm{m}}(\mathbf{x}) . \tag{17.38}$$

It should be noted that this wavefunction does not correspond to the dressed, or true, molecular state which is an eigenstate of the coupled-channel Hamiltonian and determined by (17.9). Rather, it corresponds to the bare molecular wave function. The coupling $V_{\uparrow\downarrow}(\mathbf{x}-\mathbf{x}')$ of this bare state with the continuum renormalizes it such that it contains also a component in the open channel. Moreover, as we have already seen in the previous section, this coupling also affects the energy of this bound state. Both effects are important near the resonance and are discussed in detail later on.

We are now in the position to derive the quadratic action for the quantum field that describes the bare molecule. To do this, we consider first the case that the exchange interaction $\hat{V}_{\uparrow\downarrow}(\mathbf{x}-\mathbf{x}')$ is absent. Within the above approximations, the two-point function for the pairing field is given by

$$\langle\Delta(\mathbf{k},\mathbf{K},\mathrm{i}\Omega_n)\Delta^*(\mathbf{k}',\mathbf{K},\mathrm{i}\Omega_n)\rangle = -2\hbar\frac{\langle\mathbf{k}|\hat{V}_{\downarrow\downarrow}|\chi_{\mathrm{m}}\rangle\langle\chi_{\mathrm{m}}|\hat{V}_{\downarrow\downarrow}|\mathbf{k}'\rangle}{\mathrm{i}\hbar\Omega_n - \varepsilon_{\mathbf{K}}/2 - E_{\mathrm{m}} - \Delta\mu B + 2\mu} . \tag{17.39}$$

We introduce the field $\phi_{\mathrm{m}}(\mathbf{x},\tau)$, that describes the bound state in the closed channel, i.e, the bare molecule, by considering configurations of the pairing field such that

$$\Delta(\mathbf{x},\mathbf{x}',\tau) = \sqrt{2}V_{\downarrow\downarrow}(\mathbf{x}-\mathbf{x}')\chi_{\mathrm{m}}(\mathbf{x}-\mathbf{x}')\phi_{\mathrm{m}}((\mathbf{x}+\mathbf{x}')/2,\tau) . \tag{17.40}$$

Using this we have that

$$\langle\phi_{\mathrm{m}}(\mathbf{K},\mathrm{i}\Omega_n)\phi^*_{\mathrm{m}}(\mathbf{K},\mathrm{i}\Omega_n)\rangle = \frac{\hbar}{-\mathrm{i}\hbar\Omega_n + \varepsilon_{\mathbf{K}}/2 + E_{\mathrm{m}} + \Delta\mu B - 2\mu} , \tag{17.41}$$

which shows that the quadratic action for the bare molecular field is, in position representation, given by

$$S[\phi_{\mathrm{m}}^*,\phi_{\mathrm{m}}] = \int_0^{\hbar\beta} \mathrm{d}\tau \int \mathrm{d}\mathbf{x}\ \phi_{\mathrm{m}}^*(\mathbf{x},\tau) \times \left\{ \hbar\frac{\partial}{\partial\tau} - \frac{\hbar^2\nabla^2}{4m} + E_{\mathrm{m}} + \Delta\mu B - 2\mu \right\} \phi_{\mathrm{m}}(\mathbf{x},\tau) \ . \tag{17.42}$$

In the absence of the coupling of the bare molecular field to the atoms, the dispersion relation of the bare molecules is given by

$$\hbar\omega_{\mathbf{k}}(B) = \varepsilon_{\mathbf{k}}/2 + E_{\mathrm{m}} + \Delta\mu B \ . \tag{17.43}$$

As expected, the binding energy of the bare molecule is equal to

$$\varepsilon_{\mathrm{m}}(B) = E_{\mathrm{m}} + \Delta\mu B. \tag{17.44}$$

The momentum dependence of the dispersion is due to the kinetic energy of the molecule.

To derive the coupling of this bare molecular field to the field $\phi_\uparrow(\mathbf{x},\tau)$ it is convenient to start from the effective action in (17.27) and to consider again only terms that are quadratic in the selfenergy. Integrating out the pairing fields leads to an interaction term in the action for the field describing the open channel, given by

$$\frac{1}{2}\int_0^{\hbar\beta} \mathrm{d}\tau \int \mathrm{d}\mathbf{x} \int \mathrm{d}\mathbf{x}' \int_0^{\hbar\beta} \mathrm{d}\tau' \int \mathrm{d}\mathbf{y} \int \mathrm{d}\mathbf{y}' V_{\uparrow\downarrow}(\mathbf{x}-\mathbf{x}')\phi_\uparrow^*(\mathbf{x},\tau)\phi_\uparrow^*(\mathbf{x}',\tau) \times G^{(4)}_{\downarrow\downarrow}(\mathbf{x},\mathbf{x}',\tau;\mathbf{y},\mathbf{y}',\tau')V_{\uparrow\downarrow}(\mathbf{y}-\mathbf{y}')\phi_\uparrow(\mathbf{y},\tau')\phi_\uparrow(\mathbf{y}',\tau'),$$

where the two-atom four-point Green's function is given diagrammatically in Fig. 17.7. For our purposes it is, for the same reasons as before, sufficient to neglect the many-body effects on this propagator and to consider again only the contribution that arises from the bound state in the closed channel. This gives for the Fourier transform of this Green's function

$$G^{(4)}_{\downarrow\downarrow}(\mathbf{k},\mathbf{k}',\mathbf{K},\mathrm{i}\Omega_n) \simeq \frac{\chi_{\mathrm{m}}^*(\mathbf{k})\chi_{\mathrm{m}}(\mathbf{k}')}{\mathrm{i}\hbar\Omega_n - \varepsilon_{\mathbf{K}}/2 - \Delta\mu B - E_{\mathrm{m}} + 2\mu} \ , \tag{17.45}$$

where $\chi_{\mathrm{m}}(\mathbf{k})$ is the Fourier transform of the bound-state wavefunction. After substitution of this result into the above integral, the resulting interaction term is decoupled by introducing the field $\phi_{\mathrm{m}}(\mathbf{x},\tau)$ with the quadratic action given in (17.42). This procedure automatically shows that the bare atom-molecule coupling constant is equal to $V_{\uparrow\downarrow}(\mathbf{k})\chi_{\mathrm{m}}(\mathbf{k})/\sqrt{2}$.

Summarizing, we have thus derived from a microscopic atomic Hamiltonian, a bare atom-molecule theory for the description of a Feshbach resonance. It is determined by the action

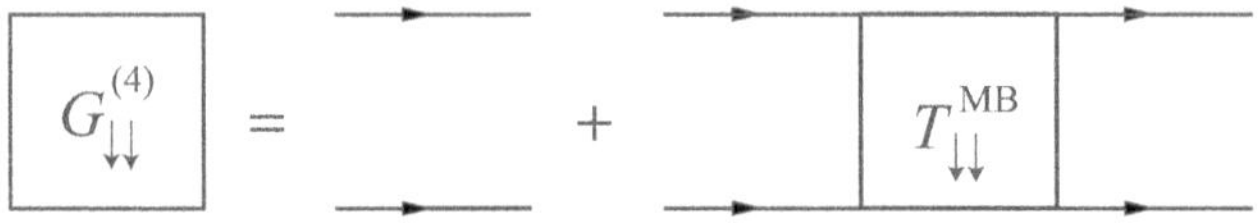

Fig. 17.7 Diagrammatic representation of the two-particle Green's function in the closed channel. The solid lines correspond to single-atom propagators.

$$
\begin{aligned}
&S[\phi_\uparrow^*, \phi_\uparrow, \phi_\mathrm{m}^*, \phi_\mathrm{m}] = \\
&\int_0^{\hbar\beta} \mathrm{d}\tau \left\{ \int \mathrm{d}\mathbf{x} \left\{ \phi_\uparrow^*(\mathbf{x},\tau)\hbar\frac{\partial}{\partial\tau}\phi_\uparrow(\mathbf{x},\tau) + \phi_\mathrm{m}^*(\mathbf{x},\tau)\hbar\frac{\partial}{\partial\tau}\phi_\mathrm{m}(\mathbf{x},\tau) \right\} \right. \\
&\left. + H[\phi_\uparrow^*, \phi_\uparrow, \phi_\mathrm{m}, \phi_\mathrm{m}^*] \right\} ,
\end{aligned}
\tag{17.46}
$$

where the bare or microscopic atom-molecule Hamiltonian functional is given by

$$
\begin{aligned}
&H[\phi_\uparrow^*, \phi_\uparrow, \phi_\mathrm{m}, \phi_\mathrm{m}^*] = \\
&\int \mathrm{d}\mathbf{x}\, \phi_\uparrow^*(\mathbf{x},\tau) \left\{ -\frac{\hbar^2\boldsymbol{\nabla}^2}{2m} - \mu + \frac{1}{2}\int \mathrm{d}\mathbf{x}' \phi_\uparrow^*(\mathbf{x}',\tau) V_{\uparrow\uparrow}(\mathbf{x}-\mathbf{x}')\phi_\uparrow(\mathbf{x}',\tau) \right\} \phi_\uparrow(\mathbf{x},\tau) \\
&+ \int \mathrm{d}\mathbf{x}\, \phi_\mathrm{m}^*(\mathbf{x},\tau) \left\{ -\frac{\hbar^2\boldsymbol{\nabla}^2}{4m} + \Delta\mu B + E_\mathrm{m} - 2\mu \right\} \phi_\mathrm{m}(\mathbf{x},\tau) \\
&+ \int \mathrm{d}\mathbf{x} \int \mathrm{d}\mathbf{x}' g_{\uparrow\downarrow}(\mathbf{x}-\mathbf{x}')\phi_\mathrm{m}^*((\mathbf{x}+\mathbf{x}')/2,\tau)\phi_\uparrow(\mathbf{x}',\tau)\phi_\uparrow(\mathbf{x},\tau) \\
&+ \int \mathrm{d}\mathbf{x} \int \mathrm{d}\mathbf{x}' g_{\uparrow\downarrow}(\mathbf{x}-\mathbf{x}')\phi_\uparrow^*(\mathbf{x},\tau)\phi_\uparrow^*(\mathbf{x}',\tau)\phi_\mathrm{m}((\mathbf{x}+\mathbf{x}')/2,\tau) ,
\end{aligned}
\tag{17.47}
$$

and the bare atom-molecule coupling is given by $g_{\uparrow\downarrow}(\mathbf{x}) = V_{\uparrow\downarrow}(\mathbf{x})\chi_\mathrm{m}(\mathbf{x})/\sqrt{2}$, where $V_{\uparrow\downarrow}(\mathbf{x})$ is the coupling between the open and closed atomic collision channel of the Feshbach problem, that has its origin in the exchange interaction of the atoms. Note also that the atom-molecule coupling is proportional to the wavefunction $\chi_\mathrm{m}(\mathbf{x})$ for the bound molecular state in the closed channel responsible for the Feshbach resonance.

Physically, the microscopic Hamiltonian in (17.47) describes bosonic atoms in the open channel of the Feshbach problem in terms of the field $\phi_\uparrow(\mathbf{x},\tau)$. These atoms interact via the interaction potential $V_{\uparrow\uparrow}(\mathbf{x}-\mathbf{x}')$. Apart from this background interaction, two atoms in the gas can also form a molecular bound state in the closed channel with energy E_m that is detuned by an amount of $\Delta\mu B$ from the open channel. This bare molecular state is described by the field $\phi_\mathrm{m}(\mathbf{x},\tau)$. The most important input in the derivation of (17.47) is that the energy difference between the various bound states in the closed channel is much larger than the thermal energy, so that near resonance only one molecular level is of importance. This condition is very well satisfied for almost all the atomic gases of interest.

In an application of the above microscopic atom-molecule action to realistic atomic gases we have to do perturbation theory in the interaction $V_{\uparrow\uparrow}(\mathbf{x}-\mathbf{x}')$ and the coupling $g_{\uparrow\downarrow}(\mathbf{x}-\mathbf{x}')$. Since the interatomic interaction is strong, this perturbation theory requires an infinite number of terms. Progress is made by realizing that the atomic and molecular densities of interest are so low that we only need to include two-atom processes. This is achieved by summing all ladder diagrams as explained in detail in the next section.

17.3 Ladder Summations

From the bare or microscopic atom-molecule theory derived in the previous section we now intend to derive an effective quantum field theory that contains the two-atom physics exactly. This is most conveniently achieved by renormalization of the coupling constants. Moreover, the molecules acquire a selfenergy. Both calculations are done within the framework of perturbation theory to bring out the physics involved most clearly. It is, however, also possible to achieve the same goal in a nonperturbative manner by a second Hubbard-Stratonovich transformation.

Because we are dealing with a homogeneous system, it is convenient to perform the perturbation theory in momentum space. Therefore, we Fourier transform to momentum space, and expand the atomic and molecular fields according to

$$\phi_{\mathrm{a}}(\mathbf{x},\tau) \equiv \phi_{\uparrow}(\mathbf{x},\tau) = \frac{1}{(\hbar\beta V)^{1/2}} \sum_{\mathbf{k},n} a_{\mathbf{k},n} \mathrm{e}^{\mathrm{i}\mathbf{k}\cdot\mathbf{x}-\mathrm{i}\omega_n\tau} \,, \tag{17.48}$$

and

$$\phi_{\mathrm{m}}(\mathbf{x},\tau) = \frac{1}{(\hbar\beta V)^{1/2}} \sum_{\mathbf{k},n} b_{\mathbf{k},n} \mathrm{e}^{\mathrm{i}\mathbf{k}\cdot\mathbf{x}-\mathrm{i}\omega_n\tau} \,, \tag{17.49}$$

respectively. The even Matsubara frequencies $\omega_n = 2\pi n/\hbar\beta$ account for the periodicity of the fields on the imaginary-time axis. With this expansion, the grand-canonical partition function of the gas is written as a functional integral over the coefficients $a_{\mathbf{k},n}$ and $b_{\mathbf{k},n}$ and their complex conjugates. It is given by

$$Z = \int \mathrm{d}[a^*]\mathrm{d}[a]\mathrm{d}[b^*]\mathrm{d}[b] \exp\left\{-\frac{1}{\hbar}S[a^*,a,b^*,b]\right\} \,, \tag{17.50}$$

where the action $S[a^*,a,b^*,b]$ is the sum of four terms. The first two terms describe noninteracting atoms and noninteracting bare molecules respectively, and are given by

$$S_{\mathrm{a}}[a^*,a] = \sum_{\mathbf{k},n} (-\mathrm{i}\hbar\omega_n + \varepsilon_{\mathbf{k}} - \mu)\, a^*_{\mathbf{k},n} a_{\mathbf{k},n} \,, \tag{17.51}$$

and

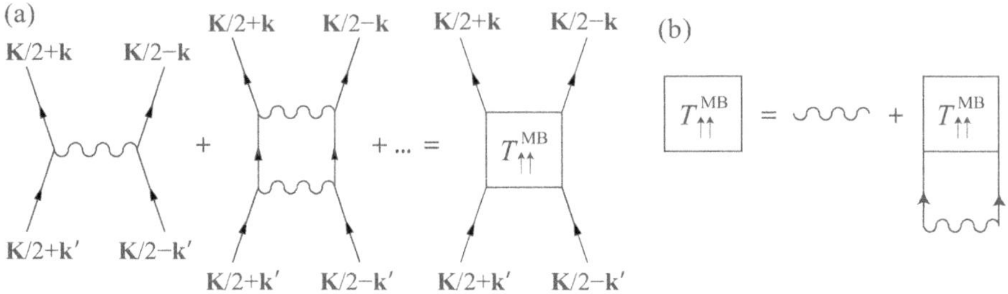

Fig. 17.8 (a) Ladder diagrams that contribute to the renormalization of the interatomic interaction. (b) Diagrammatic representation of the Lippmann-Schwinger equation for the many-body T matrix. The solid lines correspond to single-atom propagators. The wiggly lines correspond to the interatomic interaction $V_{\uparrow\uparrow}$.

$$S_{\mathrm{m}}[b^*,b] = \sum_{\mathbf{k},n}(-i\hbar\omega_n + \varepsilon_{\mathbf{k}}/2 + E_{\mathrm{m}} + \Delta\mu B - 2\mu)\, b^*_{\mathbf{k},n} b_{\mathbf{k},n} \ . \tag{17.52}$$

The atomic interactions are described by the action

$$S_{\mathrm{int}}[a^*,a] = \frac{1}{2}\frac{1}{\hbar\beta V}\sum_{\substack{\mathbf{K},\mathbf{k},\mathbf{k}'\\ n,m,m'}} V_{\uparrow\uparrow}(\mathbf{k}-\mathbf{k}')\, a^*_{\mathbf{K}/2+\mathbf{k},n/2+m} a^*_{\mathbf{K}/2-\mathbf{k},n/2-m} \times a_{\mathbf{K}/2+\mathbf{k}',n/2+m'} a_{\mathbf{K}/2-\mathbf{k}',n/2-m'} \ , \tag{17.53}$$

where $V_{\uparrow\uparrow}(\mathbf{k})$ is the Fourier transform of the interatomic interaction potential. This Fourier transform vanishes for large momenta due to the nonzero range of the interatomic interaction potential. The last term in the action describes the process of two atoms forming a molecule and vice versa, and is given by

$$S_{\mathrm{coup}}[a^*,a,b^*,b] = \frac{1}{(\hbar\beta V)^{1/2}}\sum_{\substack{\mathbf{K},\mathbf{k}\\ n,m}} g_{\uparrow\downarrow}(\mathbf{k})\left\{ b^*_{\mathbf{K},n} a_{\mathbf{K}/2+\mathbf{k},n/2+m} a_{\mathbf{K}/2-\mathbf{k},n/2-m} + a^*_{\mathbf{K}/2-\mathbf{k},n/2-m} a^*_{\mathbf{K}/2+\mathbf{k},n/2+m} b_{\mathbf{K},n} \right\} , \tag{17.54}$$

where $g_{\uparrow\downarrow}(\mathbf{k})$ is the Fourier transform of the bare atom-molecule coupling constant. This coupling constant also vanishes for large momenta since the bare molecular wavefunction has a nonzero extent.

We first discuss the renormalization of the microscopic atomic interaction $V_{\uparrow\uparrow}(\mathbf{k})$, due to nonresonant background collisions between the atoms. The first term that contributes to this renormalization is of second order in the interaction. It is found by expanding the exponential in the functional-integral expression for the grand-canonical partition function in (17.50). To second order in the interactions this leads to

$$Z = \int \mathrm{d}[a^*]\mathrm{d}[a] \left(1 - \frac{1}{\hbar}S_{\mathrm{int}}[a^*,a] + \frac{1}{2\hbar^2}S^2_{\mathrm{int}}[a^*,a] + \cdots\right)$$
$$\times \exp\left\{-\frac{1}{\hbar}S_{\mathrm{a}}[a^*,a]\right\}. \tag{17.55}$$

After the decoupling of the eight-point function resulting from the square of the action $S_{\mathrm{int}}[a^*,a]$ with the use of Wick's theorem, it gives rise to various terms in the perturbation theory which can be depicted by Feynman diagrams. As mentioned already, we only take into account the ladder Feynman diagram. This diagram is given by the second term of the Born series depicted in Fig. 17.8a, and corresponds to the expression

$$-\frac{1}{\hbar\beta V}\sum_{\mathbf{k}'',m} V_{\uparrow\uparrow}(\mathbf{k}-\mathbf{k}'')G_{0;\mathrm{a}}\left(\mathbf{K}/2+\mathbf{k}'',\mathrm{i}\omega_{n/2+m}\right)$$
$$\times G_{0;\mathrm{a}}\left(\mathbf{K}/2-\mathbf{k}'',\mathrm{i}\omega_{n/2-m}\right)V_{\uparrow\uparrow}(\mathbf{k}''-\mathbf{k}')\ ,$$

where

$$G_{0;\mathrm{a}}(\mathbf{k},\mathrm{i}\omega_n) = \frac{-\hbar}{-\mathrm{i}\hbar\omega_n + \varepsilon_{\mathbf{k}} - \mu}\ , \tag{17.56}$$

is the noninteracting propagator of the atoms. After performing the summation over the Matsubara frequencies we find that, to second order, the renormalization of the interatomic interactions is given by

$$V_{\uparrow\uparrow}(\mathbf{k}-\mathbf{k}') \to V_{\uparrow\uparrow}(\mathbf{k}-\mathbf{k}') \tag{17.57}$$
$$+\frac{1}{V}\sum_{\mathbf{k}''} V_{\uparrow\uparrow}(\mathbf{k}-\mathbf{k}'')\frac{1+N_{\mathrm{BE}}\left(\varepsilon_{\mathbf{K}/2+\mathbf{k}''}-\mu\right)+N_{\mathrm{BE}}\left(\varepsilon_{\mathbf{K}/2-\mathbf{k}''}-\mu\right)}{\mathrm{i}\hbar\omega_n - \varepsilon_{\mathbf{K}/2+\mathbf{k}''} - \varepsilon_{\mathbf{K}/2-\mathbf{k}''} + 2\mu}V_{\uparrow\uparrow}(\mathbf{k}''-\mathbf{k}')\ ,$$

which is finite due to the use of the true interatomic potential. In comparing this result with the first two terms of the Born series for scattering in vacuum in (10.22), we see that the only difference between the two-body result and the above result is the factor involving the Bose distributions. Called a statistical factor, this accounts for the fact that the scattering takes place in a medium and is understood as follows. The amplitude for a process where an atom scatters from a state with occupation number N_1 to a state with occupation number N_2 contains a factor $N_1(1+N_2)$. The factor N_1 simply accounts for the number of atoms that can undergo the collision, and may be understood from a classical viewpoint as well. However, the additional factor $(1+N_2)$ is a result of the Bose statistics of the atoms and is therefore called the Bose-enhancement factor. For fermions this factor would correspond to the Pauli-blocking factor $(1-N_2)$, reflecting the fact that a fermion is not allowed to scatter into a state that is already occupied by an identical fermion. In calculating the Feynman diagram we have to take into account the forward and backward scattering processes, which results in the statistical factor in (17.57).

Continuing the expansion in (17.55) and taking into account only the ladder diagrams leads to a geometric series, which is summed by introducing the many-body

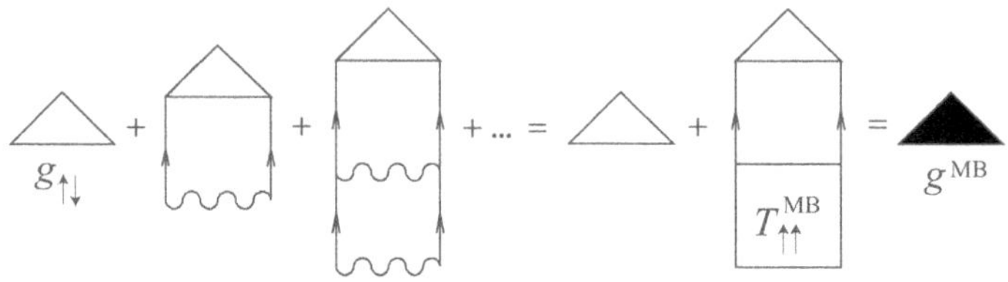

Fig. 17.9 Renormalization of the atom-molecule coupling constant by interatomic interactions. The solid lines correspond to single-atom propagators. The wiggly lines corresponds to the interatomic interaction $V_{\uparrow\uparrow}$.

T matrix in the open channel. It is given by

$$T^{\rm MB}_{\uparrow\uparrow}(\mathbf{k},\mathbf{k}',\mathbf{K},z) = V_{\uparrow\uparrow}(\mathbf{k}-\mathbf{k}') \tag{17.58}$$
$$+\frac{1}{V}\sum_{\mathbf{k}''} V_{\uparrow\uparrow}(\mathbf{k}-\mathbf{k}'')\frac{1+N_{\rm BE}\left(\varepsilon_{\mathbf{K}/2+\mathbf{k}''}-\mu\right)+N_{\rm BE}\left(\varepsilon_{\mathbf{K}/2-\mathbf{k}''}-\mu\right)}{z-\varepsilon_{\mathbf{K}/2+\mathbf{k}''}-\varepsilon_{\mathbf{K}/2-\mathbf{k}''}} T^{\rm MB}_{\uparrow\uparrow}(\mathbf{k}'',\mathbf{k}',\mathbf{K},z)\ .$$

Its diagrammatic representation is given in Fig. 17.8b. For the moment we neglect the many-body effects on the scattering atoms and put the Bose-distribution functions equal to zero. As we have discussed in the single-channel case, this assumption is valid at temperatures far below the critical temperature [110]. This reduces the many-body T matrix to the two-body T matrix $T^{\rm 2B}_{\uparrow\uparrow}(\mathbf{k},\mathbf{k}',z-\varepsilon_{\mathbf{K}}/2)$. For the low temperatures of interest to us here, we are allowed to take the external momenta equal to zero. For small energies we find, using the result in (10.38), that the effective interaction between the atoms reduces to

$$T^{\rm 2B}_{\uparrow\uparrow}(\mathbf{0},\mathbf{0},{\rm i}\hbar\omega_n-\varepsilon_{\mathbf{K}}/2+2\mu) = \frac{4\pi a_{\rm bg}\hbar^2}{m}$$
$$\times\frac{1}{1-a_{\rm bg}\sqrt{\frac{-m({\rm i}\hbar\omega_n-\varepsilon_{\mathbf{K}}/2+2\mu)}{\hbar^2}}-\frac{a_{\rm bg}r_{\rm bg}m({\rm i}\hbar\omega_n-\varepsilon_{\mathbf{K}}/2+2\mu)}{2\hbar^2}}\ . \tag{17.59}$$

Here, $a_{\rm bg}$ and $r_{\rm bg}$ are the scattering length and the effective range of the open-channel potential $V_{\uparrow\uparrow}(\mathbf{x})$ respectively. Although these could in principle be calculated with the precise knowledge of this potential, it is much easier to take them from experiment. For example, the magnitude of the scattering length can be determined by thermalization-rate measurements [41]. The effective range is determined by comparing the result of calculations with experimental data. We will encounter an explicit example of this in Sect. 17.7.

The next step is the renormalization of the microscopic atom-molecule coupling constant. Using the same perturbative techniques as before, we find that the effective atom-molecule coupling is given in terms of the bare coupling by

$$g^{\mathrm{MB}}(\mathbf{k},\mathbf{K},z) = g_{\uparrow\downarrow}(\mathbf{k}) + \frac{1}{V}\sum_{\mathbf{k}'} T^{\mathrm{MB}}_{\uparrow\uparrow}(\mathbf{k},\mathbf{k}',\mathbf{K},z) \times \frac{1+N_{\mathrm{BE}}(\varepsilon_{\mathbf{K}/2+\mathbf{k}'}-\mu)+N_{\mathrm{BE}}(\varepsilon_{\mathbf{K}/2-\mathbf{k}'}-\mu)}{z-\varepsilon_{\mathbf{K}/2+\mathbf{k}'}-\varepsilon_{\mathbf{K}/2-\mathbf{k}'}} g_{\uparrow\downarrow}(\mathbf{k}') , \tag{17.60}$$

and is illustrated diagrammatically in Fig. 17.9. Again neglecting many-body effects, the coupling constant becomes $g^{2\mathrm{B}}(\mathbf{k},z-\varepsilon_{\mathbf{K}}/2)$ with

$$g^{2\mathrm{B}}(\mathbf{k},z) = g_{\uparrow\downarrow}(\mathbf{k}) + \frac{1}{V}\sum_{\mathbf{k}'} T^{2\mathrm{B}}_{\uparrow\uparrow}(\mathbf{k},\mathbf{k}',z)\frac{1}{z-2\varepsilon_{\mathbf{k}'}} g_{\uparrow\downarrow}(\mathbf{k}') . \tag{17.61}$$

From the above equation we infer that the energy dependence of this coupling constant is the same as that of the two-body T matrix. This result is easily understood by noting that for a contact potential $V_{\uparrow\uparrow}(\mathbf{k}) = V_{\mathbf{0}}$ and we simply have that $g^{2\mathrm{B}} = g_{\uparrow\downarrow}T^{2\mathrm{B}}_{\uparrow\uparrow}/V_{\mathbf{0}}$. Hence we have for the effective atom-molecule coupling

$$g^{2\mathrm{B}}(\mathbf{0},\mathrm{i}\hbar\omega_n-\varepsilon_{\mathbf{K}}/2+2\mu) = \frac{g}{1-a_{\mathrm{bg}}\sqrt{\frac{-m(\mathrm{i}\hbar\omega_n-\varepsilon_{\mathbf{K}}/2+2\mu)}{\hbar^2}}-\frac{a_{\mathrm{bg}}r_{\mathrm{bg}}m(\mathrm{i}\hbar\omega_n-\varepsilon_{\mathbf{K}}/2+2\mu)}{2\hbar^2}} . \tag{17.62}$$

where g is the effective atom-molecule coupling constant at zero energy. The latter is also taken from experiment. We come back to this point in Sect. 17.4.1 when we discuss the two-atom properties of our effective many-body theory.

Finally, we have to take into account also the ladder diagrams of the resonant part of the interaction. This is achieved by including the selfenergy of the molecules. It is first given by the expression

$$\Pi^{\mathrm{MB}}(\mathbf{K},z) = \frac{2}{V}\sum_{\mathbf{k}} g_{\uparrow\downarrow}(\mathbf{k})\frac{1+N_{\mathrm{BE}}(\varepsilon_{\mathbf{K}/2+\mathbf{k}}-\mu)+N_{\mathrm{BE}}(\varepsilon_{\mathbf{K}/2-\mathbf{k}}-\mu)}{z-\varepsilon_{\mathbf{K}/2+\mathbf{k}}-\varepsilon_{\mathbf{K}/2-\mathbf{k}}} \times g^{\mathrm{MB}}(\mathbf{k},\mathbf{K},z) , \tag{17.63}$$

and shown diagrammatically in Fig. 17.10. We neglect again many-body effects which reduces the selfenergy in (17.63) to $\Pi^{2\mathrm{B}}(z-\varepsilon_{\mathbf{K}}/2)$ with

$$\Pi^{2\mathrm{B}}(z) = \langle\chi_{\mathrm{m}}|\hat{V}_{\uparrow\downarrow}\hat{G}_{\uparrow\uparrow}(z)\hat{V}_{\uparrow\downarrow}|\chi_{\mathrm{m}}\rangle , \tag{17.64}$$

where the propagator $\hat{G}_{\uparrow\uparrow}(z)$ is given by

$$\hat{G}_{\uparrow\uparrow}(z) = \frac{1}{z-\hat{H}_{\uparrow\uparrow}} , \tag{17.65}$$

with the Hamiltonian

$$\hat{H}_{\uparrow\uparrow} = \frac{\hat{\mathbf{p}}^2}{m} + \hat{V}_{\uparrow\uparrow} \equiv \hat{H}_0 + \hat{V}_{\uparrow\uparrow} . \tag{17.66}$$

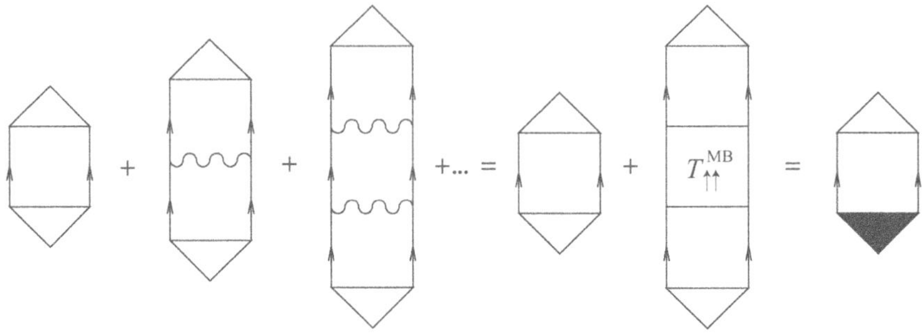

Fig. 17.10 Molecular selfenergy. The solid lines correspond to single-atom propagators. The wiggly lines corresponds to the interatomic interaction $V_{\uparrow\uparrow}$.

We insert in (17.64) a complete set of bound states $|\psi_\kappa\rangle$ with energies E_κ and scattering states $|\psi_{\mathbf{k}}^{(+)}\rangle$ that obey the Schrödinger equation for the Hamiltonian in (17.66). This reduces the selfenergy to

$$\Pi^{2\mathrm{B}}(z) = \sum_\kappa \frac{|\langle\chi_{\mathrm{m}}|\hat{V}_{\uparrow\downarrow}|\psi_\kappa\rangle|^2}{z-E_\kappa} + \int \frac{d\mathbf{k}}{(2\pi)^3} \frac{|\langle\chi_{\mathrm{m}}|\hat{V}_{\uparrow\downarrow}|\psi_{\mathbf{k}}^{(+)}\rangle|^2}{z-2\varepsilon_{\mathbf{k}}}, \qquad (17.67)$$

where we replaced the sum over the momenta $\mathbf{k}$ by an integral. Using (17.61) and the equation for the scattering states we have that

$$g^{2\mathrm{B}}(\mathbf{k}, 2\varepsilon_{\mathbf{k}}^+) = \frac{1}{\sqrt{2}}\langle\chi_{\mathrm{m}}|\hat{V}_{\uparrow\downarrow}|\psi_{\mathbf{k}}^{(+)}\rangle\ . \qquad (17.68)$$

Neglecting the energy dependence due to the contribution of the bound states since their binding energies are usually large compared to the thermal energy, we have, using the result for the atom-molecule coupling constant in (17.62), the intermediate result

$$\Pi^{2\mathrm{B}}(z) = 2\int \frac{d\mathbf{k}}{(2\pi)^3} \left|g^{2\mathrm{B}}(\mathbf{0}, 2\varepsilon_{\mathbf{k}}^+)\right|^2 \frac{1}{z-2\varepsilon_{\mathbf{k}}}\ . \qquad (17.69)$$

The remaining momentum integral yields the final result

$$\hbar\Sigma_{\rm m}^{\rm 2B}(z) \equiv \Pi^{\rm 2B}(z) - \Pi^{\rm 2B}(0) \equiv \Pi^{\rm 2B}(z) + (\Delta\mu B_0 + E_{\rm m}) \tag{17.70}$$
$$= -\frac{g^2 m}{4\pi^2\hbar^2}\left\{ -2\pi\sqrt{a_{\rm bg}-2r_{\rm bg}}\sqrt{\frac{-mz}{\hbar^2}} \right.$$
$$+\mathrm{i}\sqrt{a_{\rm bg}}\left\{\log\left(-\frac{\mathrm{i}\sqrt{a_{\rm bg}}r_{\rm bg}}{\sqrt{a_{\rm bg}-2r_{\rm bg}}}\right) - \log\left(\frac{\mathrm{i}\sqrt{a_{\rm bg}}r_{\rm bg}}{\sqrt{a_{\rm bg}-2r_{\rm bg}}}\right)\right\}$$
$$\left.\times\frac{mz}{\hbar^2}\left\{3r_{\rm bg}-2a_{\rm bg}-\frac{a_{\rm bg}r_{\rm bg}^2 mz}{2\hbar^2}\right\}\right\}$$
$$\times\left\{\sqrt{a_{\rm bg}-2r_{\rm bg}}\left\{1+a_{\rm bg}\left(a_{\rm bg}-r_{\rm bg}\right)\frac{mz}{\hbar^2}+\left(\frac{a_{\rm bg}r_{\rm bg}mz}{2\hbar^2}\right)^2\right\}\right\}^{-1},$$

where we have denoted the energy-independent shift $\Pi^{\rm 2B}(0)$ in such a manner that the position of the resonance in the magnetic field is precisely at the experimentally observed magnetic-field value B_0. This shift is also shown in the results of the calculation of the bound-state energy of the coupled square wells in Fig. 17.4.

17.4 Effective Atom-Molecule theory

Putting the results from the previous section together, we find that the atom-molecule system is described by the effective action

$$S^{\rm eff}[a^*,a,b^*,b] = \sum_{\mathbf{k},n}(-\mathrm{i}\hbar\omega_n + \varepsilon_{\mathbf{k}} - \mu)\,a_{\mathbf{k},n}^* a_{\mathbf{k},n} \tag{17.71}$$
$$+\sum_{\mathbf{k},n}\left\{-\mathrm{i}\hbar\omega_n + \varepsilon_{\mathbf{k}}/2 + \delta(B) - 2\mu + \hbar\Sigma_{\rm m}^{\rm 2B}(\mathrm{i}\hbar\omega_n - \varepsilon_{\mathbf{k}}/2 + 2\mu)\right\} b_{\mathbf{k},n}^* b_{\mathbf{k},n}$$
$$+\frac{1}{2}\frac{1}{\hbar\beta V}\sum_{\substack{\mathbf{K},\mathbf{k},\mathbf{k}'\\ n,m,m'}} T_{\rm bg}^{\rm 2B}(\mathrm{i}\hbar\omega_n - \varepsilon_{\mathbf{K}}/2 + 2\mu)$$
$$\times a_{\mathbf{K}/2+\mathbf{k},n/2+m}^* a_{\mathbf{K}/2-\mathbf{k},n/2-m}^* a_{\mathbf{K}/2+\mathbf{k}',n/2+m'} a_{\mathbf{K}/2-\mathbf{k}',n/2-m'}$$
$$+\frac{1}{(\hbar\beta V)^{1/2}}\sum_{\substack{\mathbf{K},\mathbf{k}\\ n,m}} g^{\rm 2B}(\mathrm{i}\hbar\omega_n - \varepsilon_{\mathbf{K}}/2 + 2\mu)\, b_{\mathbf{K},n}^* a_{\mathbf{K}/2+\mathbf{k},n/2+m} a_{\mathbf{K}/2-\mathbf{k},n/2-m}$$
$$+\frac{1}{(\hbar\beta V)^{1/2}}\sum_{\substack{\mathbf{K},\mathbf{k}\\ n,m}} g^{\rm 2B\,*}(\mathrm{i}\hbar\omega_n - \varepsilon_{\mathbf{K}}/2 + 2\mu)\, a_{\mathbf{K}/2-\mathbf{k},n/2-m}^* a_{\mathbf{K}/2+\mathbf{k},n/2+m}^* b_{\mathbf{K},n},$$

where $\delta(B) \equiv \Delta\mu(B - B_0)$ is the so-called detuning. From now on we use the notation $T_{\rm bg}^{\rm 2B}(z) \equiv T_{\uparrow\uparrow}^{\rm 2B}(\mathbf{0},\mathbf{0},z)$, and $g^{\rm 2B}(z) \equiv g^{\rm 2B}(\mathbf{0},z)$. Since these coupling constants are the result of summing all ladder diagrams, these diagrams should not be taken into account again. We next show that our effective field theory correctly contains the two-atom physics of a Feshbach resonance. First, we show that the correct Feshbach-resonant atomic scattering length is obtained after the elimination of the molecular field. Second, we calculate the bound-state energy and show that it has the correct threshold behavior near the resonance. We also investigate the molecular density of states.

17.4.1 Scattering Properties

To calculate the effective interatomic scattering length, we have to eliminate the molecular field from the action in (17.71). Since the scattering length is related to the scattering amplitude at zero energy and zero momentum, we are allowed to put $i\hbar\omega_n - \varepsilon_{\mathbf{K}}/2 + 2\mu$ equal to zero when performing the Gaussian integral over the molecular field. This then leads to the semiclassical result

$$\phi_{\rm m}(\mathbf{x},t) = -\frac{g}{\delta(B)}\phi_{\rm a}^2(\mathbf{x},t). \tag{17.72}$$

Substitution of this result into the action for the atomic field leads for the interaction terms to

$$\begin{aligned}&\frac{4\pi a_{\rm bg}\hbar^2}{m}\phi_{\rm a}^*(\mathbf{x},t)\phi_{\rm a}^*(\mathbf{x},t)\phi_{\rm a}(\mathbf{x},t)\phi_{\rm a}(\mathbf{x},t) + 2g\phi_{\rm a}^*(\mathbf{x},t)\phi_{\rm a}^*(\mathbf{x},t)\phi_{\rm m}(\mathbf{x},t) = \\ &\left(\frac{4\pi a_{\rm bg}\hbar^2}{m} - \frac{2g^2}{\delta(B)}\right)\phi_{\rm a}^*(\mathbf{x},t)\phi_{\rm a}^*(\mathbf{x},t)\phi_{\rm a}(\mathbf{x},t)\phi_{\rm a}(\mathbf{x},t)\,.\end{aligned} \tag{17.73}$$

From this result we observe that we have to take the renormalized atom-molecule coupling constant at zero energy equal to $g = \hbar\sqrt{2\pi a_{\rm bg}\Delta B\Delta\mu/m}$, so that we have

$$\frac{4\pi a_{\rm bg}\hbar^2}{m} - \frac{2g^2}{\delta(B)} = \frac{4\pi a(B)\hbar^2}{m}, \tag{17.74}$$

where we recall that the scattering length near a Feshbach resonance is given by

$$a(B) = a_{\rm bg}\left(1 - \frac{\Delta B}{B - B_0}\right) \equiv a_{\rm bg} + a_{\rm res}(B). \tag{17.75}$$

Since both the width ΔB and the background scattering length $a_{\rm bg}$ are known experimentally, the knowledge of the difference in magnetic moment between the open and the closed channel $\Delta\mu$ completely determines the renormalized coupling constant g. Since the open and the closed channel usually correspond to the triplet and singlet potential respectively, we often have that $|\Delta\mu| \simeq 2\mu_{\rm B}$, with $\mu_{\rm B}$ the Bohr magneton. More precise values of the difference in magnetic moments are obtained from coupled-channels calculations using the interatomic interaction potentials [17, 188, 189, 190].

From the above analysis we see that the correct Feshbach-resonant scattering length of the atoms is contained in our theory exactly. Next, we show that our effective theory also contains the correct bound-state energy.

17.4.2 Bound-State Energy

The energy of the molecular state is determined by the poles of the molecular propagator $G_{\mathrm{m}}(\mathbf{k},\omega)$. It is given by

$$G_{\mathrm{m}}(\mathbf{k},\omega) = \frac{\hbar}{\hbar\omega - \varepsilon_{\mathbf{k}}/2 - \delta(B) - \hbar\Sigma_{\mathrm{m}}(\hbar\omega - \varepsilon_{\mathbf{k}}/2)} \ . \tag{17.76}$$

For positive detuning $\delta(B)$ there only exists a pole with a nonzero and negative imaginary part. This is in agreement with the fact that the molecule decays when its energy is above the two-atom continuum threshold. The imaginary part of the energy is related to the lifetime of the molecular state. For negative detuning the molecular propagator has a real and negative pole corresponding to the bound-state energy. More precisely, in this case the poles of the molecular propagator are given by $\hbar\omega = \varepsilon_{\mathrm{m}}(B) + \varepsilon_{\mathbf{k}}/2$, where the bound-state energy is determined by solving for E in the equation

$$E - \delta(B) - \hbar\Sigma_{\mathrm{m}}(E) = 0. \tag{17.77}$$

In general this equation cannot be solved analytically, but it is easily solved numerically, and in Sect. 17.7 we discuss its numerical solution for the parameters of ^{85}Rb. Close to resonance, however, we are allowed to neglect the effective range of the interactions. This reduces the selfenergy of the molecules to

$$\hbar\Sigma_{\mathrm{m}}(E) \simeq -\frac{g^2 m^{3/2}}{2\pi\hbar^3} \frac{\mathrm{i}\sqrt{E}}{1 - \mathrm{i}|a_{\mathrm{bg}}|\sqrt{\frac{mE}{\hbar^2}}} \ . \tag{17.78}$$

Moreover, the bound-state energy is small in this regime and we are allowed to neglect the linear terms in the energy with respect to the square-root terms. This reduces the equation for the bound-state energy in equation (17.77) to

$$\frac{g^2 m^{3/2}}{2\pi\hbar^3} \frac{\mathrm{i}\sqrt{E}}{1 - \mathrm{i}|a_{\mathrm{bg}}|\sqrt{\frac{mE}{\hbar^2}}} = \delta(B) \ . \tag{17.79}$$

This equation is easily solved analytically, and yields the result

$$\varepsilon_{\mathrm{m}}(B) = -\frac{\hbar^2}{m[a(B)]^2} \ , \tag{17.80}$$

which analytically proves the numerical result in (17.18). This numerical result was obtained for the specific case of two coupled attractive square wells. The above analytic proof, which does not depend on the details of the potential, shows that the result is general.

The same result is found by noting that after the elimination of the molecular field the effective on-shell T matrix for the atoms in the open channel is given by

$$T^{2B}(E^+) = T^{2B}_{bg}(E^+) + \frac{2}{\hbar}|g^{2B}(E^+)|^2 G_m\left(\sqrt{mE/\hbar^2}, E\right) . \tag{17.81}$$

Close to resonance this expression reduces to

$$T^{2B}(E) \simeq \frac{4\pi a_{res}(B)\hbar^2}{m} \frac{1}{1 + ia_{res}(B)\sqrt{\frac{mE}{\hbar^2}}} . \tag{17.82}$$

The pole of this T matrix, which gives the bound-state energy, is indeed equal to the result in (17.80) close to resonance.

17.4.3 Molecular Density of States

The molecular density of states is obtained by taking the imaginary part of the molecular propagator, i.e.

$$\rho_m(\mathbf{k}, \omega) = -\frac{1}{\pi\hbar} \mathrm{Im}\left[G_m(\mathbf{k}, \omega^+)\right] . \tag{17.83}$$

For simplicity, we discuss here only the situation that we are close to resonance, and therefore approximate the molecular selfenergy by the square-root term resulting from Wigner's threshold law as given by $\hbar\Sigma_m(E) \simeq -ig^2 m^{3/2}\sqrt{E}/2\pi\hbar^3$. The extension to situations further off resonance are straightforward.

For the case of negative detuning, the molecular density of states is shown by the solid line in Fig. 17.11 and has two contributions. One arises from the pole at the bound-state energy, and the other from the two-atom continuum. Within the above approximation, it is given by

$$\begin{aligned}\rho_m(\mathbf{k}, \omega) &= Z(B)\delta(\hbar\omega - \varepsilon_{\mathbf{k}}/2 - \varepsilon_m(B)) \\ &+ \frac{1}{\pi}\theta(\hbar\omega - \varepsilon_{\mathbf{k}}/2) \frac{(g^2 m^{3/2}/2\pi\hbar^3)\sqrt{\hbar\omega - \varepsilon_{\mathbf{k}}/2}}{(\hbar\omega - \varepsilon_{\mathbf{k}}/2 - \delta(B))^2 + (g^4 m^3/4\pi^2\hbar^6)(\hbar\omega - \varepsilon_{\mathbf{k}}/2)} ,\end{aligned} \tag{17.84}$$

with $Z(B)$ the wavefunction renormalization factor

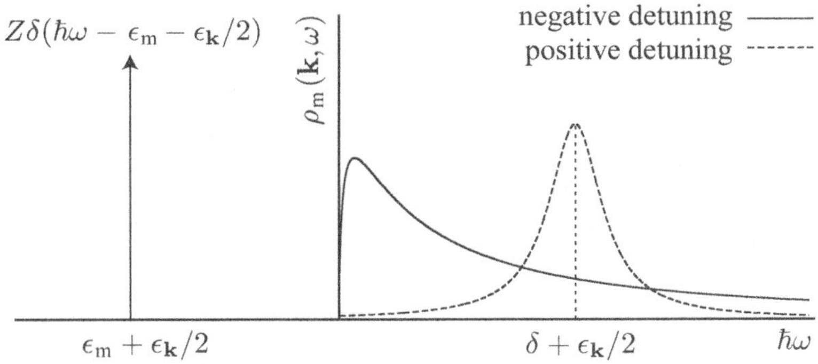

Fig. 17.11 Molecular density of states. The solid line shows the density of states for negative detuning. Since there is a true bound state in this case there is a pole in the density of states. For positive detuning the density of states is approximately a Lorentzian as shown by the dashed line.

Here, the second-quantized operator $\hat{b}_{\mathbf{0}}^{\dagger}$ creates a molecule with zero momentum. It acts on the vacuum state $|0\rangle$. The bare molecular state is therefore given by $|\chi_{\mathrm{m}}\rangle = \hat{b}_{\mathbf{0}}^{\dagger}|0\rangle$. The operator $\hat{a}_{\mathbf{k}}^{\dagger}$ creates an atom with momentum $\hbar\mathbf{k}$ and hence the part of the dressed molecule wavefunction in the open channel of the Feshbach problem is given by $\mathrm{e}^{-r/a}/\sqrt{2\pi a}r$.

For positive detuning the molecular density of states has only a contribution for positive energy. For large detuning it is in first approximation given by

$$\rho_{\mathrm{m}}(\mathbf{k},\omega) = \frac{1}{2\pi}\frac{\hbar\Gamma_{\mathrm{m}}(B)}{(\hbar\omega - \varepsilon_{\mathbf{k}}/2 - \delta(B))^2 + (\hbar\Gamma_{\mathrm{m}}(B)/2)^2}\,, \tag{17.88}$$

where the lifetime of the molecular state is defined by

$$\Gamma_{\mathrm{m}}(B) = \frac{g^2 m^{3/2}}{\pi\hbar^4}\sqrt{\delta(B)}\,. \tag{17.89}$$

As expected, the density of states is, in the case of positive detuning, approximately a Lorentzian centered around the detuning with a width related to the lifetime of the molecule. It is shown in Fig. 17.11 by the dashed line.

17.5 Bogoliubov Theory for the Bose-Einstein Condensed Phase

In this section we derive the mean-field equations for the atomic and molecular Bose-Einstein condensate wavefunctions. The mean-field equations for the atomic and molecular condensate wavefunctions are derived most easily by varying the effective action in (17.71) with respect to $a_{\mathbf{k},n}^{*}$ and $b_{\mathbf{k},n}^{*}$ respectively. Before doing so, however, we remark that an important property of this effective action is its invariance under global $U(1)$ transformations. Any transformation of the form

$$a_{\mathbf{k},n} \to a_{\mathbf{k},n}\mathrm{e}^{\mathrm{i}\theta}\,, \text{ and } b_{\mathbf{k},n} \to b_{\mathbf{k},n}\mathrm{e}^{2\mathrm{i}\theta}\,, \tag{17.90}$$

with θ a real parameter, leaves the action unchanged. The conserved quantity, the Noether charge, associated with this invariance is the total number of atoms. The appearance of the atomic and the molecular condensates breaks the $U(1)$ invariance since the wavefunctions of these condensates have a certain phase. According to Goldstone's theorem, an exact property of a system with a broken continuous symmetry is that its excitation spectrum is gapless [191]. Since our mean-field theory is derived by varying a $U(1)$-invariant action, this property is automatically incorporated in the mean-field theory as we will see shortly.

To derive the time-independent mean-field equations that describe the equilibrium values of the atomic and molecular Bose-Einstein condensate wavefunctions, we substitute into the effective action $a_{\mathbf{0},0} \to \phi_{\mathrm{a}}\sqrt{\beta\hbar V} + a_{\mathbf{0},0}$ and $b_{\mathbf{0},0} \to \phi_{\mathrm{m}}\sqrt{\beta\hbar V} + b_{\mathbf{0},0}$. Here, ϕ_{a} and ϕ_{m} correspond to the atomic and molecular condensate wavefunctions respectively. Requiring that the terms linear in $a_{\mathbf{0},0}$ and $b_{\mathbf{0},0}$

vanish from the effective action leads to the equations

$$\mu\phi_{\rm a} = T^{\rm 2B}_{\rm bg}\left(2\mu - 2\hbar\Sigma^{\rm HF}\right)|\phi_{\rm a}|^2\phi_{\rm a} + 2\left(g^{\rm 2B}\left(2\mu - 2\hbar\Sigma^{\rm HF}\right)\right)^*\phi^*_{\rm a}\phi_{\rm m}\,,$$
$$2\mu\phi_{\rm m} = \left(\delta(B) + \hbar\Sigma^{\rm 2B}_{\rm m}\left(2\mu - 2\hbar\Sigma^{\rm HF}\right)\right)\phi_{\rm m} + g^{\rm 2B}\left(2\mu - 2\hbar\Sigma^{\rm HF}\right)\phi^2_{\rm a}\,. \quad (17.91)$$

A crucial ingredient in these equations is the Hartree-Fock selfenergy of the noncondensed atoms. This selfenergy is the mean-field energy felt by the noncondensed atoms due to the presence of the atomic Bose-Einstein condensate. Taking into account the energy-dependence of the interactions, it is determined by the expression

$$\hbar\Sigma^{\rm HF} = \frac{4n_{\rm a}\left|g^{\rm 2B}\left(\mu - \hbar\Sigma^{\rm HF}\right)\right|^2}{\hbar\Sigma^{\rm HF} + \mu - \delta(B) - \hbar\Sigma^{\rm 2B}_{\rm m}\left(\mu - \hbar\Sigma^{\rm HF}\right)} + 2n_{\rm a}T^{\rm 2B}_{\rm bg}\left(\mu - \hbar\Sigma^{\rm HF}\right), \quad (17.92)$$

with $n_{\rm a} = |\phi_{\rm a}|^2$ the density of the atomic Bose-Einstein condensate. Its diagrammatic representation is given in Fig. 17.12. The overall factor of two comes from the constructive interference of the direct and exchange contributions. Far off resonance we are allowed to neglect the energy-dependence of the effective atom-atom interactions, and the Hartree-Fock selfenergy of the atoms is given by $8\pi a(B)\hbar^2 n_{\rm a}/m$, as expected. The Hartree-Fock selfenergy is essential for a correct description of the equilibrium properties of the system. The physical reason for this is understood as follows. In the Bose-Einstein condensed phase the chemical potential is positive. The energy of a condensate molecule is equal to 2μ, which is therefore larger than the continuum threshold of two atoms in vacuum. Without the incorporation of the Hartree-Fock selfenergy, the molecular Bose-Einstein condensate would therefore always decay and an equilibrium solution of the mean-field equations would not exist. However, due to the presence of the atomic Bose-Einstein condensate, the continuum threshold shifts by an amount $2\hbar\Sigma^{\rm HF}$ and the molecular Bose-Einstein condensate is stable.

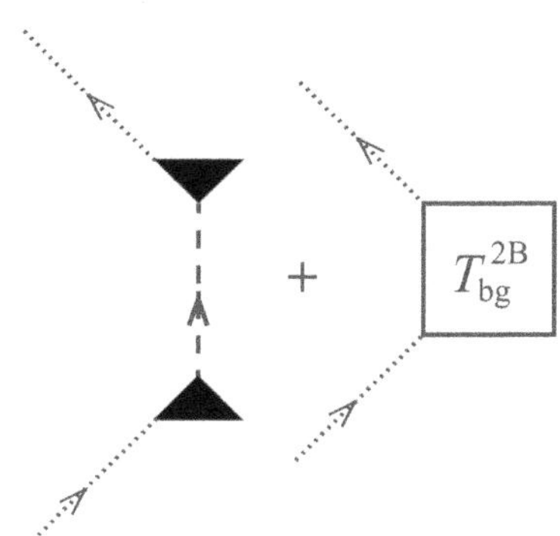

Fig. 17.12 Hartree-Fock selfenergy of the atoms. The dotted lines correspond to condensate atoms. The dashed line corresponds to the full molecular propagator.

To study the collective excitation spectrum over the ground state determined by (17.91), we consider the effective action up to second order in the fluctuations, which is known as the Bogoliubov approximation [48]. To facilitate the notation we introduce the vector $\mathbf{u}_{\mathbf{k},n}$ by means of

$$\mathbf{u}_{\mathbf{k},n} \equiv \begin{bmatrix} a_{\mathbf{k},n} \\ a^*_{-\mathbf{k},-n} \\ b_{\mathbf{k},n} \\ b^*_{-\mathbf{k},-n} \end{bmatrix} . \tag{17.93}$$

With this definition, the quadratic part of the effective action is given by

$$S_{\mathrm{B}}[\mathbf{u}^\dagger, \mathbf{u}] = -\frac{\hbar}{2} \sum_{\mathbf{k},n} \mathbf{u}^\dagger_{\mathbf{k},n} \cdot \mathbf{G}_{\mathrm{B}}^{-1}(\mathbf{k}, i\omega_n) \cdot \mathbf{u}_{\mathbf{k},n} , \tag{17.94}$$

where the Green's function of the fluctuations is determined by

$$\mathbf{G}_{\mathrm{B}}^{-1} = \begin{bmatrix} \mathbf{G}_{\mathrm{a}}^{-1} & \mathbf{G}_{\mathrm{coup}}^{-1} \\ \mathbf{G}_{\mathrm{coup}}^{-1\,*} & \mathbf{G}_{\mathrm{m}}^{-1} \end{bmatrix} . \tag{17.95}$$

The atomic part of this Green's function is found from

$$\begin{aligned} \mathbf{G}_{\mathrm{a}}^{-1}(\mathbf{k}, \mathrm{i}\omega_n) = & \begin{bmatrix} G_{0,\mathrm{a}}^{-1}(\mathbf{k}, \mathrm{i}\omega_n) & 0 \\ 0 & G_{0,\mathrm{a}}^{-1}(\mathbf{k}, -\mathrm{i}\omega_n) \end{bmatrix} \\ & -\frac{1}{\hbar} \begin{bmatrix} 2T^{2\mathrm{B}}_{\mathrm{bg}}(\mathrm{i}\hbar\omega_n - \varepsilon_{\mathbf{k}}/2 + 2\mu')\, n_{\mathrm{a}} & T^{2\mathrm{B}}_{\mathrm{bg}}(2\mu')\, \phi_{\mathrm{a}}^2 + 2g^{2\mathrm{B}}(2\mu')^* \phi_{\mathrm{m}} \\ T^{2\mathrm{B}}_{\mathrm{bg}}(2\mu')\, \phi_{\mathrm{a}}^{*2} + 2g^{2\mathrm{B}}(2\mu')\phi_{\mathrm{m}}^* & 2T^{2\mathrm{B}}_{\mathrm{bg}}(\mathrm{i}\hbar\omega_n - \varepsilon_{\mathbf{k}}/2 + 2\mu')\, n_{\mathrm{a}} \end{bmatrix} \end{aligned} \tag{17.96}$$

where $\mu' \equiv \mu - \hbar\Sigma^{\mathrm{HF}}$. Note that in the absence of the coupling to the molecular condensate, this result reduces to the well-known result for the Green's function that describes phonon propagation in a weakly-interacting Bose-Einstein condensate. We have in this case, however, also explicitly taken into account the energy dependence of the coupling constants. Therefore we know that in the limit of vanishing coupling $g^{2\mathrm{B}}$ the propagator in (17.96) has a pole that determines the gapless dispersion relation for the phonons. For energy-independent interactions this Bogoliubov dispersion is given by

$$\hbar\omega_{\mathbf{k}} = \sqrt{\varepsilon_{\mathbf{k}}^2 + \frac{8\pi a_{\mathrm{bg}}\hbar^2 n_{\mathrm{a}}}{m} \varepsilon_{\mathbf{k}}} . \tag{17.97}$$

The molecular part of the Green's function $\mathbf{G}_{\mathrm{B}}(\mathbf{k}, \mathrm{i}\omega_n)$ is determined by

$$\mathbf{G}_{\mathrm{m}}^{-1}(\mathbf{k}, i\omega_n) = \begin{bmatrix} G_{\mathrm{m}}^{-1}(\mathbf{k}, \mathrm{i}\omega_n) & 0 \\ 0 & G_{\mathrm{m}}^{-1}(\mathbf{k}, -\mathrm{i}\omega_n) \end{bmatrix} , \tag{17.98}$$

where the single-molecule propagator is given by

$$\begin{aligned} -\hbar G_{\mathrm{m}}^{-1}(\mathbf{k}, \mathrm{i}\omega_n) = & -\mathrm{i}\hbar\omega_n + \varepsilon_{\mathbf{k}}/2 + \delta(B) - 2\mu \\ & + \hbar\Sigma_{\mathrm{m}}^{2\mathrm{B}}\left(\mathrm{i}\hbar\omega_n - \varepsilon_{\mathbf{k}}/2 + 2\mu - 2\hbar\Sigma^{\mathrm{HF}}\right) . \end{aligned} \tag{17.99}$$

From the previous section we know that the Green's function in (17.99) for negative detuning has a pole at the molecular binding energy. There are now, however, mean-field effects on this binding energy due to the presence of the atomic condensate, incorporated by the Hartree-Fock selfenergy $\hbar\Sigma^{\mathrm{HF}}$ [192]. Finally, the Green's function that describes the coupling between the atomic and molecular fluctuations is given by

$$-\hbar\mathbf{G}_{\mathrm{coup}}^{-1}(\mathbf{k},\mathrm{i}\omega_n) = \begin{bmatrix} 2g^{2\mathrm{B}}\,(\mathrm{i}\hbar\omega_n-\varepsilon_{\mathbf{k}}/2+2\mu')^*\,\phi_{\mathrm{a}}^* & 0 \\ 0 & 2g^{2\mathrm{B}}\,(\mathrm{i}\hbar\omega_n-\varepsilon_{\mathbf{k}}/2+2\mu')\,\phi_{\mathrm{a}} \end{bmatrix} \quad (17.100)$$

The spectrum of the collective excitations is determined by the poles of the retarded Green's function for the fluctuations $\mathbf{G}_{\mathrm{B}}(\mathbf{k},\omega)$. This implies that we have to solve for $\hbar\omega$ in the equation

$$\mathrm{Det}\left[\mathbf{G}_{\mathrm{B}}^{-1}(\mathbf{k},\omega)\right] = 0\,. \quad (17.101)$$

This is achieved numerically in the next section to determine the frequency of the Josephson oscillations between the atomic and the molecular Bose-Einstein condensates. However, we are already able to infer some general features of the excitation spectrum of the collective modes. We have seen that in the absence of the coupling between the atomic and molecular condensate, we have that one dispersion is equal to the gapless Bogoliubov dispersion with scattering length a_{bg}. In the presence of the coupling this branch corresponds again to phonons, but the dispersion is now approximately equal to the Bogoliubov dispersion for the full scattering length $a(B)$. There is a second dispersion branch that for small coupling $g^{2\mathrm{B}}$ lies close to the molecular binding energy. At nonzero coupling this branch corresponds to coherent atom-molecule oscillations, i.e. pairs of atoms oscillating back and forth between the atomic and molecular condensate. Physically, the difference between the two branches is understood by realizing that for the phonon modes the phases of the atomic and the molecular condensate are locked to each other and oscillate in phase. Since the action is invariant under the transformations in (17.90) we conclude that the phonons are indeed gapless and, in fact, correspond to the Goldstone mode associated with the breaking of the $U(1)$ symmetry by the condensates. For the coherent atom-molecule oscillations the phases of the atomic and molecular condensate oscillate out of phase and hence the associated dispersion is gapped. As a final remark we note that we indeed have

$$\mathrm{Det}\left[\mathbf{G}_{\mathrm{B}}^{-1}(\mathbf{0},0)\right] = 0\,, \quad (17.102)$$

which shows that there is indeed a gapless excitation, in agreement with Goldstone's theorem.

17.6 Experiments

To compare the constructed many-body theory for an atomic Bose gas near a Feshbach resonance with some beautiful experimental results, we are going to look in more detail at two experiments performed in Wieman's group at JILA, namely the one by Donley et al. [193] and by Claussen et al. [194]. The Feshbach resonance of interest is located at $B_0 = 155.041(18)$ G(auss) in the $|f = 2; m_f = -2\rangle$ hyperfine state of ^{85}Rb. The width of this resonance is equal to $\Delta B = 11.0(4)$ G and the off-resonant background scattering length is given by $a_{bg} = -443a_0$, with a_0 the Bohr radius. The difference in the magnetic moment between the open channel and the closed channel is given by $\Delta\mu = -2.23\mu_B$, with μ_B the Bohr magneton [189]. In both experiments, one starts from a stable and essentially pure condensate of about $N_c = 1 \times 10^4$ atoms at a magnetic field such that the effective scattering length is close to zero. This implies that, since the condensate is in the noninteracting limit, its density profile is determined by the harmonic-oscillator ground-state wavefunction. The harmonic external trapping potential is axially symmetric, with trapping frequencies $\nu_r = 17.4$ Hz and $\nu_z = 6.8$ Hz in the radial and axial direction respectively.

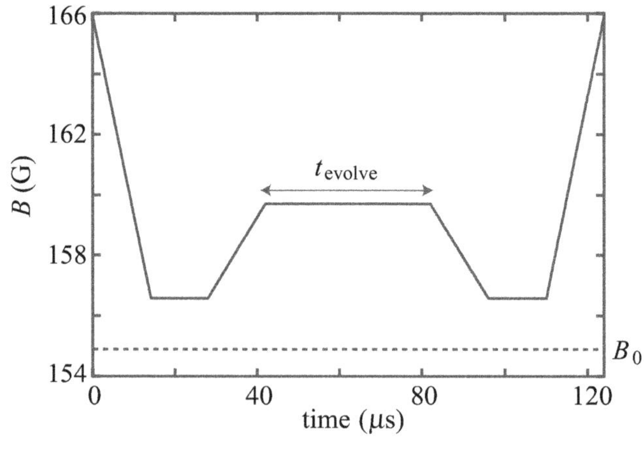

Fig. 17.13 Typical magnetic-field pulse sequence as used in the experiments of Donley et al. [193] and Claussen et al. [194].

Starting from this situation, one quickly ramps the magnetic field to a value B_{hold} close to the resonant value and keeps it there for a short time t_{hold} before ramping to a value B_{evolve}. The magnetic field is kept at this last value for a time t_{evolve} before performing a similar pulse to go back to the initial situation. The duration of all four magnetic-field ramps is given by t_{ramp}. A typical pulse is illustrated in Fig. 17.13. Both the ramp time t_{ramp} and the hold time t_{hold} are kept fixed at values of 10–15 μs. The time t_{evolve} between the pulses is variable. Such a double-pulse experiment is generally called a Ramsey experiment. Its significance is most easily understood from a simple system of two coupled harmonic oscillators. Consider therefore the Hamiltonian

$$\hat{H} = \frac{1}{2}\left[\hat{a}^\dagger \; \hat{b}^\dagger\right] \cdot \begin{bmatrix} \delta(t) & \Omega \\ \Omega & -\delta(t) \end{bmatrix} \cdot \begin{bmatrix} \hat{a} \\ \hat{b} \end{bmatrix} , \tag{17.103}$$

where $\hat{a}^\dagger$ and $\hat{b}^\dagger$ create a quantum in the oscillators a and b respectively, and Ω denotes the coupling between the two oscillators.

We consider first the situation that the detuning $\delta(t)$ is time independent. The exact solution is found easily by diagonalizing the Hamiltonian. We assume that initially there are only quanta in oscillator a and none in b, so that we have that $\langle \hat{b}^\dagger \hat{b} \rangle(0) = 0$. The number of quanta in oscillator a as a function of time is then given by

$$\langle \hat{a}^\dagger \hat{a} \rangle(t) = \left(1 - \frac{\Omega^2}{(\hbar\omega_\mathrm{R})^2} \sin^2(\omega_\mathrm{R} t/2)\right) \langle \hat{a}^\dagger \hat{a} \rangle(0) , \tag{17.104}$$

with the frequency ω_R given by

$$\hbar\omega_\mathrm{R} = \sqrt{\delta^2 + \Omega^2} . \tag{17.105}$$

We see that the number of quanta in the oscillator a oscillates in time with frequency ω_R. Such oscillations are called Rabi oscillations. Note that the number of quanta in oscillator b is determined by

$$\langle \hat{b}^\dagger \hat{b} \rangle(t) = -\frac{\Omega^2}{(\hbar\omega_\mathrm{R})^2} \sin^2(\omega_\mathrm{R} t/2) \langle \hat{a}^\dagger \hat{a} \rangle(0) , \tag{17.106}$$

so that the total number of quanta is indeed conserved.

Suppose now that we start from the situation with all quanta in the oscillator a and none in b and that the detuning is such that $\delta(t) \gg \Omega$. Then we have from (17.104) that $\langle \hat{a}^\dagger \hat{a} \rangle(t) \simeq \langle \hat{a}^\dagger \hat{a} \rangle(0)$ and $\langle \hat{b}^\dagger \hat{b} \rangle(t) \simeq 0$. Starting from this situation, we change the detuning instantaneously to a value $\delta(t) \simeq 0$ and keep it at this value for a time t_hold. During this hold time quanta in oscillator a will go to oscillator b. Moreover, if t_hold is such that

$$t_\mathrm{hold} \simeq \frac{\pi}{2}\frac{\hbar}{\Omega} , \tag{17.107}$$

on average half of the quanta in oscillator a will go to oscillator b. Such a pulse is called a $\pi/2$ pulse. The defining property of a $\pi/2$ pulse is that it creates a superposition of the oscillators a and b, such that the probabilities to be in oscillators a and b are equal, and therefore equal to $1/2$. This is indicated by the average $\langle \hat{a}^\dagger \hat{b} \rangle(t)$. At $t = 0$ this average is equal to zero because there is no superposition at that time. We can show that after the above $\pi/2$ pulse the average $\langle \hat{a}^\dagger \hat{b} \rangle(t)$ reaches its maximum value. In detail, the state after the $\pi/2$ pulse is equal to

$$\frac{1}{\sqrt{N!}}\left(\frac{\hat{a}^\dagger+\hat{b}^\dagger}{\sqrt{2}}\right)^N|0\rangle\,, \tag{17.108}$$

where the ground state is denoted by $|0\rangle$, and $N=\langle\hat{a}^\dagger\hat{a}\rangle(0)$.

We can now imagine the following experiment. Starting from the situation $\delta(t)\gg\Omega$, we perform a $\pi/2$ pulse. Then jump to a certain value δ_{evolve} for a time t_{evolve}, and after this perform another $\pi/2$ pulse and jump back to the initial situation. The number of quanta in the oscillator a, a measurable quantity, then oscillates as a function of t_{evolve} with the oscillation frequency determined by (17.105) evaluated at the detuning δ_{evolve}. The second $\pi/2$ pulse enhances the contrast of the measurement thus providing a method of measuring the frequency ω_{R} as a function of the detuning with high precision.

This is essentially the idea of the Ramsey experiments performed by Donley et al. [193] and Claussen et al. [194]. Roughly speaking, the atomic condensate corresponds to oscillator a and the molecular condensate to oscillator b. Therefore, after performing the double-pulse sequence in the magnetic field one makes a light-absorption image of the atomic density from which one extracts the number of condensed and noncondensed atoms. Since this imaging technique is sensitive to a specific absorption line of the atoms it does not measure the number of molecules. From the above discussion we expect to observe oscillations in the number of condensate atoms. In the context of particle-number oscillations between Bose-Einstein condensates, Rabi oscillations are referred to as Josephson oscillations and the associated frequency is called the Josephson frequency. Moreover, if the situation is such that the detuning between the pulses is relatively large, the effect of the coupling can be neglected and the frequency of the observed oscillations corresponds to the energy difference between the atoms and the molecules, i.e. the molecular binding energy. This is indeed what is observed, thereby providing compelling evidence for the existence of coherence between atoms and molecules.

In Fig. 17.14, the experimental results of Claussen et al. [194] are presented. Fig. 17.14a and b show the number of atoms in the atomic Bose-Einstein condensate as a function of t_{evolve} after a double-pulse sequence. Clearly, there is an oscillation in the number of atoms in both cases. In Fig. 17.14a the magnetic field between the pulses is $B_{\text{evolve}}=156.840(25)$ G. In Fig. 17.14b, we have $B_{\text{evolve}}=159.527(19)$ G which is further from resonance. This explains also the increase in frequency from Fig. 17.14a to b, since further from resonance the molecular binding energy is larger.

What is also observed is that there is a damping of the oscillations and an overall loss of condensate atoms. Experimentally, the number of atoms in the condensate is fit to the formula

$$N_{\text{c}}(t)=N_{\text{average}}-\alpha t+A\exp(-\beta t)\sin(\omega_{\text{e}}t+\phi)\,, \tag{17.109}$$

where N_{average} is the average number of condensate atoms, A and ϕ are the oscillation amplitude and phase, respectively, and β is the damping rate of the oscillations. The overall atom loss is characterized by a rate constant α. The experimentally observed frequency is equal to $\omega_{\text{e}}=2\pi\sqrt{\nu_{\text{e}}^2-[\beta/2\pi]^2}$. By defining the

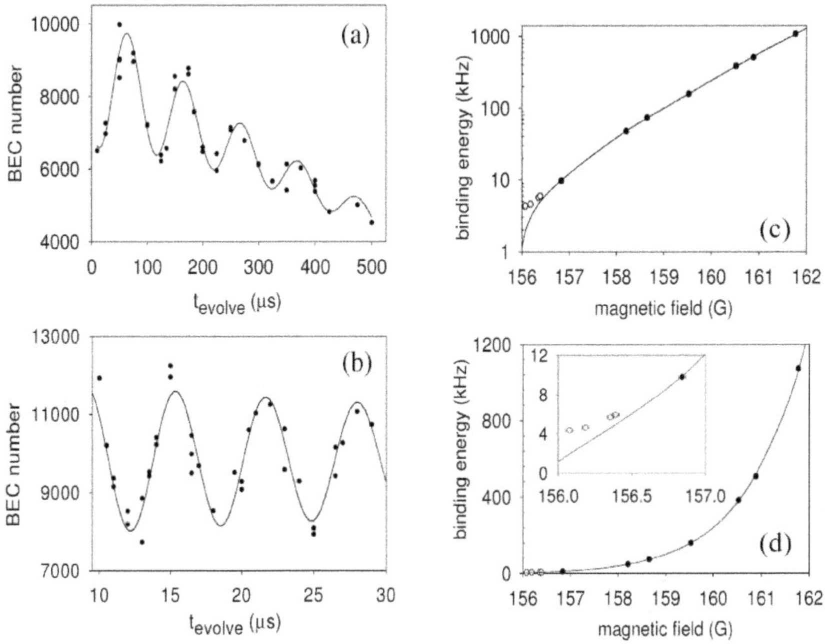

Fig. 17.14 Experimental observation of coherent atom-molecule oscillations [194]. Figs. a) and b) show the number of atoms in the atomic condensate as a function of the time between the two pulses in the magnetic field. The solid line indicates the fit in (17.109). For a) we have that $B_{\text{evolve}} = 156.840(25)$ G. The frequency and damping rates are respectively given by $\nu_e = 2\pi \times 0.58(12)$ kHz, $\alpha = 7.9(4)$ atom/μs, and $\beta = 2\pi \times 0.58(12)$ kHz. For b) the magnetic field $B_{\text{evolve}} = 159.527(19)$ G and $\nu_e = 157.8(17)$ kHz. The damping is negligible for the time that is used to determine the frequency. Note that the frequency has increased for the magnetic field further from resonance. Figs. c) and d) show the observed frequency of the coherent atom-molecule oscillations as a function of the magnetic field. The solid line is the result for the molecular binding energy found from a two-body coupled-channels calculation using the experimental results for the frequency to accurately determine the interatomic potential [194]. Only the black points were included in the fit. The inset shows that, close to resonance, the observed frequency deviates from the two-body result. Reprinted figure with permission from N.R. Claussen, S.J.J.M.F. Kokkelmans, S.T. Thompson, E.A. Donley, and C.E. Wieman, Phys. Rev. A **67**, 060701R (2003). Copyright 2008 by the American Physical Society.

frequency of the coherent atom-molecule oscillation in this way one compensates for the effects of the damping on the frequency. For the results in Fig. 17.14a, we have that $\beta = 2\pi \times 0.58(12)$ kHz and $\alpha = 7.9(4)$ atom/μs. The frequency is equal to $\nu_e = 9.77(12)$ kHz. For Fig. 17.14b, the frequency is equal to $\nu_e = 157.8(17)$ kHz. The damping and loss rate are negligible for the short time used to determine the frequency. It is found experimentally that both the damping rate and the loss rate increase as B_{evolve} approaches the resonant value.

In Fig. 17.14c and d, the results for the frequency as a function of B_{evolve} are presented. The solid line shows the result of a two-body coupled-channels calculation of the molecular binding energy [194]. The parameters of the interatomic potentials are fit to the experimental results for the frequency. Clearly, the frequency of the coherent atom-molecule oscillations agrees very well with the molecular binding energy in vacuum over a large range of the magnetic field. Moreover, in the magnetic-field range $B_{\text{evolve}} \simeq 157-159$ G the frequency of the oscillations is well described by the formula $|\varepsilon_{\text{m}}(B)| = \hbar^2/ma^2(B)$ for the binding energy, derived in Sect. 17.4.2. Close to resonance, however, the measured frequency deviates from the two-body result. The deviating experimental points are shown by open circles and are not taken into account in the determination of the interatomic potential. This deviation is due to many-body effects [195].

Although some of the physics of these coherent atom-molecule oscillations can roughly be understood by a simple two-level picture, it is worth noting that the physics of a Feshbach resonance is much richer. First of all, during Rabi oscillations in a simple two-level system one quantum in a state oscillates to the other state. In the case of a Feshbach resonance pairs of atoms oscillate back and forth between the dressed-molecular condensate and the atomic condensate. Therefore, the Hamiltonian is not quadratic in the annihilation and creation operators and the physics is more complicated. In particular the dressed molecule may decay into two noncondensed atoms instead of forming two condensate atoms. This is a contribution to the damping seen experimentally. Second, the observed atom-molecule oscillations are oscillations between an atomic condensate and a dressed molecular condensate. The fact that we are dealing with dressed molecules implies that by changing the magnetic field not only is the detuning altered, but also the internal state of the molecule itself.

17.7 Josephson Frequency

To get a quantitative description of the experimentally observed magnetic-field dependence of the Josephson frequency, we calculate this frequency in a linear-response approximation, including the energy-dependence of the atom-molecule coupling and the atom-atom interactions. With the mean-field theory derived in the previous sections, we now calculate the magnetic-field and density dependence of the Josephson frequency of the coherent atom-molecule oscillations, in a linear approximation. The only parameter that has not been determined yet is the effective range of the interatomic interactions r_{bg}. All other parameters are known for ^{85}Rb.

The effective range is determined by calculating the molecular binding energy in vacuum and comparing the result with the experimental data. We have seen that far off resonance the Josephson frequency is essentially equal to the molecular binding energy. Since the effect of a nonzero effective range only plays a role for large energies, and thus is important far off resonance, this comparison uniquely determines the effective range. As explained in detail in Sect. 17.4.2, the molecular binding

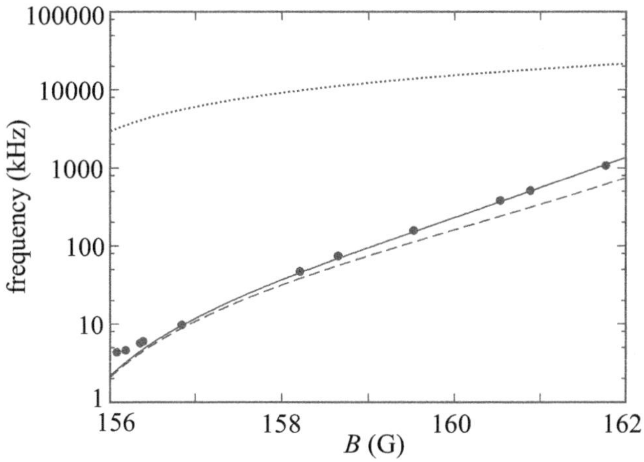

Fig. 17.15 Molecular binding energy in vacuum. The solid line shows the result of a calculation with $r_{bg} = 185a_0$. The dashed line shows $|\varepsilon(B)| = \hbar^2/ma^2$. The experimental points are taken from [194]. The dotted line shows the detuning $|\delta(B)|$.

energy is determined by solving for E in the equation

$$E - \delta(B) - \hbar\Sigma_m(E) = 0. \tag{17.110}$$

For ^{85}Rb the background scattering length is negative and the effective range turns out to be positive. The molecular selfenergy is therefore given by

$$\hbar\Sigma_m(E) = -\frac{g^2 m}{2\pi\hbar^2\sqrt{1-2\frac{r_{bg}}{a_{bg}}}} \frac{i\sqrt{\left(1-2\frac{r_{bg}}{a_{bg}}\right)\frac{mE}{\hbar^2} - \frac{r_{bg}mE}{2\hbar^2}}}{1 + ia_{bg}\sqrt{\left(1-2\frac{r_{bg}}{a_{bg}}\right)\frac{mE}{\hbar^2} - \frac{r_{bg}a_{bg}mE}{2\hbar^2}}}. \tag{17.111}$$

In Fig. 17.15, the result of the numerical solution of (17.110) is shown for $r_{bg} = 185a_0$. Also shown in this figure are the experimental data points. Clearly, far off resonance there is good agreement between our results and the experimental data points. Therefore, we use this value for the effective range from now on in all our calculations. The absolute value of the detuning is shown by the dotted line, and deviates significantly from the binding energy. The dashed line in Fig. 17.15 indicates the formula $|\varepsilon_m| = \hbar^2/ma^2$. As we have derived in Sect. 17.4.2 this formula should accurately describe the magnetic-field dependence of the binding energy close to resonance. Clearly, the solid line that indicates the result that includes the nonzero effective range becomes closer to the dashed line as we approach resonance. However, there is a significant range of magnetic field where we need to include the effective range in our calculations. Closer to the resonance, the experimental points start to deviate from the two-atom binding energy. This deviation is taken into account by considering many-body effects. Note, therefore, that the expected oscillation frequency $\hbar^2/ma^2$ never leads to a quantitative agreement with experiment.

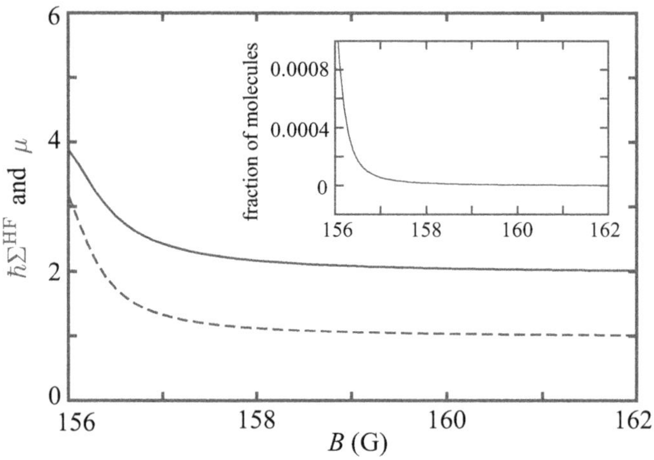

Fig. 17.16 Hartree-Fock selfenergy (solid line) and chemical potential (dashed line) as a function of the magnetic field for an atomic condensate density of $n_{\mathrm{a}} = 2 \times 10^{12}$ cm^{-3}. Both quantities are shown in units of $4\pi a(B)\hbar^2 n_{\mathrm{a}}/m$. Far off resonance, where the energy dependence of the interactions can be safely neglected we have that $\hbar\Sigma^{\mathrm{HF}} = 8\pi a(B)\hbar^2 n_{\mathrm{a}}/m$ and $\mu = 4\pi a(B)\hbar^2 n_{\mathrm{a}}/m$, as expected. The inset shows the fraction of bare molecules as a function of the magnetic field.

As mentioned previously, we calculate the many-body effects on the frequency of the coherent atom-molecule oscillations in linear approximation. Therefore, we first need to determine the equilibrium around which to linearize. In detail, the equilibrium values of the atomic and molecular condensate wavefunctions are determined by solving the time-independent mean-field equations in (17.91) together with the equation for the Hartree-Fock selfenergy in (17.92) at a fixed chemical potential μ. To compare with the experimental results it is more convenient to solve these equations at a fixed condensate density. The chemical potential is then determined from these equations.

In Fig. 17.16, we show the result of this calculation for an atomic condensate density of $n_{\mathrm{a}} = 2 \times 10^{12}$ cm^{-3}. The solid line shows the Hartree-Fock selfenergy $\hbar\Sigma^{\mathrm{HF}}$ and the dashed line the chemical potential as a function of the magnetic field, both in units of the energy $4\pi a(B)\hbar^2 n_{\mathrm{a}}/m$. Note that far off resonance, where the energy dependence of the interaction may be neglected, we have that $\mu = 4\pi a(B)\hbar^2 n_{\mathrm{a}}/m$ and $\hbar\Sigma^{\mathrm{HF}} = 2\mu$. This is the expected result. The inset of Fig. 17.16 shows the fraction of bare molecules $|\phi_{\mathrm{m}}|^2/n_{\mathrm{a}}$. Note that this fraction is always very small. This justifies neglecting the atom-molecule and molecule-molecule interactions since from this figure we see that the mean-field energies associated with these interactions are at least three orders of magnitude smaller.

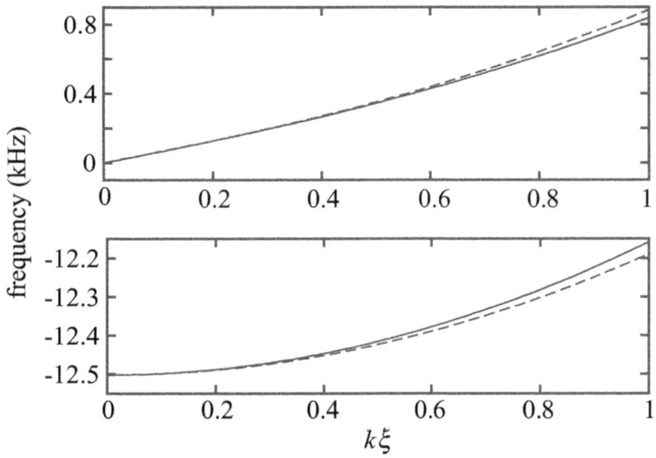

Fig. 17.17 The dispersion relation for the collective modes of an atom-molecule system for a condensate density of $n_a = 2 \times 10^{12}$ cm^{-3} at a magnetic field of $B = 157$ G. The momentum is measured in units of the inverse coherence length $\xi^{-1} = \sqrt{16\pi a(B)n_a}$. The upper branch corresponds to the gapless dispersion for phonons. The solid line is the result of the full calculation, the dashed line shows the Bogoliubov dispersion for the scattering length $a(B)$. The lower branch corresponds to the coherent atom-molecule oscillations. The solid line is the result of the full calculation whereas the dashed line shows the result with the same zero-momentum part, but with the momentum dependence determined by $\hbar^2\mathbf{k}^2/4m$.

Since the coherent atom-molecule oscillations are a collective mode where the amplitude of the atomic and molecular condensate wavefunctions oscillate out-of-phase, we study the collective modes of the system. As explained in detail in the previous section, the frequencies of the collective modes are determined by (17.101). This equation is solved numerically and yields a dispersion relation with two branches. The result of this calculation is shown in Fig. 17.17 for an atomic condensate density of $n_a = 2 \times 10^{12}$ cm^{-3} and a magnetic field of $B = 157$ G. The momentum is indicated in units of the inverse coherence length $\xi^{-1} = \sqrt{16\pi a(B)n_a}$. The upper branch corresponds to the gapless phonon excitations. For small momenta this branch has a linear momentum dependence. The upper dashed line indicates the Bogoliubov dispersion in (17.97) evaluated at the scattering length $a(B)$. For small momentum the solid and the dashed line are almost identical. For larger momenta the numerically exact result is smaller, due to the energy-dependence of the interactions that effectively reduce the scattering length.

The lower branch corresponds to the coherent atom-molecule oscillations and is gapped. The solid line indicates the result of the full calculations. For small mo-

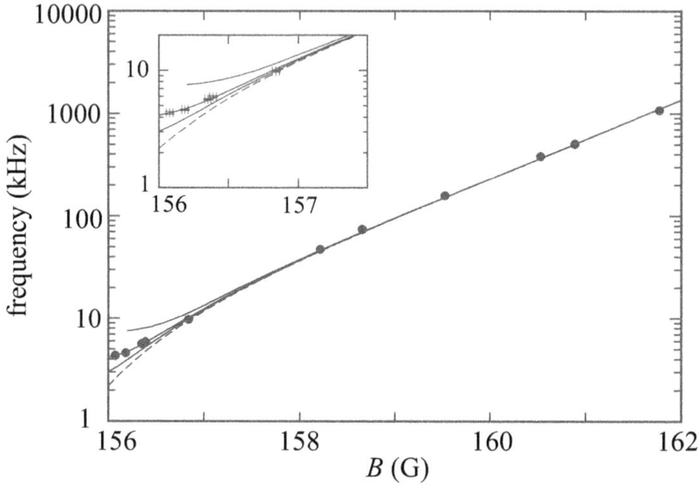

Fig. 17.18 Josephson frequency of coherent atom-molecule oscillations for various values of the condensate density. The solid lines are the results of calculations for nonzero condensate density. The different lines correspond from top to bottom to the decreasing condensate densities $n_a = 5 \times 10^{12}$ cm^{-3}, $n_a = 2 \times 10^{12}$ cm^{-3}, and $n_a = 10^{12}$ cm^{-3}. The dashed line corresponds to the molecular binding energy in vacuum, i.e. $n_a = 0$. The experimental data points, taken from reference [194], are also shown.

menta it is well described by

$$\hbar\omega_{\mathbf{k}} \simeq -\hbar\omega_J + \varepsilon_{\mathbf{k}}/2 \, , \tag{17.112}$$

where ω_J is the Josephson frequency. The dispersion resulting from this last equation is shown in the lower part Fig. 17.17 by the dashed line. This momentum dependence is to be expected since sufficiently far from resonance the atom-molecule oscillations reduce to a two-body excitation. The fact that the dispersion is negative is due to the fact that we are linearizing around a metastable situation with more atoms than molecules. Although this is the experimentally relevant situation, the true equilibrium situation for negative detuning corresponds to almost all atoms in the molecular state [196].

In Fig. 17.18, we present the results for the Josephson frequency as a function of the magnetic field, for different values of the condensate density. The solid lines in this figure show, from top to bottom, the results for an decreasing nonzero condensate density. The respective condensate densities are given by $n_a = 5 \times 10^{12}$ cm^{-3}, $n_a = 2 \times 10^{12}$ cm^{-3}, and $n_a = 10^{12}$ cm^{-3}. The dashed line shows the molecular binding energy in vacuum. The Josephson frequency reduces to the molecular binding energy for all values of the condensate density, in agreement with previous remarks. Nevertheless, sufficiently close to resonance there is a deviation from the two-body result due to many-body effects. This deviation becomes larger with increasing condensate density.

In order to confront our results with the experimental data we have to realize that the experiments are performed in a magnetic trap. Taking only the ground states $\phi_{\mathrm{a}}(\mathbf{x})$ and $\phi_{\mathrm{m}}(\mathbf{x})$ into account for both the atomic and the molecular condensates, respectively, this implies effectively that the atom-molecule coupling g is reduced by an overlap integral. Hence we define the effective homogeneous condensate density by means of $n_{\mathrm{a}} = N_{\mathrm{a}}\left[\int \mathrm{d}\mathbf{x}\phi_{\mathrm{a}}^2(\mathbf{x})\phi_{\mathrm{m}}(\mathbf{x})\right]^2 = 16\sqrt{2}N_{\mathrm{a}}m^{3/2}\nu_r\sqrt{\nu_z}/(125\pi^3\hbar^{3/2})$, where N_{a} denotes the number of condensed atoms and ν_r and ν_z the radial and axial trapping frequencies respectively. For the experiments of Claussen et al. we have that $N_{\mathrm{a}} \simeq 8\times 10^3$ during the oscillations close to resonance, as seen from Fig. 17.14, which results in an effective density of $n_{\mathrm{a}} \simeq 2\times 10^{12}$ cm^{-3}. This agrees also with the effective homogeneous density quoted by Claussen et al. [194]. The solid curve in Fig. 17.18 clearly shows an excellent agreement with the experimentally observed frequency for this density.

It is important to note that there are two hidden assumptions in the above comparison. First, we have used that the dressed molecules are trapped in the same external potential as the atoms. This is not obvious because the bare molecular state involved in the Feshbach resonance is high-field seeking and therefore not trapped. However, (17.87) shows that near resonance almost all the amplitude of the dressed molecule is in the low-field seeking open channel and its magnetic moment is therefore almost equal to twice the atomic magnetic moment. Second, we have determined the frequency of the coherent atom-molecule oscillations in equilibrium. In contrast, the observed oscillations in the number of condensate atoms is clearly a nonequilibrium phenomenon. This is, however, expected not to play an important role because the Ramsey-pulse sequence is performed on such a fast time scale that the response of the condensate wavefunction can be neglected. By variationally solving the Gross-Pitaevskii equation for the atomic condensate wave function, we have explicitly checked that after a typical pulse sequence its width is only a few percent larger than the harmonic oscillator ground state.

17.8 Problems

Exercise 17.1. Molecular Selfenergy

Rederive (17.77) by solving the scattering problem for two atoms with a coupling to a molecular state in the center-of-mass frame. The Hamiltonian for the coupled atom-molecule system is given by

$$\hat{H} = \hat{H}_{\mathrm{a}} + \hat{H}_{\mathrm{m}} + \hat{V}_{\mathrm{am}}. \tag{17.113}$$

Here, $\hat{V}_{\mathrm{am}}$ is the atom-molecule coupling potential, while the kinetic-energy part is described by

$$\hat{H}_{\mathrm{a}} = -\frac{\hbar^2\nabla^2}{m}, \tag{17.114}$$

and the molecular part equals

$$\hat{H}_{\rm m} = \delta_{\rm B}|\chi_{\rm m}\rangle\langle\chi_{\rm m}|, \tag{17.115}$$

with $\delta_{\rm B}$ the bare detuning.
(a) Derive the following equation for the eigenenergy of the coupled atom-molecule system

$$E - \delta_{\rm B} = \langle\chi_{\rm m}|\hat{V}_{\rm am}\frac{1}{E-\hat{H}_{\rm a}}\hat{V}_{\rm am}|\chi_{\rm m}\rangle. \tag{17.116}$$

(b) In the absence of the atom-molecule coupling, the energy difference of the molecules with the atomic continuum is simply equal to the bare detuning $\delta_{\rm B}$. Due to the coupling potential, the molecular binding energy is modified and the right-hand-side of (17.116) can be interpreted as the selfenergy $\hbar\Sigma(E)$ of the molecules. Since the spatial extent of the molecular wavefunction is small, we have that the atom-molecule coupling potential can be approximated by the pseudopotential,

$$\langle\mathbf{r}|\hat{V}_{\rm am}|\phi_{\rm m}\rangle = \begin{cases} \sqrt{2}g\delta(\mathbf{r}) & \text{Bosons} \\ g\delta(\mathbf{r}) & \text{Fermions, Bose-Fermi mixture} \end{cases}. \tag{17.117}$$

Here, g is the atom-molecule coupling constant that obeys

$$g = \begin{cases} \hbar\sqrt{2\pi a_{\rm bg}\Delta\mu\Delta B/m} & \text{Bosons} \\ \hbar\sqrt{4\pi a_{\rm bg}\Delta\mu\Delta B/m} & \text{Fermions, Bose-Fermi mixture} \end{cases}. \tag{17.118}$$

The eigenstates $|\phi_{\mathbf{k}}\rangle$ of $\hat{H}_{\rm a}$ are plane waves with energies $\hbar^2\mathbf{k}^2/m$. Use this observation to obtain the selfenergy in the bosonic case, namely

$$\hbar\Sigma(E) = -\mathrm{i}\frac{m^{3/2}g^2}{2\pi\hbar^3}\sqrt{E} - \lim_{\mathbf{r}\downarrow\mathbf{0}} 2g^2\int\frac{\mathrm{d}\mathbf{k}}{(2\pi)^3}\frac{m\,\mathrm{e}^{\mathrm{i}\mathbf{k}\cdot\mathbf{r}}}{\hbar^2\mathbf{k}^2}. \tag{17.119}$$

(c) Finally, show that the divergent term in the selfenergy can be written as,

$$\lim_{\mathbf{r}\downarrow\mathbf{0}} 2g^2\int\frac{\mathrm{d}\mathbf{k}}{(2\pi)^3}\frac{m\,\mathrm{e}^{\mathrm{i}\mathbf{k}\cdot\mathbf{r}}}{\hbar^2\mathbf{k}^2} = \lim_{r\downarrow 0}\frac{mg^2}{2\pi\hbar^2 r}. \tag{17.120}$$

To deal with this divergence, we have to use the renormalized detuning instead of the bare detuning. The former is defined as $\delta = \delta_{\rm B} - \lim_{r\downarrow 0} mg^2/2\pi\hbar^2 r$, where we have that $\delta = \Delta\mu(B-B_0)$, which is determined by the experimental value of the magnetic field B_0 at resonance. For the case of fermions or a Bose-Fermi mixture, the above expression for the selfenergy is a factor of two smaller and in those cases the renormalized detuning is given by $\delta = \delta_{\rm B} - \lim_{r\downarrow 0} mg^2/4\pi\hbar^2 r$.

Aditional Reading

- For a detailed account on Feshbach resonances in ultracold gases, see E. Timmermans, P. Tommasini, M. Hussein, and A. Kernan, Phys. Rep. **315**, 199 (1999), and R. A. Duine and H. T. C. Stoof, Phys. Rep. **396**, 115 (2004).

References

1. P. Kapitsa, Nature **141**, 74 (1938)
2. J.F. Allen, A.D. Misener, Nature **141**, 75 (1938)
3. J.F. Allen, H. Jones, Nature **141**, 243 (1938)
4. N.S. Bose, Z. Phys. **26**, 178 (1924)
5. A. Einstein, Sitzber. Kgl. Preuss. Akad. Wiss., 261 (1924)
6. H.K. Onnes, Comm. Phys. Lab. Univ. Leiden, Nos. 119, 120, 122 (1911)
7. W. Meissner, R. Ochsenfeld, Naturwiss. **21**, 787 (1933)
8. J. Bardeen, L.N. Cooper, J.R. Schrieffer, Phys. Rev. **108**, 1175 (1957)
9. J.G. Bednorz, K.A. Müller, Z. Phys. B **64**, 189 (1986)
10. M.H. Anderson, J.R. Ensher, M.R. Matthews, C.E. Wieman, E.A. Cornell, Science **269**, 198 (1995)
11. C.C. Bradley, C.A. Sackett, J.J. Tollett, R.G. Hulet, Phys. Rev. Lett. **75**, 1687 (1995)
12. K.B. Davis, M.O. Mewes, M.R. Andrews, N.J. van Druten, D.S. Durfee, D.M. Kurn, W. Ketterle, Phys. Rev. Lett. **75**, 3969 (1995)
13. E.L. Raab, M. Prentiss, A. Cable, S. Chu, D.E. Pritchard, Phys. Rev. Lett. **59**, 2631 (1987)
14. C. Monroe, W. Swann, H. Robinson, C.E. Wieman, Phys. Rev. Lett. **65**, 1571 (1990)
15. N. Masuhara, J.M. Doyle, J.C. Sandberg, D. Kleppner, T.J. Greytak, Phys. Rev. Lett. **61**, 935 (1988)
16. H.T.C. Stoof, M. Houbiers, C.A. Sackett, R.G. Hulet, Phys. Rev. Lett. **76**, 10 (1996)
17. E. Tiesinga, B.J. Verhaar, H.T.C. Stoof, Phys. Rev. A **47**, 4114 (1993)
18. C.A. Regal, M. Greiner, D.S. Jin, Phys. Rev. Lett. **92**, 040403 (2004)
19. M.W. Zwierlein, C.A. Stan, C.H. Schunck, S.M.F. Raupach, A.J. Kerman, W. Ketterle, Phys. Rev. Lett. **92**, 120403 (2004)
20. J. Kinast, S.L. Hemmer, M.E. Gehm, A. Turlapov, J.E. Thomas, Phys. Rev. Lett. **92**, 150402 (2004)
21. M. Bartenstein, A. Altmeyer, S. Riedl, S. Jochim, C. Chin, J.H. Denschlag, R. Grimm, Phys. Rev. Lett. **92**, 203201 (2004)
22. T. Bourdel, L. Khaykovich, J. Cubizolles, J. Zhang, F. Chevy, M. Teichmann, L. Tarruell, S.J.J.M.F. Kokkelmans, C. Salomon, Phys. Rev. Lett. **93**, 050401 (2004)
23. G.B. Partridge, K.E. Strecker, R.I. Kamar, M.W. Jack, R.G. Hulet, Phys. Rev. Lett. **95**, 020404 (2005)
24. D.M. Eagles, Phys. Rev. **186**, 456 (1969)
25. A.J. Leggett, *Modern Trends in the Theory of Condensed Matter* (Springer-Verlag, Berlin, 1980), p. 13
26. D. Bailin, A. Love, Phys. Rep. **107**, 325 (1984)
27. P.S. Jessen, C. Gerz, P.D. Lett, W.D. Phillips, S.L. Rolston, R.J. Spreeuw, C.I. Westbrook, Phys. Rev. Lett. **69**, 49 (1992)
28. G.V. Chester, Phys. Rev. A **2**, 256 (1970)

29. E. Kim, M.H.W. Chan, Nature **427**, 225 (2004)
30. K. Góral, L. Santos, M. Lewenstein, Phys. Rev. Lett. **88**, 170406 (2002)
31. M. Greiner, O. Mandel, T. Esslinger, T.W. Hänsch, I. Bloch, Nature **415**, 39 (2002)
32. R.A. Duine, Atom-molecule coherence in Bose gases. Ph.D. thesis, Utrecht University (2003)
33. A.G. Truscott, K.E. Strecker, W.I. McAlexander, G.B. Partridge, R.G. Hulet, Science **291**, 2570 (2001)
34. I.M. Gelfand, A.M. Yaglom, J. Math. Phys. **1**, 48 (1960)
35. A.L. Fetter, J.D. Walecka, *Quantum Theory of Many-Particle Systems* (McGraw-Hill, New York, 1971)
36. L. Mandel, E. Wolf, *Optical Coherence and Quantum Optics* (Cambridge, New York, 1995)
37. J.W. Negele, H. Orland, *Quantum Many-Particle Systems* (Westview Press, Boulder, 1998)
38. A.J. Leggett, Rev. Mod. Phys. **47**, 331 (1975)
39. F.J. Dyson, Phys. Rev. **75**, 1736 (1949)
40. H.J. Metcalf, P. van der Straten, *Laser Cooling and Trapping* (Springer-Verlag, New York, 1999)
41. C.J. Pethick, H. Smith, *Bose-Einstein Condensation in Dilute Gases* (Cambridge University Press, Cambridge, 2002)
42. R.A. Duine, H.T.C. Stoof, Phys. Rep. **396**, 115 (2004)
43. D.G. Fried, T.C. Killian, L. Willmann, D. Landhuis, S.C. Moss, D. Kleppner, T.J. Greytak, Phys. Rev. Lett. **81**, 3811 (1998)
44. O. Penrose, Phil. Mag. **42**, 1373 (1951)
45. O. Penrose, L. Onsager, Phys. Rev. **104**, 5767 (1956)
46. A.J. Leggett, Rev. Mod. Phys. **73**, 307 (2001)
47. C.N. Yang, Rev. Mod. Phys. **34**, 694 (1962)
48. N.N. Bogoliubov, J. Phys. Moscow **11**, 23 (1947)
49. L.P. Pitaevskii, Sov. Phys. JETP **13**, 451 (1961)
50. E.P. Gross, J. Math. Phys. **4**, 195 (1963)
51. K.G. Singh, D.S. Rokhsar, Phys. Rev. Lett. **77**, 1667 (1996)
52. M. Edwards, P.A. Ruprecht, K. Burnett, R.J. Dodd, C.W. Clark, Phys. Rev. Lett. **77**, 1671 (1996)
53. M. Lewenstein, L. You, Phys. Rev. Lett. **77**, 3489 (1996)
54. A.L. Fetter, Ann. Phys. **70**, 67 (1972)
55. V.N. Popov, *Functional Integrals in Quantum Field Theory and Statistical Physics* (Reidel, Dordrecht, 1983)
56. F. Dalfovo, S. Giorgini, L.P. Pitaevskii, S. Stringari, Rev. Mod. Phys. **71**, 463 (1999)
57. D.A.W. Hutchinson, E. Zaremba, A. Griffin, Phys. Rev. Lett. **78**, 1842 (1997)
58. R.J. Dodd, M. Edwards, C.W. Clark, K. Burnett, Phys. Rev. A **57**, R32 (1998)
59. H.T.C. Stoof, J. Low Temp. Phys. **114**, 11 (1999)
60. L. Thomas, Proc. Camb. Phil. Soc. **23**, 542 (1927)
61. E. Fermi, Mat. Natur. **6**, 602 (1927)
62. V.V. Goldman, I.F. Silvera, A.J. Leggett, Phys. Rev. B **24**, 2870 (1981)
63. S. Stringari, Phys. Rev. Lett. **77**, 2360 (1996)
64. D.S. Jin, J.R. Ensher, M.R. Matthews, C.E. Wieman, E.A. Cornell, Phys. Rev. Lett. **77**, 420 (1996)
65. D.S. Jin, M.R. Matthews, J.R. Ensher, C.E. Wieman, E.A. Cornell, Phys. Rev. Lett. **78**, 764 (1997)
66. M.O. Mewes, M.R. Anderson, N.J. van Druten, D.M. Kurn, D.S. Durfee, C.G. Townsend, W. Ketterle, Phys. Rev. Lett. **77**, 988 (1996)
67. L.V. Hau, B.D. Busch, C. Liu, Z. Dutton, M.M. Burns, J.A. Golovchenko, Phys. Rev. A **58**, R54 (1998)
68. N.R. Cooper, N.K. Wilkin, J.M.F. Gunn, Phys. Rev. Lett. **87**, 120405 (2001)
69. K.W. Madison, F. Chevy, W. Wohlleben, J.B. Dalibard, Phys. Rev. Lett. **84**, 806 (2000)
70. J.R. Abo-Shaeer, C. Raman, J.M. Vogels, W. Ketterle, Science **292**, 476 (2001)

71. S.L. Cornish, N.R. Claussen, J.L. Roberts, E.A. Cornell, C.E. Wieman, Phys. Rev. Lett. **85**, 1795 (2000)
72. For an equilibrium argument at zero temperature, see L.D. Landau and E.M. Lifshitz, *Statistical Physics*, (Pergamon, London, 1958) and P. Nozières, in *Bose-Einstein Condensation*, edited by A. Griffin, D.W. Snoke, and S. Stringari, (Cambridge, New York, 1995). The nonzero-temperature case is discussed in H.T.C. Stoof, Phys. Rev. A **49**, 4704, (1995)
73. M. Houbiers, H.T.C. Stoof, Phys. Rev. A **54**, 5055 (1996)
74. T. Bergeman, Phys. Rev. A **55**, 3658 (1997)
75. For the anisotropic generalization see, for instance, G. Baym and C.J. Pethick, Phys. Rev. Lett. **76**, 6, (1996).
76. L. Fetter, Phys. Rev. A **53**, 4245 (1996). The exact result is obtained in P.A. Ruprecht, M.J. Holland, K. Burnett, and M. Edwards, Phys. Rev. A, **51**, 4704, (1995).
77. H.T.C. Stoof, J. Stat. Phys. **87**, 1353 (1997). For a different calculation of the tunneling rate that neglects the phase fluctuations of the condensate, see E.V. Shuryak, Phys. Rev. A. **54**, 3151 (1996)
78. C.A. Sackett, C.C. Bradley, M. Welling, R.G. Hulet, Appl. Phys. B **65**, 433 (1997)
79. P. Fulde, R.A. Ferrell, Phys. Rev. **135**, A550 (1964)
80. A.I. Larkin, Y.N. Ovchinnikov, Sov. Phys. JETP **20**, 762 (1965)
81. C.A.R. Sá de Melo, M. Randeria, J.R. Engelbrecht, Phys. Rev. Lett. **71**, 3202 (1993)
82. L.P. Gor'kov, T.K. Melik-Barkhudarov, Sov. Phys. JETP **13**, 1018 (1961)
83. H. Heiselberg, C.J. Pethick, H. Smith, L. Viverit, Phys. Rev. Lett. **85**, 2418 (2000)
84. T.L. Ho, Phys. Rev. Lett. **92**, 090402 (2004)
85. J. Carlson, S.Y. Chang, V.R. Pandharipande, K.E. Schmidt, Phys. Rev. Lett. **91**, 050401 (2003)
86. G.E. Astrakharchik, J. Boronat, J. Casulleras, S. Giorgini, Phys. Rev. Lett. **93**, 200404 (2004)
87. E. Burovski, N. Prokof'ev, B. Svistunov, M. Troyer, Phys. Rev. Lett. **96**, 160402 (2006)
88. P. Nozières, S. Schmitt-Rink, J. Low Temp. Phys. **59**, 195 (1985)
89. M.W. Zwierlein, J.R. Abo-Shaeer, A. Schirotzek, C.H. Schunck, W. Ketterle, Nature **435**, 1047 (2005)
90. M. Kozuma, L. Deng, E.W. Hagley, J. Wen, R. Lutwak, K. Helmerson, S.L. Rolston, W.D. Phillips, Phys. Rev. Lett. **82**, 871 (1999)
91. D.M. Stamper-Kurn, A.P. Chikkatur, A. Görlitz, S. Inouye, S. Gupta, D.E. Pritchard, W. Ketterle, Phys. Rev. Lett. **83**, 2876 (1999)
92. G. Grynberg, C. Robiliard, Phys. Rep. **355**, 335 (2001)
93. A.G. Martin, K. Helmerson, V.S. Bagnato, G.P. Lafyatis, D.E. Pritchard, Phys. Rev. Lett. **61**, 2431 (1988)
94. I. Bloch, T.W. Hänsch, T. Esslinger, Phys. Rev. Lett. **82**, 3008 (1999)
95. C. Chin, M. Bartenstein, A. Altmeyer, S. Riedl, S. Jochim, J.H. Denschlag, R. Grimm, Science **305**, 1128 (2004)
96. M. Greiner, C.A. Regal, D.S. Jin, Phys. Rev. Lett. **94**, 070403 (2005)
97. S. Gupta, Z. Hadzibabic, M.W. Zwierlein, C.A. Stan, K. Dieckmann, C.H. Schunck, E.G.M. van Kempen, B.J. Verhaar, W. Ketterle, Science **300**, 1723 (2003)
98. C.J. Pethick, H.T.C. Stoof, Phys. Rev. A **64**, 013618 (2001)
99. C.H. Schunck, Y. Shin, A. Schirotzek, W. Ketterle, cond-mat/0802.0341
100. P. Massignan, G.M. Bruun, H.T.C. Stoof, Phys. Rev. A **77**, 031601(R) (2008)
101. H.T.C. Stoof, Phys. Rev. B **47**, 7979 (1993)
102. P.W. Anderson, Phys. Rev. **112**, 1900 (1958)
103. E. Abrahams, T. Tsuneto, Phys. Rev. **152**, 416 (1966)
104. H. Kleinert, Forts. Phys. **26**, 565 (1978)
105. For simplicity we here again neglect a topological term in the action, which essentially plays no role in the following. This is also in agreement with our discussion of the superconductor, where exactly the same topological term was not included in the effective action for the global phase.
106. K.G. Wilson, Phys. Rev. B **4**, 3174 (1971)

107. K.G. Wilson, J. Kogut, Phys. Rep. C **12**, 75 (1974)
108. J. Zinn-Justin, *Quantum Field Theory and Critical Phenomena* (Oxford, New York, 1989)
109. T. Donner, S. Ritter, T. Bourdel, A. Öttl, M. Köhl, T. Esslinger, Science **315**, 1556 (2007)
110. M. Bijlsma, H.T.C. Stoof, Phys. Rev. A **54**, 5085 (1996)
111. N. Hasselmann, S. Ledowski, P. Kopietz, Phys. Rev. A **70**, 063621 (2004)
112. M.W. Zwierlein, A. Schirotzek, C.H. Schunck, W. Ketterle, Science **311**, 492 (2006)
113. G.B. Partridge, W. Li, R.I. Kamar, Y. Liao, R.G. Hulet, Science **311**, 503 (2006)
114. M.M. Parish, F.M. Marchetti, A. Lamacraft, B.D. Simons, Nat. Phys. **3**, 124 (2007)
115. K.B. Gubbels, M.W.J. Romans, H.T.C. Stoof, Phys. Rev. Lett. **97**, 210402 (2006)
116. C. Lobo, A. Recati, S. Giorgini, S. Stringari, Phys. Rev. Lett. **97**, 200403 (2006)
117. Y. Shin, C.H. Schunck, A. Schirotzek, W. Ketterle, Nature **451**, 689 (2008)
118. K.B. Gubbels, H.T.C. Stoof, Phys. Rev. Lett. **100**, 140407 (2008)
119. R. Combescot, A. Recati, C. Lobo, F. Chevy, Phys. Rev. Lett. **98**, 180402 (2007)
120. S. Pilati, S. Giorgini, Phys. Rev. Lett. **100**, 030401 (2008)
121. R. Shankar, Rev. Mod. Phys. **66**, 129 (1994)
122. U.A. Khawaja, J.O. Andersen, N.P. Proukakis, H.T.C. Stoof, Phys. Rev. A **66**, 013615 (2002)
123. A. Görlitz, J.M. Vogels, A.E. Leanhardt, C. Raman, T.L. Gustavson, J.R. Abo-Shaeer, A.P. Chikkatur, S. Gupta, S. Inouye, T.P. Rosenband, W. Ketterle, Phys. Rev. Lett. **87**, 130402 (2001)
124. F. Schreck, L. Khaykovich, K.L. Corwin, G. Ferrari, T. Bourdel, J. Cobizolles, C. Salomon, Phys. Rev. Lett. **87**, 080403 (2001)
125. H. Ott, J. Fortagh, G. Schlotterbeck, A. Grossmann, C. Zimmermann, Phys. Rev. Lett. **87**, 230401 (2001)
126. W. Hänsel, P. Hommelhoff, T.W. Hänsch, J. Reichel, Nature **413**, 501 (2001)
127. W.J. Mullin, J. Low. Temp. Phys. **106**, 615 (1997)
128. T.L. Ho, M. Ma, J. Low Temp. Phys. **115**, 61 (1999)
129. D.S. Petrov, M. Holzmann, G.V. Shlyapnikov, Phys. Rev. Lett. **84**, 2551 (2000)
130. D.S. Petrov, G.V. Shlyapnikov, J.T.M. Walraven, Phys. Rev. Lett. **85**, 3745 (2000)
131. N.D. Mermin, H. Wagner, Phys. Rev. Lett. **22**, 1133 (1966)
132. P.C. Hohenberg, Phys. Rev. **158**, 383 (1967)
133. J.M. Kosterlitz, D.J. Thouless, J. Phys. C **6**, 1181 (1973)
134. M. Schick, Phys. Rev. A **3**, 1067 (1971)
135. D.S. Fisher, P.C. Hohenberg, Phys. Rev. B **37**, 4936 (1988)
136. H.T.C. Stoof, M. Bijlsma, Phys. Rev. E **47**, 939 (1993)
137. F.D.M. Haldane, Phys. Rev. Lett. **47**, 1840 (1981)
138. M. Olshanii, Phys. Rev. Lett. **81**, 938 (1998)
139. M.D. Girardeau, E.M. Wright, Phys. Rev. Lett. **84**, 5239 (2000)
140. H.T.C. Stoof, M. Bijlsma, Phys. Rev. B **49**, 422 (1994)
141. H.T.C. Stoof, L.P.H. de Goey, W.M.H.M. Rovers, P.S.M.K. Jansen, B.J. Verhaar, Phys. Rev. A **38**, 1248 (1988)
142. A.I. Safonov, S.A. Vasilyev, I.S. Yasnikov, I.I. Lukashevich, S. Jaakkola, Phys. Rev. Lett **81**, 4545 (1998)
143. T.D. Lee, C.N. Yang, Phys. Rev. **105**, 1119 (1957)
144. Z. Hadzibabic, P. Krüger, M. Cheneau, D. Batelier, J.B. Dalibard, Nature **441**, 1118 (2006)
145. E.L. Bolda, D.F. Walls, Phys. Rev. Lett. **81**, 05477 (1998)
146. J. Tempere, J.T. Devreese, Solid State Commun. **108**, 993 (1998)
147. V. Schweikhard, S. Tung, E.A. Cornell, Phys. Rev. Lett. **99**, 030401 (2007)
148. J.M. Kosterlitz, J. Phys. C **7**, 1046 (1974)
149. D.J. Amit, *Field Theory, the Renormalization Group, and Critical Phenomena* (World Scientific, Singapore, 1984)
150. P. Minnhagen, M. Nylén, Phys. Rev. B **31**, 5768 (1985)
151. W. Ketterle, N.J. van Druten, Phys. Rev. A **54**, 656 (1996)
152. L. Fetter, D. Rokhsar, Phys. Rev. A **57**, 1191 (1998)
153. S. Stringari, Phys. Rev. Lett. **77**, 2360 (1996)
154. D.S. Petrov, G.V. Shlyapnikov, J.T.M. Walraven, Phys. Rev. Lett. **85**, 3745 (2000)

155. H.T.C. Stoof, J. Low Temp. Phys. **114**, 11 (1999)
156. H.T.C. Stoof, M.J. Bijlsma, J. Low Temp. Phys. **124**, 431 (2001)
157. M. Bijlsma, H.T.C. Stoof, Phys. Rev. A **55**, 498 (1997)
158. D. van Oosten, P. van der Straten, H.T.C. Stoof, Phys. Rev. A **63**, 053601 (2001)
159. D. Jaksch, C. Bruder, J.I. Cirac, C.W. Gardiner, P. Zoller, Phys. Rev. Lett. **81**, 3108 (1998)
160. M.P.A. Fisher, P.B. Weichman, G. Grinstein, D.S. Fisher, Phys. Rev. B. **40**, 546 (1989)
161. M. Snoek, M. Haque, S. Vandoren, H.T.C. Stoof, Phys. Rev. Lett **95**, 250401 (2005)
162. M. Snoek, S. Vandoren, H.T.C. Stoof, Phys. Rev. A **74**, 033607 (2006)
163. J.M. Ziman, *Theory of Solids* (Cambridge University Press, New York, 1972)
164. N.W. Ashcroft, N.D. Mermin, *Solid State Physics* (Saunders College Publishing, Philadelphia, 1976)
165. W. Zwerger, J. of Opt. B **5**, s9 (2003)
166. N.F. Mott, Proc. Phys. Soc. A **62**, 416 (1949)
167. A. van Otterlo, K.H. Wagenblast, Phys. Rev. Lett. **72**, 3598 (1994)
168. E. Roddick, D. Stroud, Phys.Rev. B **48**, 16600 (1993)
169. K. Sheshadri, H.R. Krishnamurthy, R. Pandit, T. Ramakrishnan, Europhys. Lett. **22**, 257 (1993)
170. T. Stöferle, H. Moritz, C. Schori, M. Köhl, T. Esslinger, Phys. Rev. Lett. **92**, 130403 (2004)
171. C. Schori, T. Stöferle, H. Moritz, M. Köhl, T. Esslinger, Phys. Rev. Lett. **93**, 240402 (2004)
172. K. Xu, Y. Liu, J.R. Abo-Shaeer, T. Mukaiyama, J.K. Chin, D.E. Miller, W. Ketterle, K.M. Jones, E. Tiesinga, Phys. Rev. A **72**, 043604 (2005)
173. F. Gerbier, S. Fölling, A. Widera, O. Mandel, I. Bloch, Phys. Rev. Lett. **96**, 090401 (2006)
174. F. Gerbier, A. Widera, S. Fölling, O. Mandel, T. Gericke, I. Bloch, Phys. Rev. Lett. **95**, 050404 (2005)
175. P. Sengupta, M. Rigol, G.G. Batrouni, P.J.H. Denteneer, R.T. Scalettar, Phys. Rev. Lett. **95**, 220402 (2005)
176. H.A. Jahn, E. Teller, Proc. Roy. Soc. **A164**, 117 (1937)
177. S. Fölling, A. Widera, T. Müller, F. Gerbier, I. Bloch, Phys. Rev. Lett. **97**, 060403 (2006)
178. G.K. Campbell, J. Mun, M. Boyd, P. Medley, A.E. Leanhardt, L.G. Marcassa, D.E. Pritchard, W. Ketterle, Science **313**, 649 (2006)
179. D. van Oosten, D.B.M. Dickerscheid, B. Farid, P. van der Straten, H.T.C. Stoof, Phys. Rev. A **71**, 021601 (2005)
180. D. van Oosten, Quantum gases in optical lattices: the atomic Mott insulator. Ph.D. thesis, Utrecht University (2004)
181. L.D. Landau, E.M. Lifshitz, L.P. Pitaevskii, *Electrodynamics of Continuous Media* (Pergamom, Oxford, 1984)
182. L. van Hove, Phys. Rev. **89**, 1189 (1953)
183. W.C. Stwalley, Phys. Rev. Lett. **37**, 1628 (1976)
184. M.W.J. Romans, H.T.C. Stoof, Phys. Rev. Lett. **95**, 260407 (2005)
185. S. Inouye, M.R. Andrews, J. Stenger, H.J. Miesner, D.M. Stamper-Kurn, W. Ketterle, Nature **392**, 151 (1998)
186. C.A. Regal, C. Ticknor, J.L. Bohn, D.S. Jin, Nature **424**, 47 (2003)
187. H.T.C. Stoof, M. Bijlsma, M. Houbiers, J. Res. Natl. Inst. Stand. Technol. **101**, 443 (1996)
188. A. Marte, T. Volz, J. Schuster, S. Dürr, G. Rempe, E.G.M. van Kempen, B.J. Verhaar, Phys. Rev. Lett. **89**, 283202 (2002)
189. S.J.J.M.F. Kokkelmans, M.J. Holland, Phys. Rev. Lett. **89**, 180401 (2002)
190. E.G.M. van Kempen, S.J.J.M.F. Kokkelmans, D.J. Heinzen, B.J. Verhaar, Phys. Rev. Lett. **88**, 093201 (2002)
191. J. Goldstone, Nuova Cimento **19**, 154 (1961)
192. R.A. Duine, H.T.C. Stoof, J. Opt. B: Quantum Semiclass. Opt. **5**, S212 (2003)
193. E.A. Donley, N.R. Claussen, S.L. Cornish, J.L. Roberts, E.A. Cornell, C.E. Wieman, Nature **412**, 295 (2001)
194. N.R. Claussen, S.J.J.M.F. Kokkelmans, S.T. Thompson, E.A. Donley, C.E. Wieman, Phys. Rev. A **67**, 060701R (2003)
195. R.A. Duine, H.T.C. Stoof, Phys. Rev. Lett. **91**, 150405 (2003)
196. E. Timmermans, P. Tommasini, H. Hussein, A. Kerman, Phys. Rep. **315**, 199 (1999)
197. E. Shuryak, Phys. Rev. A **54**, 3151 (1996)

Index

GPSR Compliance
The European Union's (EU) General Product Safety Regulation (GPSR) is a set of rules that requires consumer products to be safe and our obligations to ensure this.

If you have any concerns about our products, you can contact us on

ProductSafety@springernature.com

In case Publisher is established outside the EU, the EU authorized representative is:

Springer Nature Customer Service Center GmbH
Europaplatz 3
69115 Heidelberg, Germany

www.ingramcontent.com/pod-product-compliance
Ingram Content Group UK Ltd.
Pitfield, Milton Keynes, MK11 3LW, UK
UKHW020226290926
15470UKWH00001B/394

9781402087622